BIOLOGY

HEW Will Study DES Link to Breast Cancer

Higher Risk Feared for Women Given Hormone

BY MARLENE CIMONS
Times Staff Writer

WASHINGTON—Acknowledging that new statistics for women who

No Ice Age Soon, Scientists Say

Stable Climate Expected for Next 20 Years

BY ROBERT GILLETTE
Times Science Writer

government policymakers a feeling for the limits of consensus among sis Strategy: Climate and Global Survival," represents one pole of that

SMOKING DRIVE

HEW Huffs, Tobaccoland Plows Ahead

BY JEFF PRUGH
Times Staff Writer

Dramatic, Surprising Decline in Rate of World's Population Growth Reported

DALLAS SCHOOLS PICK DISPUTED TEXTBOOK

Board Votes, 6 to 3, for Biology Classes to Have Access to Bible's Concept of the Creation

UP TO 11 MILLION EXPOSED
Early Death Forecast for Asbestos Workers

World population expected to be 8.08 billion by 2013

Radiation Rise in Bikini Islanders' Bodies Called Incredible; Atoll to Be Evacuated

Birth Control Study Finds The Pill Most Effective

NEW YORK (AP)—A new study of contraceptive meth-

POPULATION GROWTH RATE IS DECLINING

WASHINGTON (AP) — The rate of the world's population growth has unexpectedly peaked and is now actually declining, according

Study Links Uterine Cancer to Affluence

Incidence Found to Be Much Greater in Women From County's Wealthier Areas

U.S. Scientists Link Clay to Formation of Primitive Life

Nickel-Bearing Material May Have Attracted Chemicals That Formed Living Organisms in Primordial Oceans

Researcher says breast X-rays may cause a cancer epidemic

'LAST PLAGUE'
Flu: Age-Old Illness They Can't Control

BY ROBERT C. TOTH
Times Staff Writer

Kepone problem still not licked

RICHMOND, Va. (AP) — The

FDA Urged to Treat Cigarettes as Drugs

WASHINGTON (AP)—Two former U.S. surgeons general and 10 organizations opposed to cigarette smoking pe-

Toledo, Ohio, showed the

Bacterium Altered to Make It Produce Human Hormone in 'Scientific Triumph'

WASHINGTON (UPI)—California that Bover and coworkers had trans-

Report Cites Nuclear Waste Disposal Risk

No Safe Method in Sight, Energy Panel Concludes

Aggressive Behavior Linked to Chemicals

Study Measures Level of Three Substances in Human Brains

BY LOIS TIMNICK
Times Human Behavior Writer

ATLANTA—The difference

PUBLIC UNHAPPY
Food, Drug Tests Become a Bitter Pill

PHILOSOPHER RAPS MOVIES
'Retreat From Reason' in Pseudosciences Seen

Scientists Plunge Into Lobbying For More Medical Research Aid

Panel of Scientists Urges $1 Billion U.S. Research Effort to Ease World Hunger

Organics Not More Nutritious, Institute Says

f insect vantages in raising a ral ene- yard garden by orga in the methods.
the re-

SCHIZOPHRENIA BREAKTHROUGH CLAIMED
Scientists Find 'Sites of Craziness'

Majority in U.S. breathe harmful air, EPA reports

Doctor Hails Upset of Laetrile Conviction

SCIENTISTS PUT LIFE'S HISTORY UNDER THE MICROSCOPE

U.S. Tightens Rules for Use of Benzene

Chemical Used in Detergents and Gasoline Linked to Cancer

Poisonous Gas Cloud Still Haunts Town

SEVESO, Italy (AP)—It was just af-

Recent newspaper headlines testify to the public importance of biological problems.

BIOLOGY
Second Edition

James F. Case University of California, Santa Barbara

ILLUSTRATED BY
Larry Jon Friesen / **Maggie Day** / **Charles Parmely**
Santa Barbara City College

Earlier edition entitled
BIOLOGY: OBSERVATION AND CONCEPT   By James F. Case and Vernon E. Stiers

Macmillan Publishing Co., Inc.
New York

Collier Macmillan Publishers
London

Copyright © 1979, James F. Case, and Estate of Vernon E. Stiers

Printed in the United States of America

All rights reserved. No part of this book may be reproduced or transmitted in any form or by any means, electronic or mechanical, including photocopying, recording, or any information storage and retrieval system, without permission in writing from the Publisher.

Earlier edition entitled *Biology: Observation and Concept*, copyright © 1971 by Macmillan Publishing Co., Inc.

Macmillan Publishing Co., Inc.
866 Third Avenue, New York, New York 10022

Collier Macmillan Canada, Ltd.

Library of Congress Cataloging in Publication Data

Case, James F   (date)
 Biology

 Includes index.
 1. Biology. I. Stiers, Vernon E., joint author. II. Title. [DNLM: 1. Biology. QH307 C337b]
QH308.2.C37 1979   574   77-22613
ISBN 0-02-319980-6

Printing: 1 2 3 4 5 6 7 8 Year: 9 0 1 2 3 4 5

INTRODUCTION
To the Second Edition

Vernon Stiers and I were grateful for the acceptance of the first edition, which appeared under the title of *Biology: Observation and Concept*. Criticisms and suggestions from users of the first edition have been important in developing this new edition. In addition to the anonymous readers of the draft, to whom I am tremendously indebted, I thank my wife and Kathryn Hamilton, Linda Strause, Mark Lowenstine, and David Cook for careful reading that has saved me from many an error.

Examination of the book will instantly reveal that it has been vastly improved by the art work, which is the product of thoughtful design and execution by Dr. Larry Jon Friesen and Margaret Day, assisted by Charles Parmely.

Science, it goes without saying, has progressed between editions, providing the agreeable task of explaining new discoveries to replace old errors. However, the world situation as regards problems of biological significance has regressed. We have, therefore, not only corrected errors but have greatly increased our emphasis on those phases of biology that specifically relate to the human condition. Unfortunately, it remains a sad reality of teaching that what are plain facts to the teacher do not necessarily produce plainly required changes in the world. Thus, although it is well known that the easiest way to eliminate about half of all environmentally caused cancer is to stop cigarette smoking, the habit persists. Or, although nearly everyone knows the message written in human population data, the world growth curve continues its upward swing. Certainly there is more to such problems than simple biology, but I hope that an occasional student will appreciate the lessons retold here and, perhaps persuaded by the biology, may add his or her effort to the cause.

Vern Stiers lived long enough to take great and significant interest in preparations for the writing of this edition. He was a splendid teacher and this book suffers for lack of his masterful touch. His wife, Velma, was a major help with preparation of the manuscript.

J.F.C.

TO The Student

There is some possibility that you are not undertaking the study of biology entirely as a free agent. You may need biology for graduation or as a prerequisite for another course. If so, perhaps a way to generate interest is to view biology as a prerequisite for *survival* rather than an academic prerequisite. This may put things a little strongly as far as your own immediate survival is concerned, but it seems that the world as a whole is going to take a change for the worse if more people do not learn certain lessons from biology and rapidly put them to use.

Unfortunately, as news of various biological problems spreads, the average person doesn't know enough about biology—or about any phase of science—to respond intelligently. Test yourself by looking at the headlines in the Frontispiece. These come from recent issues of newspapers. Do you know enough about the topics cited to react intelligently? If not, are you not at the mercy of the purveyors of special interests? Are you likely to make wrong decisions, lured by the special pleading of one interest or another? Also remember that our ignorance results in a tendency to overreact, to consider all scientific problems with social impact as either all good or all bad, just as there is a tendency on the part of many to shrug off every alarm in the newspapers as the doomcrying of the crazies. In short, you have to know some science to be an enlightened citizen, to choose a safe path between conservation and exploitation or just to keep your body running well and as inexpensively as possible in a world where suddenly there is a chemical, diet, exercise, or way to meditate for every conceivable malfunction and where every day the newspapers announce new findings about the mortal effects of old habits.

There are many ways to gain enough knowledge to give you a running start on disaster. Taking a survey course is not necessarily the best approach because the time is so short and the subject so large that, usually, the format is mostly answers without reasons. This is no way to convince a rational being and probably has little lasting effect. We hope that the compromise between evidence and conclusions that we present in this text will reduce this problem and give you some sense of participation in science.

We are taking little for granted, except your intelligence. You should be able to understand nearly everything in this book just by reading it carefully a few times; a minimal high school science background should be sufficient preparation. Obviously, beginning at this level would produce an impossibly long book if we tried to cover all of biology at the same depth. So, we have been selective. For the topics we emphasize, we try to give enough detail so that you can see the connections between fundamental ideas and the conclusions presented.

In deciding what to leave out, we were guided by two ideas: what remained had to hang together and make a logical and interesting story, and it had to be of value to you as a person interested both in survival and in the way humanity fits into the world. For this purpose, we singled out three continuing themes. The first shows that life is a matter of physics and chemistry, with nothing necessarily supernatural or mystical about its beginnings or subsequent ramifications. The second theme shows that the basic human is just a very smart animal, subject to all the natural laws governing all other organisms—which does not imply that we must always act like "animals." The third theme is practical; it shows what humanity is doing to itself and to the world.

We sincerely hope that you will find this story interesting and useful and that it may help you live a little better.

Using the Book

To learn science you have to learn a new vocabulary and learn to use this vocabulary precisely. To help with this, important words in the text are printed in bold type. If a word is not defined in the text at the place of its first use, it can be found in the Index-Glossary, where the place of the text definition is cross-referenced by an italicised page number. It is also a very good idea to look up scientific terms in a good dictionary or scientific dictionary.[1] These give the origin of the term, usually a Greek or Latin word often called a "root." Learning these roots can help you remember the meaning of scientific words. In addition, you will soon begin to recognize the roots of new words, which will often give a hint as to their meaning.

Chapters end more as they might in a novel rather than as in a typical textbook; that is, they do not have summaries, questions, or references. This is not because such material is unimportant but because we wish to preserve as much as possible the flow of the argument and promote continuous reading from topic to topic. All the missing material is available in the guide that comes with the text. You should turn to the guide to consolidate your knowledge.

[1]Among the excellent biological dictionaries in print are the following: Abercrombie, M., C. J. Hickman, and M. L. Johnson: *A Dictionary of Biology*. Penguin, Baltimore, Md., 1973 (paperback); Lapedes, D. N. (ed.): *McGraw-Hill Dictionary of the Life Sciences*. McGraw-Hill, New York, 1976 (illustrated); Gray, P.: *The Dictionary of the Biological Sciences*, 2d ed. Van Nostrand-Reinhold, New York, 1970. Gray, P.: *Student Dictionary of Biology*. Van Nostrand-Reinhold, New York, 1972.

The following are books devoted to the root words of scientific terminology: Ayers, D. M.: *Bioscientific Terminology, Words from Latin and Greek Stems*. University of Arizona Press, Tucson, Az. 1972 (paperback); Jaeger, E. C.: *A Source-book of Biological Names and Terms*. 3rd ed. C. C. Thomas, Springfield, Ill., 1955.

CONTENTS

Foundations

Chapter 1
Science and Biology 1

Chapter 2
Biology 15

Chapter 3
Life and the Cell 27

Chemistry and the Beginnings of Life

Chapter 4
The Chemistry of Life 38

Chapter 5
Beginnings 57

Chapter 6
The First Ecological Crisis: Depletion of Initial Resources 77

Chapter 7
Oxygen and Life 96

Genetic Biology

Chapter 8
Continuity of Life Through Time 110

Chapter 9
Gene Expression in Simple Cells 129

Chapter 10
Cellular Reproduction: Duplication of Self 142

Chapter 11
Sexual Reproduction: The Cellular Mechanisms 151

Chapter 12
The Natural History of Genes: Transmission Genetics 163

Multicellular Life and Evolution

Chapter 13
Multicellular Organisms 179

Chapter 14
Organization of Higher Plants 210

Chapter 15
The Variety of Life 229

Chapter 16
The Great Explanation 284

ix

Human Biology

Chapter 17
Human Evolution   308

Chapter 18
Human Physiology: Regulation   333

Chapter 19
Human Physiology: Maintenance   367

Chapter 20
The Brain   411

Chapter 21
Human Reproduction   447

Chapter 22
Human Development and Fetal Health   471

Chapter 23
Problems in Human Life: Disease and Aging   489

Humanity and the Environment

Chapter 24
The Environment: Organization of the Ecosphere   519

Chapter 25
Populations   561

Chapter 26
Pollution   583

Appendices

Appendix 1
Numbers and Measurement   627

Appendix 2
Time Scale of Earth Evolution   630

Appendix 3
A Classification of Organisms   631

Appendix 4
The Darwin and Wallace Publications   633

Index-Glossary   641

A hundred times every day I remind myself that my inner and outer life depend on the labours of other men, living and dead, and that I must exert myself in order to give in the same measure as I have received and am still receiving.

Albert Einstein
The World As I See It

1 Science and Biology

THE IMPORTANCE OF SCIENCE

We live in remarkable times of our own making. Never before has one form of life had the power to destroy itself together with most other life on earth. Yet, again as never before, we have the ability to create a golden age, one in which humanity might be almost wholly free from the ravages of the natural world. These self-created alternatives are breathtaking; on one hand a long and pleasant life is potentially attainable; on the other hand there are ever-growing prospects of self-destruction.

How did such astoundingly opposite prospects arise? Major contributions to this situation have been made by **scientists,** by **technologists** who make practical applications of the discoveries of science, by the **producers** who market and by the **consumers** who use scientific discoveries and technological developments, often without fully understanding their potential effects. We are all involved, all responsible.

It may not be clear how both our perilous condition and our prospects for a good life come from the same circumstances. One of the major purposes of this book is to illustrate how the work of scientists has brought us face to face with the possibility of world devastation and also so near to the capability of living in harmony with the natural world. Many believe that a choice between these alternatives is still possible, but the way to the obviously desirable goal is far from clear and the time is probably short. Terrible problems face us, such as the nuclear threat and destruction of the resources of the planet through pollution and population overgrowth.

In a more general way the route to the good life is not plain because scientific and technological progress have gone ahead of our collective ability wisely to use science and the products and techniques it makes available. Scientific knowledge has grown so rapidly in the past hundred years that most individual scientists fully comprehend only a tiny fraction of its total span. Moreover, scientific truths, even when obvious to all, often are ignored in the face of practical requirements, political and financial "realities," and entrenched human biases.

What can be done? Should the advance of *all* science be stopped until humanity becomes more responsible? Or would a *selective* halt, designed to terminate bad aspects of science, be desirable, assuming that these are identifiable? Considering how the world works, neither alternative seems possible. Worldwide agreements are generally known by their violations. Even if attained, probably nothing would be gained by a total or selective ban on scientific research. Humanity already has enough scientific rope to hang itself, and, on the other hand, we see many problems facing humanity that can probably be solved only by further scientific and technological advances. Although such problems are usually rooted in human ignorance, at least temporary scientific solutions appear attainable. Thus, further scientific progress may buy enough time for humanity to mature sufficiently to grope its way into the golden age.

We must conclude: like it or not, our time is inescapably an age of science. If we are to survive and, as they tritely say, "build a better world," we must all understand as much as we can of the results and the limitations of science. In the phrase "we must all," *all* is to be emphasized. *Science can no longer be the business of isolated scientists.* Science is too important to let pass as something too difficult for the citizen to understand, something best left to a highly trained elite of scientists and technologists. Science is too important to be dominated by special groups, be they political, industrial, or military. *All* of us must be concerned with the quality of life; we *all* have an enormous stake in scientific progress and the manner of utilization of scientific discoveries. We must act intelligently to influence science and its applications instead of passively observing its progress. This must be done through a widespread, sound general understanding of science resulting in enlightened and effective citizenship. Although we cannot understand all of the details of science, we can learn enough about how scientific work is done and about its implications to be able to evaluate intelligently its impact upon ourselves. Good citizenship today requires intelligent awareness of the significance of science very nearly as much as it requires knowing the humanistic principles that are the cornerstone of society.

There are many ways to an understanding of science sufficient for the purposes of enlightened citizenship. For several reasons biology, the science of life, may be the most directly useful branch of science to study for this purpose: First, the logic and method of biology are like that of any other branch of science. Second, biology deals directly with critical problems affecting our lives. And finally, biology is a useful route to a general appreciation of science because its roots are in the other two fundamental sciences, chemistry and physics. Thus, the study of biology can lead to a useful general impression of the entire scope of the natural sciences together with a detailed view of many scientific problems of immediate significance.

THE DIMENSIONS OF SCIENCE

Science in the broadest sense is organized knowledge about the structure and functions of the universe gained by application of a particular way of thinking called the **scientific method.** Although the word science was coined as late as 1831, by William Whewell, the business of science began far earlier in common sense, the way of thinking that finds straightforward explanations for our experiences. The birth of science came with separation of the natural from the supernatural in the thoughts of our ancestors. They began to explain the world scientifically when events like floods, the succession of night and day, or the flowering of deserts after rain came to be seen as natural, as occurrences explainable in terms of readily observable causes, rather than as the consequences of supernatural action beyond the reach of everyday experience. We inherited this remarkable system of thought called science after, perhaps, a hundred thousand years of human contemplation, so we can best describe the basic principles of science as an organized common sense. Although the instruments and machinery of today's science are awe inspiring, science basically remains a simple art rooted in careful observation, clear thinking, and objective testing—the scientific method.

Science has many branches. Each focuses on some particular aspect of the universe and often has special tools and methods of study. Still, all the branches have their foundations in the scientific method whether they are called physics, chemistry, biology, archaeology, or whatever. The scope of science ranges from subatomic particles to the incomprehensible vastness of the astronomically detectable universe and extends over expanses of time measured in billions of years. Within a tiny fraction of these limits lie the many branches of science concerned with that unique state of matter called life. But even within the relatively restricted domain of the sciences of life, there are, for us, virtually limitless frontiers ranging from the chemistry of the molecules of the living substance all the way to the behavior of populations of organisms.

The word science comes from a Latin term meaning *knowledge.* As the broadness of the root word and our comments on common sense suggest, science as now practiced originated in a wide range of human activities. Many of these are no longer part of science. For centuries much of what we now call science was joined with philosophy in the discipline called natural philosophy, though philosophy has long since gone its independent way. Within the domain of science itself much splitting took place as knowledge grew. The term natural science was once applied to much of what we now call physics, chemistry, geology, and biology. Although this fragmentation of the body of science into a family of sciences was a necessary and practical result of the growth of knowledge, the truth remains that there is only one nature. Obviously then, any single subdivision of science, owing to its specialized way of viewing its piece of the whole, produces an incomplete picture. There is, therefore, a tendency sometimes for the fusion of parts of science to occur, particularly in areas best studied from the point of view of more than one field. Such fusions are often known by hybrid terms such as biophysics, biogeology, or biochemistry.

The historical development of a subject is not always logical. Mistakes are made in the development of even such a logical body of knowledge as science. Thus false or pseudosciences were, and are perhaps still, involved in the growth of science. The best known of these was alchemy (Figure 1-1) a medieval ancestor of chemistry. Alchemy was primarily devoted to irrational and ill-fated attempts to create precious metals from inexpensive sources and to a preposterous search for a universal cure for diseases. Even so, alchemy provided basic knowledge of chemical reactions that was essential to the development of chemistry, and even the great Newton dabbled in transmutation of the elements. When sufficient accurate scientific knowledge accumulated, the erroneous

Figure 1-1 Alchemist's workroom. In this sketch the astrological symbols illustrate the largely irrational basis of alchemy, even though some of the apparatus in view obviously has descendants in the modern chemistry laboratory.

Figure 1-2 **Hoodwinking the public.** Fake cancer diagnostic equipment sometimes fools unsophisticated victims with impressive but functionless dials, switches, and lights. (Helene Brown, American Cancer Society.)

concepts of alchemy were discarded from the accepted body of scientific knowledge. This winnowing and purifying of the body of scientific knowledge is never over. It is, indeed, one of the major characteristics of science and is made possible by the scientific method, which provides a way of testing and verification of ideas as they accumulate.

Pseudoscience still appears in many forms with serious impact. Pseudoscientific, or "quack," medicine at the least wastes its victims' money and at the worst lures the sick away from proper medical care, sometimes costing them their lives. Quack cancer "cures" are particularly tragic forms of this problem, and they are ever present. Less serious are an almost infinite array of food fads, weight control methods, and supposed cures for any number of annoying disorders. And even relatively harmless across-the-counter drugs are advertised in terms coming as close as legally possible to implying phenomenal curative powers. Sometimes practitioners of pseudoscientific medicine are ignorant believers in their own methods. More often they are charlatans who cloak their fakery with a superficial glitter of science (Figure 1-2). Although the public is protected by state and federal agencies that regulate health products and health care delivery, the first line of defense must be an aware and enlightened public. The enormity of the problem is revealed by the estimate of the American Cancer Society that the annual expenditure in the United States for quack cancer cures is about 1 billion dollars.

The Scientific Method

Science is distinguished from other things people do not so much by the objects or processes studied as by the method of study. For example, the behavior of human populations can be studied in many ways, not all of which are part of science. Human populations are studied in nonscientific systems of thought such as history or politics. That such studies now are nonscientific does not necessarily mean that they will always be nonscientific, and even now such subjects tend to merge imperceptibly with related scientific fields such as sociology and anthropology.

What is this scientific way of thought? There is nothing particularly difficult to understand about it. We all use if frequently and probably do so without thinking of it as scientific. The basic process is simply described as *making and testing ideas about how something operates*. You do this, perhaps not with perfect logic, every time you think to yourself, "Here I have something which probably works in the following way. Therefore, if I do *this* then *that* ought to happen." It is, as we have said, common sense. Formally there are three steps to the process,

1. A question arises.
2. A thought model (**hypothesis**) explaining the question is developed.
3. The explanation is tested.

The test may prove the hypothesis wrong, or the outcome of the test may well suggest an improvement in the thought model, and so on through, perhaps, many cycles of models, and tests suggested by models, until a satisfactory explanation is obtained. Here a difficulty arises: What do we mean by the phrase satisfactory explanation?

Satisfactory explanation in science has three extremely important characteristics. It is first of all *predictive*. This represents the "practical" side of science leading to control over future events or permitting their occurrence to be anticipated. It is also this aspect of scientific explanation that is the most easily tested. Thus, one can consider an explanation in science adequate if a test that it suggests turns out as the explanation says it should.

Although a satisfactory explanation may be accurately predictive, it may not be correct. There is a good example having to do with malaria. Long before the Italian scientists Bignami, Grassi, and Bastianelli proved in 1898 that malaria is transmitted by mosquito bite, the preventive was to stay away from swamps at night. The preventive was rather effective because the probability of getting bitten by mosquitoes was highest in swamps at night. However, the explanation, although leading to accurate prediction in one limited instance, was incorrect. The disease was attributed to the bad air of swamps, hence the probable origin of the name of the disease: malaria, bad air.

There are so far few, if any, explanations in science that are totally and eternally correct, but those explanations that are most likely to endure fulfill two further requirements: they are economical, and they are reductive. An explanation is **economical** if the fewest possible statements are used to explain the largest possible number and variety of facts; it is **reductive** if complex matters are explained with simpler concepts. In other words the economical and reductive nature of good scientific explanation adds up to simplicity.

AN EXAMPLE OF METHOD

As will be seen later, these steps of hypothesis *development* and hypothesis *testing* represent a vast oversimplification of the scientific method actually used by scientists to explain the physical world. Nevertheless, it is useful to follow these simplified steps through a model situation. The fictional example we have chosen concerns a question in practical chemistry which must often have occurred to our ancestors. The question is: What makes wood burn? The full answer to this question came to us only within the last century. Let us forget about that for now, however, and look at the problem through simpler eyes.

When fire came to the notice of early humans, almost immediately they must have speculated about it. Fire inflicted pain, they discovered, and it was hot. This might have led to the question: Does the heat of fire produce more fire? This might have led to an hypothesis: Something burns because heat is applied to it by already burning objects. The **experiment** used to test this hypothesis might have been to apply heat to dry wood with a hot coal (Figure 1-3). To complete this sequence of observation, question, hypothesis, and experiment, our ancestor might then have communicated the success of this primitive scientific venture to his or her companions, which might have led to the domestication of fire. Many years later, around the campfire, less primitive persons may have discussed a new question: Once heat is applied, is it necessary to continue to apply more heat in order to keep the wood burning? And thus a new experiment was born.

Later, more inquisitive members of our hypothetical campfire community would have been interested to know that their experimentally verified hypothesis that heat starts fire is only partially true. Heat causes fire only if certain other conditions are satisfied. We

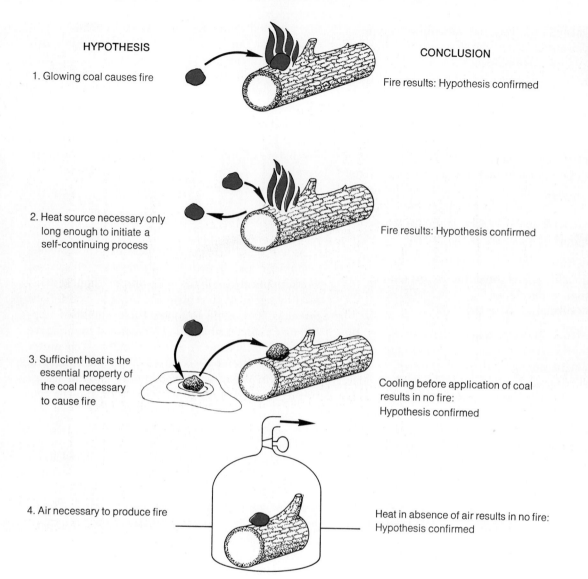

Figure 1-3 Scientific method. A sequence of hypotheses, tests, and conclusions regarding fire.

know this is true because a latter-day scientist asked himself the same question and, having a vacuum pump available, applied heat to a piece of wood in the absence of air, thereby producing a lump of charcoal but no flames. His hypothesis, of course, was that air is required for the burning of wood. This better-equipped scientist found that his hypothesis was confirmed: air is necessary to burn wood. With the advent of still better equipment for separating the components of air, scientists discovered that oxygen is the component in air required for the burning of wood.

Wood turns into a lump of charcoal without burning, when heated in another component of air, nitrogen, just as it did when heated in a vacuum. These hypotheses and experimental verifications were communicated to the scientific community, no longer in campfire conversation, but by means of reports published in scientific journals (Figure 1-4).

Building on this information, other scientists ultimately hypothesized that the burning of wood is a chemical reaction called oxidation. In this process atoms of wood substances combine chemically with

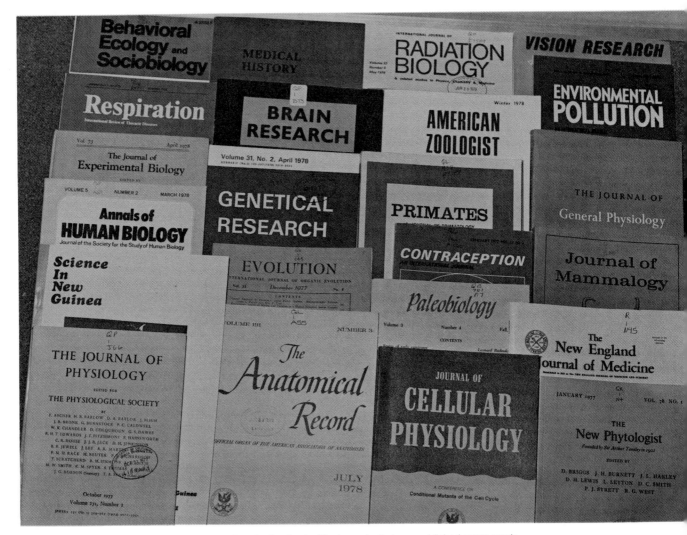

Figure 1-4 Scientific journals. A few of the hundreds of scientific journals that are published every week.

atoms of oxygen. The products of such a reaction are chemicals called oxides, as readily shown by the finding that burning wood always produces carbon dioxide and water, both of which are oxides (Figure 1-5).

Later scientists hypothesized that an initial **energy of activation,** the externally appplied heat that starts the fire, was essential in order to bring the oxygen into an initial reaction with the molecules of the wood. Once started, it was hypothesized that the reaction itself would release enough heat to keep the process going spontaneously, using up oxygen from the air, changing wood to carbon dioxide and water with release of heat, and leaving mineral ash behind. Once confirmed by experiment, other scientists were to hypothesize and confirm that the flame that excited primitive curiosity is a stream of gas molecules and carbon particles heated to incandescence by the energy released from the reaction. Still others, much later, were to point out the similarity between the burning of wood and the metabolic process that supplies energy inside the cells of all living systems, the "fire of life."

To conclude this partially fanciful stream of scientific method we see that the method involves observations that result in *testable hypotheses* about the workings of nature. The example shows how the proc-

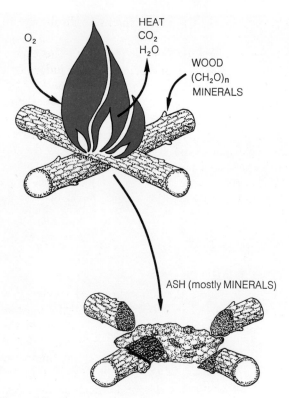

Figure 1-5 Products of combustion.

ess can lead to a long continuing series of interdependent cycles of observations, hypotheses, and tests of hypotheses, each cycle building upon the understanding developed by the previous cycle. We see also that the process not only leads to improving the *detail* of what is known, carrying us from the vague appreciation that heat produces flames to understanding of the chemistry of oxidation, but we see that it leads to *generalization*, making us aware of underlying similarities among phenomena. In this instance application of the scientific method revealed the similarity of burning and respiration, both in the final analysis being oxidative processes.

SCIENTIFIC ATTITUDE

That we have used the teacher's device of oversimplification in explaining the scientific method is evident when we read a scientist's description of how he works. Peter Medawar received the Nobel Prize for work basic to understanding how the body reacts to grafts of foreign tissues and organs. He writes:

All advances of scientific understanding, at every level, begin with a speculative adventure, an imaginative preconception of what might be true—a preconception which always, and necessarily, goes a little way (sometimes a long way) beyond anything which we have logical or factual authority to believe in. It is the invention of a possible world, or of a tiny fraction of that world. The conjecture is then exposed to criticism to find out whether or not that imagined world is anything like the real one. Scientific reasoning is therefore at all levels an interaction between two episodes of thought—a dialogue between two voices, the one imaginative and the other critical; a dialogue, if you like, between the possible and the actual, between proposal and disposal, conjecture and criticism, between what might be true and what is in fact the case.

In this conception of the scientific process, imagination and criticism are integrally combined. Imagination without criticism may burst out into a comic profusion of grandiose and silly notions. Critical reasoning, considered alone, is barren.

P. B. MEDAWAR: 1972.
The Hope of Progress.
Methuen and Co., London, p. 22

VERIFICATION AND CONTROL

The imaginative aspect of the scientific method, although essential to progress, needs regulation. This is done by two procedures that are parts of the third (or testing) step in the scientific process. One of these, **verification**, is done once scientific procedures have produced a new theory about how something works. Verification may simply involve further replications of the original steps of observations and tests by the same or other observers, or it may involve an entirely new set of observations and tests based on the new model. This is most important, and, indeed, an hypothesis, a new model of nature, is not very valuable unless there is some way to test its truth.

The second way to make certain that scientific work is proceeding accurately involves special tests generally called **controls,** or controlled experiments. These are conducted during the observation and testing period. In an ideal experiment only one important condition, or variable, is allowed to change at a time. All others are held constant so that the effect of the one variable can be observed without confusion from the effects of others. To illustrate, let us suppose that our ancestral fire researcher went about the hot coal experiment without regard to the dryness of the test samples of wood. This early experimenter probably would have discovered that sometimes the coal set fire to the test wood and sometimes it simply caused a lot of smoke and steam. False conclusions would have been forthcoming if it was not understood that the experimental situation had at least two important conditions, namely, contact with the igniting coal and the degree of inflammability of the firewood. An adequate control would have been deliberately to do experiments with firewood in various stages of drying.

Consider another possibility. Perhaps some property of the coal besides its heat causes ignition of the firewood. An appropriate control experiment could test this possibility. The coal might be allowed to cool and then tested. This experiment would then be a control with the purpose of showing that whatever properties the coal had besides heat could not cause ignition.

Controls are particularly important in experiments that involve many possible influencing factors that cannot be held constant and might actually be unknown to the experimenter. In medical experimentation upon humans this situation arises frequently. Suppose that it is necessary to test the hypothesis that a particular drug reduces the severity of headaches. It is known that people vary in sensitivity to pain. Often they are highly suggestible, easily convinced that they feel better just by taking an impressive pill. Thus the test of the headache remedy must cope with psychological variables and, of course, besides these there may be any number of physiological variables of an unknown nature. The only practical control possible for unknown physiological variables would be to make the experimental population rather large and select its members to be uniform or to have a known distribution of obvious physiological characteristics that might influence the test. These might include sex, age, weight, and general state of health.

To eliminate the effects of psychological bias, the experimental population might be divided into three groups. One would receive the drug. Members of the second group would be told they are receiving the drug but would actually receive a harmless, inactive substitute, called a **placebo.** A third group would receive no medication at all. Comparisons among these three groups can be expected to eliminate psychological bias as far as the experimental subjects are concerned.

In experiments of this sort in which the experimenter has to deliver some sort of a value judgement, the *experimenter must be considered a part of the experiment* and his own psychological biases may require controls. To eliminate this type of bias a so-called "blind" experimental procedure is often used. In a **blind experiment** another party actually decides which experimental group receives which treatment, and the person gathering the results has no idea as to their cause until the experiment is finished.

CHANCE AND THE INTERPRETATION OF EXPERIMENTS

There is always the possibility that chance may affect the outcome of an experiment, and this must be taken into account. Proper controls will reveal the presence of influences that can be anticipated, but it is obvious that there may be others that cannot be anticipated. The scientist simply lumps these together and calls them chance. Their influence on the interpretation of an experiment is then minimized by performing a sufficient number of repetitions of the experiment and evaluating the outcome of the total experimental series according to the mathematical rules of probability. These rules are the basis of the branch of applied mathematics called statistics, which provides powerful techniques for the analysis of experimental data. Statistics is particularly useful in situations when it is impossible or impractical to perform a large enough number of experimental repetitions to indicate directly the outcome of an experiment.

It must be clearly understood that probability statements apply to averages of many events and not to specific events. This is easily made clear by coin flipping. A coin flipped in an unbiased way (with proper controls) comes down half of the time heads and half of the time tails. This might suggest that the minimal number of tests required to prove this is two. Perform these tests and you will find that it is not certain that the second flip will turn up the face of the coin opposite to the one turned up at the first flip. This must mean that the probability rule does not apply to a specific event; rather, the rule predicts only what will happen on the average. The more times the coin is flipped, the closer the cumulative outcome approaches half heads and half tails. This conclusion is reinforced by brief reflection on what it would mean if the rule actually did apply to specific events. It would mean that the outcome of every flip of the coin after the first would be fixed. The probability statement would have become a law of predestination.

Let us next consider a less simple example of coping with chance in a slightly more realistic experimental situation. Suppose that our ancestral experimenter with fire was aware that a single example might not prove the point. Perhaps one experimental fire started accidentally from causes not connected with the application of the hot coal. The experimenter might then try again and again and, to be truly convincing, attempt to build 100 such fires. Among those 100 trials a few pieces of wood might be found that would not ignite at all, perhaps because the twigs were too green. Another small percentage might have seemed to ignite spontaneously, before being touched with the hot coal, because an unseen spark was blown from one of the other fires. Now, from our vantage point we know why the fire maker did not have complete success in demonstrating that hot coals start fires. That is, we know the *causes* of the variability that was observed. But without this knowledge the experimenter is able to think of the results only in terms of the intervention of chance.

As long as nonburning and spontaneously burning piles remain very small in number, the hypothesis is still clearly acceptable. But if the number of nonburning and spontaneously burning piles was larger, the conclusion from the experiment—that heat starts fire—would be less certain. When the number of results that do not confirm nearly equal the number of tests that do confirm the hypothesis, clearly the hypothesis would have had to be discarded or the obscuring effects sought out and eliminated. Had the experimenter access to statistics, its methods would aid in determining the confidence to be attached to the group of observations. (See Student's Guide)

SCIENTIFIC METHOD VARIES: ATTITUDE DOES NOT

It would be surprising, in the light of the foregoing, if any two scientists pursued exactly the same steps in research. Each scientist's work is the unique product of personal experience, knowledge, imagination, intelligence, of available facts and tools, and of the tendency to prejudice observations and results by personal desires.

Every scientist has a characteristic scientific attitude. It is typified by curiosity about nature, which leads the scientist to bring to bear on the problem under study as much intelligence, imagination, and patient analysis as can be mustered. The good scientist is unshakably honest and continually skeptical of his or her own work and that of other scientists until personally satisfied that all necessary tests of validity have been conducted.

WHERE DO THE IDEAS COME FROM?

Here lies mystery. How can the good scientist see in a set of events or a collection of facts implications about their causes and relationships that may never occur to others? To a large extent this ability is a matter of training and experience and is similar to the ability of a skilled mechanic to diagnose the ailments of a piece of machinery with a few swift observations.

Beyond this matter-of-fact level of scientific thinking, there are other levels more difficult to describe. We tend to associate these ways of thought with the poet, artist, or novelist rather than with the scientific mind. One is the sense of fantasy, that ability to force the mind out of ordinary paths of thought and to forge seemingly unrelatable facts into unexpected models. As Albert Szent-Gyorgyi put it, one must

". . . see what everybody else has seen, and think what nobody else has thought . . ."

Several other great scientists acknowledge the value of fantasy. Einstein wrote:

". . . I come to the conclusion that the gift of fantasy has meant more to me than my talent for absorbing positive knowledge."

Surprisingly, important scientific thoughts have often occurred quite irrationally, sometimes in a dream or sometimes leaping fully formed into the consciousness, often prompted by trivia not obviously related to the problem under study. There is a remarkable example in how the great physiologist Otto

Lowei got the idea for his famous experiment showing that nerves might act by releasing chemicals (see page 343). He describes it this way:

Consciously I never before had dealt with the problem of the transmission of the nervous impulse. It therefore will always remain a mystery to me that I was predestined and enabled to find the mode of solving this problem, considered for decades to be one of the most urgent ones in physiology. And like me you will find it still more mysterious when now I tell you the story of how the discovery happened.

In the night of Easter Saturday, 1921, I awoke, turned on the light, and jotted down a few notes on a tiny slip of paper. Then I fell asleep again. It occurred to me at six o'clock in the morning that during the night I had written down something important, but I was unable to decipher the scrawl. That Sunday was the most desperate day in my whole scientific life. During the next night, however, I awoke again, at three o'clock, and I remembered what it was. This time I did not take any risk; I got up immediately, went to the laboratory, made the . . . [necessary experiment] . . . and at five o'clock the chemical transmission of the nervous impulse was conclusively proved. O. LOWEI: 1953

From the Workshop of Discoveries
University of Kansas Press, Lawrence, Kansas p. 33.

Two things are evident about these unusual forms of scientific discovery. First, they do not occur in an unprepared mind. It is as if the mind that accomplishes some feat of creative imagination has been working at an unconscious level using information provided by earlier, intense conscious effort on the problem at hand. Second, although this statement implies that the more the mind knows about the matter under study the more likely it is that the creative act will occur, it is also true that there is sometimes a curious tendency for the body of scientific knowledge and opinion actually to slow the creative process. A mind long immersed exclusively in one area of science may tend to abide by the traditional habits of thought of that area, and, to the extent that these may prevent new ways of thinking, be shut off from great feats of creative imagination. Very often a stagnant area of science gets moving again by entry of new workers from other fields.

The origin of molecular biology shows this effect of one branch of science on another. By the 1940s genetics, the science of heredity, was well understood at the level of cell and organism, owing to the work of biologists such as Gregor Mendel, T. H. Morgan, and H. J. Muller. Yet the molecular aspects, the *chemistry* of genetics, remained almost totally unknown. The chemical nature of the gene and the chemistry of how the gene reproduced itself and regulated the chemistry of the cell was unknown territory. Then, prompted by the thoughts of Niels Bohr, the great physicist who was founder of the modern concept of atomic structure, another major physicist of that era, Erwin Schrodinger, wrote a small book titled *What Is Life?* In it he said that the major problem in describing life in terms of chemistry and physics is how the gene is able to transmit an unchanging message over vast periods of time even though it is constructed of chemical units which, if of an ordinary sort, ought to be too unstable to permit the uniformity that we see in heredity. In other words, how can the message be more stable than the messenger? Schrodinger put this as a challenge to physicists, asking them to determine if the stability of the gene meant that there must be new laws of nature to be discovered. This challenge, which boiled down to finding out how the genetic substance maintains stability when it is itself constructed out of unstable chemicals, was accepted. The resulting flow of new ideas and techniques of research into biology, brought largely by scientists initially trained in the physical sciences, had the remarkable effect of bringing about virtually complete understanding of genetics at the chemical level within the next 25 years (see Chapter 8).

Pure and Applied Science

Science can be classified in terms of how it affects the quality of human life. Consider as an example the recent development of new varieties of rice that have higher yields than the rice traditionally used in Asia.

This is clearly science with an immediate practical effect. Indeed, such efforts have been called the Green Revolution (but see page 552). However, this kind of science has not produced new general principles of heredity because the work to produce high yield rice or improve other crops involved application of already existing scientific knowledge to *specific practical problems*. These two characteristics of the work, (1) application of known scientific principles rather than the development of new ones and (2) immediately practical effects, mark this work as **applied science**.

Pure science by contrast is represented, in terms of this example, by all of the research that has gone into working out the general principles of heredity. These principles were certainly developed without a primary interest in rice cultivation. It is highly probable, however, that the scientists who developed these general principles were spurred on by the hope that their work would ultimately have beneficial effects in the practical realms of animal and plant breeding and in the area of human heredity. The important point is that this work was done primarily out of the desire of scientists to understand a particular aspect of nature, heredity. *Understanding* was the primary goal, not immediate profit or other benefits of a practical or humanitarian sort. This is the hallmark of pure science.

This terminology is really not very good because the word pure has more than one meaning. As used here, pure refers to the singleness of purpose that is characteristic of the activity that we call pure science. There is one goal, understanding; and that is all. Pure science is not "contaminated" by practical considerations. Unfortunately, this terminology also tends to suggest that applied science, by being contrasted with pure science, is not pure, and, therefore, is of a baser, or lesser, quality. It might be, of course, but the entire range of efforts called applied science does not deserve such a label. Each specific piece of applied science research must be independently appraised and, depending on the practical goals sought, might be truly base, in the worst sense of the word, or might be an expression of the highest qualities of mankind.

How are pure and applied science related? Actually, they are highly interdependent. Applied science would soon suffer a slowing in its rate of progress and possibly even stagnate without continual fresh input from pure science. Pure science provides the new facts and concepts that applied science fashions into advances in industrial technology, health care, and other practical benefits. Applied science reciprocates by providing the technology essential to further advancement of many areas of pure science and often uncovers ideas that initiate work in pure science. One without the other suffers. It is a matter of great importance that an appropriate balance be maintained between the two forms of science.

Everyone understands the value of applied science, but pure science has its critics. They either fail to see why it is necessary at all or they believe that it should be more closely harnessed to applied science. To such critics two things should be said:

First, they should realize that as far as we now know we are the only organisms in the universe capable of asking and obtaining answers to questions about the nature of the universe. We would, therefore, be wasting our human uniqueness if we did not ask questions in the domain of pure science. In the words of Nobelist George Wald:

Surely this is a great part of our dignity as men, that we can know, and that through us matter can know itself; that beginning with protons and electrons, out of the womb of time and the vastness of space, we can begin to understand; that organized as in us, the hydrogen, the carbon, the nitrogen, the oxygen, those 16 to 21 elements, the water, the sunlight—all, having become us, can begin to understand what they are, and how they came to be.

G. WALD: 1964.
The origins of life.
Proceedings of the National Academy of Sciences, 52:595–611

The second point is that applied and pure science must be only loosely coupled, if that, because it is impossible to predict how or when a new scientific discovery may have a practical application. If the effort that now goes into pure science were instead devoted to the immediate problems of applied science, those radical new discoveries that may now have no application, but are the raw material of applied science in the future, will cease to be made. Slowing the progress of applied science must be the result.

Michael Faraday, a founder of the study of electricity, is said to have put this point most vividly. A lady is supposed to have asked him, after one of his famous public lectures, just what all this business about electricity that he had just talked about would amount to.

This was many years before electricity ever powered a motor or lit a lamp, but Faraday unhesitatingly replied, "Madam, what good is a baby?"

The National Science Foundation, the principal federal supporter of pure science in the United States, has documented well the interdependence of pure and applied science by analyzing the complex interactive web of discovery leading to a number of major advances in applied science. The oral contraceptive pill is an important example. The first effective oral contraceptives underwent clinical trials as recently as 1953. Yet the basic knowledge essential to their development can be traced back to 1922 and the beginnings of the work of a group of physiologists interested in the physiology of reproduction—not in such practical matters as birth control. Applied science also made great contributions, even before the final push to develop the contraceptive pill. Thus, the particular drugs finally used in the pill were to a large degree already commercially available when needed, owing to the work of applied scientists who had synthesized them while searching for drugs useful in the treatment of menstrual and reproductive disorders.

SCIENCE AND MORALITY

In beginning this chapter we said that this is an age of science. Science influences our every action from birth to death. It is equally true, as we have seen from the consideration of pure and applied science, that the trail of scientific discovery so interweaves the work of unnumbered scientists that no individual scientist can ever be sure whether his or her work will someday have material effects upon humanity, no matter how far removed from the practical the work may seem to be when it is done. Of course, in addition to scientists, a host of others jointly labor to develop scientific discoveries into forces that affect our lives. These include technologists of various sorts, industrialists, financiers, and members of the regulatory and legislative branches of government. Who, if any, among these, is responsible for the consequences of scientific developments?

Perhaps scientists have no responsibility except to do their work well. Thus some scientists concerned with the development of the nuclear bomb disclaimed moral responsibility for its use. They say, as scientists, they are dedicated only to discovery. What is done with their discoveries must be the moral responsibility of the users rather than of the scientist. The scientist's obligations are met by issuing a warning that application of a new discovery may be dangerous and outlining the danger.

Scientists with this outlook maintain that pathways of discovery must not be blocked by moral considerations. They argue that science is amoral, neither good nor bad, and that it is only the *use* of science that has moral qualities. According to this concept the "middle people," the appliers of scientific discoveries bear the responsibility, the moral burden.

One good thing may be said for the philosophy of the amorality of science. If adhered to we do not risk missing important discoveries or applications of science that might have come from research terminated for reason of its adjudged immorality. Thus, if we had given up on the nuclear bomb, we would have been deprived for many years of other beneficial fruits of nuclear technology.

However, those were not the only considerations. Everyone knows much more was involved. The bomb project is perhaps the most striking example of all time of scientific research done out of political necessity. Development of the bomb was initiated and carried out as a race with the Axis Powers. The bombing of Hiroshima and Nagasaki was considered necessary by many wartime leaders in order to avoid even greater bloodshed they thought would attend an allied invasion of Japan. Reflection upon the proliferation of nuclear weapons, even after such an ominous example of short- and long-term destructive capability, gives us reason to wonder if civilization is mature enough to possess such a powerful instrument as science.

Our hope is, of course, that people everywhere will in sufficient time learn that it is to their mutual benefit to develop ways to regulate such terrible exploitations of science. Fortunately, at less immediately terrifying levels, moral responsibility for science is more readily assignable and technological advances are more readily controllable for the public good. A recent display of scientific morality occurred when it seemed likely that certain advances in molecular biology made possible creation in the laboratory of new, possibly harmful organisms. Some of the principal scientists involved met and formulated guidelines designed to prevent the creation of such material without adequate safeguards. It now appears that

Figure 1-6 The importance of thorough drug testing. A child, whose mother took thalidomide during a critical phase of pregnancy, adapts to the results. (Keystone Press.)

their recommendations generally will be respected, although debate continues among scientists and legislators.

This evidence of moral responsibility in the scientific community is heartening, but much is left to be done. Undoubtedly one reason this particular problem seems to be proceeding towards a solution is that all of the individuals initially involved were scientists. However, things are now becoming much more complicated since this discovery has been shown to have financial significance and is entering the hands of technologists.

The birth defect tragedy associated with the tranquilizer drug thalidomide particularly well illustrates this technological aspect of the problem of scientific morality because it involved, in addition to the scientists who developed and tested this drug, the manufacturers, retailers, physicians, and governmental regulatory agencies that were in one way or another responsible for its use.

Thalidomide was developed and distributed internationally by a German pharmaceutical company. It evidently underwent insufficient clinical tests because, after its use became widespread, deleterious effects began to be reported. The manufacturer took little note and neither did regulatory agencies in countries where thalidomide was in use. The drug continued to be prescribed, and it was only due to the brilliance of an Australian physician that its use by pregnant women was demonstrated to be the cause of an upsurge in the incidence of a rare and terrible congenital malformation, phycomelia. This is a condition in which the child is born with one or more limbs present only as flipperlike stumps (Figure 1-6). Even when confronted with strong scientific evidence of this effect of thalidomide, the manufacturer did not voluntarily withdraw it from the market. A long series of legal battles was necessary before justice was done, after a fashion, for the malformed infants and before the drug was banned. Unfortunately, the ban is not worldwide, even today.

The thalidomide story had one good result. It awakened us to the fact that drug testing and regulatory procedures were insufficient and promoted many reforms in these areas. But was it not an expensive way to learn a rather obvious lesson? The price is estimated to have been at least one thousand living phycomelic children and an unknown number stillborn.

2 Biology

Claude Bernard, the great nineteenth century French scientist, honored as the founder of experimental medicine, wrote:

"Nature is a unity; the frontiers in nature are erected by mankind."

This was a remarkable statement for the times because biology, the science of life, was then generally considered to be completely different from other types of science. Living things were thought to possess a unique property, sometimes called a vital spirit, essential to life and nowhere else present in nature.

VITALISM AND MECHANISM

In the long history of biology, there have been many theories supporting the view that the living differ qualitatively from the nonliving. These are vitalist theories. The essential belief in vitalism is that the complexity of organisms is associated with properties that are not found in their separate parts. Earlier vitalist theories proposed the existence of a guiding force that determined all the fundamental properties of life. This force was thought to appear complete and functional in the organism at its inception. As knowledge of biology grew, more and more properties of organisms were explained in terms of physics and chemistry. Vitalist theories, consequently, became more restricted in scope and sought accomodation to physicochemical explanations of many aspects of life. Thus, living systems were seen as regulated by two sets of laws. The ordinary laws of nature explained what they could explain, and everything that remained unexplainable was attributed to vitalist phenomena.

Although during the peak of vitalism, the eighteenth and nineteenth centuries, scientists were disposed towards vitalist concepts by the general religious climate, they also often accepted these concepts because of experiments or observations that could not be explained in terms of the science of the time. An illustration is the famous experiment of the German student of embryology, Hans Driesch. He divided a sea urchin embryo at the two-cell stage. Each half-embryo became a well proportioned but half sized, larval sea urchin (Figure 2-1). The experiment raised crucial questions: How did the half-embryo "recognize" that it was a half-embryo? How did the complex process of embryonic development undergo adjustment to produce, not the half-larva that the isolated cell would have become during normal development, but a smaller, otherwise normal larva?

Driesch concluded that the ordinary machinery of development could neither accomplish this recognition nor achieve the rearrangement of developmental events necessary to produce a normally proportioned

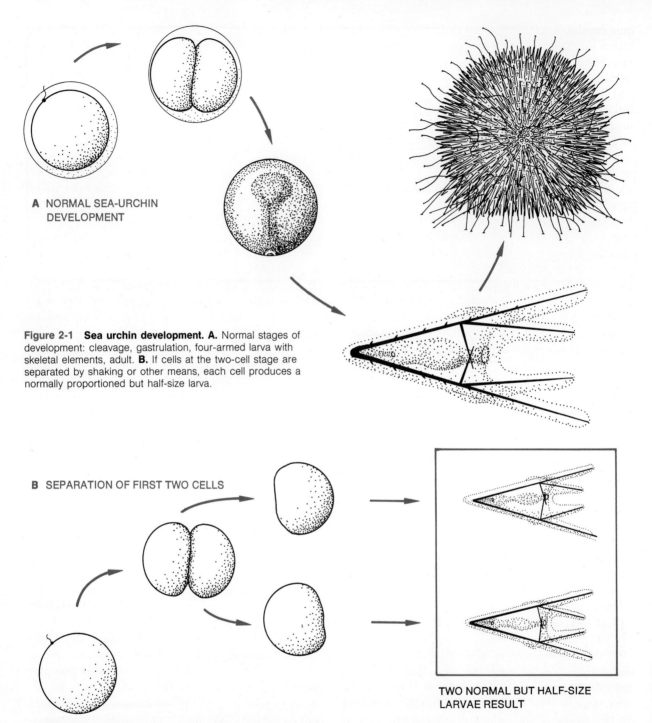

Figure 2-1 Sea urchin development. A. Normal stages of development: cleavage, gastrulation, four-armed larva with skeletal elements, adult. **B.** If cells at the two-cell stage are separated by shaking or other means, each cell produces a normally proportioned but half-size larva.

larva. He decided that each organism must possess a guiding force, which he called an entelechy, directing the functions that are so characteristic of life. Considering this force to be undecipherable by the techniques of science, he gave up experimental science and for the rest of his life was a philosopher.

However, Driesch's experiment was followed by several generations of experimental work on embryonic development that showed he gave up experimentation too soon. Today we know that the total hereditary developmental control system is repeated in the nucleus of every cell of the organism. As Robert Briggs and Robert King have shown, just the nucleus from a cell of even a very old embryo can be transplanted into an egg cell, whose nucleus has been removed, and still support normal embryonic development. This important experiment does not mean that the entire problem raised by Driesch has been solved. For example, it is still not known how the process of nuclear regulation of development is caused to go back to the beginning and start anew. In respect to vitalism, the important point is not that some things about development are still unknown; it is that a vital force overseeing life processes such as embryonic development is no longer a logical necessity. The overseeing mechanism resides in each cell as its nuclear hereditary material and, because molecular biology tells us so very much about how this material functions, few doubt that soon even the most obscure details of development controlling mechanism will be understood.

This history of a critical biological issue is typical in showing how explanation of biological problems in terms of physics and chemistry becomes more all-inclusive with the passage of time. **Mechanistic** theories of life hold that this process will continue, or is potentially capable of continuing, until all aspects of life are explained. The mechanist believes that all nature is explained by a uniform or continuous set of laws. One set explains both the living and nonliving, and there is no necessity for special biological laws other than the weak excuse of incomplete understanding. History is certainly on the side of mechanistic theory as the example of Driesch's experiment shows. Vitalist strongholds fall one by one before the advance of physicochemical knowledge.

However, a completely mechanistic explanation of life may never be achieved for two reasons which do not invalidate the mechanistic philosophy. These reasons are founded in our *technical* incompetence in the face of the great complexity of life. The first reason is that many components of life are so complex that there is little hope of fully understanding them using research techniques of the present or the foreseeable future. However, even though they are too complex to be deciphered by present day science, we realize from the behavior of related components that *are* now understood that none of them appear to involve phenomena immune to the laws of physics and chemistry. For example, the human brain is by far the most complex biological structure. It does things that are plainly beyond present understanding. Yet the nerve cells of the brain are known to function according to well-known physicochemical principles, and even today many simple whole-brain functions are understood in these terms. Are we therefore entitled to assume that the brain functions that are not now understood are somehow different from the understandable components, that they are subject to laws distinct from the ordinary laws of nature?

The second reason that a completely mechanistic explanation of life may never be achieved comes not from our human failings in the analysis of life but from our lack of adequate technical skill. Remember that ultimate proof of the mechanistic theory would be construction of a living organism from ordinary chemicals. Even though we eventually may be able to list completely the chemicals of a cell, be able to produce in the test tube every chemical reaction in which cells participate, and be able to characterize in molecular detail every structure of a cell, this does not mean that vitalism is dead. Presented with this information, the vitalist would perhaps say, "This is an interesting catalog but there seems to be missing the most important ingredient, the vital spark that makes everything work, whatever it is that makes this pile of chemicals alive. To demonstrate complete understanding of life, just put all these things together and make them live."

This challenge is unfair. Nature herself did not create a living cell from simple chemicals in one great leap, and certainly biologists have no way to bring together in a functioning whole all the constituents of a cell at one time and in the proper spatial relations. The impossibility of the task is brought home in this graphic comparison by Albert Lehninger of the chemical skills of cells and biochemists:

A bacterial cell synthesizes simultaneously perhaps 3000 or more different kinds of protein molecules in specific molar ratios to each other. Each of these protein molecules contains a minimum of 100 amino acid units in a chain; most contain many more. Yet at 37°C the bacterial cell requires only a few seconds to complete the synthesis of any single protein molecule. In contrast, the synthesis of a protein by man in the laboratory, a feat which was accomplished for the first time only in 1969, required the work of highly skilled chemists, many expensive reagents, hundreds of separate operations, complex automated equipment, and months of time in preparation and execution.

A. L. LEHNINGER: 1975.
Biochemistry, 2nd ed.
Worth Publishers, New York. p. 11

As we shall see in later chapters, complex living cells probably arose gradually, in mechanistically plausible steps, over *millions* of years. Once established, life depends for its continuation on the reproductive activities of cells. That is, subsequent to the origin of life, the only thing that can make a cell is another cell. But as far as we know this limitation is imposed only by our intellectual and technical inabilities rather than by inadequacy of the mechanistic conception of life.

Finally, there is a practical consideration about the choice between vitalism and mechanism. One may believe that mechanistic explanations will truly and finally fail at some stage in the decoding of the secrets of life and that further answers will have to be sought in vitalism or some other philosophy. Even so, it is essential that biological thinking and research up to that point be strictly mechanistic. This is because vitalism is not subject to scientific explanation. Hence science cannot progress if vitalist theory is imposed upon scientific research. When a scientist does research, then, at least, he or she must think mechanistically.

The Scope of Biology

Perhaps we now agree that life is a phenomenon of the physicochemical world of ordinary experience. At least we agree that life must be treated as though it is, until overwhelming evidence forces other conclusions. If life were not entirely part of the natural world, it would be easily defined by exclusion—as that "other" material—and we would not have to pay much attention to organizing our study of biology along the rigorous lines expected of a scientific discipline. The facts are otherwise, and we have a lot of work ahead of us.

THE PROBLEMS OF DEFINING LIFE

A common way to begin the study of biology is to define life. Unfortunately we cannot logically do this, as we might define chemistry or physics in the first chapter of a textbook, because the best definition of life is only a list of properties, themselves requiring much study before *they* can be defined. Thus, we might say that living things grow, that they have the *property of growth*. But what does that mean? Does it mean the same as get bigger, or is there more to it? Is it the same as the growth that we see in things that are definitely nonliving, such as a salt crystal (Figure 2-2). Obviously we are getting in rather deeply already and we still have to consider a long list of things life does, like exhibit irritability, carry on metabolism, and evolve. Clearly it will be a while before we can approach a rigorous definition of life and state with precision the characteristics of the living state.

For now, let us be satisfied with our intuitive knowledge of the nature of life, admitting that anyone recognizes the difference between salt crystals and people. This will be sufficient for the present. Starting from this, let us first look at the science of biology itself and learn, from consideration of the ways in which life is studied, more about its nature without the necessity of an initial precise definition.

Figure 2-2 Salt crystals (halite) from the Salton Sea grow in a predictable, regular pattern from brine. (Robert Gill.)

Central Problems in Biology

Every aspect of the study of biology is ultimately addressed to one or both of two principal themes, the *history* of life and the *mechanisms* of the living state.

PROGRESSION OF LIFE THROUGH TIME

Physics and chemistry are not governed by history. That is, to understand most problems in physics it is unnecessary to study the situations on which they are based over great spans of time. When history *is* important in the nonbiological sciences, it is usually found to be a highly logical type of history that can be explained by application of simple rules. In physics, we start the universe with a few fundamental types of particles, and their subsequent history is accounted for through application of physical laws. The evolution of the elements, condensation of matter into stars and planets, and even the future evolution of these bodies are explainable by these rules.

Biology, in contrast, is characterized by the overriding significance of historical development, a stream of events that differs from the history of the physical universe in lacking logical explanation by simple rules. Charles Darwin provided the first significant explanation of biological history. Yet Darwin and his followers teach us only that randomly occuring, inheritable differences in organisms survive into future generations in proportion to the reproductive success that their carriers experience relative to other similar organisms. "Survival of the fittest" is the rule, and the candidates for the survival test originate at random. The consequence of random origin of the raw material of biological evolution is that the details of biological evolution are not explainable, even after the fact, as the logical consequences of a few simple laws.

The laws of evolution apply only to the question of the survivability of a living form once it appears. These laws do not apply to the *variety* of living forms that do appear, except in terms of the survival of their ancestors under the evolutionary test. Thus, at each of the myriad branching points in the path from primitive to modern organisms, there is unpredictability. For example, there is no explaining why there are *five* major kinds of vertebrates—mammals, birds, reptiles, fish, and amphibians—instead of some other number. The only "certainty" is that *organisms multiply and diversify to occupy all possible environments.* However, the forms and functional capacities that they develop are explainable only most generally in terms of rules governing survival and in terms of their actual evolutionary history. This evolutionary path is not logical at the level of our understanding. The importance of history, of knowing the course of evolution, is clear. A particular organism is not a logically deducible consequence of the starting conditions of life. It is most fully understood only when we know its ancestry.

Study of the actual course of evolution is mostly based on **fossils** and on organizing them into evolutionary sequences—lines of descent—according to their structural similarities and ages. The later stages of evolution of modern organisms are well known from such evidence. Earlier phases, which cover a much greater span of time, are almost totally unknown as fossils and remain problems of great interest to biologists. In recent years sound physicochemical reasoning has been applied to the problem of understanding the environment of earth before and at the time of the origin of life. This has made possible the development of a plausible outline of the general details of the origin of life. According to this outline it appears certain that before the advent of life there was a long period of *chemical evolution* on the primitive earth. During this time most of the organic molecules necessary to primitive life were synthesized by natural processes. The earliest living forms are believed to have utilized this accumulated store while their own biochemical synthetic mechanisms developed. The beginnings of life, according to this scheme, are mechanistically satisfying as a logical extension of the evolution of matter and are explainable by the same natural rules.

Beyond this point there is much unknown at the stage when biological structure must have appeared and with it the indeterminancy that is associated with biological evolution. Only enough is known to make it

Figure 2-3 The Mars lander, Viking, tests for life on Mars. The trench from which the biology test sample was taken can be seen at the right. (Jet Propulsion Laboratory.)

clear that life arose out of the nonliving and started on the path to the immense variety of modern organisms by a series of steps that are mechanistically plausible, if not now totally understood.

Biological evolution requires a biological memory. This memory must be capable of storing the pattern of the organism and of causing its faithful replication in the offspring of each organism. As already implied, our understanding of biological memory has become virtually complete in the last few years. This feat of molecular biology complements the efforts of the Darwinian evolutionists by providing a chemical basis for the evolutionary process that they describe. Because of these advances, biologists are able to study the evolutionary relationships of modern organisms in an entirely new way, examining at the chemical level the detailed structure of the biological memory and its immediate chemical products. They find that the evolutionary history of an organism is very nearly as clearly written in the subtle variations of the chemical structure of these molecules as in the fossil record. Thus, the study of the progression of life through time is entering a final phase of study at the chemical level.

From the fact that the origin of chemicals necessary for the origin of life is a logical, inevitable step in the physical evolution of matter, there is a high probability that life exists in other places in the universe. This realization occurs as humanity develops the technology to allow direct exploration of the planets and, with the techniques of radio astronomy, listens over vast interstellar distances for signals from other civilizations. The consequence has been the emergence of a new branch of biology called **exobiology**, concerned with finding and describing life elsewhere in the universe (Figure 2-3).

THE MECHANISMS OF THE LIVING STATE

The second major theme of biology concerns (1) the development and operation of the individual organism and (2) interactions among organisms and with the environment. Biological memory is the link between these themes because its mechanisms are responsible for the construction of the individual. Biologists now understand in broad detail how the hereditary molecules evoke the synthesis of the structural and functional chemicals necessary for development and maintenance of the individual, but many questions persist about the process of development. Primarily these are of the type considered in the discussion of the experiment of Driesch, namely: How are specific components of the hereditary substance activated at the proper time to produce required products in specific parts of the organism? Since the hereditary message detailing the construction of the entire organism exists in all cells of the organism there must be a way to cause only relevant parts of this message to be activated at appropriate times and places in the organism. At the time and place for, let us say, construction of an eye, how are the necessary parts of the message dealing with eyes translated instead of the whole message or perhaps the wrong part? It is inescapable that there must be ways for at least some of the requirements of the organism to influence the activation of the hereditary message. Thus, the understanding of this problem is, indeed, presently one of the most interesting areas of biological research.

The processes that maintain the completed individual organism constitute another major family of biological problems. At the molecular level there is an intricate web of chemical reactions to be controlled and supplied with necessary raw materials and energy. Learning the details of how all of these processes are fitted together within minute cellular confines will provide questions for many future generations of scientists.

Most organisms are multicellular. Therefore, superimposed upon the activities of cells, there must be other levels of biological processes. These integrate the work of various specialized parts of the organism and provide these parts with the requirements of life that their specialization may prevent them from providing for themselves. In complex organisms such as man, we find many specialized systems to serve these functions. For example, the **endocrine system** produces **hormones** that circulate to all parts of the body to regulate processes such as growth or the overall rate with which the body consumes energy. The chemical structures of almost all major hormones are now known, and current research on hormones is focused on how they affect cellular processes.

The **nervous system** is the major remaining frontier in understanding the operation of complex organisms. The human brain is the ultimate in biological complexity. There is sincere doubt that it will ever be completely understood. Biologists do, however, understand rather well how nerve cells function, although even here the ultimate questions about nerve cell function seem to disappear into still unfathomable problems about the functional microstructure of cells. The question of how the cell membranes of nerve cells control ions to produce nerve impulses, the traveling electrical waves essential to nervous system function, reduces to difficult problems at this level. The higher functions of nervous systems are only partially understood. The broad outlines of how the nervous system receives information and how it controls the body in which it resides are known. The mystery lies in those processes known as learning and memory and in such states of the mind as emotions.

The power of the human brain, particularly as expressed in the ability that it has given man to communicate and store knowledge, is an unprecedented force in the living world. So great is this force that biologists now speak of a new process called **social evolution**. We change our environment so rapidly, and so cleverly and rapidly adapt ourselves to the changing environment by means of learned behavior, that biological evolution is being outstripped. Humans conquered fire, learned to make tools and weapons, and began to wear clothing in a split second of time as measured on the clock of biological evolution. Yet these learned behaviors accounted for more success in the evolutionary race than millions of years of accumulated beneficial changes in human heredity.

The continual acceleration of human social evolution is obvious even upon consideration of just one aspect of the rise of human societies, namely the exploitation of energy sources: Several hundred thousand years ago our ancestors learned to use fire in simple ways. It took until 1769 for fire to be put to use in the first practical steam engine of James Watt. However, from then it was only about a hundred years to the beginning of the electrical age. Humanity had barely adapted to this advance when, in 1945, explosion of the Alamogordo nuclear bomb thrust us into the age of nuclear power. Each of these advances gave humanity immense increases in useable energy and thereby greatly facilitated the opportunity of humanity to interfere with the balance of nature.

Appropriately, the goal of one of the most active areas of science today is understanding the fundamental meaning of the well-worn phrase, "balance of nature." We are now faced with the stark results of our interference with nature and are becoming more and more aware that virtually everything that society does to the environment ultimately seems to have some bad effect. Thus, we have embarked upon a crash program to understand the balance of nature. We must determine how our lives are affected by changes in this balance, and, it is hoped, we may learn how to avoid some final, irrevocably fatal tipping of the natural scales.

The interactions among animals, plants, and the environment are so reminiscent of the vital exchanges between the parts of a single organism that some biologists have gone so far as to call the total mass of life on earth a superorganism. Perhaps this idea is overdrawn, but it does reveal an important analogy. Just as an organism can be killed by destruction of some critical part, the biological world exhibits similar vital interdependences. Upon severe damage to some component of the worldwide biological system, such widespread deleterious changes may follow that biological world death might be a rather accurate description of the result.

That branch of biological science concerned with the interactions of populations of organisms among themselves and with the environment is called **ecology**. Since the concerns of ecology are at the level of groups of organisms not always of the same kind, the subject bears the full impact of biological complexity. Consequently, it has only slowly matured into an exact science, and even now ecology is far from possessing the tightly logical structure of, say, the molecular aspects of biology.

In spite of the complexity of the subject, ecologists are able to study the exchanges occurring between the members of some favorably situated groups of organisms. Such groups are called **ecosystems** and are characterized by relative isolation and independence of their surroundings or by having easily measured inputs and outputs of energy and materials. Among the first such studies was one done on an isolated coral reef in the south Pacific Ocean (Figure 2-4). Because relatively little food enters a reef of this type from surrounding waters it must produce most of its own from that unique activity of green plants, **photosynthesis.** The flow of energy and material from this source was then followed in this study by sampling representatives of the various other types of nonphotosynthetic organisms living in the reef. The result was an estimate of the "metabolism" of the reef: determination

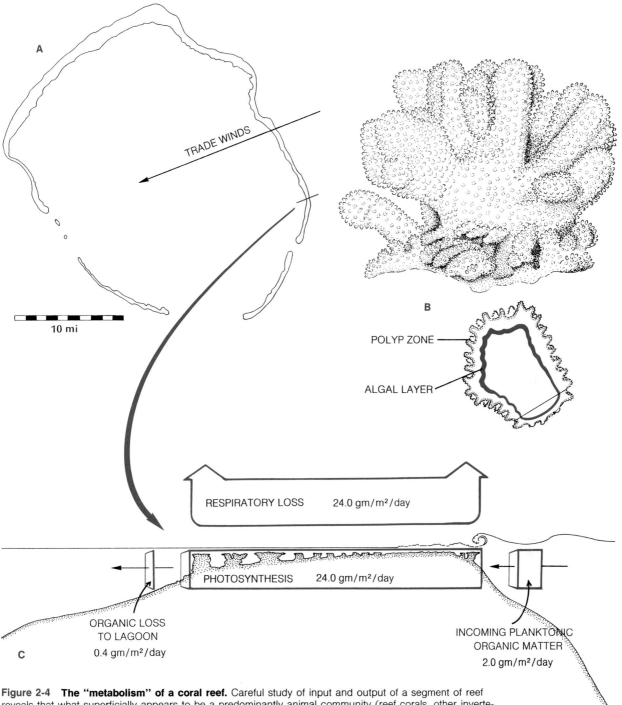

Figure 2-4 The "metabolism" of a coral reef. Careful study of input and output of a segment of reef reveals that what superficially appears to be a predominantly animal community (reef corals, other invertebrates, and fish) could not possibly maintain itself on incoming organic matter and does, like animal communities on land, depend on plant life. In this instance the plant life is algae associated with corals. **A.** Map of Eniwetok Atoll showing study site and the direction of prevailing winds. **B.** A typical reef coral with a cross-section showing the zone of associated algae beneath the living layer of coral polyps. **C.** The analysis of income, photosynthesis, and losses from the section across the reef. (After H. T. Odum and E. P. Odum, (1955) Ecological Monographs **25**:291–320.

BIOLOGY

of how much energy and material entered and left and the routes it followed through the different organisms making up the reef ecosystem. Analysis of ecosystems is valuable for providing models that aid understanding of larger and less easily characterized parts of the world ecosystem.

Ecologists are also able to follow the paths of certain easily traceable substances through very large parts of the world ecosystem and thus learn about worldwide energy and material pathways, even though the entire system is beyond characterization. The technique is analogous to the way a scientist might study the biochemistry of an intact animal by feeding it radioactive molecules and then tracking these through the various organs. The insecticide DDT was one of the first substances to be traced through the world ecosystem. By the time tracking studies on DDT began, it was in use in all inhabited parts of the world to control insect pests affecting crops, livestock, and humans, It was by far the most effective insecticide ever developed. Unfortunately its use had to be drastically curtailed because the tracer studies revealed that DDT had spread far beyond its insect targets. All manner of organisms, including humans, on all continents and in the sea were contaminated.

The way DDT accumulates in organisms graphically illustrates the ultimate interdependence of all organisms. DDT is chemically stable and is highly soluble in the body fat of animals. Whenever it enters a **food chain** (the flow of food from photosynthesizers through plant eaters to animal eaters) it tends to become concentrated at each step in the chain (Figure 2-5). This is because a plant eater accumulates a certain amount of DDT in its fat from a lifetime of eating plant material that has been sprayed with DDT. When that plant eater is eaten by another animal, much of its entire accumulated dose of DDT is retained by the predator. This is because DDT is broken down only very slowly and is retained in body fat of the predator. Therefore, the predator's lifetime accumulated dose of DDT would rise to much higher levels than those in its prey since it has, in effect, harvested the DDT accumulated by many plant eaters. Another stage of predation would raise the concentration even higher.

Some results of the spread of DDT by such means through the environment were entirely unexpected. These sometimes included outbreaks of the very insects DDT was intended to eradicate. The cause was high mortality of insect eating birds due to accumula-

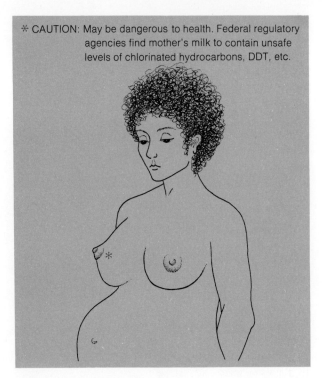

Figure 2-5 As this environmentalist poster shows, humans are directly involved in accumulation of toxic substances in many food chains. The milk, eggs, and meat that a nursing mother eats may contain toxic substances concentrated from the diet of domestic animals, such as organic insecticides sprayed on fodder. These substances are particularly lipid soluble and so are concentrated in the mother's milk.

tion of DDT from insects on which they had fed. The lesson of such occurrences is that the web of interaction among organisms is so intricate and far reaching that almost any interference with nature can be expected to produce many unpredictable and often deleterious side effects. In the years to come environmental biologists will have their hands full providing the basic scientific information necessary to allow some prediction of the effects of interfering with the environment, mapping the effects of damage already done, and suggesting remedies.

The preceding paragraphs illustrate the central problems and show the current directions of research in biology. Although the human implications are obvious from some of the examples, we still have not truly shown our great dependence upon the biological sciences. In the following section this dependence will be illustrated by questions of great human concern, which are at least to some extent answerable through the biological sciences.

BIOLOGY AND HUMAN AFFAIRS

Our species, despite hard won abilities, is not free of the inconveniences of being biological. We must be born in the usual way, and similarly we must die. In the time between we exercise our wits to avoid starvation and fatal encounters with any number of viruses, bacteria, parasitic worms, or mechanical accidents. In the time remaining we reproduce, go mad, and interfere with our metabolism by drinking alcohol, smoking tobacco, taking drugs with various effects, and breathing bad air. Clearly we need all the help we can get.

Biological science comes to our aid in ways classified in four principal categories: **medicine, agriculture, human ecology,** and **human biology.** These subjects overlap, but each has a distinctive central theme. Medicine is a practical science dealing with all aspects of the biology of the human that are related to health. Agriculture, also a practical science, is concerned with optimal production of plants and animals used as human food, clothing, and shelter. Human ecology treats specifically the subject of the environmental interactions of humanity and has inputs to practical subjects such as medicine and environmental engineering. Human biology deals with nonmedical aspects of the biological nature of humanity. A list of the major kinds of problems that these subjects treat follows, along with examples of specific questions of current interest. Many of these questions will be discussed in later chapters. As your study of biology progresses, you should return to these questions from time to time to see if your growing knowledge has changed your attitudes towards them.

Human Origins and Diversity
1. How and from what ancestral forms did modern humans originate?
2. Are there physical and mental differences between the races of humans?
3. Are certain races better adapted to certain environments?
4. Are the races amalgamating or growing more different?

Human Reproduction and Development
1. Is it possible to insure the birth of an infant of one sex or the other?
2. Why is it dangerous for a pregnant woman to take certain drugs or to have a disease such as German measles?
3. What factors initiate childbirth?
4. Do female children mature faster than males?
5. Is parthenogenetic (single parent) human development possible?

Human Mentality and the Nervous System
1. How do we remember?
2. What factors, genetic and otherwise, influence intelligence?
3. What is sleep and why is it necessary?
4. What is pain and how can it be controlled?
5. Is there a genetic basis for criminal behavior?

Health
1. What is a good diet?
2. Are "organically" grown foods significantly different from those grown by ordinary methods?
3. Do we need extra vitamins?
4. How are we to be protected from unsafe food and dangerous drugs?

Medicine
1. How do viruses and bacteria cause illness?
2. Is cancer caused by viruses?
3. How do antibiotics act, and why can they be dangerous when misused?
4. What can be done to reduce the incidence of inherited disease?
5. Why is it difficult to transplant an organ from one person to another?
6. How are we made immune to some diseases by inoculation?
7. Is medical experimentation on humans necessary?
8. Is vivisection in the study of disease ethical?
9. How does nuclear radiation affect life?
10. What are the causes of aging?

Human Populations
1. Is it necessary to limit growth of human populations?
2. What factors have regulated human population growth in the past?
3. Is it practical to move excess human populations to other planets or to space stations?

Human Effects on the Environment
1. What are the risks and advantages associated with the use of nuclear power? Are nuclear plants more dangerous than coal-fueled power plants?
2. Are new agricultural practices such as extensive fertilization and widespread use of new plant varieties likely to have only good effects?
3. Is addition of fluoride to a city water supply mass poisoning?
4. Is it important to protect economically unimportant species from extinction?

Do We Have the Answers?
It would be gratifying to be able to say that, after the reader has acquired a biological background from the subsequent chapters, questions of this type will have been answered. Unfortunately, many of these questions are not yet answerable. Others may be only partially answered. Nevertheless, with the information made available here, you will be able to follow subsequent discussions of these questions at a useful level of awareness. When you make known your opinions on these and similar questions, in voting or otherwise, your decisions will be fortified by some knowledge of the biology underlying the fundamental issues.

SCIENTIFIC QUESTIONS WITHOUT SCIENTIFIC ANSWERS

It is important to realize that there are not always scientific answers to questions that may seem to be about scientific matters. There are several reasons for this.

Some questions involving science have moral aspects that must be resolved by nonscientific means. Even so, the scientific aspect of such questions is important and needs to be carefully worked out to provide clear alternatives and an accurate evaluation of the consequences of the alternatives so that the ultimate moral decision is not clouded by false information. The desirability of medical experimentation on humans is a much debated example of this type of question—one with both scientific and moral elements. The moral aspect lies in the question: Under what conditions is it proper that the few should suffer for the many? Before this can be answered, it is necessary to know certain things that can be determined scientifically. These include (1) alternatives to human experimentation that might provide the necessary information, (2) the risks to the experimental subjects and the degree of discomfort they may suffer, and (3) the importance of the data that human experimentation provides. With this information in hand the moral questions are at least seen in clear perspective.

Another category of scientific question that cannot be answered scientifically involves problems about which it is simply impractical to obtain a scientifically valid answer. These include medical questions concerning the minimum safe dose of some environmental pollutant. Usually such substances have obvious effects at the relatively high concentrations used in laboratory toxicity tests, but it is often impossible to extrapolate from the effects of high concentrations down to the effects that might be expected at the concentrations found in the environment. Since so few test animals would show a measurable effect at very low concentrations of the pollutant, unrealistically large numbers of test animals would have to be used to obtain reliable results. The effects of low level atomic radiation, such as might be expected in the environment if nuclear technology continues to increase, is an example. One authority calculates that to determine the effects of radiation at such low levels would require the use of about *eight billion* experimental mice, clearly an impossibility. There are also other difficulties that make reliable answers to such a question difficult to obtain. For example, in this instance there is uncertainty as to the degree of similarity of human and mouse responses to radiation.

3 Life and the Cell

How does one begin the study of a subject as immense as biology? By now you know about the importance of biology and something of how it is related to other branches of science. This, together with practical knowledge unavoidably accumulated out of your own experience of the subject, provides a clear impression of the job we are about to undertake. Building upon this and appreciating that biology today is reasonably well understood at all levels of complexity, the best approach probably is to proceed from first principles. We begin with the fundamental question: What is life?

A question as general as this has many answers and many routes to answers. Our approach will be to first sketch a working definition of life. Then we will fill in this sketch by describing some of the specializations in living systems in terms of how they may have developed during the origin and early evolution of life. Very early in this phase of our study, in Chapter 4, we shall find it necessary to take up briefly the chemistry of life. Finally we will shift our focus to the biology of our own species.

THE FUNCTIONAL CHARACTERISTICS OF LIFE

Life has many manifestations of form and habit. It is estimated that about ten million living species of animals, plants, and microorganisms now exist on earth. Therefore, we must be certain that the characteristics we pick to define life are truly representative of all life and not just of particular forms of life. There are six fundamental characteristics common to all organisms (Table 3-1 and Figure 3-1). Let us consider them briefly.

Living Organisms are Dynamic.

Living organisms are often called **open systems.** This means that they are not walled-off, or isolated,

Table 3-1 **Basic Properties of Life**

Living organisms	
1. are dynamic	They interact with the environment by exchange of matter and energy.
2. are self maintaining	They replace and repair their structure and obtain a steady supply of energy.
3. grow	Biological growth involves production of new living substance from dissimilar raw materials.
4. reproduce their kind	Life on earth no longer appears from inanimate sources.
5. evolve	Interaction of variant offspring with the environment improves the chances of survival of living systems through long spans of time.
6. are excitable	They make adaptive, short-term responses to environmental changes.

SIX PROPERTIES OF LIFE

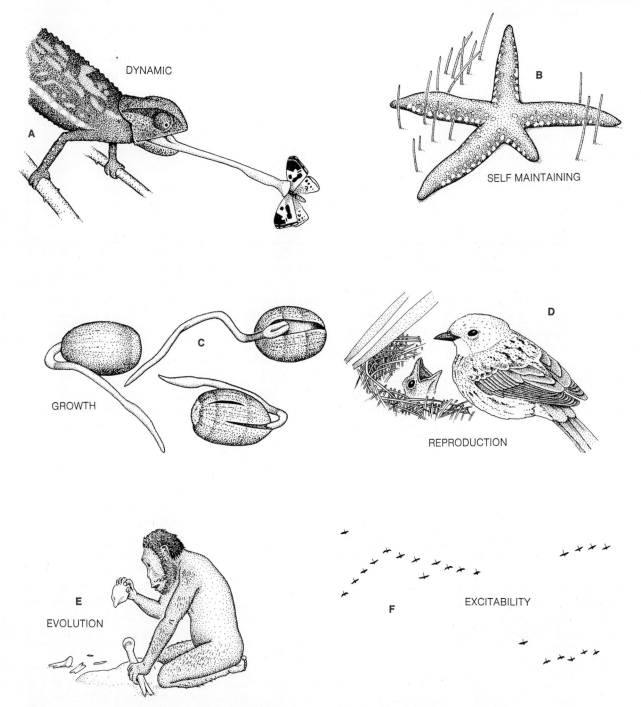

Figure 3-1 The characteristics of life as illustrated by the activities of higher organisms. A. A chameleon illustrates the dynamic nature of life as it maintains itself by taking matter and energy from its surroundings. **B.** A starfish regenerates an arm, demonstrating the self-maintaining nature of life. **C.** Seed development illustrates growth. **D.** A bird attends to the reproduction of its kind. **E.** The tendency to evolve is evoked by a sketch of a human ancestor. **F.** Excitability of living organisms is illustrated by the migratory flight of birds, which is triggered by the detection of the progression of the seasons.

from the surrounding environment. They must interact with it. Organisms obtain from the environment (1) *matter*, those chemicals necessary for sustenance, growth, and reproduction, and (2) *energy* to operate vital processes, this being either chemical energy obtained from certain types of matter or light energy from the sun. To the environment organisms return matter and energy as waste products and heat, or in the form of their own substance at death. This interaction between organisms and the environment is called dynamic because the exchanges that occur are continuously varying, influenced by the requirements of the organism. They do not take place passively. The exchanges vary, with direction (in or out of the organism) and rate, changing from instant to instant, as dictated by the requirements of the organism. Thus, a person running needs more oxygen and must get rid of more carbon dioxide than when sitting.

Living Organisms Are Self-maintaining.

All matter and energy harvested by the organism is used in a complex network of chemical processes. These convert energy into utilizable forms and synthesize molecules necessary to maintain the structure of the living system itself. Most organisms replace virtually their entire structure, molecule by molecule, many times over during their life span. Some have the ability to replace major parts by a process called regeneration, and a few are even able to restore themselves completely from just a few cells.

Living Organisms Grow.

Living things increase in size in a way that is different from the growth of nonliving systems. A nonliving crystal grows by accretion or accumulation of identical parts. Thus a table salt crystal (Figure 2-2) grows by accumulating more and more molecules of sodium chloride, nothing else. In contrast, not only do living things increase in size as they grow but they also increase in complexity, producing more and different kinds of parts. These are all constructed from materials that are different from the final parts. The organism takes in molecules of various types. These are broken down and formed anew into different molecules characteristic of the organism. We eat the protein of cows, break it down into its components, and reform these into human proteins.

At the cellular level growth of all living things starts from a single, simple source, often one cell. Many organisms then grow and become more complex through cell division and differentiation, resulting in the formation of a variety of different tissues. Muscle tissue, for example, is easily observed to be quite different from skin, and both are vastly different from tissues such as those found in the liver or kidneys.

Rates at which living systems grow are sometimes phenomenal. A blue whale at birth, after a gestation period of 2 years, weighs 8 tons. It grows at a rate of 240 pounds per day for the next several months! At the other end of the size spectrum almost any bacterium, if allowed to multiply at its maximum rate, would result in bacteria overgrowing the earth in a few weeks.

Living Organisms Reproduce their Own Kind.

Although it is certain that life had to arise from nonliving chemicals at least once in the remote past, under present conditions on earth this does not happen. The demonstration that all life now comes from other living organisms was one of the landmarks of biology (see Note 3-1). Not only does life on earth now come only from life, but each organism gives rise to offspring highly similar to itself. Reproduction may be as simple as the division of a cell, as in bacteria or single-celled plants or animals, or it may be extremely complex, with the offspring formed by budding or embryo formation. Commonly reproduction is sexual, with two organisms of different sex participating in the production of offspring.

Living Organisms Evolve.

Although each organism gives rise to offspring that are very similar to the parental type, there are inherited variations, often of an almost imperceptible nature. Over long spans of time these variations accumulate or disappear in populations of organisms in accordance with their effect on survival. The result is that each kind of organism tends to undergo change through time, to *evolve*. This process makes possible the survival of organisms in the face of changing environments, as has frequently occurred in the history of the earth. Although it is proper to speak of the evolution of nonliving systems such as the earth or a galaxy, the term as applied to life implies more than the predictable unfolding that is the evolution of a nonliving system. In effect as the living system evolves, it continually presents a variety of alternative types of offspring to environmental test. Some may survive,

Note 3-1 *Life from life: disproof of spontaneous generation*

The concept of spontaneous generation, or the origin of complex life from nondescript, nonliving materials, was widely accepted into the sixteenth century, by scientists and public alike. Their evidence that spontaneous generation occurred we now consider laughable, but it was not easily dispelled in those days because the life cycles of many organisms were unknown and the concepts of controlled experimentation were not highly developed. Thus, "formulae" for the production of organisms, such as one calling for putting some grain and old rags in a warm corner to produce mice, were considered validated when such a concoction did, indeed, yield mice after a week or two.

Disproof of the theory of spontaneous generation was accomplished in two stages. First, the spontaneous generation of complex, macroscopic (visible to the naked eye) forms of life was disproven; later the same was done for microorganisms. The Italian, Francisco Redi (1626–1697) provided the model demonstration for macroscopic life by showing that rotting meat does not turn into maggots. He left samples of rotting meat about, some covered with guaze and some completely exposed. Although the gauze-covered meat rotted as promptly as the exposed meat, no flies appeared, whereas the exposed meat rapidly became covered with maggots. He observed that flies attempted to reach the gauze-covered meat and, when foiled, laid their eggs on the gauze. Thus it was obvious that in this instance there was a necessary connection of the appearance of maggots with the life cycle of the fly. Of course, as in all situations in which a disproof is required, it is virtually impossible to provide a universal disproof. Believers in spontaneous generation persisted long after Redi's experiment, bringing forth example after example. Even today not a few among us believe that the little worms that sometimes appear in vinegar—vinegar eels—appear by spontaneous generation.

However, after experiments like Redi's the scientific believers in spontaneous generation shifted to the microbial level, holding that the growths that appear in nutrient broths represent spontaneous generation. It fell to the great French scientist Louis Pasteur (1822–1895) to undertake this second phase of the disproof, nearly 200 years after Redi. Pasteur's experimental genius is illustrated by the experiment shown in Figure 3-2. He had already shown that nutrient broths never became cloudy with microbial growth if they were sterilized by boiling and immediately sealed. This did not satisfy the believers in spontaneous generation who argued that a vital force necessary to evoke spontaneous generation out of inanimate matter pervades the air. Boiling and sealing the containers of broth would, therefore, be expected to prevent the appearance of life because the vital substance had been damaged or destroyed by heating. Pasteur replied to this criticism by exposing sterile nutrient bottles to air at various elevations on a mountainside and then resealing them. As he expected, those opened at high altitude more frequently remained sterile than those opened lower down, since, he reasoned, the air at higher altitudes is freer of bacteria and contaminated dust particles. But his clincher was the experiment with U-necked flasks. After boiling, these flasks could stand indefinitely with the neck open and yet not turn cloudy. Contaminating matter from the air was trapped by the U, as shown by the fact that growth began immediately if a flask, after standing for some time, was tipped so that some of the broth contacted the lower part of the U. Although they did not give up immediately, the vital force faction was effectively discredited because the open U-flask neck certainly ought to have given an airborn vital force a free entry.

Figure 3-2 Disproving the spontaneous origin of life at the macroscopic and microscopic levels.

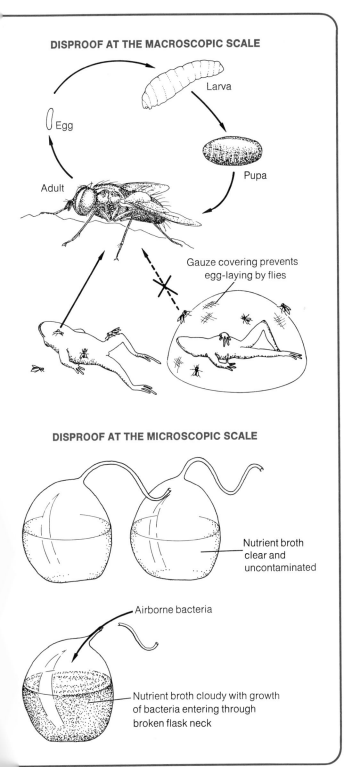

allowing life to continue in the face of change. The exact course of the process, although it obeys natural law, depends on so many factors that it is unpredictable.

Living Organisms Are Excitable.

Living things are excitable by appropriate stimuli. **Excitable** has a special meaning in biology, namely, that organisms respond to a stimulus in ways which are not simply immediate consequences of the stimulus. A billiard ball is not called excitable if it moves when hit by another billiard ball. Water is not excitable because it boils when heated. Although excitability as exhibited by living things varies greatly with the type of organism, it has one uniform characteristic. It is usually adaptive. This means that the response of the organism improves its chances of survival in a changed environment. Thus, a bacterium may undergo changes in the operation of its internal chemical systems in response to a change in the chemical composition of its environment, perhaps enabling it to destroy an antibiotic drug. Or a plant, by slow growth changes, may achieve an arrangement of its leaves so that they receive sunlight optimally. Or migratory birds may depart for summer nesting grounds as a response to changing day length. All are examples of excitability in living systems.

THE FUNDAMENTAL STRUCTURE

Now that we have a list of six functional characteristics of life the question becomes: How are these functions accomplished? One way to approach this question is to ask whether life is always associated with any characteristic structure. If we use only the evidence of the naked eye, we would say it is not. Considering the variety of forms that we see in living things it would seem too much to expect that all organisms would have common structural elements. On the evidence of the naked eye, would you expect bread molds, elephants, and roses to have many common features?

Beginning in the seventeenth century, scientists attacked the great problem of finding universal structural elements in animals and plants. Ultimately they were successful. From their work came the **cell theory**,

Note 3-2 *Microscopy*

The quality of a microscope is generally defined in terms of two characteristics, magnification and resolving power. Magnification refers to how much the image of the object viewed is enlarged, whereas resolving power denotes the distance between two objects that can just be discriminated as separate. Resolving power increases as the wavelength of light used in observing decreases. The resolving power of the best light microscopes enables discrimination of objects 2000 Å (Ångstroms, see Appendix 1) apart. The electron microscope, using a shorter wavelength electron beam, has a resolving power of 5 Å or better, permitting viewing of such cell components as ribosomes and even DNA molecules. The maximum useful magnification of the light microscope is about $2000\times$. Electron microscopes achieve magnifications in excess of $100,000\times$.

The topmost instrument in Figure 3-3 represents the first microscope used for extensive observations of microorganisms, including bacteria. This is the simple microscope (with a single, biconvex lens) of Anton van Leeuwenhoek (1622–1723). His instruments had lenses ground and polished by techniques he never divulged, and some of them attained magnifications in the neighborhood of $300\times$. The tiny lens was held between two flat plates of metal and objects were viewed through holes in the plates. On one side of the plate a screw arrangement permitted objects, often held in tiny glass tubes, to be moved into focus. Leeuwenhoek also never revealed how he made his most critical observations, such as seeing the transparent flagellae of living bacteria. Some believe that he must have discovered dark-field illumination. This technique permits seeing otherwise transparent objects by illumination with light that is markedly off the viewing axis. The result is that the observer sees only light that is deflected into the viewing axis by striking the object viewed, which thereby appears bright against a dark background.

Leeuwenhoek corresponded for years with the Royal Society of England, revealing a remarkable scientific acumen. He undoubtedly saw microbes as small as 9 μ and made many precise measurements using various quaint yardsticks, thus " 'Twas of a length anigh that of a louse's eye.'' He knew the objects seen were alive because they moved erratically in convection currents set up by the warmth of his fingers, whereas inanimate particles moved smoothly with the flow, and because the organisms ceased their peculiar motion after contact with a drop of vinegar.

From such beginnings, the compound microscope has achieved a high degree of sophistication. Special combinations of glass are used in its lenses to reduce chromatic abberation caused by wavelength dependent variation in the diffraction of light by the medium, the lens glass in this instance. But in its essentials the compound microscope consists of a light source to illuminate the specimen, an objective lens which forms a magnified image of the specimen, and an ocular lens which further magnifies the objective image and projects it onto the viewer's retina. A typical student's microscope is shown in the central figure.

The electron microscope, shown at the bottom of the figure, can be viewed as an upside down analog of the light microscope in which the light beam is composed of electrons and the lenses are magnetic or electrostatic devices that direct the flight of the electrons. The light source is an electron gun, an electrically heated filament. Electrons from the filament are accelerated down an evacuated tube by an electrical potential gradient of up to -100 Kv. Focussed by the electrical lenses, the beam of electrons passes through an extremely thinly sliced specimen, is magnified by other lenses and then forms an image on a fluorescent screen, which is viewed through glass ports at the bottom of the evacuated tube. Alternatively, the screen may be replaced by a photographic plate.

The accelerated pace of modern science is illustrated by the fact that several hundred years were needed to perfect light microscopy while the first practical transmission type electron microscope appeared in 1938 and by 1945 resolving powers of better than 10 Å had been obtained, that is in only 7 years.

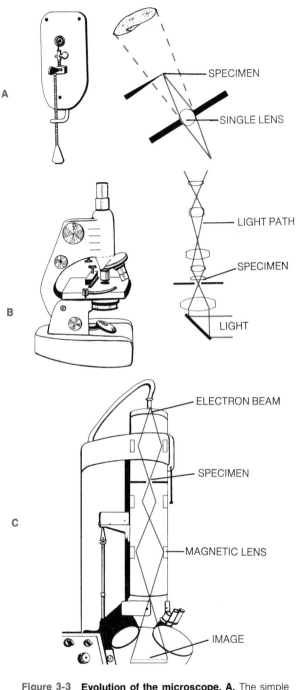

Figure 3-3 **Evolution of the microscope. A.** The simple (single-lens) microscope of Leeuwenhoek. **B.** A modern student's compound microscope. **C.** An electron microscope.

one of the most fundamental generalizations of biology. This theory tells us that all living things are constructed of highly similar microscopic units called cells (except for the special case of the viruses and relatively minor variations among the bacteria). We now know that, in turn, cells are built of components, cell organelles, that are themselves highly similar throughout the living world. Finally, in recent years it has become known that this uniformity of microscopic structure is based on a uniform chemistry of life.

The cell is the basic structural unit of life. Although some early Greek naturalists believed that living things were made of simple parts repeated numerous times, there was no possibility of scientific verification of this idea until the development of the microscope (Note 3-2).

In about 1663, Robert Hooke, an Englishman, examined thin slices of cork with a simple (single lens) microscope that magnified about 100 times. In the slices he saw tiny compartments. Since cork is the dead bark of a Spanish oak, he saw the dead walls of plant cells. With their contents gone, the remaining honeycomb structure reminded him of cells in the architectural sense and thus he named them. Following Hooke there was a period of great activity by the early microscopists who turned their new instruments on every imaginable biological material. The greatest of these early workers was Anton van Leeuwenhoek, a Dutch merchant, not scientifically trained, who pursued microscopy as an all-consuming hobby. He built remarkable fine single-lens microscopes and with them observed for the first time many structures that we now know are single cells (Note 3-2).

Although such observations showed the way, it required another 150 years of accumulated information about the microscopic structure of organisms before the critical generalization was made. More or less jointly, by 1845, two German scientists, M. J. Schleiden and T. Schwann, proposed that all animals and plants were built of cells. Besides identifying cells as the basic structural elements of organisms, they saw them as the basic functional unit of life as well. As they put it, "The cells are organisms, and animals as well as plants are aggregates of these organisms. . . ." Thus, what might be called a modular concept of both biological structure and function was born. According to this concept, the most diverse forms of organisms are built up of basically similar components, cells.

Inside Cells.

It was already evident when the cell theory was formulated that the cell was not the minimal structural component of living systems. Things could be seen inside cells, and certain of these structures seemed to occur in virtually all cells. Then perhaps the cell was not the ultimate, the irreducible unit of structure and function. Perhaps the cell would suffer a fate similar to that experienced later by the atom of physics and undergo further subdivision. In a sense this did happen. Cells were found to contain a characteristic family of **organelles.** As the term implies, organelles are to cells what organs are to organisms. Although cell organelles have been found to carry out functions essential to life, the important thing to remember is that the cell is still the minimal independently living unit. A cell organelle does not survive alone. *The cell is the minimal unit that is potentially capable of performing all life functions,* that is potentially capable of life independently of other living material. This is what Schleiden and Schwann meant when they wrote that cells are organisms.

There were opposing ideas for a while. In the late 1800s biologists thought that cells were made up of a uniform and basic substance in which the qualities of life resided. The name applied to this substance was **protoplasm** and at the time it was a serious candidate for replacing the cell as the fundamental living unit. Today we understand that protoplasm is simply a collective term for a large variety of components of the living parts of cells and thus it has no special theoretical significance.

However, when protoplasm was very much on the minds of scientists it is amusing to note, in illustration of the sometimes crooked path to scientific truth, that great excitement was caused when some gelatinous and amorphous ooze was dredged up from the ocean floor during one of the earliest oceanographic studies. Found in many places on the bottom, this ooze evoked visions of a great, ocean-wide, living, protoplasmic mass, getting along without resorting to compartmenting itself into cells. In his book, *The Depths of the Sea,* the pioneer oceanographer Sir Wyville Thompson wrote about this substance:

"To this organism, if a being can be so called which shows no trace of differentiation of organs, consisting apparently of an amorphous sheet of a protein compound, irritable to a low degree and capable of assimilating food, Professor Huxley has given the name of Bathybius haeckelii. . . . The circumstance which gives its special interest to Bathybius is its enormous extent: whether it be continuous in one vast sheet, or broken up into circumscribed individual particles, it appears to extend over a large part of the bed of the ocean. . . ."

Unfortunately, Huxley, the same Thomas Henry Huxley later to be the great defender of Darwin, had to eat his words this particular time. *Bathybius* was found to be an artifact created by the reaction of sea water and mud with the alcohol used in preserving specimens for study, and the cell theory was secure again.

Cell Organelles.

Today a long list of cell organelles is recognized, and we have a rather clear understanding of the role each plays in the life of the cell. To give you some appreciation of cellular complexity, we shall quickly list the organelles with the promise that we will return to some of them many times in the remainder of the text. A typical cell in a higher organism (Figure 3-4) contains the following:

The cell **membrane** is the functional boundary between cell and environment, although it is only a few molecules in thickness. More than simply a passive barrier or sieve, the cell membrane actively regulates entry and exit of many kinds of molecules into and out of the cell. In plants, a nonliving **cell wall** of cellulose lies immediately outside the cell membrane and provides it with rigid support. **Plasmodesmata** provide channels of communication through the plant cell wall.

Cells are filled with a complex water suspension and solution of many substances. The suspended particles are so small that electrical charges upon them keep them in solution so that we may speak of this watery medium as a **colloidal suspension.** All of this material together, essentially all the cell water and everything dissolved or suspended in it that is not otherwise designated as an organelle, is called **cytoplasm.** It serves as the storehouse and circulatory system of the cell.

Well-defined organelles are suspended in the cytoplasm. The **nucleus** is the most conspicuous of these. It is relatively large among organelles and contains nucleoplasm, similar to cytoplasm. The nuclear boundary is a pore-containing nuclear membrane that

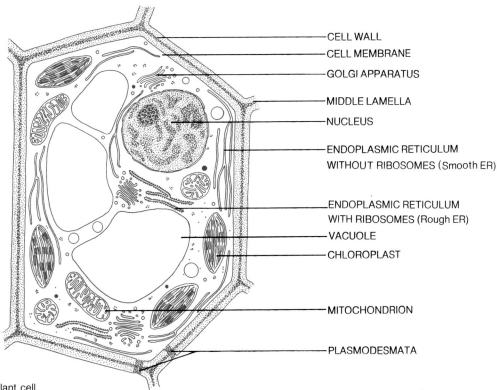

Figure 3-4 A typical plant cell.

assists in regulating interchange between nucleus and cytoplasm. The hereditary apparatus lies within the nucleus. This includes the primary genetic units, the genes, arranged along the length of several chromosomes, and one or more nucleoli. The genetic apparatus regulates virtually all aspects of cell life. It is so important for each cell to have a complete complement of genes that a remarkably complex process of gene replication and precise distribution of genes between daughter cells occurs each time a cell divides. The centriole, which lies just outside the nucleus in animal cells, plays an important role in this division process.

At magnifications attained by the electron microscope, fine canals may be seen wandering through the cytoplasm. These are the **endoplasmic reticulum**. Small bodies called **ribosomes** are attached to the reticulum. Together these structures play a major role in cell chemistry, particularly in the synthesis of proteins. **Mitochondria** are bean shaped structures containing folded surfaces called cristae on which are arranged many of the enzymes and other molecules responsible for providing energy for cellular activities.

A third membrane-rich structure in the cytoplasm is the **Golgi apparatus**. It plays an important role as the site of assembly of many cell products.

Even at this early stage in our study it is evident that cells are membrane-rich. Surfaces are obviously important to cell function. A profusion of surfaces is necessary to allow physical coordination of many of the complex chemical processes that go on in cells. If these processes occurred only in solution in the cytoplasm, undoubtedly the cell would be less efficient. Some membranes serve to isolate materials, others selectively to pass or interrupt the movement of materials, whereas still others serve as frameworks upon which chemical processes are conducted by attached enzymes (biological catalysts).

Vacuoles are storage sites within cells. A common type contains fat. Often, especially in plant cells and in cells with large amounts of stored material, vacuoles occupy most of the cell volume. Lysosomes are special, vacuole-like structures that contain protein destroying enzymes. Normally lysosomal enzymes are used to destroy foreign protein, such as from invading bacteria, but under some conditions they may attack the

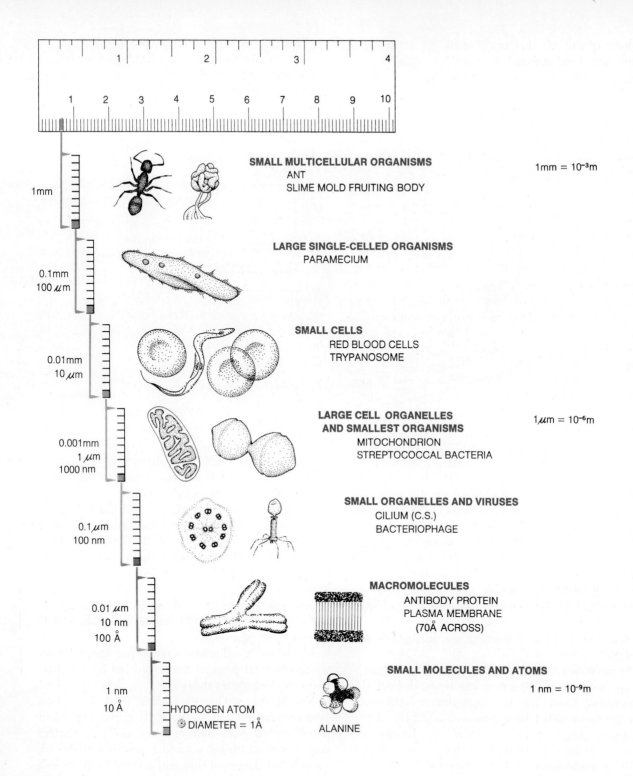

Figure 3-5 Measurement at the microscopic level.

proteins of the cell that contains them. Hence lysosomes are kept isolated in vacuoles. **Plastids** are membrane-rich structures of various types, depending on the principal substance they contain. The color of green plants is due to a pigment, chlorophyll, which is contained in **chloroplasts.** These are among the most important of biological structures because chlorophyll captures light energy for the synthesis of the foodstuffs on which, essentially, all life on this planet depends.

Locomotor organelles are characteristic of some cells. They enable cells either to move about or to move small objects that are in contact with them. Among protozoa, the ameba and its relatives, and in man, the white blood cells, are examples of cells with **pseudopodia**. The properties of the cell surface and underlying cytoplasm of ameboid cells are such that the cell can rapidly throw out from any point on its surface a narrow extension, a foot or pseudopod, into which the rest of the cell flows and thus moves. Other kinds of cells, including some protozoa, unicellular plants, and nearly all kinds of sperm, move by means of flagella. A **flagellum** is a long, hairlike extension of the cell surface with a central core of fine tubules. By a mechanism that is not well understood, the flagellum is able to beat, making a sinuous wave motion that propels the attached cell. Cilia are smaller versions of flagella found in large numbers on the surfaces of some types of cell. An important example is in the human upper respiratory tract where the coordinated beating of the ciliated lining cells helps to remove foreign particles from the lungs.

Motion by means of contractile proteins is common in animal cells. Contractile proteins are long molecules which specific chemical reactions cause to move along each other, thus doing mechnical work, causing movement of whatever is attached. The most highly developed example is found in the muscle cell, but similar contractile proteins are involved in such widespread processes as the constriction of the cell membrane that takes place during animal cell division.

MEASUREMENT AT THE CELLULAR LEVEL

Cell organelles and even cells are so small that to see them light microscopes, magnifying up to about 2000 times, or even electron microscopes, magnifying up to about 100,000 times, must be used. It is not easy to comprehend the smallness of such objects. One way to do so is to imagine how large some everyday object that you can see with your naked eye would appear after such magnification. For example, a housefly magnified 2000 times would have an apparent length of 10 meters (m). At 100,000 times magnification it would seem to be 500 m long.

To facilitate measurement of objects at the cellular and subcellular level we substitute for Leeuwenhoek's louse's eye a downwards extension of the metric system from the familiar millimeter (mm, 10^{-3} m) to the micrometer or micron (μm or μ, 10^{-6} m), nanometer (nm, 10^{-9} m), and finally the Ångstrom (Å, 10^{-10} m). These measures are illustrated by application to cell parts in Figure 3-5 (see also Appendix 1), which shows that they cover a very wide range of object sizes down to the molecular level. The smallest dimensions of cellular components are only large enough to include a few molecules. For example, the cell membrane is so thin that across its width there may be only room for four or five of the protein and lipid molecules that it is known to contain. Yet other organelles of animal and plant cells are many times larger than an entire bacterial cell, which may have a diameter of about 0.5 μ. Actually the smallest structures that we consider among the living are viruses. These may be only 200 Å in diameter, but it must be remembered that they are able to attain this small size only by doing without nearly all of the cellular apparatus essential to a free life. Most of the tiny volume of the virus particle is occupied by the genetic apparatus, a single giant molecule of nucleic acid that diverts the chemistry of the host cell to the task of making more virus. The smallest of free-living cells are many times larger, on the order of 5 μ in diameter. At the other end of the cellular scale, the largest of cells are probably giant nerve cells in squids. These may be as much as 2 mm in diameter and nearly a meter in length in large squids. There are also some very large plant cells, but these have actually a rather small amount of cytoplasm since most of their volume is occupied by a large vacuole. Similarly, the eggs of birds are technically single cells, but most of their volume is occupied by yolk and the amount of cytoplasm present at the time of fertilization is far less.

4 The Chemistry of Life

It is obvious from the last chapter that discussion of the basis of life must center on chemistry. We saw that organisms exchange matter and energy, that they replace and repair their structure, that they grow. These are all chemical processes. Moreover, we saw that discussion of cellular organization immediately involved chemical concepts and terminology, words like enzyme, protein, and chlorophyll. Clearly our path of inquiry requires an immediate introduction to chemistry as it relates to life.

FOUR FUNDAMENTAL CHEMICALS

The central core of the chemistry of life is based on just four types of organic chemicals: carbohydrates, lipids (fats), proteins, and nucleic acids. There are many other types of molecules involved, but understanding the roles of these four is an excellent start towards appreciation of the chemistry of life.

These four centrally important kinds of molecules have many diverse functions. **Carbohydrates** are primary energy providing molecules. **Lipids** are secondary sources of energy and play other important roles, such as contributing to the structure of cell membranes. **Proteins** are of prime importance in the structure of cell membranes and most other cell organelles. As **enzymes,** proteins are the catalysts of innumerable chemical reactions. As **antibodies,** they are essential in resistance to disease. **Nucleic acids** carry information necessary for cell replication and for control of virtually all aspects of cellular function. The genetic code, which is based on the structural arrangements of nucleic acids, is the chemical language in which this information is written.

Much is known about the synthesis and breakdown of these four kinds of chemicals. What follows is a great simplification of the most important aspects of their chemistry. To make the explanation easy to understand, in each case the "building-block" of which the chemical is made will be described, as well as the manner in which these blocks are built up. Then something will be said about the mechanism of their breakdown and function in cellular operations.

ATOMS, MOLECULES, AND CHEMICAL REACTIONS

Before discussing biologically important chemicals specifically, it is necessary to provide a foundation from general chemistry. We start with the **atom,** which everyone in this age of nuclear power knows is the smallest particle of matter that still has definite chemical properties. It was long thought that there were only 92 kinds of atoms in the universe. This number is now increased to about 103 by the discovery of rare, often short-lived radioactive atoms. These radioactive atoms play no part in the normal operation of living systems, so we may disregard them at

Table 4-1 **Elements Found in Living Systems***

I The Big Four			II Essential			III Occasionally Found	
Hydrogen	(H)	9.9	Sodium	(Na)	0.10	Vanadium	(V)
Oxygen	(O)	63.0	Magnesium	(Mg)	0.07	Molybdenum	(Mo)
Carbon	(C)	20.2	Phosphorus	(P)	1.14	Lithium	(Li)
Nitrogen	(N)	2.5	Sulfur	(S)	0.14	Fluorine	(F)
			Chlorine	(Cl)	0.16	Silicon	(Si)
			Potassium	(K)	0.11	Arsenic	(As)
			Calcium	(Ca)	2.50	Bromine	(Br)
			Iron	(Fe)	0.01	Tin	(Sn)
			Manganese	(Mn)		Iodine	(I)
			Cobalt	(Co)		Barium	(Ba)
			Copper	(Cu)			
			Nickel	(Ni)			
			Zinc	(Zn)			

* The elements in column I, the Big Four, are found in all organisms and together represent 96% of the chemicals in them. The 13 elements in column II are found in most organisms, but together constitute less than 5% of the total. Thus, only 17 of the 103 elements are found in nearly all living systems. Those elements in column III are found only occasionally and at concentrations less than 0.001%. Despite their low concentrations (for which they are known as **trace elements**) several are important to life. Following the chemical symbol for each element the per cent by weight in soft tissues is given when that value is at least 0.01%.

Figure 4-1 Various simple combinations of carbon, hydrogen, and oxygen produce molecules with markedly different properties. Space-filling models and formulae are shown.

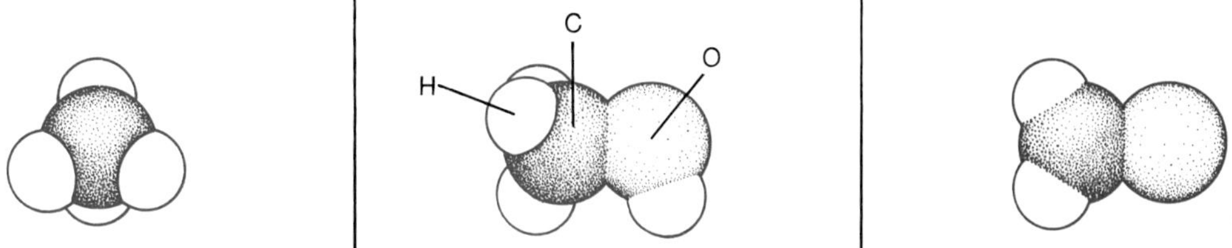

CARBON 1	1	1
HYDROGEN 4	4	2
OXYGEN 0	1	1
METHANE	METHYL ALCOHOL	FORMALDEHYDE
CH_4	CH_3OH	CH_2O

THE CHEMISTRY OF LIFE

present. In fact, most of the original 92 elements do not play any major role in the function of living systems (Table 4-1). Only four **elements** make up the larger part of the chemical units found in living systems. These are carbon (C) hydrogen (H), oxygen (O), and nitrogen (N).

At ordinary temperatures, such as at the surface of the earth, carbon is a solid (charcoal, graphite, or diamond). Hydrogen, oxygen, and nitrogen are colorless gases. Oxygen may combine with hydrogen to produce water (H_2O), or it may form carbon dioxide (CO_2, also a colorless gas) by reacting with carbon. Theoretically, since they contain the most important atoms, water and carbon dioxide, with a little nitrogen added, could serve as the raw material for nearly all the complex structures in a living system. As we shall see, plants use these raw materials very much in this way. When carbon, hydrogen, oxygen, and nitrogen are chemically combined in different ways, they produce the many hundreds of chemicals whose interactions result in what we see as a living system.

Each time they are rearranged, a new chemical results (see Figure 4-1). For example, when carbon and hydrogen are combined in the ratio of one carbon atom to four hydrogen atoms, methane (CH_4) is formed. Methane is a colorless, highly inflammable gas. Sometimes called marsh-gas, methane is an important part of the natural gas used for industrial and domestic heating. It also occurs in the gaseous wastes from the intestines of higher animals. Although no more than a simple combination of carbon and hydrogen, its properties are vastly different from those of either carbon or hydrogen alone. If now an oxygen atom is combined with the methane molecule, still another kind of chemical, methyl alcohol (CH_3OH, wood alcohol), is produced (Figure 4-1). This kind of oxidation is different from "burning" in that the oxygen combines with the chemical without breaking it all the way down to carbon dioxide and water. Methyl alcohol is a clear, inflammable, poisonous liquid. Adding an oxygen atom to methane has changed its properties dramatically. Adding another oxygen atom, again without burning, changes the chemical to formic acid. The loss of two hydrogens and one of the oxygens (that is, the removal of a water molecule) converts formic acid to formaldehyde. Formaldehyde is the chemical used in preserving biological specimens and formic acid is a toxic substance found in a bee sting. Remarkable changes in the properties of molecules accompany changes in the number of oxygens and hydrogens.

Definition of Chemical Units.

If you haven't already done so, you should now take special note of the terminology used by chemists, biologists, and physicists, namely, that the simplest chemical particles are called atoms and that atoms combine to produce **molecules** of chemical compounds. Iron (Fe), for example, is a metal and oxygen is a colorless gas. When two atoms of iron combine with three atoms of oxygen, one molecule of the reddish compound ferric oxide (rust) is produced.

Chemical compounds made up primarily of carbon, hydrogen, and oxygen, are called **organic** compounds. Chemists once thought that organic compounds were found only in living systems, but it is now clear that innumerable organic chemicals can be made and used outside living systems. Nearly all plastics, for example, are organic compounds. Nearly all molecules characteristic of life have been artificially synthesized, except for very large proteins.

The Carbon Backbone of Organic Molecules.

We have not yet mentioned a most important property of the carbon atom, namely, that carbon atoms may link to each other forming long chains. These chains are backbones to which hydrogen, oxygen, and nitrogen atoms are attached in various ways to form complex organic molecules. Although these molecules have highly similar atomic compositions, the varied arrangement of atoms within the molecule produces markedly different chemical properties (see Figure 4-2). To understand precisely why this is so would require a long excursion into the details of how atoms bond together. At the level of our present considerations this knowledge is unnecessary, but we will need to consider it later.

PROTEINS

Proteins are the most common organic molecules found in organisms, where they exist in tremendous variety. Even though they are large and complex molecules, specialized to perform many different tasks, the structure of proteins is easily understood (Table 4-2, Figure 4-3). This is because proteins are assembled in a uniform way from a few kinds of small organic

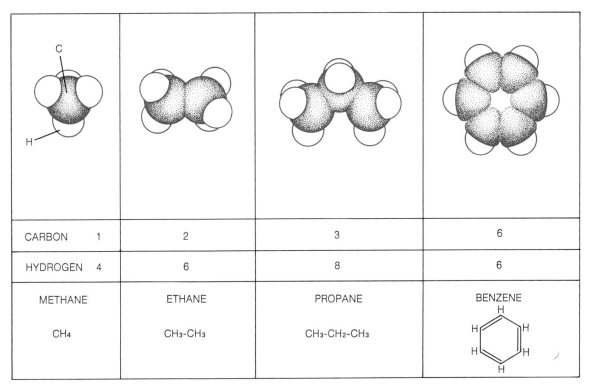

Figure 4-2 The ability of carbon atoms to link together, forming chains and various ring structures, almost infinitely increases the variety of carbon compounds. When a ring structure is drawn, carbon atoms are frequently omitted for simplicity, as for benzene, C_6H_6.

molecules called **amino acids.** The special properties of each kind of protein are *exclusively* determined by the sequence in which its constituent amino acids are attached together in a linear string.

Amino Acids: Protein Building Blocks

About 20 different amino acids are important as protein building blocks. They all have similar structures although they differ from one another in detail. The structure of an amino acid is simply a carbon chain backbone containing an **amino group** and a **carboxyl group.** The amino group and the carboxyl group are always found separated to some extent from one another and form regions of the molecule that have distinctive chemical properties.

To understand protein structure it is necessary to

Table 4-2 Some Typical Proteins

Name	Biological Function	Molecular Weight	Number of Amino Acids
Insulin	Hormone	11,466	104
Myoglobin	Oxygen storage in muscle	17,000	154
Hemoglobin	Oxygen carrier in blood	64,500	586
Catalase	Enzyme	232,000	2109
Myosin	Muscle contraction	468,000	4254

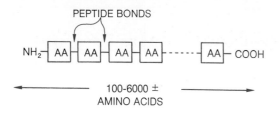

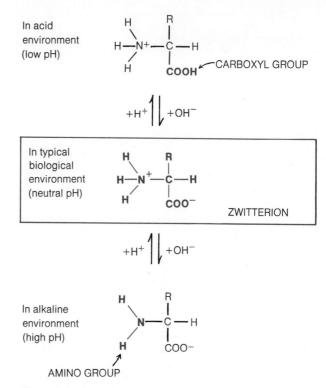

Figure 4-4 An amino acid has both basic and acidic properties.

PROTEIN TYPE	EXAMPLE	LOCATION OR FUNCTION
STRUCTURAL	KERATIN	FINGERNAIL
PROTECTIVE	FIBRIN	BLOOD CLOTTING
TRANSPORT	HEMOGLOBIN	OXYGEN CARRIER IN BLOOD
ENZYME	AMYLASE	CONVERT STARCH TO SUGAR
MOTILITY	MYOSIN	MUSCULAR CONTRACTION
STORAGE	FERRITIN	IRON STORAGE
HORMONE	INSULIN	REGULATE BLOOD SUGAR
POISON	BOTULINUM TOXIN	FOOD POISON

Figure 4-3 Protein functional types.

see how carboxyl and amino groups function. A carboxyl group (Figure 4-4) is that part of an organic molecule which predominantely determines whether or not that molecule behaves as an **acid**. An organic molecule, generally speaking, is more or less acidic depending upon whether it has many or few carboxyl groups. Acidity also depends upon the extent to which the carboxyl groups are able to release hydrogen ions (protons), which are hydrogen atoms that have left their electron behind when separated from the parent molecule.

It is the hydrogen ion that makes acidic substances taste sour. Vinegar, which contains acetic acid, tastes very sour because it readily releases hydrogen ions when dissolved in water. On the other hand, an amino acid such as glycine does not readily release hydrogen ions and so does not taste sour.

The amino group consists of a nitrogen and two hydrogen atoms linked to a backbone carbon atom in much the same way that the carboxyl group consists of a hydrogen and two oxygen atoms linked to a backbone carbon atom. The amino group behaves oppositely to the carboxyl group. Whereas the carboxyl group gives up protons, the amino group accepts them and thus is said to have **basic** properties. Molecules like amino acids with both acidic and basic groups are called **zwitterions**.

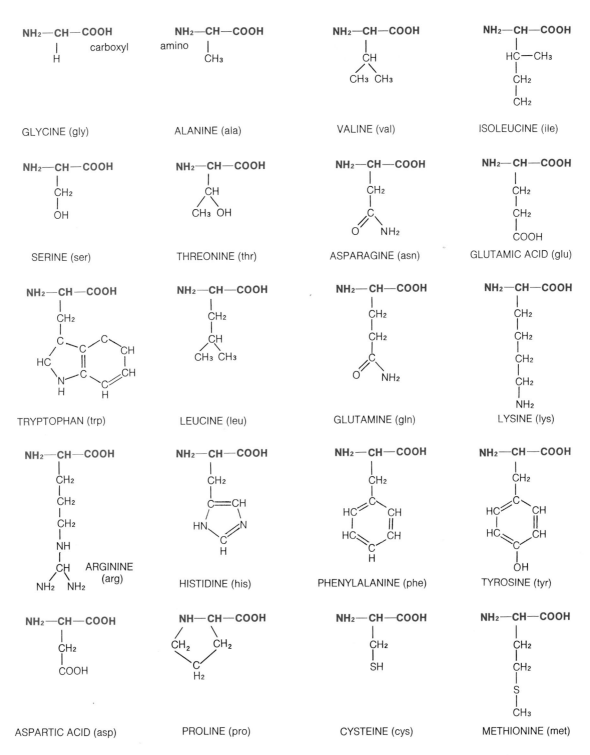

Figure 4-5 **The major amino acids.** Each of these is common in proteins. Note that aspartic and glutamic acids have an extra carboxyl and that lysine and arginine have an extra amino group. Proline differs from the rest in having its amino group in a closed ring; it is called an imino acid. The abbreviation beside each name is used in diagramming protein structure.

THE CHEMISTRY OF LIFE

The major amino acids that appear in proteins are shown in Figure 4-5. All of these are found in nearly every kind of protein.

Peptide Bond Formation

In the chain of amino acids that forms a protein, the carboxyl group of each amino acid is linked to the amino group of the next amino acid. This arrangement continues throughout the protein, and each such linkage is called a **peptide bond**. If the resulting molecule contains only two amino acids, it is called a dipeptide. If there are more amino acids, the molecule is called a polypeptide until it reaches a size of about 100 amino acids, when it is a full-fledged protein. An example of dipeptide formation is shown in Figure 4-6. During the formation of a protein molecule, each time a carboxyl group links to an amino group a water molecule is removed. This is called **dehydration synthesis**, or, to put it more simply, "putting together by water removal." When a protein breaks down, this process is reversed. A water molecule splits and is inserted between a carboxyl and an amino group, breaking the chain at that point. This kind of molecular breakdown is called **hydrolysis**, or "water splitting" (see Figure 4-6).

Dehydration synthesis and hydrolysis, then, are a "matched pair" of processes: Putting together? Take water out. Taking apart? Put water in. As will be seen, this same simple pair of reactions is responsible for the synthesis and breakdown not only of proteins but also of polysaccharides, fats, and nucleic acids. Dehydration is an almost universal mechanism for synthesis of important complex molecules within a cell.

Protein Structure

The sequence of peptide linked amino acids in a protein is called its **primary structure**. The chemical properties of the individual amino acids in the sequence cause it to assume a **secondary structure**, usually a helix (screwlike coil) or pleated sheet. **Tertiary structure** is the shape of the entire chain of amino acids and involves elaborate folding of the entire molecule. Often a protein is formed of more than one amino acid chain, and the structure that these subunits assume together is called the **quaternary structure** of the protein. Levels of protein structure are diagrammed in Figure 4-7. Although these various levels of structure are ultimately dependent on the primary structure, that is, the amino acid sequence, the *biological function* of the protein depends on its higher levels of structure, its overall shape. Denaturation, which is the disruption of protein structure by various chemical procedures or by heating, completely changes the properties of protein without changing primary or secondary structure. To illustrate, consider the effect of boiling highly proteinaceous egg white.

The total possible number of proteins is staggering. Since the substitution of even one amino acid for another results in a different protein, the number of possible different kinds is determined by the number of different combinations of the principal 20 amino acids that is possible in chains of hundreds of amino acids. This creates so many possibilities that an assemblage of just one molecule of each kind would fill the volume of the visible universe!

CHEMISTRY IN THREE DIMENSIONS

In order to explain how biological molecules such as proteins are constructed and work we must now consider more carefully their actual three-dimensional nature. In most of the remainder of the text we will continue to use the conventional two-dimensional notation because of its convenience, but at least you will know what it represents. Figure 4-8 is a series of

Figure 4-6 Peptide bond formation by dehydration synthesis.

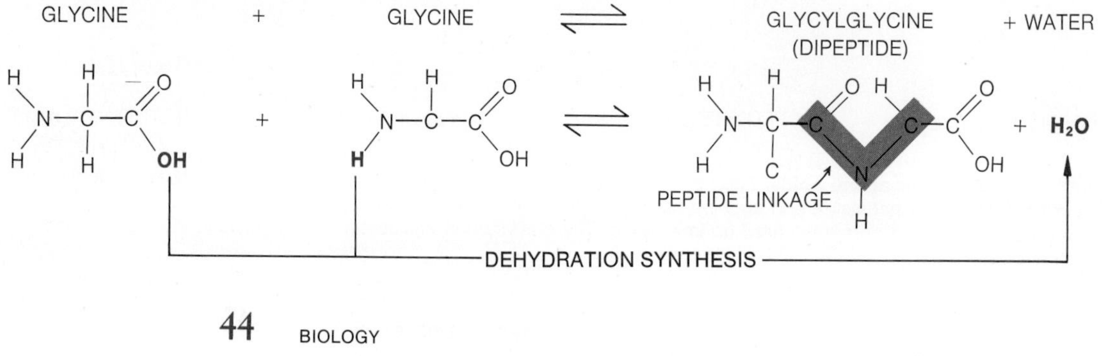

A PRIMARY STRUCTURE

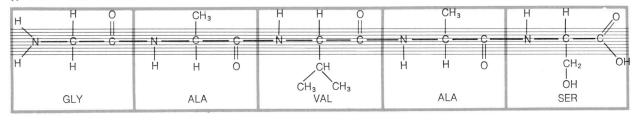

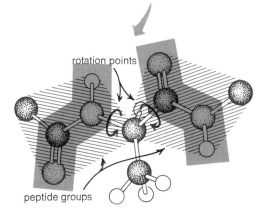

rotation points

peptide groups

B SECONDARY STRUCTURE

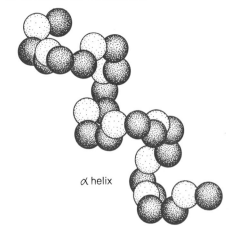

α helix

C TERTIARY STRUCTURE

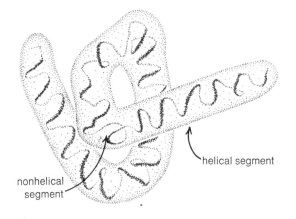

nonhelical segment

helical segment

Figure 4-7 Levels of protein structure.

Figure 4-8 Three-dimensional chemistry: methane.

CH_4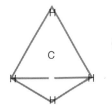

THE CHEMISTRY OF LIFE 45

representations of the methane molecule. You are already familiar with the two ways of representing the structure shown on the left. Two three-dimensional representations are on the right. The first of these shows that the four bonds of the carbon atom project towards the corners of a regular tetrahedron, which has the carbon atom in its center. No two of these bonds lie in the same plane, so it is obvious that a typical biological molecule with several carbon atoms bonded together is going to do anything but lie in one plane. The last representation is a space-filling model rotated 90° with respect to the tetrahedron model.

Stereoisomerism

Another structural complexity, stereoisomerism, appears if we convert the methane carbon tetrahedron into part of an amino acid. In fact, all amino acids except glycine are stereoisomeric. To understand this compare glycine with another amino acid, as shown in Figure 4-9. Glycine can be rotated 180 degrees around the axis indicated by the arrow and still look the same. This is not true of any other amino acid, as represented in the figure by alanine. Alanine has to be shown three-dimensionally by two structures. Both structures have the same **empirical formula**; the alpha (α) carbon in each instance carries one carboxyl, one hydrogen, one methane, and one amino group. However, with four unlike groups attached to the α carbon, there are two ways to draw the three-dimensional structure. These are truly different as you can see by trying to rotate them as you did the glycine and finding that one does not look like the other after rotation. The two molecules are mirror images of each other in the same sense that your hands are mirror images. Such molecules are called **optical isomers**. Whenever a molecule contains a carbon with four unlike groups bonded to it, that molecule has two optical isomers. If there is more than one such carbon, the molecule may have more than two isomers.

All amino acids except glycine have optical isomers. If an amino acid is synthesized in the laboratory, a mixture of equal quantities of the two kinds of molecules is produced. One of the remarkable things about life is that it deals almost exclusively with only one type of isomer. By convention this is called the L- form (left) whereas the companion, nonbiological isomer, is the D- form (dextro-, right). The biological form of alanine is called L-alanine and a mixture of the two isomers is called DL-alanine. Except for rare instances, such as the antibiotics produced by some bacteria, L-amino acids are the only ones found in organisms. The chemical machinery of life easily tells D from L isomers. Feed a bacterial colony a DL mixture of an amino acid and only the L form will be metabolized. Even your tongue can tell the difference. Many D-amino acids taste sweet whereas their L isomers are usually tasteless, except for L-glutamic acid (used as seasoning) which has a unique "meaty" taste.

There is no generally accepted explanation for this biological preference for one isomer over the other, although it is easy to see that a mixture of the two would disturb protein structure. Recent experiments on destruction of D- and L-amino acids by various types of radiation have shown preferential breakdown of one or the other isomer, depending on the nature of the radiation. Possibly solar radiation conditions at the time life arose influenced the first "choice" and, once made, it is easy to see why the usage would persist to prevent mixups in protein structure.

SHAPING PEPTIDE CHAINS

From the tetrahedron model it is evident that a chain of peptide linked amino acids does not lie flat. What shape will it assume? Although this is an extremely complex question, it is easy to show the types of factors that are important in determining shape at the level of secondary protein structure. These include:

1. *Rotation of atoms around bonds.* Some types of bonds permit free rotation and others do not. Parts of molecules attached by carbon-to-carbon single bonds (C—C) rotate freely (Figure 4-7). There is little rotation about the peptide bonds in the peptide backbone (C—N) and none about unsaturated (double) carbon-to-carbon bonds (C=C). However, even if rotation around a bond is theoretically possible, it may not occur if the structure of the rest of the molecule would cause parts of the molecule to come together upon rotation. The "stiffness" that is caused in this way is called steric hindrance.

2. *Chemical linkage.* The linkage between amino acids in different parts of the same peptide chain or between two separate peptide chains is important in determining structure, particularly tertiary structure. A cysteine molecule in one part of the chain may react with another lying some distance away in the same or

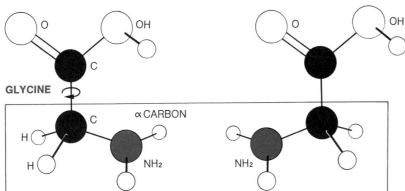

Figure 4-9 Sterioisomerism illustrated with amino acids.

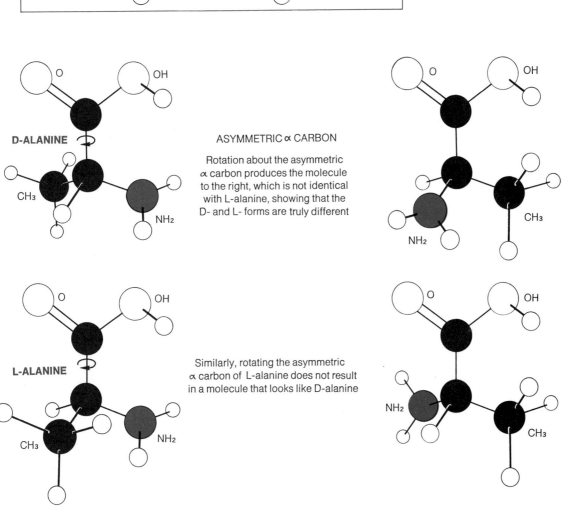

in a different chain to make a disulfide (two sulfur atom) link that firmly bonds the two components together. For example, hair contains a large amount of cysteine in its constituent proteins (keratins). Perma-

nent waves are made by breaking and reforming the disulfide bridges between cysteine molecules. When treated with a reducing agent, (hydrogen donor), disulfide bridges are broken and the hair becomes flexi-

THE CHEMISTRY OF LIFE

ble. It may then be set in a new pattern and the disulfide bridges reformed by oxidation, making the pattern permanent.

3. *Hydrogen bonding.* Although weaker than the kind of linkage just described, hydrogen bonding is extremely important because there are many opportunities for it to occur in and between peptide chains. Hydrogen bonding occurs when two atoms with negative charges share a hydrogen loosely between them. Hydrogen bonding is important in determining the properties of water (see Figure 5-1) because the oxygen of each water molecule forms hydrogen bonds with four other water molecules. Hydrogen bonds occur in proteins between the double bonded oxygen of the carboxyl group (C=O) and properly positioned nitrogens (N—H) of the amino group, either within the same chain or between chains in close contact.

Secondary Protein Structures: Coils and Sheets

Two hints led to the fundamental work of L. Pauling and R. B. Corey in determining the major type of protein secondary structure, known as the α-helix (alpha helix). First, since there are many opportunities for hydrogen bonding in a protein chain, it seemed likely that the chain would assume a shape that satisfied as many of these opportunities as possible because this would be the most stable condition. The more hydrogen bonds, the more stable the molecule will be.

The second hint came from x-ray diffraction studies of highly purified samples of protein. When an x-ray beam is passed through a substance some x-rays are bent by collision with the atoms within. This makes a pattern that can be photographed and is characteristic for the arrangement of the atoms in the x-rayed sample. The x-ray diffraction pattern can, in other words, be used to understand the spatial distribution of atoms. When keratin, a protein characteristic of fingernails and hair, was examined, the diffraction pattern showed that its structure involved large numbers of similar units repeating regularly at intervals of about 0.05 nm.

Pauling and Corey then experimented with scale models of peptide chains to see how they might be arranged so as to maximize hydrogen bonding and simultaneously satisfy the requirement for regularly repeating units. They found two basic types of structure, ture, the α-helix and β-helix (beta helix). These are shown in Figure 4-10, from which you can see that the α-helix structure is a tight spiral and the β-helix is a more extended, zig-zag arrangement. The spirals of the α-helix structure satisfy the requirement for regularly repeated similar structures while providing maximum opportunity for hydrogen bonding within the amino acid chain. Groups of α-helix chains are held together by many sulfur bonds (—S—S—) between cysteine molecules. Proteins of the β-helix type include substances like silkworm silk. In such proteins there is a great excess of small amino acids and this

Figure 4-10 Protein tertiary structure: coils and sheets.

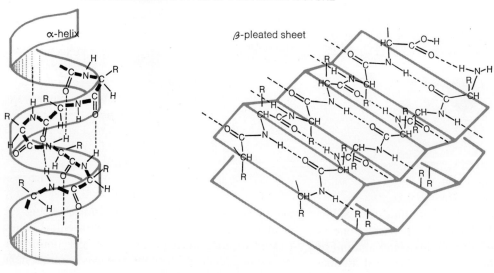

TWO FUNDAMENTAL TYPES OF PROTEIN STRUCTURE

allows the zig-zag chains to pack together into flat sheets held together by hydrogen bonding between chains.

Tertiary and Quaternary Protein Structure

Proteins are synthesized by stepwise addition of amino acids to form long chains. The sequence and number of these additions is all that the biochemical mechanisms of the organism specify about the structure of the final protein molecule. The remainder of the structure (secondary, tertiary, and quaternary) follows automatically from the amino acid sequence. You have just seen how amino acid composition determines whether the secondary structure will be α or β in form, and so it seems plausible that further determination of shape would automatically follow from secondary structure.

At first glance this might seem improbable for one of the principal tertiary structural types, the globular proteins (globulins), since their roughly spherical form seems to be consequent upon irregular and complex tight folding of α-helix chains. However, after such a protein has been denatured into a loose chain by gently breaking the (—S—S—) bonds that hold regions of the chain in contact, allowing the (—S—S—) bonds to reform leads automatically to reformation of the original tertiary structure. It is in fact thought to be theoretically possible to predict the final structure of a protein just from a knowledge of its amino acid sequence. In practice, tertiary structure is deduced both from x-ray data and knowledge of amino acid sequences. The first such structure to be determined was that of myoglobin, a small protein consisting of a single chain of 154 amino acids. By now several protein structures including enzymes, hormones (see Note 4-1 for the amino acid sequence of insulin), and antibodies have been determined. In all of these the tertiary and quaternary structure is of the greatest importance to biological function.

THE FUNCTIONS OF PROTEINS

Proteins appear everywhere in the cell as both structural and metabolic molecules. They are fundamental components of cell membranes and of most other organelles in the cell. Proteins may also be broken down to serve as energy sources and raw materials for the synthesis of important molecules. They may act as carriers of critical smaller molecules. In many animals protein antibodies play a major role in defense against disease. We shall return to these roles of proteins later. First we must briefly discuss an extremely important role of proteins, their catalytic function.

Catalysis

A **catalyst** is a molecule that speeds up chemical reactions, the joining or splitting apart of other molecules. Only a very tiny amount of a catalyst is needed compared with the quantity of products formed, and the catalyst is not altered or used up in the process. From these two observations it must follow that each catalyst molecule must be doing the same work over and over again, and the energy for this work must be coming from the other molecules involved in the chemical reaction the catalyst promotes. There is evidence that many catalysts do their work by promoting the bending and shaping of molecules being put together or taken apart so that the points at which they must bond match up more readily and precisely.

Catalysts Are Essential to Organic Reactions. Although there are catalysts to influence almost any chemical reaction, catalysis is essential to the organic reactions that take place inside cells. This is true because nearly all of the chemical reactions of molecules formed mainly from carbon, hydrogen, oxygen, and nitrogen—that is, organic molecules—are extremely slow reactions. Types of interactions that took minutes or seconds to occur in the inorganic world are found to take hours, or perhaps days, in the biological environment unless they are accelerated by catalysts. Many of these reactions proceed so slowly without catalysis that it may be said that they are not proceeding at all for purposes of any useful cellular function.

The extremely slow rate of organic chemical reactions is in part a reflection of the large complex molecules involved. These have scattered bonding points that must be aligned before reaction can occur. The slowness is also partially a reflection of the small energy differences that generally exist between "before" and "after" in an organic chemical reaction.

It is because the reactions are extremely slow that catalysts become all-important to living systems, for without them the organic reactions of a cell would not be fast enough to support life. Conversely, the absence of a catalyst can slow an undesirable chemical process to the point where it does not interfere with

Note 4-1 *Deciphering Protein Primary Structure*

The amino acid sequence of the protein hormone insulin, which regulates carbohydrate metabolism, is shown in Figure 4-11. The protein is formed of two chains with a total of 51 amino acids held together by two —S—S— links.

How are such complex structures worked out? Once the substance is isolated in pure form, the procedure is to break up the molecule in a systematic way so that each amino acid and its place in the chain may be identified. Identification is accomplished by chromatography.

Chromatography. If a bit of oil is shaken in a bottle containing two immiscible (meaning not dissolving in each other) solvents, let us say water and benzene, the oil will be found in the benzene after shaking and reseparation of the solvent layers. Sugar treated similarly would be found in the water layer. Some other compound might be present in both layers. In other words, chemicals partition themselves (separate) between pairs of immiscible solvents according to the extent of their solubility in each solvent, this being determined by the electric charge and other aspects of the structure of the chemicals. In chromatography advantage is taken of this fact by placing chemicals to be separated from each other in a system where one solvent flows over another. In paper chromatography the nonmoving solvent is water trapped in the fibers of the paper; the moving solvent may be any of several fluids immiscible with water.

Now a mixture of amino acids may be placed at one end of a sheet of wet paper and a moving solvent caused to move across the sheet. Those amino acids that are relatively the most soluble in the moving solvent will be carried across the paper more rapidly than amino acids that are more water soluble. After a time the amino acids will be strung out across the paper according to their relative solubilities in the two solvents. If necessary the paper may then be turned at right angles and a second kind of moving solvent may be used to achieve further separation of the amino acids. Then the spots where the amino acids lie are located by staining and identified by comparison with the position of spots generated by known samples of amino acids (Figure 4-12). This procedure has now been automated in a machine, the amino acid analyser, in which the moving solvent changes over time in an apparatus arranged so that the unknown sample flows through and is collected as a series of identified components (Figure 4-13). The process is called chromatography because it was first used to separate colored pigments.

Breaking Down the Protein for Analysis. First phase: separate the amino acid chains of the protein into like kinds. The proteins are denatured, splitting the —S—S— bonds, causing them to open up into random coils without fragmentation. Then the chains are separated into like kinds by a form of chromatography.

Second phase: break up the chains into large polypeptide fragments and purify each type. The protein is gently treated chemically so as to break a few peptide bonds at random. This produces a large number of polypeptides, essentially all possible combinations of segments of several amino acids from the parent chain. These polypeptides are next separated into like kinds by chromatography.

Third phase: determine the amino acid sequence in each polypeptide. Sequencing is accomplished by an ingenious technique in which a marker chemical can be made to combine with the amino acid at the amino terminal end of the polypeptide, after which that amino acid and marker are removed from the polypeptide and identified by chromatography. The next amino acid is similarly treated and the procedure continues until the entire sequence for that particular peptide is known.

Fourth phase: determine how all the polypeptide sequences fit together in the parent protein chain. This is simply done because enough of the polypeptides will have portions of the parent sequence in common to allow their positions relative to each other to be deduced.

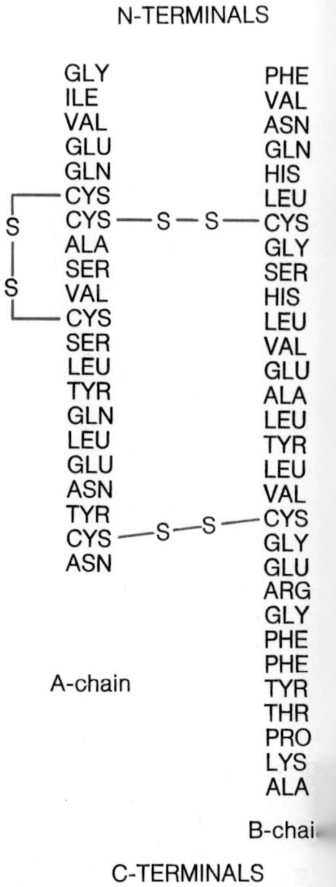

N-TERMINALS

A-chain	B-chain
GLY	PHE
ILE	VAL
VAL	ASN
GLU	GLN
GLN	HIS
CYS —S—————S—	LEU
CYS	CYS
ALA	GLY
SER	SER
VAL	HIS
CYS	LEU
SER	VAL
LEU	GLU
TYR	ALA
GLN	LEU
LEU	TYR
GLU	LEU
ASN	VAL
TYR —S—S—	CYS
CYS	GLY
ASN	GLU
	ARG
	GLY
	PHE
	PHE
	TYR
	THR
	PRO
	LYS
	ALA

C-TERMINALS

Figure 4-11 The amino acid sequence of bovine insulin.

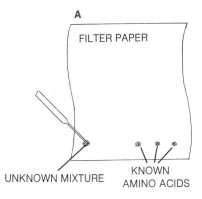

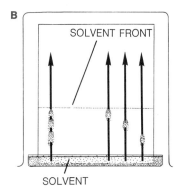

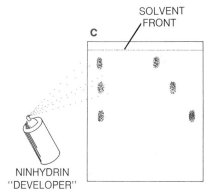

Figure 4-12 Paper chromatography. A. Unknown and reference amino acids are applied to filter paper. **B.** As the solvent front advances, the samples are carried along (see text). **C.** At completion of the solvent run, the paper is dried and a developing agent applied. The constituents of the unknown are identified according to their positions relative to the reference amino acids.

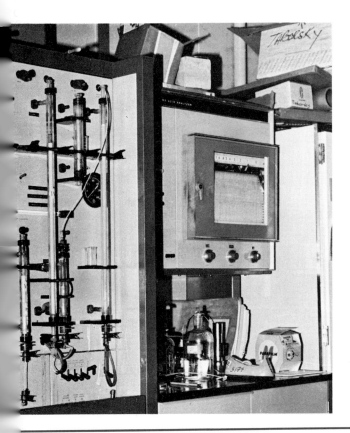

Figure 4-13 An amino acid analyser automates chromatographic processes and provides quantitative and qualitative data on the amino acid content of proteins. Separation of amino acids is effected as they flow through the chromatographic columns on the left, and a photometric device signals their identity and concentration on the recorder, right.

THE CHEMISTRY OF LIFE

the operation of the cell. Therefore we may say that, for all practical purposes, cellular reactions are turned on or off by the activity of catalysts. Their importance to life is reflected in their numbers. There may be as many as 5000 different catalysts at work in a single living cell, perhaps more. In living systems such catalysts are called **enzymes**. Some enzymes operate only inside cells; others move from cell to cell or act extracellularly, as in digestion. All are proteins. If cellular reactions will not proceed without enzymes and if

A   MONOSACCHARIDES EXIST IN TWO CLASSES

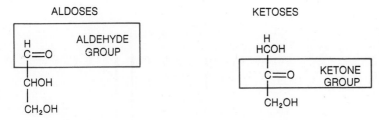

IN SOLUTION FEW ALDEHYDE OR KETONE GROUPS ARE FREE. THEY REACT WITH HYDROXYL GROUPS TO FORM CYCLIC MOLECULES. FOR EXAMPLE, D-GLUCOSE FORMS TWO CYCLIC MOLECULES.

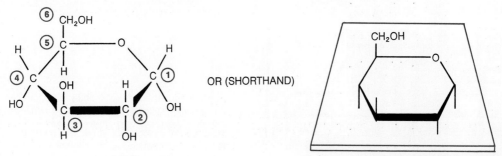

The molecule is considered to be in a plane, with the thick end bonds toward the observer, and with hydroxyl groups projecting above and below that plane.

enzymes are all proteins, then it must be true that whatever it is that controls protein synthesis also controls the cell. As we shall see, the information for protein synthesis ultimately comes from nucleic acids found in the control center of the cell, the nucleus.

CARBOHYDRATES

Carbohydrates play two principal roles, as energy sources and as structural components in organisms. They include the common substances that we call

B DISACCHARIDES ARE FORMED FROM TWO MONOSACCHARIDES JOINED BY A GLYCOSIDIC LINKAGE

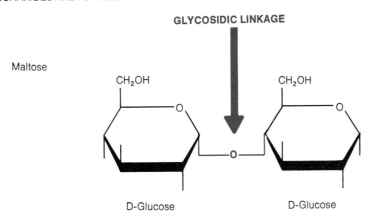

POLYSACCHARIDES CONTAIN MORE THAN 10 MONOSACCHARIDE UNITS
The monosaccharide units are connected in one or both of two ways, allowing branching

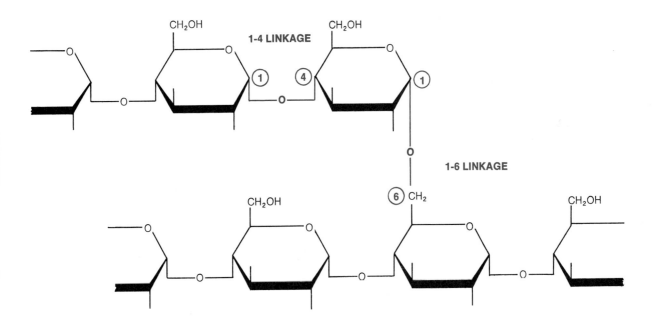

Figure 4-14 **Carbohydrates. A.** Monosaccharides. **B.** Di- and polysaccharides.

sugar, starch, and cellulose. Carbohydrates typically have empirical formulae that consist of repeating units of CH_2O, which implies that they are "hydrates" of carbon and hence their name. The structures of some carbohydrates are shown in Figure 4-14 A and B. Monosaccharides (from Greek, sakcharon, sugar) are the simplest carbohydrates. All of the more complex carbohydrates, the oligo- and polysaccharides, are **polymers** of various monosaccharides. These are built up of repeating units of the same or alternating pairs of monosaccharides. Glucose is biologically the most important monosaccharide, since it is the principal energy source for life processes and it is the basic unit of three important polysaccharides. These are **starch,** the plant energy storage molecule; **glycogen,** an energy storage molecule for animals; and **cellulose,** the principal plant structural molecule. These are very large molecules. Glycogen, with as many as 20,000 glucose units per molecule, has a molecular weight of several million.

LIPIDS

Lipids are of varied structural types but all share water insolubility owing to the possession of water repelling molecular regions. A typical example is the fat tristearin (Figure 4-15), which is formed of glycerol and the fatty acid stearic acid. The very long hydrocarbon "tail" of the fatty acid molecule makes the fat molecule water insoluble. The many types of lipids play a number of important roles. They serve as parts of cell membranes and as protective surfaces of organisms, for example, the waterproofing of insects. Some vitamins and hormones are lipids. Lipids are important fuel reserves for organisms.

NUCLEIC ACIDS

Nucleic acids are of two types: ribonucleic acid (RNA) and deoxyribonucleic acid (DNA). Although the chemical differences between the two are slight, they have markedly different biological functions. However, both DNA and RNA are concerned with storage, transfer, and application of information in cellular operations. *The information that nucleic acids carry determines form and function in all living systems.* The discovery in 1944 by O. T. Avery and his coworkers that DNA could transfer inherited characteristics from one strain of bacteria to another was a landmark in the history of biology. This discovery provided the first definite link between the mechanisms of inheritance and cell chemistry. There followed a remarkable 30 years in which the chemistry of information handling in living systems has been nearly completely worked out by molecular biologists.

TRIGLYCERIDES: GLYCEROL + 3 FATTY ACIDS
tristearin:

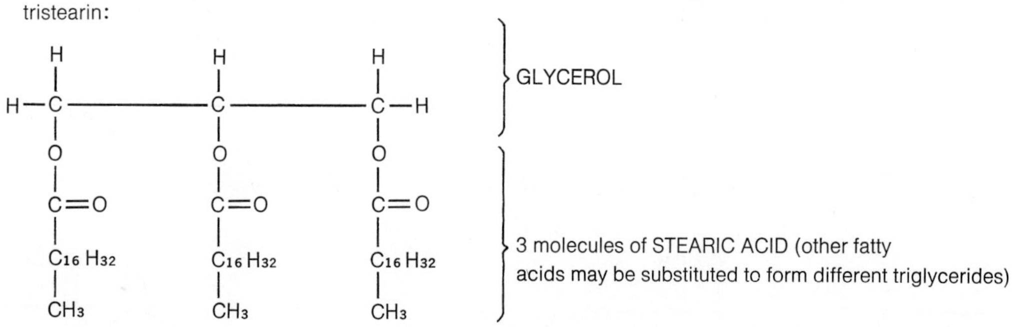

Figure 4-15 Lipids.

GLYCEROL

3 molecules of STEARIC ACID (other fatty acids may be substituted to form different triglycerides)

SATURATED vs. UNSATURATED

A lipid such as tristearin is termed saturated because its fatty acids lack double bonds; they carry the maximum possible number of hydrogen atoms. If the fatty acids have many double bonds, and therefore carry fewer hydrogens, the lipid is polyunsaturated, as would be a lipid containing this fatty acid with 4 double bonds,

Arachidonic Acid: $CH_3(CH_2)_4(CH{=}CHCH_2)_3 CH{=}CH(CH_2)_3 COOH$

NUCLEIC ACIDS: FUNDAMENTAL STRUCTURE

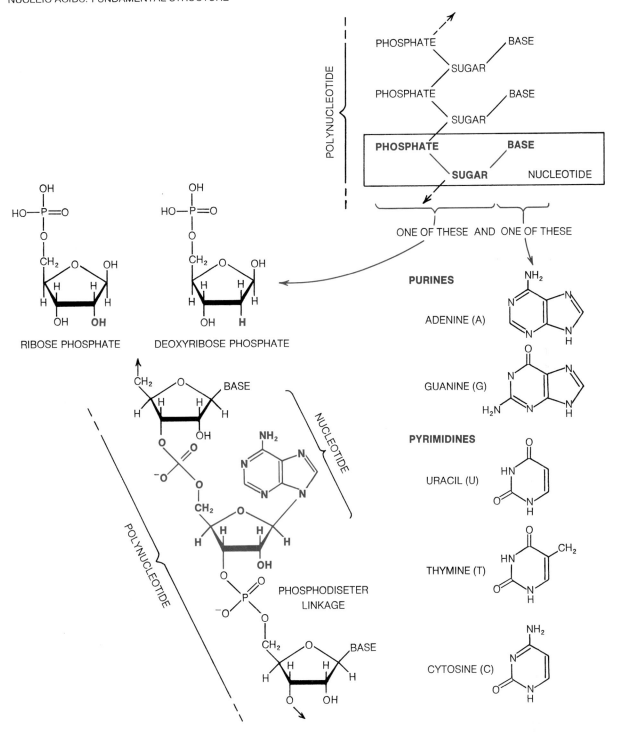

Figure 4-16 The components of nucleic acids.

The Structure of Nucleic Acids

Nucleic acid structure is analogous to protein structure because the sequences in which the building blocks are attached together determines the properties of the resultant molecule.

The Building Blocks of Nucleic Acids. The synthesis of nucleic acids, illustrated in Figure 4-16, involves four kinds of subunits: two kinds of purines and two kinds of pyrimidines. When one of these subunits is linked with ribose, a simple five-carbon sugar readily made from glucose, a molecule called a **nucleoside** is produced. If a phosphate group (one phosphorus and four oxygen atoms, PO_4) is then linked to this nucleoside, a three-part molecule called a **nucleotide** results. To summarize, the three parts of a nucleotide are the original purine or pyrimidine subunit, the ribose sugar, and the phosphate.

Nucleotides are building blocks that are strung together in a chain to form short **polynucleotides**. As the chains grow longer, they become fully formed **nucleic acids**. The links between nucleotides in this chain form between the phosphate of one building block and the sugar of the next (see Figure 4-16). Linking takes place by the usual dehydration synthesis with loss of a water molecule. Replacement of the water molecule under the influence of a breakdown enzyme (called a nuclease) reduces a nucleic acid to its nucleotide units or even to its more fundamental subunits.

DNA and RNA differ chemically in two ways. They differ in the nature of one of their pyrimidine subunits, and the ribose sugar in DNA (deoxyribose) has one less oxygen atom than the ribose in the RNA. These rather small chemical differences produce large and significant differences in the properties of DNA and RNA. One result is that DNA always tends to string out in very long, complex molecules involving thousands of nucleotides, whereas some of the RNA molecules may be much smaller. Indeed, DNA molecules in higher organisms may have molecular weights of 80 million and thus are the largest molecules known.

5 Beginnings

The understanding that we have thus far developed about the fundamental properties of living systems can now be profitably applied to the question of our ultimate "roots," the origin of life on earth. Besides being inherently fascinating, an understanding of origins is essential if the biologist is ever to understand fully biological structures and processes.

Evolution is the central theme of biology and the origin of life is its first chapter.

When and Where to Look for What

We now know what we must seek as critical conditions for the emergence of the first organisms. We must identify conditions that would lead to the production of a structure somewhat like a cell and which would satisfy certain chemical requirements. Since the chemistry of life takes place only in water solution, we conclude that life is essentially a water phenomenon (see Figure 5-1). Therefore, we presume that *life originated in water*, probably in the primitive seas, where sources of protein building blocks in the form of amino acids, glucose molecules or other sources of energy, basic subunits and phosphates to build nucleic acids, and components of lipids must have occurred.

If the assumption is made that life began here on earth, we must examine conditions as they existed on the planet during its youth, for scientists have long been convinced that life does not appear spontaneously under *present* earth conditions (see Note 3-1). Thus we must enlist the aid of the geologist and the astrophysicist to discover what is known of the history of the earth and of the record left by living systems.

THE TIME SCALE

It has been about 4.6 billion years since the earth condensed out of a cloud of primordial matter circling the sun. Time on this scale is extremely difficult to comprehend. To get a feel for the great age of the earth it might help to imagine those roughly 5 billion years compressed into 12 hr of a normal 24-hr day (Figure 5-2). On this scale 1 sec would equal about

57

WATER AND LIFE

CERTAIN PROPERTIES OF WATER MAKE IT ESSENTIAL TO LIFE

PROPERTIES	EFFECTS ON LIFE
High heat capacity, high heat of vaporization	Protects organism against thermal change. Promotes efficient cooling of terrestrial organisms.
Volume increases on freezing	Prevents complete freezing of deep bodies of water, thereby allowing overwintering of freshwater organisms. Prevents tying up of much of the earth's water in polar sea ice.
High surface tension	Assists in formation of biological membranes. Aids water circulation in plants.
Low density and viscosity	Permits rapid diffusion of molecules in cytoplasm.
High solvent capacity	Almost the universal solvent, water transports all life-essential molecules in cytoplasm and circulatory fluids.
Enters many chemical reactions	Water is also an almost universal catalyst. It is central to many biological reactions which involve adding or removing water.

THE STRUCTURE OF WATER EXPLAINS THESE PROPERTIES

Water is a **polar molecule** which forms hydrogen bonds.

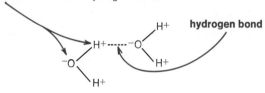

The polar nature of water makes it a good solvent. Thus, electrolytes in solution are surrounded by water molecules and their reassociation is hindered.

Hydrogen bonding causes water to form **super molecules.** These must be broken up to cause vaporization, requiring more heat than otherwise; and, as water freezes, the super molecules molecular structure dominates—increasing volume.

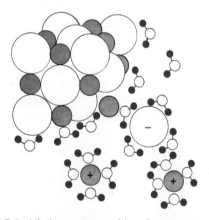

Figure 5-1 Life is a water-requiring phenomenon.

116,000 years. Of that 1 sec, roughly 3000 years of human civilization would occupy only 0.03 sec.

The earth as it formed became extremely hot due to gravitation, which caused compressional heating of gases and solid matter, and from radioactivity. Consequently, chemical processes occurred on a tremendous scale during the first third of earth history, setting the scene for the appearance of life. A time for the origin

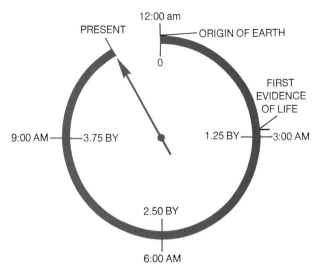

Figure 5-2 Earth history condensed to the scale of a 12-hour clock.

of life is difficult to assign because the earliest living organisms, undoubtedly small and fragile, were unlikely to leave recognizable traces over the long span of time. There may have even been *several* successive origins and failures. However, the marked similarity of present day organisms at the molecular level is taken by many scientists to indicate that a single original type is ancestral to all modern organisms.

The earliest traces of life date to around 3 billion years ago. That would be about 3:00 A.M. in our 5 billion year day. Most fossil remains of living organisms are less than 500 million years old, and the history of humans dates back only about 2 million years, with forerunners extending back as long as 14 million years. The time scale of life thus shows great acceleration over time. In the beginning little happened for vast periods except for accumulation of "raw materials." Then life processes began and continued almost as slowly for further ages. Finally, there occurred a great increase in rate, with all of the modern forms of life developing within the most recent tenth of the total period of earth history.

THE PRIMITIVE EARTH ENVIRONMENT

The physicochemical environment of the planet was markedly altered by the appearance of life. To illustrate this recall that the present-day atmosphere contains 21% oxygen. This has major effects on the types of chemical compounds that are present and the reactions they may undergo. Oxygen also limits the amount of ultraviolet radiation from the sun that reaches the earth surface because the ozone (O_3) layer, formed from oxygen at high altitudes, absorbs the most biologically harmful of these rays. Prior to the appearance of photosynthetic organisms (primarily green plants), which release oxygen in the process of capturing light energy, atmospheric oxygen amounted to less than 1%.

Important conclusions follow from this critical fact. Owing to the absence of oxygen many types of chemicals produced by nonbiological processes, which would break down, or oxidize, under the influence of oxygen, were stable and probably accumulated to high concentrations during the long ages before the appearance of life. Further, in the absence of the ozone layer (see page 527), high intensity ultraviolet radiation reaching the earth's surface served to increase still more the amount and variety of these accumulating chemicals because ultraviolet promotes many chemical reactions. The primitive earth atmosphere probably contained large amounts of water vapor, carbon dioxide, ammonia (NH_3), and methane. The primitive seas undoubtedly contained much sulfur, phosphorus, and other heavy elements in various compounds. These substances were exposed for very long periods of time to high temperatures, electrical discharges (lightning), and high energy radiation (ultraviolet light).

The Russian A. I. Oparin, and the Englishman J. B. S. Haldane, independently suggested a theory of the origin of life on the basis of these probable early earth conditions. This new theory dispelled at one stroke the critical problem that had crippled all earlier theories. Before Oparin-Haldane it was commonly thought that no significant amount of organic molecules existed on earth before the appearance of life. If true, this produced an impossible dilemma, since the only known life forms that can survive in a purely inorganic environment are photosynthesizers and **chemoautotrophs**. Yet the biochemical machinery of such organisms is so complex that there is no way imaginable for them to arise in an inorganic environment except by special creation, which of course is not a theory open to scientific analysis.

It is highly probable, according to Oparin-Haldane, that under plausible early earth conditions, *before* the

Figure 5-3 A laboratory experiment modelling early earth conditions produces many types of molecules essential to life.

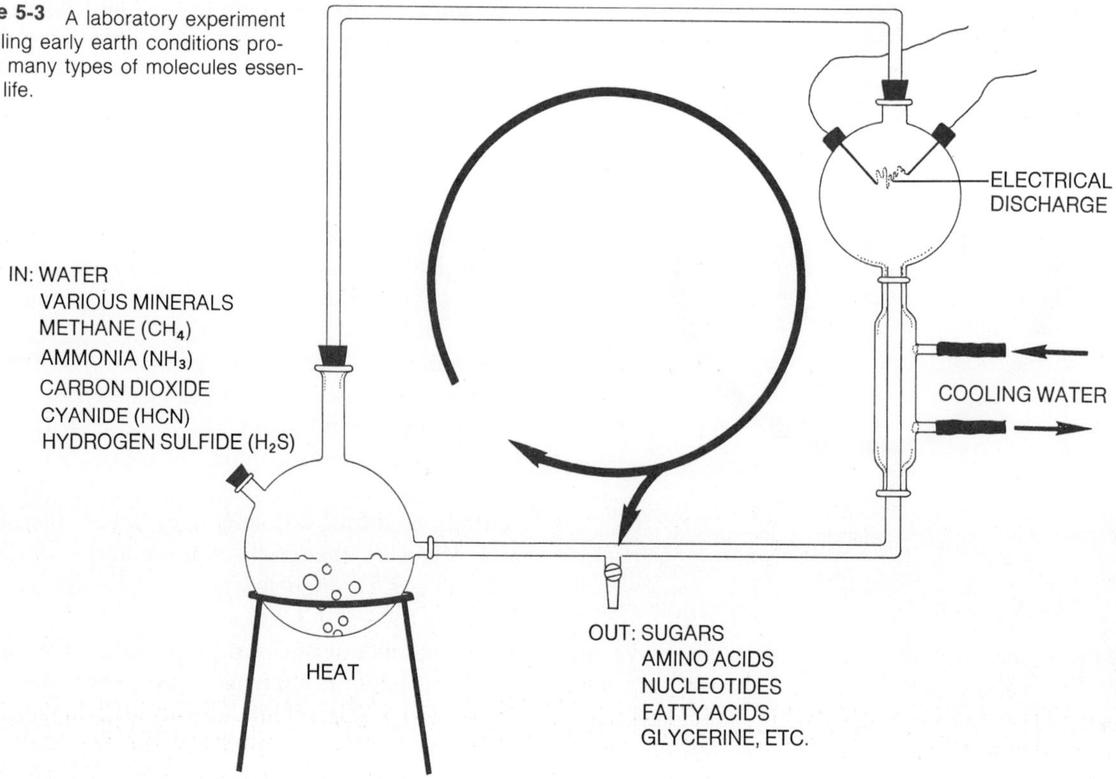

IN: WATER
VARIOUS MINERALS
METHANE (CH$_4$)
AMMONIA (NH$_3$)
CARBON DIOXIDE
CYANIDE (HCN)
HYDROGEN SULFIDE (H$_2$S)

ELECTRICAL DISCHARGE

COOLING WATER

HEAT

OUT: SUGARS
AMINO ACIDS
NUCLEOTIDES
FATTY ACIDS
GLYCERINE, ETC.

advent of life, the inorganic substances originally present were formed into a large variety of organic molecules. If true, organic molecules could accumulate over time to very high concentrations because there would have been no living organisms present to consume them. Thus, the absence of life both promoted the formation of organic molecules and helped to insure their accumulation. The oceans eventually must have become a thick solution, a soup, of organic molecules. Recent calculations suggest that the organic content of the prebiological seas could have been 10%. Oparin writes of this period as a time of *chemical evolution*. It ended, in his view, with the appearance of very simple protoorganisms (first organisms) capable of using molecules from the organic soup as energy sources and structural components.

This theory leaves unsolved many problems regarding the origin of life, but it obviously is better than earlier theories because it allows a more gradual and chemically plausible development of living systems. The theory does not require the protoorganism to have the sophisticated chemical machinery necessary to live on inorganic materials.

Is the Oparin-Haldane theory correct? How can it be tested? Obviously it is not practical to set up an experiment to create life. Our laboratory would have to be a prebiological earth, and we would have to wait a billion years for the results. But, short of such a test, there are many indications that the theory is close to the truth. Geologists and astrophysicists confirm that the inorganic chemicals essential to the Oparin-Haldane theory were undoubtedly present on the primitive earth. Further, it has been shown by experiment that, when these chemicals are exposed to primitive earth conditions, organic molecules are formed (Figure 5-3). Experiments of this type, starting with chemicals probably present on the young earth, have produced all of the basic chemical components of living systems: polypeptides, nucleotides, sugars, and lipids.

Extraterrestrial input

For the initial supply of organic molecules we are not entirely limited to earth sources. Even today about 100 metric tons/day of meteorites and micrometeorites enter the earth's atmosphere, and this rate of accumulation was far higher in the period following the primary accretion of the earth. Scientists now believe that much of the volatile material in the superficial layer of the earth accumulated in this way. Spectroscopic analysis of interstellar dust and direct analysis of fallen meteorites show an abundance of organic material. Spectroscopy of interstellar clouds (Figure 5-4) demonstrates the presence of at least 25 carbon compounds, including formaldehyde, formic acid, and methylamine. Amino acids have been found in meteors. This material at the least could contribute to the stockpile of organic molecules used in the evolution of life on earth.

Some authorities go even further and suggest that primitive biological systems could have evolved in interstellar space and could have been carried to earth or other planets by meteorites. Thus the "cosmozoa theory" of the great physical chemist, Svante Arrhenius, is born again. According to this theory the spores of life move through interstellar space from sites of origin until they find fertile new planets. This idea, although it anticipated by many years the universal distribution of the organic molecules that we believe are precursors to life, suffered from the defect that any relatively complex life form, even as metabolically

Figure 5-4 A nebulous cluster, 4600 light years distant, shows dark clouds of gas and particles against a background of luminous gas and stars. (Lick Observatory, University of California.)

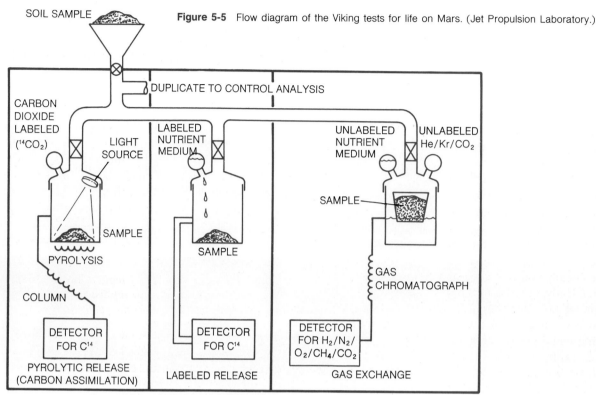

Figure 5-5 Flow diagram of the Viking tests for life on Mars. (Jet Propulsion Laboratory.)

inactive "spores," would in all probability be destroyed by radiation during long transits in space.

Another exciting way to obtain ideas about how life might have evolved on earth is to examine other planets that are accessible to us. Our moon is so small that it has not developed an atmosphere and evidently never had any significant amount of free water. Hence it cannot serve as a model for what we believe took place on the young earth. Mars appeared to be a more likely place to look for life or its chemical forerunners; thus, tests for the presence of life, or for the necessary conditions for life, were a major task for the two Viking flights to Mars in 1977.

The logic of the Viking search for life or life conditions on Mars is illustrated in Figure 5-5. The basic premise was that Mars life would be based on carbon chemistry. First, Mars soil was heated and the resulting volatilized material tested for the presence of organic matter by a gas chromatograph–mass spectrometer. None was detected, which was an astounding observation because, even if there was insufficient water for prebiological synthesis on the surface of the planet, Mars must have been exposed to the same meteorite rain of organic molecules that the earth experiences. The explanation seems to be that organic matter is destroyed by reactions occuring on the surface of Mars. Destruction is undoubtedly favored by strong ultraviolet radiation, since there is essentially no atmosphere or ocean depths to reduce its intensity.

Other experiments to detect metabolic activity by organisms that might be in the soil samples were also negative. These tests involved detection of photosynthesis or other forms of biosynthesis by measuring the assimilation of radioactively labeled carbon dioxide and detection of consumption of carbon labeled nutrient molecules. Although the analysis of the Viking experiments is continuing, it appears that Mars most probably did not start on the path to evolving life because it never developed an atmosphere capable of providing the right amount of shielding from ultraviolet and did not retain water for any significant period in its evolution, even though flowing water appears to have created some surface features of the planet.

The Problem of Structure

BEFORE THE PROTOORGANISM

Even though we may be confident that many biochemicals essential for life were present on the prebiological earth, we still have the *problem of organization*, that is, how these molecules formed living *units* able to survive and to give rise to similar "offspring." Even the little that we have presented so far about cells makes it obvious that all the complex structure of a cell could not have come into being all at once. This is a problem to our theorizing, since we have already said that the cell is the minimum living unit. Clearly, in terms of the special conditions of the origin of life, this statement cannot be true. Cells must have once emerged from simpler structures possessing some essential attributes of life. No direct evidence exists on the nature of these structures. The geological record is no help in trying to imagine what they might have been like because the earliest geologically detectable life forms look pretty much like modern single celled plants and bacteria (Figure 5-6).

One way to approach the problem of the nature of the earliest living organisms is to ask whether any basic chemical constituent of living systems—carbohydrate, protein, lipid, nucleic acid—might be able, *singly*, to sustain essential life processes. If so, it would be much more probable that life began with one of these rather than by several of them simultaneously beginning to cooperate in carrying out life functions. A clue as to which type of molecule might be such an initiator was provided by the American geneticist H. J. Muller. He wrote

. . . given a system capable of replication (reproduction), mutation (change) and replication of mutant forms, life and evolution automatically follow.

Muller's concept provides one way to fill the conceptual void between the chemical evolution of Oparin and evolution of "true" organisms. His statement thus warrants careful examination.

Suppose some kind of protoorganism arose by accident in the primitive world. Perhaps a droplet of some

Figure 5-6 Ancient microfossils. A. Filamentous procaryotic algae about 2 billion years old, from the Gunflint Formation on the North Shore of Lake Superior, Canada. **B.** Unicellular, eucaryotic algae, about 0.8 billion years old, from the Bitter Springs Formation, near Alice Springs, central Australia. 10 μm. (Preston Cloud.)

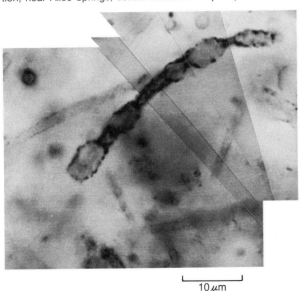
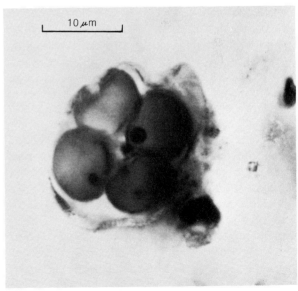

BEGINNINGS

sort existed that had fortuitously trapped within itself a simple group of chemicals able to break down organic molecules from the surrounding organic soup. Suppose that these breakdown products aided in sustaining the existence of the droplet. Already we have strained probability to the limits. The droplet had to form in the first place, had to contain the proper chemicals, and the products of their reactions with chemicals from the environment had to be of use in continuing existence of the droplet, and so on. Even so, such a protoorganism could not have led to present systems of life for two reasons: (1) it had no way to reproduce itself, and (2) it had no way usefully to adapt to change in the environment.

Note that reproduction without useful adaptation would serve for little. The world would simply fill with duplicates of organisms with similar capabilities. These would go on reproducing until some unfavorable environmental change disposed of them. For example, the population of nonadaptive protoorganisms might use up the available supply of some essential organic molecule from the environment. On the other hand, the ability to change without reproduction, without being able to pass on beneficial modifications to offspring, would also have no impact on the future of life. Clearly, what Muller tells us is that, of all of the attributes of life, the essentials, which cannot be dispensed with from the very beginning, are those of *replication* and persistence of change, or *inheritability*. Everything else can come later.

Taking this lesson to heart, we again consult our list of essential biomolecules. It is immediately obvious that nucleic acids, the information molecules, must have been primarily involved in the initiation of life. As we shall see later in more detail, nucleic acids arranged in sequences are able to direct the synthesis of more identical nucleotide sequences. Perhaps the appearance of this process was the watershed between chemical evolution and life. *Chemical evolution must have been a random process in which all possible chemical reactions took place, whereas life is a process in which chemical systems direct their own reproduction with great precision.*

The next great advance along the path to modern organisms would logically have been the transition from unimolecular to multimolecular systems. So long as the cycle of replication and change involved only nucleotides, there could be no buildup of supportive local environments for nucleotide reproduction, such as a cell provides. Various theorists have speculated about how this local environment could have first come into existence, but it remains a very serious problem. It has, for example, been suggested that lipid droplets might have trapped critical molecules, thereby serving as physical precursors of cells.

In whatever way it occurred, the next important chemical advance on the evolutionary path must have been the development of the ablity of nucleotides to direct the synthesis of polypeptides and proteins. This remarkable chemical process, unbelievably complex among chemical reaction systems of modern organisms, must have arisen in its essentials at the very threshold of life because it exists in virtually identical form in all organisms, from viruses to man. Once possessed of this ability to direct protein synthesis, nucleotides would have the potential for evolving life as we know it. The ability to direct the synthesis of proteins, would make ultimately possible the development of complex enzyme(protein)-regulated cell chemistry and the development of the complex protein-based structure of cells. As Sol Spiegelman puts it:

Cells, as we know them, can be looked upon as inventions of nucleic acids to provide themselves with a local environment optimally suited to provide the materials and conditions required for nucleic acid replication. Similarly, the evolution to the multicellular plants and animals can be interpreted as devices evolved to permit DNA to exploit all terrestrial space, including the land, the seas, and the air.

S. SPIEGELMAN: 1971.
An Approach to Experimental Analysis of Precellular Evolution.
Quarterly Reviews of Biophysics, 4:213-253.

Obviously this scheme is sketchy and is not presented as the only possible origin of life. Rather, its purpose is to show how, commencing with conditions believed to have existed on the prebiological earth, it is possible to sketch a chemically plausible sequence of events to bridge the gap between the nonliving and the living.

STRUCTURAL ESSENTIALS OF THE PROTOORGANISM

Order: in Time and Space.

The next step to take in developing an evolutionary scenario is to try to decide which *structural* components of modern cells are so essential that they must have appeared very early in cellular evolution. Suppose that a stage in evolution of life has been reached in which large chains of nucleotides control synthesis of proteins with functions that promote continued survival of the nucleotide chain. The next obvious requirement would seem to be a container, a boundary, to keep these collections of molecules together. Although this boundary might initially have functioned only as a purely physical barrier, a boundary structure with the properties of modern cell membranes would have been required eventually. Only such structures allow the necessary spatial organization and, at the same time, permit vital exchange of materials between environment and cell.

We have seen that one of the essentials of life is replication. The term has the implication of exactitude. This is one of the characteristics of life that challenged Schrodinger, and it is clearly an essential one if complex organisms are to evolve and increase their dominance over the environment. Nothing would be accomplished if the protoorganism could not transmit to its progeny improvements in function that develop by chance. Thus, we suspect that there was very early development of chemical processes and structures promoting orderly replication and transmission of nucleotide chains to daughter cells. In the simplest organisms this was probably not difficult to accomplish. Even today the hereditary material of bacteria is confined to a single long nucleotide chain, and perhaps that was how things were when the first cell divisions took place. Perhaps the nucleotide chain duplicated itself many times so that, whenever some accident caused the protoorganism to break up, each fragment usually included a copy of the essential nucleotide chain. Only when evolution had progressed to a stage in which so much hereditary material was necessary that it had to be stored in more than one nucleotide chain would the complex process of mitosis (page 147) be necessary. Mitosis insures that each daughter cell receives a duplicate of each kind of nucleotide chain represented in the parent cell.

These matters of generating order in time, of cellular reproduction, we must set aside for later chapters. Now our concern is with order in space, especially the nature of the cell boundary and what it accomplishes.

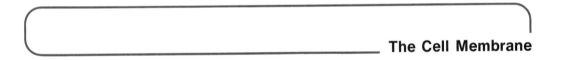

The Cell Membrane

The cell membrane is more than a cage for cell contents. It does play such a role, but it must do more. It must let some types of chemicals in, and it must let others escape. Actually it does even more than this. The cell membrane actively promotes the transfer of many types of chemicals into and out of the cell. The details of how these functions are accomplished are still under intensive study, but much is already known.

DIFFUSION

What makes chemicals in solution move from one side of a membrane to the other? Diffusion is the simplest mechanism. **Diffusion** is the tendency of molecules and ions to disperse from where they are concentrated by moving to where they are less concentrated. Every molecule, if its motion is not restricted, tends to move increasingly farther away from every other molecule. Diffusion is not confined to movements of molecules in water, as is easily demonstrated by the diffusion of odorants, for example, a perfume or ammonia, from a small open bottle to all parts of a closed room.

If you put salt crystals into a glass and then pour water very carefully over them until the glass is full, the water at the top of the glass, if tasted gently with a straw, will not be salty at first. However, as the salt crystals dissolve at the bottom of the glass, dissolved salt particles (ions) will diffuse from that region of higher concentration until, eventually, a sample taken from anywhere in the glass will taste just as salty as a sample taken from any other place in the glass.

Diffusion, considered on a molecular scale, is quite rapid. A hydrogen molecule at 0°C travels at an average speed of about 4000 mi/hr between collisions with other molecules. Molecular speed is temperature dependent and is also influenced by molecular size and electrical charge. Distance travelled between collisions and consequent changes in direction depend on the concentration of all molecules of all types present. Scaled up to the dimensions of everyday experience, the net motion of diffusing molecules is slow because of the frequency of collisions and changes in direction.

Diffusion Across Membranes

If we repeat the salt diffusion experiment with a thin barrier, a membrane separating the top half of the glass from the bottom half, the salt in solution will have a tendency to penetrate the membrane as it moves along the diffusion gradient, that is, down the concentration gradient, towards the top half of the glass. If the membrane is permeable to dissolved salt, these particles in solution will move across the membrane into the top compartment.

To keep essential chemicals in, to keep other harmful chemicals out, and at the same time to admit those molecules from the outside which are required by the system, membranes enclosing living cells must have the property of being closed to certain molecules. A membrane of this type is said to be **semipermeable**. One simple way to attain semipermeability is for there to be holes in the membrane of a size to admit small molecules and exclude larger ones. Many nonliving membranes have this quality. However, if the membrane admits one kind of molecule but not another kind of about the same size or smaller, the membrane is said to be **selectively permeable**. Membranes of living systems are the best examples of this kind of control.

OSMOSIS

Diffusion of Water Through Membranes

Osmosis is the movement of water across a membrane down its own concentration gradient. Problems caused by this movement of water are serious for living cells and have been solved in a variety of interesting ways in the course of evolution. Osmosis is best understood if we first consider an example of osmosis in a nonliving system. If a small amount of sugar-water is sealed inside a cellophane bag, we have an osmotic

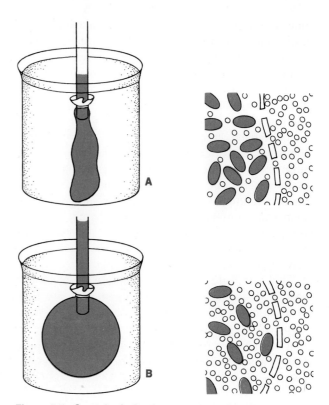

Figure 5-7 Osmosis. **A.** Starting conditions: flaccid, semipermeable bag containing sugar solution is attached to glass tube and immersed in water. **B.** With passage of time, water enters bag along its concentration gradient, causing bag to swell, and rises in tube, indicating osmotic pressure. Sugar molecules = large ovals; water molecules = small circles; semipermeable membrane = broken line.

model of a cell (Figure 5-7). The bag represents the cell membrane and the sugar-water represents the internal substance. If this model is immersed in a pan of tap water, representing a fresh-water environment, we observe:

1. The bag swells as water enters.
2. The rate of water entry diminishes as the bag fills.
3. The bag eventually becomes taut (turgid) and either ruptures or ceases taking on water.

The explanation of these events is simple. The assumption can be made that water molecules, in ceaseless random motion, are capable of passing through tiny holes in the membrane and that this membrane has no control over their passage or direction of passage. This assumption makes cellophane a useful analog for the cell membrane because cells are unable directly to control water movement. In other words, living cells, like the cellophane bag in the model, live in a water environment under the influence of the diffusion gradient of water. In the model, the water molecules striking a particular outside section of the bag, are more numerous than those striking the inside of that same section. This is because the sugar molecules trapped on the inside get in the way of the water molecules and separate them from one another (Figure 5-7). It follows that water is actually more dilute inside the bag than outside and the result is diffusion of water from a region of higher concentration (outside) to a region of lower concentration (inside).

That the water molecules are constantly striking against the membrane from both sides is shown by observation of Brownian movement under the microscope. **Brownian movement** is the constant random motion of visible particles suspended in water, caused by the impacts of water molecules. The phenomenon was first described in detail by the English botanist Robert Brown in 1827. According to the kinetic theory of gases, atoms and molecules of gases and liquids are in continual motion. Their motion becomes visible as the movement of small visible particles upon which they impact.

Dynamic Equilibrium Across the Membrane

To review, in the cellophane bag model of a cell, more water molecules hit the outside than the inside of the same surface because, on the inside, the water molecules are diluted, separated from one another, by sugar molecules, which are unable to pass through the membrane. There is a diffusion gradient of water molecules across the membrane from outside to inside. Although water molecules move both ways through the holes in the membrane, more will enter the bag of sugar-water than will leave it. This accounts for the first observation: water enters the bag. The situation could be reversed, of course, by putting tap water in the bag and sugar water in the pan, in which case water would move out of the bag since the pores through which water moves are nondirectional.

But why did the entry of the water into the bag gradually fall to zero? Remember that as water enters the bag the sugar solution inside is progressively diluted. If the bag were large enough and if there were enough water outside the bag, eventually so much water would enter the bag, that there would be no essential difference between osmotic concentrations inside and outside the bag. The osmotic gradient of water molecules would virtually be eliminated.

In the model the bag is not that large, and eventually fills. But, while it fills, the water molecules become more concentrated inside, and the difference between the rate of water movement in and the rate of water movement out becomes less and less. This must happen as the osmotic gradient becomes less steep. As observed previously, the rates of movement in and out would eventually be virtually the same, and osmosis would cease if the bag were big enough. So it is that there must always be a decline in the rate of osmotic movement of water into the model cell as it fills.

Turgor Offsets Osmosis

Now, what about the observation that the model cell becomes rigidly inflated and osmotic swelling stops altogether? Here, another factor comes into play. This is the effect of the pressure inside the cell on the movement of the water molecules. As the enclosed bag becomes turgid, the water molecules tend to collide more often. This is the same as saying that the pressure inside the bag has increased. The pressure of the water inside the bag (hydraulic pressure) results in more water molecules moving outwards through the membrane against the osmotic gradient. Finally, when the pressure is high enough in the bag, the number of water molecules moving out equals the number moving in, even though an inward osmotic gradient still exists. At this point the model cell remains turgid and water molecules move both ways at equal rates, an example of *dynamic* equilibrium. Although the system may appear to be in *static* equilibrium, water molecules continually are passing, unobserved, both ways across the membrane. Nothing appears to be going on because two unseen processes are exactly balanced. The remaining concentration gradient (outside to inside) is exactly offset by the pressure gradient (inside to outside), and the cell is in dynamic equilibrium.

Osmosis and Environmental Change

The earliest cells probably lived in and were nearly in osmotic equilibrium with sea water. Even today the body fluids of simple marine organisms are osmotically

Table 5-1 Ion Concentrations in Body Fluids as Compared with Sea Water

	Ion concentrations, mM/l*			
	Na	K	Ca	Mg
Modern sea water	470	10	10	54
Simple marine animals				
Jellyfish (*Aurelia*)	454	10	10	51
Starfish (*Asterias*)	428	10	12	49
Freshwater animals				
Frog (*Rana*)	104	2	2	1
Lamprey (*Petromyzon*)	139	6	2	2
Terrestrial animals				
Rat (*Rattus*)	145	6	3	2
Human	147	6	3	1

*See Appendix 1
From C. L. Prosser: 1973. *Comparative Animal Physiology*, 3rd ed. Saunders, Philadelphia, Pa.

and ionically similar to sea water. Fresh-water animals, by contrast, have markedly reduced concentrations of the major inorganic cations in their body fluids, which is advantageous because of the consequent reduction of ion gradients between body fluids and ion-poor fresh water. Similar changes are seen in terrestrial animals (Table 5-1). You should note that these data do not indicate equilibrium of all ions between sea water and body fluids. Ionic equilibrium, equal concentrations of all ions on both sides of the membrane, is obviously impossible because the internal fluids of living cells must contain a large number of complex biomolecules that are rare in sea water. If at the same time internal fluids contained identical concentrations of the ionic constituents of sea water—principally sodium, potassium, chloride, calcium, and magnesium—they would of necessity be more concentrated than sea water and out of osmotic balance. Thus, the attainment of osmotic balance must mean that the organism is able to reduce the concentration of some ion or group of ions to less than the external concentration. (Figure 5-8). This is not an easy trick because the reduction in concentration must be made against the natural concentration gradient, and this requires some sort of energy consuming metabolic ion pump. Despite its complexity, this ability must have arisen very early in the evolution of life because the problem that it solved would have occurred as soon as the primitive cell membrane became semipermeable. Lacking such a mechanism, the primitive cell would swell to the bursting point.

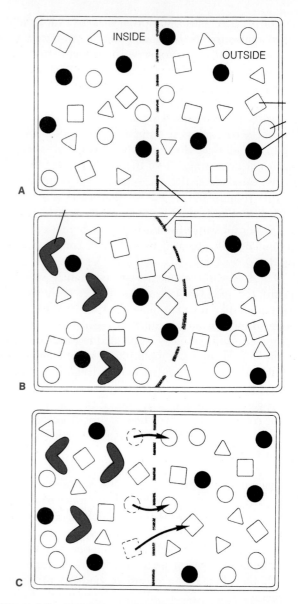

Figure 5-8 Osmotic and ionic equilibrium. **A.** Two compartments are in osmotic and ionic equilibrium across a semipermeable membrane. **B.** When biomolecules are present, inside compartment is no longer in osmotic equilibrium with outside compartment and tends to swell as water enters. **C.** Osmotic equilibrium is restored by metabolism-powered "pump," which moves ions to restore osmotic equilibrium at cost of loss of ionic equilibrium.

Change in the osmotic environment causes serious difficulties for many types of cells and organisms. When they are placed in fluid of lower osmotic concentration, there is a tendency for water to move in. The result might be disruption of cellular processes or even bursting of cell membranes. Changes that place a cell in solutions of higher osmotic concentration have the opposite effect. Water moves out and the cell becomes flaccid. In plant cells, the result is **plasmolysis** in which the cell contents shrink away from the rigid cell wall.

Two Directions of Osmotic Evolution

Subsequent to the first appearance of cells, probably very nearly in osmotic balance with the primeval sea, life seems to have followed two basic evolutionary paths in terms of osmotic problems. In one instance, the cells and organisms remaining in the ocean as it became more concentrated with salts from land runoff, found themselves fighting loss of water. In a sense, they had entered an arid environment, as barren as a desert, even though surrounded by water. On the other hand, organisms that migrated to fresh water, rivers and lakes, found themselves facing an opposite situation with water constantly entering.

This problem of water relations, or water balance between organism and environment, as will be discovered in Chapters 13 and 18, must have been a prime factor in the evolution of organ systems of multicellular animals. Even so, it was a problem that must have demanded a solution even among the earliest single celled animals. Observation of the internal organization of the single celled animal *Paramecium* demonstrates one effective answer to the problem of life in a dilute environment. This organism possesses vacuoles that contract regularly. Their behavior when *Paramecium* is placed in solutions of different concentrations indicates that these contractile vacuoles remove water, which enters constantly down the osmotic gradient and which the cell membrane cannot block effectively (Figure 5-9). Other simple fresh water organisms, *Hydra* (see page 186), for example, lack obvious water removing organelles. They regulate their water content in ways involving the ability of their cell membranes to take up salt ions from the dilute solution of such ions in the fresh water about them. This type of membrane function must involve more than

Figure 5-9 *Paramecium* rides itself of excess water through the contractile vacuole.

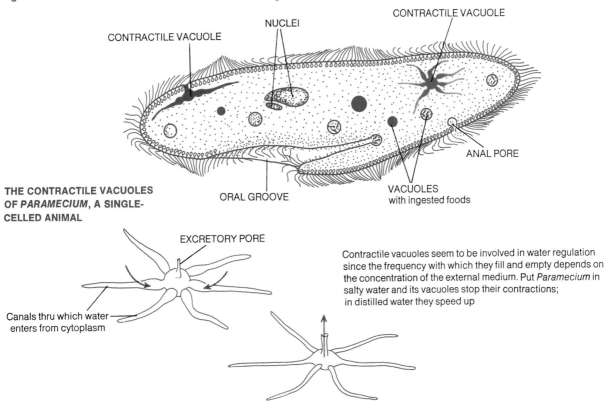

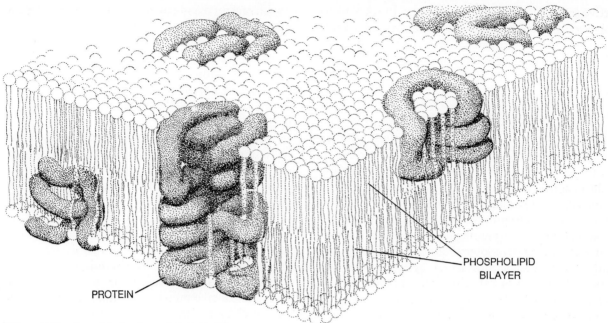

Figure 5-10 A model of cell membrane structure.

selective admittance of ions through pores, since the diffusion gradient is in the opposite direction—from the cell to the environment. The membrane under such circumstances acts as a "pump" for the required ions. Once again the fact that the cell cannot directly control water movement is illustrated. Control is achieved by changing ion concentrations, thereby causing water to flow along its concentration gradient.

ACTIVE TRANSPORT BY CELL MEMBRANES

Besides promoting uptake of inorganic ions against the concentration gradient, cell membranes can select specific biomolecules from the environment and promote their entry into the cell. This ability must have evolved early in the history of life. Primitive cells are thought initially to have relied upon the biochemically rich sea for necessary molecules. These chemicals would begin to disappear with continued exploitation by ever increasing populations of organisms. Those organisms best able to harvest this diminishing supply would have the advantage and be the more likely to survive. Thus, the membranes of primitive cells probably specialized early for this function.

The details of such specific uptake mechanisms are known only generally, for they require explanation in terms of the molecular organization of the cell membrane, itself an obscure subject. In general, an **active transport** mechanism requires a specialized region on the cell membrane that is capable of recognizing and temporarily binding the molecule that it transports. From this point, various mechanisms have been postulated. Perhaps the binding process itself may cause the membrane receptor site to change its shape in such a way that the bound molecule is drawn to the inside of the cell. According to another theory a receptor molecule transfers the trapped molecule to a carrier molecule, which then conveys it to the interior.

CELL MEMBRANE STRUCTURE

Chemical analysis of cell membranes show that they contain about 40% lipids and 60% proteins. Precisely how these chemical constituents are arranged is not known, but the best evidence is that the lipid molecules are lined up in a double layer with protein molecules penetrating all the way through or attached to one side or the other. Some of these are receptor protein molecules and other proteins concerned with active transport. The membrane is also thought to be traversed by pores whose shape, size, and electrical charge regulate the passage of some kinds of molecules (Figure 5-10).

Energy for the Protoorganism

We have used our knowledge of the conditions on early earth and of contemporary organisms in sketching a plan for the construction of a protoorganism. However this organism still lacks a major ability: it has no way to extract energy from the environment and put it to work in cellular operations. Since all life processes that we have identified as essential to the earliest organisms are energy requiring, this is an urgent problem. What is the source of this energy and how is it put to use?

CHEMICAL ENERGY

Life has two primary energy sources. They are the energy of sunlight and chemical energy, the energy stored in the bonds holding atoms together in molecules. Utilization of light energy, as in photosynthesis, although it is now essential to life, was probably, as indicated, not the earliest way organisms obtained energy. We will, therefore defer discussion of photosynthesis until we have considered the earlier utilized energy source, chemical bonds.

Burning releases energy from organic matter in the form of heat, as when wood, gasoline or sugar are burned. That heat can be harnessed to do work in a variety of machines. Burning is an example of an oxidation-reduction chemical reaction. All such reactions require the interaction of two kinds of molecules: **reductants** (reducing agents) and **oxidants** (oxidizing agents). The reductants transfer electrons to the oxidants. When electrons are transferred, in other words, when chemical bonds are broken, energy is released. The organic constitutents of wood, the hydrocarbon chains of gasoline, and the sugar molecule are reductants; they give up electrons to an oxidant. In the example of burning, the oxidant is molecular oxygen. In this kind of oxidation the electrons are transferred with hydrogen atoms and thus the oxygen is reduced to water.

The Calorie

It is useful to quantify the energy available in molecules, and this is done in terms of its heat equivalent. The unit of measure is the calorie (cal), which is the heat required to raise the temperature of 1 gram (g) of pure water by 1°C, from 14.5° to 15.5°C. Usually a more convenient unit of 1000 cal, the kilocalorie (kcal), is used. (See also Appendix 1). The energy content of a pure substance is usually expressed as kilocalories per mole (kcal/mole), the mole being the weight in grams numerically equivalent to the molecular weight of the substance. In nutrition studies the caloric value for various foodstuffs is the value obtained by complete combustion. Caloric value may also be expressed in terms of specific chemical bonds, rather than for entire molecules. For example, a rough figure for the energy bonds characteristic of biomolecules is 90 kcal/mole, whereas the total combustion of glucose releases 686 kcal/mole.

HARVESTING CHEMICAL ENERGY BIOLOGICALLY

Organisms are not machines in the sense of steam engines that use the heat from combustion to do work. Living things oxidize foodstuffs in a more gradual fashion, harvesting the available energy at each step and transferring it to sites within the cell or the organism where it may be used to power energy requiring processes. Enzymes accomplish the breakdown, bond by bond, in a highly specific way, which is under elaborate control at every step. Available molecules of foodstuffs are broken down only as required. The energy liberated is transferred with relatively

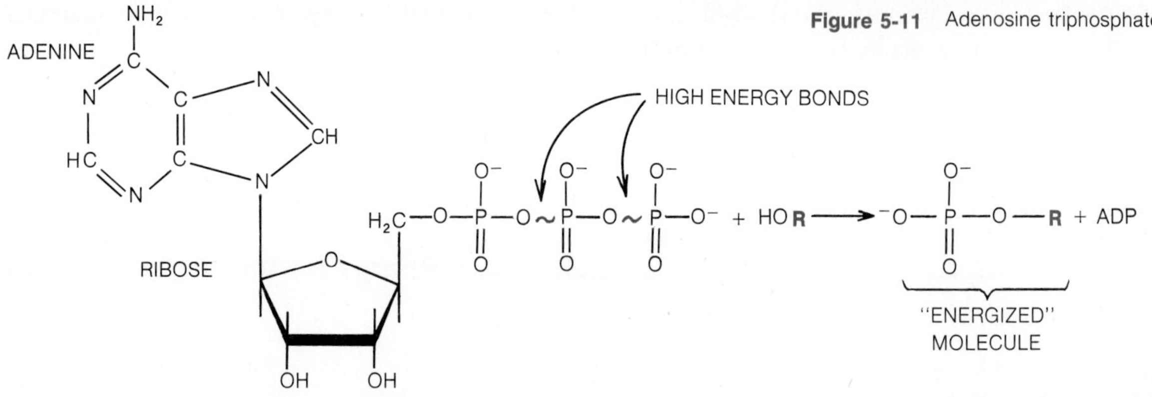

Figure 5-11 Adenosine triphosphate (ATP).

little loss to energy carrier molecules that convey it to energy requiring chemical reactions.

To initiate the study of the chemistry of the living state, we are going to study the process of **glycolysis.** This is the series of chemical reactions by which organisms obtain energy by partially breaking down glucose without the use of oxygen. Breakdown is only partial because oxygen is not used; the result is that not as much energy is released as when glucose is completely broken down.

GLYCOLYSIS

Glycolysis is probably an ancient biochemical pathway that evolved before there was free oxygen in the atmosphere. This idea is supported by the fact that glycolysis persists in all organisms as an alternate pathway for glucose utilization, even if the organisms are able to break glucose down more efficiently using oxygen. Further evidence is that one of the enzymes of glycolysis, found in such distantly related organisms as rabbits, lobsters, yeast, and bacteria, has a structure so uniform that mixtures of parts of the enzyme from any of these species work normally together. The enzyme must be built of highly similar amino acid sequences, irrespective of the organism from which it comes. This is strong evidence that the enzyme arose early in the evolution of life when organisms yet to evolve, such as rabbits and bacteria, had a common ancestor.

The Process of Glycolysis

The series of enzymatic reactions in glycolysis converts one molecule of glucose to two molecules of a three-carbon acid, pyruvic acid, with the net production of two energy-rich molecules of adenosine triphosphate (ATP) (Figure 5-11). The process merits close attention because it illustrates many important principles of the operation of cellular chemistry.

Glycolysis is divisible into two phases:

Preparation Phase. Glucose is **phosphorylated** forming glucose-6-phosphate. Glucose-6-phosphate is broken down into two molecules of glyceraldehyde-3-phosphate (Figure 5-12).

ATP Production Phase. Glyceraldehyde-3-phosphate is broken down into lactic acid with net production of energy-rich adenosine triphosphate (ATP) (Figure 5-12).

The first thing to note about this sequence of reactions is that all of the intermediate compounds between glucose and lactic acid (lactate)[1] are phosphorylated. This occurs for at least three reasons:

1. Because all of the reactions in glycolysis occur floating free in the cytoplasm, the small, intermediate molecules of the reaction sequence might diffuse out of the cell and be lost. The negative charge of the phosphate group insures that molecules carrying it will be repelled by the

---

[1]Usually weak acids of this type are dissociated in biological systems so that in structural formulas they are not indicated as $-C{\begin{smallmatrix}O\\OH\end{smallmatrix}}$ but $-C{\begin{smallmatrix}O\\O^-\end{smallmatrix}}$ with the H⁺ playing the field. In this instance, acid would be dropped from the name of the chemical and the ending -ate added, thus pyruvate, indicating that dissociation of the molecules is understood.

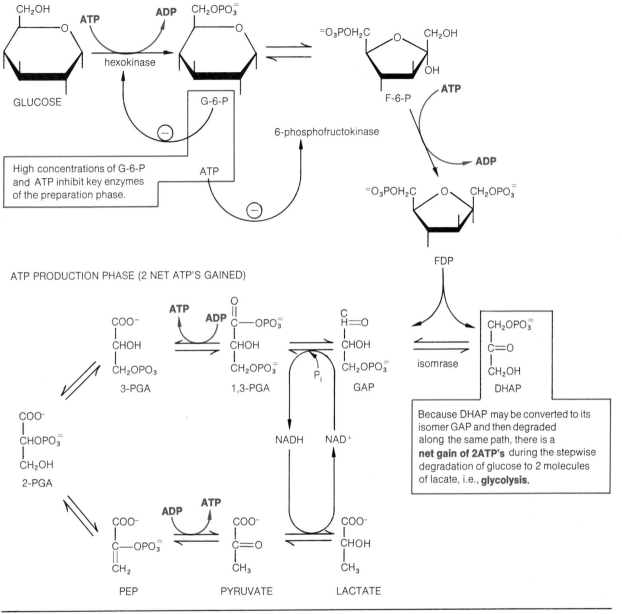

Figure 5-12 Glycolysis

G-6-P	glucose-6-phosphate
F-6-P	fructose-6-phosphate
FDP	fructose-1,6-diphosphate
DHAP	dihydroxyacetone phosphate
GAP	glyceraldehyde-3-phosphate
1,3-PGA	1,3-diphosphoglycerate
3-PGA	3-phosphoglycerate
2-PGA	2-phosphoglycerate
PEP	phosphoenolpyruvate

electrical charges on the inner surface of the cell membrane.

2. Because enzymes of glycolysis recognize the molecules that they act upon (their substrates) by, among other things, the presence of phosphate groups.

BEGINNINGS 73

3. Because, most importantly, the bonds formed when phosphate groups attach to the intermediates are able to retain much of the energy liberated in the several steps of glycolysis. They are able to carry over much of this energy to the primary energy transferring molecule of the organism, ATP (Figure 5-11). The importance of this function of ATP cannot be overemphasized because only ATP and a few similar molecules provide a way for energy to be transferred between chemical reactions in biological systems. Nothing at all is gained by oxidizing energy-rich molecules such as glucose unless the energy liberated can be transferred eventually to energy-requiring reactions.

Next consider the facts that (1) many different reactions are carried out within the cell and that (2) many of these reactions use the same raw materials. Obviously it would not do to have all possible reactions running independently at whatever rates the amount of available substrates could support. In actuality, cellular chemical processes are controlled in many ways, and these cellular controls are well coordinated with higher level controls exerted over the whole organism and its parts.

The first steps of glycolysis provide good examples of how cellular chemistry is controlled. For example, although glucose and glucose-6-phosphate are interconvertible in a single step, the reaction in each direction is controlled by *separate* enzymes. This provides many ways to control the relative concentrations of these two compounds, including relatively long term (hours to days) regulation by changes in the amounts of the two enzymes present. Also, the enzymes themselves can be made to change their rates of activity—without changes in their concentrations—and such changes make possible short-term (seconds to minutes) control of cell chemistry.

Enzymes at key points in reaction sequences may be inhibited, that is caused to become less active, by their own reaction products. Undue depletion or buildup of products is thus prevented. For example, the first enzyme in glycolysis, hexokinase, is inhibited by the end product of the reaction it catalyzes. If a cell contains sufficient glucose-6-phosphate, the product of the hexokinase reaction, hexokinase shuts down and the glucose that it would otherwise convert remains available for other uses, such as conversion into the sugar storage molecule glycogen or diffusion into other cells where the need might be greater.

The rate-changing molecule does not have to be a direct product of the reaction it affects. An example of this is seen farther along in glycolysis where the enzyme 6-phospho-fructokinase, which promotes the phosphorylation of fructose-6-phosphate, is inhibited, not by its own reaction product but by ATP. This is logical because one of the results of glycolysis is synthesis of ATP. With plenty on hand there is no point in making more. Remember that ATP is a *working form* of biological chemical energy. The organism stores energy in other molecules like glycogen or fats and doles it out into ATP only when necessary.

The result of the preparation phase of glycolysis is the production of two molecules of glyceraldehyde-3-phosphate (g-3-p) for each molecule of glucose entering the reaction sequence. Reference to Figure 5-12 shows that two molecules of ATP were required to supply energy to the reaction. The net result so far in terms of energy is the use of 3.6 kcal of energy per mole of glucose converted to g-3-p. No energy has been gained so far, but the g-3-p produced is used in the ATP production phase of glycolysis and this does result in a net energy gain.

The g-3-p produced by phosphorylation of glucose is next converted to 3-phosphoglyceryl phosphate. The German biochemist Otto Warburg worked out the details of this reaction, which was the first demonstration of conservation of the energy of oxidation of a biomolecule by the formation of ATP. The reaction begins as an oxidation of the aldehyde group (—CHO) of g-3-p. What might be expected to occur would be an ordinary oxidation with the formation of 3-phosphoglyceric acid and the energy of oxidation going up the flue. Instead the oxidation goes in two steps. First, a hydrogen is removed from the aldehyde and is replaced by a phosphate (PO_4^{2-}), whose bond is left in the energy-rich state. The fate of the hydrogen is also important in this regard and will be mentioned later. In the second step, the energy-rich phosphate is transferred to a molecule of adenosine diphosphate (ADP) to generate a molecule of energy-rich ATP, leaving a molecule of 3-phosphoglycerate.

A similar trick happens again before the end product, lactate, is reached. In this instance the oxidation-reduction reaction essential to energy production is not as obvious as in the previous sequence. What happens is a shifting of the relative number of hydro-

NAD AND NADP TRANSFER HYDROGEN ATOMS IN MANY ENZYMATIC OXIDATION-REDUCTION REACTIONS

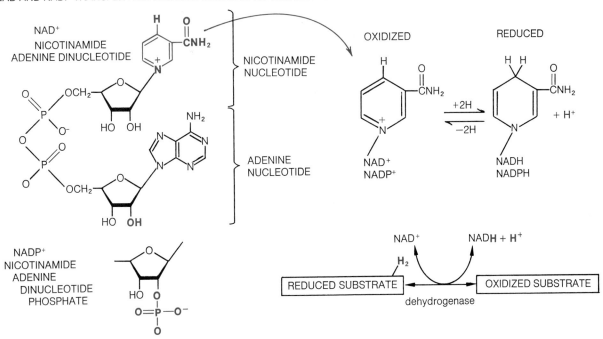

Figure 5-13 Nicotinamide adenine dinucleotide (NAD$^+$).

gens between two carbon atoms in the production of phosphoenolpyruvate, with the effect of accumulating energy in the remaining bond. This bond is then used to make another molecule of ATP, leaving pyruvate.

The reduction of pyruvate to lactate is the final step in glycolysis. At this point we have to consider the hydrogen removed from g-3-p as it becomes phosphorylated. This hydrogen can be thought of as eventually returning to the pathway of glycolysis as the hydrogen used in the last step. The transfer is accomplished by another type of carrier molecule, which has the formidable name nicotinamide adenine dinucleotide (NAD$^+$) (Figure 5-13). This molecule and others with similar function transfer hydrogen (and electrons, but don't worry about this now) from molecules to be oxidized to those to be reduced. Many biosynthetic processes involving reduction, for example, conversion of carbon-carbon double bonds to single bonds, depend on this type of carrier. A "bucket brigade" of these transfer molecules ultimately completes the oxidative processes of the cell by carrying hydrogen to oxygen to form water.

The result of glycolysis may be written as shown in Figure 5-14 and from this we see that the process has a net yield of two ATPs for each molecule of glucose oxidized to lactate. Glycolysis is not particularly effi-

Figure 5-14 The results of glycolysis.

SUMMARY OF GLYCOLYSIS

I. PREPARATORY PHASE   GLUCOSE + 2ATP + 2P$_{inorganic}$ + 2NAD$^+$ ⟶ 2(1,3-PGA) + 2ADP + 2NADH

II. ATP PRODUCTION PHASE   2(1,3-PGA) + 4ADP + 2NADH ⟶ 2(LACTATE) + 4ATP + 2NAD$^+$

III. NET REACTION (by cancelling terms and adding)   GLUCOSE + 2P$_{inorganic}$ + 2ADP ⟶ 2(LACTATE) + 2ATP

cient because it stops with the formation of two lactates per starting molecule of glucose. However, in organisms that use oxygen (**aerobic** organisms), lactate is never more than a temporary end product. Muscle, for example, may accumulate a high concentration of lactate when its oxygen supply cannot keep up with demand during hard work. The result is an *oxygen debt* that the muscles pay off by continued elevated use of oxygen after exercise, and much of this oxygen is used in further oxidations of accumulated lactate. In aerobic organisms lactate eventually is oxidized to pyruvate, which then enters other chemical pathways that are capable of extracting still more energy (see page 101).

WHERE WE STAND

We are now beginning to assemble a picture of the requirements and basic functions of the first organisms and to form an impression of the immense complexity that has developed in the process of living as life prevailed through time. Regarding this picture, the important thing to remember is that we are constructing our primitive organism and theorizing about the course of its evolution in rather simple steps, all of which are scientifically plausible, if not yet completely understood. *No impediment to a mechanistic explanation of the origin and evolution of life has yet appeared.*

To have survived through the long ages of earth history, organisms must have undergone modifications in structure and function that enabled them to exist under changing conditions of life. Next we begin to look at how organisms accomplished this. To begin we will consider one of the first serious problems that faced primitive organisms, the depletion of the initially rich supply of biomolecules in the primitive seas by the activities of living organisms. We might well call this the first ecological crisis.

6 The First Ecological Crisis: Depletion of Initial Resources

A BIOCHEMICAL GARDEN OF EDEN

Life may have continued without environmentally stimulated change for many millions of years once organisms reached the evolutionary state indicated in the previous chapter. Created by such agents as heat, ultraviolet light, and lightning, organic compounds probably accumulated by prebiological synthesis over the vast extent of time preceding the origin of life. L. E. Orgel observes that, even if the rate of synthesis and accumulation was as small as 1 kilogram per year per square kilometer, (1 kg/yr km^2), this could cover the earth with a 3-ft deep layer of organic solids in 1 billion years. To this must be added organic material from meteorites. Since a very large fraction of these chemicals would have dissolved in the oceans and other bodies of water, it is evident that there was little chance of organisms running out of material and energy sources in the first millions of years of life.

Except for the relatively restricted parts of the earth where evaporation or freezing and thawing cycles would be expected to concentrate solutes, it is likely that the waters of the earth contained a uniform concentration of organic matter, estimated by some to have been about 1% by weight of carbon. Although by no means all of the compounds formed from this carbon would have been of direct utility to primitive organisms, it is still instructive to compare this value of 1% concentration in the primitive seas with the 5% carbon concentration of one of the most sophisticated of fluid life support systems, human blood. Viewed in this perspective the primitive sea becomes impressive as a substrate for life. It does not seem too great a flight of fancy to call the primitive seas a worldwide circulatory system supporting a worldwide "organism" of widely dispersed and similar cells. The stabilizing physical properties of water (see page 58) augmented by the immense mass of the seas would have insured thermal and chemical stability over vast stretches of time, making the oceans a protective cradle for the first organisms. Certainly the ample initial supplies of organic compounds would have relieved them of many of the problems that face modern organisms.

BEGINNINGS OF THE BIOLOGICAL EVOLUTIONARY PROCESS

Under these conditions what could have led to inherited change and evolution in organisms? Once having attained some minimally functional state, would the first organisms have simply gone on without change, perhaps not even reproducing, until destroyed by some accident? Obviously, inheritable change must have occurred in at least some early organisms if we are to trace our beginnings to them. The causes of such change in the earliest organisms would appear to be rooted in two properties that these organisms most probably had, namely:

1. Their hereditary machinery made occasional inherited mistakes.
2. They had an optimal size, approximately that of modern single cells.

With regard to the first property, the occurrence of inherited mistakes in the operation of the hereditary apparatus almost certainly existed from the beginnings of the replication processes characteristic of life. After all, these chemical processes, were undoubtedly without the sophisticated mechanisms of modern organisms for repair of damage to hereditary molecules and were, moreover, exposed to the chemically destructive effects of radiation. The majority of changes would probably have resulted in death or inability of the organism to replicate. However, a sufficient few must have produced nonlethal variations of small degree in any of a number of characteristics of the primitive organism. These variations would be expected to affect the biological efficiency of their possessors and thus influence the composition of subsequent generations. This seems plausible enough and is simply a description of what happens in modern organisms. The problem in translating the modern situation into the context of the primitive world comes in deciding what the shaping forces (or as we shall say later, selective factors) would be like. What possible changes might occur in primitive organisms that would increase their chances of survival under early earth conditions?

A clue lies perhaps in the second property that we have specified for these organisms, that they would have a rather small optimal size. If there was no optimum size for early organisms one might imagine all sorts of strange consequences. For example, they might just continue growing without breaking up into progeny until destroyed by their own physical disruption. Actually, there are many reasons of a functional nature that the first, or any, cells should be rather small. Cells today rarely have a volume greater than 1 microliter (μl). The principal reason for this size limit is that 1 μl is nearly the maximum volume in which unaided diffusion can efficiently serve the metabolic requirements of a cell. Even considering that the first cells would undoubtedly not have been as metabolically active as modern cells, it still seems unlikely that they were more than an order of magnitude larger than modern cells.

The consequences of these two premises seem obvious. First, there would ensue an evolutionary competition in which the survivors would be those varieties of organisms that produced viable offspring most rapidly. Whenever by accident an inherited change occurred that favored rapid reproduction, the descendants of that organism would increase in numbers generation by generation relative to other organisms. Limited size before reproduction would itself tend to increase the rate of reproduction and would insure that reproduction occurred frequently enough so that inherited changes would be represented in offspring.

The second consequence is that a race for rapid reproductive ability would sooner or later have filled the seas with teeming hordes of cells. What would these cells have been like? There is no way of knowing anything with certainty about them. We can guess however, that they remained highly similar to each other in terms of how they gained matter and energy. All probably still obtained their requirements from the bounty of the organically rich seas.

Two closely linked factors would have eventually brought an end to this uniformity. First, rates of reproduction would have become limited by the rates at which necessary raw materials could be obtained from the sea, molecule by molecule, by diffusion. Second, reproductive rates would also have been slowed by reduction of concentrations of raw materials in the sea. This reduction would have come about as the increasing numbers of organisms used up the accumulated supply from the prebiological era and finally exceeded the restorative abilities of new nonbiological synthesis.

In modern terms, we would say that the world had experienced a population explosion of organisms so successful that they had depleted their natural resource base. Perhaps this happened many times. We have no measure of the time required for life to evolve to this stage. It might require only a few million years. If so, there would be ample time for repeat performances since the organic remains of previous die-offs would be a rich beginning for accumulation of materials necessary for the next origin of life. Obviously, since we exist, organisms eventually found a solution to this limitation.

Development of New Ways to Obtain Raw Materials

As supplies grew limiting, survival required development of new sources of supply and increased efficiency in extracting remaining available supplies from the environment. Both methods have become elaborately developed.

THE ORIGIN OF METABOLIC PATHWAYS

Suppose that at some time in their evolution early organisms required a naturally available and initially plentiful molecule A. Under natural conditions molecule A breaks down by simple and reversible stages through compounds B, C, and D. When biological utilization finally made A rare in the environment, any change in the chemistry of organisms that allowed them to make A out of B would have obviously increased their chances of survival. B utilizers would outbreed A- requiring organisms. Furthermore, since A converts naturally to B, reversal of the process might only require a simple modification of cell chemistry, perhaps a slight change in an enzyme already in existence for the purpose of acting on A. As further depletions of environmental chemical resources occurred, reversal of subsequent steps by similar means would each in turn confer immediate advantages. The result would be an organism with a metabolic system or pathway that transformed D into A in a sequence of three reactions involving two intermediate compounds, C and B.

Now suppose that biochemists undertake to study this metabolic pathway from D to A and, in particular, try to understand how it evolved. By the time this study was being made the enzymes involved in the pathway had no role other than cooperating in the production of A. This would make it seem that there could be only two ways that these enzymes could have appeared in the course of evolution so as to confer an immediate advantage on their possessors: (1) The enzymes would have had to appear simultaneously in the same organism so that A could be produced immediately or (2) each step from D through B must have been individually advantageous in the past, that is D, C, and B must be compounds that were once valuable but which now have lost their individual functions and serve only as precursors for A. As for case (1), considering the improbability of a single inheritable change, the chances of three of a specific type occurring at the same time in the same organism are not worth bothering about. Today mutations occur at a rate of perhaps one mutation per 10,000 opportunities. Mechanism (2) might actually have played a role in evolution, but it is extremely unlikely that a large part of the vast complexity of metabolic pathways in modern organisms could have arisen in this way.

Problems of this type are laid to rest by the splendidly simple idea that metabolic sequences arose in an order reverse to the way they proceed in modern organisms (Figure 6-1). As N. W. Horowitz observed, this hypothesis is made possible by the Oparin-Haldane theory of the origin of life in a sea rich with complex organic molecules. In one stroke the hypothesis makes a gradual and plausible evolutionary development of the complex metabolic network of modern organisms all the way back to their most remote beginnings in ancient seas. Assuredly it is logical to postulate a long period of evolution of metabolic pathways as the oceans were gradually depleted of their organic resources and organisms turned to previously unused molecules to use as precursors of depleted essential molecules.

GENERATION OF REDUCING POWER

If primitive organisms were to engage in extensive chemical synthesis, that is building up useful molecules from less complex organic precursors, they had to have a source of hydrogen, of reducing power. We did not see this happen in glycolysis (page 72). Although ATP is formed in glycolysis, there is no net

IN A MODERN ORGANISM THIS 3-STEP REACTION SEQUENCE EXISTS

← THIS IS THE USEFUL MOLECULE

← THESE HAVE NO VALUE EXCEPT AS PRECURSORS OF D

PROBLEM: If A has always been the ultimate precursor of D, then reactions

$A \longrightarrow B$
$B \longrightarrow C$
$C \longrightarrow D$

would need to come into existence simultaneously to have survival value. **THIS SEEMS UNLIKELY**

A SOLUTION: If, under primitive conditions, all 4 were present in the environment

AND

If D was initially used by organisms, as it was consumed the advantage would be to the organism that became capable of making **D from C**. This is a single step, and a probable one to evolve since it confers immediate survival value.

Similarly, as C became limiting, the advantage would go to organisms able to convert **B to C**, and so on.

Thus by postulating **reverse order of evolution** of complex biochemical sequences, **each step has survival value** and evolution of the sequences is seen to be plausible.

Figure 6-1 An hypothesis for the origin of metabolic pathways. It has the advantage of assigning adaptive value to each step at the time of its origin.

gain in reducing power because the hydrogen produced early in the breakdown of glycogen is used up at the end. There is, however, another pathway of **anaerobic** glycogen utilization that does quite the opposite. The hexose monophosphate pathway uses up ATP but produces hydrogen as glucose is broken down into a five-carbon sugar, ribose-5-phosphate (Figure 6-2). This pathway, like glycolysis, is thought to be an ancient one because it is anaerobic and because it is found throughout the living world. Its early development would have insured maintenance of biosynthetic capability well into the period of depletion of prebiological chemical stores.

IMPROVEMENTS IN HARVESTING EFFICIENCY

New Developments in Cell Membranes

Presumably as long as the environment contained sufficient amounts of required substances, the external membranes of primitive organisms could have been relatively simple structures permeable to the requisite molecules. However, as soon as an essential molecule became scarce, there would have been an advantage conferred on cells that developed ways to prevent the loss of essential molecules by outward diffusion. Phosphorylation of glucose is a good example of a possible

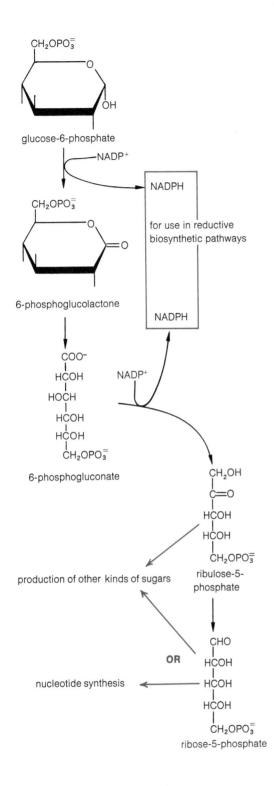

Figure 6-2 The hexose monophosphate pathway.

early way to do this since it requires no structural changes in the cell membrane (see page 72). Eventually cells must have been faced with such scarcity that inward diffusion was no longer adequate. Then there must have ensued the evolution of active transport mechanisms, membrane specializations able to recognize and transport specific molecules to the interior. This stage in the evolution of cells would seem to represent the limits of plausible evolutionary development of mechanisms serving to improve the abilities of single celled organisms to harvest molecules uniformly distributed in solution in water.

The Switch to Concentrated Sources of Essential Materials

In modern organisms there are two ways for materials to be taken inside cells besides the molecule by molecule mechanisms of diffusion and membrane transport. The first method is illustrated by the free-living, single celled ameba and by the **phagocytes** of vertebrates. These engulf small organisms and fragments of biological materials whole by flowing around and collecting them in a vacuole newly formed from the cell membrane. The second method, **pinocytosis** (cell drinking), is a smaller scale engulfment process found in many kinds of cells. In pinocytosis small vacuoles form from the surface membranes of cells and take in small volumes of the outside medium. Pinocytosis has been experimentally induced by the stimulus of protein outside the cell.

Pinocytosis might have developed from an already present membrane recognition and transport ability. Perhaps filling many molecular recognition sites by a much higher than ordinary concentration of some essential molecule would sufficiently change membrane structure to allow it to undergo pocket formation. When the environment contained localized and highly concentrated food sources, such a process would have been sufficiently useful to have been an obvious next step in the evolution of food getting.

THE FIRST FOOD CHAINS

When that step was taken the environment must have contained concentrated food sources in the form of organisms alive and recently dead. It follows that evolution of the ability to recognize and engulf other organisms and their products would place engulfers at a decided advantage over those organisms still limited to accumulating food a molecule at a time from solution. Using the terminology applied to modern organisms, we would say that **scavengers** (consumers of the remains of dead organisms) and **predators** (consumers of living organisms) have appeared. At this juncture the primitive seas would have been supporting a simple, two-step **food chain,** or flow of energy and materials. This consisted of primary harvesters of the prebiological organic soup, and these first organisms supplied a second level of scavenger-predators. The chain might be viewed as having a third level since the scavengers undoubtedly would have had little compunction at dining on either the remains of primary harvesters or of predators. Although this postulated early appearance of predation was important in signalling the first appearance of animal life in the modern sense, this simple food chain offered no long term solution to the basic problem of the dwindling food resources. It was only a redistribution mechanism and created no new supply of energy or materials.

NEW BEGINNINGS: BIOLOGICAL UTILIZATION OF SOLAR ENERGY

The crisis was clear cut: unless new energy and material resources appeared to replace those supplied by prebiological chemistry, life would suffer a serious curtailment. Without new resources, the most optimal outcome would be a die-back to population levels sustainable by new nonbiological chemical synthesis. This would have been far less than peak population levels prior to the onset of severe exhaustion of resources, and probably no further evolutionary steps of significance could have been taken in such an impoverished world.

The first phase of the crisis must have been using up the energy-rich molecules necessary to power biological synthesis. Reduced carbon compounds would have remained relatively plentiful as long as there was no free oxygen to complete the inefficient oxidations that were probably characteristic of primitive metabolism. These compounds could be built up again into useful molecules, using the reducing power generated by mechanisms similar to the hexose monophosphate pathway. Thus, the first factor to become limiting was likely to have been useful chemical energy. It appears likely that the first step towards a solution of the crisis was the use of solar energy in a process called **photophosphorylation** to generate ATP or its ancient equivalent. Subsequently organic molecules would have become limiting, whereupon a mechanism for creating new organic material from carbon dioxide emerged. Together we call these two processes **photosynthesis.**

The advantages of direct biological harnessing of solar energy are obvious. The immense energy of the sun was directly or indirectly the driving force in all prebiological organic syntheses. However, these syntheses were inefficient and *random* with respect to the requirements of life. Thus, if life itself could utilize solar energy directly in the synthesis of its *particular* requirements, resource limitations would be removed, conceivably until the final burn-out of the sun, when earth life will be snuffed out for reasons more impelling than nutrition.

The ultimate success of photosynthesis as conducted in green plants is phenomenal. Today it is estimated that the annual accumulation of energy from photosynthesis over the whole earth corresponds to the formation of 10^{11} tons of organic carbon. E. I. Rabinowitch, a major worker in photosynthesis research, wrote about this remarkable fact:

The reduction of carbon dioxide by green plants is the largest single chemical process on earth. To make clear what a yield of 10^{11} tons/year means, we may compare it with the total output of the chemical, metallurgical, and mining industries on earth, which is on the order of 10^9 tons annually. Ninety percent of this output is coal and oil, i.e., products due to photosynthesis in earlier ages.

E. I. RABINOWITCH: 1945.
Photosynthesis and Related Processes, Vol. 1. p. 9.
Interscience Publishers, New York.

Photosynthesis

Ideas regarding the evolutionary beginnings of photosynthesis are sparse. Indeed, the modern form of the process is so complex that much about it is still to be learned. The following pages summarize some general aspects of photosynthesis in modern organisms and provide ideas for speculations about its beginnings.

First it is necessary to say something about the interactions of light and matter.

LIGHT FROM THE SUN

Deep within the sun, under great pressure and at temperatures approximating 10 million °C, atomic nuclei undergo thermonuclear fusion, or nuclear burning. The fuel is hydrogen nuclei from a source so immense that utilization of only 1% of the supply is enough to maintain the solar furnace for nearly a million years. Gravity allows only radiation from the solar fires to escape. About 10^{25} kcal/day escape, and about 500 kilocalories per square meter per day (kcal/m² day) of this radiation reaches the surface of the earth.

Solar radiation is part of a continuum of radiation collectively known as the electromagnetic spectrum (Figure 6-3). Other parts of this spectrum with familiar names are cosmic rays, x rays, ultraviolet light, visible light, microwaves, and radiowaves. All such radiation moves through space at the same speed, 186,000 mi/sec (3×10^8 m/sec). All types of radiation are treated as having a dual nature. Radiation moves as though it consists of waves. However, in making energy exchanges with matter, radiation behaves like a stream of particles. These particles are called **photons.** They carry specified and indivisible quantities of energy.

The energy of a photon is described in electron volt units (eV), where 1 eV = 3.8×10^{-23} kcal. Photon energy varies inversely with the wavelength of radiation (distance between successive crests in the radiation wave), ranging from 10^{17} eV for the most powerful cosmic rays to 10^{-9} eV for long radio waves. Pho-

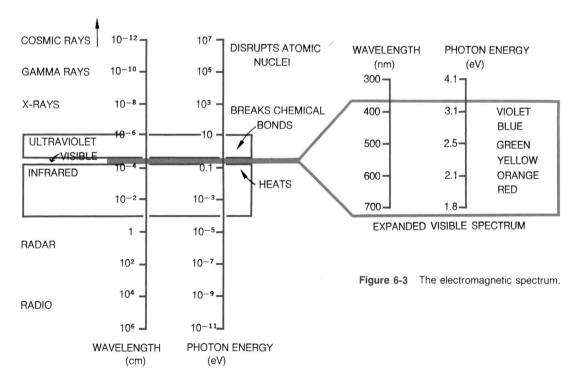

Figure 6-3 The electromagnetic spectrum.

NATURAL HISTORY OF ATOMS

ATOMIC STRUCTURE

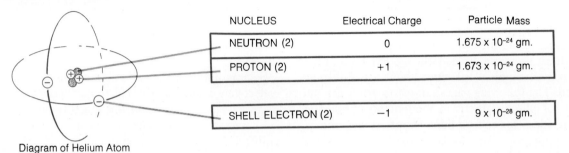

Diagram of Helium Atom

NUCLEUS	Electrical Charge	Particle Mass
NEUTRON (2)	0	1.675×10^{-24} gm.
PROTON (2)	+1	1.673×10^{-24} gm.
SHELL ELECTRON (2)	−1	9×10^{-28} gm.

ENERGY LEVELS AND CHEMICAL REACTIONS

Three "simple" rules:
(1) The outermost energy level never has more than 8 electrons.
(2) If the outer energy level has few electrons, the tendency is to give up electrons in chemical reactions
(3) If the outer energy level is nearly full, the tendency is to accept electron in chemical reactions

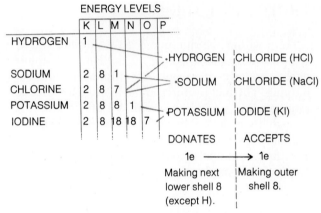

ENERGY LEVELS

	K	L	M	N	O	P
HYDROGEN	1					
SODIUM	2	8	1			
CHLORINE	2	8	7			
POTASSIUM	2	8	8	1		
IODINE	2	8	18	18	7	

HYDROGEN CHLORIDE (HCl)
SODIUM CHLORIDE (NaCl)
POTASSIUM IODIDE (KI)

DONATES	ACCEPTS
1e ⟶	⟶ 1e
Making next lower shell 8 (except H).	Making outer shell 8.

TWO KINDS OF CHEMICAL REACTIONS

A. IONIC — Electron(s) move from one atom to another leaving the two electrically unbalanced

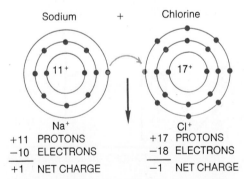

Na+	Cl+
+11 PROTONS	+17 PROTONS
−10 ELECTRONS	−18 ELECTRONS
+1 NET CHARGE	−1 NET CHARGE

The 2 ions thus formed are attracted to each other but may dissociate in water. In solid form a "supermolecule" is formed with each ion surrounded by others of opposite charge.

B. COVALENT — Electron(s) do not permanently leave or join atoms but are shared, orbiting both atoms in the molecule thus formed

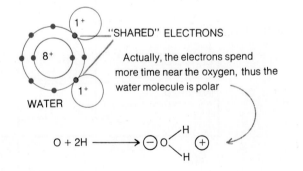

"SHARED" ELECTRONS

Actually, the electrons spend more time near the oxygen, thus the water molecule is polar

Figure 6-4 Atomic structure and chemical reactions.

ton energy is about 30 eV for a typical wavelength in the visible spectrum. The photon energy of visible light is more than enough to effect great changes in the energy and material balance of living organisms. Thus of the 500 kcal/m^2 day of solar energy reaching the surface of the earth, about 200 kcal/m^2 day are at visible wavelengths suitable to support photosynthesis. When the efficiency losses of photosynthesis and related metabolic processes are taken into account, this is still enough energy to form about 700 g of carbohydrate per square meter of photosynthetic surface per day.

ATOMS, MOLECULES, AND LIGHT

In order to understand how light energy is utilized in photosynthesis, it is necessary to examine a little more of the behavior of electrons, atoms, and molecules (Figure 6-4). The atom is conveniently pictured, in terms of the well-known Bohr planetary model, as a nucleus surrounded by electrons orbiting at specified distances in zones called electron shells, or energy levels. The nucleus contains protons with positive charges and neutrons, which have no charge. The number of protons is the same for all atoms of an element and is known as the **atomic number.** Thus all carbon atoms have six protons, all hydrogen atoms have one, and so on. The total number of protons and neutrons in the nucleus is called the **mass number.** Elements may have atoms with different mass numbers caused by variation in the number of neutrons; a different mass number represents an **isotope** of the element. Isotopes that have more neutrons than protons may be radioactive, as is the isotope of carbon which has eight neutrons (mass number 14, containing eight neutrons and six protons, called carbon-14) in contrast with the common form of carbon, which contains six neutrons (carbon-12).

Atomic nuclei occupy only a small fraction of the space filled by the atom, perhaps as much as 1/10,000 of a typical atomic diameter of 10^{-12} cm. This is roughly the same ratio that the diameter of the earth bears to the distance to the moon. Although the nucleus occupies a very small space in the volume of the atoms, it is dense beyond comprehension. A 1 cm diameter sphere formed of pure atomic nuclei would weigh over 130 million tons! What holds the nucleus together is poorly understood, but the examples of nuclear reactions in the sun and nuclear weapons show that atomic nuclei harbor great forces. Such forces do not contribute to the matters that concern us now since we are only interested in the chemical properties of atoms that reside in their electrons. Nuclear forces are considered in Chapter 26.

Electrons each have a negative charge of 1 and are equal in number to the number of positively charged protons in the nucleus in an electrically neutral atom. Because of the Bohr model, we tend to think of electrons as solid objects orbiting the nucleus along a defined, eliptical path at a predictable speed, like the moon around the earth. Actually the electron is somewhat more mysterious: Although it does have definite charge and mass, it also has the wave properties of electromagnetic radiations such as light. Its orbits are only generally defined, and their energy content may vary. Electrons orbit the nucleus at various distances or energy levels. Levels nearest the nucleus have electrons with the lowest energy, and electrons farthest out have the highest energy and are the most easily lost by the atom. Each energy level may be occupied by only a specified maximum number of electrons (Figure 6-4). The outermost energy level is our principal interest because it contains **chemical bonding** or **valence electrons** that are involved in chemical reactions.

CHEMICAL REACTIONS

When an atom naturally has the maximum allowable number of electrons in its outer shell, as with helium (He), neon (Ne) and argon (Ar), it does not enter into chemical reactions except under rare circumstances. If the valence shell has less than the maximum allowable number of electrons, then the atom is able to enter into chemical reactions. The basic rule of such reactions is that the valence shells of participating atoms should be filled. This is accomplished by giving up or receiving electrons or by sharing electrons among participating atoms. When a valence shell contains few electrons, the atom tends to give up electrons until the shell is empty. Sodium is an example of this type of atom. If the valence shell is nearly full, the tendency is to accept sufficient electrons to fill the shell, as is illustrated by chlorine.

Chlorine can accept one electron to make the maximum allowable number of eight in its valence shell.

If in a chemical reaction an electron is transferred completely from one atom to another, the donor is left with a positive charge, the recipient becomes negatively charged, and the two atoms are held together by the unlike charges. Should the reaction take place in solution, the charged atoms may drift apart, that is, dissociate and exist as ions, which are charge-bearing atoms. Reactions of this type, characterized by complete change in ownership of electrons, are called **ionic reactions** and the bond between the atoms involved is called an **ionic bond.** Sodium and chlorine combining to make common salt is an example (Figure 6-4).

The biologically important atoms, carbon, nitrogen, and oxygen can form **covalent bonds,** which are bonds formed without completely giving up electrons. Instead, electrons are shared by the atoms participating in the bond, with the electrons forming a new orbit around both nuclei. Such bonds are strong, and dissociation of covalently bonded atoms does not occur in solution. Carbon represents a particularly important and complex form of covalent bonding since it may use any or all of its four valence electrons to form bonds with other carbon atoms and thereby build up the complex molecules characteristic of life.

EFFECTS OF LIGHT ON ATOMS AND MOLECULES

If a photon with appropriate energy passes sufficiently near an atom, all of its energy may be transferred to a valence electron. The photon goes out of existence as the electron attains a higher energy level. If the new energy level is high enough, the electron escapes from the atom, which then is ionized. Smaller energy absorptions elevate an electron from its minimal energy state, called the **ground state,** to one of several possible **excited states** short of the ionization level. If raised only to the lowest of these excited

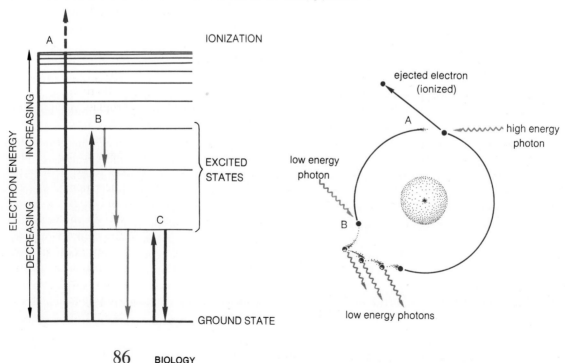

Figure 6-5 Photon interactions with atoms. Graphical and diagrammatic representations of three types of interactions. A. A high-energy photon imparts sufficient energy to an electron to cause ionization. B. A lower energy photon raises an electron through two possible excited states (intermediates impossible) to a third, followed by a "cascade" of successive low-energy photon emissions as the electron returns to the rest state through intermediate excited states. C. A still lower energy electron raises an electron to the next higher excited state, from which it immediately returns to rest state with emission of a photon of the same energy as the exciting photon.

states, the electron may immediately return to the ground state by emitting a photon of the same energy as the one first absorbed. If an electron is raised to some higher energy level, the electron cannot return in one step to the ground state; it does so in smaller steps through each intermediate level, giving off a less energetic photon at each jump downwards (Figure 6-5).

Instead of being reemitted as photons, some of the energy absorbed by the atom may be utilized in other ways. By far the commonest usage is for the energy to be degraded into increased atomic and molecular motion which is detected as heat. Of particular interest to us is the fact that *energy from absorbed photons may have direct chemical effects.* Absorbed photon energy may break chemical bonds. It may be stored, as by causing a structural change in a molecule that may later revert to the initial state with release of energy. Absorbed energy also may be transferred to some distance from the absorbing atoms, both within and between molecules and there produce chemical effects.

What determines whether or not a photon is absorbed by atom or molecule? The single requirement is energy matching between the photon and an electron. A photon is an indivisible unit of energy and must give up its energy in an all or none fashion. On the other hand an electron can only accept energy in packages that are precisely large enough to raise it to one of several allowable energy levels or that are large enough to cause ionization, that is ejection of the electron from the atom. The more valence electrons an atom has, the finer is its net for catching photons; the smaller the number of valence electrons, the coarser is its net. Thus hydrogen, with one valence electron, absorbs photons at only four wavelengths in the visible region of the spectrum. Similarly, energy matching of photons to electrons is much easier in molecules than in atoms because the bonds holding molecules together provide additional energy levels beyond those existing in the isolated atomic components of the molecule.

PHOTOSYNTHESIS IN MODERN PLANTS

If we apply these ideas about the interaction of light and matter, we find that photosynthesis consists of (1) utilization of light energy to elevate electrons from oxidation of a substrate molecule to sufficiently high energy levels to permit synthesis of ATP and the reduced form of nicotinamide adenine dinucleotide phosphate (NADPH), and (2) utilization of these energy-rich compounds in production of new carbohydrate molecules by reduction (hydrogen addition) of carbon dioxide.

In photosynthesis by green algae and higher plants, the substrate molecule oxidized is water and the process of photosynthesis involves:

1. Oxidation of water with liberation of oxygen and use of hydrogen in photosynthetic reactions.
2. Light absorption by chlorophyll with captured energy used to raise electrons to energy levels sufficient for ATP and NADPH formation.
3. Utilization of ATP and NADPH to support incorporation of carbon dioxide into carbohydrate (carbon dioxide fixation).

Light energy is required in two steps of the electron energizing process, whereas the remainder of photosynthesis may occur in the dark. Conventionally the processes of step 3 are called the **dark reactions,** whereas steps 1 and 2 are called the **light reactions.**

In some bacteria and algae the substrate oxidized is an organic molecule, such as an alcohol, or a reduced sulfur compound, such as hydrogen sulfide.

Chlorophyll and Chloroplasts

Both the light and dark reactions of photosynthesis in green algae and higher plants take place in **chloroplasts** (Figure 6-6) whose structure is critically important to the process; isolated chloroplast contents do not photosynthesize. Chloroplasts are saucer shaped, about 6μ in diameter and 1μ thick. Seen with electron microscope (Figure 6-7) they are enclosed in a double membrane whose inner part forms thin partitions or **lamellae** that partition the chloroplast and unite at intervals with piles of membranes known as **grana.** The grana and lamellae contain the pigment chlorophyll, whereas the rest of the chloroplast contains the enzymes and other factors necessary for carbon dioxide fixation and the synthesis of sugar, starch, fats, and proteins from the new carbohydrate.

The chlorophyll molecule has two principal parts, a tetrapyrrole ring bearing a magnesium atom and a

A STARCH ACCUMULATES IN ILLUMINATED PARTS OF GREEN LEAVES

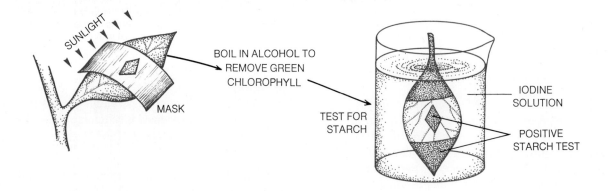

B OXYGEN IS LIBERATED BY ACTION OF SUNLIGHT ON CHLOROPLAST

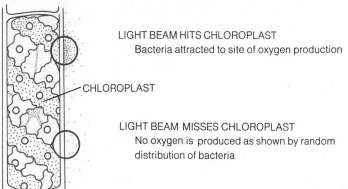

LIGHT BEAM HITS CHLOROPLAST
Bacteria attracted to site of oxygen production

CHLOROPLAST

LIGHT BEAM MISSES CHLOROPLAST
No oxygen is produced as shown by random distribution of bacteria

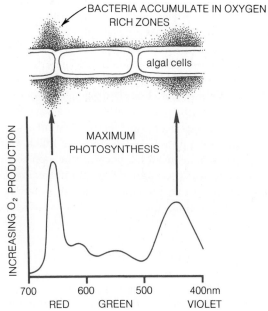

BACTERIA ACCUMULATE IN OXYGEN RICH ZONES

algal cells

C CHLOROPHYLL ABSORPTION SPECTRUM IS SIMILAR TO PHOTOSYNTHESIS ACTION SPECTRUM

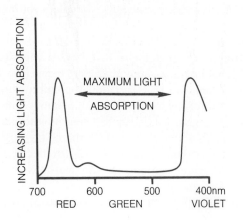

Figure 6-6 Fundamental experiments demonstrating photosynthetic role of chloroplasts. **A.** First performed by Julius Sachs (1832–1897), this experiment takes advantage of the fact that during intense photosynthesis sugar is converted to starch, which remains near the point of origin and can be detected chemically, showing that photosynthesis occurs only in the illuminated part of the leaf. **B.** Theodor Englemann, in 1880, used the chemotaxis of certain motile bacteria toward oxygen and the very large chloroplast of the green alga, *Spirogyra*, to show, by bacterial aggregation at sites of oxygen evolution, that oxygen is evolved when the chloroplast is illuminated. **C.** Later experiments show that the **absorption spectrum** of chlorophyll is identical with the **action spectrum** of photosynthesis, that the wavelengths of light most strongly absorbed by chlorophyll are the wavelengths that most strongly promote photosynthesis.

Figure 6-7 A chloroplast lies in a thin layer of cytoplasm adherent to the cell wall in a leaf mesophyll cell of tobacco. ×24,000. (Robert Gill.)

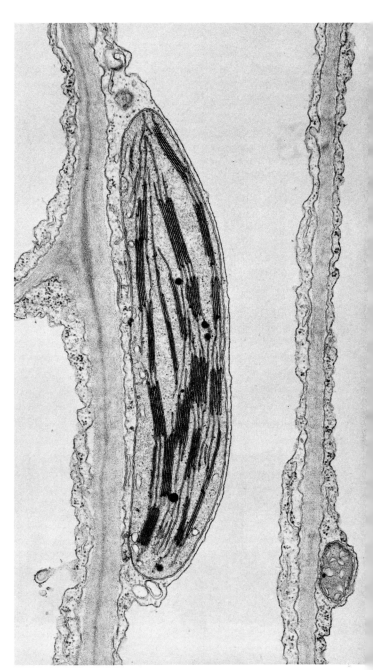

long chain alcohol, phytol (Figure 6-8). It is an ancient molecule, as evidenced by two facts. First, the basic component of its ring structure was probably readily synthesized under prebiological conditions. Second, various molecules incorporating the tetrapyrrole group are found in all organisms where they perform important functions having to do with electron transport and respiration.

Chlorophyll molecules are located in the membrane structures of the chloroplast grana. Their fat soluble phytol tails are held in the lipid component of the membranes, and their tetrapyrrole rings are lined up in the immediately adjacent protein components of the grana. If chloroplast membranes are broken apart with sufficient care, particles are released that can perform some of the steps in photosynthesis. Each particle consists of about two hundred chlorophyll molecules plus a few other molecules that have important cooperative functions with the chlorophyll. There appear to be two types of these units, and both are required to achieve the entire process of photosynthesis. These observations, showing that a very large number of chlorophyll molecules must work together in photosynthesis, fit well with the demonstration that several photons are required for the oxidation of each water molecule.

Now, it would be altogether impractical to expect photosynthesis to work as well as it does if the process initially required several photons simultaneously to hit a single site of the size of a few chlorophyll molecules. Instead it appears that each unit of several hundred chlorophyll molecules transmits the energy received from photons (which can be absorbed anywhere among them) to a central reaction center where the basic reactions of photosynthesis take place. This transport of photon energy is facilitated by the tight, almost crystalline, packing together of chlorophyll and

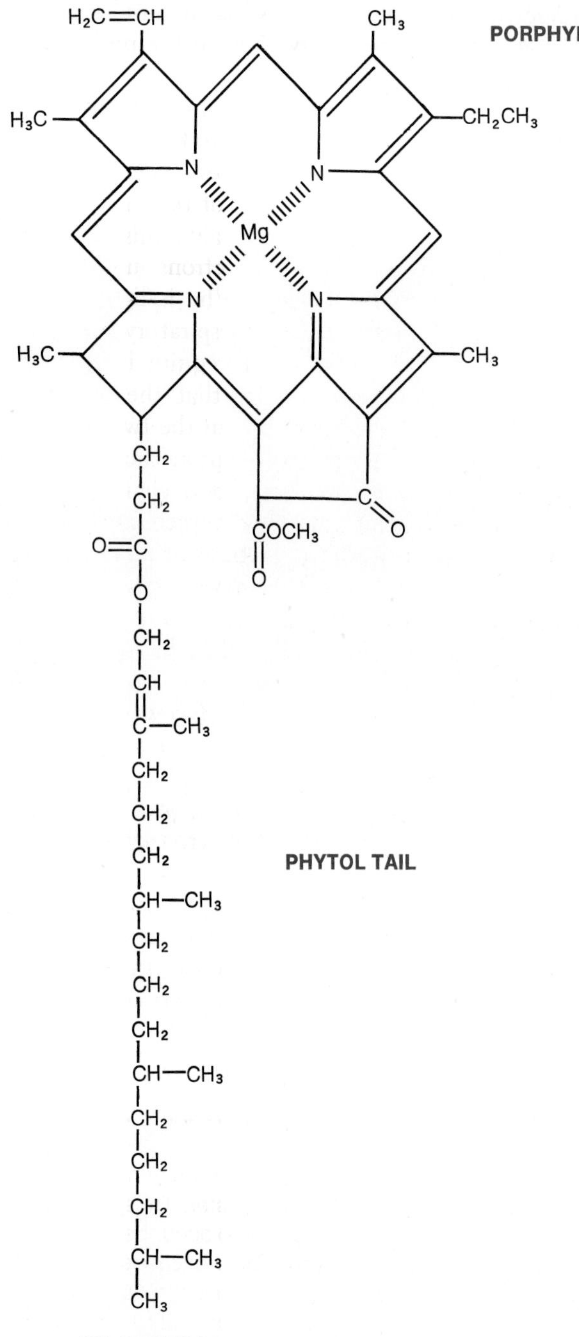

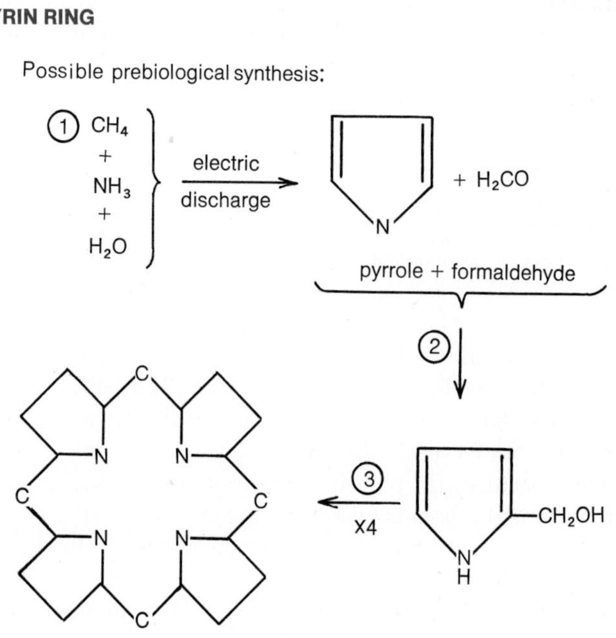

PORPHYRIN RING

Possible prebiological synthesis:

CHLOROPHYLL a

Figure 6-8 The structure of chlorophyll.

reaction center molecules. Probably photon energy moves through the molecules of the photosynthetic unit by resonance transfer, a mechanism by which energy is transmitted over very short distances from an excited electron of one atom to another atom without movement of an electron between the two. This is somewhat like the way radio waves are able to transfer energy from a transmitting antenna to a receiving antenna.

Electron Transport

An instructive way to write an overall equation for the steps of photosynthesis just described is as follows:

$$H_2O + CO_2 \xrightarrow{light} (CH_2O) + O_2$$

That is, water and carbon dioxide in the presence of light form carbohydrate (CH_2O) and oxygen. Described this way, photosynthesis is the reverse of the burning (oxidation) of organic matter, a reaction that, once started, proceeds without input of energy and releases energy as heat.

The equation for burning is a gross description of how organisms extract energy from oxidation of carbohydrates. Remember that it was said in connection with anaerobic glycolysis (page 71) that the oxidation could not usefully proceed in one step, as in burning, if the organism is effectively to harvest a substantial amount of the energy available in the overall reaction. Consequently, glycolysis proceeds in a series of steps, allowing energy units of manageable size to be used in the formation of ATP. The example of glycolysis also revealed an identical problem in respect to conservation of reducing power. There was a similar solution, namely addition of steps between reductant (CH_2O) and oxidant to allow stepwise flow of hydrogen (protons) and electrons in a manageable way, with biologically useful results. In anaerobic glycolysis only one link in this series of steps was seen because, in the absence of oxygen, glucose oxidation does not proceed to the final step of reduction of oxygen to water. That link was the important hydrogen carrier molecule NAD^+, which in glycolysis cycled hydrogen (becoming NADH) between glyceraldehyde-3-phosphate and lactate.

In the metabolism of most organisms, in the presence of oxygen, electrons and hydrogen flow through a complex series of carrier molecules to oxygen, giving up energy along the way to form ATP. This series of carrier molecules is collectively known as the **respiratory chain**. In oxidative metabolism, the role of the respiratory chain is to draw off the energy liberated when the hydrogens of carbohydrates reduce oxygen by allowing the oxidation to proceed in gradual steps. In some of these the energy released is of a suitable magnitude for conservation as chemical energy in ATP. A good analogy, returning to our campfire example of the first chapter, would be to imagine the difficulty of cooking over a great bonfire in which all available fuel for cooking a meal is burned at once as compared with the ease of cooking by burning fuel a stick at a time, as necessary to keep the pot boiling nicely.

Since photosynthesis is the formal reverse of oxidative metabolism, we would expect to find an electron transport chain at work in the process. There is such a chain, and it is precisely the role of the light energy required in photosynthesis to enable this chain to operate, to transport hydrogen (electrons and protons) from water to carbon dioxide. Why is light energy necessary, since, after all, the respiratory chain of oxidative metabolism does not require light? The answer is indicated by the fact that the transport chains are functioning oppositely in the two systems. Oxygen is an avid hydrogen acceptor; thus, it is no great problem to operate a transport chain carrying hydrogen down the energy hill to oxygen, as occurs in oxidative metabolism. The reactions of photosynthesis are quite the opposite; they have to pry hydrogen away from the grip of oxygen in water and carry it up the energy hill away from oxygen. *Light supplies the energy to do this.*

There is a very useful and simple way to make precise such fuzzy descriptions of pushing hydrogens up and down energy hills. It is possible to indicate quantitatively the relative tendencies of chemicals to attract or give up hydrogens or electrons through the use of oxidation-reduction potentials (Note 6-1). From the oxidation-reduction potential of a molecule, measured in electron volts (eV) you can tell at a glance whether a certain molecule can supply or receive electrons with regard to any other molecule

Note 6-1 *Oxidation-Reduction Potential*

In an oxidation-reduction reaction electrons are transferred from a donor (reductant) to an acceptor (oxidant). The electron transfer may be made by hydrogen transfer. Thus dehydrogenation is the equivalent of oxidation. The tendency of such reactions to occur between various molecules is measured as the oxidation-reduction potential difference between them. Oxidation-reduction potentials are determined by making the chemicals in question one electrode of an electrochemical battery whose other electrode is a standard oxidation-reduction reaction and measuring the magnitude and sign of the electromotive force (emf) that the battery develops. The more positive the emf the more readily the molecule accepts electrons and the more negative the greater the tendency to give up electrons. By a series of such measurements comparing various molecules with the standard a table may be constructed showing the relative tendency of chemicals to be donors or acceptors of electrons. Such tables are useful in predicting chemical reactions.

whose oxidation-reduction potential is known. Using this system, we can make a graph in which the energy relationships among the molecules involved in photosynthesis (or any other phase of metabolism) is shown, making it much easier to grasp the chemical logic of the process (Figure 6-9).

Electron Transport in Photosynthesis

The oxidation reduction diagram in Figure 6-9 shows that the electron transport problem in photosynthesis involves getting hydrogens (protons and electrons) away from water and raising them to a high enough oxidation-reduction potential so that they will be useful in reducing carbon dioxide. The energy hill to be climbed has a "height" of about 1.2 eV. The photosynthesis system accomplishes this and also generates ATP, which is used in carbon dioxide reduction. To do this, photon-capturing units work in series with each other. One of these units energizes electrons, ultimately derived from water, sufficiently so that they can participate in ATP formation and then be used by another photon capturing unit. This second unit, starting from a higher energy level, gives electrons sufficient energy to rise to an oxidation-reduction level allowing participation in formation of NADPH. This and the energy of ATP produced in the process are then used to reduce carbon dioxide.

Light energy is necessary at only two specific points in the entire process. At each of these points, the energy from an absorbed photon allows a photosystem (many chlorophyll and other types of molecules) to supply an energized electron to the high energy end of a carrier chain. From there the process runs downhill without further energy input.

Water enters the picture because new electrons have to come from somewhere to resupply the system. These are provided by water to the photosystem that provides the first uphill kick. Light, therefore, has no direct effect in splitting water. It simply creates an attractive place for electrons from the hydrogens of water to go by making the initial photosystem strongly electropositive. How the water is split is unknown, beyond the facts that it is a process that takes several steps and requires manganese.

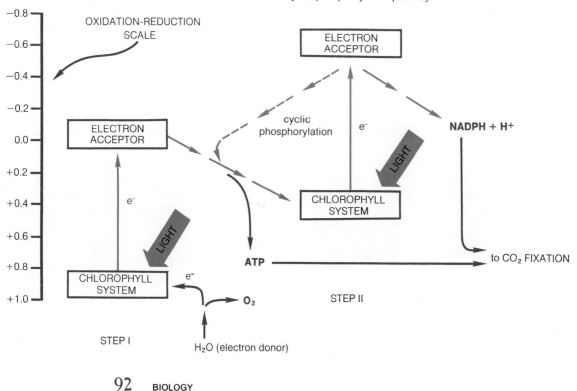

Figure 6-9 In photosynthesis light acts at two sites to sequentially raise electrons from water to high enough oxidation-reduction levels to generate ATP and NADPH. From the electron acceptor of Step II, high-energy electrons may flow downhill to either produce NADPH or additional ATP via the cyclic phosphorylation pathway.

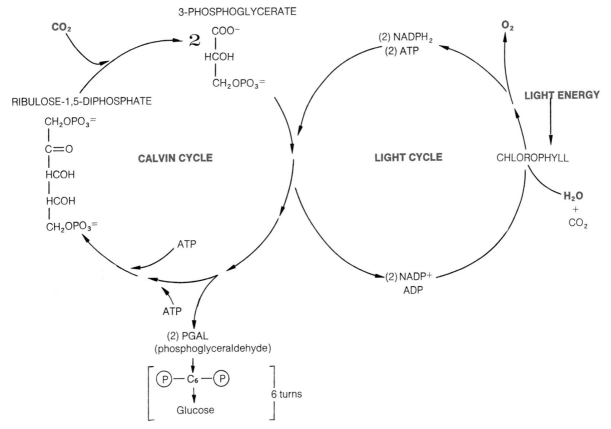

Figure 6-10 Carbohydrate is produced by CO_2 fixation in the Calvin Cycle using energy obtained in the light-dependent reactions of photosynthesis. The arrows between 3-phosphoglycerate and ribulose-1,5 diphosphate represent at least 10 reactions. Some of the carbohydrate products accumulate within the chloroplast, for example, starch grains (see Figure 6-6).

Carbohydrate Production in Photosynthesis

With the light reactions of photosynthesis providing reducing power and ATP, the next step is the utilization of these materials in synthesis of *new* carbohydrate. By new we mean that inorganic carbon in the form of carbon dioxide is utilized, thereby adding to the total supply of organic carbon on earth. This process, *carbon dioxide fixation*, is not exclusive to photosynthetic organisms, but when it occurs in others there is no net production of new organic carbon.

In photosynthesis the dark reactions do not proceed by direct construction of organic molecules whose entire carbon backbones are newly formed from carbon dioxide. Instead carbon dioxide is added to a five-carbon sugar, ribulose-1,5-diphosphate, which immediately breaks down into two molecules of 3-phosphoglycerate. One of these bears the carbon atom derived from carbon dioxide. Still within the chloroplast, the 3-phosphoglycerate enters into a cycle of conversions among several kinds of phosphorylated sugars. In each turn of the cycle one six-carbon glucose molecule is released and one ribulose-1,5-diphosphate that is necessary for turning the cycle over again, is made. Six rounds of the cycle is the minimum necessary to insure that the glucose molecule released on that sixth turn could come directly from carbon dioxide fixation in the cycle. This cycle, called the **Calvin cycle** after its principal discoverer, is shown in Figure 6-10. All of the carbon of the green plant ultimately is produced in this way.

The Calvin cycle is quite efficient since it provides all of its own molecular requirements except for carbon dioxide, ATP, and NADPH. This is made evident by summing up the cycle, cancelling out all of the components that stay in, and listing only input and output, as shown in Figure 6-11. The equation is

NET* RESULT OF 6 TURNS OF THE CALVIN CYCLE:

$$6CO_2 + 18\ ATP + 12\ NADPH + 12H^+ + 12O_2 \longrightarrow SUGAR\ (6\text{-carbon}) + 18P_{inorganic} + 18\ ADP + 12\ NADP^+$$

*ribulose-1,5-diphosphate is made and used up at the same rate and so cancels out.

Figure 6-11 Results of the Calvin Cycle.

summed over six turns of the cycle, since one carbon at a time is added to make the hexose output. Energy input to the cycle is about 140 kcal/mole for the NADPH, making a total of 760 kcal/mole of glucose produced. The energy available from 1 mole of glucose is 670 kcal, which marks the Calvin cycle as highly efficient among complex organic reactions. The energy loss, 90 kcal, or about 11% of the input energy, is the cost of the cycle. Efficiency of the dark reaction is thus 89%. The efficiency of the light reactions of photosynthesis is also high, with about 35% of the actual energy input (photons trapped usefully by chlorophyll) being converted into ATP and NADPH. That would make the efficiency of the entire photosynthetic process, light plus dark reactions leading directly to glucose, about 31% (0.35 × 0.89 × 100). Of course, only a small amount of the light intercepted by a plant is put to use in photosynthesis and so the efficiency of photosynthesis on the basis of total light absorbed by the plant is 1% or 2%.

In actuality the Calvin cycle may supply raw materials to other biosynthetic pathways rather than exclusively manufacturing glucose. This is possible because some intermediate compounds in the photosynthetic dark reactions are also intermediates in other metabolic cycles. For example, 3-phosphoglyceric acid is familiar as a part of the glycolysis cycle and, as we shall see in Chapter 7, this cycle is linked with pathways that produce amino acids. Similarly, reduction of glyceraldehyde phosphate can lead to glycerol formation, and this to the synthesis of lipids. Such examples show how the total array of biomolecules found in the green plant are ultimately attributable to inputs of carbon dioxide, water, nitrogen, and several inorganic chemicals.

Evolution of Photosynthesis

Our scheme of evolution indicates that photosynthesis had to originate early in the evolution of life to offset depletion of organic molecules in the primitive seas. It must be obvious that the complex process just described for modern green plants, involving two photosystems interacting with electron transport chains to effect the very large oxidation-reduction step from water to NADP, could not have suddenly come into being without simpler antecedents. Is there evidence as to what these might have been?

There are existing anaerobic bacteria that conduct photosynthesis with a one-step photosystem by the use of already reduced compounds as the primary hydrogen donor, rather than water. Green sulfur bacteria use hydrogen sulfide (H_2S) as the hydrogen donor, liberating free sulfur and forming high energy compounds necessary for carbohydrate synthesis. Certain other bacteria use simple organic compounds such as alcohols and fatty acids for the same purpose. These are organisms of great interest, providing us with models of what photosynthesizers might have been like in times when a significant amount of organic matter still existed in the sea and while there was still a primitive earth atmosphere containing hydrogen sulfide but no oxygen. Although this pattern of photosynthesis is simplified, because only one photosystem is required, it is still highly complex and can at best represent only a model of an intermediate in the evolutionary route to photosynthesis based on water as the hydrogen donor.

Two other considerations suggest still earlier forms of photosynthesis. The first is the fact that many organisms are able to conduct photophosphorylation independently of the synthesis of carbohydrate. Since phosphorylation takes place in the electron transport chain, it is worth speculating that the beginnings of photosynthesis lie in the electron-hydrogen transport chains of early organisms. A slight modification of one of the transport molecules might have rendered it capable of collecting light energy and using that energy to facilitate phosphorylation of ADP (adenosine diphosphate). This, of course, would have been a most useful development because, as already suggested, energy sources for building up required molecules would probably have become limiting before the supply of organic molecules was exhausted.

The second point is that certain common molecules in electron carrier chains are highly similar to chlorophyll, making development of photochemical abilities in carrier chains chemically rather probable.

Before there was oxygen in the atmosphere, there could have been no ozone shield in the stratosphere to ward off high energy ultraviolet photons. Reference to Figure 6-3 will show that these are so much more energetic than the green and red wavelength photons that are used in two-stage photosynthesis that one might wonder why an early photosynthetic system based on such photons might not have developed. It might have been capable of flipping electrons to most respectable energy levels in a single step. The truth of the matter is that ultraviolet radiation of wavelengths only slightly shorter than those now admitted through the atmosphere are extremely destructive to living material, producing chemical changes that break down nucleic acids, proteins, and other critical components of the cell. In order to survive, the earliest forms of life must have had to avoid direct exposure to short wavelength ultraviolet, perhaps by living under a few meters of water. They could not have taken advantage of this rich but destructive energy source. Thus, as is often the case, a biological process, seeming needlessly complex at first, turns out to be about the only conceivable way to achieve a particular function within the limitations imposed by biological materials and the environment.

7 Oxygen and Life

An immensely important, life-induced change came over the earth during the first 1000 million years or more after its formation. The change began as the earliest photosynthesizers utilized all readily available hydrogen donor molecules, making the switch to water as the hydrogen donor for photosynthesis a final necessity. The result was *liberation of free oxygen* into the environment as a by-product of photosynthesis. This ultimately brought about a revolution in the chemistry of the earth's surface, changing it from a *reducing* to an *oxidizing* environment. Although free oxygen is also produced by the action of ultraviolet light on water vapor, it is thought that photosynthesis was the major source.

Oxygen buildup in the atmosphere to the present day level of 21% probably took on the order of 2000 million years after appearance of the first oxygen releasing photosynthesizers, with the most rapid and significant increase occuring in the last 600 million years. Thus, the concentration of oxygen in the atmosphere was only about 80% of present levels as recently in earth's history as the time of appearance of land plants, about 400 million years ago. Rise of oxygen concentration to present atmospheric levels was coincident with the great flourishing of plants that produced our oil and coal deposits (Figure 7-1). It is perhaps fortunate that oxygen buildup to present levels took as long as it did. Although the ability to use oxygen confers great advantages, it is nonetheless readily capable of oxidizing and, thereby, destroying virtually all biomolecules. Thus organisms had many difficult adjustments to make before they could take full advantage of free oxygen.

HISTORY OF THE DEVELOPMENT OF OXYGEN IN THE ATMOSPHERE

The facts regarding the atmospheric buildup of oxygen and its consequences for life, although written indirectly, are plain in the geological record. For nearly 2000 million years after the first known single celled algae and bacteria appeared, very little happened insofar as the development of higher forms of life is concerned. Then, only about 600 million years ago, there seems to have occurred a great acceleration of evolution in which all of the principal types of plants and animals appeared. Many authorities suggest this sudden flowering of life was due in part to the rapid increase of atmospheric oxygen to present levels, since these events happened more or less at the same time.

The periods of these events are known in geological terminology as the Precambrian and Cambrian ages. In the Precambrian, which began about 3500 million years ago, the concentration of oxygen in the atmosphere was probably less than 0.1% of present levels. This had two important consequences: first, metabolism of organisms was obviously limited to anaerobic pathways and was thereby inefficient; second, since

Figure 7-1 Artist's impression of the luxurious vegetation existing at the time when the earth's major oil, gas, and coal deposits were laid down.

there was no oxygen-produced ozone layer of any significance, the earth's surface was exposed to intense and lethal ultraviolet radiation. Ozone (O_3), produced by reaction of ultraviolet light with oxygen, strongly absorbs high energy ultraviolet. At least for photosynthetic organisms lack of the ozone layer meant that the only suitable environment was beneath a layer of water deep enough to reduce ultraviolet to nonlethal levels and yet shallow enough to transmit sufficient visible light to support photosynthesis. Initially the required water depth would have been about 10 m.

Stromatolites (Figure 7-2), algae similar to modern blue-green algae, are examples of the plant life of this period. They appear to have existed as mats, which collected sediments in their filamentous strands. These mats evidently grew in shallower and shallower water as the ozone layer built up and also because the superficial, sediment filled layers of the mat would protect underlying organisms from lethal exposure to ultraviolet rays.

When the atmospheric oxygen level reached 1% of present values, the ozone layer probably would have been sufficiently developed to allow plant life to exist in very shallow waters. This increased area of the earth suitable for photosynthesis would have promoted rapid increase in atmospheric oxygen concentration. As a consequence, ultraviolet radiation intensity was soon reduced sufficiently by the ozone layer to permit plants to colonize the land, and it is thought that this raised the oxygen concentration to modern levels. These events would have culminated at the beginning of the Cambrian period, 600 million years ago. By that time the switchover to aerobic metabolism in both aquatic and terrestrial organisms would have been well established, and it is likely that this, together with the great increase in inhabitability of the land made possible by reduction of incident ultraviolet radiation, contributed greatly to the vast diversification of living forms which then took place. Of equally great importance were new developments in cellular organization that occurred in the late Precambrian. These led to sexual reproduction which, as we shall see later (Chapter 11), greatly accelerated the evolutionary process.

OXYGEN AND LIFE

Figure 7-2 Ancient and modern stromatolites are constructed of sediments trapped or otherwise bound, principally by blue-green algae and also by bacteria and green algae. **A.** Fossil stromatolites, *Katernia africana*, about 2200 million years old, from South Africa. (Preston Cloud.) **B.** Recent stromatolites from intertidal zone, Shark Bay, Western Australia. Scale is given by the 2m shovel at right. (S. Awramik.)

Major deposits of iron ore (principally hematite, Fe_2O_3), such as the vast iron mines in the Lake Superior region of the United States, appear to have been laid down as biological oxygen production began and peaked. Some authorities believe that these deposits originated as the consequence of reaction of biologically produced oxygen with water soluble ferrous iron, thus:

$$2\,Fe^{2+} + 3\,O_2 \rightarrow 2\,Fe_2O_3$$

According to this line of reasoning, the initial toxic effects of oxygen would have been offset to some extent as long as the waters of the earth contained sufficient ferrous ions to react with oxygen released during photosynthesis.

METABOLISM IN THE PRESENCE OF OXYGEN

Presumably even the earliest photosynthesizers, and ultimately almost all organisms, experienced sufficiently high oxygen concentrations to make development of ways to channel oxygen into useful biochemical reactions advantageous. These developments, seen in the light of the biochemistry of modern organisms, included (1) protective antioxidant molecules and protective enzymes, and (2) aerobic metabolic pathways allowing the use of oxygen to provide energy in excess of that generated by the incomplete oxidations of anaerobic metabolism.

Metabolism of this complexity could no longer satisfactorily be carried out free-floating in the cytoplasm of cells and so we find that the many necessarily closely coordinated steps in aerobic metabolism are carried out by enzymes attached to membranes such as the inner membranes of mitochondria.

Antioxidants and Protective Enzymes

Several common biomolecules are strong reducing agents. Two well known vitamins, vitamin C (ascorbic acid) and vitamin E (alpha-tocopherol), have this property and thus are able to protect against oxidation (Figure 7-3). For example, vitamin E protects the lipids in biological membranes from oxidations that destroy their effectiveness in membrane function. The toxic superoxide anion, O_2^-, and hydrogen peroxide,

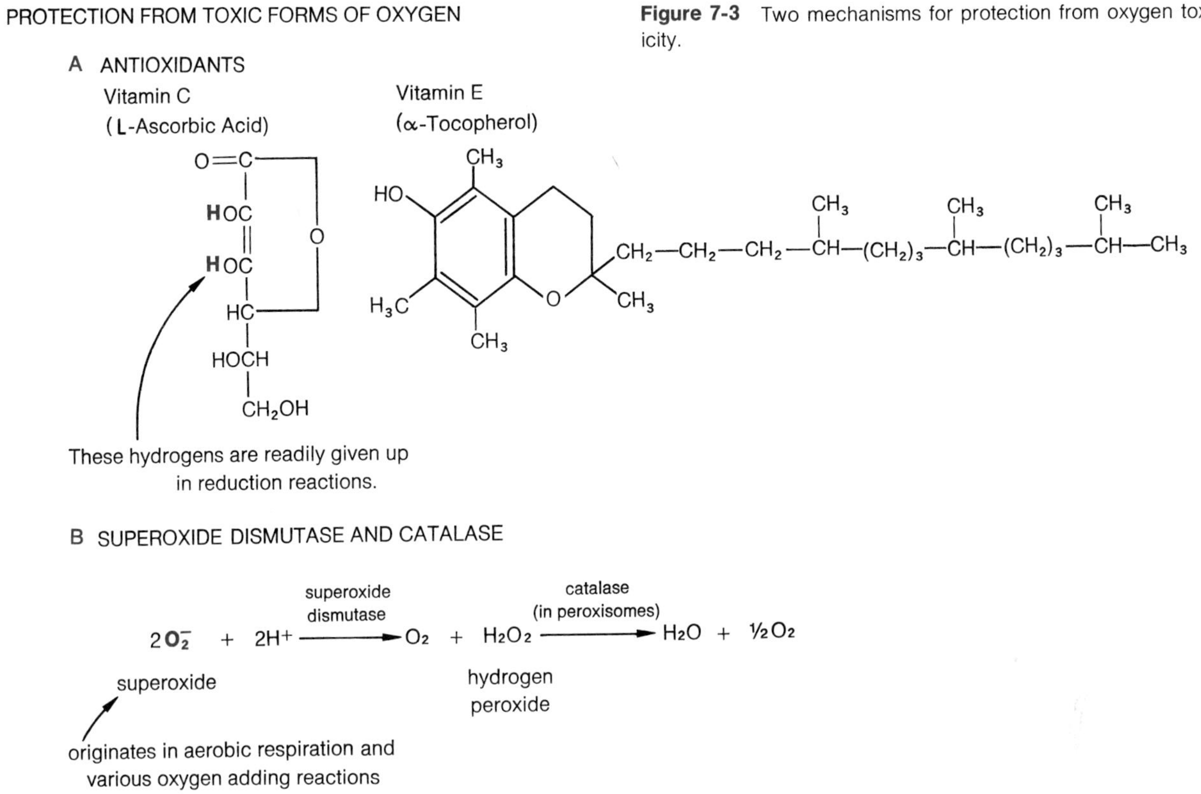

PROTECTION FROM TOXIC FORMS OF OXYGEN

Figure 7-3 Two mechanisms for protection from oxygen toxicity.

A ANTIOXIDANTS

Vitamin C (L-Ascorbic Acid)

Vitamin E (α-Tocopherol)

These hydrogens are readily given up in reduction reactions.

B SUPEROXIDE DISMUTASE AND CATALASE

$$2 O_2^- + 2H^+ \xrightarrow{\text{superoxide dismutase}} O_2 + H_2O_2 \xrightarrow{\text{catalase (in peroxisomes)}} H_2O + \tfrac{1}{2} O_2$$

superoxide — originates in aerobic respiration and various oxygen adding reactions

hydrogen peroxide

H_2O_2, are common products of the metabolism of aerobic organisms. Organisms are protected from the effects of these oxidants by two types of enzymes—superoxide dismutases, which convert superoxide anions to hydrogen peroxide, and catalase, which breaks down hydrogen peroxide to water and oxygen. Completely anaerobic organisms lack the dismutase enzyme; thus, it may be that this enzyme is essential for life in the presence of oxygen.

Aerobic Metabolic Pathways

The chemistry of life is complexly interwoven, making it difficult to discuss any part of it in isolation. This is particularly true of the aspects of metabolism that now concern us. The primary goal of this section is to show how carbohydrates are broken down all the way to carbon dioxide and water and how oxidations in general are usefully conducted in cells. To do this we must widen our view to consider two central mechanisms of cellular metabolism, the *tricarboxylic acid cycle* and the *oxidative phosphorylation* that occurs in the chain of electron transport molecules. These reactions are crucial because proteins and lipids, not only carbohydrates, enter the tricarboxylic acid cycle, making it the essential pathway for generation of energy from breakdown of carbohydrates, lipids, and proteins as well as for the interconversions of these several types of chemicals. Figure 7-4 shows these relationships. In general the central core of these reactions, involving the production of energy by oxidation of biomolecules by oxygen, is called **cellular respiration**.

THE TRICARBOXYLIC ACID CYCLE

You will remember that, when glucose moves through the pathway of anaerobic glycolysis, the end product is lactate. Breaking down glucose to lactate makes available 47 kcal of energy per mole. However, the total energy available from complete oxidation of glucose is many times that amount. The first step in harvesting this remaining energy from glucose is the tricarboxylic acid cycle. The tricarboxylic acid cycle does this by oxidizing two-carbon units (acetyl groups, CH_3COO^-), which come from pyruvate and ulti-

Figure 7-4 Tricarboxylic acid cycle.

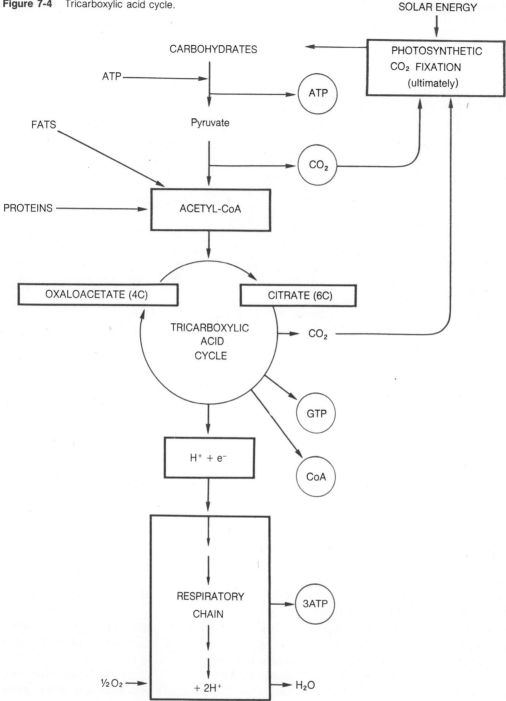

mately from carbohydrates, proteins, and lipids, by splitting off carbon dioxide and hydrogen. The latter is fed into the chain of electron carrier molecules, which uses them to generate ATP and finally reacts them with oxygen to form water, thus completing the oxidation. In the instance of oxidation of glucose by the tricarboxylic acid cycle the total potential energy yield is 686 kcal whereas energy losses associated with

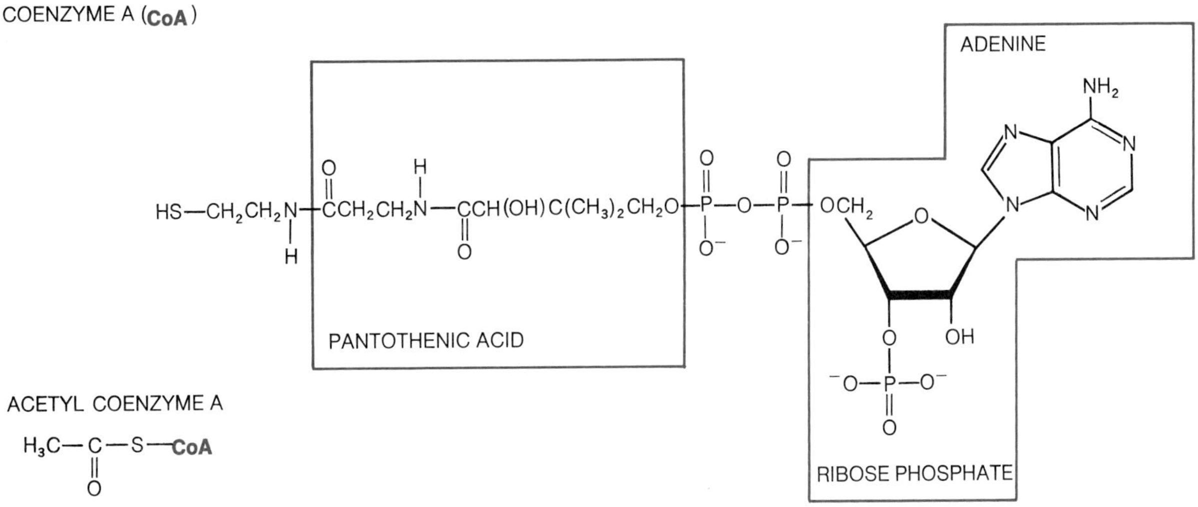

Figure 7-5 Coenzyme A (CoA).

the process are approximately 280 kcal. The net energy yield of about 406 kcal is available to support energy-requiring metabolic reactions.

Acetyl Coenzyme A

Our consideration of enzymes neglected the important fact that most enzymes require more than just their substrates to function. They may require **cofactors,** which are small molecules or even ions like magnesium. **Coenzymes** are often necessary. These are nonprotein organic molecules that serve essential roles in supplying hydrogen, electrons, or small groups of atoms necessary in a reaction promoted by an enzyme. Coenzyme A (CoA) plays the latter role by serving as a donor of acetyl groups. CoA (Figure 7-5) has a familiar look because, like ATP, it is an adenine ribose phosphate. Like ATP it carries a high energy group, not as part of a terminal phosphate, as in ATP, but in the thiol (sulfur) linkage with the acetyl group that it carries. When CoA picks up an acetyl group, ATP is used to establish a high energy bond. Thus primed, CoA uses this bond energy in transferring the acetyl group to some other appropriate molecule.

Acetyl CoA, that is a CoA molecule carrying an acetyl group, plays a central role in metabolism because it is the principal input molecule to the tricarboxylic acid cycle. Carbohydrates enter the cycle by way of pyruvic acid; proteins and lipids similarly use CoA for access to the tricarboxylic acid cycle. *Acetyl CoA is thus the central molecular link in both the oxidation and synthesis of most of the biomolecules of the cell.*

The Path of Pyruvate Through the Tricarboxylic Acid Cycle

The priming step for the oxidation of pyruvate in the tricarboxylic acid cycle is the enzymatic attachment to CoA of an acetyl group derived from pyruvate. The reaction is very complex, requiring joint action by three enzymes and several cofactors. Acetyl CoA then enters the tricarboxylic acid cycle by reacting enzymatically with the four-carbon product of the cycle, oxaloacetic acid, to form the tricarboxylic acid, citric acid. As citric acid proceeds around the cycle, the acetyl group is removed in two steps, each of which releases carbon dioxide and hydrogen, the latter being used to reduce NAD^+ to NADH (Figure 7-7). The cycle also generates one molecule of guanosine triphosphate (GTP), another energy-rich molecule like ATP.

After some ten enzyme-controlled steps, oxaloacetic acid is reformed and ready for another turn of the cycle. The cycle is thus self-replenishing as far as its major components are concerned. Each turn of the cycle converts one pyruvate molecule to three molecules of carbon dioxide (one having been generated during the formation of acetyl CoA) and ten hydrogens. Two turns of the cycle would then account for one glucose molecule. Of course, the process of oxidation of pyruvate is not yet complete at this point

A FLAVIN MONONUCLEOTIDE (FMN)
(riboflavin phosphate)

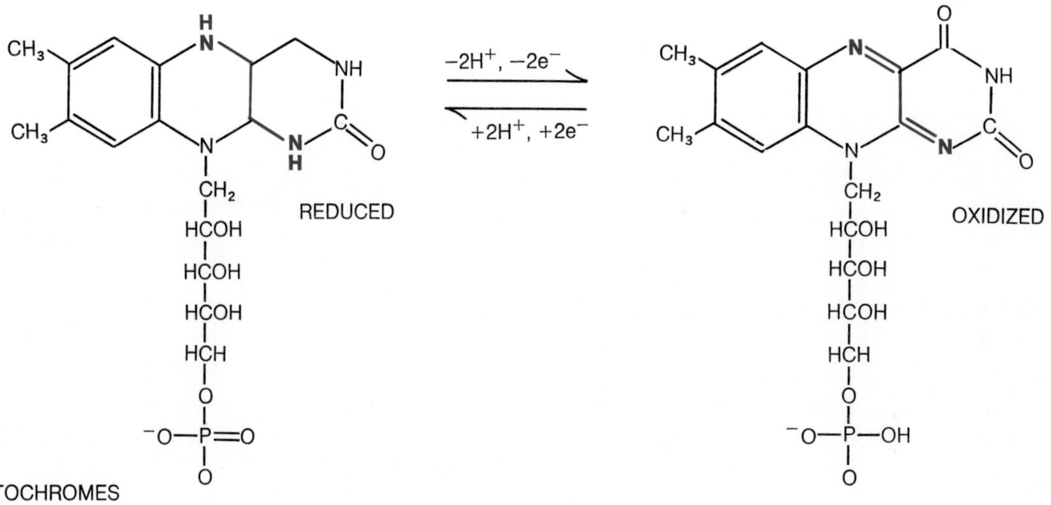

B CYTOCHROMES

Heme component of type α cytochromes. There are more than 30 different types of cytochromes differing in the side chains of the heme (a tetrapyrrole) and in the amino acid sequence of the protein (not shown) to which the heme is attached.

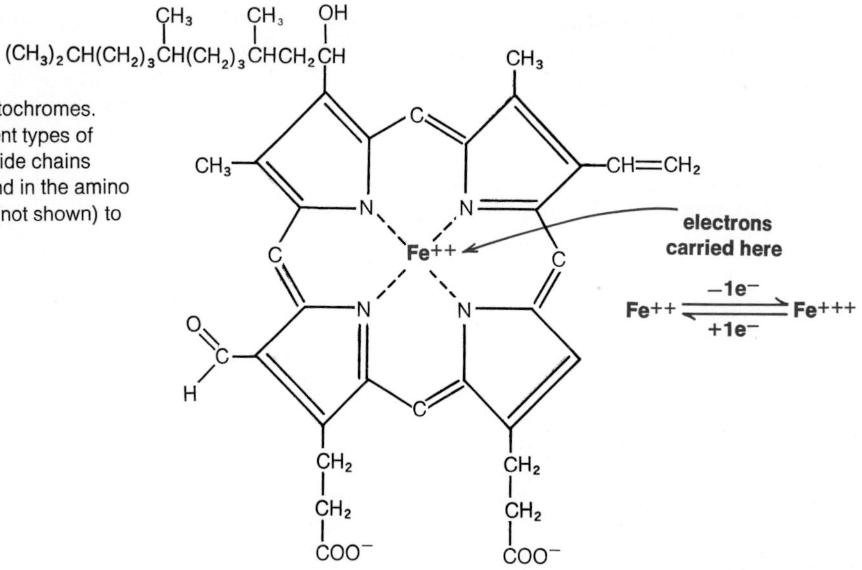

C COENZYME Q

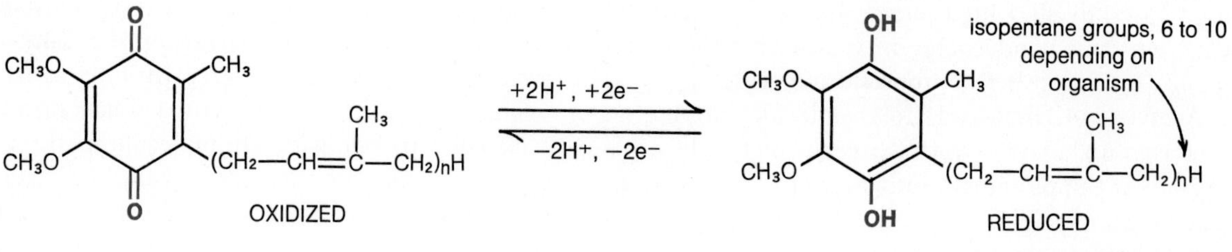

Figure 7-6 Respiratory carrier molecules.

because the hydrogens used to reduce NAD^+ have not been carried all the way to water. To see how this occurs with the generation of high energy ATP bonds, we must look at the closely linked electron transport chain and the process of oxidative phosphorylation that it supports.

THE RESPIRATORY CHAIN: FINAL PATHWAY OF BIOLOGICAL OXIDATION

The story begins with the pyridine nucleotide hydrogen acceptor molecules, NAD and NADP. To some extent you are already familiar with these in their role of accepting hydrogens in biological oxidations and giving them up in biological synthesis. They are, in other words, hydrogen receiving or supplying coenzymes. They function in many different reactions with many different enzymes, but always the basic process is addition or removal of hydrogen from organic molecules. In general NADP acts in syntheses, the building up of biomolecules, as in photosynthesis. NAD is primarily concerned in reactions breaking down molecules, and particularly in those leading to supplying hydrogen atoms to the carrier chain leading to oxygen. However, the two coenzymes are linked by an enzyme, NAD-NADP transhydrogenase, which exchanges hydrogens between them so that hydrogen made available by oxidations can be used for syntheses.

After NAD accepts hydrogen atoms, it supplies them to the first carrier molecule in a chain of hydrogen and electron carrier molecules. The hydrogen atoms then pass in bucket brigade fashion, from one carrier molecule to another, until the last one passes them to oxygen. Thus water is formed, and the biological oxidation process is complete. Some of the energy made available during this carrier process is used to generate ATP by a mechanism called *oxidative phosphorylation*. In this way every two hydrogen atoms entering the carrier chain results in the production of three molecules of ATP.

A chain of carrier molecules of this nature is generally called an **electron transport chain**, even though its early steps transport hydrogen. When such a chain terminates with the reduction of oxygen (addition of hydrogen) it is called a **respiratory chain**.

The respiratory chain works by conveying hydrogen and electrons down an electrical potential gradient; a carrier high in the chain can transfer its load only to the next downstream carrier and not to one several steps away. In addition to this close matching of carrier oxidation-reduction potentials with each carrier being reduced by the one preceding and oxidized by the one following, the efficiency of the respiratory chain depends on the precise location of carrier molecules in the mitochondrial membrane.

At least three kinds of carrier molecules are found in the respiratory chain. One is a flavin mononucleotide (FMN), which has a structure somewhat like NAD or NADP. Several types of cytochromes are present (cyt a, b, c, and so on) and these again are familiar structures, having as their principal functional component a tetrapyrrole ring similar to that seen in chlorophyll but containing iron instead of magnesium. The third type of carrier molecule, coenzyme Q (CoQ), has a structure that you have not seen yet. It is a quinone with a long, lipid-soluble tail. Examples of these carrier molecules and some of the details of their modes of action are shown in Figure 7-6.

With this introduction to its components, let us now see how the whole system operates. The first three steps involve in succession NADH, FMN, and CoQ with two hydrogens being transferred from NADH (which got them from the oxidation of some biomolecule) to FMN, reducing it to $FMNH_2$, which in turn reduces CoQ to $CoQH_2$. At this point it should be noted that when NAD^+ is reduced it gains two hydrogen atoms from the substrate molecules being oxidized, that is two protons (H^+) and two electrons (e^-). Actually only one of these protons is taken up by NAD^+ while the other goes into solution (Figure 7-7). Thus, the conventional symbol for reduced NAD, namely NADH, really symbolizes NADH + H. The presence of this unattached proton is understood. FMN and CoQ behave differently in that they both accept two hydrogens at a time, so when FMN is reduced by NADH it picks up this free floating proton, becoming $FMNH_2$.

At the level of CoQ there is a fundamental change in the transport mechanism. When $CoQH_2$ reduces the next element of the chain, cyt b, it transfers to it only the two electrons that originally entered the chain. The two protons are ejected. In Figure 7-7 you can see that these protons or their equivalents are

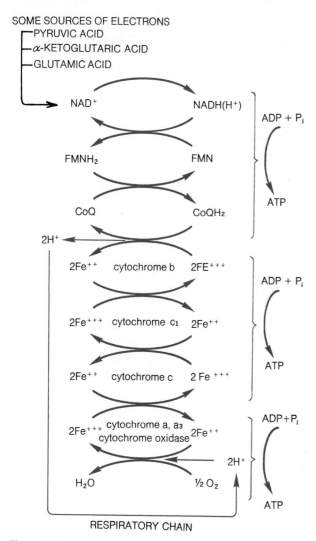

Figure 7-7 The respiratory chain is a series of linked oxidation-reduction reactions which allow stepwise energy release and its conservation as ATP.

from $CoQH_2$ is in the form of pairs of electrons, it is thought that the cytochromes accomodate them by functioning in linked pairs. The process of electron transport continues to cyt c in a similar manner. Electrons are removed from cyt c and transferred to oxygen by a complex consisting of cyt a,a_3, and two copper atoms, collectively known as cyt c oxidase. At the same time two protons are added and water is formed.

OXIDATIVE PHOSPHORYLATION

As shown in Figure 7-4 the production of three ATP molecules from ADP takes place during the movement of two electrons down the respiratory chain. Energetically this is a reasonable number, since about 7.3 kcal are required to convert ADP into ATP whereas the total energy drop through the entire chain is about 526 kcal. In three of the steps along the chain, the energy drop is sufficient to support ATP synthesis, namely:

$$
\begin{aligned}
\text{NAD to FAD} &= 12.0 \text{ kcal} \\
\text{cyt b to cyt c} &= 8.3 \text{ kcal} \\
\text{cyt a},a_3 \text{ to } O_2 &= 24.4 \text{ kcal}
\end{aligned}
$$

A very interesting problem remains as to how this energy is actually used to promote ATP synthesis. One possibility is that some molecule in the mitochondrion serves to couple chemically the electron transfer events in the respiratory chain with oxidative phosphorylation. To illustrate: this intermediate, or coupling molecule (I) might intervene in the ATP synthesis that occurs between cyt b and cyt c according to the following scheme:

$$\text{cyt b-}e^- + \text{cyt c} + I \rightleftharpoons \text{cyt b} \sim I + \text{cyt c-}e^-$$
$$\text{cyt b} \sim I + P + ADP \rightleftharpoons ATP + \text{cyt b} + I$$

Remember that $\sim$ means a high energy bond. Although coupling reactions of this type are of importance in metabolism, it actually appears that, in this instance, no such reaction is involved. For one thing, no coupling intermediate has ever been isolated from mitochondria, and, perhaps most relevant to an eventual explanation of what happens, the coupling idea does not account for the fact that oxidative phosphorylation can only occur when the mitochondrial mem-

used at the end of the cycle when, along with the electron pair transported through the chain, they reduce oxygen to water.

The remainder of the sequence involves electron transfer along a series of cytochromes. $CoQH_2$ initiates transport into the cytochrome part of the chain by giving up two electrons to cyt b and, at the same time it ejects the two protons (H^+) from the chain. Iron is present in oxidized cyt b and in the other cytochromes, as the ferric ion (Fe^{3+}). When cytochromes accept an electron, it is taken up by the iron which is thereby reduced to Fe^{2+}. Since the input

brane exists as an intact vesicle, or bag, isolating the interior from the exterior.

A theory fitting the structural requirements for oxidative phosphorylation in a most interesting way has been put forward by P. Mitchell. According to his **chemiosmotic theory** of oxidative phosphorylation, the transport molecules of the respiratory chain are placed in the mitochondrial membrane so that they may work like pumps to carry protons from the inside of the mitochondrion to the outside, as suggested in Figure 7-8. This would cause buildup of OH^- on the inside and H^+ on the outside. The gradient could be put to use in the synthesis of ATP in the following way: ATP synthetase, an enzyme in the mitochondrial membrane, promotes ATP formation from phosphate and ADP, a reaction that can be written in the familiar water-removal fashion, thus:

$$ADPH + POH \rightleftharpoons ATP + H^+ + OH^-$$

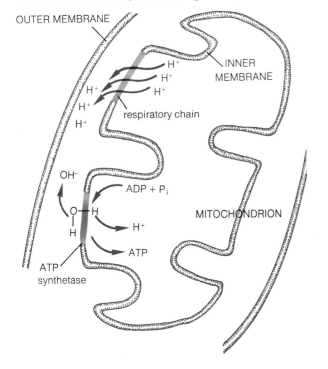

Figure 7-8 A theory of mitochondrial function. In the chemiosmotic theory the physical orientation of components of the respiratory chain in the outer mitochondrial membrane causes accumulation of H^+ outside and OH^- inside. This gradient would then favor a membrane-contained ATP synthetase thought to be structured and oriented so that the water produced in ATP synthesis is "trapped" (H^+ to inside and OH^- to outside of mitochondrion), thus favoring ATP synthesis.

According to the chemiosmotic theory, this reaction system must be arranged in the mitochondrial membrane so that the OH^- can only move to the outside of the mitochondrion and the ATP and H^+ can only move to the inside. If so, the gradients produced by the respiratory chain, being opposite to this flow, would favor the ATP synthetase reaction proceeding in the direction of ATP formation by "trapping" the other reaction products, H^+ and OH^-. Thus the OH^- would tend to move outside and be trapped by H^+, forming water, and the reverse would be true of the H^+ produced in the ATP synthetase reaction. H^+ would move inside the mitochondrion and be trapped by the excess of OH^- created by respiratory chain transport of H^+ to the outside. The result would be ATP synthesis since, according to the chemical law of mass action, a reversible reaction tends to proceed in the direction in which products are in lowest concentration.

The basic idea of the chemiosmotic theory is that energy available from the work of the respiratory chain is invested in producing a proton concentration gradient across the mitochondrial membrane. This fits the observation that oxidatitive phosphorylation is abolished by rupture of the membrane. The energy stored in this chemical gradient is then used up by allowing elimination of the gradient by H^+ and OH^- from one specific reaction, ATP synthesis in the mitochondrial membrane. In sum, the chemiosmotic theory holds that coupling between respiratory chain transport and ATP synthesis is achieved through osmotic gradients, rather than by an energy carrier molecule bridging the two processes. If the theory proves correct, it will be an elegant demonstration that cellular chemistry is far more than events transpiring in a bag of randomly mixed chemicals.

EXPERIMENTS WITH MITOCHONDRIA

We have been spinning out theories in explanation of biological processes so freely that it is time to consider their experimental basis. This is a particularly interesting task in regard to mitochondria because studies on these organelles exemplify the power of modern methods of biological research.

The first group of observations demonstrates what may be learned by a combination of electron microscopical observations and biochemical technqiues for

SPECTROPHOTOMETRY

A spectrophotometer measures the absorption of specific wavelengths of light

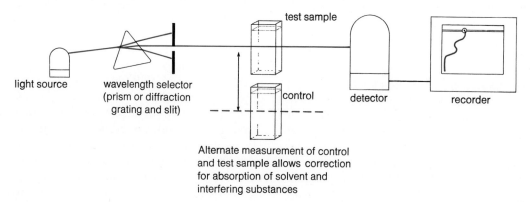

Alternate measurement of control and test sample allows correction for absorption of solvent and interfering substances

The spectrophotometer may be used to:

identify light absorbing molecules by their characteristic absorption spectrum.

follow chemical reactions if a change in light absorption occurs as an indicator of the progress of the reaction.

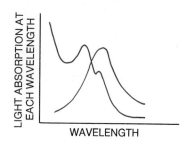

In this example when NAD+ goes from the oxidized to the reduced form a new absorption peak (340 nm) appears.

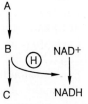

Since this occurs in many biochemical reaction sequences, change in absorption at 340 nm as a function of time can be used to follow a great many biochemical processes, thus

Any reaction sequence that can be coupled with NAD+ ⇌ NADH

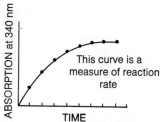

This curve is a measure of reaction rate

Figure 7-9 Spectrophotometry.

fractionation of cell components and assay of their functional capabilities. In this instance the focus is on the mitochondrial inner membranes whose properties figure importantly in development of theories of electron transport and oxidative phosphorylation. With high resolution electron microscopy, the inner membranes of mitochondria are seen to be lined with many spheres carried on stalks. To determine if specific functions are localized in the spheres in comparison with the membranes proper, mitochondria may be broken up in such a way that the spheres are isolated from inner membranes. Upon isolation the membranes seal up to form wrong-side-out sacks, or vesicles. Tests on these two components show: (1) the vesicles formed from membranes contain the components of the respiratory chain, and they are able to transport H^+ ions since they have resealed into closed containers as demanded by the chemiosmotic theory; and (2) the spheres contain the ATP synthetase enzyme which, presumably due to the absence of traps for OH^- and H^+, breaks down ATP rather than synthesizing it. When the two components are put back together, they are able to conduct oxidative phosphorylation.

Since the carrier molecules of the respiratory chain absorb light at different wavelengths (have different colors) when oxidized or reduced, their functional state in a suspension of living mitochondria can be determined by **spectrophotometry** (Figure 7-9). This technique may be used to determine the sequence in which the carrier molecules of the respiratory chain pass electrons to each other, even in the intact mitochondrion. Thus if all of the oxygen is removed from a suspension of living mitochondria, the components of the mitochondrial respiratory chain can be shown by spectrophotometry to be all in the reduced state. The whole chain is clogged with electrons with no place to go. When oxygen is suddenly admitted, the sequence in which the carriers become oxidized can be determined. The first to be oxidized is the carrier that transfers electrons directly to oxygen. The second carrier to oxidize would be the next one up the line. The outcome of the experiment is as you would expect. The first carrier to oxidize is cyt a,a_3 which confirms that it is next in line to oxygen. After this last cytochrome complex begins to oxidize, it can accept electrons from the next carrier and so it is found that cyt c is the next to oxidize, and so on up the chain.

THE IMPORTANCE OF OXYGEN BASED METABOLISM

We have already mentioned one widespread benefit of the presence of free oxygen to all life, namely the reduction by the ozone layer of ultraviolet radiation to levels permitting organisms to inhabit shallow waters and the land. This benefit accrues to organisms whether or not they use oxygen metabolically. Let us next review the more specific advantages that the use of oxygen in metabolism confer on the organism. These are readily shown by the example of glucose utilization. You recall that complete oxidation of glucose yields 686 kcal/mole. When glucose is utilized anaerobically, the end product is two molecules of lactate per molecule of glucose with a net yield of two molecules of ATP. Since the energy stored in ATP is 7.3 kcal/mole, the net yield of energy in glycolysis is 14.6 kcal/mole which, in terms of the total available energy in glucose, represents an efficiency of only 2%, or 31% of the 47 kcal/mole released during the breakdown of glucose to lactate. In contrast, converting lactate to carbon dioxide and water by way of the tricarboxylic acid cycle and the respiratory chain nets a further 36 molecules of ATP, or its equivalent, per initial molecule of glucose. The overall net yield of ATP for the aerobic breakdown of glucose to carbon dioxide and water is then 38 ATP, or 277 kcal/mole (38×7.3), representing an efficiency of 40% and a 19-fold increase in available energy as compared with anaerobic glycolysis. Finally, it is important to note that this major improvement is attained with no loss in available reducing power. The hydrogen carriers utilized in the process (NAD, NADP, FAD) are all regenerated (Figure 7-7).

Besides the energetic advantages of respiration, another mark in its favor is the fact that the respiratory end products are carbon dioxide and water, wastes posing no serious disposal problem. This may be contrasted with the accumulation during glycolysis of lactic acid, which builds up to toxic levels unless rapidly removed. A similar example of practical significance occurs in **fermentation** by yeast, which is the anaerobic metabolism of sugar to ethyl alcohol. One of the factors terminating the fermentation process during wine production is the buildup of alcohol within the confines of the fermentation containers to levels that inhibit further growth of the yeast organisms.

PROGRESS, OR MERE SURVIVAL?

In these last two chapters detailing the origin of photosynthesis and respiration (metabolism utilizing oxygen), life is seen to become more and more chemically complex. Although we theorize that this complexity was initiated by two successive crises, depletion of prebiologically accumulated organic molecules followed by the appearance of free oxygen in the atmosphere, it is important to understand that the biological responses to these crises, photosynthesis and respiration, were far more significant than merely immediate solutions to immediate problems.

This is made especially clear by examining the results of photosynthesis. This process provided ultimately such a rich excess of energy and raw materials over the requirements of the earliest forms of life that only now, due to the excesses of humans, are we beginning to forsee limitations in this regard. Similarly, respiration made possible great improvements in the efficiency of resource utilization and, together with the metabolic processes upon which it was superimposed, laid down the fundamental patterns of metabolism that have persisted as the basis of life throughout the long course of evolution. Far more than immediate solutions of immediate problems, the appearance of photosynthesis and respiration firmly established the energy and material base for all remaining events of biological evolution.

The next chapter considers the means by which organisms direct these resources into the generation of the living substance.

A TURNING POINT

At this point in our development of a plausible, mechanistic scheme for the appearance of life on earth, we pause for a brief recapitulation. Thus far the story carries life from its initiation out of the nonliving to a state in which its indefinite survival is assured, owing to the switch from heterotrophy, based on "random" nonbiological chemistry, to a photosynthetic basis. Photosynthesis assured a supply of energy and material and provided the opportunity for further perfections in reproductive mechanisms and other

Table 7-1 Hypothetical Stages in the Early Evolution of Life

Initiating Condition	Response	Result
A. Prebiological Phase		
1. Primitive earth chemistry; absence of life	Accumulation of organic molecules	Appearance of self-replicating molecules, molecular systems
B. Biological Phase		
2. Competition among self-replicating molecular systems	Development of favorable local environments supporting replication—boundary membranes	Cells
3. Competition among cells	Improvements in hereditary mechanisms—mitosis	Depletion of prebiologically accumulated biochemical resources
4. Competition for survival amidst declining chemical resources	Switch to alternate molecules	Development of metabolic pathways
	Processes to generate reducing power	Biosynthesis
	Improvements in cell membranes	More effective uptake of materials from environment; stability of internal environment of cell
5. More efficient cells further deplete environment of organic molecules	Photophosphorylation; photosynthesis	Abundant supply of life supporting molecules
6. Presence of heterotrophs and autotrophs	Appearance of food chains	Diversification of life
7. Increased atmospheric oxygen from photosynthesis; ozone layer	Aerobic respiration	Increased metabolic efficiency
	Invasion of shallow water, then land	Diversification of life
8. Competition among organisms	Meiosis and sexual reproduction	Increased pace of evolution; further diversification

processes that were essential for the ultimate successes of life. The early crises are past. If the evolutionary scheme thus far is sound, mechanistic origins of life deserve the most serious consideration.

To assist you in thinking about the progression of our argument up to this turning point, examine Table 7-1 with care. It presents the argument as a series of cause-response-result statements to emphasize the gradualness and plausibility of each stage in a process, the origin of life, which might seem scientifically incomprehensible if viewed as a whole. The final stage, development of sexual reproduction, is included here for completeness, although we will not consider it fully until Chapter 11. Note also that phases such as the appearance of food chains or the appearance of sexual reproduction might well have occurred earlier than indicated.

8 Continuity of Life Through Time

Biochemical answers to the great chemical problems faced by early living organisms have held our attention up to this point. Now we turn to another great problem. This one was certainly as difficult as those created by the ecological crises just discussed. Its solution was essential to the persistence of life. In analyzing the responses of organisms to the first ecological crisis, we sidestepped this problem. No attention was paid to anything but the machinery of the cell. Thus we ignored how this machinery comes to be produced in each cell and how the information necessary to guide its production is handed along from cell generation to cell generation. This is our subject now: how the directions for building a cell are generated.

WHY DIRECTIONS ARE NECESSARY

A cell comes into existence by no single accident, nor by several. This is evident from the long list of cell components and their intricate structure. Directions for the construction of a cell are necessary, and it is clear that these directions must be very accurate. The easiest way to illustrate this is to consider what is necessary to make just one protein, out of the hundreds in a cell, accurately enough to perform its required function. You will recall from page 41 that the primary structure of a protein, namely, the order in which amino acids attach to each other to form the peptide chain, determines its properties. At the time nothing was said about how critical that amino acid sequence is. In fact, one misplaced amino acid out of hundreds can result in a protein that functions poorly or perhaps not at all.

Sickle-cell anemia demonstrates this remarkable fact. In this inherited human disease, oxygen carrying red blood cells do not function properly because their oxygen transporting protein, hemoglobin, crystallizes instead of staying in solution. The difference between normal and the sickle cell hemoglobin is exactly one out of 574 amino acids that constitute the hemoglobin molecule. Hemoglobin, of course, is only one of thousands of different macromolecules that might occur in a cell, so it is obvious that a large number of detailed instructions must exist and must be followed exactly if adequately functional cells are to be produced.

The magnitude of the instructions necessary to direct the assembly of a complex organism is perhaps best appreciated if we imagine the required instructions were to be set forth in print as a simple code of some sort. In this form the instructions for building a human would fill over 1000 very large books. Yet, as we shall see in this chapter, biological systems are so adept at miniaturization that all of these instructions for one human are stored in only 10^{-15} g of that most remarkable of biomolecules, DNA. Indeed, all of the DNA that has gone into the gametes used in the construction of humans since the origin of the species would amount to far less than 1 g.

ACCURATE REPRODUCTION IS ESSENTIAL TO EVOLUTION

Beyond those just stated, there is another reason that the directions for constructing cells and organisms must be accurate. Besides functioning correctly, offspring must also resemble parents.

Until now we have considered evolution without discussing its mechanisms in any formal way. We have speculated on how life originated and on what changes it experienced as it persisted on the earth. We have accepted the obvious concept that, when confronted by sufficiently severe environmental change, only those organisms that are able to cope with the change are going to survive. Now all organisms, even of the same "kind" can never be exact copies of each other. This is obvious because we know that the offspring of sexually reproducing organisms possess a mixture of hereditary instructions from each parent. It is therefore easy to imagine a change in the environment in which some copies of a kind of organism do better than others. The survivors in such a situation are said to have been better adapted to the new environment, or, according to the classical statement, there has been survival of the fittest.

For this to have any long term effect, for organisms to continue to survive under the new conditions, the survivors must possess the ability to transmit to their offspring whatever characteristics made their survival possible. If an organism better able to survive could not transmit directions for that improvement to its offspring, then there could be no evolution, no adaptation of organisms to change in the environment, and life would at best have persisted at a very low level.

DNA PROVIDES THE DIRECTIONS

Nucleic acids, DNA and RNA, were listed among the constituents of the cell in Chapter 4. They were described as information molecules. Now is the time to discuss them in more detail for they are responsible for the continuity of life through time. Deoxyribose nucleic acid (DNA) (Figure 8-1) carries virtually all of the information necessary for the construction of cell and organism from generation to generation. It achieves this by *control of protein synthesis* through the formation of ribose nucleic acid (RNA), which actually directs protein construction, following in-

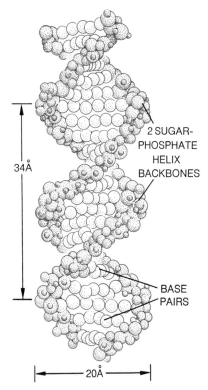

Figure 8-1 The structure of DNA is depicted in a space-filling model of a very small segment of this immense molecule.

structions supplied by DNA. This statement summarizes some of the most widely influential facts of biology. It represents the coming together of the study of *development, inheritance,* and *evolution* at the biochemical level. Such important information was not easily obtained and even now is not fully understood in all of its immense complexity. Perhaps the best way to arrive at an understanding is to make a series of general statements about the role of DNA and RNA and then present some of the evidence supporting them.

The Flow of Information

The information directing the manufacture of the cell is contained in the sequence of purine and pyrimidine bases found in DNA molecules. DNA directs the making of highly exact copies of itself in a process called **replication** and complete sets of these copies are transmitted to each daughter cell. DNA produces copies of its store of information in a form useful in the synthesis of protein by transcribing its information

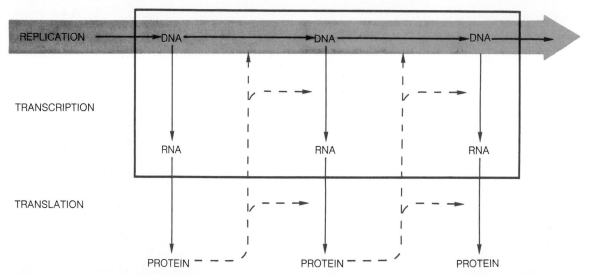

Figure 8-2 The central concept of molecular biology. DNA supplies information for its own reproduction (replication) and for production of RNA (transcription), which in turn supplies information for protein synthesis (translation). Dotted lines indicate that protein synthesis leads directly or indirectly to all materials necessary for further cycles of replication, transcription, and translation.

into RNA molecules, the process of **transcription**. The RNA then supplies information to guide protein synthesis in the process of **translation** (Figure 8-2).

DNA and the Flow of Information between Cells

The major part of our knowledge about the role of DNA is quite recent. By 1941 the rules of heredity were well understood (see Chapter 12). The unit of heredity, the **gene,** had been defined in terms of its effects, its location in chromosomes within the cell had been defined, and the means by which it is transferred from cell to cell had been determined. But neither the chemical nature of the gene nor the chemical processes by which it acted were known.

George W. Beadle and Edwin W. Tatum in that year took an important step towards understanding these two great problems by *linking genes with enzymes,* the protein cellular catalysts. They worked with a simple organism which was especially favorable for combined genetic and biochemical studies, the mold *Neurospora* (Note 8-1).

The studies of Beadle and Tatum showed convincingly that the presence or absence of a given enzyme in *Neurospora* could be correlated with a change in a particular gene. Their **one gene:one enzyme hypothe**sis expressed the idea that each gene in an organism achieves its effect by somehow controlling the presence or absence, or some vital property, of one enzyme. With relatively minor modifications this theory has stood the test of time. The important fact that it tells us is this: whatever the nature of the gene, it must allow the gene to act by influencing the synthesis of protein, since enzymes are, of course, proteins.

The first important clues to the chemical nature of the gene came before the work of Beadle and Tatum in a study of what was called the **transforming principle** in one of the bacteria that cause pneumonia, *Streptococcus pneumoniae.* Transforming principle extracted from a virulent, or lethal, strain of *S. pneumoniae* causes a nonvirulent form of the pneumonia bacterium to become virulent. A smooth, polysaccharide coat is secreted outside the cell wall of the virulent form, giving it immunity from host defenses. This coat is lacking in the nonvirulent form. In 1928, F. Griffith discovered that *dead,* virulent and *living,* nonvirulent bacteria could be injected into a mouse with the result that the nonvirulent bacteria would become virulent, secrete smooth coats, and kill the host mouse.

This was a true hereditary change because descendants of the newly virulent bacteria remained virulent.

It seemed that some chemical, a transforming principle, from the killed bacteria was able to cause a permanent hereditary change from nonvirulent to virulent. What was its chemical nature? It could not have been just the enzyme necessary to make the polysaccharide coat because that would have eventually been lost in subsequent generations of bacteria unless there was some means of making more enzyme. Thus, it must be the enzyme maker itself, that is, the *gene* regulating formation of an enzyme essential to the formation of the polysaccharide coat. This was a remarkable fact, not entirely appreciated at the time. Ultimately this fact suggested an opportunity to catch a gene out in the open, away from its hiding place in the living machinery, and to find out how it is made.

It was soon found that a cell-free extract of virulent bacteria contained the transforming principle and could cause transformation of nonvirulent to virulent bacteria in cultures grown outside a living host. Working with such extracts, a group headed by Oswald Avery conducted chemical isolation procedures on extracts of virulent bacteria to show that the active component of the transforming principle is DNA. It, alone among the constituents of the extract, caused formation of the enzyme necessary to form the polysaccharide coat in nonvirulent bacteria and make them virulent. One of the most convincing of these experiments involved exposing the active DNA extract to deoxyribonuclease, an enzyme that specifically breaks down DNA. Extracts treated in this way lost the ability to cause transformation. The final steps in this sequence of discovery were made much later, in 1970, when H. G. Khorana and his associates achieved the complete laboratory synthesis of a functional gene and, in 1976, when such an artificial gene was incorporated into a bacterium and shown to function. It was, of course, made of DNA.

How DNA Carries Information about Protein Synthesis

Finding that the gene is composed of DNA did not immediately answer the question about how it carries information because the essential facts about the structure of DNA were unknown at the time. Although it was known that DNA contained four kinds of bases (Figure 4-16), it was originally thought that these were repeated in monotonous fashion along the length of the DNA molecule. Thus DNA was considered to be the same wherever it occurred. This made it difficult to see how DNA could direct the specific types of protein synthesis that the genetic substance must direct. Biochemists soon discovered that, far from containing a monotonously repeating sequence of bases, DNA contained bases in different ratios to each other, depending upon the organism. With this information available, it was proposed that the DNA molecule carries genetic information coded in the *sequence* of its bases, that is *the gene is DNA and its information is expressed in a code in which the "words" are sequences of bases.* If true, there is more than enough DNA in a cell to provide constructing and operating directions for that cell, and, indeed, for the entire organism in which it resides. Thus each nucleated human cell contains enough DNA to make a strand two meters long.

The Genetic Code

Knowing the size of the DNA book and the alphabet on which it is based is not enough to understand how it works. We need to know the language in which it is written. In other words, what is the genetic code? To answer this question, remember that the critical thing about building a protein is the linear order of its constituent amino acids. Obviously the genetic code must somehow, directly or indirectly, specify that linear order with precision. Something about the base sequence of DNA must determine how amino acids are strung together to form proteins. The simplest possibility would be that each of the four bases specifies a particular amino acid. This is obviously wrong because it would mean that proteins could be made of only four amino acids, and we know that 20 amino acids are common in proteins. So the "words" of the code must consist of groups of bases. Forming the words from combinations of two bases allows only 16 combinations, which still is not enough. If the words are formed of three bases, that would allow specification of 64 different amino acids, which is more than enough. In fact this is the way it is: successive sets of three bases specify the sequence of amino acids in the protein. The long sequence of bases necessary to specify the amino acid sequence of a polypeptide chain is simply read as a message beginning with the first amino acid, and then each successive set of three bases is read in turn specifying each successive amino acid.

Note 8-1 *Studies on Bread Mold Facilitated Understanding of the Chemical Action of the Gene.*

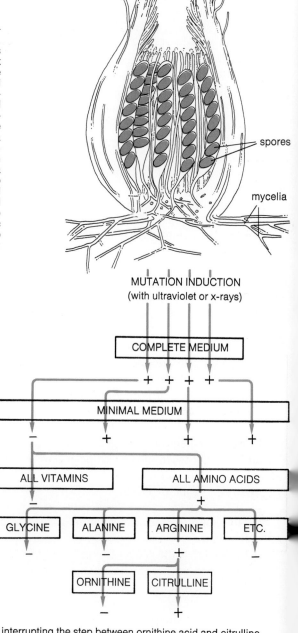

Figure 8-3A Steps in the identification of *Neurospora* mutants.

STUDIES ON BREAD MOLD FACILITATED UNDERSTANDING OF THE CHEMICAL ACTION OF THE GENE.

Before 1940 George W. Beadle and Edwin L. Tatum were trying to discover the chemical action of the gene using the fruit fly, *Drosophila,* the subject of much earlier work in genetics. Various difficulties caused them to seek a better organism for their work and they tried the bread mold, *Neurospora crassa.* The choice was good because *Neurospora* (1) is easily grown on a minimal medium containing only sugar, the vitamin biotin, and a few inorganic salts and, (2) it has a very simple life cycle, well suited both to the biochemical studies that Beadle and Tatum proposed, and also suitable to the genetical studies that would be necessary to back up their work. They reasoned as follows: If a mutation is caused in *Neurospora* (they used UV and X rays) which affects either the ability to produce a necessary vitamin or amino acid, then the mutated organism will grow only in a medium that has the particular vitamin or amino acid added to the minimal medium. By collecting enough of such mutants and determining what molecules had to be added to the minimal medium to restore growth to normal they hoped to be able to map out the relationship of the genes to the important metabolic processes regulating the production of vitamins and amino acids. Their procedure in simplified form is outlined here.

STEPS IN MUTANT IDENTIFICATION

1. Isolate single reproductive cells (spores) and grow a colony from each in complete medium. All grow even if mutated, since complete medium supplies all required vitamins and amino acids.

2. Detect mutants by growing sample from each culture in minimal medium. No growth means a mutation requiring something present in complete medium but not in minimal.

3. Determine if the mutation involves vitamins or amino acids by adding either all vitamins or all amino acids to minimal medium.

4. Determine the specific amino acid required by tests with single amino acids.

5. Determine where in arginine biosynthetic pathway mutation has occurred by supplying other possible intermediates.

THEREFORE: The mutation affects arginine synthesis by interrupting the step between ornithine acid and citrulline (see fig. 8-3B).

Using the technique illustrated in fig. 8-3, Beadle and Tatum worked out many complex biosynthetic pathways. Here we see part of the pathway for the production of the amino acids proline and arginine from glutamic acid. The diagram also shows the outcome of their tests with various amino acid supplements to the basic medium which identified the site of action of each of the mutations in this series. Thus the mutant Proline-1 could grow with no replacement other than proline itself; therefore, the mutated gene must govern the final step in the synthesis of proline. In contrast Arginine-5 in their series could grow on either arginine, ornithine, or citrulline, but not on glutamic acid, thus locating the site of action of the mutated gene at the step from glutamic acid to ornithine. Since each step is controlled by a single enzyme, Beadle and Tatum saw very quickly that in some way the gene probably acted by interfering with the synthesis of or action of specific enzymes. Thus the "one gene: one enzyme" hypothesis was born, and there was started a trail of research leading to our modern understanding that the genetic substance, DNA, actually carries specific instructions for the synthesis of the chains of polypeptides that form proteins.

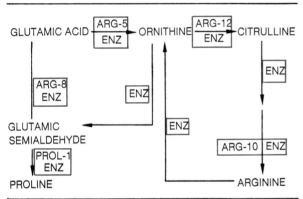

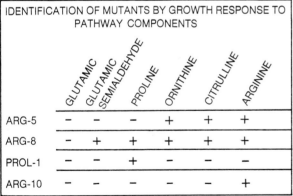

Figure 8-3B Use of *Neurospora* mutants to reveal synthetic pathways.

The "dictionary" of the nucleotide code is shown in Figure 8-4.

DNA STRUCTURE AND REPLICATION

We have just said that a DNA base sequence is read in the process of synthesizing a protein. Figuring out what "read" means in chemical terms has proven to be no small problem. There must in fact be two kinds of "reading." One results in the production of more DNA to supply the genetic machinery of newly forming daughter cells—**DNA replication**. The other results in the formation of amino acid sequences in proteins. It involves two processes, *transcription* and *translation*, which involve RNA. To understand how reading is accurately carried out, we must first consider more details of the structure of DNA.

It is perhaps already obvious, by analogy with protein, that we have been discussing only the primary structure of DNA. Just as the amino acid chain of a protein coils and twists to form secondary and tertiary structure, the chain of nucleotides forming DNA coils to generate a secondary structure. Several kinds of information were necessary before the secondary structure of the DNA molecule could be worked out. X-ray diffraction studies, exactly similar to those conducted to determine protein structure, revealed that DNA has a coiled or helical structure like many types of protein. The x-ray data also told how tightly the DNA is coiled. Chemical studies provided at least two other important clues about the three-dimensional structure of the DNA molecule. One was that there had to be more than one chain of nucleotides per molecule, and the other was that in each nucleic acid molecule there is always as much adenine as thymine and always as much guanine as cytosine. This was all the information necessary for James Watson and Francis Crick in 1953 to figure out the three-dimensional structure of DNA and to suggest how this molecule performs the remarkable feat of directing the synthesis of exact copies of itself.

The Double Helix

If it were not for the informational part of the DNA molecule, that is, the sequence of bases strung along its phosphodiester-linked sugar backbone, it would not be particularly hard for the cell to build new DNA molecules by enzymatic processes of the

A.

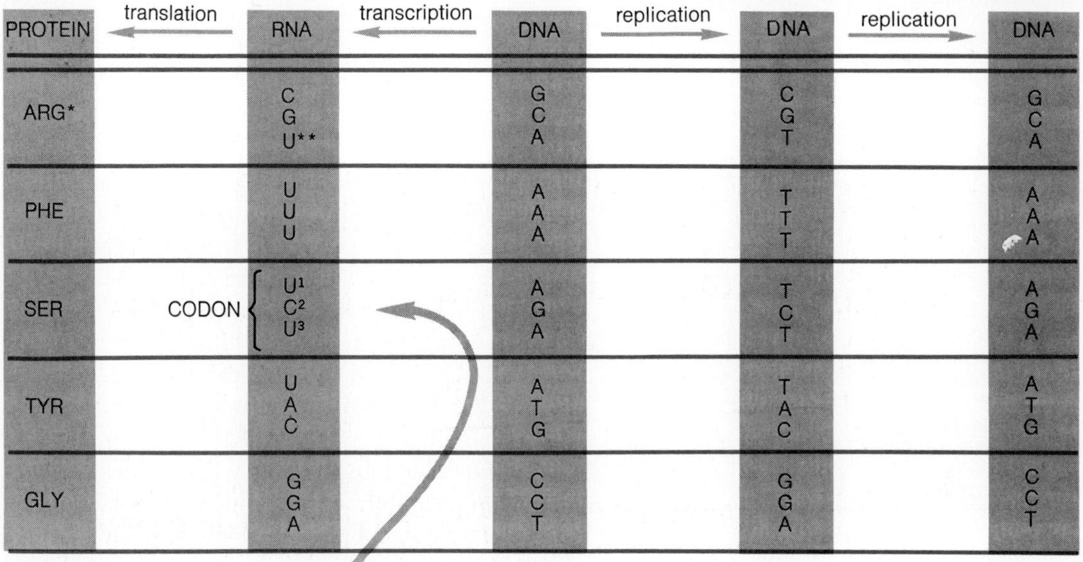

PROTEIN	← translation	RNA	← transcription	DNA	replication →	DNA	replication →	DNA
ARG*		C G U**		G C A		C G T		G C A
PHE		U U U		A A A		T T T		A A A
SER	CODON {	U¹ C² U³		A G A		T C T		A G A
TYR		U A C		A T G		T A C		A T G
GLY		G G A		C C T		G G A		C C T

B.

NUCLEOTIDE ②

NUCLEOTIDE ①		U		C		A		G		NUCLEOTIDE ③
U		UUU UUC UUA UUG	PHE PHE LEU LEU	UCU UCC UCA UCG	SER SER SER SER	UAU UAC UAA UAG	TYR TYR (T) (T)	UGU UGC UGA UGG	CYS CYS (T) TRP	U C A G
C		CUU CUC CUA CUG	LEU LEU LEU LEU	CCU CCC CCA CCG	PRO PRO PRO PRO	CAU CAC CAA CAG	HIS HIS GLN GLN	CGU CGC CGA CGG	ARG ARG ARG ARG	U C A G
A		AUU AUC AUA AUG	ILE ILE ILE MET	ACU ACC ACA ACG	THR THR THR THR	AAU AAC AAA AAG	ASN ASN LYS LYS	AGU AGC AGA AGG	SER SER ARG ARG	U C A G
G		GUU GUC GUA GUG	VAL VAL VAL VAL	GCU GCC GCA GCG	ALA ALA ALA ALA	GAU GAC GAA GAG	ASP ASP GLU GLU	GGU GGC GGA GGG	GLY GLY GLY GLY	U C A G

* see Fig. 4-5
** U replaces T as partner of A in RNA
(T) signals termination of polypeptide chain

Figure 8-4 The genetic code in replication, transcription, and translation. A. Right side of diagram, two successive DNA replications showing that alternate replicates are identical. Left side shows transcription of the same segment of DNA into RNA, whose condons specify amino acids in a peptide chain. **B.** A list of RNA codons and the amino acids that they specify. Note that the code is degenerate, meaning that all amino acids (except methionine) are coded by more than one codon.

kind you have already seen. Nucleotides (deoxyribose sugar plus one of the four purine and pyrimidine bases: adenine, cytosine, guanine, and thymine) are readily synthesized, and it is simple enough to hook them together with phosphodiester linkages (Figure 4-16) to form DNA. That would be straightforward as long as there is no need to produce a specific base sequence. The problem of forming a specific base sequence is quite different from what one sees in ordinary enzyme reactions, where an enzyme recognizes and promotes the reaction of a very limited variety of substrates or often only one. In the case of DNA, the enzyme DNA polymerase, which assembles the DNA molecule, has the additional problem of determining which of four fairly similar kinds of molecules has to be fitted at a particular position in the new DNA under construction. What happens is that the DNA polymerase itself does not determine the sequence of base additions. It takes instructions from the DNA that it is copying. Thus the process of DNA synthesis goes as follows:

Old DNA + nucleotide precursors $\xrightarrow{\text{DNA polymerase}}$

Old DNA + New DNA

The old DNA molecule, the one being copied, may be called a substrate in the synthesis of new DNA, but in the new sense that it provides *information rather than material or energy* to the reaction.

To understand how this occurs we must look more carefully at the structure of DNA. Certain important new facts must be comprehended. One is that the DNA molecule, in all but a very few viruses, consists of two polynucleotide strands coiled about each other just as you might twist two wires symmetrically together, so that they cannot be pulled apart without uncoiling (Figure 8-5). The two polynucleotide chains twisted in this way have a very precise structural relationship to each other. They are lined up, in the chemical sense, head to toe with the front end of one polynucleotide chain at the tail end of its partner. They are said to be in **antiparallel** arrangement rather than parallel. (Examine Figure 8-5 to see how direction of the polynucleotide chain is determined from the nature of the phosphodiester linkages between the deoxyribose sugars of the DNA backbone.) The second important fact is that the bases of the two antiparallel polynucleotide chains occur in a very exact relationship to each other. When adenine occurs in one chain, exactly across from it and linked to it by hydrogen bonding, lies a thymine in the other chain; similarly, when guanine occurs in one chain, its partner in the other is cytosine. The rule is: A(adenine) pairs with T(thymine) and G(guanine) pairs with C(cytosine), A + T : G + C. This **pairing rule** is satisfying to the chemist because it accounts for the long unexplained fact, first discovered by biochemist Irwin Chargaff, that in DNA the concentration of A is always equal to the concentration of T and the same is true for G and C. This, of course, would always be true if the pairing rule applies and the DNA studied exists as molecules of paired polynucleotide chains.

The pairing rules depend on the relative sizes of the four bases and on the space within which they must fit in the helically coiled DNA molecule. There is accurate data to work from in determining these matters. The sizes of the four bases are exactly known, and the x-ray data show the dimensions of critical parts of the coiled DNA molecules. In fact, this information plus Chargaff's chemical data was essentially all the information needed by James Watson and Francis Crick to make their two-strand model of the DNA molecule and to understand what the model implied about the nature of the DNA replication process.

With this data the structural logic behind the pairing rules becomes obvious. Pairing of T with G cannot occur because the bridge that they would make between the two backbones of the polynucleotide chains would be too wide and force other parts of the molecule out of alignment. We actually know this pairing does not occur because the x-ray data show the double helix is not as wide as would be required to fit T + G base pairs. Pairing of A with C will not work because they would form too short a bridge. The only pairing combinations that fit the x-ray data, Chargaff's biochemical data, and which can be shown to allow the large amount of hydrogen bonding that is necessary to hold the polynucleotide chains together are A with T and G with C.

Having the chains related by the pairing rules allows any one chain to have any base sequence and still allow a close enough fit between chains to permit necessary hydrogen bonding. This freedom must exist if DNA does actually carry hereditary information in the sequence of its bases, according to a triplet code. It is easy to see why this is so. Just suppose that the two chains had to be exactly alike rather than being constructed according to the base pairing rule. Then

Figure 8-5 The coiling relationship of the two DNA strands in the DNA molecule.

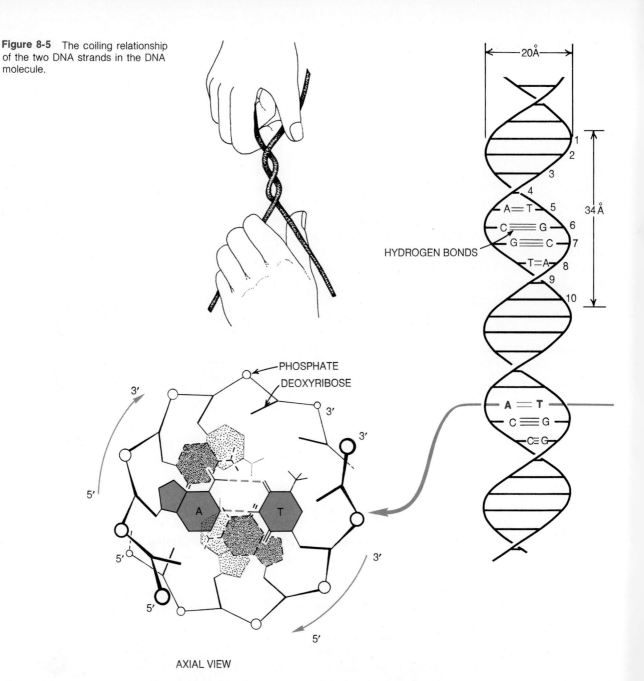

AXIAL VIEW

the only way that the two chains could fit together would be for the bases across from each other in the two chains to be similar; in other words we would be restricted to a two letter code, either involving A and G or T and C, that is, the equal size pairs, since that would be the only way the two strands could fit together closely. The way it actually is, a long base on one chain pairs with a short base on the other, pre-

serving the fit of the two strands and allowing any base sequence in one chain, but at the cost of exact duplication of base sequence in each strand. As we shall see this causes no trouble.

DNA Replication

Although the double helix structure of DNA has now been described sufficiently to show that it fits the

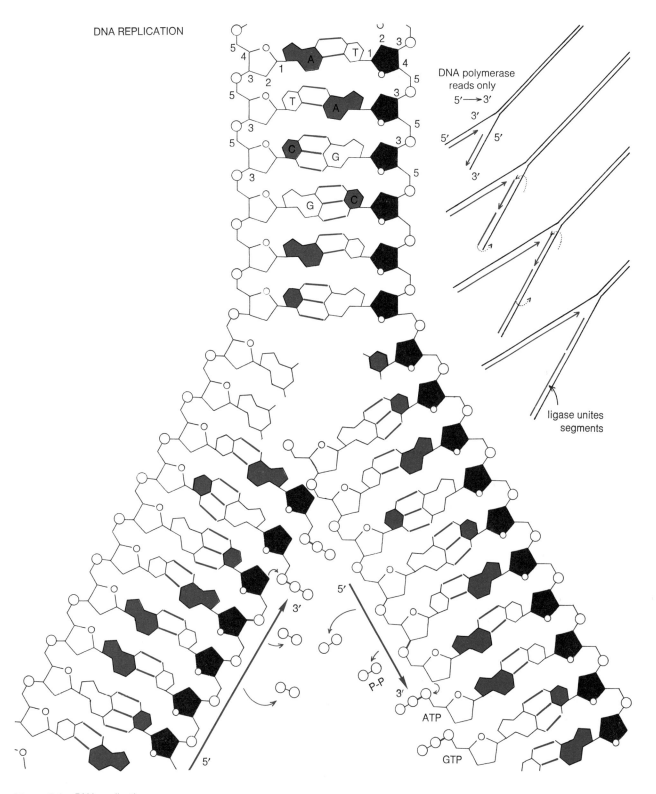

Figure 8-6 DNA replication.

x-ray and chemical data, we have not yet seen how it undergoes replication. Unless the proposed structure of the molecule lends itself to this remarkable process we are missing vital information. Perhaps you have already seen a problem. DNA as described here consists of two polynucleotide strands that have base sequences that are related by the pairing rule, but the sequences in each strand are not the same. In other words the message carried in the base code of one segment of a strand is not the same as that carried by the complementary part of the partner strand. That is, if one strand reads ACTGCAAGG the comparable part of the other reads TGACGTTCC.

The problem is which one is the genetic message? Or are they both genetic messages, and, if not, what is the role of the other one? The answer is simply this: there are two kinds of messages carried by each region of the paired DNA polynucleotide strands. One is the message that directly controls cellular activity. We would call this the gene in the classical sense. The other is the message that directs the formation of the gene in the next generation of cells. The two messages are *complementary* copies of each other because one message can be deduced from the other, if one knows the base pairing rule. Which of the two is the replicating fragment and which is the gene fragment is not obvious, since both strands differ only in their base sequences as far as is now known. Nonetheless, the cellular machinery that links the gene fragment with cellular processes somehow can tell the difference between the two strands.

The logic of the replicative process now becomes clear. If we label the two strands in the DNA I and II and pay no attention to which parts of I and II are genetic and which are replicative, in the sense just given, it is clear that, if there is some process that can use I as a template to make II and that can use II as a template to make I, then DNA replication is assured from cell generation to generation. Each DNA molecule carries its own replicative pattern.

This is exactly what takes place (Figure 8-6). The DNA molecule uncoils enough to allow DNA polymerase to use each strand as a template to guide the construction of a new strand. If I is the template, II is produced and the old I strand and the new II strand coil about each other in the usual way as II is synthesized. The same process takes place on the other strand so that the result of DNA replication is the formation of two new DNA molecules, each containing one old strand and one new strand. It is important to understand that DNA polymerase itself does not determine the base sequence. This is done by the strand of DNA that is being copied. How the DNA controls the process is unknown, but the mechanism must be related to the same kind of structural considerations that determine the base pairing rules. Perhaps as the DNA polymerase enzyme moves along the strand being copied it has a functional site that resembles the backbone of the DNA being copied so that only bases that satisfy the pairing rule would fit well enough and long enough to be enzymatically attached to the growing backbone of the new DNA polymer.

TWO EXPERIMENTS ELUCIDATE DNA REPLICATION

DNA Polymerase Copies any DNA

You have been expected to take a lot on faith and now some proofs of such unusual processes are in order. First, let us consider the idea that the DNA strand is the guide to the actions of DNA polymerase, that DNA polymerase simply tacks onto the strand of DNA being copied whatever base satisfies the pairing rule for that particular point. This is easily demonstrable by DNA synthesis in the test tube. To do this, a functioning DNA polymerase system is extracted from cells and provided with a supply of single strand DNA of the experimenter's choice. It might even be a synthetic DNA that the polymerase could never have come in contact with in all of its evolutionary history. Even so, the polymerase goes to work and faithfully fills the test tube with copies of the DNA strand that was provided.

This fact, indeed, makes the simple life of the virus possible. The infective part of many viruses is just a large DNA molecule. Yet soon after it has been introduced into a susceptible host cell it is readily shown that *host cell DNA polymerases* are hard at work making *virus DNA*.

DNA Copying Is Semiconservative

The second experiment pertains to the statement that the new DNA molecule formed by DNA polymerase contains one old strand and one new strand. Even if we agree that the experiments just described tell us that DNA does direct its own copying by DNA polymerase, we have to admit that there are several ways the copying might be done. Each DNA strand might break up and be copied as described but old and new parts might be mixed so that any given strand of DNA would contain some old and some new DNA. Or, in the course of replication, the two old strands might rejoin and the newly formed strands might form an entirely new DNA molecule. However, experimental evidence assures us that the process takes place as originally described. Each newly formed DNA molecule contains one old polymer chain and one new one. The process is, therefore, called **semi-conservative** since the process in which the two old DNA strands stay together would be called conservative.

Experimental proof of semiconservative replication is obtained by growing cells in a supply of DNA precursor molecules containing the heavy isotope of nitrogen, ^{15}N, rather than the ordinary light isotope, ^{14}N. After incubation in heavy ^{15}N, the cells containing newly made heavy DNA are shifted to light ^{14}N-containing culture medium and allowed to duplicate. The weight of DNA molecules subsequently formed during growth in the ordinary ^{14}N can be determined by using a technique employing centrifugation to accent the small differences in weight due to the presence of the two isotopes. You can guess what the outcome would be. If replication is conservative, there would be two weight classes, the heavier, old DNA composed of two strands each containing the heavy ^{15}N isotope, and the lighter new DNA, made of two strands, each containing the light ^{14}N isotope. If replication is not conservative, all of the DNA would be of a weight intermediate between the heavy and light forms, with each of the two strands composed of random ^{15}N- and ^{14}N-containing segments.

The results after two generations of replication clearly demonstrate semiconservative replication (Figure 8-7). After one generation there appears a single weight class intermediate between the light and the heavy forms. The appearance of this single weight class eliminates the possibility of conservative replication but is not, as yet, proof of semiconservative replication. However, the results after a second generation of duplication shows two weight classes, DNA of a weight intermediate between light and heavy, and DNA of a weight corresponding to the light ^{14}N isotope. Therefore, each newly formed DNA molecule has one old strand and one new one.

Further Problems in DNA Replication

The story of DNA replication is still intensely studied. In filling in the details of this remarkable process, fascinating insights into the incredible complexity of cellular chemistry emerge. Here are just two examples:

Two-way DNA Copying. You will recall that the DNA chains were described as being *antiparallel*. Reference to Figure 8-5 will show that the antiparallel nature of the DNA pair involves the backbone internucleotide bridges in which the phosphodiester linkages are between the 3 and 5 carbons of the deoxyribose sugars. One polynucleotide is said to run in the 3,5 direction, and the other runs in the 5,3 direction. This fact might cause a great deal of trouble at the molecular level because DNA polymerase can only copy DNA by running along it in the 5,3 direction. Since the two DNA strands open up in opposite directions as the molecule uncoils for replication, only one of the strands can be copied continuously, namely the one that opens in the 5,3 direction.

How is the other strand copied? Copying occurs in very short segments starting at the just-opening end of the uncoiling site and running backwards, thus in the proper, or 3,5 direction, to where the strand starts to coil about its newly formed partner, where replication must stop (Figure 8-6). These short segments are readily detected by biochemical techniques. It is also possible to show that there is an enzyme present, DNA ligase, whose function is to hook these short fragments together into a perfect DNA strand. Presumably the ligase uses some feature of the parent DNA as a guide in this task.

DNA Copying Rate. The second example has to do with how fast DNA replication takes place. If we consider a typical bacterial cell, which may divide every 20 min and which has a single DNA molecule of 240,000 bases, we calculate that it must undergo reproduction at a rate of 16,000 bases per minute! Since there are 10 bases per turn of the helical DNA molecule, this must mean that, if replication goes on at one site, called a **replication fork**, the DNA molecule must be coiling and uncoiling at a rate of 1600 turns per minute.

Astonishing as this is, it seems to be exactly what happens. It is even possible by electron microscopy to see that there is only one replication fork per DNA molecule in bacteria (Figure 10-2). Such tremendous rates of turning could be expected to build up great forces on the DNA. This seems to be so because there

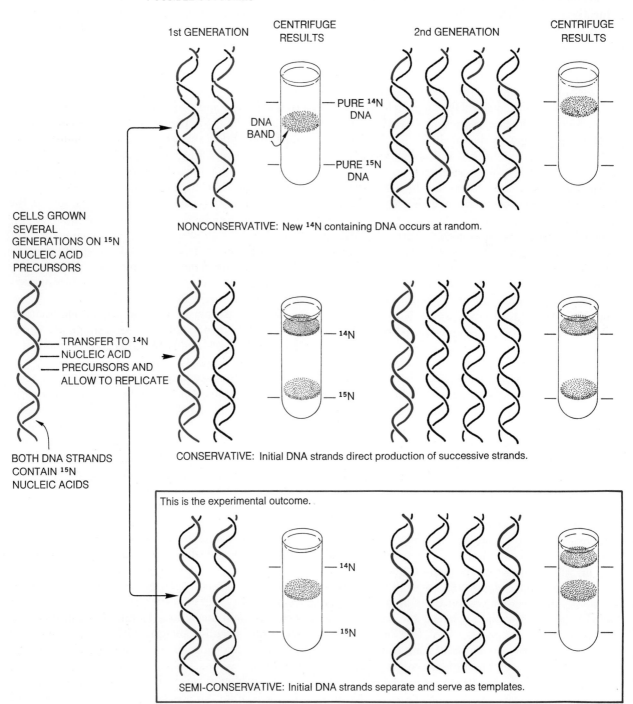

Figure 8-7 An experiment demonstrating semiconservative DNA replication.

is yet another enzyme, whimsically called swivelase, which seems to function to break too tightly twisted DNA molecules, let them unwind a bit, and then seal them back the way they were. There seems not only to be a gene for every enzyme but an enzyme for every problem.

FROM BACTERIA TO HIGHER ORGANISMS

Most of what is known about DNA replication has been learned from viruses and bacteria. Their simplicity, rapid rates of reproduction, and the ease with which immense numbers of them are cultured are essential to much of the research in molecular genetics. The ultimate hope is to use the clues derived from these simple systems to unravel the greater complexity of the genetic apparatus of higher organisms. For example, as we shall later see in more detail, the DNA of higher organisms exists as a number of independent molecules, each one of which is contained in a chromosome.

Even so, the replication process seems to be about the same as in bacteria except that the replication fork moves considerably more slowly, and there are many replication forks instead of one or a few. One complex system in which this has been studied, the developing egg cell of the fruit fly *Drosophila*, provides an interesting comparison with bacteria. A fruit fly cell contains a great deal more DNA than is in a bacterium, yet it replicates DNA at only about 2600 bases per minute per replication fork. If there were only one replication fork, it is estimated that replication of all of the DNA of the fruit fly cell would require about 2 weeks, which would be awkward since that is about how long a fruit fly lives. *Drosophila* solves the problem by having some 6000 replication forks in action at once and does the whole job in 3 min.

MUTATION

The relationship between DNA and the life processes that it directs, through the control of protein synthesis, is largely a one-way street. Although events taking place in the cell can affect the *magnitude* of DNA activity, they do not affect the *qualitative* nature of what the DNA produces. To illustrate this important fact suppose that a cell comes upon a larger than usual amount of some sugar, which it is capable of metabolizing, but that it has an insufficient supply of the enzyme necessary to handle the increased work load. There are ways (page 136) for this fact to be signalled to the DNA that controls production of the required enzyme, resulting in increased production. In this way it can be said that the environment directly influences the DNA, but only in the very specific and limited way of affecting the magnitude of its effort. The environment, we have found, does not affect DNA's product. Thus, if the cell comes upon a sugar for which it has no enzymatic machinery at all, there is no way for this information to be presented to the DNA; thus, DNA cannot be instructed to undertake production of an enzyme capable of metabolizing the new sugar. Even if a protein produced by the DNA underwent an environmentally induced change, perhaps just the loss of a few terminal amino acids, this information could not be communicated to the DNA in a way that would lead to the synthesis of this modified protein. This is true even if the new protein improved the chances for survival of the cell. All this being so, we are faced with a surprising fact. *Inherited changes in an organism cannot come about as a direct response to environmental influences.*

This has been a hard lesson for biologists to learn. Knowing nothing about the molecular nature of inheritance, early students of evolution, quite naturally believed there ought to be a direct way for environmental requirements to be translated into inherited changes if evolution were to occur. Some very cumbersome theories were proposed in explanation of what was believed to happen. Even the great Charles Darwin (page 286) fell for one of these, the theory of Pangenesis. According to this theory, every cell of the organism contributed particles called gemmules to the hereditary message carried by its reproductive cells. Each gemmule was believed to communicate not only the details of how to make its part of the organism but also to communicate changes induced in that part of the organism by environmental effects. Pangenesis was, in other words, a theory of a mechanism for the *inheritance of acquired characteristics.* Thus if a certain set of muscles was used to excess, that fact would be expressed in the gemmules representing those muscles and result in more robust muscles in the next generation.

Today all such ideas may be dismissed because we know that

1. The basic fabric and function of the organism is determined by its genes, with the environment only serving to influence the extent to which the genetic message is expressed.
2. The only way to change the genetic message is to change its words, and environmental effects can only produce such changes *randomly*.

These changes in the wording of the genetic message are called mutations. Wording changes are of two general types. The first is **gene mutation** which is a change in the sequence of bases in DNA. The second involves addition, deletion, or change in position of entire sections of the DNA sequence. This second class of mutation is generally called **chromosomal mutation** (see page 176).

If mutations occur in ways that are not directly related to new environmental or other requirements for survival, how is it possible for organisms to adapt, to cope with change? How can inherited changes appear that improve survival? The briefest possible answer is that mutations occur at random, and the organisms that possess them survive on the basis of whether or not the change caused by the mutation assists or hinders the organism's ability to survive. There will be much more on this in Chapter 16. Our intent now is to focus on the mutation process itself, and for now exclusively on the process of gene mutation with emphasis on what gene mutation can tell us about what the gene is and how it works.

Change Reveals the Reality

Now that we know a little about how DNA replicates itself, let us begin to look at how it works to build the organism. The first step that we must take is to develop a clearer idea of the relationship between DNA and those factors that we have been glibly calling genes. The several statements already made about this relationship may be summarized as follows: Beadle and Tatum gave us the idea that there was a direct relationship between genes and enzymes (proteins), and over the past 20 years the ideas of Griffith and Avery and many others came to fruition in the Watson-Crick structure for DNA, representing the physical embodiment of the genes. The linear sequence of bases in the DNA has been described as a code in which each successive set of three bases designates through RNA, another nucleic acid, a specific amino acid in a particular spot in the protein for which that segment of DNA codes.

How Much DNA Makes a Gene?

It must be obvious that one DNA molecule must represent more than one gene. Consider bacteria; each bacterium contains only one DNA molecule, but it contains many proteins. Then the question becomes, how much DNA is required to make a gene? If one gene makes one protein, then the number of nucleotide pairs in the gene (considering both strands) must be three times the number of amino acids in the protein for which it codes. An average protein has a molecular weight of about 30,000. If we allow 100 as the average molecular weight of an amino acid, an average protein is composed of about 300 amino acids. Since we claim that it takes three base pairs to code for one amino acid, that would make the size of the gene about 900 base pairs, or about 90 turns of the double helix (Figure 8-1).

There is a very interesting way to confirm this estimate of the size of the gene. Basically it involves determining how many different ways a gene can be mutated, changed in an inheritable way. Because of the very large number of organisms required for such a study, it could not be done until it became possible to work with bacteriophages, viruses that attack bacteria. These are favorable subjects for such experiments because each virus has only one DNA molecule containing, perhaps, only a few dozen genes, and because techniques are available by which immense numbers of viruses may be grown, treated with agents that cause mutations, and the mutations identified. When Seymour Benzer performed such an analysis of the genetic fine structure of a gene of bacteriophage T4, he found about 500 sites at which an identifiable mutation of that one gene could occur. That suggests that the gene has at least 500 parts. This in turn leads to the thought that the only informationally critical parts of the DNA present in such numbers are the bases. Thus by two radically different tests we come to about the same figure for the number of bases in a gene.

The Action of Mutagenic Chemicals

Another way to study the nature of the gene involves studying the action of chemicals that react rather specifically with the chemical subunits of DNA. Many such chemicals are known, and some may be significant causes of cancer (see Chapter 23). Nitrous acid (HNO_2) is a particularly good example. It reacts directly with the nucleotide bases, converting one base into another by replacing an amino group with oxygen (Figure 8-8). This alteration should produce

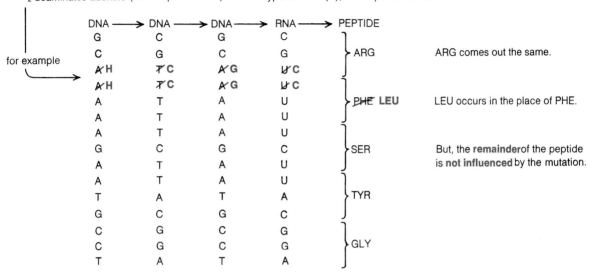

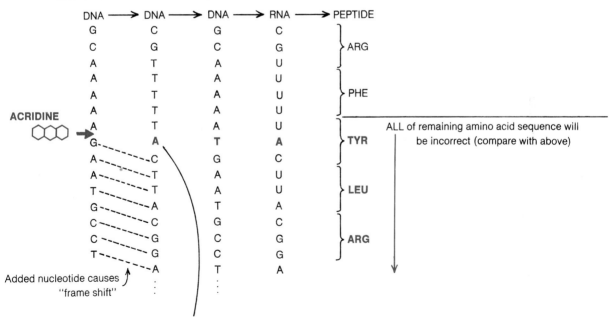

Figure 8-8 The effects of mutagens.

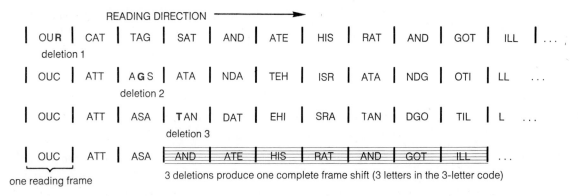

Figure 8-9 In a three-letter (triplet) code, sense is restored to the message after three deletions of letters.

"single word" (one amino acid) changes if the genetic code is as we have described it. These would be changes in which one base triplet would be misspelled per molecule of nitrous acid reacting with a DNA base with the result that one amino acid would be out of place for each such reaction. The effect on the cell may not be observable, since proteins often function relatively normally with such single substitutions. However, as we have seen in the instance of sickle cell anemia, if single amino acid substitutions occur in critical parts of the protein, the results can be devastating.

Acridine dyes (Figure 8-8) produce mutations by causing the addition or deletion of a base in the DNA chain. The results of this action confirm the idea that the genetic code is a triplet code, requiring reading along the base sequence, three by three. Since the base sequence is read continuously, removal or addition of one base acts to change the meaning of each following triplet in the DNA molecule, Figure 8-8. The result is a totally nonfunctional protein and, probably, a nonviable cell, unless the site of action of the dye is very close to the end of the DNA molecule. Proof that the code is read in groups of three bases comes from experiments in which more than one dye-induced deletion or insertion occur close together in the DNA. For example, if there are two deletions of bases very close together, the effect is still likely to be serious because the reading frame is still shifted, with nonsense triplets occurring all the way to the end of the DNA molecule. However, three deletions close together result in a comparatively normal protein because only the triplets between the three deletions would be frame shifted. After three deletions the reading is returned to normal if the code is a triplet code (Figure 8-9). If it were to be a four-unit code, four deletions would be necessary in this experiment to make things come out right again.

The Connection between Mutation and the Continuity of Life

At first glance two characteristics of DNA seem at odds with its role as the link between cell generations. First, DNA is changeable in an inherited way, and, second, these changes do not occur in ways that directly result in adaptation. This dilemma is resolved by the appreciation of two facts. The first is that there must be some inherited change experienced by the DNA of living things if evolution is to occur. Continual perfect replication of the original forms of DNA would result in no change, no adaptation to changing conditions, and would ultimately be fatal to life. That is obvious enough. The second fact is somewhat harder to appreciate, namely, that adaptive change in DNA does occur as the result of increased survivability of certain random mutations.

With the knowledge that you now have of the chemical basis of heredity and mutation, it is not hard to grasp why there is not likely to be a very direct relationship between (1) an environmental change, (2) the changes necessary in various proteins to adapt the organism to the new environmental condition, and (3) the changed sequence of DNA nucleotide bases that are necessary to produce the required change in properties of the proteins. It would be convenient if this did happen, but, since such direct and adaptive effects of the environment on the DNA are impossible, the more cumbersome but ultimately effective method of selection of random mutations is utilized.

Aside from chemical arguments, it is difficult to provide experimental proof that there is no direct and adaptive effect of the environment on DNA in most organisms because of the technical problems caused by their long life cycles and the difficulty of dealing with large enough numbers of individuals to provide the necessary data. However, in bacteria it is possible to do such experiments, and these show clearly that the environment cannot induce adaptive changes directly in the bacterial DNA. An experiment by Joshua and E. M. Lederberg illustrates this.

The bacterium *Escherichia coli* is normally susceptible to bacteriophage T1, but it can mutate to T1 resistance. Resistant mutations in a culture of *E. coli* may be detected by growing samples in culture dishes uniformly covered with bacteriophage. The *E. coli* colonies that appear must have started from a resistant mutant bacterial cell that can live in the presence

Figure 8-10 The replica plating experiment supports the random nature of mutation.

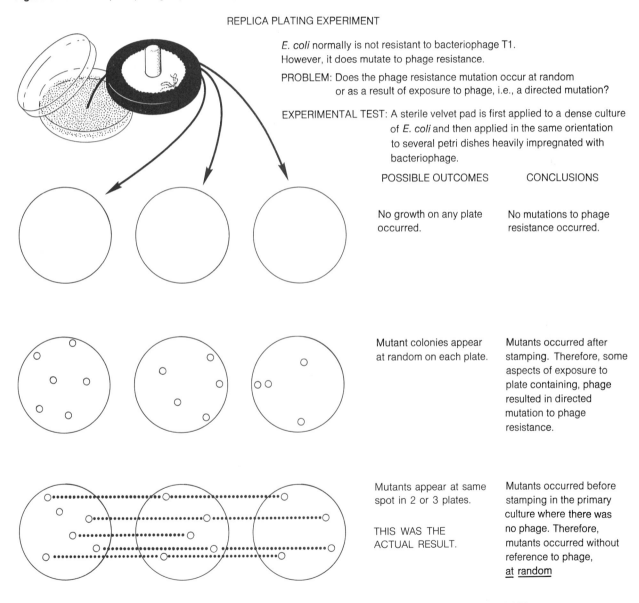

of T1. The question is, did the mutation arise because of contact with the phage, which would be a direct environmental effect, or did it already exist in the culture from which the sample was taken?

The answer was obtained by the technique of replica-plating (Figure 8-10). In this technique the original colony of *E. coli* before exposure to T1 can be thought of as a stamp pad "inked" with bacteria. A stamp, in this instance a sterile velvet pad, is pressed against the surface on which the bacteria are growing and some adhere. The stamp is then applied in the same orientation to several culture dishes containing bacteriophage. These are incubated until phage resistant colonies are identifiable by their growth. Their positions are plotted in the several dishes. If there are resistant colonies in the same relative positions on several dishes, then the stamp must have picked them up already mutated from the original dish, for there is no reason for mutations to occur at the same place on several plates. If mutations are found in different sites on each plate, then there is a strong possibility that they arose *after* the stamping process, and, therefore, possibly as a result of the effects of the phage containing environment. The experiment showed that the latter did not happen with anything like the frequency of appearance of place-related resistant colonies. Thus the experiment demonstrated selection of mutants already in existence.

Since, as such experiments show, new candidates for survival must be chosen from random mutations, there obviously must be an appreciable mutation rate to provide sufficient raw material for the evolutionary process. Yet this mutation rate must not be so high as to engender the risk of loss of hard-won hereditary characteristics. Mutation rates will be discussed later but we can say as a rule of thumb that there is, perhaps, one mistake made per 10 million nucleotide replications. This figure has undoubtedly not been constant through time and was probably much higher during the early phases of the evolution of life. In very early organisms the replicative machinery must have been more poorly developed, and the incidence of natural mutagenic agents may have been higher than it is now.

The fact that it is important to maintain mutation rates at a low level is emphasized by the presence in cells of elaborate enzymatic machinery for detecting mutated DNA and repairing it. A set of enzymes is able to recognize abnormal DNA sequences, cut them out, and weave in new, correct strands. This process illustrates another advantage of the double helix. If a mutation occurs, the partner strand provides a way to identify the error.

Ultraviolet light is a powerful mutagen because it is strongly absorbed by DNA with the production of chemical changes that result in coding errors during subsequent replications. We have seen that the first organisms were limited in their conquest of the shallow waters and land by the strong ultraviolet radiation impinging upon the earth before the ozone layer developed. It is, therefore, interesting to note that many organisms have a repair mechanism specifically tailored to repair ultraviolet effects on DNA. Although the process is not fully understood, what happens is that the ultraviolet mutated DNA is detected by a special protein. By attaching to the mutated site, the protein can transfer visible light energy to the DNA and repair the ultraviolet damage, a process called **photoreactivation.** Still important in organisms without the protection from ultraviolet provided by large body size, photoreactivation has perhaps been with us since the earliest days of life.

The importance of such DNA repair mechanisms in the human is illustrated by the hereditary disease xeroderma pigmentosum. Affected individuals develop cancerous lesions of the skin upon exposure to light and die young. The fundamental defect appears to be failure of the normal mechanisms for repair of light-damaged DNA.

9 Gene Expression in Simple Cells

We have seen how DNA replicates to insure distribution of faithful copies of itself to daughter cells. Next we will look at a far more complex process, the role of DNA in the life of the cell. We know that the DNA code is transcribed into another message in the form of molecules of RNA and that these molecules control the steps of protein synthesis, called **translation.** *Protein synthesis is the key step in the translation of the plan of the organism from the base sequences of DNA into the substance of a functioning cell.* Protein synthesis produces both structural protein and enzyme protein, and the latter generates the remainder of the cell.

In this chapter we will explain these bare statements and then consider (1) how protein synthesis is regulated to insure production of proper amounts of the many required gene products to bring about economical and effective operation of the cell and (2) how the activity of the gene products may be directly controlled. This chapter emphasizes bacteria because they illustrate the basic processes very well. Indeed, they are the organisms in which most of these processes were discovered. But you should be aware that regulation is much more complex in multicellular organisms. For the present, let us be grateful for the relative simplicity of the bacterial cell and reserve some of the complexities of gene expression in higher organisms for later chapters where the development and physiology of higher organisms are discussed.

IS PROTEIN SYNTHESIS A DIRECT TRANSLATION OF DNA NUCLEOTIDE SEQUENCES INTO AMINO ACID SEQUENCES?

Two facts among many show that this simplest hypothesis of protein synthesis cannot work. First of all, DNA is confined to the cell nucleus in higher organisms, yet it is readily demonstrated that protein synthesis takes place in the cytoplasm. Second, amino acids do not have sufficient structural complementarity to the DNA nucleotide bases that code for them to allow amino acids to be simply accumulated and formed into polypeptide chains directly on a DNA template. These facts require us to search first for additional kinds of molecules participating in protein synthesis. We need a molecule that carries the message of the DNA out into the cytoplasm. Then, in the cytoplasm, we must identify molecules that act as intermediates between amino acids and the triplet code of nucleotide bases; that is, we must find molecules that supply the correct amino acid at the correct position during polypeptide synthesis to satisfy the instructions of the DNA code. Three kinds of RNA perform these important functions. All three are made as complementary copies of nuclear DNA. They are messenger RNA (mRNA), ribosomal RNA (rRNA), and transfer RNA (tRNA).

The RNAs and Protein Synthesis

MESSENGER RNA IS THE WORKING TEXT OF THE GENETIC CODE

With simple variations, all types of RNA are constructed like DNA. RNA has a backbone of phosphate-linked sugar molecules bearing nucleotide bases. The sugar is ribose, rather than deoxyribose as in DNA; and the bases are the same as in DNA except for the substitution of uracil for thymine. This does not interfere with base pairing because the structure of uracil allows it to form uracil: adenine pairs, just as adenine pairs with thymine in DNA. Unlike DNA, RNA does not exist as a dimer—as a double helix, a pair of helically coiled complementary molecules. Although RNA can form a helical coil with DNA, or even coil upon itself, it is fundamentally a single stranded polynucleotide molecule (Figure 9-1).

Transcription

RNA is synthesized upon a DNA template in the process called **transcription**. One strand of the DNA serves as a template for RNA polymerase, which synthesizes long segments of RNA as complementary copies of the template DNA. RNA cannot serve as

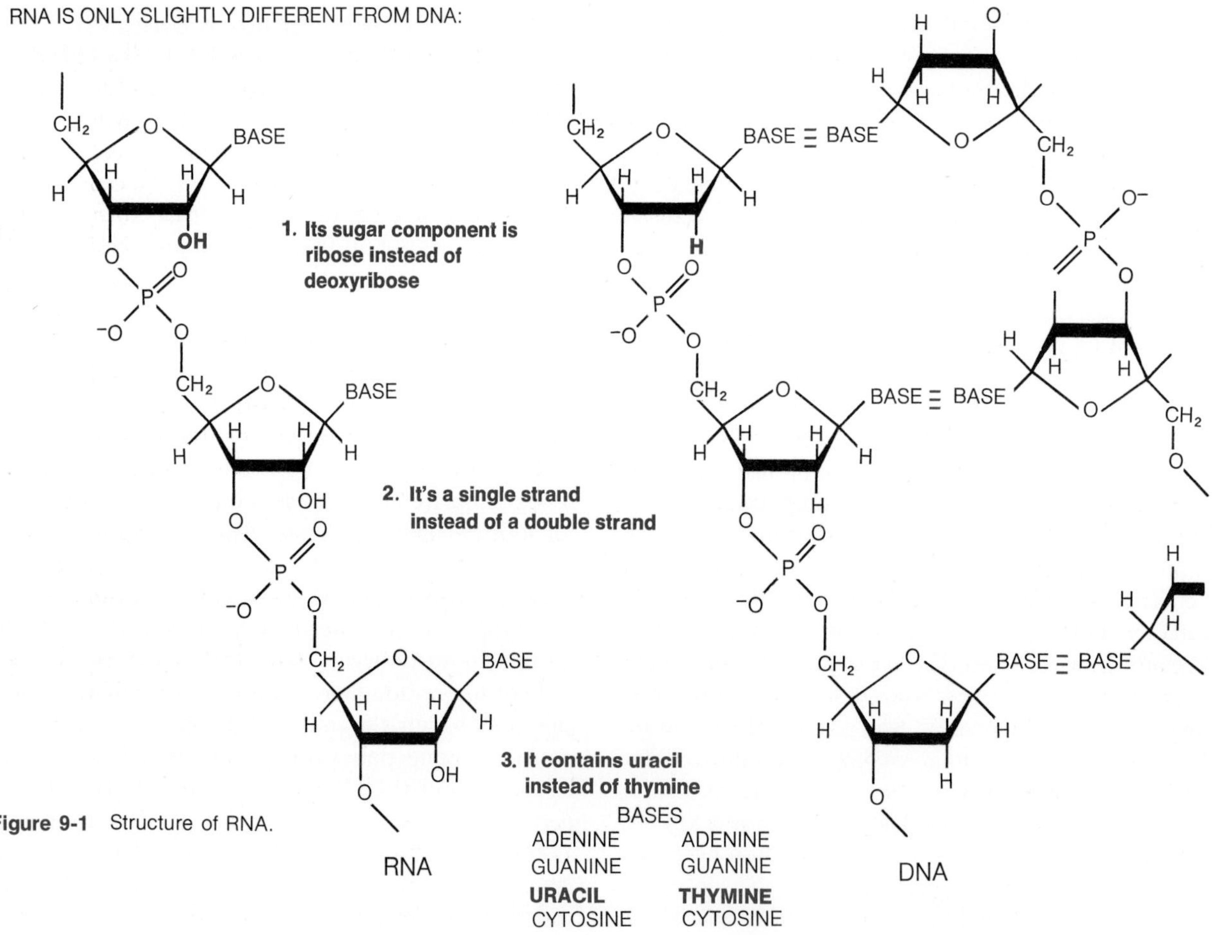

RNA IS ONLY SLIGHTLY DIFFERENT FROM DNA:

1. Its sugar component is ribose instead of deoxyribose
2. It's a single strand instead of a double strand
3. It contains uracil instead of thymine

BASES
ADENINE   ADENINE
GUANINE   GUANINE
URACIL **THYMINE**
CYTOSINE   CYTOSINE

RNA   DNA

Figure 9-1 Structure of RNA.

130 BIOLOGY

template for more copies of itself because RNA polymerase functions only with DNA as a template. The enzyme identifies specific initiation and termination points on the DNA by special base sequences so that complete genetic units are copied. In this way the directions contained in the DNA are copied out into a form that may be directly utilized in the synthesis of protein.

In bacterial transcription a protein coding segment of RNA requires no modification before serving as a guide for protein synthesis. However, in higher organisms, modification of the RNA is sometimes necessary because the DNA that codes for a particular protein has been found to contain 500 to 1000 nucleotide segments of "spacer" DNA. Spacer DNA is transcribed along with the rest of the DNA, but it is removed before protein synthesis begins. Excision is accomplished by special enzymes, which also join together the RNA that was on each side of the spacer segment. This must be accomplished with absolute precision, since an error of just one nucleotide would ruin the coding sequence. If spacer DNA proves to be common among organisms above the level of bacteria, it may be that it plays a role in the permanent switching on or off of parts of the genome and probably occurs as a part of the cellular specialization process in multicellular organisms.

Transcription is Exact

The technique of **molecular hybridization** permits demonstration that the copying process is an exact, base by base, complementary copying, with A on the DNA being represented by T on the RNA, and so on. It is reasoned that, if a segment of RNA is an exact, complementary copy of a particular segment of DNA, the RNA should be able to make a satisfactory substitute for the complementary DNA strand of that particular DNA segment and, with it, form a double helix molecule. The molecular hybridization test is done by gently heating DNA until it uncoils into single strands. Strands of the presumed complementary RNA are added and then cooling is allowed to take place with resultant double helix formation. Among these will be hybrids in which one strand is DNA and the other is RNA if the RNA is complementary to the DNA. When this occurs it demonstrates that the base sequence in complementary RNA is enough like that in the template DNA to allow normal pairing, a delicate process requiring precise molecular positioning to facilitate hydrogen bond formation between bases in the two polynucleotide strands.

That part of the cell's RNA known as mRNA contains the instructions for the assembly of specific sequences of amino acids into proteins, whereas tRNA and rRNA, copied from DNA in the same way as mRNA, have more generalized functions. They support the synthesis of *all* proteins rather than the synthesis of specific ones. They are the machinery that mRNA directs.

THE SITE OF PROTEIN ASSEMBLY IS THE RIBOSOME

Ribosomal RNA is found in structures called ribosomes, which consist of 40% to 60% rRNA with the remainder being protein. The molecular weight of ribosomes is about 2.8 million, and they are visible in the electron microscope as numerous small bodies attached to the endoplasmic reticulum or lying free in the cytoplasm. Ribosomes synthesize protein from amino acids, even when isolated from the cell, if they are provided with essential cofactors. *No other cellular component is capable of protein synthesis.* Thus, ribosomes are the crucial link in the process of information flow from DNA to protein.

Much is unknown about ribosome assembly and function. It is possible that rRNA works initially like mRNA to direct the synthesis of some of its own ribosomal protein, but thereafter rRNA does not function to direct the synthesis of specific protein. Instead the ribosome produces whatever protein is specified by the mRNA with which it happens to be associated. It is useful to think of the ribosome as a sort of molecular sewing machine in which the mRNA instructions and the many components necessary to construct a polypeptide chain are brought together in a physical alignment that allows sequential stitching of amino acids to the growing polypeptide chain.

The ribosome is formed of several subunits that aggregate to form two major units, one large and one small, with a groove between. The small subunit in bacterial ribosomes contains one kind of RNA and 21 different proteins, whereas the large subunit contains two kinds of RNA and 34 proteins. The groove between the subunits is thought to serve as a binding site for mRNA. Conformational (shape) changes in the molecules lining the groove are believed to move the ribosome along the mRNA. At any one time, several mRNA **codons** (base triplets specifying amino acids) lie within the groove and, in this position, they control the sequence of adding amino acids to the growing polypeptide chain, which is attached to the larger of the two ribosome subunits.

TRANSFER RNA MATCHES CODONS WITH AMINO ACIDS

The critical role in this step of specifying the addition of amino acids belongs to tRNA because it is specialized for two functions, the recognition of a particular codon and recognition of the amino acid which that codon represents. There are at least 20 types of tRNA, at least one for each amino acid involved in protein synthesis. Transfer RNA is a relatively small molecule containing up to 90 nucleotide units. The nucleotide sequences of several tRNA molecules have been determined. These data plus other information indicate that tRNA exists as a cloverleaf shaped molecule in which about half of the structure is in double helix form, produced by winding the single polynucleotide chain (monomer) upon itself. Transfer RNA differs from other kinds of RNA in that about 10% of its bases are of unusual types, several of which are formed by addition of methyl groups to the usual four bases (UAGC) after their insertion into the chain. These methylated bases are thought to be responsible in part for the secondary structure of the molecule since they would interfere with coiling and thus could cause the characteristic four armed structure (Figure 9-2).

Two of the tRNA arms are essential to the role of tRNA as translator of the mRNA code into specific amino acid additions to a growing polypeptide. The **anticodon** arm of tRNA contains a specific base triplet that is complementary with a codon triplet of mRNA, hence the name anticodon. Elsewhere on the molecule there is an as yet unidentified site which insures that the enzyme attaching an amino acid to the tRNA will attach only the amino acid for which the anticodon site codes. Amino acids are attached to an arm that bears no specificity and, in all tRNAs, terminates identically with the three bases, CCA. It is

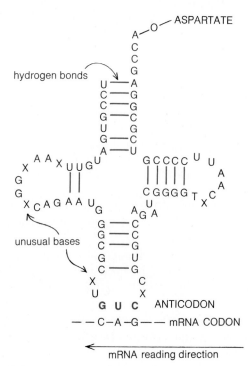

Figure 9-2 Structure of a tRNA for aspartic acid.

to this last base, adenine, that the amino acid is attached by an amino acid activating enzyme specific for each amino acid and its specific tRNA. Finally, there is a fourth site on the rRNA that controls attachment of the tRNA to the ribosome.

Addition of an amino acid to its proper carrier tRNA occurs as follows. First, the amino acid is activated by combination with an enzyme-ATP complex, and next it is attached to the terminal adenine group of tRNA (Figure 9-3). This process must be highly specific for the kind of amino acid because, once attached to tRNA, the amino acid is identified by the protein synthetic machinery only in terms of the anticodon. This can be proven by chemically converting one amino acid into another after attachment to tRNA and observing the effects on protein synthesis. For example, cysteine is readily converted into alanine by removal of its SH group. When this is done to cysteine already attached to its specific tRNA, incorrect coding results because the anticodon says it is carrying cysteine when it is actually carrying alanine. If this tRNA participates in protein synthesis, the protein synthesized contains alanine where there should have been cysteine.

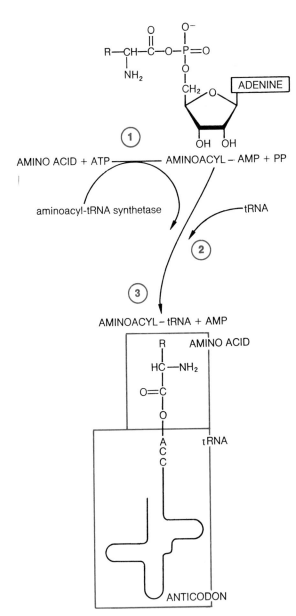

Figure 9-3 Steps in binding of an amino acid with its specific tRNA.

1. Amino acid reacts with ATP to form aminoacyl-AMP, which remains bound to AA-specific enzyme.
2. Enzyme recognizes proper tRNA for amino acid.
3. Transfer of amino acid from AMP to tRNA.

THE STEPS OF PROTEIN SYNTHESIS

With the basic components of protein synthesis identified, let us see how they cooperate to produce protein molecules. The process involves the formation of peptide bonds between amino acids (see page 44). Synthesis proceeds from the free amino end of the protein, amino acid by amino acid. Great complexity is built around this simple reaction because of the necessity of producing a specific amino acid sequence. To facilitate describing the many events that are involved, we shall divide the process of polypeptide synthesis into three phases, (1) initiation, (2) peptide formation, and (3) completion.

Initiation

The starting point for reading mRNA is always the codon AUG, the code for methionine. In bacteria this initial codon recognizes only a special aminoacyl-tRNA that carries formylmethionine (Figure 9-4). Presence of the formyl group prevents the amino group from participating in peptide bond formation, and this limits formylmethionine to the initial position in peptides. The same initial molecule is used in mitochondria of higher organisms, but in their cytoplasmic protein synthesis the initiator tRNA carries

Figure 9-4 Formylmethionine is the initiator amino acid in bacterial peptide synthesis.

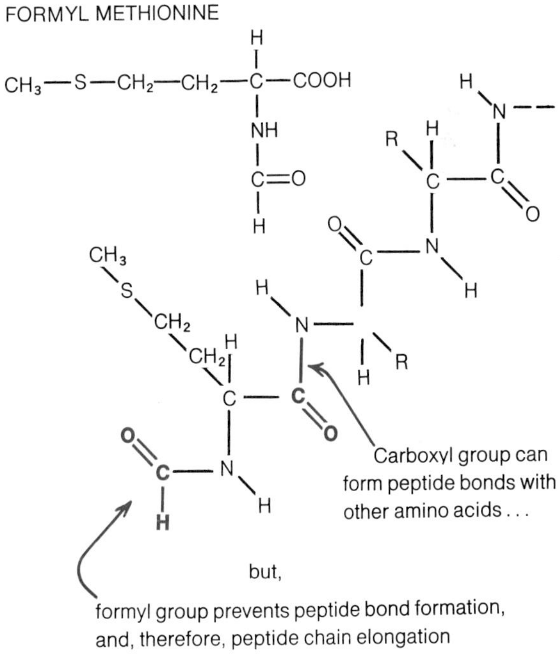

FORMYL METHIONINE

Carboxyl group can form peptide bonds with other amino acids...

but,

formyl group prevents peptide bond formation, and, therefore, peptide chain elongation

an unmodified methionine, although the tRNA is itself specialized for its initiator role.

At the start of the initiation phase the ribosome is in two parts, the small (SRU) and large (LRU) ribosome units. The SRU binds with a protein initiation factor (IF-3) and then it can bind with an initiator complex consisting of initiator tRNA, another protein initiation factor (IF-2), and guanosine-5'-triphosphate (GTP). The resulting complex (IF-3, SRU + IF-2, initiator tRNA, GTP) then binds to a strand of mRNA, a process that probably involves another initiation factor (IF-1). In bacteria the SRU has an exposed short nucleotide sequence that is complementary with a short segment of mRNA that seems typically to preceed the initiation codon. It is believed that this serves to facilitate proper starting orientation of mRNA to the SRU. At this stage, the LRU binds to the SRU complex, the initiation factors are released for reuse, and the GTP high energy bond is utilized (Figure 9-5). The completely assembled ribosome is ready to translate the attached mRNA strand.

Peptide Formation

The assembled ribosome contains two sites for aminoacyl tRNAs. In the initial events, the special initiator tRNA comes to lie in one of these two sites. The P (peptidyl) site has its anticodon hydrogen bonded to the initiator codon of the mRNA strand. In this orientation the second codon of the mRNA strand lies so that it determines which variety of aminoacyl-tRNA (aa-tRNA) can bind at the second, or A (aminoacyl) site. Figure 9-6 illustrates the steps that follow the attainment of this orientation, steps that result in formation of the peptide by successive addition of amino acids as specified by the mRNA.

In order to accomplish the assembly of the first two amino acids in the peptide chain, the A site must be occupied by the aa-tRNA specified by the second codon. This is not accomplished directly. The aa-tRNA first combines with GTP and an elongation factor protein. Then the aa-tRNA anticodon hydrogen bonds with the codon defining the A site, GTP is utilized, and the elongation factor is released for recycling (Figure 9-6, Step 1). Peptidyl transferase, an

Figure 9-5 The initiation phase in protein synthesis.

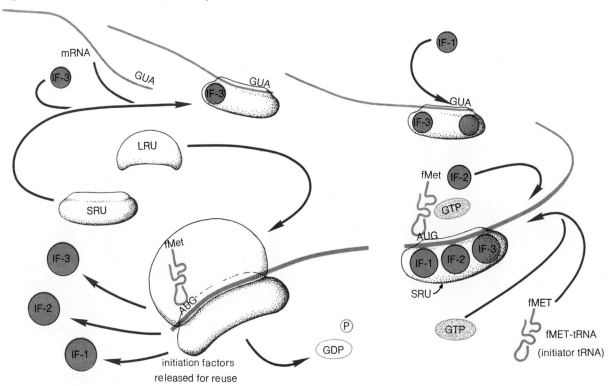

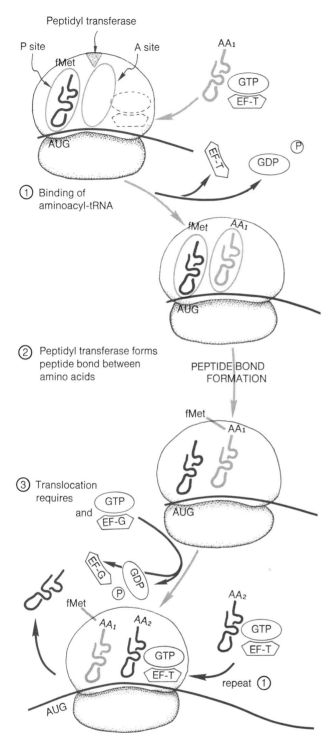

Figure 9-6 Assembly of peptide units.

enzyme that is part of the LRU, is then able to promote peptide bond formation between the amino acids occupying the P and A sites. As this takes place, the amino acid occupying the P site is released from its tRNA and remains attached to the ribosome only by the peptide bond with the aa-tRNA occupying the A site. The unloaded tRNA at the P site dissociates from the ribosome for recycling (Figure 9-6, Step 2).

To prepare for forming the second peptide bond with the third amino acid specified by the mRNA, the aa-tRNA at the A site moves to the P site. This energy requiring process also requires another elongation factor. As a consequence of this shift between sites, the entire ribosome also shifts along the mRNA so that the codon specifying the third amino acid controls the A site (Figure 9-6, Step 3). The correct aa-tRNA then attaches at the A site, and peptide bond formation occurs. The cycle then repeats until the peptide is completed. Cycling rates vary between 1 and 20 peptide bonds per second. Utilization of the mRNA is made more efficient by the fact that more than one ribosome may occupy it at once. Such aggregations of several ribosomes connected by a single strand of mRNA are called polyribosomes. Ribosomes may also initiate translation of mRNA before it is completely transcribed from DNA.

Completion

Examination of the genetic code tabulation in Figure 8-4 shows that there are several codons that signal stop, or the end of the peptide chain. These codons are not recognized by tRNA but by special protein release factors. When the ribosome reaches a stop codon the release factor causes the last aa-tRNA to shift to the P site and evidently modifies the specificity of the peptidyl transferase to allow it to break the bond between the last amino acid of the peptide chain and the tRNA. This frees the newly formed peptide, which in most cases spontaneously assumes its tertiary structure. In some instances the formylmethionine is modified or removed, and sulfhydryl bonds may be enzymatically formed to crosslink folds in the peptide chain. Instances are also known in which internal segments of peptide chain must be removed before the newly synthesized peptide is functional.

Once the peptide is free, the last tRNA and the mRNA come free of the ribosome. The ribosome then dissociates into SRU and LRU, which are then ready to commence another sequence of peptide synthesis.

Finally, it should be noted that protein synthesis actually consumes more energy than any other biosynthetic process, four high energy groups being used per peptide bond formed. This large amount of energy is a necessary expenditure to operate the elaborate machinery that insures faithful copying of the mRNA message because peptide bonds could be formed with a much smaller expenditure of energy if coding were not a consideration.

Regulation of Gene Action

The system of gene action as so far detailed lacks controls; if it were to work as described, each gene would promote the synthesis of its product without regard to the needs of the cell, a wasteful situation not long to be tolerated by nature. To function optimally, a cell must use its genetic machinery with that minimum expenditure of energy allowing production of *just sufficient amounts* of gene products necessary for survival. The cell also must be able to *vary the proportions* of gene products to meet changing environmental conditions. Particularly in higher organisms, cells must produce *specific gene products* as dictated by cell specialization or the stage in the life cycle.

These and other regulatory processes of cells and organisms are matters of the greatest complexity and will concern us throughout the remainder of the text, for assuredly one of the dominating requirements of the organism is maintenance of a stable and well regulated internal environment. In the following pages we shall begin to study this problem by examining two of the three major levels of regulation. The first of these is regulation of the construction of proteins, that is, regulation at the level of DNA transcription into mRNA and at the level of mRNA translation into protein. The second level is regulation of action of the protein enzyme, itself. As before, we will be primarily concerned with these processes as they occur in bacteria.

TRANSCRIPTIONAL AND TRANSLATIONAL REGULATION IN BACTERIA

Most of the uncertainty in the life of a bacterium comes from outside. The bacterium is a single cell with little protection from the direct action of the environment upon its protoplasm. Also it depends continuously on the environment to supply essential molecules which, because of its small size, it cannot easily stockpile against future shortages. These facts, together with the short life span typical of bacteria, indicate that survival in bacteria is favored by rapid adaptation to environmental change—with rapid being defined in terms of a fraction of a life cycle that often occupies less than an hour.

Adaptation of *Escherichia coli* to Different Carbon Sources

Escherichia coli, which we have met before, illustrates well the problem of environmental adaptation in the life of bacteria. *E. coli* requires one of several sugars. It has genes for producing the various enzymes essential to utilization of each of these sugars. However, the relative concentrations of these various enzymes in the bacterial cell depend on which sugar is available in the environment. If glucose is the only sugar available, *E. coli* cells have many copies of the glucose metabolizing enzymes present and at work. At the same time there may be very few copies of the enzymes necessary for the utilization of other sugars, for example lactose, which the bacteria is also capable of utilizing. Sudden transfer of cells from glucose to lactose is followed by a brief pause in sugar metabolism, but within a few minutes the ability to use lactose begins to increase and soon the bacteria are able to use it as well as they formerly did glucose. This change is accompanied by a rapid increase in the number of copies of an enzyme, β-galactosidase, which breaks down lactose to glucose and galactose. Such an effect is called **enzyme induction**. The cause is not selection of different mutants because the change occurs within only a few cell division cycles and can be shown to involve all cells in the experimental population.

THE OPERON CONCEPT

To explain situations such as this, in which gene products vary in concentration with the requirements of the cell, F. Jacob and J. Monod argued that the DNA had to contain factors that sensed cell requirements so that genes could be activated or inactivated as needed. Their concept is *not* a violation of the well-accepted idea that the environment cannot *directly* cause adaptive change in the genes. It postulates only that there is a way to change the *activity* of existing genes; the genes themselves are not changed.

Jacob and Monod propose that there are actually three kinds of genes, and these are packaged in functional units of DNA segments called **operons.** The first type of gene we have been discussing all along. It is the **structural gene,** a gene that actually codes for a particular peptide chain. Associated with the structural gene, or groups of functionally related genes, are **regulator** and **operator genes.** They work to regulate the activity of the associated structural gene or genes in the following way: The **regulator gene** codes for a protein that is called the **repressor protein,** synthesized in the usual way by the ribosomes. In the ordinary course of events, when the repressor protein is released by the ribosome, it finds its way to the operator gene site in the DNA and is able to bind with it. When this happens, RNA polymerase cannot copy the associated structural gene or genes because the mass of the repressor protein prevents the RNA polymerase from attaching to what is called the promotor site on the DNA, a necessary step lending to transcription. Thus the structural genes are inactive, and few or no copies of them are transcribed into mRNA. The genes are said to be **repressed** by the repressor protein.

The structural genes may be activated, or **derepressed,** by removing the repressor protein from the operator gene. Actually the operator gene is only a dozen or so bases in length and bonds to the repressor protein with easily broken hydrogen bonds. Consequently, the repressor protein continually cycles on and off the operator gene, but most of the time it is attached and the associated structural gene is inactive.

However, the repressor protein is constructed so that it can combine with a molecule that is significantly related to the activity of the structural genes that it controls. For example, if the structural gene codes for an enzyme that breaks down a particular sugar, that sugar may be able to combine with the repressor. In doing so, it produces a change in the structure of the repressor protein that prevents it from attaching to the operator gene, thus derepressing the structural gene or genes and allowing formation of their products (Figure 9-7).

Let us examine how this actually works in the case of the genes involved in lactose breakdown in *E. coli*. These lie together in one region of the bacterial DNA called the **lactose (lac) operon.** There are in the lac operon three structural genes, one for β-glucosidase, which breaks down the disaccharide lactose into glucose and galactose, a second for β-galactose permease, which promotes entry of the lactose into the cell, and a third, galactoside transferase, of unknown but necessary function in the utilization of lactose. The operator gene and the regulator gene are nearby (Figure 9-7). In the absence of lactose, the repressor occupies the operator gene and mRNA for the three enzymes is not produced. Lactose and a few related molecules are able to combine with the repressor protein, derepressing the three structural genes and allowing them to produce mRNA for the three enzymes necessary for lactose utilization. To this scheme we must add another control element to account for the fact that, even when lactose is present, if glucose is also present the lac operon remains repressed.

As a matter of fact, in the presence of glucose the enzymes for utilization of several sugars besides lactose are repressed. Presumably there is an advantage to using glucose exclusively when it is present. Precisely how this action occurs is not yet clear, but it appears to involve cyclic AMP (c-AMP) (page 363). The intracellular concentration of c-AMP appears somehow to be controlled by glucose. When glucose is low, c-AMP concentrations rise. This allows c-AMP to react with a protein, which is then able to derepress the lac operon, allowing production of lactose enzymes.

Thus adaptively useful control of enzyme levels in the cell is attained by action at the transcriptional level. These mechanisms of control are well documented in bacteria, and there is every likelihood that they also occur in the cells of higher organisms. In-

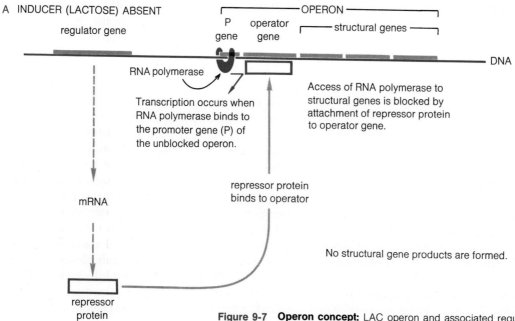

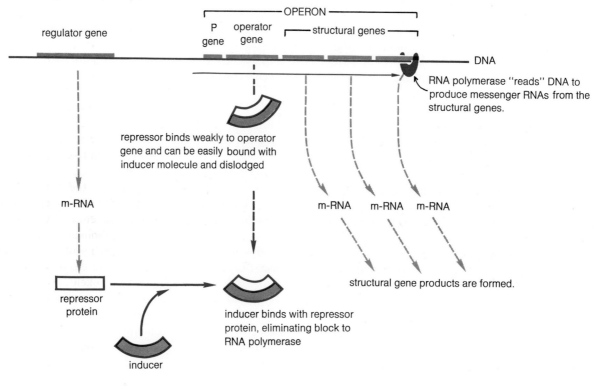

Figure 9-7 Operon concept: LAC operon and associated regulatory gene. Regulatory gene causes production of protein repressor able to bind to a particular DNA region (operon), blocking transcription of DNA to mRNA. Repressor protein binds to operator gene, blocking RNA polymerase, which is bound to the P gene (promoter region). Presence of an inducer (lactose) favors inactive form of repressor, resulting in normal transcription and subsequent translation of enzymes for substrate utilization. **A.** Repression; **B.** Induction.

deed, c-AMP has for some time been known as a "second messenger" mediating many of the effects of hormones upon the metabolism of cells in higher organisms.

END PRODUCT REPRESSION

In essence the type of repression just considered for the lac operon is one in which the repression is relieved by the substrate of an enzyme. This, of course, is not the only situation in which it might be necessary to control an enzyme. For example, with sufficient substrate present, an enzyme might work away, utilizing precious resources without limit and flooding the cell with far more of its product than needed. Since what we already know of the complex and interdependent web of chemical reactions in cells tells us that this would be disastrous, we may expect some form of control to prevent it from happening.

A particularly good example of this type of control is seen in the *E. coli* system of enzymes that manufacture the amino acid tryptophan. This molecule is produced by the sequential action of five enzymes, yet when the concentration of tryptophan in the medium rises sufficiently to make biosynthesis unnecessary, all five of these enzymes cease to be produced. The explanation is simply that all five structural genes for these enzymes constitute a single operon controlled by one regulator gene. The difference between this situation and the lactose operon mechanism is that the tryptophan operon repressor substance is nonfunctional unless it is in the presence of the end product, tryptophan. With tryptophan present, the repressor is activated and can prevent synthesis of the entire sequence of enzymes.

Thus it appears that transcriptional level enzyme induction and repression both occur and both represent relatively minor modifications of a process in which a repressor protein blocks transcription. The repressor is rendered inactive by the substrate molecule in the case of enzyme induction, whereas it is made active by the end product in the case of end product repression. In addition to these two processes, another mechanism involving c-AMP operates at the transcriptional level to allow interaction among different systems of enzymes when this is biochemically logical. Thus, when the cell has sufficient glucose, it does not need to bother with other enzyme systems for the metabolism of other sugars; these are then turned off at the transcriptional level by c-AMP mediated corepression.

Transcriptional level control is the most saving of cellular resources, but controls at other levels are possible and do occur. Translational control, involving rRNA and tRNA is possible but it can only affect the overall level of protein synthesis since these entities are active in the synthesis of all proteins.

It is important to note that the life of mRNA in bacterial cells is quite short, often only a few minutes, and this insures that transcriptional level control is rapid and conforms to the requirements of the moment. In higher organisms mRNA may have a much longer lifetime, so that the actual concentrations of enzymes present in the cell may only change slowly in response to transcriptional control. In such instances controls that act directly upon the functioning enzyme come to be of great importance.

REGULATION AT THE ENZYME LEVEL

Once synthesized, enzymes are regulated in several ways. An enzyme may be synthesized in an inactive form, called a **zymogen**, which must be enzymatically or otherwise converted into an active form. This stratagem is useful if an enzyme, for example, a protease, might do damage at the site of synthesis and must be kept in the inactive zymogen form until it reaches the proper site for its action (see page 376).

In another type of regulation, an enzyme may have two forms, one more active than the other, and be switched back and forth between the two by other enzymes in accordance with the requirements of the cell. Glycogen phosphorylase, the enzyme that breaks down glycogen (the storage carbohydrate of animal cells) into glucose-1-phosphate, is a well-known example. The most active form of the enzyme is a four-peptide chain subunit. This is broken down into less active two-chain subunits by one enzyme and reconstituted into the four-chain more active form by another.

Finally, there is a large group of regulatory enzymes that are affected by **modulator molecules** that may be their own substrates or substrates of related reactions. These enzymes have been named **allosteric enzymes** by J. Monod, J. P. Changeux, and F. Jacob, who first proposed their mechanism of action. The name allosteric was given because it means "another structure," reflecting the idea that the enzyme changes its activity by means of changes in its structure. Allosteric enzyme molecules possess at least two critical sites. One is the enzymatic site where the reaction the enzyme mediates takes place. A different site, the allosteric site, allows the modulator molecule to bind in a reversible way. This binding of the modulator with the enzyme serves to influence the properties of the enzymatic site, perhaps by a change in the shape of that part of the molecule. Very commonly the allosteric enzyme is the first in a series of related reactions so that changes in its level of activity may control the entire sequence. Alternatively, allosteric enzymes may occur at branch points in complex reaction systems, so as to control the flow of molecules among the various possible alternative routes. Modulator molecules may either increase or decrease the level of activity of the enzymes on which they act.

NUCLEIC ACIDS, PROTEINS AND THE ORIGIN OF LIFE

In the last two chapters we have been caught up in the story of a remarkable process that is central to being alive, self-replication. This is the work of DNA; its nucleotide code directs construction of more identical DNA, and through RNA, it produces the living molecular environment, serving in the ultimate analysis to insure replication and persistence of more DNA. We have only seen how this process takes place in the simplest of organisms. Our data has been selected almost exclusively from the world of bacteria and viruses. More complexity will appear in future chapters when we consider how higher organisms manage this task. But for the moment we need to attend to a problem in the other direction. We ask how such complexity as we have already seen might have evolved in the simple circumstances that we believe attended the origin of life.

There is a serious dilemma associated with this question. The replication process found in even the simplest modern organisms involves molecular machinery based on *two* types of complex molecules, nucleic acids and proteins. The system does not work without both. Without nucleic acids there is no replicative message; without proteins there is nothing to act on the message. Yet it is highly implausible that such a system, even in its barest form, with nucleic acids and proteins serving their mutual needs, could have arisen in a *single* evolutionary event. If we agree, then we must reexamine the replicative process with an eye to simplifying it, to seeing what components may be discarded, so that what is left might show us what the primitive replicative process could have been like.

Perhaps replication began only with proteins. They at least seem to be favored over nucleic acids in their enormous functional capacity, in the art of acting on the environment. However, valuable as they are, proteins alone get us nowhere unless they can reproduce themselves. Since this seems to be an impossibility, we must turn to nucleic acids for a plausible answer. It does appear that there is a tendency for the codelike self-assembly of nucleotides and that this tendency occurs in no other molecular type.

During the long ages before the origin of cells, perhaps chemical evolution of nucleotides took place with selection favoring some factor, such as rapidity of self-assembly from prebiotically synthesized sugars and nucleotide bases. During this phase of evolution, the construction of nucleotides into polynucleotide chains would have involved hydrogen bond formation, holding nucleotide groups in position on the template of an already formed polynucleotide, until **polymerization** occurred. Molecules that attained the ability to accomplish this process the most rapidly, and which could accurately transmit this characteristic to their molecular descendants, would come to be the best represented in the primitive environment.

Probably, a limit was soon reached beyond which self-replicative activity by single nucleotide molecules could not be improved. Perhaps the next evolutionary step would have been selection for those molecules that could obtain assistance in replication from other kinds of molecules. This step is difficult to conceptualize, but we presume it began with the nucleic acid molecule somehow keeping the assisting molecule in close proximity. Possibly in the beginning the cooperative molecule was a polypeptide whose contact with the nucleotide it assisted was by hydrogen bonding between roughly complementary structures. Indeed, some theorists believe that there must have been some preferential relationship of a template nature between certain amino acids and certain nucleotide bases and that these were important in the establishment of the genetic code that now exists.

In whatever way this cooperative relationship between nucleic acids and polypeptides came to be established, it was undoubtedly the final great hurdle to be gotten over in the precellular, chemical phase of evolution. The next phase was at a supramolecular level and involved the formation of some sort of an enclosure, a boundary membrane, to keep together the nucleic acid and the molecules that assisted in its replication. Thus freed of the necessity of carrying these assisting molecules bonded to the nucleic acid, the machinery of life could become more diversified.

ONLY ONE REPLICATIVE SYSTEM SURVIVED

There might have been many ways to accomplish the replicative process up to the level of complexity in which nucleic acids directly specified the structure of their necessary cooperative polypeptide molecules. However, once the evolutionary process reached a point at which amino acids were being specified by certain nucleotide bases, only one code survived. One of two things must have happened: either the postulated template relationship between given pairs of amino acids and nucleotide bases was sufficiently efficient to cause the present triplet code to come out as it is, or one of several codes that developed was for some reason so successful that it overwhelmed all others. Once established, by either means, deviations from the code would have been lethal. This is obvious because, if the code triplet for a particular amino acid changed by some accident, all proteins of the organism would experience amino acid substitution. With such widespread effects, it is difficult to see how a change in the code of this type could be other than lethal. This does not mean that the entire 64 word code had to arise at once, since there are ways in which it could have evolved gradually, but with the proviso that there could be no new development that modified the meaning of any previously developed code word.

EVIDENCE FOR UNIVERSALITY OF THE GENETIC CODE

The strongest evidence that the genetic code is universal comes from the ready interchangeability of components of the protein synthetic apparatus—mRNA, tRNA and ribosomes—among widely diverse organisms. In a typical experiment, mRNA obtained from rabbit blood-forming cells is added to a protein synthetic system extracted from *E. coli*. Rabbit hemoglobin is produced, showing that *E. coli* tRNA reads the code the same way that rabbit tRNA does. Since rabbits and bacteria lie almost at the extremes of the variety of life, it is most likely that the remainder of living things also use the same code.

The fact that all organisms so far tested do read the code the same way has opened the new field of recombinant DNA research in which hybrid DNA molecules are created. For example, segments of DNA from virtually any organism can be introduced permanently into the *E. coli* genetic apparatus. The bacterial cell is then theoretically able to produce whatever gene product is coded by the foreign DNA. The technique has great practical significance because it holds promise for production of valuable and difficult-to-obtain natural products in easily grown *E. coli* if the genes controlling the synthesis of such products can be hybridized with the *E. coli* DNA.

Recently this has been accomplished with an artificially produced gene for the newly discovered hormone **somatostatin.** The gene could be synthesized relatively simply since somatostatin is a small peptide containing only 14 amino acids, or, at least, synthesis was easier than isolating the gene from the mammalian genome. When this gene was coupled with a natural component of the *E. coli* DNA, which serves to trigger gene copying, and inserted into *E. coli,* somatostatin was synthesized in useful amounts. In the initial experiments about 2 gal of bacterial culture is reported to have yielded as much somatostatin as could be isolated from nearly a half million sheep brains, the source until now.

If this work is an indication, medicine is again about to reap great profits from pure science. However, such advances are not without risks. This technique might also inadvertently produce new and harmful forms of bacteria. Development of safety guidelines for this type of recombinant DNA research is currently in progress.

10 Cellular Reproduction: Duplication of Self

As primitive cells became more complex their DNA information packages grew larger. Rough and ready division could no longer suffice to produce equally viable daughter cells. These obvious deductions suggest that the most likely to survive among early organisms would have been those able to produce good copies of themselves upon division. To survive, each daughter cell must have a complete set of DNA and of whatever other materials are necessary to establish living machinery at least as good as that in the parent cell. How these critical goals are achieved in cell division is the primary subject of this chapter. Secondarily, we shall use facts appearing in this discussion to speculate on how the cell division process might have arisen in the course of evolution. The subject of the subsequent chapter is a radically different process: the cellular basis of sexual reproduction, or the production of new individuals unlike the parent by mixing DNA from different ancestries.

WHY SHOULD A CELL DIVIDE?

Let us consider the necessity of cell division before discussing mechanisms. For single celled organisms there are two reasons for cell division, evolutionary and functional. The simplest way to state the evolutionary reason is to say that the only way a line of organisms can make long-term adaptation to environmental change is by the production of descendants whose inherited variations may increase their ability to survive in changed conditions. A cell's offspring are the gamble it takes for survival of its kind in the future.

The functional reason for cell division has many aspects, some will reappear in later chapters. For now we consider them only in terms of two functional problems of cells. The first is that a cell must maintain a favorable ratio between its surface area and the cell mass served by that surface. The surface of a cell is a bottleneck through which essential materials enter and wastes leave the cell. As a cell grows, its volume increases more rapidly than its surface (see page 196). The result is that beyond a certain size a cell is inefficient, as a large factory with doors too small would be. Increasing the ratio of cell surface to volume by cell division is one solution to this problem. Next we should inquire why a cell should get big rather than growing to some optimum size and then stopping. For many cells this is what happens. Instructions that precisely limit size are programmed in their DNA.

Even so, many cells that have size limitations divide, so there must be functional reasons for cell division that we have not yet considered. Among these is the second functional problem, the matter of growing old at the cellular level. Although there is a constant tearing down and renewal of many cell components, so that these cannot be said to age in the sense that the cell as a whole ages, there are others that are not renewable and which, through chemical accident—wearing out—become more and more nonfunctional as time passes. Cell division provides an opportunity to dilute aging molecules with fresh ones produced in the great spurt of biosynthetic activity that takes place as new cells grow.

Looking beyond the horizon of the single cell to multicellular organisms, a final argument appears: Cell division is the obvious prerequisite for construction of multicellular organisms since these originate from one or a very few cells.

Two Methods of Cell Division in Two Major Forms of Life

Although they achieve similar ends, cell division and sexual reproduction differ greatly in detail according to the type of organism in which they occur. Consequently we must now abandon the terms "higher" and "lower" that we have been loosely applying to the two fundamental divisions of living organisms and define them precisely.

There are two basic cell types in the living world, representing the greatest evolutionary discontinuity between organisms. The simplest cells are those of bacteria and blue-green algae. They are called **prokaryotes** because they lack a nuclear membrane. Although prokaryote cells have a nuclear region, it is not clearly separated from the cytoplasm. All other organisms, plants and animals, are **eukaryotes.** Their cells have a well-defined nucleus clothed in a nuclear membrane. There are many other differences between the cells of Prokaryotes and Eukaryotes, and some of these are shown in Figure 10-1 and Table 10-1. In general the prokaryotic cell is simpler than the eukaryotic cell, as emphasized by the fact that it lacks specialized organelles. In addition to lacking a nucleus, prokaryotes lack mitochondria and chloroplasts, those membrane-rich organelles specialized for respiratory processes and photosynthesis. Instead, prokaryotes perform these activities with enzyme systems associated with the cell membrane. Finally, and of most immediate interest, prokaryotes package their

Figure 10-1 Diagram of prokaryote and eukaryote cells.

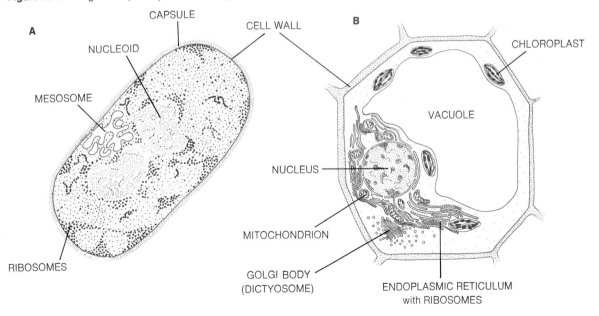

Table 10-1. Some Distinguishing Characteristics of Prokaryotic and Eukaryotic Cells

Characteristics	Prokaryote	Eukaryote
Structure		
Size	Usually smaller than 6 μm	Usually larger than 6 μm
Organelles:		
nucleus	None	Present
mitochondria	None	Present
ribosomes	Present	Present: 2 types; mitochondrial, resembling prokaryote ribosomes; cytoplasmic
chloroplasts	None, but analogous structures, **chromoplasts**, present in bacteria.	Present
endoplasmic reticulum	None	Present
Golgi apparatus	None	Present
vacuoles	Rare	Present, with many diverse functions
locomotor	Flagellae and unexplained gliding motion	Flagellae, pseudopods, contractile mechanisms based on contractile proteins
Cell division	Nonmitotic fission	Mitosis and meiosis
Sexuality	Partial exchange of DNA only	Usually present
DNA	Single molecule, attached to cell membrane or **mesosome**, specialized invagination of cell membrane; replicates from single replication site	As several chromosomes; replication from multiple sites
Metabolism	Anaerobic and aerobic; heterotrophic or photosynthetic	Usually aerobic; heterotropic or photosynthetic

principal store of DNA in one single, immense molecule, whereas eukaryotes have at least two and usually many more units of DNA in the form of **chromosomes**. In keeping with this difference the process of cell division in prokaryotes is quite simple, whereas in eukaryotes the mechanism for equally dividing the several chromosomes at cell division is an elaborate process called **mitosis**.

CELL DIVISION IN PROKARYOTES

Cell division in bacteria and blue-green algae is called **fission**. The cell typically elongates by growth, and a septum consisting of new cell membranes and cell wall material appears and grows across the cell. Since many of the critical enzymes of prokaryotes are attached to the cell membrane, an equitable sharing of these elements between the daughter cells automatically occurs. Cytoplasmic materials are also equally shared since they are present either in solution or as multiple small particles that are randomly distributed throughout the cell as the dividing wall forms.

DNA Sharing Mechanism in *Escherichia coli*

The process by which DNA is shared between the daughter cells in prokaryote fission is typified by the situation in *E. coli*. As is evidently the case in all bacteria, the DNA of *E. coli* exists as a single, closed loop. There is genetic evidence for this, but the most convincing evidence comes from the direct visualization made possible in an experiment by J. Cairns. *E. coli* DNA was made radioactive by supplying the bacteria with tritium-labeled thymine. Then, using the technique of **radioautography**, the DNA was caused to photograph itself. This was accomplished by gently breaking up the bacteria so that their DNA might be spread out to ease visualization, and then bringing them into contact with a photographic emulsion with results as shown in Figure 10-2. Wherever there was a radioactive thymine molecule in the DNA, if it gave off radiation during the period of exposure, there was a good chance of a visible black grain appearing in the photographic emulsion. All such events throughout the DNA molecule provide a photograph giving its dimensions and shape. It proves to be an extremely large molecule, 1300 μ (1.3 mm) in length. Note that the bacterium from which it came is only 2 to 3 μ long! A DNA molecule of this size is

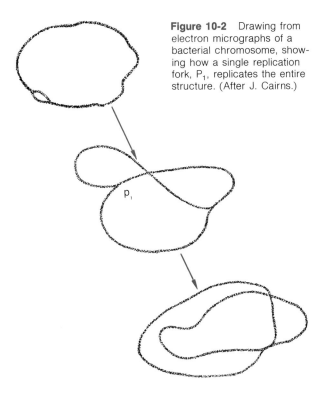

Figure 10-2 Drawing from electron micrographs of a bacterial chromosome, showing how a single replication fork, P_1, replicates the entire structure. (After J. Cairns.)

the grains in the photograph can be interpreted to show that replication was semiconservative (see page 120). Each DNA molecule resulting from the replication contains one old strand and a newly synthesized strand.

Once formed, how do the new DNA molecules separate and move into the daughter cells? If they simply float free in the cytoplasm, chance would sometimes result in unequal segregation. One daughter cell might get two DNA molecules and the other none. Electron microscopical evidence shows how this is prevented. The parental DNA molecule is attached to the membrane of the cell at the site of the replication fork (Figure 10-2). As the daughter DNA molecule forms it also attaches to the membrane. Somehow this attachment always occurs on the other side of the site of elongation of the cell from the attachment point of the parental DNA. Since the division plane of the cell then lies between these two attachment sites, in the center of the zone of elongation, equal segregation is assured.

made up of about 4×10^6 base pairs. You can figure this out by reference to Figure 8-5, from which it is evident that there are 10 base pairs per turn of the DNA molecule, a distance along the DNA axis of 3.4 nm. The 4 million base pairs are considered enough to code for all the genes that *E. coli* might possess so it seems certain that this single loop represents the entire genetic complement except for a few factors that are carried on very small DNA units in the cytoplasm.

The DNA radioautograph has caught the molecule replicating. The best interpretation of the labeling experiment shown is that replication of the molecule began at p_1, that replication is complete and that there are two new molecules of DNA. The density of

CELL DIVISION IN EUKARYOTIC CELLS

As noted, the DNA of the eukaryotic cell occurs in a number of chromosomes instead of in a single molecule. There is a simple quantitative reason for this additional complexity. If the 6.4×10^{-12} g of DNA in a eukaryote cell (human) could exist as a single molecule of the type found in *E. coli* it would be about 2 m long. This would undoubtedly cause difficulties at mitosis. Thus, eukaryote DNA is broken up into smaller units—chromosomes. These are rendered even more linearly compact at cell division by having various forms of secondary coils superimposed upon the familiar primary helical coil.

Chromosome Organization and Duplication

When the cell is not dividing, the chromosomes are elongate and thin. As the time for division approaches, they become greatly shortened and quite thick, largely due to coiling. Much of the mass of the chromosome consists of two kinds of molecules, basic lysine- and arginine-rich **histones,** and acidic proteins. The function of these constituents of chromosomes and the manner in which they are combined with chromosomal DNA is largely unknown. However, the histones figure in theories concerning the regulation of DNA transcription.

The DNA of eukaryote chromosomes differs from bacterial DNA in that up to 30% of its bases exist as repeated nucleotide sequences, many of which code for histones. These DNA sequences probably play some structural or regulatory role in the function of the eukaryote chromosome. In addition, there are "spacer" DNA segments within some sequences of DNA (see page 135).

Another unique feature of eukaryote chromosomal DNA is that on each chromosome there is a special region, the **centromere.** Within it lies the **kinetochore.** This may be a genetic unit that functions during mitosis, when the rest of the chromosome is genetically inactive. The kinetochore plays an important role in the movements of the chromosome into the daughter cells during division. These movements involve **microtubles** and the kinetochore may be concerned either with the assembly of microtubules from already formed subunits or possibly with their synthesis during or just before mitosis.

Aside from complications introduced by the structure of the chromosome, chromosomal DNA replication proceeds very much as in bacteria except for the presence of multiple replication forks in each DNA molecule. Replication is semiconservative (see page 120), as may be demonstrated by brief application of radioactive thymine (**pulse labeling**) and following the distribution of radioactively labeled DNA in descendants of the pulse-labeled cell.

As shown in Figure 10-3, when the chromosome divides the daughter units formed are called **chromatids.** The two newly formed chromatids remain attached at the centromere although they each contain a complete DNA molecule. If the radioactive thymine is present during replication of DNA then the two daughter chromosomes will each have one strand each of nonradioactive and radioactive DNA, corresponding to the one old, one new pattern of semiconservative replication. This is confirmed in the next cell generation in which half of the chromosomes are entirely nonradioactive and half are radioactive, and in the third generation when the ratio is one radioactive chromosome in four. It is important to examine Figure 10-3 carefully and avoid confusing the paired nature of the DNA molecule (a double helix of DNA monomers) and the paired nature of the replicated but not yet separated chromosome (two chromatids, each containing a DNA double helix). The chromosome exists in this paired state only when the cell is preparing to divide. In the nondividing cell, each chromosome consists of one double helix of DNA.

DIPLOIDY AND HAPLOIDY

Another complexity of eukaryote heredity is that there may be either one or two of each kind of chromosome. The cell is called **haploid,** if one of each kind of chromosome is present. If there are two of each, the cell is said to be **diploid.** The cell, or organism, may shift between diploidy and haploidy in different phases of its life cycle, as we will see in the next chapter. In some instances more than two chromosome sets may be present, a condition known as **polyploidy.**

Figure 10-3 Replication of DNA in eukaryote chromosomes.

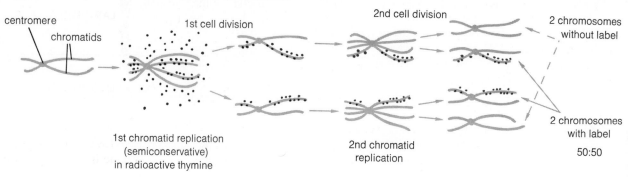

MITOSIS

Mitosis is principally a matter of coordinating the division of the chromosomes with division of the rest of the cell so that each daughter cell receives a normal complement of chromosomes.

The details of the process in a plant cell, as shown in Figure 10-4, are as follows: When a cell is not undergoing mitosis, it is said to be in **interphase**. Upon completing mitosis the cell may enter a state in which it is inactive as far as multiplication is concerned for a variable period of time, from hours to years. Or it may continue the multiplication cycle by immediately entering the **S phase** (synthetic phase) of the interphase in which there is extensive synthesis of DNA and chromosomal proteins leading to doubling of the chromosomes. The chromosomes double only before cell division and remain single in the mitotically inactive period, no matter how long it lasts. In rapidly dividing cells the S phase takes from 30% to 50% of the total cycle time.

After the chromosomes have doubled, the cell enters the actual process of mitosis. This goes quickly, occupying only 5% to 10% of the cycle. Mitosis is initiated with **prophase**, which is characterized by coiling of the chromosomes. In this process each chromatid coils upon itself in such a way that there will be no mechanical difficulty when it comes time for chromatids to separate. As prophase ends the nuclear membrane breaks up and the chromosomes lie free in the cytoplasm.

Preparations for cell division have been under way in the cytoplasm during these nuclear events. In animal cells two **centrioles**, complex self-reproducing protein structures, move to opposite poles of the cell. They become surrounded by radiating aggregations of microtubules called **spindle fibers**. Each centriole with its collection of spindle fibers is termed an **aster** (Greek, star). Many spindle fibers extend between the two asters. The two asters, together with the spindle fibers, are known as the **spindle apparatus** (Figure 10-5). In plant cells neither asters nor centrioles are present. In all cells it is probable that the spindle fibers are constructed from preformed microtubular subunits since they may be disrupted only to form again after a brief treatment with the drug **colchicine**.

As these events are completed the cell enters **metaphase**. The defining characteristic of this stage is alignment of the chromosomes in the metaphase plane. The mechanism by which chromosomes move during this and later stages of mitosis is still obscure. It is known that each chromosome is attached by its

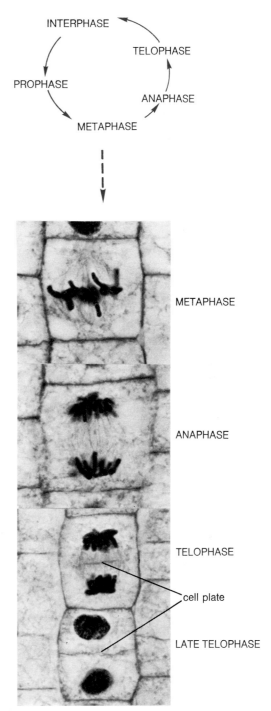

Figure 10-4 Mitosis in the onion root tip. (Robert Gill.)

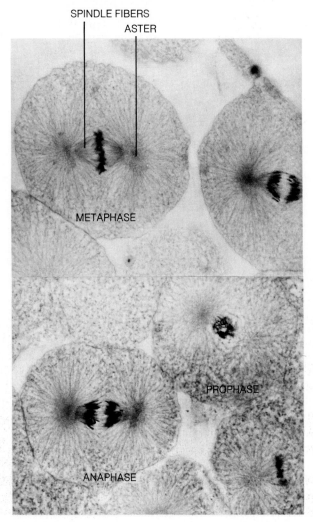

Figure 10-5 Mitosis in whitefish embryo. (Robert Gill.)

relatively normally in subsequent divisions in the absence of colchicine.

With the end of metaphase all daughter chromatids separate at the same time by breaking free of their connections in the centromeric region. The next stage is **anaphase,** during which the daughter chromatids continue separation and move to opposite poles of the cell. Thus the normal chromosomal complement is established for each daughter cell, which is formed in the next and last phase of mitosis, **telophase.** The principal event in telophase is the forma-

Figure 10-6 *Drosophila* salivary gland chromosomes. The four paired chromosomes are united in a single chromocenter. An inversion results in the looped pairing seen at the bottom of the figure. (Robert Gill.)

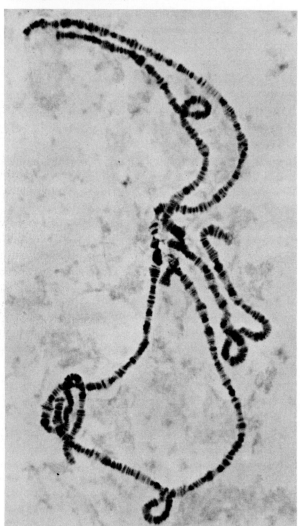

kinetochore to microtubules similar to the spindle fibers. It appears that chromosomal movement involves interaction of the chromosomal spindle fibers with the fibers extending between asters and may also involve movements caused by rapid assembly and disassembly of the chromosomal spindle fiber at the kinetochore. If colchicine is applied throughout mitosis, the spindle fibers remain disassembled and, although the daughter chromatids separate, they do not move from the metaphase plane. This procedure may be used to produce cells that have more than the normal number of chromosomes, since a nuclear membrane may form around the colchicine-produced double set of chromosomes, which may then behave

tion of the daughter cells, either by constriction of the cell membrane, as in animal cells, or by the rapid synthesis of a new cell membrane and wall, as in plants. In either case the plane of the division between the two new cells is approximately the metaphase plane. During these events a nuclear membrane forms about the newly divided chromosomes in each daughter cell. Then the chromosomes once more elongate in preparation for DNA replication and transcription, and the cell is in interphase again.

SYNCHRONIZATION OF NUCLEAR AND CYTOPLASMIC EVENTS

At least two types of evidence show that synchronization of nuclear and cytoplasmic division during mitosis is under positive control by both cytoplasmic and nuclear factors. The role of the cytoplasm is shown by D. M. Prescott in an experiment in which an amoeba was prevented from undergoing mitosis for 6 months by regular amputation of cytoplasm. During the same period a control amoeba divided 65 times. Thus it appears that an insufficient amount of cytoplasm serves to suppress nuclear division. Nuclear influences are revealed in experiments by P. N. Rao and associates in which an extra nucleus was placed in a cell whose own nucleus was ahead of the introduced nucleus in preparation for mitosis. The effect of the introduced nucleus was to delay the more advanced nucleus so that the two nuclei were able to enter mitosis together.

Endomitosis

Division of chromosomes and cytoplasm does not always occur together. Sometimes nuclear division occurs without cytoplasmic division. There may also be normal duplication of chromatids without separation. A good example is found in the fruit fly, *Drosophila*, in which the chromosomes of salivary gland cells are huge due to, perhaps, thousands of replications of chromatids without separation. This condition was valuable in the early days of genetics because the *Drosophila* chromosomes thus enlarged show distinctive banding patterns, which can be correlated with the position of genes (Figure 10-6). It is believed that this condition serves to increase the transcriptional potential of the chromosome, which is probably important in these actively secreting cells.

Extrachromosomal Inheritance

All genetic mechanisms that we have considered involve chromosomes. All of the chromosomal genetic factors of an organism can be conveniently termed the **genome** of that organism. Evidence from many sources tells us that there are genetic factors of several types that occur outside the nucleus. These may be collectively termed **episomal factors.** Episomes are quite common in bacteria, for example one type controls sex, as we will see in the next chapter. In eukaryotes there are two remarkable examples, mitochondria and chloroplasts.

The Genetic Independence of Mitochondria and Chloroplasts.

In the mold *Neurospora* administration of radioactively labeled DNA precursor thymine can be followed by isolation of labeled DNA from both nucleus and mitochondria. This indicates that the mitochondria produce DNA since nuclear DNA does not transfer to cytoplasmic organelles. In this same organism the DNA of mitochondria regulates synthesis of at least some mitochondrial proteins and the mitochondria are to a large part independent of nuclear control. These facts are illustrated by a mutant of *Neurospora* called "poky," Poky *Neurospora* grow slowly because their mitochondria have defective respiratory enzymes. When mitochondria from the poky strain are injected into the normal strain, very high concentrations of poky mitochondria may be isolated from the normal cells after a period of growth. Since the experiment involves no change in the genome of the normal cells and since more poky mitochondria were isolated than were injected, the only plausible conclusion is that the mitochondria are self-reproducing. The injected mitochondria perpetuate the poky defect, which must be carried in their own mitochondrial DNA, since nothing else was injected.

The Commensal Theory of Mitochondrial and Chloroplast Origins

The suggestion had been made that mitochondria and chloroplasts might have originated as guest organisms taking up residence in the cells of organisms ancestral to eukaryotes. Presumably the host organisms provided something of value to the guests so that the situation was one of **commensalism**. This is a relationship between organisms in which both are benefited. The idea has much appeal, not only because of the undoubted presence of DNA in these organelles but because they also contain unusual ribosomes of a type not found in cytoplasm and closely resembling the ribosomes of prokaryotes. Also, even in modern organisms there are undoubted examples of commensalism involving entire cells. Unicellular algae are intracellular commensals in some animal cells, endowing them with the advantages of photosynthesis while themselves accruing benefits from the host cells.

However appealing such ideas might be, no decision can now be made regarding their validity because there are "conventional" explanations for the origin of such organelles, particulary mitochondria. In prokaryotes, as exemplified by bacteria, many of the enzymes characteristic of mitochondria are localized in the cell membrane where they perform mitochondria-like respiratory functions. Perhaps as a response to heightened respiratory demand in the course of evolution, the surfaces bearing respiratory enzymes were increased by infoldings of the cell membrane. These may eventually have become membrane-surrounded vesicles free to wander through the cytoplasm, in other words mitochondria in the modern sense, except for the absence of an independent, DNA-controlled protein synthetic system. Even this hurdle is not hard to clear because we now know that there is frequent exchange of genetic material between the bacterial chromosome and the cytoplasm, where the DNA dislodged from the chromosome functions as episomes. Since many of the proteins required by the mitochondria could not get inside them if synthesized in the ordinary way in the cytoplasm, there would be a substantial advantage derived by mitochondria having their own essential DNA, allowing them to be self-replenishing and, according to this idea, eventually self-reproducing.

11 Sexual Reproduction: The Cellular Mechanisms

The ultimate significance of sex is transfer of functional DNA between cells by some means other than cell division. In its simplest form, sexuality is expressed in prokaryotes by bacteria that transfer DNA by injection. In eukaryotes the sexual process is obscured by many complex events of cellular reproduction and by changes between diploidy and haploidy, which may occur contemporaneously. In the following pages we will try to avoid these difficulties by studying the sexual process in an evolutionary way, first examining sexual reproduction in bacteria as a model of how it might have arisen in primitive organisms, before undertaking its study in eukaryote cells.

THE VALUE OF SEX

As discussed so far, cellular reproduction has been strictly asexual. All DNA in the asexually reproducing cell is exclusively obtained from that cell's single parent cell, except for viral DNA that might intrude. This has been true in all cell division described previously, whether it involved fission in bacteria or mitosis in eukaryote cells. Such asexual cell multiplication has one purpose, the production of exact copies of parental cells. As explained earlier, exactitude in this process is important. Although change of a heritable nature is essential to evolution, for life to persist change must, nonetheless, be rare. But now let us look at the other side of the coin and analyze the difficulties of achieving change, when it might be desirable, in cells that lack sexuality.

Consider the hereditary relationships existing in a population of asexually reproducing cells. All descendants of any single cell within that population constitute a **clone** (Figure 11-1). Within the entire population the members of any clone are, except by lines of descent from common ancestors, as hereditarily isolated from the members of any other clone as they are from organisms of a completely different kind. The significant point is if a favorable mutation occurs in one clone, there is no way to get it into the cells of other clones, except by the independent occurrence of the same mutation in those clones. Likewise, if different favorable mutations occur in different clones—even if these several mutations are more advantageous together than singly, there is no way in asexual cells to bring them together in the same genome. Only if mutation rates are high enough might the several mutations occur relatively often in the same clone. As is probably more often the case, with large populations and low mutation rates, the only effective way to bring such mutations together in the same genome is by some means that allows DNA exchange between cells in other than the strict line of descent by asexual reproduction, in other words by a *sexual process*. This is basically an argument for a long term evolutionary advantage of sex: it speeds up the evolutionary process by producing a greater variety of genomes to try the test of survival than does asexual reproduction.

In addition to the long-term effect, sex is also important because it results in greater genetic diversity of offspring than does asexual reproduction, a point which will become obvious as we proceed further in the study of heredity in eukaryotes. Increased diversity is thought to provide a short-term advantage since some of the genetically varied offspring generated in sexual reproduction might be better adapted than the parents. This would be especially true in rapidly changing environments that vary from conditions to which the parental organisms are adapted by the time progeny appear. This role of sex in the production of diversity in offspring is similar to bet spreading in gambling. That is, if you intend to make several bets on the ball games some Saturday, your chances of winning at least some money are improved if you bet on more than just one team.

Just as some bets will be lost in this instance, it is true that some of the genetic recombinations produced by sexual reproduction will have low adaptive value and not survive. It seems, however, that the opportunity to produce good new combinations of DNA generally results in more benefit to the population than the harm caused by production of deleterious combinations. The appearance of a seriously defective genome would be presumed to result in only a relatively few maladapted organisms before the deleterious genome died out, whereas the appearance of an improved genome would eventually result in production of large numbers of progeny with the new and advantageous genome.

ORIGINS OF SEX

Predation was discussed in Chapter 6 as one of the possible adaptations of life to dwindling resources of prebiological organic molecules. It should be obvious that predation, a process by which one organism consumes another, might provide more than energy-rich and essential building-block molecules. Informational molecules (DNA, RNA) would be taken in by the predator along with these other substances. Although in modern predators DNA and RNA would be immediately broken down by specialized enzyme systems, it is altogether likely that such mechanisms would not

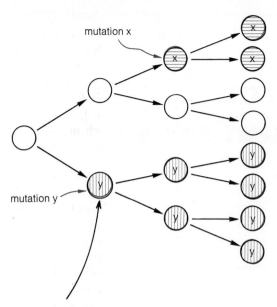

Only the descendents of this mutated cell will receive copies of mutation y occurring in the ancestral cell. Similarly, mutation x will be found only in descendents of the cell in which it occurred. Even if x and y are advantageous together, the only way then they may occur together is by the unlikely event of their occurring as mutations in the same line of descent, or clone, as shown below:

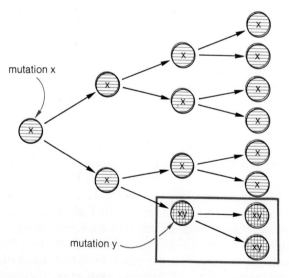

Figure 11-1 Inherited modifications cannot be exchanged between clones of asexually reproducing organisms, except under special conditions mentioned in the text. Ordinarily, even if modification x would be advantageous in combination with modification y, the two cannot occur together in the same organism unless they appear in the same line of descent, as shown for clone x. Since such single changes occur but rarely, their occurrence together would have a vanishingly small probability.

have evolved until after the appearance of predation. Therefore, until organisms developed sophisticated digestive equipment, it seems likely that there must have been extensive transfer by predation of genetic material among primitive organisms. Thus, the early predator might have profited by the capture of new genetic material as well as building-block and energy-rich molecules with each successful capture of prey. Such transfer satisfies the basic definition of sex. Perhaps from such simple beginnings the complex mechanisms associated with sex in higher organisms had their start.

In modern organisms no process of this type occurs since random accumulation of new genes would wreak havoc on a finely tuned genetic apparatus. However, viruses clearly show how foreign genetic material can be accepted and its instructions followed by a host cell. Viruses incorporate and transmit genes of their host organisms and can even establish them in totally unrelated cells. To illustrate, it has been shown that a virus carrying the *Escherichia coli* gene for the enzyme galactose transferase can establish this gene in a functional state in human cells in tissue culture. The cells used were from a patient with a hereditary disease, galactosemia, in which galactose transferase is absent, with serious results. After growth in the presence of the virus the human cells were shown to be transcribing viral DNA into mRNA and producing the galactose transferase which they lacked. Such experiments are basic to the concept of **genetic "engineering"** whose proponents suggest that human diseases caused by genetic defects might be someday cured by techniques like this, which might allow introduction of normal genes to overcome the effects of defective ones.

While we do not know how early viruses arose in the history of life, observations such as these suggest that viruses, once they appeared, must have, and may still, play an important role in evolution by promoting widespread transfer of DNA among organisms, even of the most diverse types, as the galactose transferase example shows. Viruses are everywhere in immense numbers, constituting a large fraction of all DNA, and thus their role in promoting redistribution of DNA among living organisms is probably of significance. If viruses arose before sexual processes became well developed, there may have been a time when viruses were the only agents transferring DNA between clones of organisms to augment DNA transfer by predation.

Bacterial Sex

The simplest form of sexuality that we know of in prokaryotes or eukaryotes occurs in bacteria that have a mechanism for transfer of part or all of the single bacterial chromosome from one bacterial cell to another. In the early days of molecular biology it was really impractical to look for bacterial sex until some early steps in understanding bacterial genetics were taken. What had to be done to prove that sex occurs in bacteria? The most direct proof would be to mix together two bacterial cultures, each possessing a distinguishable trait determined by one gene, and obtain bacteria possessing both characteristics. This would demonstrate a mixing of the genomes of the two kinds of bacteria. Such mixing of DNA from two parental organisms is called **recombination.**

Recombinants were immediately detected when this experiment was first done, by J. Lederberg and E. W. Tatum on *E. coli*. In order to eliminate the chance that the presumed recombinants might instead be mutants, they did the experiment somewhat more complexly, as shown in Figure 11-2. The next obvious question has to do with how the bacterial genetic exchange is accomplished. Perhaps exchange might be as in bacterial transformation (see page 113) in which no direct contact is required between the participating cells. In the pneumoccocal transformation, nonvirulent (rough) cells need only to be exposed to extracted DNA from virulent (smooth) cells to convert them into virulent forms. This possibility was ruled out by showing the necessity of direct contact between cells, as demonstrated by an ingenious experiment in which the two *E. coli* cell types were grown separated by a porous barrier preventing passage of cells but through which the culture medium

DEMONSTRATION OF SEXUAL REPRODUCTION IN BACTERIA

1. Two strains of bacteria with mutations causing different amino acid requirements are chosen:

	STRAIN 1	STRAIN 2
Amino Acids required in medium for growth	A,B	C,D
Growth in minimal medium (no amino acids)	NONE	NONE
Growth in complete medium (all required amino acids present)	GROWTH OCCURS	GROWTH OCCURS

2. Then strains 1 and 2 are tested on minimal medium separately and mixed together.

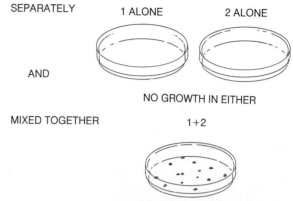

SEPARATELY

AND

MIXED TOGETHER

NO GROWTH IN EITHER

Growth occurs, resulting in visible colonies, each originating from an initial bacterial cell capable of growth on minimal medium.

3. What happened?

Since no growth occurred in cultures of 1 and 2 alone, the chance of mutations restoring the missing abilities to produce the required amino acids (A and B in strain 1, B and C in strains 2) must be very low.

In each two specific mutations in the same clone would have been necessary. Yet, many colonies appeared in the mixed culture. To explain them as originating from mutation would require believing that such pairs of mutations occurred for each colony that appears. It is far more probable that some consequence of strain 1 and 2 cells coming into contact produced the results seen.

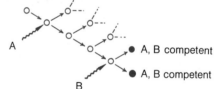

There are two possibilities:
    I. Strains 1 and 2 might exchange diffusion substances that allow growth.
    OR
    II. Direct contact might be required between the two types of bacteria.

4. Possibility II was tested by arranging for exchange by diffusion without direct contact, as shown:

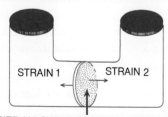

BARRIER ALLOWING CHEMICAL PASSAGE BUT NOT BACTERIAL PASSAGE

5. Growth did not occur under these conditions. Therefore it was proven that direct contact between strain 1 and 2 bacteria is necessary if bacteria able to live on minimal medium are to be produced. This supports the hypothesis that bacteria may exchange genetic material on contact, that is, reproduce sexually.

Figure 11-2 A demonstration of bacterial genetic recombination.

could readily be pumped (Figure 11-2). No recombinants were detected, showing that cell contact is required.

BACTERIAL "MALES" AND "FEMALES" AND THE MECHANISM OF DNA TRANSFER

With the knowledge that contact between bacteria is necessary to achieve recombination, it was soon possible to work out the details of the transmission process. Certain strains of E. coli were found to behave as donors, whereas others behaved as recipients of DNA. This was shown in experiments in which one strain was rendered incapable of fission by treatment with the antibiotic streptomycin and then mixed with another strain. If each strain in the mixture could serve as both donor and recipient of DNA, this treatment should only result in a lowering of the number of recombinants. Actually what happened was that when one strain was treated with streptomycin, no recombinants appeared; when the other strain was treated, recombinants appeared in the normal numbers. The interpretation was that in this experiment only one strain can be a recipient. If cells of this strain are prevented from dividing by pretreatment with streptomycin, all genetic transfers to them are of no avail because the recipient cells can have no progeny. It is believed that streptomycin treatment does not affect the ability of a donor strain to transfer DNA; thus, pretreating the donor strain with streptomycin would have no effect on the appearance of recombinants. By analogy with higher organisms, we call the DNA donor strain male and the recipient strain female.

However, unlike the situation in higher organisms, maleness in E. coli can be transmitted to recipient, or female, cells. They then behave as donors. The agent that produces this effect is called the **F factor**. It is a short segment of DNA with control of the essential processes in sexual transfer. The F factor controls the appearance of hollow tubes that extend from the donor bacteria and are thought to be the route by which donor DNA enters the recipient bacterium. The factor also contains a DNA replication initiation site, which is thought to be activated when a recipient cell is contacted. Evidently this F replication site is located so that, when it initiates replication of the donor chromosome, one of the resulting DNA strands is directed down the hollow connecting tube into the recipient cell. It is certainly true, as this theory implies, that the donor DNA enters the recipient cell in a linear sequential way, and contact between the two cells is usually disrupted before a copy of the entire donor DNA is transferred. Sequential transfer is readily shown by breaking apart connected bacterial pairs at various times after the initiation of the transfer process. The longer the contact, the more DNA is transferred.

THE SEX FACTOR AND BACTERIAL EVOLUTION

Bacterial resistance to the antibiotics used in medical practice is a remarkable example of the adaptive value of sexuality in bacteria. Sex factors of the type just discussed are readily transferred between bacteria of medical importance, and these may carry with them segments of bacterial DNA, much like the case of the virus-transmitted galactose transferase gene. Other independent segments of bacterial DNA called plasmids contain genes responsible for bacterial resistance to antibiotics. This makes for extremely rapid spread of bacterial resistance. Not only is this process effective for members of a particular species of bacteria but it has been shown that the sex factors and resistance genes may even be transferred between different species. Although this is an ominous matter for humanity (Note 11-1), it beautifully illustrates the importance of the sexual process in enabling an organism to respond to an unprecedented demand from the environment.

Sexual Reproduction in Eukaryotes

Sex is not related in any necessary way to cellular reproduction in prokaryotes, but there is an obligatory relation in eukaryotes. This is because DNA exchange between eukaryote cells occurs only when a special-

SEXUAL REPRODUCTION: THE CELLULAR MECHANISMS 155

Note 11-1 *Penicillin-resistant Venereal Disease*

In November 1976, the U. S. Public Health Service Center for Disease Control reported the presence of penicillin-resistant gonorrhoea. Although rare, there being about 150 cases out of the possibly 3 million cases of gonorrhoea in the United States during the first year after its discovery, this development is disturbing because it has been independently reported in England. Fortunately, the new form of the disease is controlled by another antibiotic, spectinomycin, but this is more expensive and means that the disease will be difficult to control in poorer parts of the world where it evidently already exists in strength. The United States cases, for example, appear to have originated in the Philippines, where a large fraction of the prostitutes have the resistant form of the disease.

Resistance is caused by a penicillin-attacking enzyme coded by a gene on a **plasmid,** a small circular piece of DNA independent of the main bacterial DNA strand (See Figure 11-3). Resistance carrying plasmids evidently have arisen twice in the current outbreak because the plasmids in the United States and British organisms are of different sizes. Spread of resistance plasmids is favored by presence of sex factors in a high percentage of resistant gonorrhoea organisms. Origin of the penicillin-resistant gene may have been from another species of bacteria that causes severe infections of the respiratory tract. Authorities fear that the gene may be similarly transferred to the dangerous meningococcus organism, which is related to the gonorrhoea bacillus.

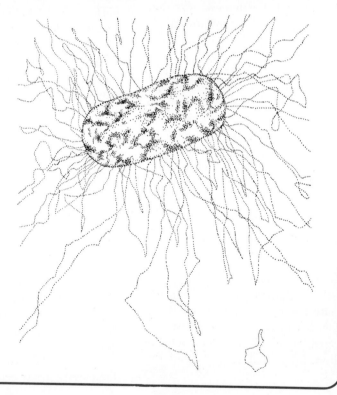

Figure 11-3 The small ring structure amidst the dispersed DNA of a bacterial cell is a plasmid.

ized reproductive cell called a **gamete** from each of two individuals merge in the act of **fertilization** to produce a single celled **zygote.** This is either the new individual or the forerunner of all the cells of the new individual. Unlike bacteria, in which there is usually incomplete transfer of DNA between male and female, the fertilization process in eukaryotes brings together two complete sets of DNA, one from each parental organism.

DIPLOIDY AND ITS SIGNIFICANCE

In the last chapter we learned that a cell might have only one copy of each kind of chromosome, in which instance it is called haploid; or it might have two copies of each chromosome and be diploid. Commonly an animal is diploid throughout its life history, except for the gametes, which are reduced to the haploid condition by two specialized cell divisions, the **reduction divisions**, or meiosis. Consequently, fusion of two haploid gametes at fertilization produces a diploid zygote with one complete set of chromosomes from each parent (Figure 11-4).

The pairs of chromosomes in the diploid genome (one member of each pair from each parent) are said to be alike, or **homologous**, but what is actually meant is that homologs are *potentially* rather than *actually* alike. Homologous chromosomes might have the same genes in the same order, but two kinds of processes can produce changes. These are *chromosomal mutations* and *gene mutations*. Chromosomal mutations may involve actual *loss* (**deletions**) of segments of a chromosome, may result in *rearrangements* of segments of chromosomes, or may result in *translocation* of part of one chromosome to another (Figure 11-5). Gene mutations, as you already know, affect the functions of specific genes by altering the sequence of nucleotide bases in the chromosomal DNA. An important consequence of these processes of change is the fact that in the diploid genome each gene may be present in the following permutations:

1. As only one copy.
2. As two copies that differ because of gene mutation, in which case the two copies are called **alleles** of each other.
3. As two exact copies.
4. As more than two copies.

When two exact copies of a gene are present, the organism is **homozygous** for that particular gene. When two differing copies are present, the organism is **heterozygous** for that particular gene. Since both members of the gene pair can, if functional at all, participate in cellular processes, it must be clear that homo- or heterozygosity for the various possible alleles of a particular gene can have significant effects on the cell and organism containing them.

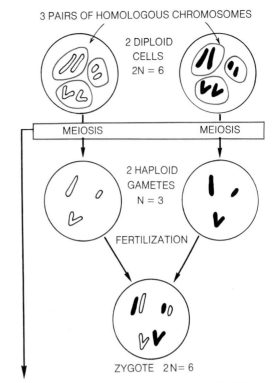

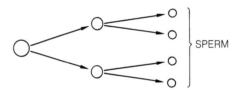

IN MALES 4 SPERM ARE PRODUCED PER SPERMATOCYTE

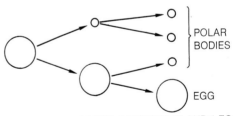

IN FEMALES 3 POLAR BODIES (NON VIABLE) AND 1 EGG ARE PRODUCED PER OOCYTE

Figure 11-4 Zygote formation.

The various outcomes possible from the situation just described allow *four important effects of the diploid condition.*

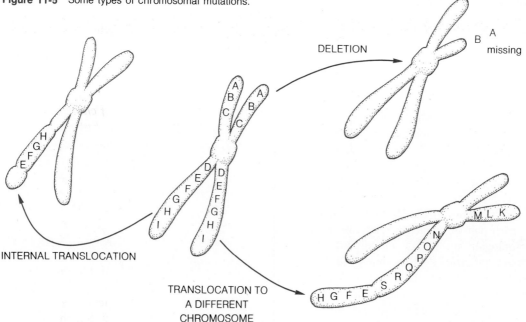

Figure 11-5 Some types of chromosomal mutations.

1. Because the diploid state results from receipt of one entire chromosome set from each parent, diploidy allows for maximal expression of the essential nature of sex. The maximum amount of recombination occurs, and the short-term evolutionary advantage of sex, the establishment of genetic variation in progeny, is maximally attained.
2. Storage of mutant genes in the populational gene supply (gene pool) is favored. To appreciate this, consider the fate of a gene mutation in a haploid organism. Since there is only one functional copy of the gene per cell, if the gene plays a critical role, little mutational change may be tolerated without a severe effect on viability. This means that the mutant allele will probably be rapidly eliminated. However, if the same mutant appeared in a diploid, the normal form of the gene will probably also be present in each cell, usually permitting normal cellular function. Consequently the mutant gene is not immediately lost by death of its carrier organism. The mutant allele thus has opportunity at some future time to contribute usefully to the genome, possibly by undergoing further mutational changes or by finding itself in a different genome or a different environment, in which it can be useful. In short, *diploidy* favors an increase in the supply of genetic raw material on which the organism may draw in the attempt to cope with changed requirements for survival.
3. Frequently the presence of two different alleles of various genes produces a markedly improved organism, a condition described as **hybrid vigor.** Inbreeding, the breeding together of close relatives, has the opposite effect of producing homozygosity, which usually results in inferior organisms. Agriculture has profited immensely from the practical application of these principles, as the huge annual hybrid corn crop in the midwest attests.
4. Finally it is likely that diploidy protects many-celled organisms from the deleterious effects of **somatic mutations.** These are mutations that occur in the course of the many cell divisions of somatic, or body cells, necessary to construct a large organism. Should a deleterious mutation occur early in the growth of a haploid organism it would, of course, be expressed in all of the descendants of the cell (clone) in which it occurred, and major defects would surely result. This outcome might be averted by the presence of the normal allele in a diploid organism.

In sum, *diploidy is valuable because it makes possible extensive genetic recombination, provides a storage mechanism for mutant genes, makes possible hybrid vigor, and provides a protection against expression of deleterious somatic mutations.*

MEIOSIS

In the absence of a mechanism to reduce the chromosome complement to the haploid number, either during gamete formation or at some other stage in the life cycle, there would be an ultimately impossible increase in chromosome number. Meiosis accomplishes the necessary reduction (hence its common name, reduction divisions). Additionally, meiosis is of great significance as a source of hereditary variation because, during one stage of meiosis, homologous chromosomes are able to exchange segments of DNA in a process called **crossing over**.

The Stages of Meiosis

Meiosis is a getting out of step of cell and chromosomal replication so that, during *two cell divisions*, there occurs only *one highly specialized cycle of chromosomal replication* (Figure 11-6).

When meiosis commences and the cell enters the **prophase** of the first meiotic cell division, each chromosome has replicated, forming two daughter chromatids. These show no sign of separation. Contraction and thickening of the chromosomes takes place, as in mitosis, and as metaphase approaches the chromosomes move to the metaphase plane. At this juncture, the meiotic chromosomes behave differently from mitotic chromosomes. In diploid cell mitosis homologous chromosomes move independently of each other, but in meiosis a most remarkable pairing of homologous chromosomes takes place. Pairing occurs not only between homologous chromosomes but between homologous *regions* of the homologous pair. The two centromeres are aligned and the remainder of the pair is in register, virtually gene for gene (Figure 11-7). The mechanism of this precise pairing is not understood, but it seems to be a general characteristic of chromosomes, being seen to some extent even between homologous chromosomes in nonmeiotic cells.

As pairing proceeds, visible evidence of crossing over, DNA exchange, is seen in the form of chiasmata. Each **chiasma** appears as a crossconnection (crossover) between one chromatid of one member of a chromosome pair and one chromatid of the other member, representing actual exchange of the involved regions of the two chromatids (Figure 11-7). Once these events occur, cell division proceeds normally *except there is separation of homologous chromosomes at anaphase rather than separation of chromatids*, as in mitosis. Each daughter cell of the first meiotic division consequently receives at random one chromosome of each homologous pair.

Following the first meiotic division, each of the two daughter haploid cells divides once more without an intervening interphase, making a total of four haploid cells for each diploid cell that enters meiosis. This second division has the appearance of a normal mitosis, but it actually differs in two important ways: there is no interphase chromatid replication, since this occurred in interphase of the reductive division, and there may have been a rearrangement of genetic material if crossing over has occurred. So, this division may not produce identical daughter cells as in ordinary mitosis (Figure 11-7).

The Genetic Consequences of Meiosis

In summary, the two successive divisions of meiosis do three things:

1. They reduce the genome to the haploid state.
2. They allow crossing over between chromatids of homologous chromosomes.
3. They sort out the four chromatids of the paired homologous chromosomes that entered meiosis into four haploid cells that are usually genetically diverse. Since these four cells are potentially the gametes that fuse with other gametes similarly produced by a mate, the basis for the diversity of offspring produced in sexual reproduction is evident.

When we think not of one chromosome pair but of all of the chromosomes present in the genome, the diversity of gametes generated by meiosis is awesome. In the human, with 23 pairs of chromosomes, the number of different types of gametes produced by the meiotic process is 2^{23}, or 8,388,608. That is, there are over 8 million different sets of 23 chromosomes that can be selected from 23 pairs. Remember also that this very large number does not reflect the number of

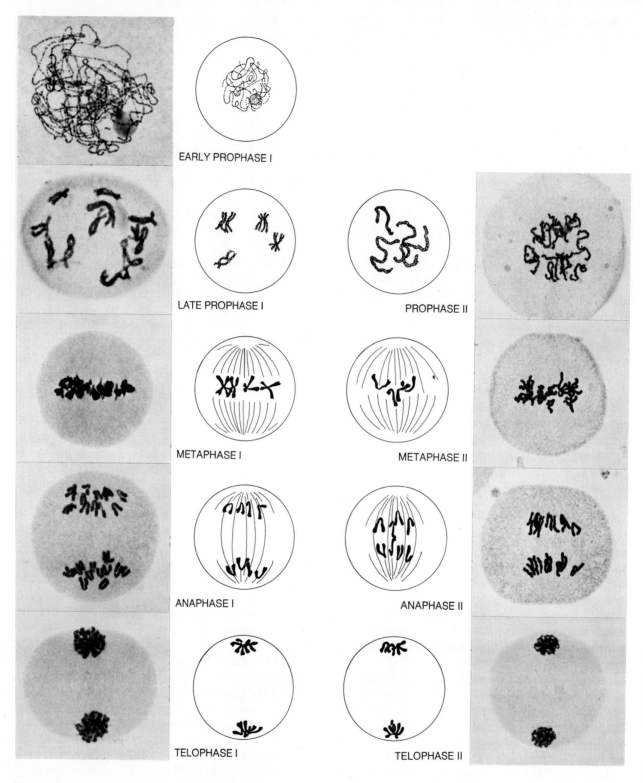

Figure 11-6 Meiosis in pollen formation in the lily. (Robert Gill.)

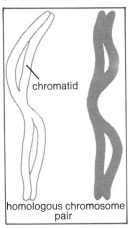
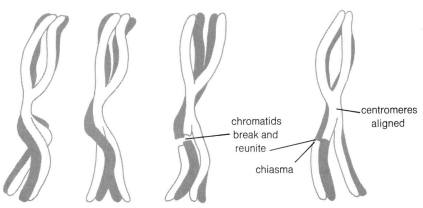

FIRST MEIOTIC DIVISION

chromatid
homologous chromosome pair
homologous chromosomes pair during Prophase I of meiosis
chromatids break and reunite
chiasma
centromeres aligned

additional types that would be caused by crossing over. Therefore, do not wonder should your children not look very much like you.

The Crossing Over Mechanism Is Not Understood

We see in Figure 11-7 that crossing over involves breaking and repair of chromatids so that two chromatids, one from each of the two chromosomes of a homologous pair, may exchange segments of DNA. Genetic investigations suggest that the breaks leading to DNA exchange probably occur at the same nucleotide position in each of the two involved chromatids. Such precision would not be too unexpected if, when the exchange occurred, the chromosomes were fully extended, as in the interphase. Extended DNA strands might effect complementary pairing, even when encumbered with the other constituents of the eukaryote chromosome. Once aligned, it is plausible that enzymes already known to cut and heal DNA molecules might be able to accomplish the required delicate molecular surgery.

However, the fact is that the chromosomes are already strongly condensed when crossing over begins. Crossing over is thus even more difficult to understand because there must be, rather than a relatively plausible mechanism to align extended double helix molecules of DNA, some way to align the supercoils that the DNA double helix is thrown into as chromosomes condense.

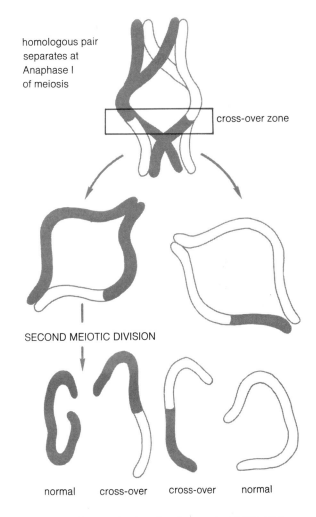

homologous pair separates at Anaphase I of meiosis

cross-over zone

SECOND MEIOTIC DIVISION

normal   cross-over   cross-over   normal

Figure 11-7 The mechanics of meiosis and crossing over.

GAMETES

All four cells produced in meiosis may become functional gametes or they may not. Some organisms, particularly among lower plants, are **isogametic,** which means that male and female gametes are identical. In these organisms all cells generated in meiosis by both sexes become gametes; none are wasted. Many organisms, particularly animals, are **heterogametic;** the two sexes produce dissimilar gametes. The male gametes, sperm, are modified for seeking out and penetrating the female gamete, the egg or ovum. **Sperm** consist of little more than a haploid chromosome set attached to an engine consisting of a flagellum and a few mitochondria sparsely enclosed in cytoplasm. In the male all cells produced in meiosis become functional sperm.

The situation is different in the heterogametic female because the **egg** is specialized by having the largest possible amount of cytoplasm, containing supplies for the early stages of development of the new individual. This cytoplasmic accumulation is brought about in part by sacrifice of three of the four cells produced by meiosis in the female, as follows: Each of the two meiotic divisions are unequal. They produce one very large cell and one very small cell that contains virtually no cytoplasm and is known as a **polar body** (Figure 11-4). The polar body produced in the first meiotic division may divide again so that the overall result of meiosis is a total of three polar bodies and one functional gamete, the egg. The logic of this process is that it is better to give one gamete an improved chance of development at the cost of the remaining three than to produce four gametes, each with only an indifferent chance of completing development upon fertilization. However, we must admit that it is not clear on such grounds why this meiotic "cost" is borne exclusively by the female. That is, why should all male cells resulting from meiosis have a chance to be represented in the next generation, whereas only one in four female meiotic products has that opportunity? Perhaps there are as yet undiscovered advantages for sperm to be as motile as possible and, therefore, to carry essentially no reserves beyond those needed to seek out an egg, thus making necessary the large reserves characteristically found in the egg.

For details of sperm and egg formation turn to Chapter 21, Human Reproduction.

12 The Natural History of Genes: Transmission Genetics

Until now our approach to genes and gene action has been largely at the chemical level. To continue exploration of the central biological problems directly concerned with heredity, we must next discuss genetics on a broader scale, one not always amenable to the magnifying lens of chemistry. However, in what follows we remain confident that the chemical basis of heredity, as developed in studies on simple organisms, still applies; but we shall for the most part consider genes as particles or units of heredity without having immediate concern for their chemical mechanisms of action. Instead, our attention will be on how genes flow from individual to individual, from parents through descendants. We will deal with how genes are related, singly and as interacting groups, to the perceivable characteristics of the organisms in which they are present.

INHERITANCE IS BIPARENTAL AND FROM LIKE PARENTS

Historically, genetics did not begin with chemistry but with the problem of transmission of physical appearance in family lines. In the beginnings of the study of heredity there was no awareness of the relationships between the observable characteristics of an organism, the **phenotype**, and the controlling hereditary factors, which we call genes. Before the study of inheritance at this level could be effectively launched, certain important preliminary facts had to be learned. It was first necessary to comprehend that in the modern world living organisms were produced only from similar living things (recall Note 3-1). Clearly there could be no logic to the science of heredity as long as mice might be generated from grain and dirty clothes, or certain kinds of trees might spawn geese, as some early naturalists thought. Next it was necessary to demonstrate the biparental nature of inheritance, with both male and female parent contributing equally in sexual reproduction to the inheritance of their offspring (Note 12-1).

With the principles established that life only comes from similar life and that both parents contribute to the heredity of the young, early geneticists were left with the more subtle question as to whether hereditary traits behaved as discrete entities or whether they blended and changed upon coming together with similar traits in a newly conceived organism. Even by the time of Darwin, the essential point that hereditary traits are not modified by transmission from generation to generation was not generally appreciated. Darwin believed in Pangenesis. He thought of the hereditary material as a set of **gemmules** generated in the tissues of the body and, carrying the essential nature of these tissues, accumulated in the gametes to blend in formation of the new individual. At least one contemporary of Darwin saw the logical flaw in this

Note 12-1 *Biparental Inheritance*

To understand inheritance in organisms with sexual reproduction it is necessary to know the contribution of the two sexes in initiation of development. Long before either plant or animal gametes had been observed, Aristotle (384–322 B.C.) believed that both parents contributed to the development of the offspring, the male providing form, motion, and soul and the menstrual blood providing the incubator and nutrients. The male contribution he compared with the work of the carpenter, the female's with the wood from which a structure is made. Aristotle had difficulty in believing that the guiding forces of development could come equally from both parents because, he wrote: "Further, if it comes equally from all of both parents, two animals are produced; for they will have every part of each parent." [*De Generatione Animalium,* Tr. D. M. Balme, 1972, Oxford, Clarendon Press] Although he was wrong about the unequal contribution, Aristotle was on the right track, in believing that the parental contributions interact to bring about development of the new individual from simpler and more general initial elements; this was the concept of **epigenesis.**

Once gametes were identified microscopically, it was not immediately concluded that they represented the physical basis for equivalent hereditary contributions from the parents. In fact, there ensued one of the more amusing controversies in the history of science. During the seventeenth century, a theory of development quite the opposite to epigenesis was born. This was the theory of **preformation;** its basic tenet was that development was a process in which there was expression of preexisting parts. The adult is in the embryo and development is simply unfolding and quantitative growth without the construction of new parts. Aristotle's problem of putting one organism together out of two sets of parts immediately returned in a severe form, and the attempt to resolve it set the scientific community at odds with some, the *spermists,* arguing that the adult-in-miniature resided in the sperm, whereas *ovists* backed the egg. In no time at all suggestive eyes saw tiny people, homunculi, in sperm (Figure 12-1). Quite soon the logical extreme of preformationist philosophy was attained in the *theory of encasement,* according to which one of the ovists held that the tiny homunculus must contain within itself the homunculi of all of its descendants, a theory with remarkable implications.

Once fertilization was clearly observed and the structure of the gametes understood, equality of parental contributions to the embryo was admitted and supported by the many clear examples from practical plant and animal breeding that had been heretofore ignored or tortuously interpreted to avoid the obvious. Curiously enough, preformation and epigenesis still cast their shadows. The developments of modern biology make it proper to say that development is both preformist and epigenetic, preformistic in that the genes come to the embryo as preformed elements, and epigenetic in that all gene products are newly formed out of simpler precursors.

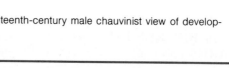

Figure 12-1 The homunculus in a sperm cell, a seventeenth-century male chauvinist view of development.

blending type of inheritance, namely that it should result in the smoothing out of inherited variation, with all interbreeding organisms ultimately merging into an average form.

Even before Darwin, the eighteenth century French scientist Maupertuis had recorded the transmission of polydactyly (excess fingers and toes) through four generations of the Rhue family (Figure 12-2). This was a very clear indication of the independence of hereditary traits, since in each recorded generation a carrier of the condition gave rise to the next generation after marrying a normal person.

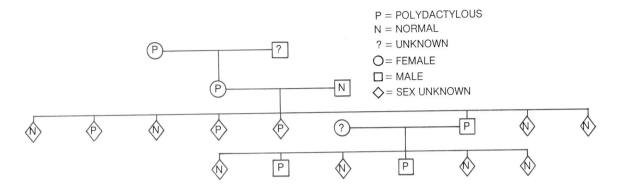

About this pedigree Maupertuis says,

"A great physician proposes... to perform some experiments on this question (of biparental inheritance)... M. de Reamur proposes to mate a hen with five toes with a four-toed cock, a four-toed hen with a five-toed cock... and (he) regards these experiments as able to decide whether the foetus is the product solely of the father, solely of the mother, or of the one and the other together.

"I am surprised that the skillful naturalist, who has without doubt carried out these experiments, does not inform us of the result. But an experiment surer and more decisive has already been entirely completed ... (as follows):

"Jacob Ruhe, surgeon of Berlin... Born with six digits on each hand and each foot, he inherited this peculiarity from his mother Elisabeth Ruhen, who inherited it from her mother Elisabeth Horstmann, of Rostock. Elisabeth Ruhen transmitted it to four children of eight she had by Jean Christian Ruhe, who had nothing extraordinary about his feet or hands. Jacob Ruhe, one of these six-digited children, espoused, at Dantzig in 1733, Sophie Louise de Thungen, who had no extraordinary trait: he had by her six children; two boys were six-digited."

Maupertuis *Lettres* (1752) Transl. Glass, B. (1947) Quart. Rev. Biol. 22: 196–210.

Figure 12-2 An early demonstration of biparental inheritance and of the independence of hereditary traits, an investigation of polydactyly in the Ruhe family, prior to 1745.

THE CONTRIBUTIONS OF GREGOR MENDEL TO GENETICS

After these beginnings the person who truly got genetics moving was an Augustinian monk, Gregor Johann Mendel. Modestly educated in the sciences, Mendel spent his adult life in a monastery in Brünn, Austria. He was active in local scientific circles, taught school, and later became abbot of his monastery. We remember him for a remarkable series of plant breeding experiments serving to establish solidly the groundwork of the science of genetics. Strangely, this work went unnoticed until 1900, 16 years after Mendel's death and some 35 years after the work was done.

Where others failed, Mendel was successful in demonstrating certain fundamental rules governing inheritance because, fortunately, he chose a favorable experimental subject, the garden pea (Figure 12-3) and because of his experimental methods. Instead of trying to follow inherited characteristics that *vary gradually*, as had so many before, he chose traits in his pea plants that were readily identified and that *varied qualitatively*, in an all-or-none fashion, rather than *quantitatively*. Thus, he could be immediately certain whether a particular trait was or was not expressed in an experimental plant. In selecting traits for study, he

Figure 12-3 The garden pea is ideal for hybridization experiments because the petals of its flowers completely enclose male and female reproductive parts until after fertilization, thus insuring self-fertilization unless the experimenter intervenes. (After J. A. Moore.)

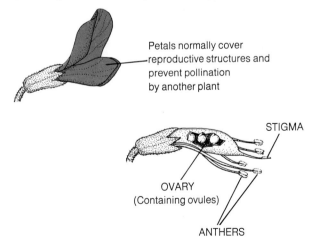

was fortunate in choosing those that we now know are controlled by genes on different chromosomes. Presently we shall see why this was an advantage.

Mendel's straightforward experimental method was greatly facilitated by the reproductive behavior of the garden pea. Left alone, the plant fertilizes itself, or "selfs," having both male and female reproductive organs. This is advantageous because a strain of plants becomes largely homozygous after a few generations of selfing and can be counted on to breed "true," so the experimenter will not be confused by the appearance of unexpected traits in the course of breeding experiments. The pea can also be crossed with other pea plants by simply transferring male gametes (pollen) from one flower to another in which the pollen forming organs have been removed. Thus the reproductive behavior of experimental plants can be totally controlled by the experimenter.

After making crosses of plants carrying the traits under study, Mendel kept accurate records of the frequency of their appearance in all progeny through several generations. Then, by application of simple mathematics, he was able to develop the basic rules governing the transmission and interaction of the traits under study.

Mendel's experiments with garden peas provided us with two landmark rules, namely that

1. Hereditary traits are transmitted between generations as independent units unchanged by the organisms in which they reside.
2. Hereditary traits may or may not be expressed in the phenotype, depending on the properties of other traits present.

To understand these important facts and to see how Mendel discovered them we must examine some of his experiments.

HEREDITARY FACTORS PASS UNCHANGED FROM GENERATION TO GENERATION

This principle is evident in an experiment in which Mendel crossed peas bearing either yellow or green seeds. Since the parental stocks had bred true for several generations, the seed color of the progeny could with confidence be attributed to the experimental cross rather than to chance appearance of some latent trait. The result of this experiment (Figure 12-4) is appearance of the yellow seed trait in all progeny. Alone this tells us little; one cannot yet determine whether the yellow progeny are the result of a blending of yellow and green hereditary factors or whether the two traits still exist unchanged in the yellow offspring.

The answer to this question comes in the next step of the experiment in which a large number of plants from the first generation of the cross, which we may call the first filial, or F1 generation, were selfed. In the next, or F2, generation the green seed trait reappears; a count of all members of the F2 showed that both yellow and green seed plants were present in a ratio of 3 yellow to 1 green seed producing plant. Reappearance of the green trait immediately demonstrated the independence of the traits for yellow and green. Had blending type inheritance occurred, there would be no way for the parental green and yellow traits to be recovered; had blending occurred in the F1, green and yellow traits would no longer exist as separate entities, and, therefore, the F2 and subsequent generations should look like the F1, that is, always have yellow seeds. Since the results came out as they did, the trait for green seeds—although not expressed in the F1— had to be present in its members independent of and unaffected by the trait for yellow seeds.

Finally, this experiment demonstrates the phenomenon of **dominance** of one gene over another. Green is **recessive** to the dominant yellow trait when the two are present together. Thus it is shown that, *although two traits may not directly affect each other, they may interact in terms of their action on the phenotype.* We shall later see that there are other types of gene interactions of which this hybridization is one example.

Mendel interpreted this experiment as follows: Each F1 hybrid must have contained a factor for yellow and a factor for green. When the F1 plants were selfed, these factors appeared in two kinds of gametes, namely gametes containing the factor for yellow and gametes containing the factor for green. Thus, when selfed, eggs with yellow or green factors would be fertilized by pollen with yellow or green factors, and the result would be three kinds of F2 plants: homozygous (pure) yellow, homozygous green, and hybrids that would appear yellow but actually

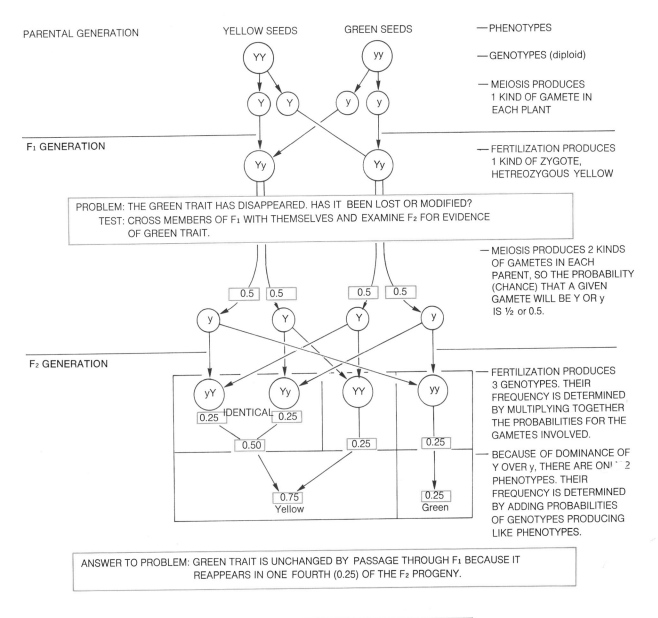

Figure 12-4 One of Mendel's experiments with yellow (Y) and green (y) seed traits shows that hereditary traits pass unchanged from generation to generation.

contain both the yellow and green factors (Figure 12-4). In modern terminology we would say that the F2 contained two **phenotypes** (external appearance) and three **genotypes** (genetic composition).

Enumerating the possible genotypes that appear in the F2, as in Figure 12-4, one can see why the phenotypic ratio is 3 yellow to 1 green, rather than 2 yellow to 1 green. As the diagram indicates, with respect to

THE NATURAL HISTORY OF GENES: TRANSMISSION GENETICS

the yellow and green traits, for every four zygotes formed there will on the average be one homozygous yellow (0.25), one homozygous green (0.25), and two heterozygotes of yellow appearance (0.50), thus producing the observed phenotype ratio (0.75:.25).

These interpretations are confirmed by observing the progeny from selfing members of the F2. When enough F2 plants are selfed to insure that all F2 genotypes are represented, the results are as follows: the green seed producing plants of the F2 produce

Figure 12-5 A hybridization involving two traits demonstrates the phenomenon of independent assortment.

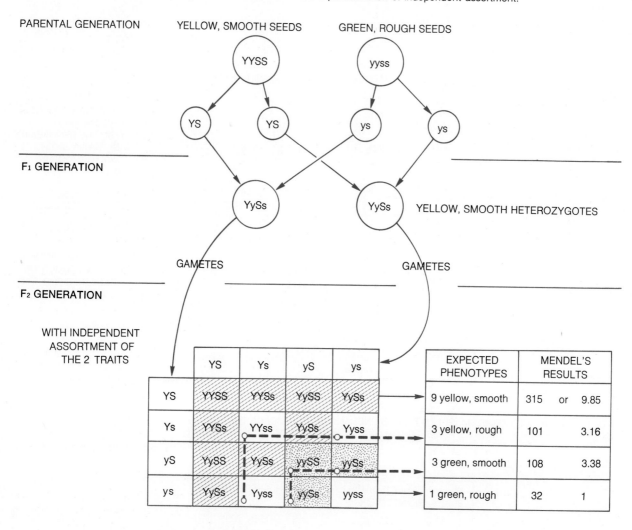

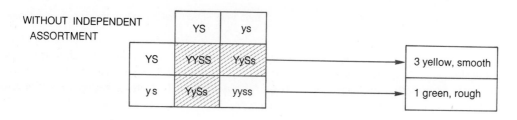

only green progeny; of the yellow seed F2, one third breed true and produce only yellow seed progeny whereas two thirds produce both yellow and green seed progeny. The fact that some of the yellow seed plants of the F2 gave rise to all yellow seed plants in the F3 and some gave rise to a mixture of yellow and green proves that the yellow plants in the F2 did actually consist of plants with two different genotypes. They were, of course (1) plants homozygous for yellow and (2) plants heterozygous for yellow, carrying the recessive factor green.

Inspection of Figure 12-4 reveals a final important conclusion from Mendel's experiment. To make things come out right, Mendel had to postulate that the progeny received one factor for each trait from each parent. Thus, although the gametes carry one factor for each trait, the organism that produces them must carry two factors for each trait. Without knowing anything about the physical basis of the traits under study–about chromosomes or meiosis–Mendel in effect stated that the adult pea plants are diploid and their gametes are haploid.

GENETIC TRAITS MOVE INDEPENDENTLY OF EACH OTHER FROM GENERATION TO GENERATION

In this next experiment Mendel did hybridizations as before but with two traits instead of one. The results were typical of his experiments with the seven selected traits that he studied in showing that all seven assorted independently, which is to say that the presence or absence of any one of them in a gamete had no effect on the presence or absence in that gamete of any of the others. The experiment illustrating this point, a two-trait cross, is shown in Figure 12-5. Plants breeding true for yellow and round seeds are crossed with plants breeding true for green and wrinkled seeds. The F1 is a single phenotype, yellow-round, indicating that wrinkled is recessive to round just as we already know that green is recessive to yellow.

The important result appears in the F2 resulting from selfing the F1. There are four classes of phenotypes in the ratio: 9 yellow-round, 3 yellow-wrinkled, 3 green-round, and 1 green-wrinkled. Obviously the yellow and round traits are inherited independently of each other, as is further illustrated by contrasting these data with the F2 that would occur if, for example, yellow and round always segregated together, that is 3 yellow-round to 1 green-wrinkled (Figure 12-5).

It is a matter of curiosity that all the traits Mendel studied segregated independently. As you realize, it is *chromosomes* that segregate independently; genes do so only if they are on different chromosomes. Strangely enough, the garden pea has seven pairs of chromosomes and Mendel was lucky enough to pick seven traits, one on each chromosome. Perhaps he did some preliminary observations before choosing these particular seven traits. There is also the possibility that he may have "improved" his data somewhat, once the underlying principles were evident. At least R. A. Fisher calculated that the chances of getting certain of Mendel's precise numerical results are on the order of 10,000 to 1.

GENES AND THE PHENOTYPE

Mendelian traits in the original sense are either dominant or recessive, they seem either to produce an effect on the phenotype or to have no effect; and this was undoubtedly true of the seven characters studied by Mendel. However, consider what happens in the F1 of a cross between red and white snapdragons. According to what has just been said, we would expect to find either all white or all red flowers, depending on which trait is dominant; actually the outcome is pink flowers. Is this evidence for the blending type of inheritance that we thought Mendel had disproved? Examination of the F2 shows that this is not so. The original genes have plainly survived (Figure 12-6). However, we still have to explain the peculiar phenotype of the F1: pink, neither red nor white.

Incomplete Dominance

If we return to the principle that one gene produces one protein–part or all of an enzyme molecule or a structural protein–an immediate solution to the problem of pink snapdragons suggests itself. It might be that the gene for red flowers produces an enzyme necessary for the production of red pigment but that two doses of the gene are necessary to provide enough gene product to synthesize sufficient pigment to produce the full red color. In the heterozygote F1 there would be only half a dose, resulting in a diluted red, that is, pink.

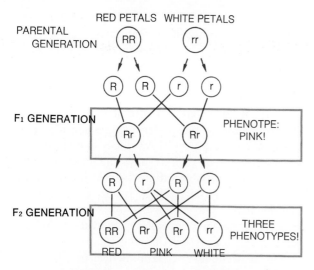

INCOMPLETE DOMINANCE IN SNAPDRAGON

Figure 12-6 Incomplete dominance of red petal color in the snapdragon. As is typical in many instances of incomplete dominance, heterozygotes for the dominant trait are phenotypically distinguishable from homozygous dominants, resulting in a phenotypic ratio of 1:2:1 in the offspring of a heterozygous cross.

Figure 12-7 Variable penetrance of a gene as shown by a family history of human polydactyly. The pedigree shows that the gene is dominant, but that it sometimes fails to be expressed (*) and produces varying degrees of the trait when it is expressed. (Adapted from Guy's Hospital Reports, 1881.)

This type of mechanism is well documented when the products of the two alleles are identifiable, as in the human genetic disease, sickle-cell anemia. The normal allele of the sickle-cell gene produces normal blood pigment molecules, hemoglobin, whereas the mutant allele produces a biochemically identifiable, abnormal hemoglobin. Although the heterozygote does not show the phenotypic characteristics of the sickle-cell trait except under unusual circumstances, the blood of such persons does contain both the normal and abnormal types of hemoglobin. In this instance both genes are working independently to produce their own type of hemoglobin, and it just so happens that the heterozygotes can get along reasonably well with only half of their hemoglobin being the normal type. This type of gene effect is called **incomplete dominance.**

Gene Interactions Affecting the Phenotype

The complexity of biochemical pathways in the organism makes it understandable that a gene seldom acts alone in producing its phenotype. Some genes must have extremely widespread effects. Consider, for

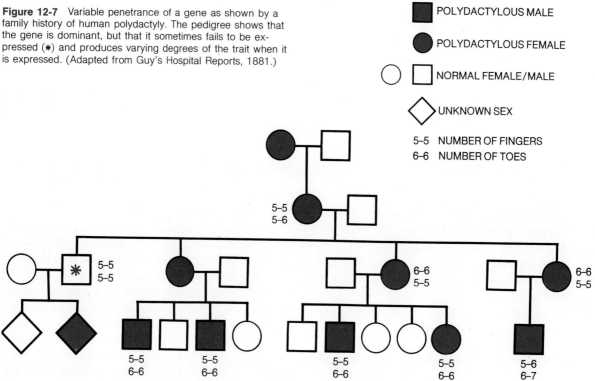

170 BIOLOGY

example, a gene that controls an important step in the synthesis of ribosomes. A defective mutation in such a gene would undoubtedly be lethal because, in the final analysis, it affects the action of *all* genes. A less drastic example of genic interaction might be the interaction of two genes controlling successive steps in a biochemical pathway; the phenotype obviously could not be normal without both genes functioning. Finally, there are many less well-defined examples of gene interaction in which genes are said to have **variable penetrance**, or **variable expression**, depending upon their total genetic environment. A gene that is variably penetrant may, although dominant, not achieve expression at all in one individual but be completely expressed in one of its progeny owing to the general properties of the two genetic environments. Polydactyly is a good example in the human. The human pedigree in Figure 12-7 shows clearly that expression is variable because the number of abnormal fingers and toes varies within this family. In one instance an entirely normal individual transmitted the trait to offspring in which it was expressed.

These examples of exceptions to simple Mendelian dominance are only a few of the many types of gene interactions that take place, and such are almost expected to be the rule rather than the exception when one considers the extent of cooperative interaction that goes on at the biochemical level in cells. Simple Mendelian dominant and recessive genes are, indeed, the exception, and generally govern processes that may be said to be at the periphery of the biochemical systems of the cell, where their actions have only a limited effect.

Widespread Phenotypic Effects by Single Genes

Just as several to many genes may act to produce a single phenotypic trait, it is also possible for a single gene to affect several traits. Imagine a gene involved in the formation of bone. Undoubtedly a mutation affecting its action would be expected to produce a whole series of defects, all traceable to this initial action. The human genetic disorder known as phenylketonuria is a particularly clear example because a single defect at the biochemical level results in the inability to produce the enzyme necessary to convert the amino acid phenylalanine to tyrosine (Figure 12-8). A consequence of this single enzyme defect is invariable mental defectiveness with a number of associated symptoms. Such conditions with multiple effects of genes are known as **pleiotropism**.

Continuously Varying Traits

Although Mendel wisely avoided them in his studies, many important traits of the organism are not of an either/or nature but are continuously varying, as are skin color, weight, height, and intelligence in the human. By continuously varying we mean that, if a

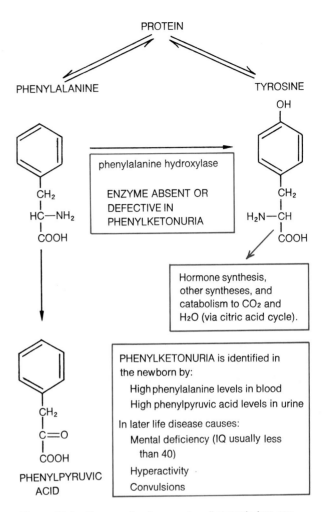

Figure 12-8 The genetic disease phenylketonuria has one primary molecular defect, the inability to convert phenylalanine to tyrosine, but this leads to many problems, primarily involving the nervous system. The controlling gene is an autosomal recessive. Heterozygotes are normal except that they may be identified by inability to properly metabolize a large dose of phenylalanine. The disease occurs in about 1 in every 10,000 live births, but perhaps 1 in 50 persons are heterozygotes.

sufficiently large sample of humans is measured with respect to such a trait, the graph of the number of persons measured against the degree of expression of the trait in each person would be a bell-shaped curve. A bimodal or multimodal curve would be observed for the expression of a trait determined by only one or a very few genes (Figure 12-9). Perhaps six or more genes are involved in the determination of human skin color. It is virtually impossible to measure how many are involved in such a complex trait as intelligence, principally for two reasons. First, it is impossible to measure intelligence accurately, and second, intelligence is markedly influenced by conditions of life such as childhood nutrition and cultural environment. Even so, studies of the interaction of environment and heredity, as in identical twins reared apart, show that intelligence has a high degree of heritability (but see Chapter 20).

FROM MENDELIAN TRAITS TO GENES ON CHROMOSOMES

Once Mendel's experiments were rediscovered and confirmed in 1900 there ensued a most exciting period in which the behavior of Mendelian traits was shown to correspond exactly with the behavior of chromosomes. These observations, resulting from work in cytology and embryology, gave genetics the firm physical basis that led in rapid sequence to the peaking of classical transmission genetics studies in the 1930s. This in turn led to the biochemical genetics of the 1940s and ultimately to the analysis of the fundamental structure of the gene and the present remarkably complete understanding of its actions at the chemical level. From the discovery of chromosomes by W. Flemming in 1882 to the seminal paper by Watson and Crick on the structure and a possible mode of replication of DNA in 1953 was only 71 years, or from virtually complete ignorance to virtually a total grasp of a major field of knowledge in the span of one human lifetime.

Two Links Between Hereditary Traits and Chromosomes

How was the connection between chromosomes and hereditary traits established? Two types of investigation provided the necessary evidence: (1) embryological observations on the effects of differing numbers and kinds of chromosomes on development and (2) cytological observations on the behavior of chromosomes in cell division, meiosis and fertilization.

Different Chromosomes Are Associated with Different Hereditary Traits

The embryological studies of T. Boveri (1862–1915) elegantly demonstrated *qualitative* differences in the effects of chromosomes. In his experiments Boveri produced sea urchin embryos with varied numbers and types of chromosomes by various techniques including fertilization by more than one sperm (polyspermy). The results of his experiments clearly showed that *one of each kind* of chromosome was necessary for normal early development to proceed and laid to rest a competing theory of the time that each chromosome carried a complete set of genetic information. As an example of the second type of observation, it was shown only a year after chromosomes were discovered that the number of chromosomes in the zygote is double that in the sperm or egg. This was done in the intestinal roundworm, *Ascaris,* an easy subject for such observations because its diploid chromosome number is only 2. Indeed, observations of this type actually led two of the early giants of embryological research, W. Roux and A. Weismann, to infer, well before 1900, that the chromosomes had to be the carriers of the hereditary material, but they, of course, lacked the necessary knowledge of Mendel's work, already published, but lying unappreciated in an obscure scientific journal.

Chromosome Behavior Parallels Behavior of Hereditary Traits

Similar behavior of chromosomes and hereditary traits throughout the life cycle has been demonstrated by many cytological observations. In early chromosome studies there was uncertainty as to whether chromosomes had continuity throughout the life of the cell or whether they formed anew at each cell division, an uncertainty which we now know was due to the difficulty of observing the dispersed chromosome during mitotic interphase. Interestingly enough, it was Boveri, in studies on *Ascaris,* who discovered the first strong evidence for continuity of the chromosomes. He found that *Ascaris* chromosomes disappeared in telophase, with their ends in characteristic outpocketings of the nuclear membrane, and then found them to first become visible at the beginning of

MULTIPLE FACTOR INHERITANCE

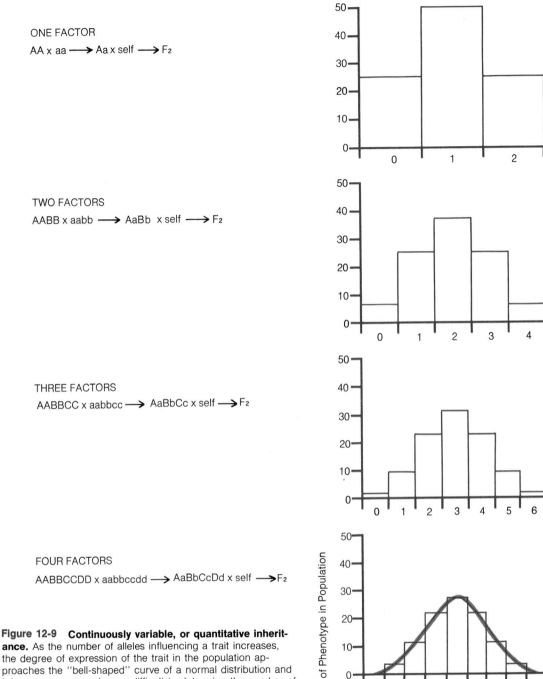

ONE FACTOR

AA x aa ⟶ Aa x self ⟶ F_2

TWO FACTORS

AABB x aabb ⟶ AaBb x self ⟶ F_2

THREE FACTORS

AABBCC x aabbcc ⟶ AaBbCc x self ⟶ F_2

FOUR FACTORS

AABBCCDD x aabbccdd ⟶ AaBbCcDd x self ⟶ F_2

Figure 12-9 Continuously variable, or quantitative inheritance. As the number of alleles influencing a trait increases, the degree of expression of the trait in the population approaches the "bell-shaped" curve of a normal distribution and it becomes more and more difficult to determine the number of alleles involved. In the figure, if it is assumed that each dominant allele has an equal and independent effect on the trait, and that the recessive alleles have none, then the expression of the trait varies directly with the number of dominant alleles in each individual. Thus in the 4-Factor F_2, the category 8 would represent the maximum expression of the trait.

the next division in the same pockets. The inference is that they had been there all the time. Meiosis, of course, provides the strongest cytologically observable parallel between hereditary factors and chromosomes because the chromosomes are clearly reduced to half their number in gamete formation, only to be restored to the diploid number at fertilization. This is exactly what Mendel theorized about the behavior of hereditary traits.

If the chromosomes do carry the Mendelian traits, there must be other similarities in their behavior beyond the numerical changes associated with gamete formation and fertilization. In parallel with Mendelian traits it should be demonstrable that chromosomes occur in pairs, with the members of each pair coming one from each parent, and it should be demonstrable that the chromosomes segregate at random in cell division. These demonstrations were readily made when organisms were found with chromosomes few enough in number and different enough in appearance to permit the cytologist to identify them all and record their movements during cell division.

One simple example illustrates the several investigations of this type that were carried out in the early 1900s. It was found that sex determination in certain insects depends upon the relative numbers present of two chromosomes called X and Y, the sex chromosomes. If the insect has two X chromosomes it is female, and if it has one X and one Y it is male. Since these chromosomes are identifiable under the microscope, it is easy enough to determine in male gamete development that (1) the X and Y segregate, never appearing in the same gamete, and (2) the sex chromosomes segregate randomly with reference to other identifiable chromosomes.

Linkage

Pursuit of the parallelism between chromosomes and Mendelian hereditary traits leads to an illogical conclusion, showing that important facts have been overlooked. According to Mendel, all traits segregate independently of each other. Since chromosomes do the same, we are led to conclude there ought to be one chromosome for each trait; however, there plainly cannot be enough chromosomes in a cell to allow each to serve as a carrier of a single trait. Obviously, then, if the chromosomes are carriers of hereditary traits, each must carry many more than one trait. If true, then all traits *cannot* segregate independently as Mendel thought; all those on the same chromosome must be transmitted together and are said to be in the same **linkage group.**

Although linkage was already known in a general way, the great school of *Drosophila* geneticists founded by T. H. Morgan made the most significant early contributions to the study of this problem of linkage between genes. Morgan's experimental animal was remarkably suitable for the study of transmission genetics in diploid organisms. *Drosophila* has a life cycle of less than 2 weeks, and is easily raised in large numbers, several hundred per pint milk bottle. The Morgan group quickly identified many mutant alleles and worked out their manner of inheritance. As the work progressed, it became apparent that all alleles fell into four groups and that the members of each group were *almost* always inherited together. Today there are at least 2000 traits known for *Drosophila*, and they still fall into one of four linkage groups. It so happens that *Drosophila* has just four pairs of chromosomes, a pair of sex chromosomes known as X and Y and three other pairs of chromosomes, called **autosomes** to distinguish them from the sex chromosomes. This correlation of linkage group with chromosome number, as it developed from the work of the Morgan laboratory, was a powerful demonstration that the chromosome must carry many genes (Figure 12-10).

Crossing Over and Chromosome Maps

We have already seen at the molecular level that chromosomal linkage groups must occasionally be disrupted by crossing over during meiosis. Evidence for this quickly developed at the level of transmission genetics during the early days of work with *Drosophila*. A typical experiment from the work of C. B. Bridges is shown in Figure 12-11 where a three-factor cross is shown. The three factors were originally on the same chromosome, and in most F2 flies they remained so. However, in about 16% of the F2, the three factors were separated. Although experiments of this type were immediately recognized as best explained by the phenomenon of crossing over, a cytological confirmation was necessary. For both plant and animal chromosomes confirmation came from experiments in which structurally abnormal, hence microscopically identifiable, chromosomes were used to correlate exchange of chromosome parts with phenotypic evidence of crossing over.

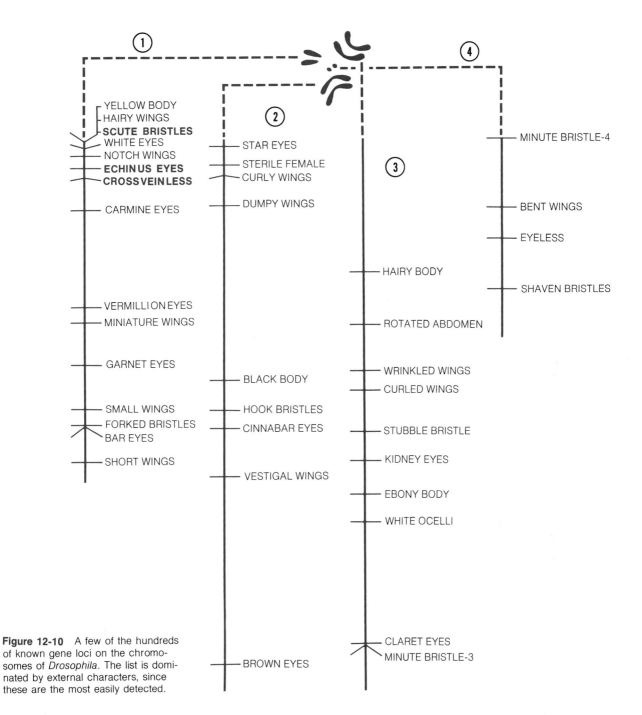

Figure 12-10 A few of the hundreds of known gene loci on the chromosomes of *Drosophila*. The list is dominated by external characters, since these are the most easily detected.

It was apparent to the early *Drosophila* workers and to other geneticists that the frequency with which crossing over occurred between genes could be used to construct a map of the linear order of genes within the chromosome for the following reason: if the chance of crossing over is the same at any point in the chromosome, then the frequency of crossing over between any two genes ought to increase the further apart they are. The greater the distance between genes, the greater the opportunity for crossing over. By measuring these crossover frequencies between successive groups of genes having one gene in common, it is possible to

DETECTION OF CROSSOVERS

If any three genes are all on different chromosomes, they should segregate completely randomly.
If all the three genes are on the same chromosome and cannot be separated, then they must always appear together.

The actual result is often an intermediate condition with most individuals possessing all three genes but with a small percentage showing "segregation", or the appearance of only one or two of the genes. This is attributed to crossing-over, exchange of homologous parts of chromatids during meiosis. For example, consider the results of a cross involving the Drosphila X chromosome genes: scute (sc), echinus (ec), and cross veinless (cv). Their chromosome positions are shown in Figure 12-10.
+ means wildtype, or "normal" allele.

$$\female \; \frac{+\;+\;+}{sc\;ec\;cv} \; \times \; sc \; ec \; cv \; \male \quad (\male \text{ has 1 X chromosome and does not crossover})$$

	PHENOTYPES	GENOTYPES ♂			GENOTYPES ♀			Number Observed	
NON-CROSSOVERS	sc, ec, cv	sc	ec	cv	sc ec cv / sc ec cv			934	Percent noncrossovers = 80.0
	wild type	+	+	+	+ + + / sc ec cv			1,174	
CROSSOVERS	sc	sc	+	+	sc + + / sc ec cv			140	Percent crossovers = 20.0
	ec, cv	+	ec	cv	+ ec cv / sc ec cv			99	
	cv	+	+	cv	+ + cv / sc ec cv			124	
	sc, ec	sc	ec	+	sc ec + / sc ec cv			164	
								2,635	

Figure 12-11 Demonstration of crossing over between 3 genes on the X chromosome of *Drosophila*. (From experiments of C. B. Bridges.)

proceed through the entire linkage group for a given chromosome and construct a linkage map showing the correct linear order of the genes on the chromosome even though various complications prevent conversion of cross over frequencies into an absolute measure of distance between genes. A linkage map of this type for *Drosophila* is shown in Figure 12-10.

CHROMOSOMAL MUTATIONS

Not surprisingly, considering the complexity of mitosis and meiosis, various accidents may intervene to cause either gross changes in chromosome structure or changes in the number of chromosomes. Because they often cause large changes in the number of genes, chromosomal mutations frequently have marked effects on the phenotypes of cells and organisms possessing them.

For convenience, the normal haploid chromosome number of an organism is termed N; thus in the human $N = 23$ and in the fruit fly $N = 4$. If an organism has this number, that is one of each of the haploid (N) chromosome complement, or some multiple of it, such as $2N$ (the diploid chromosome complement), or $3N$, or $4N$, and so on, the chromosome complement is said to be **euploid**, or balanced. If the chromosome complement varies from the balanced condition by either having more or less than the normal number of *some* of the chromosomes, it is said to be unbalanced, or **aneuploid**.

Changes in the Entire Chromosome Complement

If an organism is $3N$, it is called triploid; if $4N$, it is tetraploid, and so on. The general term applied when multiples of the normal chromosome complement are present is **polyploidy**. The condition of polyploidy may arise in various ways. For example, suppression of a division in meiosis may result in a $2N$ gamete, which, on fertilizing a normal gamete, results in a $3N$ cell, or it may arise from **polyspermy**, fertilization by more than one sperm. Polyploidy in plants is often a beneficial condition, resulting in desirable agricultural characteristics. Thus the most commonly cultivated wheat is hexaploid (6N). In animals, polyploidy is almost universally detrimental. The commonest chromosomal abnormality in aborted human fetuses is triploidy (3N = 69), and no instances of prolonged survival of polyploid humans have ever been reported. The reasons for the marked difference in effects of polyploidy in plants and animals are largely unknown except for being to some extent related to the mechanism of sex determination in animals.

Changes in the Number of Individual Chromosomes

Events, such as lagging behind of a chromosome in a gametic cell division or the failure of a pair of chromosomes to separate after meiotic pairing, termed **nondisjunction,** result in germ cells either lacking the chromosomes involved or having one more than the normal number. On fertilization with normal gametes, the result would be either nullosomy or trisomy. In **nullosomy** one chromosome is missing (2N − 1), and in **trisomy** one chromosome is present in excess (2N + 1). Although these conditions in lower organisms may result in viable individuals, they seem invariably harmful to humans. They are probably lethal in humans when the chromosomal variation involves one of the larger autosomes and result in pronounced abnormality when the smaller chromosomes are involved. Trisomy of the small human chromosome 21 was, in fact, the first demonstrated chromosomal abnormality to be correlated with a human abnormality, Down's disease (sometimes called mongoloid idiocy). A common disorder, it is found in about 1 in every 700 births, with occurrences strongly correlated with advancing age of the mother (Figure 12-12).

Changes in the Gross Structure of Chromosomes

Pieces of chromosomes may be lost or duplicated, moved from one chromosome to another, or inverted within the same chromosome. (See Figure 11-5). If higher organisms are homozygous for chromosomes with lost parts, they usually do not survive and heterozygotes are usually defective. Duplications of parts, on the other hand, usually cause no trouble and may be important to evolution as a source of new genes. Since the genes of a duplicated segment of chromo-

Figure 12-12 The characteristic features of a child with Down's syndrome along with her **karyotype,** showing trisomy of chromosome 21.

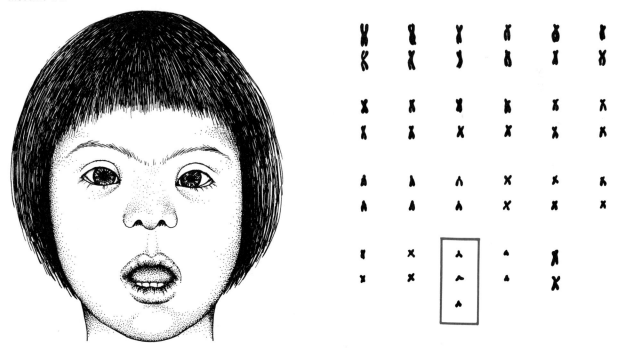

THE NATURAL HISTORY OF GENES: TRANSMISSION GENETICS

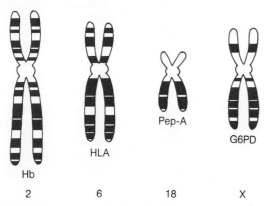

Figure 12-13 Diagrams of some of the 22 pairs of human chromosomes, illustrating the banding patterns that are revealed by staining with fluorescent dyes. Some known gene loci are indicated by abbreviations for the proteins which they specify. Thus HLA on chromosome 6 specifies certain of the human leucocyte antigens (see Chapter 23), and Hb on chromosome 2 specifies hemoglobin.

some are unnecessary for survival, they may undergo mutations without necessarily deleterious effects on the organism because the normal alleles present continue to function. Thus freed from their usual tasks, it is thought duplicate genes may mutate many times and in the evolutionary game of chance possibly develop new and valuable functions.

Translocation of part of a chromosome to another, *inversions, duplications,* and *deletions* may cause great difficulty with normal segregation of homologous chromosomes during the meiotic divisions. This is especially so if the chromosomal change involves a part of the chromosome containing the kinetochore attaching the chromosome to the spindle fibers. In the long run these difficulties have the effect of suppressing gene recombination in meiosis and, in extreme situations, suppress it totally. Thus they have a significant evolutionary effect.

MAPPING HUMAN GENES

In theory the technique described for mapping genes in *Drosophila* is applicable to humans, but in practice it is extremely difficult to apply sufficiently well to produce a map of any degree of completeness. Since there are 23 pairs of chromosomes in the human, any two genes studied would be much more likely to be in different linkage groups than in an organism with only a few chromosomes. The work is further hampered by the necessity of relying upon the essential randomness of human mating, rather than upon rigorously controlled experimental matings. Even so, some progress was made by this method, and a number of human genes were mapped. Yet, by 1970 probably not more than ten pairs of linked genes were known in the human.

Human Cell Hybridization

Recently a new family of techniques based on tissue culture has been brought to bear on the problem of mapping human chromosomes and rapid progress has resulted in the mapping of human genes that produce biochemically identifiable products. Although it has been known for many years that human and other cells grow more or less indefinitely in tissue culture, it has only recently been discovered that cells from different kinds of organisms, in the presence of certain killed viruses, frequently fuse together. Moreover, a significant number of these fusions, termed somatic cell hybridizations, result in viable cells whose nuclei contain chromosomes from both cells and are therefore called **heterokaryons.** When mouse-human heterokaryons are formed, the human chromosomes are progressively lost in subsequent divisions, but one may remain if it contains a gene necessary for survival of the heterokaryon cell clone. Such cells containing a human chromosome can be biochemically tested for the presence of certain human genes by detecting the enzymes they produce. Thus it becomes possible to test for linkage in a large number of biochemically favorable human genes by direct assay of their products.

A second powerful technique that has recently come to the aid of workers investigating human chromosomes is a straining procedure that reveals characteristic banding of the chromosomes, which otherwise are almost indistinguishable (Figure 12-13). These two techniques promise much hope for mapping of human genes. They will undoubtedly be important aids in the identification and perhaps ultimate treatment of many more human genetic defects than are now accessible to medical science. Despite these new techniques, the mapping of human genes remains a formidable task since it is estimated that the genome contains at least 50,000 structural genes.

Figure 13-2 *Acetabularia* is a multinucleate, unicellular plant. The umbrella is a reproductive structure.

differentiation. In complex organisms, similar differentiated cells are organized into **tissues,** which perform a specific function. Ultimately, tissues join to form **organs,** which are groups of tissues cooperating in some particular function.

THE PRICE OF MULTICELLULARITY

To summarize, the two great advantages of multicellularity are *size* and *specialization*. The two are necessarily linked, for multicellular organisms are unable to increase substantially in size without developing specializations in structure and function.

Although multicellularity confers many advantages, they do not come free of charge. Certain difficulties appear, and indeed much of the great variety of structure and function seen in higher organisms may be interpreted as evidence of a never ending series of compromises between the advantages of multicellularity and the difficulties generated simultaneously. The costs of multicellularity come under three principal headings.

The Complexity of Reproduction

In the world of the single celled organism, reproduction, by fission or even sexual reproduction of single celled eukaryotes, is a relatively simple matter. In higher organisms, although sometimes asexual reproduction by budding off of part of the organism does occur, sexual reproduction is usually highly complex. Gametes are produced by specialized reproductive structures, often associated with other organs and with behavior that insures fertilization and survival of the offspring.

The Complexity of Development

Development of the unicellular eukaryote is complete with zygote formation, except for relatively minor changes not requiring further cell division. In the multicellular organism the zygote is just a single step in an often long and hazardous process of cell division and shaping of the organism before it is fully able to cope with life. Much energy, structure and process is devoted to development.

The Complexity of Adult Maintenance

The single celled organism lives or dies largely by its own efforts. A constituent cell of a multicellular orga-

several centimeters long with an elaborate umbrella, long stalk, and attachment organ at its base.

Imagine the effectiveness of a colony, a cooperating group of such cells sharing the work of life so that each cell is responsible for only that function for which it is specialized: protection, digestion, reproduction, and so on. Cells could evolve in the colony with even more highly restricted functions such as the carrying of water, detecting light, or making movement possible—jobs which no jack-of-all-trades single cell could do as well. The production of such specialized cells, in a cooperating group of cells, is termed

nism, in return for the investment of its specialized labor for the good of the whole organism, has many of its requirements of life met by other cells. Thus, a cell that may be deeply buried within a multicellular organism can itself no longer directly carry on essential exchanges with the environment and must depend on other cells and tissues of the organism to provide such necessary exchanges. To accomplish this life-supporting trade between cells requires development of elaborate systems and processes that often themselves necessitate further developments of housekeeping mechanisms. Not only are these systems expensive to build and operate, but they are also vulnerable to external destructive forces; the multicellular organism works very well and is usually much better off than a unicellular organism in any given situation, but when it malfunctions *all* of its component cells are immediately in a desperate state for lack of the functional abilities that they have sacrificed to the organism. This fact in itself leads to still more complexity as organisms must invest in more systems to protect from malfunction the ones they already have.

POSSIBLE ROUTES FROM CELL TO MULTICELLULAR ORGANISM

Lacking direct evidence, we must speculate about the transition from single celled to multicellular organisms. Our speculations must conform to one rule, namely, that *each step in the transition must be of adaptive value*. We came across this rule in Chapter 6 when the manner of origin of biochemical pathways was under discussion, and it is just as applicable here as there. It must be admitted that invoking the adaptive value rule immediately creates a problem, for it is not easy to see what advantage there would be for adequately functioning single celled organisms to take the first steps towards multicellularity.

Multicellularity from Aggregation of Daughter Cells

Colonies of organisms formed by adhesion of daughter cells are common in bacteria (Figure 13-3). Substantial advantages may accrue to cells that behave in this way for several reasons. Particularly in the most primitive organisms, it is reasonable to suppose that cell membranes were not fully perfected and probably valuable molecules would have been

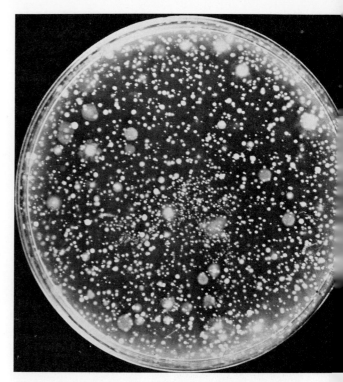

Figure 13-3 Bacterial colonies growing in a culture dish inoculated by a sneeze. (Soc. Amer. Bacteriologists.)

lost by accidental outward passage across the cell membrane. Adherence of two like cells would, of course, diminish this problem to the extent that the two cells have joint membranes; what one loses the other gains. Thus, one would assume it to be advantageous for cells to aggregate up to some maximal number, which would be determined by the trade-off between two opposing factors, the necessity for efficient chemical exchange with the outside environment and the advantages of reducing the loss of useful metabolites to the outside.

Another advantage, particularly in bacteria, is that certain activities carried on by individual cells are undoubtedly more effective when done jointly by many similar cells in close contact. An example with which you are already familiar is secretion of the polysaccharide coat that protects virulent pneumococcus bacteria from outside attack. Cells deep within a colony are substantially more protected than isolated cells, not only because of the coat but simply because the colony is too large a bite for a predator. Furthermore, bacteria may secrete toxins (poisons) or enzymes that break down substrate molecules to prepare them for ingestion, and these activities carried on collectively must create a more suitable local environment for a group of bacteria than a single bacterium could manage for itself.

Although such examples show substantial advantages for the first stages of a route to the multicellular organism by adhesion of daughter cells, the next steps along this path are difficult to picture. What we need to see is development of coordination between cells and the appearance of specialization of function. We therefore ask, how does one cell coordinate with another? How do two cells mutually influence each other? The principal means evident in modern multicellular organisms cover a wide spectrum from electrical, to chemical, to mechanical. We will fully consider these modes of cellular interaction later. For now it is obvious that the cells of a bacterial colony might easily be in highly effective chemical communication across contiguous cell membranes. Yet many who have thought about the problem of origins of multicellularity feel that it is too much to expect for originally completely separate cells to establish sufficiently good communication to lead ultimately to multicellular organisms as we know them today.

Multicellularity from Compartmentalizing a Multinucleate Cell

There is, however, another model for the origin of multicellularity that avoids the problem of establishing communication between cells. There are many instances of cells, often free living, that contain several to many nuclei; cells of this kind are called **syncytial**. In single celled green plants such as *Acetabularia* (Figure 13-2) this is a well-known condition during at least part of the life cycle, as it is in many single celled animals.

It is appealing to suppose that the origin of multicellularity involved division into separate cells of such an already well-integrated unit of cytoplasm and nuclei. It is well substantiated that many single celled animals are so remarkably compartmentalized into specialized functional units without benefit of cell membranes that they may be fairly said to be "single celled multicellular" organisms (Figure 13-4). In place of this cumbersome phrase the term **acellular** is often applied. Compartmentalization of such a going concern would not necessarily be plagued with the problem of establishing communications within the developing organism that we saw in the daughter cell model; rather, the problem would be the easier one of preserving preexisting routes of communication.

One is, of course, entitled to ask what advantages would result from compartmentalization of such a cell. This question is difficult for, in fact, an immediate disadvantage seems to appear. This is in regard to *reproduction*. Typically, the single celled animal reproduces by a division of the entire cell, and obviously cellular compartmentalization would put a kink in that process. The theorist is fortunate, however, in being able to look about among existing single celled animals and find that some reproduce by the formation of true gametes and that the sexual part of reproduction may be confined to transfer of nuclei between cells. Presumably, if multicellularity arose by compartmentalization of a multinucleate cell, it had to wait until the reproductive process was able to accommodate to the change by not involving the entire organism.

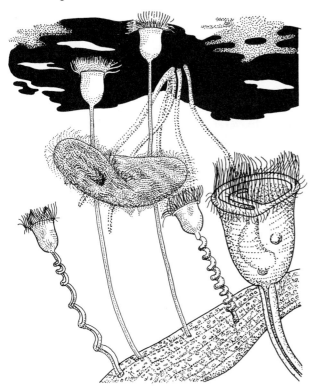

Figure 13-4 Specialization in ciliate protozoa. *Vorticella* contains within its stalk a longitudinal myoneme which functions like a muscle to retract the stalk. The motile ciliate in the center is covered with locomotor cilia that are coordinated to allow change in direction of movement.

Perhaps the most obvious advantages to attainment of multicellularity by a syncytial, single celled animal would be as follows: First, because of compartmentalization, damage to the organism would tend to be limited to the affected cell; and, second, the limits to size increase would be raised. The most immediately useful result of dividing up the cell would probably be the first, damage limitation. Although simple animals are adept at stopping cytoplasmic loss due to cell rupture, the interposition of one or more cell membranes in the organism would provide further limitation of damage. There might be a similar limitation placed upon bacterial attack. However, once multicellularity was established, advantages associated with the ability to increase size would undoubtedly soon predominate.

What is a Multicellular Organism?

By now you should, without formal definition, have a good working idea of the nature of a multicellular organism. Still, it is useful to formulate as precise a definition as possible and then to apply it to a few instructive borderline cases to see if it holds up. By such an exercise we may gain new insight into the nature of the multicellular state.

MULTICELLULARITY DEFINED

We know now that there are three types of organisms in terms of cellularity: single celled, multicelled, and an in-between group with more than one nucleus per cell, the syncytial organisms. For our immediate purposes we will include the latter among single celled organisms and point out that all in this category have one clearly identifiable attribute—they fullfill all the requirements of living within the confines of one cell membrane, except for sexual reproduction, when it occurs. In other words, the single celled organism requires no other cell to complete its life cycle except during sexual reproduction when another cell is required as a donor of genetic material.

With this in mind we come easily to a definition of a multicellular organism as an organism composed of cells which, except for gametes, *cannot* under ordinary conditions complete their life cycles in isolation from the organism of which they are a part. We add the proviso "under ordinary conditions" because the cells of many multicellular organisms may survive and multiply under the artificial conditions of tissue culture and because very often small parts of organisms are able to survive and produce a new individual by the process of regeneration. These exceptions do not seriously detract from the rigor of the definition because tissue culture is, in effect, the replacement of the organism with an artificial organism, that is, the tissue culture equipment and nutrients, and because, in the process of regeneration, the regenerating tissue can be viewed as a return to an earlier developmental stage and the undertaking of a variant of embryogenesis, the development of a new individual. This process, of course, must always begin with one or a few cells, and the aggregate of cells does not possess the full array of attributes of a multicellular organism until construction of the individual is complete.

THE BORDERS OF MULTICELLULARITY

Let us briefly look at four examples that illuminate this definition. The first is a bacterial colony, which fails the test; the next two are simple plants, which pass but raise some interesting questions; the last is a very simple animal, very clearly a multicellular organism, but which raises a most interesting question about the nature of organism themselves.

The Bacterial Colony

As we have seen, bacterial colonies may have quite definite form, and it is advantageous to the individual bacterium to be a colony member. However, each bacterium in a colony is like every other member of the colony, except for the occasionally appearing

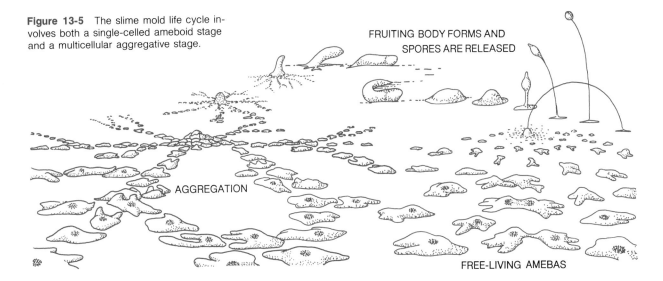

Figure 13-5 The slime mold life cycle involves both a single-celled ameboid stage and a multicellular aggregative stage.

spores, reproductive or protective modifications of the bacterial cell, which virtually any member of the colony may form. There is no cellular specialization within the bacterial colony. Each cell is capable of independent existence, and whatever contribution it makes to community welfare is by means of processes all members are equally capable of performing. For this reason the bacterial colony cannot be called a multicellular organism.

Cellular Slime Mold

The slime molds are members of a group of simple plants called fungi, which include familiar organisms like toadstools and bread mold. Examination of Figure 13-5 immediately shows that there is at least more complexity in the slime mold than in the bacterial colony. The slime mold has a well-defined life cycle beginning with haploid spores that, upon release from a structure called a fruiting body, become free moving amebas. Some of these fuse in fertilization making diploid amebas. The amebas multiply and they feed on organic matter, since fungi are plants that do not carry on photosynthesis. Ultimately they aggregate into a sluglike mass, called together by a chemical signal sent out by the members of the aggregating group. This mass then constructs the fruiting body, which is obviously structurally complex and contains cellulose, the typical supporting molecule of plants. Within the fruiting body some amebas undergo meiosis and produce haploid spores, completing the life cycle.

Now, if we did not know the complete life cycle of the slime mold, whether it is called a multicellular organism or not would depend upon the stage in the life cycle being considered. If only the free-living amebas were under consideration, the decision would undoubtedly be that they were individual, single celled organisms. On the other hand, examination of the fruiting body would clearly indicate that it is a multicellular organism. The fruiting body has a complex structure, resulting from specialization of function of particular cells, some secreting cellulose, some forming spores, and so on. Most particularly, we see that not all cells of the slime mold are reproductive. Only those cells that form spores contribute directly to the next generation. All the rest contribute only indirectly by insuring survival of the spore forming cells and by providing a safe housing for the spore forming process.

Viewed as a whole, the life cycle of the slime mold is convincingly that of a multicellular organism. There is cellular specialization not only in terms of the production of structure but in terms of activities. In particular, we see that not all cells of the organism may play an equivalent role in the sexual reproductive process; *germinative cells* are isolated from *somatic cells*. Clearly the slime mold is a multicellular organism in the sense of our definition. It is unusual only in that it passes through a stage of anarchic, single celled life, but it is a stage which carries with it the essential factor promoting return to organismic multicellularity, the chemical signal to aggregation.

Volvox

In this green plant (Figure 13-6) the most immediately striking characteristic is structural order. The living unit is a nearly perfect sphere constructed of as many as 40,000 cells, all of similar form and all similarly oriented. The typical *Volvox* cell possesses a pair of flagella and a light-sensitive eye spot; it is, in fact, not appreciably different from many types of free living, single celled plants. This alerts us to the possibility that the spherical structure may not actually, represent a multicellular organism. However, further examination shows that it meets the basic qualifications.

The cells that compose the fabric of *Volvox* are all interconnected by fine protoplasmic branches, and it appears that this direct channel of communication between cells is the means by which functional order and cellular division of labor is established. Careful study of the cells forming the spherical surface of *Volvox* shows that at one pole of the sphere the eye spots of the cells are larger than elsewhere and, at the opposite pole of the sphere, cells are specialized for reproduction. We thus see an unmistakable sign of the delegation of reproductive power that was evident in the slime mold, and again it seems to indicate advancement to multicellularity.

Volvox clearly demonstrates the advantages of multicellular organization over either the independent life of the single cell or the loose association of colonial life. The structure of *Volvox* seems adapted to two primary functions. First, since the young individuals arising from sexual reproduction are vulnerable during the early phases of their development, they are protected by being held within the living sphere. Second, orientation and movement in the direction of life-sustaining light is facilitated by the arrangement of cells and their behavior. Each eyespot controls the action of the flagella of its own cell, causing them to become more active when the light is either too bright or too dim, thus tending to keep the organism in optimum lighting for photosynthesis.

This last behavior however is, nothing more than the additive effects of strictly unicellular processes and alone would not suffice to define *Volvox* as a multicellular organism, rather than a colony. The important characteristic is the specialization of function among cells and here, as in the slime mold, this specialization is directed towards reproduction. These two examples lead us to an important corollary of our definition of multicellular organism, namely, that perhaps the earliest specialization of function to appear in simple multicellular organisms is differentiation of reproductive cells and the appearance of two cell lines, *reproductive* and *somatic*.

Hydra and the Sea Pansy

If we examine the simplest common multicellular animal generally conceded to be an organism rather than a colony, there is no difficulty at all in confirming its compliance with our definition; however, in examining some of its relatives, we discover the remarkable fact that there is a level of organization beyond that of the organism.

Hydra is a fresh water representative of a group of simple animals known as coelenterates. They include such well known organisms as jellyfish, sea anemones, and corals. Although early biologists commonly believed many of these animals were plants (thus their early name, zoophyta), a careful examination reveals their animal nature and demonstrates that they are unmistakably multicellular organisms. *Hydra* (Figure 13-7) captures, kills, swallows, and digests living prey. These acts are accomplished by cells so highly specialized for particular functions that they are totally dependent upon the remainder of the organism for survival. In addition, sexual reproduction is a delegated function of specialized cells, although *Hydra* is also capable of asexual reproduction by budding.

Even though *Hydra* contains many types of specialized cells and performs many functions requiring their cooperative action, it still reflects its descent from single celled organisms, a point well illustrated in the way it feeds. Even though the process of getting prey into the digestive cavity and the initial stages of digestion are the cooperative acts of many different kinds of cells, no single class of which could accomplish these processes alone, the final stage of digestion consists of engulfing particulate matter by digestive cells much as a free living ameba would. In other words, advancements provided to *Hydra* by multicellular organization are largely devices to improve processes already in existence in the free living cell. No new property of life is added by multicellularity to the list that we have already developed and identified in simpler organisms.

As multicellular animals go, *Hydra* is very simple. Figure 13-7 shows it is basically a digestive cavity

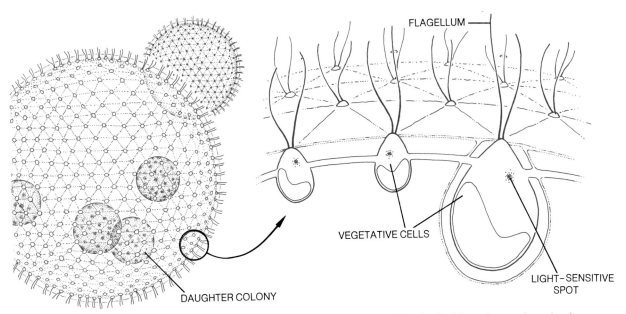

Figure 13-6 ***Volvox.*** On the left a colony containing daughter colonies. On the right, detail of the colony surface showing vegetative cells bearing paired flagellae.

Figure 13-7 ***Hydra.*** A cut away view of the digestive cavity containing captured prey shows cellular differentiation, as does the enlarged figure of the tentacle surface showing nematocysts, some discharged by the prey.

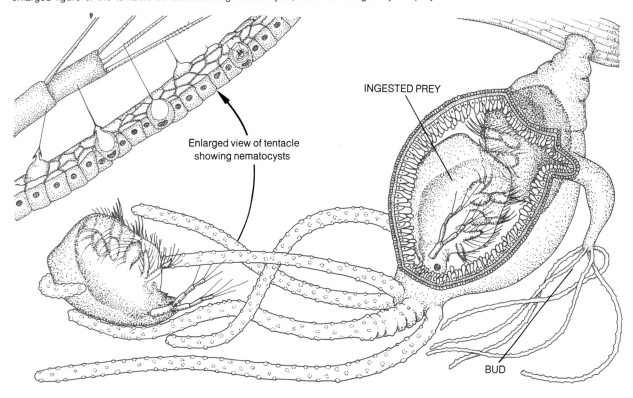

MULTICELLULAR ORGANISMS 187

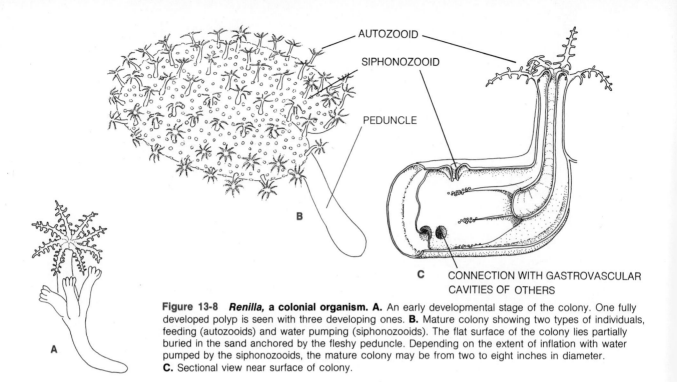

Figure 13-8 *Renilla,* **a colonial organism. A.** An early developmental stage of the colony. One fully developed polyp is seen with three developing ones. **B.** Mature colony showing two types of individuals, feeding (autozooids) and water pumping (siphonozooids). The flat surface of the colony lies partially buried in the sand anchored by the fleshy peduncle. Depending on the extent of inflation with water pumped by the siphonozooids, the mature colony may be from two to eight inches in diameter. **C.** Sectional view near surface of colony.

surrounded by two layers of cells, many of which are specialized. There is no evidence of organ formation, little sign of aggregations of cells of different kinds specifically addressed to some particular type of function; there is no liver, no heart, no brain. It seems that no coelenterate has passed much beyond the state of organization seen in *Hydra,* but it is a testimony to the resourcefulness of these simple organisms that what they have not been able to attain in the way of organization as *individuals* they have frequently been able to manage as *groups of individuals.* Coelenterates have taken a step beyond the level of multicellular organisms by forming many individual organisms into colonies. Members of the colony are able, by reason of specialization beyond that which is feasible by the isolated individual, to compensate for the absence of organs. In the colonial coelenterate, the individual organism becomes an organ of the colony.

There are many examples of colony formation among the coelenterates but one of the most clear-cut is the sea pansy, *Renilla* (Figure 13-8). It begins life as a single, hydralike organism, called a **polyp,** which asexually buds others like it. These continue multiplying their own kind and also produce other types of polyps and tissues until finally a colony is formed into which all polyps are fully integrated. For example, all of the polyps have interconnecting digestive cavities and interconnecting nervous tissues, and the entire colony shares in control of muscular tissues that enable it to move about slowly. During growth of the colony some of the polyps become specialized for the role of pumping water through the colony, so specialized, in fact, that they are unable to capture food themselves and must rely on the unspecialized polyps of the colony for food.

Organisms constructed like the sea pansy represent a new level of organization in the hierarchy of life and create a difficulty with our definition of multicellular organism. Clearly the unspecialized polyp of the sea pansy is truly a multicellular organism in structural terms as well as in many aspects of function. Thus, its nervous system functions quite independently of the nervous systems of the surrounding polyps while it feeds. Yet, when the colony is threatened, a signal sweeps through the entire colony, gains control of the nervous systems of the individual polyps, and brings about a concerted, colonywide protective withdrawal of polyps. We also find that the individual polyp is so modified that it cannot survive in isolation from the colony. It is an unfortunate matter of terminology that this level of organization has the same name as is applied to the aggregations of bacterial cells that we also call colonies. Perhaps colonies of the type represented by the sea pansy and other colonial coelenterates, in which many of the functions of undeniably multicellular organisms are irrevocably given over to the group, deserve a distinguishing name, perhaps **collectives** or superorganisms.

In coelenterates formation of collectives allows functional specialization to occur at the level of the organism, offsetting the fact that in these animals specialization does not take place to a significant degree *within* the individual organism. In a sense the individual becomes an organ of the collective. Subordination of these organ-organisms to the collective of which they are part is abundantly evident because of their permanent structural interconnections.

DO GROUPS OF HIGHER ORGANISMS FORM COLLECTIVES?

Although an obligatory physical interconnection is not evident in interrelationships among more complex organisms, it is nonetheless evident that some higher organisms do form stable groups with many of the attributes of the superorganism. Social insects such as ants and bees are universally accepted examples (Figure 13-9). The insects making up a colony exhibit specializations parallel to those we have seen in the coelenterate colony. There may be reproductive individuals, food collectors, housekeepers, soldiers, and so on. These individuals do not survive away from the environment created by other colony members. The salient difference between a coelenterate superorganism and an insect colony is the structural and functional sophistication of the insect as compared with the coelenterate polyp and the fact that integration within the insect colony or collective does not depend upon permanent physical attachments between individuals. However, this does not mean that integration within the insect colony is nonexistent; it is an extremely powerful force even though its avenues of control are the less physically tangible instinctive behavioral patterns written into the nervous systems of the insects and chemical and other cues that when

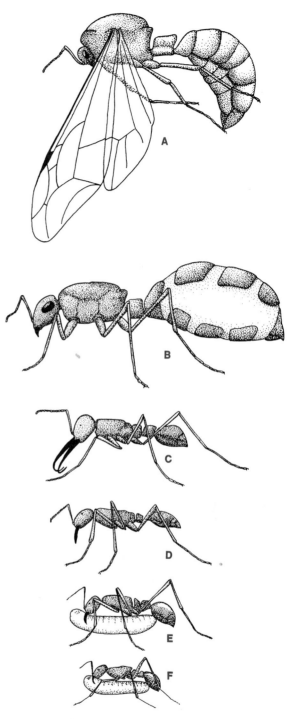

Figure 13-9 Social insects. The example of the castes of the army ant, *Eciton*. **A.** Male. **B.** Queen. **C.** Soldier. **D–F.** Workers. **B–F.** Adult females that are specialized for specific functions in the colony. Their activities, and even their production, are regulated by a variety of signals including chemical messages between the various types of individuals.

MULTICELLULAR ORGANISMS

exchanged turn patterns of behavior on and off between individuals of the colony.

It is a favorite exercise of behavioral scientists to seek signs of the collective organism in even higher levels of life, herds of mammals and even in human social behavior. Certainly the principles of the collective seem to be present, although somewhat obscured by the highly varied patterns of integration between individuals. Humans, as we well know, do band together into groups that achieve effects unattainable by the individual acting alone. In large part this is due to the specialization of functions by the individual human which this banding together permits.

Thus we see that the multicellular organism is not necessarily the apex of the continuum of biological organization that begins with the free living, single celled organism. Instead there seems to be a still more complex level in which the independence of the multicellular organism may be to a greater or lesser extent invested in group organization allowing attainment of goals otherwise unavailable to individuals acting alone.

Organ Systems

From these examples the nature of the multicellular organism emerges: it is to capitalize upon the abilities of the single cell, bringing these abilities in cooperating groups of specialist cells into a state of fuller expression than is possible in any free living generalist cell. This demands a high degree of organization of the parts of the multicellular organism; the tune is no longer produced by a lone whistler but by an orchestra, which may perform better *only* if properly directed. We have seen the beginnings of this in the way the simple *Hydra* is built as a digestive cavity surrounded by two cell layers containing specialist cells in positions appropriate to their functions. In higher plants and most multicellular animals, even more complex organization is necessary, as is illustrated in Figure 13-10 which shows the levels of organization found in a complex animal. The lower levels of this hierarchy are familiar because they have been our concern in most of the preceeding parts of this book. The middle zone, organization of cells into **tissues,** is the level attained by *Hydra*. Higher levels—organs, organ systems, and organism—are what we must now study.

STRUCTURE AND FUNCTION OF AN ORGAN

Consider the human stomach as an example of an organ (Figure 13-11). First of all, what does it do? Its functions are similar to those of the digestive cavity of *Hydra*, namely, to accept food and prepare it for complete digestion. In *Hydra* this process requires essentially the total resources of the entire organism, whereas in the human it is carried on by this single organ, accounting for only about 1% of the total body weight. How does it operate? The stomach is essentially a mixer in which food is stirred with acid to get it into a chemical and physical state suitable for further processing in other digestive organs.

The work of the stomach represents an early phase of a long process by which food is broken down into its molecular constituents, which are then taken up for use by the cells of the body. Since the stomach is a mixer, we look for mechanical devices to accomplish the mixing and find them in layers of muscle cells that crisscross in the stomach wall; their rhythmic contractions stir the ingested food. Since the stomach acts chemically on the food, we look for sources of chemicals and find them in the form of secretory cells lying among the inner lining, or epithelial cells, which line the stomach inner wall. Thus, the primary functions of the stomach are accomplished by two specialized kinds of cells, and we note that these cells are organized into two types of tissues, muscular and lining tissue, or mucosa. *We therefore define an organ as a structure composed of tissues collaborating in a common function.*

These two tissues, the stomach muscles and mucosa, include only the barest essentials of the cells

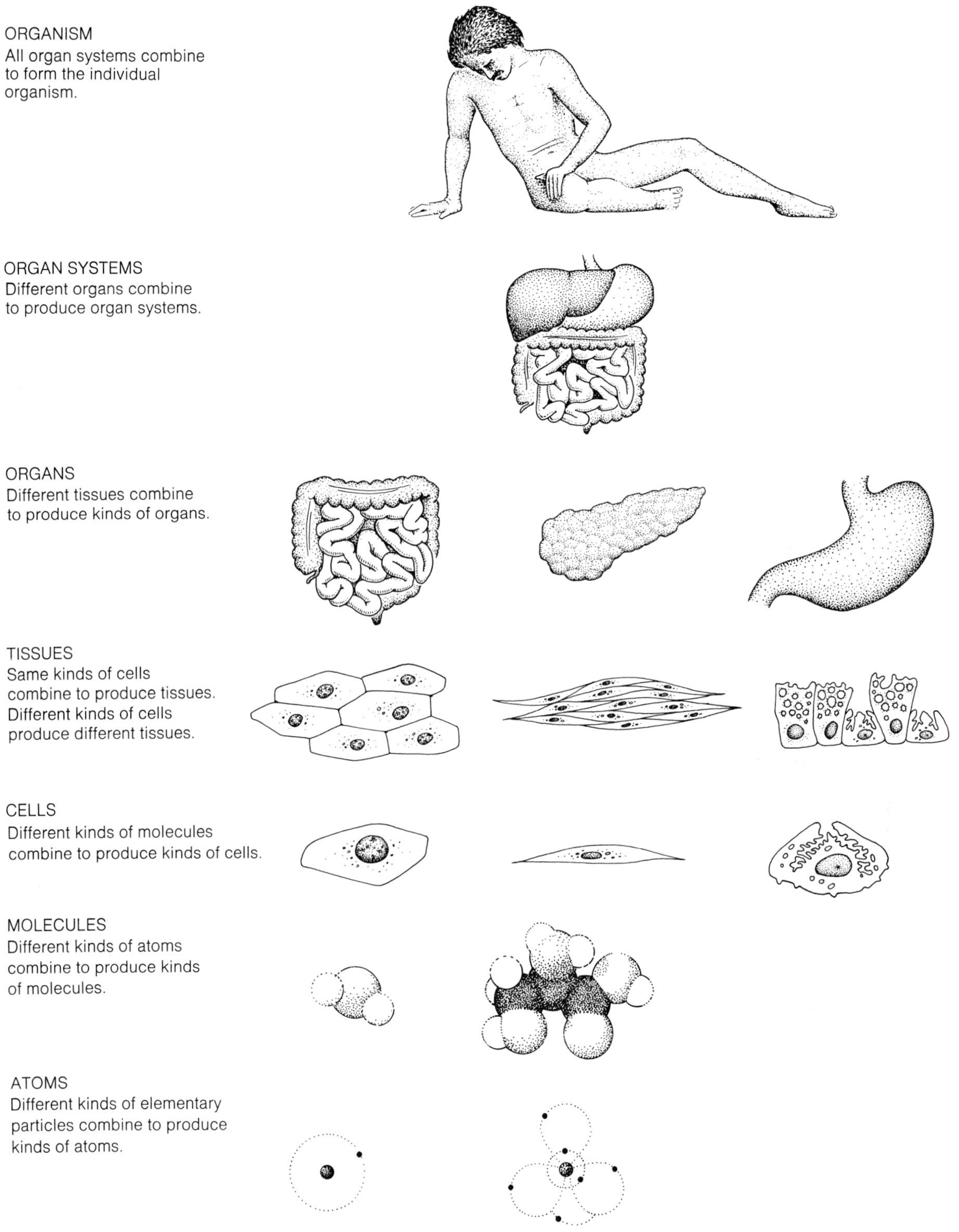

ORGANISM
All organ systems combine to form the individual organism.

ORGAN SYSTEMS
Different organs combine to produce organ systems.

ORGANS
Different tissues combine to produce kinds of organs.

TISSUES
Same kinds of cells combine to produce tissues. Different kinds of cells produce different tissues.

CELLS
Different kinds of molecules combine to produce kinds of cells.

MOLECULES
Different kinds of atoms combine to produce kinds of molecules.

ATOMS
Different kinds of elementary particles combine to produce kinds of atoms.

Figure 13-10 Levels of organization in complex animals.

MULTICELLULAR ORGANISMS

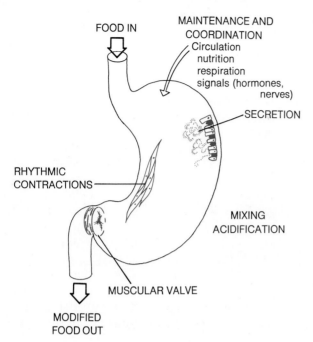

Figure 13-11 The stomach as an example of an organ. Specialized cells are organized into tissues to perform specific tasks of utility to the entire organism. In return the organ has a variety of services provided by other organ systems.

making up the stomach. Their actions are made possible and controlled by many other types of specialized cells. The stomach cells themselves must feed, obtain oxygen, and dispose of wastes; these functions are supported by blood vessels, constructed of specialized cells that serve to convey a vital stream of fluids and specialized blood cells to and from the stomach. The stomach must be controlled by other parts of the body and must be able to signal its state of health back to these other parts; such functions are served by other specialized cells, nerve cells, and by various chemical messengers produced by still other cells. Thus we add to our concept of the nature of an organ the idea that, in addition to the tissues performing its primary function, it contains others devoted to its own maintenance and integration with the organism as a whole.

ORGANS WITH COMPLIMENTARY FUNCTIONS FORM ORGAN SYSTEMS

A human stomach alone is not worth much; it cannot obtain food from the environment, and it cannot complete the digestive process. It is of value only when it is a part of a *system* of organs devoted to the common task of digestion. It must receive food from one organ, a mouth, which does chemical and physical work on it. Then, after doing its work, the stomach passes the food on to the small intestine, where further processing occurs and the process of uptake into the body principally occurs. Again, as with the individual organ, the component organs of the **organ system** require the services of coordinating and supply systems.

The Kinds of Organ Systems

From our knowledge of the requirements for cellular life and of the circumstances resulting from the housing of cells within a multicellular organism, it is a straightforward task to enumerate the classes of functions that organ systems must serve. There are five, as follows:

1. Reproduction. Sexual reproduction must have been among the first processes restricted to specialized tissues or organs when cells first became parts of organisms. Early in multicellular life, two lines of cells came into existence: **reproductive**, or germinative cells, form the link between generations and are housed and supplied by **somatic cells.**

2. Environmental Exchange. Only in the simplest organisms can the constituent cells obtain essential supplies and dispose of wastes by direct exchange with the environment. In all others, there are organ systems devoted to these processes. In complex organisms, in addition to the organ systems serving the primary functions of environmental exchange, there are others that serve as intermediators between the exchange process and the cells that are ultimately served. These include, for example, organs for storage and modification of ingested food, such as adipose (fat) tissue and the liver in the human.

3. Internal Transport. In multicellular organisms continual traffic is necessary between all cells. Each cell must be connected with sources of supply and sources of waste disposal, and each must exchange specialized products with other cells. In a substantial sense, this medium of internal transport is a substitute for the primitive ocean that bathed the earliest cells.

4. Internal Communication. While all operations conducted by multicellular organisms have antecendents in the activities of single free living cells, perhaps no group of multicellular functions are more elaborate derivatives of basic cellular functions than those serving internal communications and those associated with the next family of multicellular functions, adjustments to the environment. Both categories have, as their primary task, communication dedicated to maintaining on the one hand an optimum internal environment and on the other maintaining the entire organism in an optimal situation in the external world. The internal communication system involves collecting information regarding the state of the various parts of the organism and communicating this to other parts that require this information to guide their own activities. In many instances the internal communication routes are the same as those used by the internal transport system since chemical signals are often involved. In addition, these functions in animals are served by the nervous system.

5. *Adjustment to the Environment.* In animals the systems serving this function are, of course, directed by the nervous system; it compares what is required by the organism with what is available in the environment and then directs appropriate actions of various **effectors** such as muscles or cilia. We see these systems working in the lethal beauty of a cheeta overtaking its prey in a 45 mi/hr burst of speed; but we see them also in the steady growth of a plant towards the sun, or the searching downwards of its roots for water, processes without need for a nervous system.

THE TEN ORGAN SYSTEMS OF ANIMALS

As the functions of internal communication and adjustment with the environment suggest, we have reached a level of organization in which it is no longer practical to speak of plants and animals as one. Although their similarities at the cellular and molecular level are so extensive that they undoubtedly had a common evolutionary origin, plants and animals since their origins have travelled two different roads leading to extreme differences in organization at the levels of tissues, organs, and organisms.

As a result, although the five primary processes just enumerated certainly occur both in multicellular plants and animals, they are supported in each by radically different systems. Consequently we must divide the discussion of multicellular organ systems at this point. First, we will survey the organ systems supporting the basic five processes in multicellular animals, and then we will seek in higher plants evidences of similar systems.

Animals, beyond doubt, have a greater investment in organ system complexity than plants. However, it is equally important to study these matters in plants, not only because of the great importance and inherent interest of plants but for the less obvious reason that it is always instructive to compare radically different solutions to the same problems. The comparative approach shows what is truly fundamental about the organization of multicellular systems in contradistinction to idiosyncrasies associated with a particular style of life. For example, we have already noted the obvious, namely, that internal communication and environmental adjustment in plants is accomplished without a nervous system. Is then a nervous system not really essential to multicellular life, and, if it is not, what kinds of systems are truly essential to internal communication and adjustment to the environment?

In animals ten organ systems are recognized as follows:

1. **Circulatory system**—the circulatory fluid and cells, the containing vessels, and the muscular circulatory pump.
2. **Integument**—the outer covering, the nonliving shell it often produces, and associated structures, such as hair, feathers, and scales.
3. **Skeletal system**—internal or external supporting structures on which muscles act to effect movement.
4. **Excretory system**—structures that remove soluble wastes.
5. **Respiratory system**—gas exchanging surfaces that mediate between the circulatory fluid and the aquatic or aerial external environment, structures that promote flow of oxygen and carbon dioxide across these surfaces.
6. **Digestive system**—organs that prepare food for absorption by cells and dispose of solid wastes.
7. **Nervous system**—organs that detect environmental stimuli, monitor internal conditions, control effectors such as muscles and certain glands, and determine behavior.
8. **Endocrine system**—specialized cells and glands manufacturing and releasing regulatory chemicals called hormones.

Table 13-1 Participation of the Ten Organ Systems in the Primary Functions of Multicellular Organisms

Organ System	Primary Functions				
	Reproduction	Environmental Exchange	Internal Transport	Internal Communication	Adjustment to Environment
Circulatory	S	P	P	P	S
Integument	S	P	S	S	P[1]
Skeletal	S	S	S	S	P
Excretory	P[2]	P	S	S	S
Respiratory	S	P	S	S	S
Digestive	S	P	S	S	S
Nervous	S	S	S	P	P
Endocrine	P	S	S	P	P
Reproductive	P	NE	NE	NE	NE
Muscular	S	S	P	S	P

KEY: P = primary role; S = supportive role; NE = nonessential
[1] In its sensory capacity.
[2] Since reproductive systems commonly utilize excretory ducts.

9. **Reproductive system**—gamete forming tissues and related organs that facilitate fertilization and often development of the new individual.
10. **Muscular system**—organs made up of contractile cells that act to perform mechanical work under control of the nervous system.

The roles of these ten systems in the primary functions of multicellular animals are shown in Table 13-1. Although such a tabulation is useful for obtaining an initial impression of what an organ system does, it is misleading to the extent that all organ systems except the reproductive system are actually essential to maintenance of life.

Outline of Organ System Functions in Animals

CIRCULATORY SYSTEM

It has already been said that every cell requires its ocean, reflecting its evolutionary origins. Everything entering or leaving a cell moves by diffusion to and from the membrane of that cell. Since diffusion is dependent upon concentration gradients, in the final analysis, the role of the circulatory system is to promote optimal concentration gradients of critical molecules. It does this by bathing each cell in a benign solution, whose inorganic composition is much like that of sea water, and by rapidly circulating this solution between the cell and other structures, such as the digestive apparatus, lungs, or excretory organs, thus removing wastes and providing essential molecules. Not all multicellular organisms have a circulatory system. *Hydra* does not and neither do other simple organisms such as flatworms even though they have other complex organ systems (Figure 13-12). The physiological reason for this lies in the surface-to-volume ratio.

Surface-to-volume Ratio
Consider what happens when a balloon is blown up. As the balloon increases in size, both the volume of air inside the balloon and its outer surface area increase. However, reference to the two simple formulas for volume and surface of a sphere reveals that, as the diameter of the balloon is doubled, its volume will

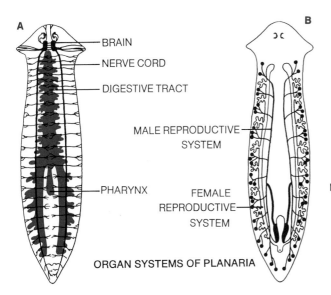

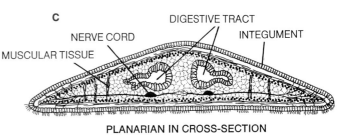

Figure 13-12 Although the flatworm is well endowed with organ systems (**A** nervous; **B** reproductive) it has no defined circulatory or respiratory system. Its flattened structure (**C**) insures that no cells are more than a mm from the exterior and makes these organs unnecessary in an organism with a low metabolic rate.

increase eight times, whereas the surface of the balloon becomes only four times greater (Figure 13-13A). In other words, the volume of the balloon increases much more rapidly than the surface. This tends to happen to every object that increases in size. Another way of expressing this difference in surface-volume growth rate is to say that the surface-to-volume ratio becomes smaller as objects become larger.

Surface-to-volume Ratio Limits Size

How is surface-to-volume ratio a matter of concern to cells and organisms? This question is answered easily if you will imagine a cell that grows larger and larger. A cell is a compartment through which material and energy must flow. That material and energy must flow in and out across the surface of the cell, thereby supplying the volume of living substance within. In addition each unit area of surface of the cell membrane is capable of a limited amount of flow. Thus, a growing cell inevitably would reach some size at which the vastly increased volume of living substance within could no longer be supported by flow across a membrane that is increasing in area less rapidly. Such a cell cannot survive because it has too much living substance and too little membrane.

This would not be true if the surface-to-volume ratio could remain constant as the cell grows larger. As we shall see, this is possible only in a very limited way. We have already seen that this situation is one of the causes of cell division. However, this solution when cells are parts of colonies or multicellular organisms has its limits because only the outer membranes of cells on the outside of the multicellular object are in contact with the environment. It is clear that becoming multicellular does not alter the limitations imposed by the surface-to-volume ratio.

How to Increase the Surface-to-volume Ratio

There are some ways of getting around this limitation—for example, by becoming very flat. If a lump of pie dough is rolled out with a rolling-pin on a breadboard, its total volume does not change much. However, as it becomes increasingly flat, its surface-to-volume ratio increases substantially. You can demonstrate this without resorting to the kitchen by visualizing a 1-cm cube sliced into four slices that are joined end-to-end; 6 cm² of surface have now changed to 10.5 cm² of surface. At the same time the volume remains unchanged (see Figure 13-13B). The multicellular flatworm of Figure 13-12 is sufficiently flat so that size does not introduce a serious surface-to-volume ratio problem; sufficient surface is exposed to the environment so the material and energy exchange necessary to support life can be maintained for every cell.

Even though the flatworm has gotten around the surface-to-volume problem by means of a severe restriction on form, it has been able to do so in part because it is limited to a watery environment. If removed from the water it dies from dehydration because its entire surface covering is water permeable, owing to the necessity of using the entire body surface for respiratory exchange.

Figure 13-13 **A.** The geometry of surface-volume is illustrated by measurements on a balloon blown up from 10 to 20 cm diameter. The graph shows how the surface-volume ratio falls as diameter of the sphere increases. The formulae show that this is because surface area increases only as the square of the diameter while the volume increases as the cube of the diameter. **B.** Certain shapes alleviate the surface-volume problem.

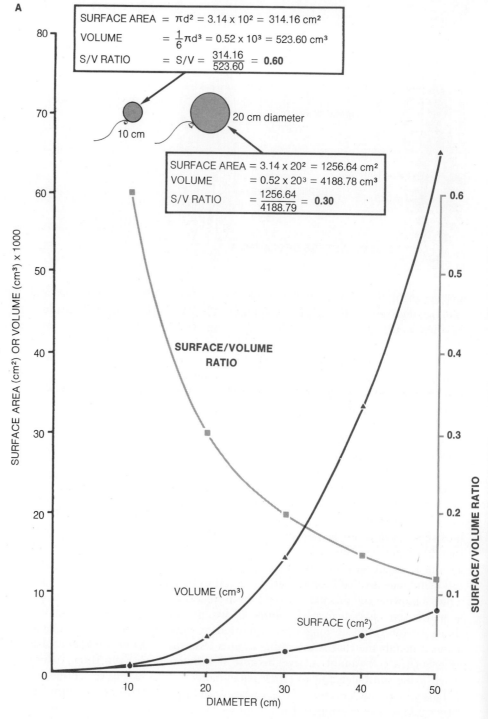

A Waterproof Container is Necessary

Even if enclosed in a waterproof coat, the planarian *still* cannot survive in a dry environment because the fundamental flow-through requirement of living systems is denied by the waterproof shell. If one now makes a hole in the waterproof surface to permit food input and another hole for waste output, and if food and water are made available in the dry environment,

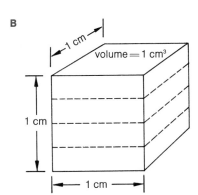
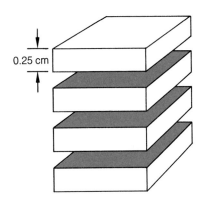

1. 1 cm cube:
   surface area = 6 cm²
   volume = 1 cm³
   S/V ratio = 6.0

2. Sliced into 4 equal parts, the total volume is still 1 cm³

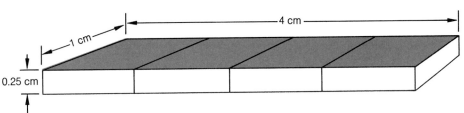

3. If the four parts are joined end-to-end, making a new and flatter shape, there is more surface relative to volume.
   surface area = 10.5 cm²
   volume = 1 cm³
   S/V ratio = 10.5

this difficulty can be met. However, the organism still will not survive because adding food at one point does not allow for distribution to all the cells. Similarly, waste removal at another point on the surface does not allow wastes to be removed from cells distant from that point. To put it another way, each individual cell in such a waterproofed, multicellular organism is no longer in direct contact with its "ocean" environment, one of the most important requirements of life.

Providing a Circulating Ocean

To get around this objection it would be possible to provide a tiny ocean inside the waterproof covering, but outside the cells, so that they would be bathed in it constantly. This small internal ocean would constitute an important element of a new internal environment—the *milieu interieur* described by Claude Bernard. In 1865, he wrote that animals

have really two environments: a milieu exterieur *in which the organism is situated, and a* milieu interieur *in which the tissue elements live.... The living organism does not really exist in the* milieu exterieur, *but in the liquid* milieu interieur *formed by the circulating organic liquid.*

This internal environment must be a relatively small volume of liquid. If very large, a prohibitively heavy outer waterproof coating would be required to contain it. The small volume of the liquid internal environment requires that it must circulate continuously and be continuously replenished and purified by digestive, respiratory, and excretory organs.

A Further Increase in Surface-to-volume

Circulation of this type would not solve all problems, for the organism, as we have designed it, could exist only so long as the multicellular mass inside the

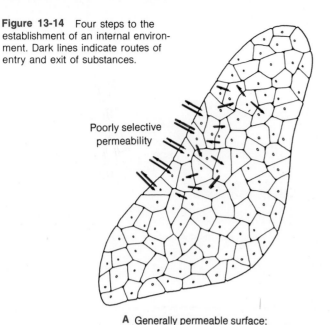
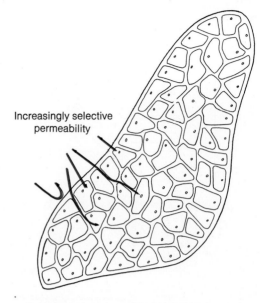

Figure 13-14 Four steps to the establishment of an internal environment. Dark lines indicate routes of entry and exit of substances.

A Generally permeable surface; cell-to-cell internal communication.

B Integument regulates exchange with environment; extracellular spaces improve internal communication.

covering maintained a suitable surface-to-volume ratio, that is only so long as it remained flat. If for some reason the internal mass of cells became bulky, then the internal environment could not adequately serve deeper lying cells of the internal cell mass. This problem is met by making channels, blood vessels or sinuses, through the cell mass so that the fluids of the internal environment can reach the surfaces of the innermost cells. The *milieu interieur* now bathes every cell of the organism (Figure 13-14A and B). Oxygen, food, and other necessary chemicals are brought to every cell and wastes are removed.

THE INTEGUMENT

The importance of the integument as a barrier to movement of materials in and out of the organism became clear in discussing the circulatory system. However, its presence introduces further problems. Thus the whole organism tends to behave osmotically, just like a single cell or like the osmometer already discussed (page 66). If the outer surface of the organism is a semipermeable membrane, then the organism will take up water when it is in fresh water or lose water when it is in a medium more concentrated than its own body fluids (Figure 13-15). As you might expect from the marine origin of life, the integuments of the simpler types of marine organisms do not have this problem because their body fluids are osmotically similar to sea water.

It must be obvious that one difficulty faced by organisms as they migrated from the sea to fresh water and to land must have been osmotic stress. Thus, on entering fresh water there would be a strong inward flow of water serving to dilute the internal environment to an impossible level. One way to offset this effect is evolution of an integument impermeable to water, but this incurs the disadvantage of making impossible the uptake of oxygen and disposal of carbon dioxide. These processes require permeable membranes. The result has been the appearance of a number of adaptations of the integument to maximize respiratory gas exchange while minimizing osmotic stress. Principally these have been the evolution of respiratory organs, which, in the final analysis, are specialized parts of the integument, and the establishment in respiratory surfaces of ion pumps. These help restore osmotic and ionic balance at the expense of metabolic energy by trapping essential ions from the dilute external medium.

The integument plays many other important roles. It may, in many organisms, be *both* integument and skeletal system, providing attachment points for the muscles used in movement. In others, it participates in processes that are primarily or in part the function

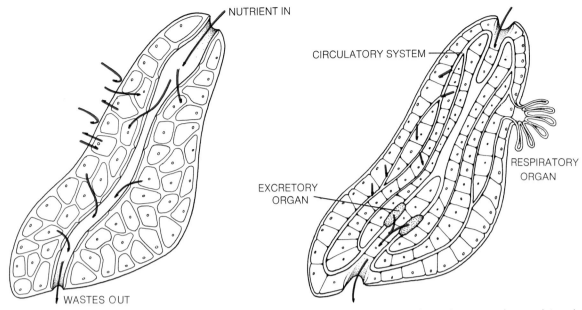

C Provision of specialized channels for nutrient.

D Circulatory, respiratory, and excretory organs improve internal communication and allow further reduction of exchange across integument.

Figure 13-15 A marine worm, a sipunculid (related to the annelids) is virtually a passive osmometer due to the properties of its integument. In **A** it is shown in a natural pose, feeding on bottom detritus with tentacles that surround the mouth. **B.** In a hypotonic (less concentrated than normal) medium water enters and the animal becomes distended while the opposite occurs (**C**) in hypertonic (more concentrated than normal) sea water. The sipunculid can survive both experiments if they are not prolonged.

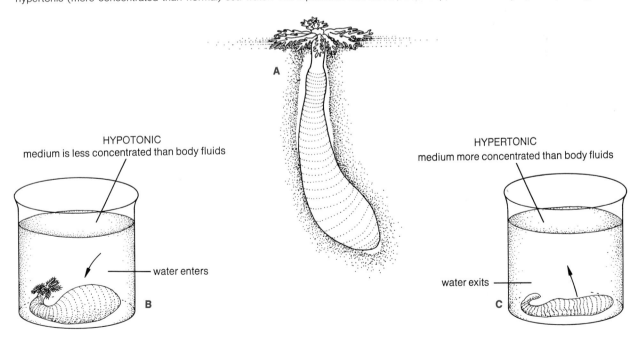

MULTICELLULAR ORGANISMS

of other organ systems. Excretion or temperature regulation in warm blooded animals are examples. Finally, the integument is the initial line of resistance to invasion by disease causing organisms.

RESPIRATORY SYSTEM

Although the simplest multicellular animals such as *Hydra* or the flatworm are small enough, and conduct their metabolism at a sufficiently low rate, to permit their generally permeable integument to serve in gas exchange, larger organisms usually have specialized gas exchange organs such as gills, lungs, or trachea. Gills and lungs are structures that promote gas exchange by exposing a large permeable surface on one side to the environment and on the other to circulating body fluids. Quite often gills and lungs have a muscular apparatus to promote rapid flow of water or air over the exchanging surface. This greatly improves the efficiency of gas exchange since, both at the surface of the respiring cell and at the surface of the gas exchanging organ—gill or lung, the limiting process is diffusion of oxygen and carbon dioxide (Figure 13-16).

The tracheal respiratory system of insects is a remarkable exception to the concept of a respiratory system in which the body fluids transport respiratory gases between the respiratory organ and the cells. The tracheal system consists of largely air-filled tubes that run from the outside directly to virtually every cell of the insect body. The circulating body fluids are not involved in the respiratory process except to a small degree in carbon dioxide transport since this gas is highly soluble in body fluids.

THE EXCRETORY SYSTEM

As we are beginning to appreciate, the tactics of many of the organ systems are directed to maintaining a stable internal environment. This is necessary because cells can tolerate very little change in the fluid in which they are bathed, perhaps, as we have suggested, as a consequence of the way life probably originated in the primeval sea. One of the great principles governing the existence of living multicellular systems is the constancy of the internal environment.

Opposing this necessary stability are four processes:

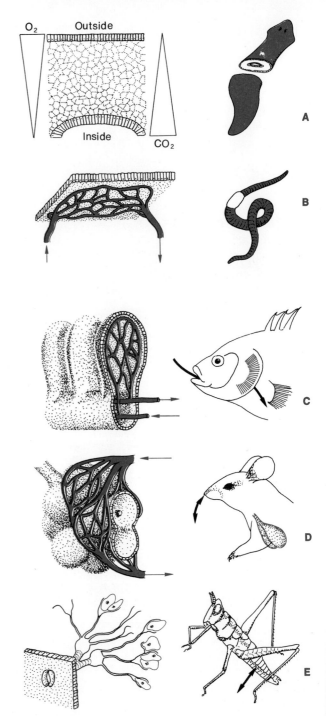

Figure 13-16 A sample of ways to promote respiratory gas exchange. **A.** Simple diffusion, as in the flatworm. **B.** Simple diffusion aided by a subintegumentary system of blood vessels which steepen the oxygen and carbon dioxide concentration gradients across the integument and aid gas exchange with deeper lying tissues, as in the earthworm. **C.** Gills. **D.** Lungs. **E.** Trachea.

1. Water access to the internal environment as part of, or mixed with, food.
2. Loss of water with waste elimination, where accumulation of wastes would also change the osmotic concentration of the internal fluid.
3. Gain or loss of water through respiratory organs.
4. Gain or loss of water owing to lack of perfection of the integument.

Since all four of these processes involve gain or loss of water, there is consequent danger of osmotic instability of the internal environment. Thus, the requirement for an osmoregulatory function is evident. There must be a means of controlling water relations with the outside environment so that the internal environment remains osmotically stable.

The Link To Excretion

Of the four processes just enumerated, only one offers much opportunity for control of water relations. An active multicellular animal has little chance to control the water content of its food and usually has little influence on water loss or gain through the integument or across gas-exchange surfaces. However, most organisms can exercise some control over the amount of water released with waste products. An organism that takes in too much water, and therefore needs to eliminate water to maintain the osmotic concentration of the internal environment, may restore balance by secreting a large volume of dilute urine. On the other hand, an organism that needs to retain water may eliminate wates in the water-saving form of a highly concentrated, even semisolid, urine, as do birds and insects.

Thus in most animals, the function of **osmoregulation** is intimately tied to the function of excretion, and the organs of excretion turn out to be also the principal organs of osmoregulation. These are known as kidneys in vertebrates, malpighian tubules in insects, green glands in lobsters, and nephridia of various

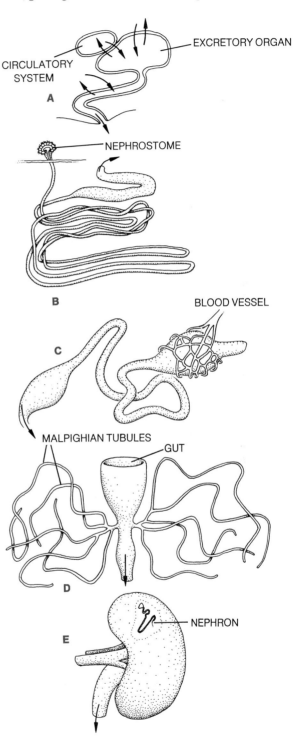

Figure 13-17 **Excretory systems. A.** The general principle: exchange is made either between the circulatory system or body cavity and the excretory organ which may further modify the excretory product before elimination. **B.** An earthworm nephridium. Two or more of these lie in each body segment and collect fluid from the body cavity through a ciliated nephrostome and convey it to the exterior through a convoluted duct which further modifies the urine before elimination. **C.** One of the pair of green glands of a crustacean in which the exchange is between the blood and the excretory organ. **D.** The malpighian tubules of insects collect wastes from the body cavity and empty it into the gut. **E.** The vertebrate kidney is composed of thousands of nephrons which exchange materials with the blood.

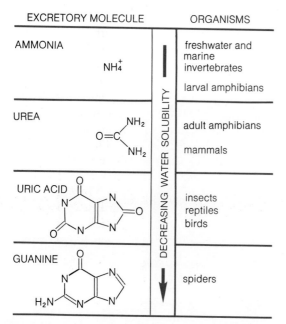

Figure 13-18 Molecules utilized in nitrogen excretion. The less toxic and the more water insoluble the excretory form of nitrogen, the less water must be wasted in its elimination.

kinds in other animals (Figure 13-17). There are exceptions to this rule. Osmoregulation is accomplished substantially by the respiratory organs of many aquatic animals, such as the gills of fish or the skin of frogs.

A major task of the animal excretory system is disposal of nitrogen compounds which appear as ammonia from breakdown of amino acids. Ammonia is poisonous and very soluble in water. To avoid self-poisoning, the organism must use a large amount of water in excretion. Fortunately there are less toxic and less water soluble molecules that can be synthesized from ammonia, and these are used in nitrogen excretion by animals that must conserve water (Figure 13-18).

SKELETAL SYSTEM

The animal that we might build using the organ systems so far described, would have no means of expressing that most characteristic of animal traits, reacting to environmental change with movement. The skeletal system, and the next two systems to be described, the muscular and nervous systems, provide that capacity. Simply the provision of muscles and nerves to control them is not enough to produce the capacity for movement of a useful sort; there must be a supporting framework to which the muscles can attach and, by pulling against, move parts of the organism. The skeletal system is that framework.

Three Kinds of Skeletal Systems

There are three types of skeletal systems.

1. High fluid pressure may be produced within cells and intercellular spaces of the organism with the result that the integumentary system will be distended like a balloon when injected with water under pressure as shown in Figure 13-19A. Such a means of stiffening is a hydraulic, or **hydrostatic, skeleton.** Many kinds of worms preserve their shape by this means alone, and by working their muscles against the pressure of their body fluids are able to move.
2. A rigid, armorlike external coating may be produced by the integumentary system. Such a structure is an **exoskeleton**, and its parts are usually jointed to allow movement (Figure 13-19B). Exoskeletons are characteristic of arthropods—insects, crabs, spiders, and their relatives.
3. Internal stiffening elements may be provided as a frame on which to hang the multicellular organism. Called an **endoskeleton** (Figure 13-19C), its stiffening members may be rigid, as are the bones in higher vertebrates, or they may be somewhat flexible, as in the cartilaginous skeleton of sharks. In either case internal joints provide for free movement.

Endoskeleton versus Exoskeleton

As you will see in Chapter 15, endoskeletons are characteristic of the great assemblage of animals called the **chordates,** which includes our species and such distant relatives such as the sharks, whereas exoskeletons are the trademark of the **arthropods,** which include our very serious competitors, the insects. The two groups of organisms are of interest because of the different ways in which they have solved the problems of life. To a large extent the type of skeletal system each group has evolved has influenced the nature of these solutions. For example, the skeleton must accomodate as an arthopod or a chordate grows. The internal skeleton of the chordate enlarges continu-

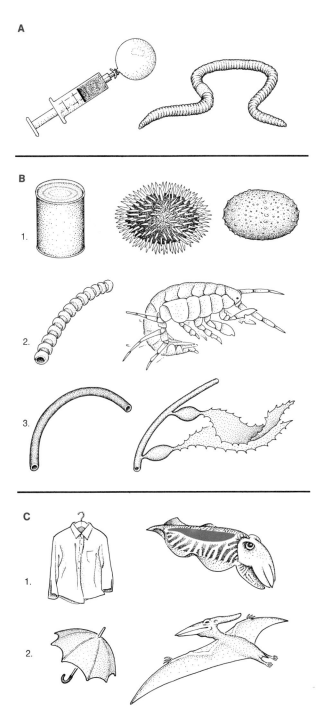

Figure 13-19 Skeletal systems. Left hand sketches illustrate the underlying physical principles found in the examples shown on the right. **A.** Hydraulic skeleton in which the body muscles work against each other *via* the internal fluids. **B.** Articulated and non-articulated skeletons in (1) sea urchin, (2) arthropod and (3) kelp. **C.** Endoskeletons, (1) non-jointed in cuttlefish and (2) jointed in a vertebrate.

ously during body growth by new growth and replacement of cartilage and bone. An arthopod cannot do this because the outer part of its skeleton is dead. Consequently it must grow in spurts, and, at the end of each spurt throw off the old, dead skeleton and form a new one. This process, called **molting**, is a delicate phase in life because the newly molted animal is soft and defenseless for a time. The exoskeleton also limits the size of an animal. This is another facet of the surface-to-volume problem; as the arthropod grows, its volume increases faster than the area of exoskeleton that must support it. The largest arthropod known, a 12-ft long scorpionlike eurypterid, lived in the sea, where bouyancy partially offsets gravity (Figure 13-20). Fortunately, this was 400 million years ago; we have nothing to fear from the giant flies and preying mantids of science fiction.

MUSCULAR SYSTEM

Animals are characteristically organisms in motion, whereas plants are only rarely capable of rapid movement, except of course for small algae, like *Volvox*, which move by means of flagella. Larger plants are poorly motile because their cells are typically encased in unyielding cellulose cell walls. These become firmly attached together in multicellular plants to produce structures too inflexible for rapid changes in shape. When plants are capable of movement, the causal agents are comparatively slow, either differential growth or osmotic swelling or shrinking of cells. Animals are readily capable of movement because of their pliable cellular structure and because they possess a remarkable set of muscle proteins capable of doing mechanical work using energy provided by ATP.

A diagram of the organization of these proteins as they occur in the most specialized of movement producing cells is shown in Figure 13-21. These are **striated muscle** cells, or, as we call them in the human, **voluntary muscle** cells because they are under conscious control of the nervous system. The same or similar proteins are present in **smooth muscle** cells, which, in the human, function in such places as the walls of blood vessels and the intestinal tract. Contractile proteins similar to those in muscle are found even in dividing animal cells, where they produce the constriction dividing the cell in telophase.

How such proteins produce rapid and controllable

MULTICELLULAR ORGANISMS 203

Figure 13-20 A Eurypterid, an extinct arthropod relative of the horseshoe crab. Members of this group approached the limits of size imposed by possession of an exoskeleton.

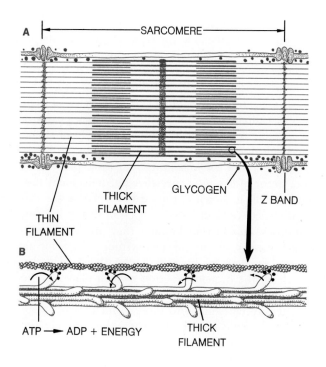

Figure 13-21 The chemical machinery of muscle contraction. **A.** One sarcomere, the basic unit of the muscle fiber. Thin filaments are rooted in the Z-bands, which are terminal structures of the sarcomere, and interdigitate with thick filaments. **B.** Detail of thin and thick filaments to show the cross bridges of myosin that connect them. Under control of calcium ions the myosin heads shift their attachment points on the thin filaments. This bending motion causes movement of the thick and thin filaments relative to each other, pulling the Z-bands together and shortening the muscle. ATP is used by myosin head ATP-ase to provide energy necessary to restore the cross link to its initial state for another contraction cycle.

movement is one of the most interesting chapters of molecular biology. If large scale motion is to be produced by molecules, without resorting to some kind of fuel burning engine, we must limit ourselves to such forces as the attraction and repulsion of electrical charges and chemical changes that alter the shape of molecules. Acting without coordination, these processes are ineffectual at moving any large mass, but, when they are properly harnessed in large numbers of contributing units, very significant amounts of work can be done. In muscle the coordination of work-producing molecules is readily apparent in the microscopic structure (Figure 13-21), which shows long interlocking thick and thin filaments of protein. The **thick filaments** are made up of many molecules of the protein myosin, whereas **thin filaments** contain three kinds of protein, troponin, tropomyosin, and, in the largest amounts, actin. Each myosin molecule has an end that can attach to actin, swivel slightly, break contact, reattach at a new point, and repeat the cycle. The result is to move the thick and thin filaments relative to each other. Since many such sets of thick and thin filaments work together and cooperate with other cells containing the same system, appreciable work can be done. Thus, in the final analysis, the work that we do when we move our muscles is produced by the bending of myosin molecules, in other words by a reversible change in the tertiary structure of protein.

The architecture of muscle systems is highly varied. In lower organisms, such as earthworms, sets of muscles work in opposing groups by contracting against the hydrostatic skeleton (Figure 15-33). A similar arrangement of smooth muscle in the human gut produces the peristaltic waves that propel the food along. In animals with exoskeletons, controlled movements of the body parts are produced by muscles that run from one body part to another. Similarly, in vertebrates with internal skeletons, muscles attach across joints to separate elements of the skeleton so that movement is caused by their contractions.

In these instances it is particularly apparent that muscles only contract; they cannot push. There is no biochemical reason why muscles cannot push. If you look at the diagram of how myosin is arranged in striated muscle (Figure 13-21) you will see that a very slight rearrangement would cause the muscle to extend rather than contract. The reason that muscles always pull, of course, lies at a higher level of organization; a muscle has no longitudinal rigidity so that trying to do work by extending a muscle would be analogous to trying to push something with a rope. This limitation causes no particular difficulty because muscles are usually arranged in opposing pairs to produce opposite motions. An interesting exception occurs in spiders whose leg joints are provided with only one muscle each. This muscle causes movement of the limb towards the body. Extension, as in jumping, is caused by sudden pumping of body fluid into the limb. Thus, the exception is only superficial because the single leg joint muscle actually works against whatever muscles produce the sudden increase in body fluid pressure.

NERVOUS SYSTEM

The nervous system and the endocrine system have somewhat overlapping functions. Together they are responsible for nearly all coordination of the activities of the various organ systems. In some of these coordinative processes, parts of the two systems work so closely together that we speak of a neuroendocrine system. In addition to internal coordination, the nervous system is responsible for providing information about the environment to the organism and for directing the behavior of the organism.

The basic component of the nervous system is the nerve cell, or **neuron.** As shown in Figure 13-22, the neuron is specialized to receive information from another neuron, or from the internal or external environment, and to transmit this information to other

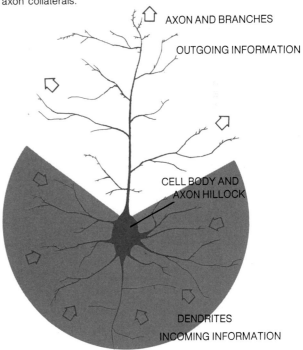

Figure 13-22   A neuron. Information gathered from many dendrites is channelled to other cells through the axon and axon collaterals.

MULTICELLULAR ORGANISMS

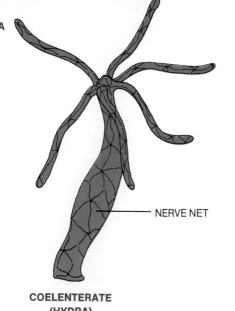

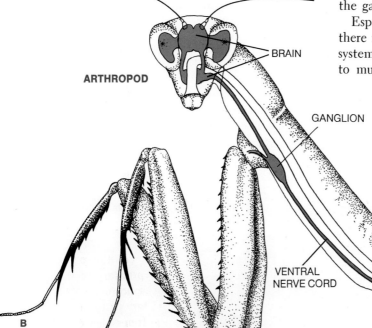

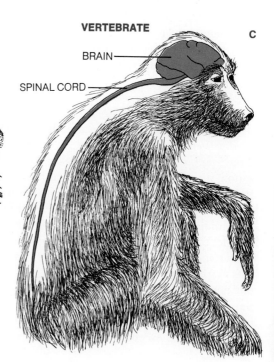

neurons or to control the action of effectors such as muscle. The detection of information from the internal or external environment is called a **sensory function**, and neurons specialized to do this are called **sensory neurons**. In some instances cells other than neurons are the sensory cells, and they communicate their information to nerve cells.

Typically, neurons transmit information, sometimes over long distances within the organism, by electrical signals traversing long, thin processes of the cell called **axons**. At the end of the axon there is a specialized structure called a **synapse** where the axon contacts either another nerve cell or an effector cell. The synapse is characterized by a narrow gap between cells, and the nerve signal usually crosses this gap by means of a **transmitter chemical** released on the arrival of the nerve signal. When a nerve controls an effector, for example, a muscle cell, there is a similar synaptic structure called a **motor end plate** interposed between the nerve and the muscle. As in the synapse, the arrival of the nerve impulse allows a chemical to cross the gap between cells and activate the muscle cell.

Especially in higher organisms, it is obvious that there must be more to the operation of the nervous system than the flow of nerve signals from sensory cells to muscle cells. Between sensory input and motor

Figure 13-23 Types of nervous systems. A. A diffuse nerve net, lacking a primary center, as in *hydra*. **B.** The highly advanced nervous system of an insect is organized into brain and nerve cord and provided with effective sense organs. **C.** The primate nervous system differs most fundamentally from all other nervous systems in the extreme development of the brain.

206 BIOLOGY

output in most animals lies an immense collection of nerve cells in what is termed the **central nervous system**. In the human this includes the **brain** and **spinal cord**.

The simplest kind of nervous system is that which occurs in *Hydra*. There is no particular concentration of nerve cells at any spot, no brain in other words. A diffuse network of nerve cells, called a **nerve net**, extends throughout the animal and suffices for its simple behavior (Figure 13-23). Among higher animals, the nervous systems of arthropods and vertebrates are similar in function but markedly different in structure. As shown in Figure 13-23, the arthropod nervous system runs along the ventral surface of the body and consists of a solid chain of clumps of nerve cells, **ganglia**, connected by longitudinal bundles of axons. Basically, each ganglion operates the part of the body in which it resides, with the central part of each ganglion serving as a switchboard to connect the nerve cell bodies lying in the outer zones of the ganglion. By contrast, the central nervous system of the vertebrate is hollow, lies on the dorsal side of the body, and is immensely specialized into a series of huge ganglia at one end, the brain. However, along the spinal cord there are additional ganglia that contain the cell bodies of sensory cells. Thus, although the two systems look rather different, they work in a similar fashion.

ENDOCRINE SYSTEM

Control by the nervous system is principally short term and is usually directed towards a limited number of effectors, a particular muscle or a single organ such as the heart. Obviously there are other phenomena, such as overall growth of the body or intensity of metabolism, that require regulation of a very large fraction of the cells of the organism over very long periods of time. Processes of this type are controlled by the endocrine system. In the human the endocrine system consists of ten recognized **endocrine glands**, which achieve their effects by releasing special chemicals called **hormones** to the body by way of the circulatory system (Figure 13-24).

Regulation of hormone release is a most complex matter. In some instances a gland such as the thyroid, whose hormone thyroxin regulates metabolic rate, is itself controlled by a hormone from another gland. In

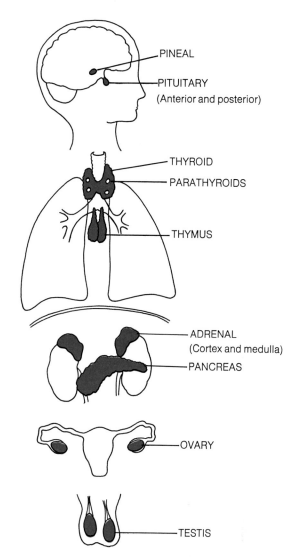

Figure 13-24 The location of the elements of the human endocrine system.

the instance of the thyroid gland, the controlling endocrine gland is the anterior pituitary, which regulates the thyroid by releasing a hormone known as thyrotropin. The anterior pituitary performs so many endocrine functions that it is often called the master gland. It is particularly important as the principal point of interaction between the nervous system and the endocrine system because the pituitary gland itself is regulated in large part by a nearby region of the brain called the hypothalamus.

A most interesting example of how hormones control the development of form occurs in insects (Figure

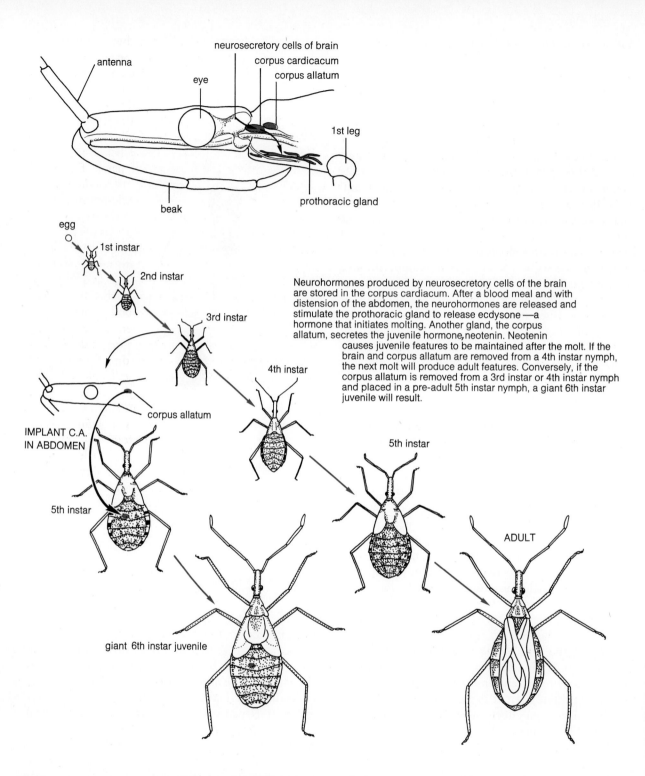

Figure 13-25 Experiments showing how the roles of two hormones in determining the character of a particular molt are ascertained. Upper figure shows the positions of the relevant organs in the anterior region of the experimental insect, a blood-sucking bug, *Rhodnius prolixus*. (From the experiments of V. B. Wigglesworth, Cambridge University.)

13-25). As an insect grows it **molts** a number of times. Finally it undergoes a molt in which it becomes sexually mature, as demonstrated by the appearance of functional sex organs and wings. Each time the insect molts, up until this final molt, the concentrations of two hormones, a growth hormone and a juvenile hormone, determine the nature of the molt, that is, whether it will be one producing another immature stage or whether it will be one producing a mature insect. This is readily demonstrated by cutting out the gland that produces the juvenile hormone, whereupon the insect matures at the next molt, even if it is far too small to be a normal adult. Conversely, providing several extra sets of the juvenile hormone producing gland to an insect about to become an adult causes it to molt into a juvenile form once more.

REPRODUCTIVE SYSTEM

Cells destined to serve in reproduction, as male and female gametes, are set aside early in the embryonic life of multicellular animals. The remaining somatic cells generally play no direct role in reproduction. However, they do form the reproductive system that nurtures the gametes, assists them in coming together in the act of fertilization, and oftentimes supports the developing embryo. Some organisms may have both male and female reproductive systems together in the same individual, in which case they are said to be **monecious,** or the sexes may be separate, the **dioecious** condition. In a few instances, an organism may initially function as one sex and later as the other.

The primary organ of the reproductive system is the **gonad**—testis or ovary—in which the gametes develop and which may also serve to produce sex hormones. These hormones influence sexual behavior and may even determine the structure of parts of the body related to sexual reproduction or to sexual behavior. The structure of the remainder of the reproductive system is related to the manner of delivery of gametes and the extent to which the embryo is assisted in its development by the female parent. In many simple marine organisms, the gametes are merely liberated into the sea, although this process is frequently synchronized by some external event, such as the phase of the moon, to increase the likelihood of male and female gametes finding each other. Terrestrial organisms, on the other hand, must resort to direct fertilization in which the sperm are directly introduced into the reproductive system of the female to fertilize the eggs before they are liberated. The result is the development of elaborate structures in both male and female to accomodate this process. The final stage in the evolution of reproductive systems was marked by the appearance of structures in the female reproductive system to shelter the developing embryo and later even to nourish it directly.

14 Organization of Higher Plants

The last chapter considered the problems and advantages of multicellularity in *animals* and examined the organ systems that evolved in response to the challenges of the multicellular state. We now ask, how did *plants* respond to these challenges? This question is of interest because it is already clear that plants differ markedly from animals in important ways: in their nutrition, in their lack of a nervous system and in their mechanisms for rapid movement. So we ask, do such differences free plants from any of the problems of multicellular life as seen in animals, or do they create new ones? To seek answers we shall study the most complex plants, the familiar seed plants of land—trees, grasses, vegetables, and so on. We know that these plants have solved the problems associated with multicellularity because even the most superficial examination shows their great complexity, their ability to attain great size, as attested by the sequoias of coastal California, and their abundance in terrestrial and fresh water environments. Thus in the seed plants we expect to see the most definitive solutions to the problems of multicellularity of which plants are capable.

INPUT AND OUTPUT RELATIONS OF A SEED PLANT

The first step in our analysis is to determine the kinds of transactions a land plant has with the environment. From Chapter 6 we know that a plant carries on photosynthesis, and thus we know it requires sunlight and carbon dixoide, from which it produces sugar and oxygen. It is immediately evident that other transactions with the environment must take place. Since a plant must have enzymes to conduct photosynthesis, as well as the multitude of other enzymes and proteins essential to life, we must look for environmental exchanges involving nitrogen, phosphorous, and other elements. Since the protoplasm of the plant contains as much water as an animal cell, we must look for transactions involving water. Where do these substances come from? How are they accumulated from the environment? How does the plant prevent excessive loss of these valuable substances to the environment?

EARLY OBSERVATIONS ON WHOLE PLANT METABOLISM

Around the year 1600, a Belgian, Jean Baptiste van Helmont, made the first truly scientific steps towards understanding such problems. With great care he weighed a young willow tree and planted it in a tub with 200 lb of dry soil (Figure 14-1). For 5 years he supplied the tree with rain water, and at the end of that time weighed the tree and dried soil again. The soil still weighed almost 200 lb, less about 3 oz, but the willow had increased from its initial weight of 5 lb

Figure 14-1 An early experiment in plant metabolism.

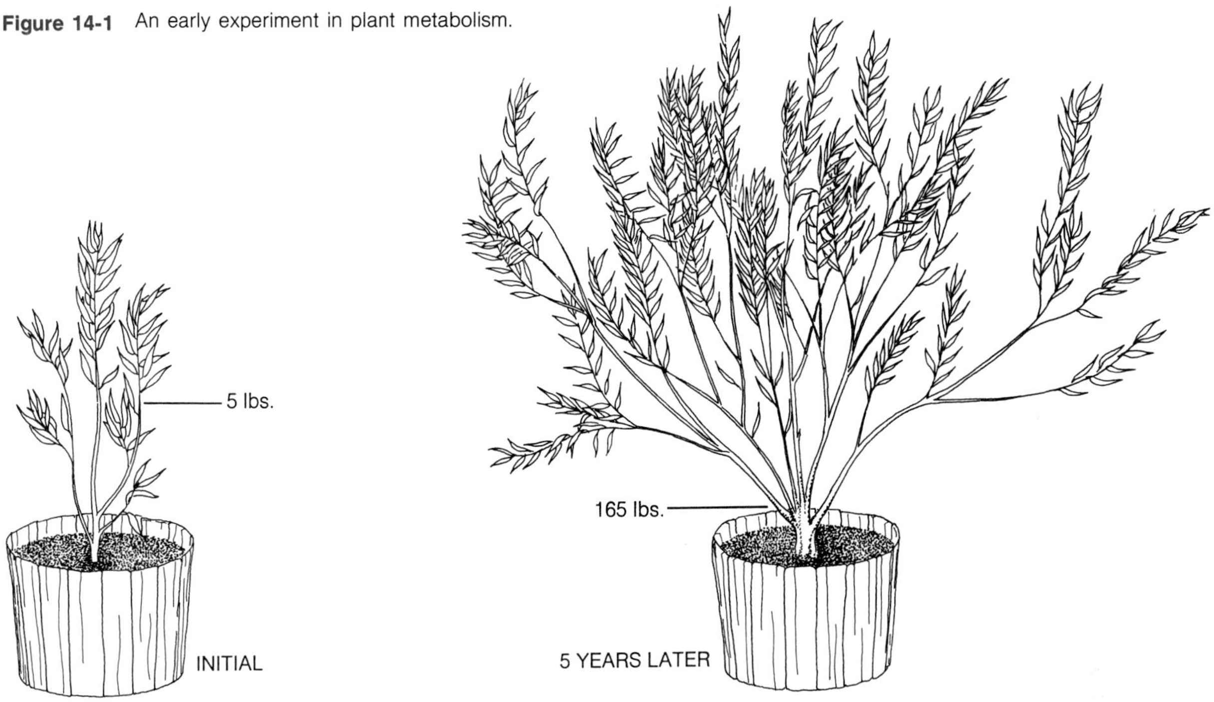

INITIAL — 5 lbs.

5 YEARS LATER — 165 lbs.

to 164 lb. Van Helmont was of an unusual philosophical turn of mind. He believed that water was the basis of all matter and took his experiment as proof of his point, that the rain water supplied to the tub had produced the substance of the willow. Although his theory was wrong, we recognize his experiment as the first to show quantitatively that the substance of a green plant must largely come from sources other than the soil in which it is rooted.

More than a hundred years passed before the next important steps were taken in unravelling the mystery of how a plant gets its food. Then the English chemist, Joseph Priestly, showed that air rendered incapable of supporting animal life by the burning of a candle could be restored to normal by a green plant. A Dutchman, Jan Ingen-Housz, refined Priestley's experiments, showing how the purification process required the plant to be in sunlight and by showing that the purified (reoxygenated) air came from the leaves. He did this cleverly by collecting and testing the gas given off by leaves submerged in water.

A few years later, when the nature of oxygen and carbon dioxide were more fully understood, a Swiss botanist, N. T. de Saussure, completed this chain of fundamental investigations by growing plants and plant parts in weighed and analyzed quantities of air. He determined the relationship between carbon dioxide taken in and the oxygen given off in light and measured the oxygen consumed in darkness. He was able to show quantitatively that most of the mass of the nutrition of the plant came from the air, but he also established the necessity of mineral constituents from the soil.

Other investigators soon demonstrated that sugar was synthesized and rapidly converted to starch in the illuminated parts of leaves and that there had to be routes for transportation of sugar to other parts of the plant and for water and other materials to be transmitted from the roots to the remainder of the plant (recall Figure 6-6).

These early investigations clearly showed that the plant, insofar as its metabolism was concerned, exhibited regional specialization of function accompanied by specialization of structure. Photosynthesis takes place in leaves, transport in stems, and uptake of water and minerals in roots. It is unmistakable that plants have taken advantage of the opportunity for cellular specialization of function that comes with multicellularity and that they have solved one of the principal problems resulting from such specialization, namely, the necessity for transport between regions of specialized function.

The Organs of Plant Metabolism

THE PLANT CELL

Compared with animal cells, plant cells assume a more limited variety of forms and tend to have a larger array of functions. Certain facts regarding the structure and water relations of plant cells must be stated in order to understand the variety and organization of the tissues and organs that they form. You recall that the plant cell characteristically is bounded just outside the plasma membrane by a nonliving cellulose cell wall. This is composed of innumerable cellulose fibers of tremendous strength, equal, in fact, to the strength of steel fibers of the same diameter. This wall allows the plant cell to exist in contact with water of lower ionic content than could be tolerated without an energy requiring compensatory mechanism.

In place of a mechanism to maintain a reasonable osmotic equilibrium by, in effect, pumping out water as it enters, the plant cell puts to use the pressure that builds up as the accumulated volume of entering water swells the cell against its inelastic wall. This pressure is called **turgor pressure**. You will recall from the discussion of osmosis in Chapter 5 that, when the turgor pressure equals the osmotic pressure (that is, mechanical push out equals diffusional push in), there is no further net flow of water molecules across the osmotic membrane, in this instance the plasma membrane of the cell. Exchange of water molecules across the membrane in both directions will continue, but the net flow will be zero until either the ionic concentrations on one side of the membrane change or until the volume of the restraining cell wall changes. Under ordinary conditions this pressure is quite high in plant cells, about 75 lb/m^2 of cell surface. The collective turgor pressure of the cells in a plant contributes in a major way to its structural strength, as all cells push out to distend their cell walls, giving the plant its rigidity. This accounts for the crispness of fresh celery; when it wilts, the cause is loss of cell turgor pressure.

Thus the cell wall eases the osmotic problems of plants in fresh water or on land and contributes to the physical support of the plant. The cell wall also limits the kinds of structural elements that a plant can build. Most obvious is the fact that the plant is highly restricted in the rapidity with which it can perform movements. In fact, we shall see that most plant movements are accomplished by the slow process of cellular growth rather than by rapid changes of the shape of cells, the technique used by animals. Consequent upon this, we find that plants cannot be "mechanical" in design. For example, they cannot have pumps, as in the circulatory systems of animals, or have bellows-type lungs, or have any kind of food processing system, for example, to move soil *en masse* through organs to extract nutrients. The result has been a most artful reliance on chemical and physicochemical ways to accomplish many of the same purposes by actions at the sub-cellular and cellular level.

PLANT CELLS INTO PLANT ORGANS

As a plant grows, at least in the early stages, new cells are produced at two kinds of sites, the growing tips of shoots and root tips (Figure 14-2). At the growing shoot tip, the youngest multiplying cells form the **apical meristem**. As apical meristem cells multiply and grow, the shoot increases in length and the cells that are left behind undergo changes, or differentiate, into the characteristic cell types of the shoot. A little farther down in an older part of the shoot, a cross section (Figure 14-3) shows these cell types. At the periphery there is an epidermis, which produces a waxy outer surface to limit water loss and provide mechanical protection. In the center there are relatively undifferentiated cells, which form the cortex or pith. Around the periphery, beneath the epidermis, there is a ring of tissue still able to divide and increase the diameter of the plant. This tissue is called the **vascular cambium**; in dividing it produces two extremely important types of cells that form the conducting system of the plant, **xylem** cells, which transport water and inorganic ions, and **phloem** cells, which transport food. Cells from the meristem also at intervals produce leaves, which contain other specialized cell types. These include **parenchyma** cells, in which

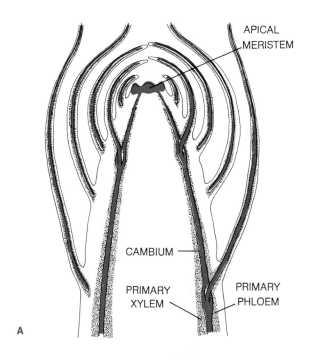
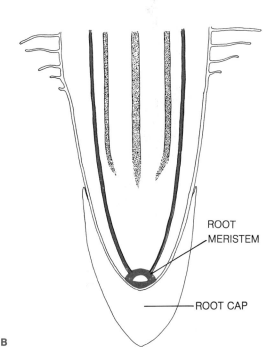

Figure 14-2 Primary growth zones in a seed plant.
A. Longitudinal section of a leaf bud. **B.** A root tip. In both, meristematic cells originate all the major cell types found in more mature regions of the plant.

nearly all of the photosynthesis of the plant occurs, and specialized **epidermal** cells, which, in pairs, form **stomata,** valve-like openings governing the access of air to the inside of the leaf.

As plant cells differentiate, they typically develop thicker cell walls by laying down secondary wall materials, often of lignin, an extremely tough waterproofing molecule. Cells destined to become purely supportive elements of the plant fill almost completely with lignin and die, forming wood. In the remainder of the plant's cells, secondary wall formation does not result in isolation of cells from each other. They retain the cytoplasmic connections that were left behind by incomplete cell wall development at cell division or form new connections. Called **plasmodesmata,** these serve as important avenues of exchange between cells.

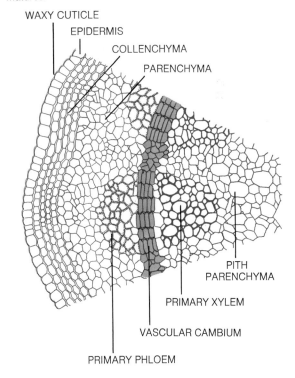

Figure 14-3 Tissue organization in a cross section of a young plant stem. Cell division in the thin cylinder of vascular cambium produces secondary phloem and xylem (not shown), additions to the primary conducting tissues. As girth of the stem increases, rupture of the epidermal region evokes "bark" formation by divisions of cells in the parenchyma. Thick walled, supportive collenchyma cells also appear as the stem matures.

ORGANIZATION OF HIGHER PLANTS

Xylem and phloem cells form long conducting tubes that run continuously from the leaves into the roots. Xylem cells form long tubes composed of stacks of xylem cells piled one atop the other, connected by perforated walls. The cells then die, but the tubes retain their conductive function. Phloem cells behave somewhat similarly, forming elongate stacks of cells that communicate by **sieve plates,** although they retain some cytoplasmic functions. The nucleus disappears and the volume of the phloem cell is nearly filled with a vacuole with remnants of cytoplasm around the periphery. Phloem cells are metabolically supported through plasmodesmata by adjacent **companion cells** with normal cytoplasmic contents. The central space of the phloem cell contains linear strands of fibrillar protein, the P-protein, which may play a role in its conductive function.

Development of the root structures is highly similar to what we have seen in the shoot. At the growing tip of the root there is a zone of meristem cells undergoing rapid cell division. Some of these form a protective cap for the root as it pushes through the soil. Others, divided off in the direction away from the tip, grow by elongation, providing the force that pushes the tip ahead, and undergo differentiation into epidermal, xylem, and phloem cells. Behind the zone of elongation, epidermal cells send out **root hairs** into the surrounding soil. The root hair is a delicate and highly permeable extension of a single epidermal cell that survives only a few days.

Leaves form as specialized outgrowths of the apical parts of the plant. When fully differentiated, they contain xylem and phloem elements that are readily visible as the veins of the leaf. Smaller subdivisions of these conductile elements extend so completely through the leaf that no cell of the photosynthetic parts of the leaf is more than a few cell diameters away from them. The photosynthetic part of the leaf is a central zone of loosely packed **mesophyll** cells, filled with chloroplasts and having free communication with the outside air by way of air paths that ramify among them and connect with the stomata.

Operation of Plant "Organs"

We now have a sufficient picture of the organization of a higher plant to consider how it works. Our principal theme will be to see how the basic functions of a multicellular organism are accomplished by a typical plant, using the animal as a comparative guide.

THE PRODUCTION, UPTAKE, AND DISTRIBUTION OF MATERIALS

Because plant cells contribute to so many different functions, it is difficult to define precisely the anatomical boundaries of specific functions, as we may often do in complex animals. In an animal, for example, the circulatory system is precisely locatable and easily distinguished from other systems, such as the excretory or digestive systems. In the plant we shall see that the analog of the animal circulatory system is clearly evident as xylem and phloem vessels, but the "pump" that drives fluid through this system consists of cells whose primary functions are not transport. Thus, it is better to discuss plant functions in somewhat broader terms.

Let us begin with a simple experiment. In a tree the phloem is situated around the periphery of the trunk with the xylem towards the center (Figure 14-4). This provides the opportunity to block transport through the phloem selectively by girdling the tree and thus to determine what xylem, as well as phloem, transports. After girdling, the tree ultimately dies, but before it does sugar accumulates above the cut and becomes depleted below. However, water continues to move upwards to the leaves until the root tissues starve. Since the phloem was the only transport channel interrupted in this experiment, we conclude that the upward flow of water from the soil is through the xylem and that sugar synthesized in the leaves moves through the phloem.

From such observations, the organization of the

Figure 14-4 A shallow cut around the trunk of a tree kills by interrupting nutrient flow in the phloem.

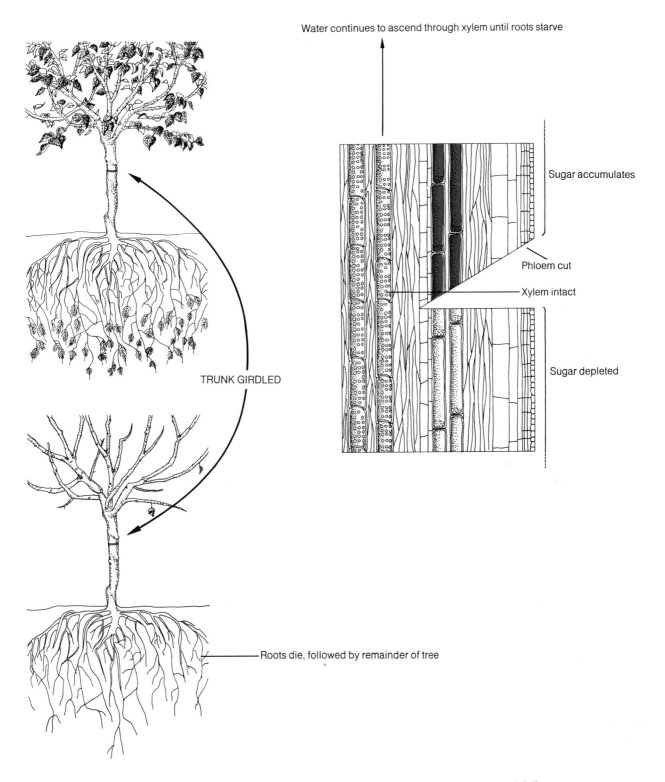

ORGANIZATION OF HIGHER PLANTS

plant transport system begins to emerge, but there remain two principal difficulties: (1) the "pump" or driving force moving fluid in the system is not yet evident, and (2) showing that fluids flow in one direction in one set of channels and oppositely in the other does not demonstrate that the two sets of channels are joined in a circulatory system. The experiment does not show that the bulk of the transported fluid undergoes a complete circuit through the system, as in animal circulatory systems. That is, flow in xylem and phloem may not be linked, so for the moment it is prudent to refer to the two pathways as transport systems, a term free of the connotation of circulation.

ACTIVE TRANSPORT AND OSMOSIS POWER PLANT TRANSPORT SYSTEMS

As we have said, by reason of its cellular structure the plant cannot construct mechanical pumps; instead it relies on the power resident in two physicochemical processes characteristic of cell membranes, osmosis and active transport. Two groups of cells at opposite ends of the plant provide the necessary power. They are the photosynthetic cells of the leaves and the cells in the root hair zone.

Pressure generated by the root is readily demonstrated by the flow of water from the cut stem of a decapitated plant when its roots are watered. Examination of a cross section of a root (Figure 14-5) shows how this flow probably comes about. The root can be thought of as two concentric cylinders of cells. The inner cylinder contains xylem and phloem and is separated from the outer cylinder by a ring of tightly connected endodermal cells. In both the inner and outer rings of cells, water can flow through extracellular spaces as well as through the cells, except at the endodermal ring. There the cells are tightly connected by an impermeable material that limits water flow to the route through the endodermal cells.

Given this arrangement, let us see how water moves from the soil into the xylem. The overall process clearly depends upon the osmotic concentration of the soil water as compared with the osmotic concentration of the root cells. This is experimentally shown by supplying the decapitated plant with water of sufficiently high salt content, whereupon water flow from the decapitated stem ceases. Under ordinary conditions the cytoplasm of the root hair cells has a higher concentration of solutes than does the surrounding soil water and, since the cell membranes of root hair cells are semipermeable, water enters them, flowing down its concentration gradient. It follows that the same would occur between peripheral cells of the root and deeper lying cells; as peripheral cells accumulated water from outside, they would become more dilute than deeper lying cells and the inward migration of water would continue. Finally, water would cross the endodermal barrier by the same means and be available to enter the conducting elements of the inner ring. This process is augmented by the fact that certain ions also reach the inner ring by active transport (see page 70) which has the effect of increasing the concentration of solutes in the inner ring, causing more water to follow. Active transport

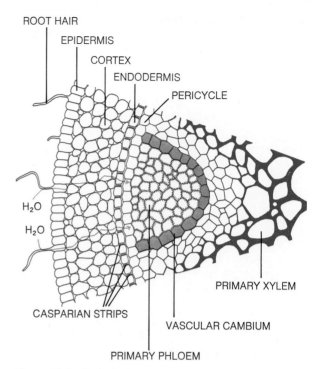

Figure 14-5 Organization of an immature root. The path of water entry shows that water must cross the membranes of at least two cell types to reach the xylem channels. These are the root hairs, epidermal cell processes that remain functional from a few days to years, depending on the species, and endodermal cells. The Casparian strips that ring each endodermal cell form a waxy, waterproof layer, that limits water and solute passage to the living part of the epidermal cell. Much of the cortex is given over to storage and cells of the pericycle give rise to lateral roots as the root matures.

also causes the accumulation of molecules specifically required for life from the dilute soil water solution outside the plant.

Moving in this way, water may enter the xylem with sufficient force to generate a pressure of from 2 to 3 atmospheres (atm). Considering the height of many trees and also the resistance to water movement created by the minute xylem channels, this is far too little pressure to account for the water flow actually encountered in the xylem. A 300-ft sequoia probably requires pressures as high as 10 atm to move water from its roots to its topmost branches. The missing power is provided at the other end of the xylem tubes by the cells of the leaves. These are, of course, water laden and in direct communication with the outside air by way of air channels and stomata. The result is **transpiration,** evaporation of water from leaf tissues.

Although transpiration involves movement of water molecules from solution at the surface of cells into the air, it can be viewed as water movement from a solution of high water concentration into a solution of low water concentration. Since the concentration of water in air is low, water evaporates from the surface of leaf cells. The consequent increase in solute concentration results in water movement out of nearby terminals of the xylem elements. Thus, at the lower end of the xylem tubes we find water is "pulled" in osmotically, whereas at the upper end it is "pushed" out osmotically. Except for the metabolic work of active transport of ions in the root, no work is done to promote this flow because, in the final analysis, it is caused by water flowing down its concentration gradient from the soil to the air through the plant.

One difficulty with this model for water movement in xylem comes from the fact that, in ordinary mechanical pumping systems, water cannot be raised more than 32 ft by suction without the water column breaking. This is not a problem in xylem because the diameter of the water column is so small that the attraction of water molecules to each other, **cohesiveness,** holds the tiny column together. Support is also provided by the attraction, or **adhesiveness,** of the xylem tube walls for water. The power of the forces lifting water in the xylem and the aggregate strength of the water columns being raised is shown by measurable shrinkage in diameter of tree trunks during active transpiration. This effect is produced by the narrowing in diameter of the rising water columns as the evaporative pull upon them exceeds the rate at which water can rise in the xylem. The magnitude of the transpirational air flow is not generally appreciated but is made evident by the fact that a corn plant, during the 4 months of active growth and maturation, passes about 54 gal of water through its xylem elements.

Movement of materials in phloem is more complex than the ringing experiment indicates. That experiment implied a tendency for sugar made in the leaves to flow *towards* the roots. However, under appropriate conditions, materials may flow *from* storage sites in the roots to the rest of the plant, as occurs in sugar maples when the sap rises during the sugar season. It is actually possible for materials to flow in two directions at once in phloem. This complexity is reflected in the more elaborate structural organization of the phloem elements as compared with xylem; presence of living cytoplasm as well as tubular flow channels seems to make several methods of transport possible. For example, phloem cytoplasm undergoes regular streaming movements in each cell and, these could bring about movement of dissolved substances, even promoting the movement of different molecules in opposite directions.

A transport scheme akin to that occuring in xylem also appears to occur in phloem. The osmotic force driving it comes from the photosynthetic activity of the leaf. As sugar builds up in photosynthetic cells, water flows into them from nearby xylem elements. The consequent increase in pressure forces water and solutes, principally sugar, through plasmodesmata into phloem cells. Thus there is a flow away from the active sites of photosynthesis, usually towards the roots. In the roots, solutes are removed from the phloem by active transport and diffusion. Sugar is converted to starch for storage, and the net effect is to free water from the phloem for another ascent to the leaves through the xylem.

The summary diagram in Figure 14-6 shows the workings of the combined system of xylem and phloem. The important feature is that transport of essential water and of essential materials both from outside and from internal manufacture is effected in two linked streams of water. The mechanism is powered by active transport and by utilization of the diffusion of water along a natural concentration gradient from the soil to air. Although more than 90% of the water that ascends to the leaves is lost as transpired water vapor, some of it does return to the lower parts of the plant carrying essential substances. Thus, it is legitimate to speak of the plant as having a circulatory system, albeit a very leaky one. This circulatory system is of critical importance in the uptake and transport of essential materials.

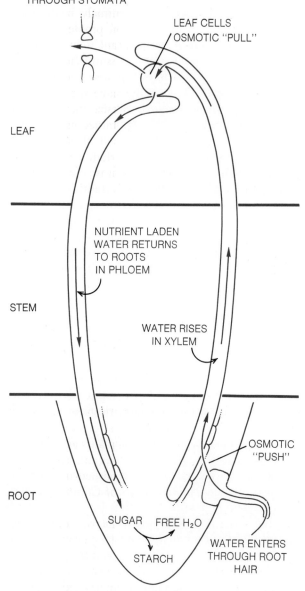

Figure 14-6 Plant internal transport system. A leaky transport of water between roots and leaves, driven by a combination of osmotic flow and gradients produced by metabolic activity and transpirational water loss, effects nutrient transport.

THE PLANT INTEGUMENT, VENTILATION, AND WATER CONSERVATION

Except in the leaves, where demand for carbon dioxide is high during photosynthesis, gas exchange in plants goes on at a low rate as compared with that in animals. Typical plant cells rarely require more than 10% of the oxygen supply of an animal cell, and they rarely undergo anything approaching the great variation in oxygen demand seen as animal cells switch from inactive to active states. As a result, plants lack elaborate specializations for gas exchange, except in the leaves. The stems of woody plants have **lenticels**, openings through the impermeable integument that admit air to the spaces between cells. Roots usually carry on sufficient gas exchange through root hairs, except in situations where they are adapted to grow in air-deficient situations. Thus, the swamp dwelling cypress has knees, which are gas exchanging parts of roots raised above the water level.

Although most of the remainder of the integument of plants is protected from both water loss and gas exchange by an impervious waxy secretion, there are stomata, valve controlled apertures in the leaves which serve as major routes of transpirational water loss and carbon dioxide uptake in support of photosynthesis. The valves controlling stomata are pairs of kidney shaped cells that are derived from the epidermis (Figure 14-7). Unlike typical epidermal cells, stomatal cells contain chloroplasts and carry out photosynthesis. Their intense metabolic activity is basic to their ability to accomodate properly the often competing demands for increased admission of carbon dioxide for photosynthesis and for decreased loss of water as the soil water supply diminishes during drying periods. Operation of stomata is illustrated in Figure 14-7.

There are many adaptations of plants to limit transpirational water loss. The stomata tend to be closed at night when they need not be open in support of photosynthesis. They are present in greater numbers on the undersides of leaves than on the upper sides where direct exposure to sunlight would greatly accelerate water loss. Desert plants often have stomata sunken within deep cavities, which further limit water loss. The familiar cactus has completely dispensed with leaves and gets by with limited transpiration and gas exchange through reduced numbers

STOMATA: DEVICES THAT REGULATE GAS AND WATER EXCHANGE

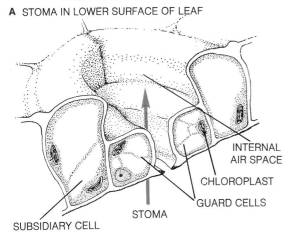

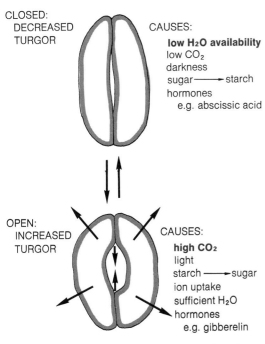

As turgor increases, thin outer wall expands outward more easily than wall at stomal side. The effect is to bow the stomal walls outward, increasing the opening.

Figure 14-7 Stomata open and close in response to the competing demands of photosynthesis and water conservation. **A.** Diagram of section of leaf undersurface showing a stoma and its two guard cells. **B.** The mechanism of stomatal opening.

of stomata on thick, leaflike branches (see Figure 15-12).

Excretion

Clearly the plant is open to difficulty in maintaining water balance since water loss is essential to photosynthesis and transport. However, the mechanisms of nitrogen metabolism in plants relieves them of the necessity of using water as a solvent to effect excretion of nitrogenous wastes. Animals, in the final analysis, have a nitrogen excretion problem because they are unable to use a significant amount of the ammonium (NH_4^+) liberated in assimilation of their protein diet for the synthesis of new protein; thus they must get rid of it. Plants have the advantage in this matter because they synthesize their own protein from ammonium which is obtained usually from nitrate (NO_3^-) derived from the soil. Because plants have this ability, the ammonium that appears in the course of normal plant metabolism is actually a valuable resource and is immediately reutilized in synthetic operations. The only situation in which there could be a toxic buildup of ammonium would be during starvation, as during prolonged darkness, when the plant would utilize its protein as food. Even then, the large vacuoles characteristic of plant cells would provide a margin of safety by providing an inert storage site for wastes.

COORDINATION IN PLANTS

The geometrical regularity of growth patterns, the synchronously timed appearance of leaves and flowers, and orientation to light and gravity show that plants regulate their activities. The first significant work leading to understanding of how plants achieve coordination was a series of observations and experiments done by Charles Darwin and published in his book, *On the Movements and Habits of Climbing Plants*, which appeared in 1865.

The critical experiment that Darwin describes is one showing how the turning of the growing shoot of a plant towards the light occurs only if the growing tip of the plant is illuminated. If just this region is masked off from the light by a tiny opaque cap, there is no response to light (Figure 14-8). As you would expect from what you already know about plants, the actual basis of the turning movement is differential

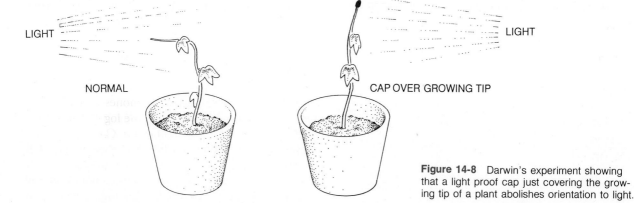

Figure 14-8 Darwin's experiment showing that a light proof cap just covering the growing tip of a plant abolishes orientation to light.

growth of cells in the region of elongation just below the tip; cells on the dark side elongate more than cells on the light side.

It took another 75 years to work out the relationship between illumination of the growing tip and cell elongation. First it developed that the tip of the plant could be sliced off and then put back in place with a small block of agar interposed (Figure 14-9). Since the agar contains water, if the signal to elongation produced by the growing tip is a water soluble chemical, it might diffuse through the agar and have a normal effect on the tissues below. This did happen, telling investigators that they were probably dealing with a plant hormone. Finally, this hormone was isolated and its chemical structure determined (Figure 14-9). It was given the name **auxin**, from the Greek word meaning to increase. Finally, it was shown that light causes auxin to move away from the illuminated side of the tip of the growing plant and from there to move directly downwards with little lateral movement, thus causing cells on the dark side to be exposed to more auxin than those on the light side.

Auxin produces the typical orientation of the plant to gravity in a similar way. If a vertically growing shoot is bent, auxin accumulates on the lower side, promoting greater elongation and thereby a correction of the bend.

The mechanism of action of auxin is still not entirely clear. Its site of action in causing cell elongation appears to be the nonliving cell wall, softening it by bringing about loosening of crosslinkages among cellulose fibrils. Cell turgor immediately swells the living

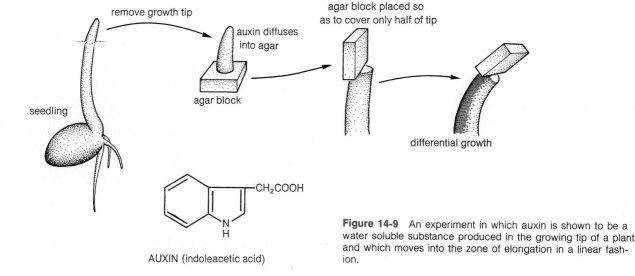

Figure 14-9 An experiment in which auxin is shown to be a water soluble substance produced in the growing tip of a plant and which moves into the zone of elongation in a linear fashion.

part of the cell to fit the larger container thus produced.

Auxin has other important effects at sites far removed from the zone of elongation. It is the means by which one primary growth point in the plant suppresses growth of lateral buds. Auxin is also produced by plant embryos, and in this instance serves to stimulate growth of the fruit. The seedless, that is, sterile, navel orange is able to develop fruit because its *ovary* produces auxin.

Although auxin can cause enlargement of plant cells, it alone does not cause growth. **Cytokinins**, another group of hormones produced in rapidly growing tissues, must be present, along with auxin, for cell division to occur. Produced in actively growing roots, cytokinins are transported to the remainder of the plant through the xylem. In addition to promoting cell division, they slow the aging of leaves, induce flowering, and promote auxin formation. Chemically the cytokinins are related to adenine (Figure 14-10) and it is remarkable that in many plants, animals, and bacteria, they are one of the bases adjacent to the anticodon region of tRNA. Not all tRNAs have cytokinins, and their significance in this site is not understood.

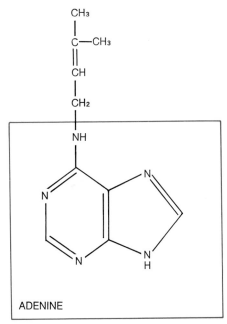

Figure 14-10 Tricanthine, a cytokinin plant hormone.

The **gibberellins** are a third group of hormones that promote cell division and growth. They are synthesized in the young leaves forming around a growing shoot tip and are also produced in the embryo, where they initiate the utilization of the food stores of the seed.

In these three classes of hormones—auxin, cytokinins, and gibberellins—we see the logic of coordination of growth activities of the plant. Generally speaking, these hormones are produced at the primary sites of growth—shoot tip, developing leaves, and root tissues. One hormone from one of these sources is not enough to promote substantial growth. This is important because it results in coordinated growth; a shoot tip cannot grow out of proportion to the root tissues, which must support the shoot's mineral and water requirements, because it requires cytokinin from the roots for cell division. These plant hormones also illustrate a way in which the plant coordinates its activities with the surrounding environment. The light and gravitationally sensitive movements of auxin insures that the plant will have an optimum chance to obtain its share of sunlight.

Several other hormones, proven or suspected, are known in plants. The gas ethylene is produced in many rapidly growing plant tissues, particularly those of fruits. At a concentration of 1 part in 10 million parts of air it promotes the dropping of fruit and the withering of flowers. Ethylene is of great practical advantage, since it can be applied to cause synchronized fruit maturation to facilitate marketing.

Environmental Coordination

In the influence of light and gravity on auxin we see clear evidence that the plant coordinates its activities with environmental conditions. The direction of incident light, although of critical importance, is only one of many important environmental phenomena to which the plant must react. The most obvious of these are related to the passage of time, as expressed in the day-night cycle and the annual succession of seasons.

Virtually all organisms that have been properly tested exhibit internal time keeping mechanisms, **biological clocks,** which are set by environmental cues but which keep time for varying periods even when completely isolated from all evident external cues. Many of the rhythms of activity thus demonstrable are called **circadian rhythms** because they approximate a standard 24-hr day when running independently of external cues. Such rhythms are easily demonstrated in the daily activities of many plants. The folding of leaves at night is a good example. If a plant with such a rhythm is placed in continuous darkness, the leaves continue to open and fold in a normal daily cycle. If, while in darkness, the lights are switched on when the plant is in its night phase, it is possible to reset its rhythm quickly; then the plant will continue in the new rhythm for an indefinite period. Both the nature of such rhythms and the means by which they are attuned to the environment, as illustrated by the resetting experiment, are unknown. They are the subject of much study because they have great practical importance in agriculture and many other areas. In humans, for example, the phenomenon of jet-lag is a well known example of interference with the circadian rhythm by resetting because of rapid travel across time zones.

Plants show a response, termed **photoperiodism**, to the relative duration of the light and dark parts of the day-night cycle. As one proceeds north or south from the equator, the length of the light period of the day varies more and more on a seasonal basis. Thus, in the northern parts of the United States there is daylight for slightly better than 15 hrs out of 24 hrs in June, whereas in December there are only about 9 hrs of daylight. This is an important environmental characteristic for the plant to be able to monitor because it provides a way to determine the progression of the seasons. Because of photoperiodism, **perennial plants**, such as trees and rosebushes, are able to cease operations, become dormant, and avoid danger before the first frosts; and annual plants, like corn or wheat, are able to complete their life cycles in one growing season. In a general way, plants may be classified into long-day plants and short-day plants in that they require either long or short days to flower. The Christmas poinsettia is an example of a short-day plant.

Surprisingly, it is not the duration of the light period that the plant measures; rather, it is the duration of the dark period. This is shown by exposing a short-day plant to a short day cycle, that is a short light period and a long dark period. Ordinarily it would proceed to flower; however, if during the dark period the lights are turned on for only a few minutes, flowering does not take place and the plant behaves as if exposed to a long day period. As in resetting of circadian rhythms, the molecular basis of photoperiodism remains unknown.

Neuroid Communication

The rapid drooping of leaves of the sensitive plant or the snapping shut of the Venus fly trap on an insect (Figure 14-11) is evidence of more rapid coordinative mechanisms than the chemical processes that we have just been discussing. Examination of a Venus fly trap reveals the presence of several hairs on the inner sides of the modified leaves that form the arms of the trap. Bending one of these twice, or two of them once within 30 sec is followed, within less than a second, by rapid shutting of the trap requiring only 100 msc. If the trap succeeds in catching an insect, it remains tightly closed for up to 2 weeks and enzymes are secreted that reduce the prey to a form utilizable by the plant. If the trap misses, it opens again in a few hours. Here is behavior worthy of an animal, yet it is generated by an organism without a nervous system. However, these processes are so similar, even at the physicochemical level to the operations of the animal nervous system, that they have been given the name **neuroid**, which means similar to nerve.

To illustrate, the drooping of the sensitive plant is caused by an electrical signal that travels in certain cells of the transport system, which thus behave in a remarkably similar way to nerve cells. This plant **action potential** (see page 341 for details of action potentials in animal cells) moves much more slowly than a nerve action potential, only a few millimeters per second, and the ions that must flow to produce it are different from those operating in nerve signalling, but the principles are identical. However, we see a major difference between animals and plants when it comes to the matter of how the plant moves in response to the neuroid signal. As we have said, plants have no muscles, no systems of contractile proteins extensive enough to move large parts of the plant. What happens is that the neuroid signal, instead of activating a contractile system, triggers a rapid change in the turgor pressure of groups of cells called **motor cells** that hold up leaves or stems. When the pressure falls, the leaves or stems fall. At present there is no clear understanding of how this rapid pressure change is brought about. In the sensitive plant, at least, the motor cells contain a large number of contractile vacuoles that suddenly dump their contents to the outside upon receiving the neuroid signal.

Figure 14-11 **Venus fly trap. A.** The trap structure is the modified distal portion of a leaf. **B.** Enlarged view showing 3 trigger hairs on the inner surface of each half of the trap.

Why No Nervous System in Plants?

These examples of neuroid behavior are clearly the exception. They do, however, demonstrate the capability of far more elaborate development of such systems in plants, making it interesting to inquire why it did not happen. The basic reason, of course, has already been stated: the nutritional processes of plants make locomotion unnecessary and would, indeed, hinder water and mineral uptake from the soil. All of the necessary movements of plants in the search for nutrition are satisfied by growth movements, by the growing of roots through the soil in search of minerals and water, and by the upward growth and spreading of the shoot and leaves. Thus, another way to explain why plants do not have nervous systems is because they are unnecessary. In the face of these facts our human gullibility is delightfully displayed by interest in certain quarters in the emotions and psychology of plants (Note 14-1).

REPRODUCTIVE SYSTEM

Of all organ systems of the plant the only one that is clearly specialized for a single purpose is the reproductive system, as exemplified in the flowers of flowering plants (Figure 14-13). Under appropriate hormonal and environmental influences, typical growing points cease elongation and production of normal leaves and instead produce flowers. Although complex and highly varied in form throughout the variety of flowering plants, a flower is basically four sets of modified leaves. The two innermost sets produce the male and female reproductive elements whereas the two outermost pairs form **sepals** and **petals** which protect the developing reproductive parts and help to attract pollinators.

Reference to the diagram in Figure 14-14 will help in understanding the complexity of the reproductive structures of the flower. First, it is important to realize that the tissues from which the reproductive structures form are diploid. Diploid plants are called **sporophytes** and all such plants have complex life cycles that involve alternation of sporophytes with haploid

Note 14-1 Plant Psychology

Plants are remarkable enough at face value. After all, they are responsible for our existence. Yet our basic human love of the bizarre forces itself on the gentle vegetable and builds up an image of unfathomable powers beneath its bland, green exterior. Not so long ago, extrapolating vigorously from the insectivorous plants and possibly building on twice-told native tales, plants with a taste for human flesh were conjured up. A good example is the legendary Madagascar person eater shown at work in Figure 14-12. At present, this form of amiable eccentricity has attained a higher plane of sophistication in the rapidly growing cult of plant psychologists. The most recent cycle of activity in this area probably had its beginnings in the fad of growing plants to music. Originally, there was a modestly plausible scientific basis to the concept, namely, that sound energy might facilitate the growth of root hairs, perhaps by vibrating soil particles. Phonograph records with selections tailored to this end were sold in quantity.

Recently, a quantum leap has been made from the idea that sounds have direct effect on cellular processes in plants to the theory that plants perceive not merely sounds but human thoughts. This great conceptual leap forward was attained by the use of lie-detector technology. Electrodes that detect changes in electrical resistance were attached to plants, and great changes in leaf resistance were obtained when the experimenter, for example,

Figure 14-12 Artist's conception of a man-eating plant, rumored to have lent excitement to rural life in Madagascar.

forms called **gametophytes**. In lower plants these two phases of the life cycle are frequently quite obvious, but in the flowering plants the gametophyte is visible only as a few cells in the reproductive structures of the flower. Sexes in the gametophyte are separate, with the male gametophyte producing pollen and the female gametophyte producing eggs. Flowers may possess both gametophytes or only one.

Figures 14-13 show the details of reproduction in male and female gametophytes. The male gametophyte forms in the **stamen**; at its tip specialized cells undergo meiosis to form spores. Each spore divides by mitosis to form a two-celled haploid plant, which is all there is of the male gametophyte generation. This two-celled plant (Figure 14-14) develops a tough wall around itself and becomes a **pollen grain**. The female gametophyte forms in a complex structure called the **carpel**. The base of the carpel is the **ovary** in which the female gametophyte develops, whereas the upper structures, **stigma** and **style**, function in the fertilization process. Within the ovary, structures called the **ovules** contain the cells that will form the seed. They

thought bad thoughts with reference to the subject plant. It seems to make little difference that these experiments have not been confirmed (for a recent example see J. Kmetz: 1977. A study of primary perception on animal and plant life. *Journal of the American Society of Psychical Research* **71:** 157), because this love affair with perceptive plants continues.

Although it is sometimes dangerous to use common sense to refute modern science, its application here might be useful because experimentation in this field cannot at present have a sound basis. First of all, we note with suspicion a strong element of anthropomorphism in the ideas of the plant psychologist who reads fear and loathing into plant responses to thoughts about killing them by burning or even to the act of dumping living brine shrimp into boiling water in the presence of the horrified plant—part of the standard methodology in this field. Although we might not think that extermination of animals by boiling is nice, the plant, victim of innumerable animal depredations down through time (Just think of what cows and grasshoppers alone have done to plants.) might very well feel that boiling is just the thing for a brine shrimp. Thus the signals reputedly evoked by shrimp boiling might well represent the highest plane of vegetable glee. This immediately leads to a problem when we are informed that the signals evoked by shrimp boiling are the same as those evoked by thoughts of cooking the plant itself. We are left to conclude that either the apparatus is not yet refined enough to differentiate among the various classes of vegetable emotions or that the experimental subject cannot be fooled by the experimenter whom it knows, despite all his ominous flicking of a cigarette lighter, is not about to set fire to a favorite *Antirrhinum*.

Secondly, we are led to inquire as to the psychological makeup of crop plants as opposed to the generally loved and pampered ornamentals upon which most studies so far have been conducted. If the perceptual skills under discussion extend to cabbage, spinach, and carrot, it is difficult to see why they would participate at all in the sad exercise of being carefully tended and growing robustly for weeks only to be plucked in the prime of life, sliced, and boiled. They should be tested as soon as possible, along with the crab grass that clearly prospers in spite of the combined ill will of at least 70 million homeowners in North America alone.

Finally, we come to more biophysical questions such as how naked thoughts are transmitted between organisms of any kind, but this is the proper subject of another interesting field of science.

At this juncture one is reminded of the quotation from Peter Medawar (page 8) to the effect that "Imagination without criticism may burst out into a comic profusion of grandiose and silly notions."

surround the one cell that will form the female gametophyte. That cell, known as the **megaspore mother cell**, undergoes meiosis to produce four haploid cells. Only one of these survives and, undergoing mitosis, produces a gametophyte consisting of seven cells. One of these seven is the egg cell and the other six form supporting elements for its benefit.

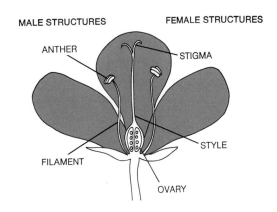

Figure 14-13 Section through a flower showing male and female reproductive parts.

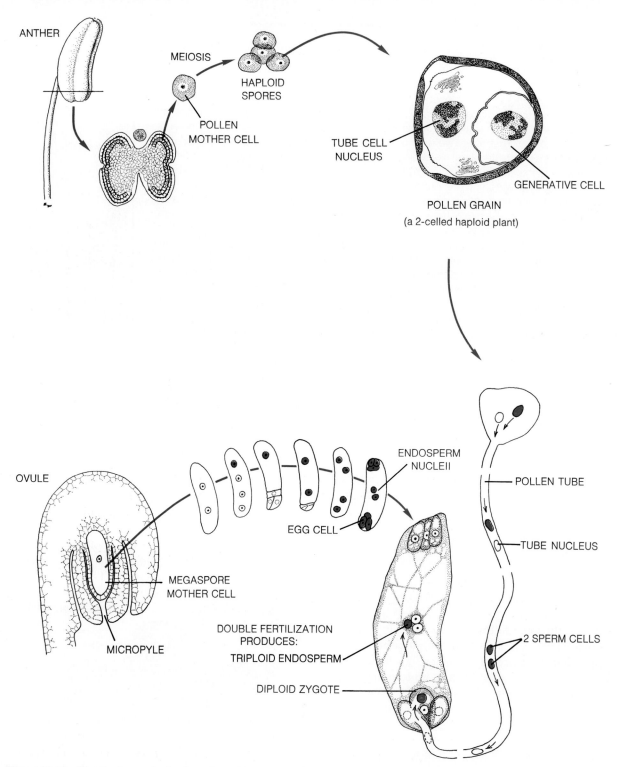

Figure 14-14 Events of reproduction in male and female gametophyte of a higher plant. See text for details.

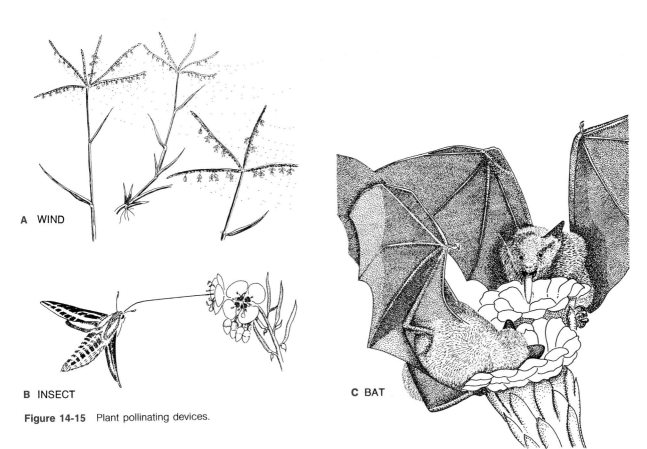

Figure 14-15 Plant pollinating devices.

When a pollen grain reaches the stigma of a mature flower it absorbs water, swells, and the protective wall cracks allowing a slender protoplasmic extension, the **pollen tube**, to grow down the stigma and approach the egg. One of the two cells of the pollen grain divides during this process to form two sperm cells. When these reach the ovule, one of them fertilizes the egg and the other fertilizes one of the supporting cells called the **endosperm mother cell**. This particular cell is formed during development of the female gametophyte from two nuclei so that the resulting endosperm mother cell is triploid after fertilization. It goes on to form the food storage tissues for the **seed** that develops around the fertilized egg or zygote, which in turn forms the new sporophyte plant.

The tactics that flowering plants employ to insure transport of pollen to achieve fertilization form one of the most amazing chapters of natural history. In many instances pollen transport is by physical agencies such as the wind, but more commonly animals do the work.

Insects are attracted to flowers by the color of their petals and, while searching for nectar, are brushed by pollen, which they transport to other plants. There are many variants of this scheme, including even some night blooming flowers that are fertilized by nectar feeding bats. Some orchids attract pollinating insects, not with nectar but with flowers shaped to look like females of the pollinating insect. Thus, by their art plants are able to compensate for the lack of a roving male in the reproductive competition (Figure 14-15).

ANIMALS AND PLANTS: COMMON PROBLEMS WITH SIMILAR SOLUTIONS

In this cataloging of the ways in which plants meet the problems of multicellularity, we have seen that, in spite of radically different nutritonal mechanisms and restrictions imposed by the rigid cell structure of plants, there is a fundamental similarity among widely diverse forms of multicellular life. Indeed, it is quite possible to list all of the organ systems of animals and list beside them comparable systems in plants or, at least to be able to specify quite exactly a logical reason for a difference between the two, as in Table 14-1.

Table 14-1 Organ System Functions in Plants and Animals

Primary Function	Organ System	Representation In Plants	Representation In Animals
Environmental Exchange	Nutritive	**Autotrophism:** photosynthetic tissues requiring only light, CO_2, water and inorganic ions.	**Heterotrophism:** digestive tract capable of assimilating complex organic foods.
	Integumentary	Cuticle, epidermis, cortex	Surface epithelia, secreted coverings
	Respiratory	Spongy tissue of leaves, stomata, root hairs	Integument, gills, lungs, tracheal system
	Excretory	Storage, metabolic recycling of nitrogen	Integument, excretory organs
Internal Transport	Circulatory	Conducting tissues—xylem and phloem, powered by transpiration and osmotic forces	Blood vessels or sinuses with fluid moved by muscular action
Internal Communication	Endocrine	Many hormone sources in tissues; hormone movement by cell-to-cell movement or through conducting tissues	Many hormone sources in tissues and specialized glands; movement through circulatory system, sometimes by intracellular transport
	Nervous	Non-existent	Elaborately developed system of sensory, conducting, integrating and effector controlling cells
Adjustment to Environment	Nervous	Rare examples of "neuroid" systems (insectivorous and "sensitive" plants)	Elaborately developed (see above)
	Skeletal	Bark, cuticle, cortex	Internal and external skeletons, often jointed; hydrostatic skeletons
	Effector	Secretory cells, cells capable of rapid turgor change; pseudopodia, cilia and flagellae in lower plants	Secretory cells, cilia and flagellae, pseudopodia, muscle, nematocysts
Reproduction	Reproductive	Flowers, seeds and analogous structures in lower plants	Gonads and accessory structures

15 The Variety of Life

Although we have developed, step by step, a plausible scheme for the progression of life to the stage of multicellular existence, it was necessary to leap ahead and discuss very complex animals and plants in order to describe multicellular organization fully. In this chapter the goal is to fill this gap in the logical progression of our study of life by showing how organisms diversified, once multicellularity became established.

A SYSTEM OF CLASSIFICATION IS NECESSARY

From the beginnings of life to the present it is estimated that there have lived or are living about 500 million distinct kinds of organisms. Perhaps 10 million are alive today on earth. Of these, scientists have described about 1.7 million, comprising about 1.2 million animals and 500,000 plants. It is clear from these large numbers that biology has a very special problem not experienced in such magnitude by other sciences, the problem of classification. Moreover, if this immense task of classification is to be of any value beyond the basic utility of giving a recognizable name to everything, it must tell us about the *relationships* of the organisms classified and help us to understand the route through time by which modern organisms came into being.

This broader function of classification is called **systematics**. The systematist is concerned with the relationships between organisms and with all the factors that shape these relationships. The basic process of systematics, the process that must be done first, is **taxonomy**, by which organisms are described and placed in a plan of classification according to the best estimate of how they are related to all other organisms. Although there is a tendency among modern biologists to look upon taxonomics as dull and dated, it is obvious that without the continued efforts of taxonomists much of biology would be in chaos. The probability that far less than half of the total number of kinds of organisms alive today have been adequately described shows very clearly that there is still room in biology for taxonomists.

THE SEARCH FOR ORDER

A natural classification of organisms, one in which organisms are arranged according to genetic relationships, was a primary goal of early biologists, and further perfection of the scheme of classification still receives much attention. Even without knowledge of genetics or of evolution, early biologists were able to develop a taxonomy that was useful. Their major guides to classification were morphological, principally the comparative anatomy of adult organisms.

Binomial Nomenclature

We are indebted to the early Swedish taxonomist, Carolus Linnaeus (Note 15-1 and Figure 15-1) for the establishment of an essential feature of taxonomy, a

Note 15-1 *Linnaeus and the Establishment of Order in Biological Classification.*

Carolus Linnaeus (1707–1778) made two memorable contributions to the subject of classification of organisms. He advanced the idea of the stability of the species, and he provided a useful system of classification based on natural relationships among organisms. As long as it was accepted that the species were changeable, that life could appear from inanimate sources, and so on, classification of such ephemeral entities could hardly be expected to be logical. Linnaeus so vigorously defended the idea of the fixity of the species that it was a problem long afterwards to the establishment of modern evolutionary theory. He initially held that each species was established by God but later came to believe that it was the genus that enjoyed that distinction.

The Linnaean scheme of classification is said to be Aristotelian because, following Aristotle, Linnaeus established three kingdoms—animal, plant, and mineral—and within these kingdoms classification was based on the fundamental characteristics of the things classified. This latter point might seem simple and obvious, but it was not obvious at the time, and some very strange taxonomies had resulted. Many classification schemes of those times were like classifying the books in a library by their external appearance rather than by their subject matter. Thus, there was a plant taxonomy based on leaf shape. By contrast, Linnaeus tried to get at the heart of the matter. He characterized three kingdoms: stones grow; plants grow and live; animals grow, live, and feel.

Linnaean taxonomy was simple compared to the huge edifice of modern taxonomy. Between species and kingdom he had only three categories—genus, order, and class. Today zoologists recognize 13 categories and botanists 17. The first edition (1735) of the *Systema Naturae* recognized 549 species, but this number had grown to 4387 by the tenth edition, in no small part due to Linnaeus' own efforts.

His classification of animals (Table 15-1) obviously had problems. The Amphibia included both the amphibians and reptiles, and the Worms included representatives of most of the major invertebrate groups except for the insects. He specifically included among worms such distantly related organisms as jellyfish, sea urchins, and cuttlefish. The scheme was particularly remarkable for its inclusion of the human among the quadrupeds, in an order Anthropomorpha, recognizing *Homo sapiens* and *Homo troglodytes,* the chimpanzee. For *Homo sapiens* the species designation was ''Nosce te ipsum,'' or ''Know thyself.''

Table 15-1 **The Major Animal Groups According to Linnaeus**

Class	Characteristics
Quadrupeds	Hairy body, four feet, female viviparous and milk producing
Birds	Feathered body, two wings, two feet, bony beak, females oviparous
Amphibians	Body naked or scaly, no molar teeth but others always present, no feathers
Fishes	Body footless, possess real fins, naked or scaly
Insects	Body covered with bony shell, head equipped with antennae
Worms	Body muscles attached at a single point to a semisolid base

Figure 15-1 Linnaeus, painted in Lapp dress a few years after his expedition of 1732 to Lapland.

system of classification that permits naming each kind of organism in a unique way, that is, with one universally accepted name for each organism, a way that also indicates the relationship of that organism to others. Linnaeus assigned two Latin names, using a system of **binomial nomenclature**, to each described kind of organism. These were set forth in a scheme of classification in his book entitled *Systema Naturae*. The tenth edition of that book, published in 1758, is the primary reference point of animal taxonomy. His *Species Plantarum* of 1753 serves the same role for plants.

Latin names are used in binomial classification for historical and practical reasons. Virtually all scholarly writing in the Western World in the time of Linnaeus was in Latin, and the habit persisted in taxonomy, not only in the names assigned to organisms but in the short, concise descriptions that are prepared for each kind of organism. Until relatively recently no one modern language was dominant in scientific writing, and so Latin continued to serve as a universal scientific language in this one specialized area of science.

A typical binomial name in this system is *Felis leo*, the African lion. The first term, *Felis*, denotes the **genus** to which the African lion belongs. Any other animal assigned this same generic name is one deemed closely related. For example, *Felis concolor*, the puma, or mountain lion of North America, is lionlike even at first glance. The second term, *leo*, or *concolor* in these examples is the **species** name. Note that the genus name is capitalized, whereas the species name is not and that both names are italicized.

The Species

Organisms of the same species are what the layman would call "of the same kind." Members of the same species are interfertile; they give rise to normal offspring which are capable of propagating. Matings between members of different species either are infertile or the progeny are sterile, as is usually the mule, the issue of mare and ass. Thus, members of a species are described as being reproductively isolated from all other species, even of the same genus.

The species is by no means clearly definable in every instance. The system of classification, it must be understood, is not applied to a static, unchanging array of different organisms. Evolution is continuous, with gradual appearance of new kinds of organisms. Since classification systems tend to describe only a particular stage in these processes, confusion is often caused by incipient species formation. Kodiak and polar bears are an interesting example (Figure 15-2). No one would confuse these giant carnivores with

Figure 15-2 Kodiak (left) and polar bears (right) do not look closely related but are interfertile.

Figure 15-3 Taxonomic investigation allows arrangement of organisms in groupings which suggest phyletic relationships. In the two parts of this Figure this is shown by arranging sketches of organisms around an arbitrarily designated animal (deermouse) or plant (sunflower) in two rings. The innermost ring shows "close" relatives, from within the same Order or Family, while the outer ring illustrates the variety of relatives within the next larger taxonomic grouping.

Part A. Deermouse **(A)** surrounded by an inner ring of members of the Order Rodentia, to which the deermouse belongs: rat **(B)**, beaver **(C)**, pocket gopher **(D)**, red squirrel **(E)**, eastern chipmunk **(F)**, Canadian porcupine **(G)**. The outer ring shows members of other mammalian orders: **(H)** Catacea, bottle-nose dolphin, **(I)** Marsupalia, opossum, **(J)** Insectivora, mole, **(K)** Chiroptera, bat, **(L)** Carnivora, stoat, **(M)** Pinnipedia, seal, **(N)** Primates, monkey, **(O)** Lagomorpha, rabbit, **(P)** Perissodactyla, tapir, **(Q)** Artiodactyla, guanaco, **(R)** Sirenia, Manatee.

Part B. Sunflower **(A)** surrounded by an inner ring of flowers of the Family Compositae, to which the sunflower belongs: fleabane **(B)**, thistle **(C)**, pearly everlasting **(D)**, dandelion **(E)**, yarrow **(F)**, eriophyllum **(G)**. The outer ring shows the flowers of members of some other plant families: **(H)** Primulaceae, shooting star, **(I)** Hydrophyllaceae, phacelia, **(J)** Labiatae, sage, **(K)** Scrophulariaceae, monkey flower, **(L)** Convolvulaceae, convolvulus, **(M)** Araceae, skunk cabbage, **(N)** Liliaceae, Mariposa lily, **(O)** Caryophyllaceae, pink, **(P)** Magnoliaceae, magnolia, **(Q)** Ranunculaceae, columbine, **(R)** Papaveraceae, California poppy, **(S)** Rosaceae, wild rose, **(T)** Papilionoideae, lupine, **(U)** Cactaceae, cactus, **(V)** Ericaceae, manzanita.

each other and would agree at first glance with the decision of the taxonomist to place them not merely in different species but in different genera, *Thalarctos* for the polar bear and *Ursus* for the Kodiak bear. Yet in captivity they breed and the cubs prove fertile. Since the environments preferred by these two bears are sufficiently different to keep them apart in nature, it is likely that geographical isolation is permitting the accumulation of sufficient inheritable changes in the two forms ultimately to prevent interbreeding when they do come together. In situations such as this, the two kinds of animals are considered to be different species if in the wild they do not interbreed for whatever reason, be it environmental isolation or behavioral trait, even though fertility is demonstrable in captivity.

Evidence of the objective reality of the species comes from two very interesting comparisons of the classification system of primitive natives of New Guinea with the taxonomy of modern science. In both instances, two biologists, Ernst Mayr and Jared Diamond found that the natives classified the birds found in their respective territories into almost exactly the same kinds, with different names, of course, as did the scientists. Clearly the objective reality of a species is strongly supported when it is derived by two totally different classification systems, one generated out of the utilitarian interests of primitive hunters and the other based on exhaustive scientific study.

In addition to assigning organisms to genera and species, the taxonomist arranges organisms in an array of successively more inclusive categories so as to show as well as possible the interrelationships of all. Members of such family trees are shown in Figure 15-3A and B.

Criteria for Determining the Relationships of Organisms

In classifying organisms according to the Linnaean scheme, we assume that the organisms we see in the world today were not deposited here by some capricious fate or created by a supernatural force. Today's organisms are related because they descended from organisms now extinct by gradual evolutionary processes, compatible with natural law.

As we have seen, the fundamental way to determine whether or not animals belong to the same species is to see if they are reproductively isolated in nature. Obviously, this is not a practical way to determine species relationships. Organisms used for taxonomic research are rarely maintained alive. Even if they were, the logistics of conducting breeding studies would be totally unmanageable. Other criteria must be employed at both the species level and at higher levels of classification. Some of these follow:

COMPARATIVE ANATOMY OF ADULT ORGANISMS

Anatomical comparisons were virtually the exclusive method used by early taxonomists, and even today this method predominates. The principle is obvious: the greater the physical similarity, the closer the relationship. At higher levels—phylum, class, order—the principle is easy enough to apply. Thus, if an animal has a backbone, it is a vertebrate. If it has feathers it is a bird, and so on. But as one proceeds towards the species level, more minor anatomical criteria must be invoked in order to discriminate among forms. Finally, such relatively minor differences must be considered that classification tends to become subjective. In general, these subjective classifications of organisms are accurate and probably represent biological reality, although it must be remembered that if it were possible to view a species from the perspective of geological time it would blend imperceptibly with ancestral forms.

COMPARATIVE ANATOMY OF FOSSIL ORGANISMS

To restrict the study of the relationships of organisms to living species is like trying to work out the complete branching pattern of a tree from the arrangement of the outermost twigs. Fortunately, this is not altogether necessary because an immense variety of extinct species is now known from fossils (Figure 15-4). These may consist of imprints of the body, footprints or tracks, or more commonly bones, teeth, or shells, which are usually mineralized. In exceptional circumstances soft tissues, which ordinarily would not be preserved, are found, as in desert caves so dry that mummification occurs or in bogs where putrifaction is arrested by acidity and lack of oxygen. In a few instances an entire prehistoric, elephantlike mastodon carcass has been preserved by freezing in the arctic.

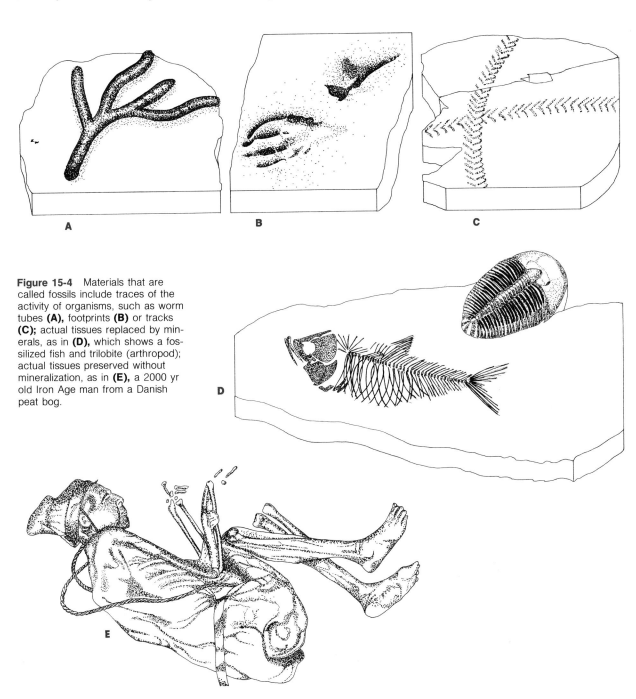

Figure 15-4 Materials that are called fossils include traces of the activity of organisms, such as worm tubes **(A)**, footprints **(B)** or tracks **(C)**; actual tissues replaced by minerals, as in **(D)**, which shows a fossilized fish and trilobite (arthropod); actual tissues preserved without mineralization, as in **(E)**, a 2000 yr old Iron Age man from a Danish peat bog.

ROCK CLOCKS: RADIOISOTOPE DECAY MEASUREMENT OF GEOLOGICAL AGE

Radioactive elements spontaneously decay at rates that are characteristic of each element and virtually independently of external conditions.

Imagine that radioactive decay is represented by the fall of sand in an hour glass.

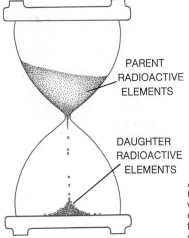

IF the fall of each grain (each decay of a parent atom) is independent of the number of grains in the upper chamber, the passage of time is a straight line function of the number of grains accumulated in the lower chamber.

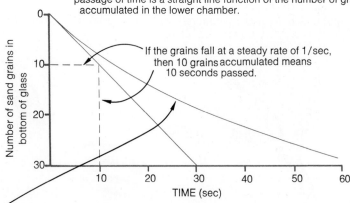

If the grains fall at a steady rate of 1/sec, then 10 grains accumulated means 10 seconds passed.

ACTUALLY, the relationship is not linear since every time a decay (fall) occurs there is one less atom (grain) left. Hence starting with a finite number of units in the upper chamber, the number of events seen decreases with time. This type of curve corresponds to the decay curve of a radioisotope (see Chapter 26 for more details)

Using the decay curve it is possible to determine the age of a rock (the time since it solidified) if it contains a radioactive element whose daughter element (decay product) would not start to accumulate—start the clock—until solidification. The decay of Potassium-40 to Argon-40 (gas) is an example since Argon is retained by the crystalline structure of many minerals when they are solidified.

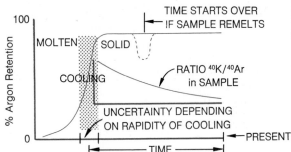

CARBON-14 is used in a similar way to determine the time since death of a biological specimen. ^{14}C is made in the atmosphere at a steady rate by the action of cosmic ray neutrons on nitrogen. The ^{14}C is incorporated into living organisms at a steady rate during life. It also decays to ^{12}C at a steady rate. The result is that carbon from a living organism gives off about 15 disintegrations/min/gm carbon. At death, incorporation of ^{14}C stops. Therefore, the disintegration rate falls with (and is a measure of) time.

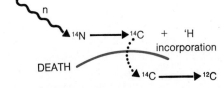

Figure 15-5 Two radioisotope methods for determining the age of ancient geological and biological material.

Indeed, the New York Explorer's Club once dined on mastodon, and sledge dogs in the Russian Arctic have been similarly fed. A most complete fossil series, demonstrating the derivation of modern horses and their relatives from a single ancestral type, *Eohippus,* illustrates the powerful assistance fossils lend to these studies (see page 303).

Regrettably the fossil record is far from complete for several reasons: First, the farther back in time one goes, the more fragmentary the record; the older a

Figure 15-6 Living fossil: an embryo of a lobe-finned fish. American Museum of Natural History and C. Lavett Smith.

geological level, the more likely it is to have undergone changes destructive to fossils lodged within it. Second, only hard structures are commonly preserved, and it would be desirable to know about soft tissues. Moreover, many invertebrates have no hard parts and thus are only rarely known as fossils. For these to be preserved, special conditions are required, for example, preservation in very fine silt such as forms the Burgess Shales from which some of the best invertebrate fossils are known (Figure 15-4). Finally, a serious difficulty with the fossil record stems from the fact that the nature of the evolutionary process tends to make it unlikely that the organisms representing "links" between groups will come to light as fossils.

This is regrettable because these links are the very specimens we would like to see. Obviously, population size influences the commonness of fossils. Linking forms, it appears, usually existed in small populations. This is probably due to the fact that linking forms tended to occur during periods when environmental changes had taken a heavy toll, leaving only small populations of organisms, possibly ill adapted to the new environment. Moreover, rates of evolutionary change tend to be more rapid in small than in large populations, and this, again, would contribute to the tendency for there to be few specimens of any given variant preserved.

Dating Fossils

If fossils are to be useful, their ages must be known. Two methods of dating are in use. The **stratigraphic method** is based on the principle that the oldest zone in a fossil deposit will be at the bottom with later zones layered serially above it. Even if the various zones have been disturbed by earth movements, they may often still be correlated with each other and with zones of similar age in other parts of the world by means of their characteristic fossil content—index fossils. Stratigraphic methods principally indicate relative age, whether one fossil is older or younger than another, but are not accurate as far as determinations of absolute age are concerned.

Since radioisotopes decay at known rates (see Chapter 26), it is possible to use the ratio of the concentration of certain isotopes to their decay products to measure the age of geological formations and fossils (Figure 15-5). These ratios are measured by detection of radioactivity from the remaining radioactive isotope or by mass spectroscopy, which directly determines the amount of the parent element and its decay products present by taking advantage of their differing mass numbers. The cyclotron has recently been put to use for this purpose and greatly increases both the sensitivity of the measurements and the time span over which measurements are useful. Using all the practically available isotopes, it is possible to measure time back to the most recent solidification of the rocks of the earth and, using carbon-14, the time of death of biological specimens is measurable from the present back 40 to 70,000 yrs.

The Study of "Living Fossils"

Extremely rarely an ancient organism persists unchanged into the present, perhaps because its habitat is an environment stable over long geological periods. An example is the lobe-finned coelocanth fish. Coelocanths are considered to have been in the path of development of terrestrial vertebrates and were assumed to have become extinct 250 million years ago. Consequently, in 1938 biologists were astounded by the capture of *Latimeria*, a representative of this group of lobe-finned fishes. By now over 80 specimens have been captured and studied—an unparalleled opportunity to look into the past (Figure 15-6).

Among plants the maidenhair tree, *Ginko biloba* (Figure 15-7), is the surviving species of a genus that has existed for as long as 150 million years. One of its remarkable primitive features is the production of motile sperm by the male tree.

COMPARATIVE ANATOMY OF EMBRYONIC STAGES

There is a tendency towards increasing anatomical resemblance between organisms at earlier and earlier stages of embryonic development. Early embryos of reptiles, birds, and mammals are superficially indistinguishable, which reinforces the strong evidence from fossils and from adult anatomy that these vertebrates had common ancestry. Among invertebrates, there are analogous examples, such as in the pattern of early cell division of annelid and mollusc embryos. This pattern is characteristic of these organisms and, at the same

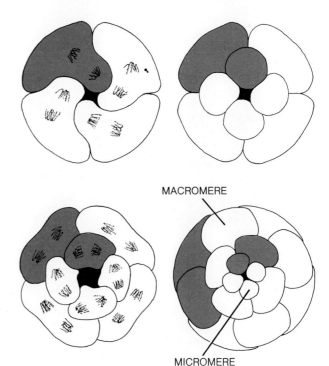

Figure 15-8 **Characteristic stages in cleavage of early annelid and mollusc embryos.** Asymmetric positioning of successive mitotic figures results in spiral disposition of micromeres at the animal pole of the embryo, atop the macromeres.

Figure 15-7 **Living plant fossil: leaves and mature seeds of the ginko tree.** This plant is known from fossils 150 million years old.

time, is so different from other invertebrates that common ancestry of the two phyla seems likely (Figure 15-8).

BIOCHEMICAL "MORPHOLOGY"

Obviously if organisms are morphologically different then they are chemically different, for as Wald puts it, "Living organisms are the greatly magnified expressions of the molecules that compose them."[1] Over the years several methods have been developed to reveal these chemical differences. The techniques of comparative serology permit rough comparison of the proteins of animals. If **antisera** to tissues of a particular animal are prepared, they will also react with tissues of related organisms with the intensity of

---

[1]Wald, G.: 1963. Phylogeny and ontogeny at the molecular level. In A. I. Oparin (ed.): *Evolutionary Biochemistry*. Macmillan, New York.

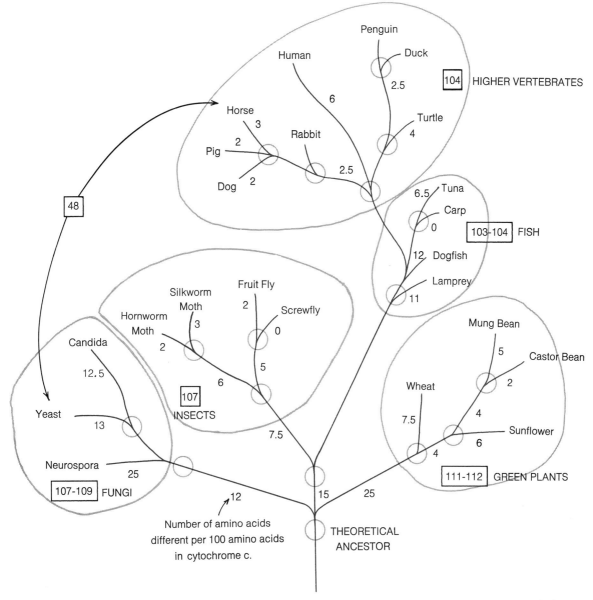

Figure 15-9 An evolutionary tree for cytochrome c. Each circle represents a computer estimation of the theoretical ancestral cytochrome c for organisms farther out on the tree. Numbers are the number of amino acids different per 100 amino acids in cytochrome c between successive ancestor molecules or between ancestor molecules and the named organisms. Numbers beside names of major groups indicate the total number of amino acids per cytochrome c in that group. The diagram shows that the largest difference in amino acid sequences is 48, between yeast and horse. Adapted from Dayhoff, M. O., et al., eds (1972) Atlas of Protein Sequence and Structure. National Biomedical Foundation, Washington, D.C.

the reaction diminishing the more distant the relation. Such comparisons can be carried a step further by actually comparing the chemical composition of selected proteins from various organisms. By protein "fingerprinting"—determining amino acid sequences—it is possible to follow the viscissitudes of evolution only a few chemical steps away from the primary changes of evolution, namely those in the genetic mechanisms controlling protein synthesis (Figure 15-9).

Genetic Analysis

In a few favorable organisms it is possible to work out evolutionary relationships by examination of chromosome pairing patterns. You recall from Chapter 11 that homologous chromosomes or parts of chromosomes pair in meiosis, and these may be identified by microscopic examination in organisms with suitable chromosomes. This method has, for example, been used in working out the evolution of New World cottons, *Gossypium*, showing their origin as natural hybrids of Asiatic and American species.

As we have seen earlier (page 175) some flies, particularly the fruit fly, *Drosophila*, have extremely large salivary gland chromosomes, which bear patterns of bands making possible identification of even very small regions of chromosomes, even if they are translocated from one chromosome to another or to another site on the same chromosome. These chromosomal mutations of course contribute to species formation and, moreover, they may be followed from one natural population of *Drosophila* to another as markers of the path of evolution.

Look Alikes May Not Be Alike, and Vice Versa

Parts of organisms may reflect common ancestry, in which case they are said to be **homologous;** or they may look or function alike but actually have different ancestry, in which case they are said to be **analogous.** Figure 15-10 shows the upper limb bones of man, mole, and whale. These structures superficially look rather different because of the different means of locomotion to which they are adapted. Closer inspection shows underlying similarities. There is a single upper limb bone (humerus) in each, followed by a pair of bones (radius and ulna), followed by a complex set of hand or flipper bones. Although these are structurally different in each form, they still have the same underlying organization; for example, each has five metacarpal bones. This basic organization can be traced back to a common ancestor similar to the lobe-finned fish, and so we are entitled to call the limb structures of these several organisms homologous and use their similarity as an argument for concluding that they are related. If we added to the sequence of limbs the legs of a crab, these could only be called *analogous* because they perform the same function but do not develop in the way that the limbs of the three vertebrates do.

More difficult versions of the crab leg versus verte-

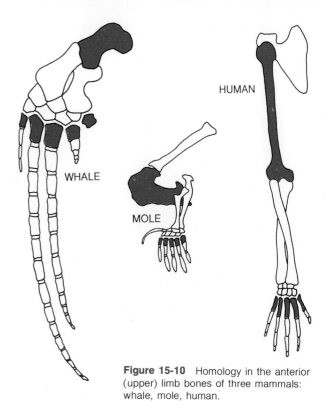

Figure 15-10 Homology in the anterior (upper) limb bones of three mammals: whale, mole, human.

brate leg problem are recognized in the phenomenon of adaptive **convergence.** A well-known instance is the adaptive convergence of body shape that has taken place in vertebrates inhabiting water. Fish, reptiles, birds, and mammals that have taken up a swimming existence have come to bear superficial resemblance to each other (Figure 15-11). This particular convergence is not complete enough to suggest anything other than the adaptation of representatives of four distinct groups of organisms to an environment in which successful ways of life are strictly limited. That is, the ability to swim rapidly is valuable to an aquatic predator, and the most energetically economical body form for rapid swimming is spindle shaped. Thus, the convergence does not indicate close evolutionary relationship, a conclusion confirmed by the fossil record. However, if we do not know the adaptive significance of similar traits, then we cannot decide whether similarity signals evolutionary relationship or whether it only signals a common solution by unrelated forms to problems of existence unknown to us. The phenomenon of convergence is not limited to animals as illustrated by Figure 15-12 which shows the effects of a dry habitat on plant structure.

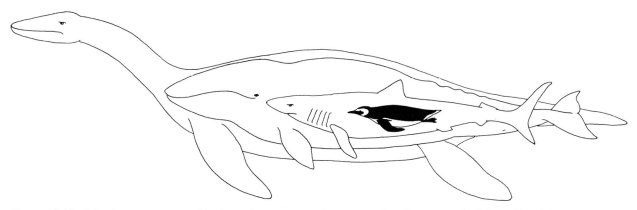

Figure 15-11 **Adaptive convergence.** Members of four classes of vertebrates (reptiles, mammals, fish, and birds) inhabiting the marine environment have highly similar body form.

Figure 15-12 **Adaptive convergence.** Two unrelated plants, the cactus and the jade plant, have undergone an increase in the ratio of volume to surface, an adaptation to a dry environment, resulting in similarity of general form.

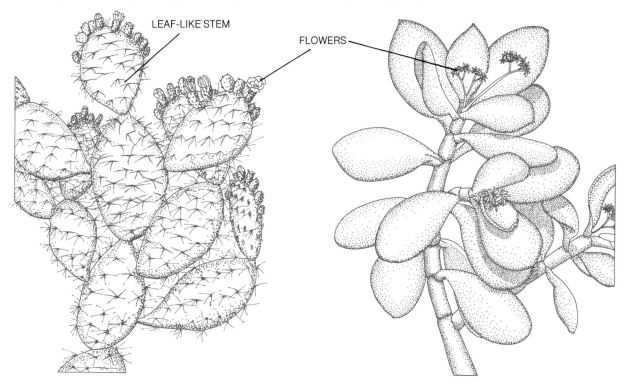

The Significance of Molecular and Biochemical Similarity

At biochemical and ultrastructural levels of organization, it is extremely difficult to determine if similarity denotes common ancestry of organisms because we do not know how many possible solutions there are to the chemical ultrastructural, and cellular requirements of life. Similarity may be because there is no other way to solve a particular biochemical or ultrastructural problem. We know, for example, that all organisms employ an identical nucleic acid coding system for directing the synthesis of their proteins.

THE VARIETY OF LIFE 241

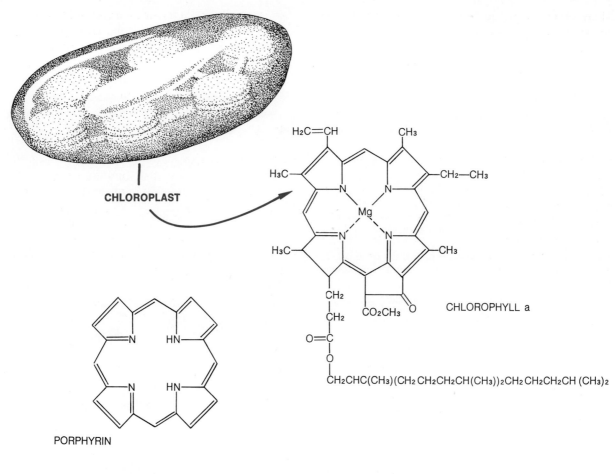
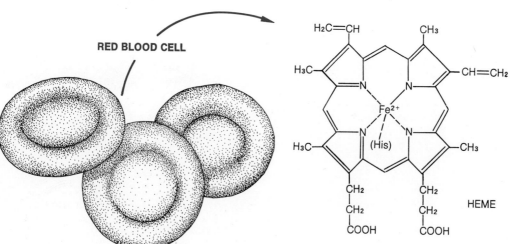

Figure 15-13 Universality of the tetrapyrrole group suggests its very early evolutionary origin. The basic tetrapyrrole structure (porphyrin) appears in such widely diverse environments as the chloroplast, where it is found in chlorophyll, and red blood cells, where it is part (heme) of hemoglobin. In the heme molecule, (His) is the amino acid histidine, attachment point of heme to the protein part of hemoglobin.

Many other less ubiquitous biochemical similarities are known, which are still remarkably widespread. Among these is the tetrapyrrole group (Figure 15-13). It occurs as the iron-carrying prosthetic group in animal blood pigments; it is part of the universal cellular respiratory pigment, cytochrome, and with magnesium substituted for iron, it is found in chlorophyll. Such biochemical similarities are taken by many as an argument for a common, single, origin of all life and to this argument is added evidence for widespread similarities at the ultrastructural and cellular level. Thus cilia and flagella, which occur throughout the living world, are always remarkably similar in structural details (Figure 15-14).

However, the very universality of the biochemical and ultrastructural traits just described actually constitutes an argument *against* their indicating origins from a common ancestor. For example, if there is only a *single* chemistry of heredity, is its uniqueness due to its having been the most successful of various possible systems or is it due to being the *only possible system*? If it is the former, then a single origin of life becomes a plausible hypothesis; if it is the latter, then more than one evolving system may have arrived at the hereditary mechanism we now recognize as universal. Such universality would be attributed to adaptive value rather than to domination of a single variety of organism in which our present hereditary system first appeared.

Turning to the example of the tetrapyrrole group, similar arguments are possible. At first glance this structure appears sufficiently unique to suggest that its present commonness in living systems represents the preservation and spread of a rare chemical accident. But consider that the precursors of the tetrapyrrole molecule have been experimentally synthesized under primitive earth conditions and that even the isolated iron atom exhibits some degree of the properties of iron ensconced in the tetrapyrrole of hemoglobin or cytochrome. Thus, the gradual evolution by plausible, adaptive steps from universally available precursors seems likely, and to this extent

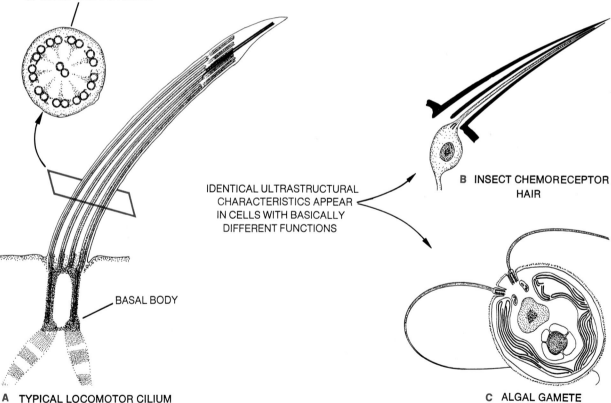

Figure 15-14 Cytostructural similarity, the universal cilium.

THE VARIETY OF LIFE

the appearance of tetrapyrrole-metal catalytic systems becomes considerably more probable in evolution than as the result of unique accident. Finally, convergence at the molecular level must be admitted to be considerably more probable than at the level of cell, tissue, or organ owing to the simple fact that fewer genes are involved in the specification of a given molecule than of a supramolecular structure.

Similar arguments are applicable to organizational details at ultrastructural and cellular levels. Consider mitosis: there would be immense adaptive value to development of a system for equitable sharing of the genetic substance between daughter cells in cell division. Moreover, how many simple mechanisms for attaining this end are possible in addition to the mitotic system with linear arrangement of genetic material that now predominates? To continue, perhaps, since all cilia are essentially of the same structure, their uniformity demonstrates only that, given the materials available to living systems, the most adaptive way to construct a whip-like locomotor organelle is according to the universal pattern that we see.

The Variety of Life

FIVE KINGDOMS: THE MAJOR KINDS OF ORGANISMS

It is a tribute to the skill of systematists and to the generally wide extent of biological knowledge that the millions of living and fossil organisms known to science can be logically organized under only five major headings, or **Kingdoms**, as shown in Figure 15-15.

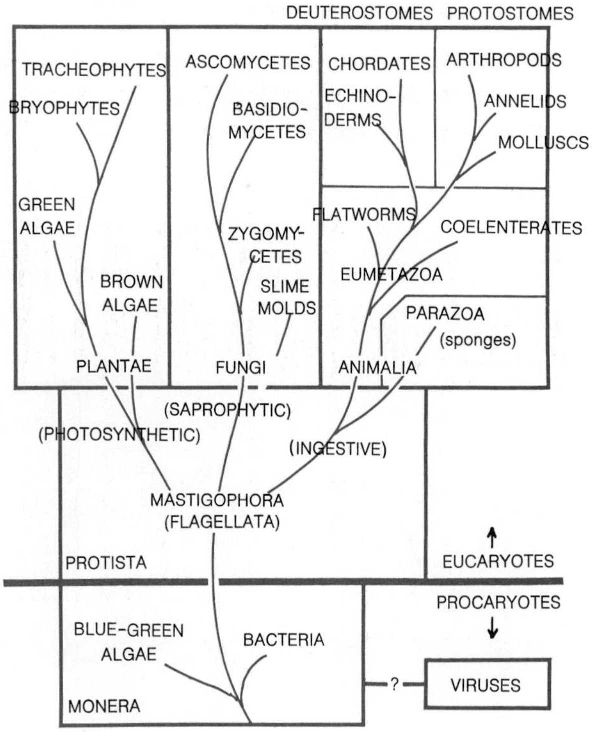

The first **Kingdom** is called the **Monera** and you are already familiar with part of it, for it includes the prokaryotes: the bacteria and blue-green algae. The remaining four Kingdoms, exclusively eukaryotes, are named **Protista**, **Fungi**, **Plantae**, and **Animalia**. These five categories include all living things except viruses. It must be confessed that there is indecision as to where these should be placed. Some authorities suggest viruses are truly primitive forms of life and warrant a Kingdom of their own; others argue that they are derived from more advanced organisms by extreme simplification. Certainly as they now exist, viruses do not belong to any of the five Kingdoms since they are noncellular in organization.

Figure 15-15 shows a very general scheme of how the Kingdoms are related to the presumed ancestral forms of life that first appeared on earth. The Protista, which are single celled eukaryotes, appear to represent the first step beyond the prokaryotes. In particular one group of protistans, the flagellates or Mastigophora, seem well suited as candidates for the ancestors of the three remaining Kingdoms: the Fungi, Plantae, and Animalia. The flagellates include *Euglena*, which is photosynthetic and yet has the motility that one would expect of an animal, and includes

Figure 15-15 Five Kingdoms encompass all known organisms, except viruses.

multicellular organisms such as *Volvox*, which some take to be a model for multicellular organisms. Thus, the present day flagellates appear sufficiently generalized to suggest that their ancestors might have given rise to the three Kingdoms of multicellular organisms.

Single Versus Multiple Roots to the Living Kingdoms

Evolutionary schemes such as we have just described are called **monophyletic**. They presume that each major group of organisms had a single ancestral type. According to this concept, the Monera are thought to have given rise to the Protista by the evolution of one restricted group of monerans. Unfortunately, there is insufficient evidence to determine whether the Protista originated in this way or from more than one source within the Monera. The reason is that homology and analogy in such simple organisms must be examined at such elemental levels of biochemistry and ultrastructure that decisions are almost impossible because there are so few plausible alternatives (recall page 243). Even at more recent levels of evolution, the problem of mono- versus **polyphyletic** origins occasionally arises. For example, the largest group of the Kingdom Animalia, the Arthropoda, comprises a vast assemblage of organisms with many important structural features in common. Yet the arthropods have so confounded attempts to work out the branchings of their evolutionary tree that sometimes the phylum is suggested to have arisen polyphyletically and to owe what common characteristics it expresses to convergent evolution rather than to common ancestry.

These are questions that must be acknowledged. However, they do not detract from the great principal before us, namely that all organisms, however different they may appear, are genetically related. Generally, our guiding principal will be to assume monophyletic connections between groups of organisms because this is the most simple hypothesis and therefore should be adopted unless we are faced with good evidence to the contrary.

THE KINGDOM MONERA

The Monera are all unicellular organisms at the prokaryote grade of organization. As you know this means that they lack nuclei, that their genetic material is contained in one strand of DNA, and that they lack mitochondria, chloroplasts, and endoplasmic reticulum. They represent a very diverse group of organisms that can roughly be divided into the blue-green algae and the bacteria. Together they play very important roles in the economy of the planet. Some bacteria and all blue-green algae are able to carry on photosynthesis. Some members of the Monera are able to fix atmospheric nitrogen, that is, incorporate it into biomolecules. The bacteria are particularly adept at the process called mineralization, in which many types of biomolecules are broken down to their ultimate constituents, carbon dioxide, ammonia, sulfate ion (SO_4^{2-}), and so on. Since the supply of such molecules available to living things is limited, it is of obvious importance to have organisms such as the bacteria, which are able to complete the cycle of materials through the natural world and, in a sense, restore chemical resources for further biological use. Bacteria, as you know are, with viruses, the principal agents of disease.

Blue-green Algae

Blue-green algae are widely distributed in the seas, in fresh water, and on land. Some of them share the ability with some bacteria of living in extremely hostile environments, for example hot springs with temperatures as high as 75°C. They are single celled but often occur in short filaments of adhering cells and in simple colonies. The cell is usually enclosed in a mucilaginous secretion and, most curiously, it is capable of a gliding motion of unknown cause. Although blue-green algae carry on photosynthesis, they do so with a chlorophyll that is somewhat different than the chlorophyll of higher plants since it is not housed in chloroplasts. The cell membrane serves both in bacteria and blue-green algae as a mechanical base for the complex enzyme systems that, in higher organisms, would be housed in mitochondria and chloroplasts.

Bacteria

The major kinds of bacteria are spirochaetes, rickettsias, and true bacteria. Spirochaetes always have helical shaped cells that move in a curious fashion, by contractions of the helix caused by the motion of cilia that lie between the cell membrane and the flexible cell wall. Although there are not many kinds of spirochaetes, they include a serious human disease agent, *Treponema pallidum*, the syphilis organism.

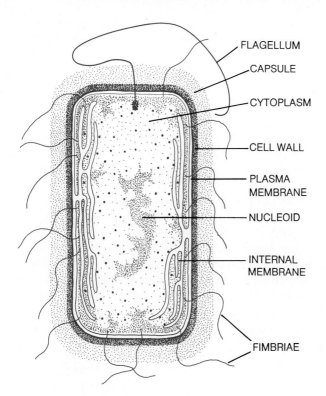

Figure 15-16 Diagram of a generalized bacterial cell.

are not to be confused with viruses because they have a definite cellular organization and possess cell wall materials, various proteins and enzymes, and both DNA and RNA. They are named in honor of their discoverer, H. T. Ricketts, who died of the rickettsial disease, typhus.

The true bacteria are a large and very diverse group. However, all have rigid cell walls and, if they move, do so by means of flagella. Figure 15-16 and Table 15-2 illustrate some of the diversity of form and metabolism in bacteria. This metabolic diversity together with the immense numbers of bacteria—the mass of bacteria probably equals the mass of all animal tissues on the planet—makes them extremely important in completing biological cycles. They obviously play central roles in human disease, as will be described in Chapter 23.

KINGDOM PROTISTA

The unicellular, eukaryotic organisms that make up the Kingdom Protista have been a cause of confusion

The rickettsias are the smallest of bacteria and they survive only in living cells; although they are very small and are obligate parasites of living tissues, they

Table 15-2 Bacterial Biochemical Diversity
These examples show that bacteria obtain energy and organic molecules in more different ways than do all other organisms.

Energy Metabolism	*Synthetic Metabolism*

Fermentation—anaerobic breakdown of organic molecules.
     glucose to lactic acid—lactic acid bacteria
 Respiration—aerobic breakdown of organic molecules
     glucose to methane—marsh gas bacteria
	Photoautotrophism—photosynthesis: in bacteria the H donor is never water and oxygen is not released.
	     H donor H_2S—green sulfur bacteria
	     H_2—hydrogen bacteria
	     organic molecules—purple bacteria
	 Chemoautotrophism—organic carbon obtained from CO_2 using energy from reduced inorganic compounds.
	     Energy source
	     H_2S—sulfur bacteria
	     Fe^{+2}—iron bacteria
	     NH_3—nitrate bacteria
	 Heterotrophism—one to many organic molecules required
	     Includes many disease causing bacteria with metabolic requirements such that they can survive only in other living organisms or in special culture conditions. For example, *Streptococcus* requires both lactic acid and porphyrins.

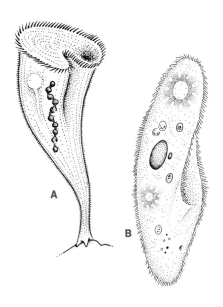
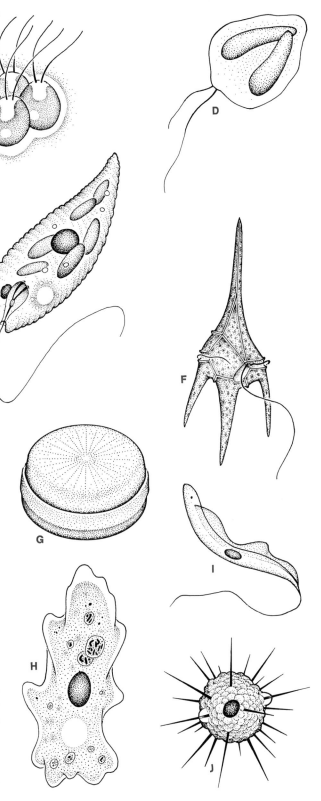

Figure 15-17 **Some protistans. A.** *Stentor*, **B.** *Paramecium*, **C.** *Gonium*, **D.** *Ochromonas* (gold alga), **E.** *Euglena* (a "plant-animal"), **F.** *Ceratium*, **G.** diatom, **H.** ameba, **I.** trypanosome, **J.** heliozoan.

among taxonomists because many of them possess characteristics of both plants and animals. Most are microscopic, live in water, and are motile, either by pseudopodia, as in the ameba, or by flagella or cilia, as in *Euglena* or *Paramecium*. Some are typical heterotrophs, obtaining organic food by ingestion or absorption, whereas others are photosynthetic, carrying on photosynthesis in chloroplasts indistinguishable from those of higher plants. *Euglena* illustrates the problem of classification quite well: if the organism is kept in the dark long enough, it will lose its chloroplasts permanently but will survive by taking up organic matter like an animal.

The Kingdom Protista was defined largely to avoid having to make decisions as to whether such organisms are animals or plants and also in recognition of the fact that certain protistans appear to be closely related to ancestral forms of animals and plants. The Kingdom may be further divided into animal-like and plantlike forms, as shown in Figure 15-17. The relationship between the plantlike protistans and the multicellular algae among modern plants is, in fact, so strong that it is most difficult to decide whether the unicellular algae should be placed in the Protista or among the true plants, the Kingdom Plantae. Such problems illustrate the continuous nature of the evolutionary process. Indeed, if it is true that all organisms are evolutionarily related, it would be surprising if situations of this kind did not frequently come to light.

The Animal Protists

The Sarcodina are protozoans that move by means of pseudopodia. Besides the familiar ameba of ponds and ditches, the group includes important parasites of man, such as the causes of amebic dysentery, *Entamoeba histolytica,* and of the several types of malaria, *Plasmodium.* Although the cytoplasmic structures of the sarcodines appears to be quite simple, some of them form remarkably complex and often beautiful skeletal structures (Figure 15-17). Heliozoans (sun animals) get their name from the fact that their needlelike pseudopods radiate from the spherical central body like the rays of the sun. Each pseudopod is supported by a stiff central rod of protoplasmic fibers that can shorten or extend. Sticky protoplasm covering the axial rod serves to capture food.

External skeletons are formed of calcium carbonate by foraminiferans and of silica by radiolarians. These skeletons can become large, several millimeters in diameter in solitary forms and even larger in colonial forms. The skeleton is always perforated by numerous holes through which pseudopods are extended to capture food. How these complex and precisely formed skeletal structures are developed is completely unknown and their adaptive significance is obscure. They may be to some extent protective and may serve to aid flotation by increasing the resistance of the organism to sinking in water. Radiolarians and foraminiferans are, and have for ages been, successful forms of life in the sea. They form such a major component of the plankton that their skeletons accumulate on the ocean floor to form geologically significant deposits. Great limestone and chalk deposits such as the white cliffs on the English Channel are formed of foraminiferan skeletons.

The most structurally complex of the animal protists are members of the Ciliata, which includes the familiar *Paramecium.* Members of this group are covered with cilia, structurally similar to but shorter than flagella. Coordinated beating of cilia is used in locomotion; in ciliates such as *Paramecium* ciliary action can be regulated so as to produce forward or backward movement to avoid noxious substances or to approach food or oxygen. A permanent mouth is present. In forms like *Paramecium,* specially placed cilia automatically sweep small particles into the mouth where they are taken up in food vacuoles for digestion by enzymes secreted into the vacuoles. At the completion of digestion, the remaining contents of the digestive vacuoles are extruded through an anal pore. Some ciliates are predatory upon other protozoans and are assisted in this activity by greatly distensible mouth structures, which are even equipped with muscle-like contractile bands. Contractile vacuoles are present in ciliates as well as in other protistans, where they serve in maintaining water balance (recall page 69). These and other cytoplasmic specializations argue that the ciliates are the most complex of single celled organisms and probably represent the culmination of evolution without multicellularity.

The protistan group most closely related to the single celled organisms that gave rise to higher organisms appears to be the Flagellata. The name means whip-bearer and reflects the fact that members of the group have as locomotor organelles one or more flagella.

Reproduction in Animal Protists. Protists may reproduce asexually by straightforward mitotic cell division, which is sometimes complicated by the presence of specialized organelles that must also be reproduced. Sexual reproduction also occurs, sometimes by fusion of two cells or sometimes by exchange of nuclear material in a process called **conjugation** (Figure 15-18). In some forms a specialized form of asexual reproduction, called **spore formation,** occurs by the sudden formation of many nucleated individuals after a period of nuclear divisions without cytoplasmic divisions.

The Plant Protists

Photosynthetic protists provide a fundamental source of food for marine organisms as well as much of the planet's oxygen. Diatoms and golden algae, the Chrysophyta, are plentiful enough to be called the grasses of the lakes and oceans. Diatoms live in cases of secreted siliceous material, elaborately sculptured but fundamentally constructed like a box with an overlapping lid (Figure 15-19). Diatomaceous earth consists of deposits of the shells of diatoms accumulated over the ages and now put to practical use in the purification of liquids and as polishing agents. The

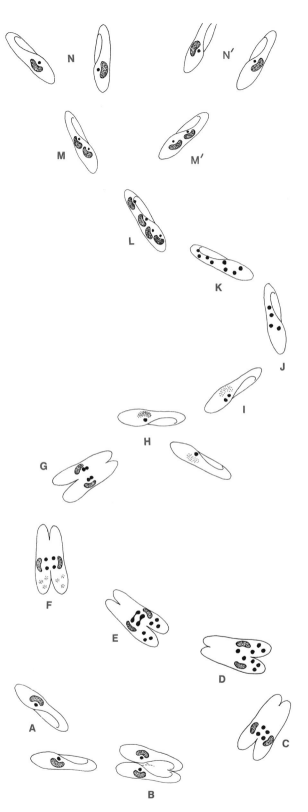

Figure 15-18 Conjugation in *Paramecium caudatum*. A–B. Paramecia of different mating types ("sexes") contact and fuse together. **C–D. Micronuclei** divide meiotically, producing four haploid micronuclei. **E–F.** One haploid micronucleus of each individual divides mitotically across the plane between them, resulting in reciprocal exchange of micronuclei. The other three micronuclei in each degenerate. **G.** Donor and recipient micronuclei fuse, restoring the diploid state. **H.** The pair separates and the **macronuclei** disintegrate. **I–K.** The diploid micronucleus divides mitotically three times, producing eight micronuclei. Four become macronuclei. **M,M'** and **N,N'.** Two binary fissions result in four paramecia, each with one macro- and one micronucleus.

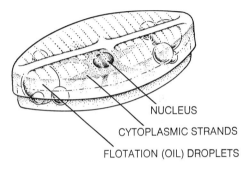

Figure 15-19 A diatom, showing its box-like silicaceous encasement. Each half box is a frustule. At cell division each daughter cell gets one old frustule and synthesizes another.

diatoms and their relatives, the golden algae, which lack silicaceous cases, together carry out about 90% of the photosynthesis that occurs on earth.

The dinoflagellates are a second major group of plantlike protists contributing heavily to photosynthesis in the seas. Their structure is unusual; characteristically they are armored by cellulose plates and they may bear two flagella beating at right angles to each other in grooves in the external surface. Some members of this group are responsible for the bioluminescence seen in breaking waves or ship's wakes. Others produce an extremely toxic poison when they occur in large enough numbers to produce the well-known red tide of coastal waters. Red tides are lethal to fish and their predators.

Finally, among the plantlike protists, the euglenoids are confined to fresh waters. Although most can carry on photosynthesis, in structure and behavior they have many animal characteristics (Figure 15-17). Instead of a cellulose or silicaceous housing, the euglenoids typically have a flexible, internal protein sheathing. Some have a gullet or oral groove capable of taking up particulate food. Locomotion is by one to several flagella and, associated with these, is an eyespot that mediates orientation to light.

THE PLANT KINGDOM: ALGAE

Ranging from single celled, microscopic individuals to the largest marine plant, the giant kelp, *Macrocystis* (Figure 15-20F), algae span the entire gamut from the simplest of plantlike protists to complex, multicellular plants. All possess the typical plant characteristics of cellulose cell walls and of photosynthesis conducted within chloroplasts. They differ from higher plants in that most of them have flagellate gametes requiring an aquatic medium to effect fertilization and in that most do not have the complex tissue organization characteristic of higher plants. Although the greatest variety of algae are marine, there are fresh water forms as well as algae that live in moist terrestrial environments, for example the "moss" that grows on tree trunks in damp forests.

There are three primary divisions of algae, the Chlorophyta (green), Phaeophyta (brown) and Rhodophyta (red) algae. As the names indicate, the brown and red algae contain colored molecules in addition to chlorophyll. Phycobilins, or linear tetrapyrrols, are the molecules that produce the red color of red algae. These molecules contribute significantly to the photosynthetic capacity of the plants for the following reason: when light descends through sea water the red wavelengths, which chlorophyll absorbs best, are filtered out early so that below about 20 m, depending on water clarity and latitude, photosynthesis is no longer maximally efficient. However, below these depths and down to about 100 m, there is sufficient blue-green light still penetrating to support photosynthesis if it is captured, and this is what the phycobilins of the red algae do. Phycobilins capture blue-green light (and hence *look* red) and transfer its energy to chlorophyll, thus allowing the red algae to live in water of greater turbidity or at depths otherwise forbidden to photosynthetic organisms.

Brown Algae: Anticipation of Terrestrial Life

Brown algae are dominant plants of the shallow and surf zones of cold seas: for example, the *Fucus* of the North Atlantic shores or the kelps of cool Pacific waters. They represent the highest level of structural complexity among algae, and in some respects so clearly evoke the organization of land plants that it is natural to suggest an evolutionary link, although this is not documented by fossil evidence.

The growing tip of a brown alga such as *Fucus* lays down a cap of small, photosynthesizing cells. Unlike the situation in simpler algae, the cells are able to divide both longitudinally and transversely to produce a strong, interlocking cellular column, rather than the thin sheets or filaments of other algae. Cells on the outer surface of the column retain photosynthetic capacity and the ability to exchange water and inorganic nutrients with the surrounding sea; in a sense they are almost as unspecialized as any free-living single celled alga, except for being anchored to the plant body. Interior cells, shaded from light, lose photosynthetic capacity and elongate, providing strength to the extending column. They are supplied with nutrient by the outer, photosynthetic cells. Some interior cells undergo structural modifications similar to those seen in vascular tissues in higher land plants, rendering them able to transport nutrients. Water transport longitudinally is unnecessary because the whole plant is submerged in water and is permeable over its entire surface.

Since the most advantageous site for photosynthesis is at the well illuminated water surface, the brown algae have developed regional specializations to facilitate the reach for the surface that foreshadow the tripartite organization of higher land plants into root, shoot, and leaves. A kelp such as *Macrocystis* forms a large holdfast of intertwining, tough rhizoids that grip the bottom and anchor the cablelike stipe, which may reach 50 ft to the surface where it flattens out into a long, leaflike blade that is the principal site of photosynthesis. Such a plant seems almost ready to stand erect on land.

Algal Reproductive Cycles. In the algae there are many examples of the alternation of generations characteristic of higher plants. As demonstrated in Figure 15-20 the life cycle may include two kinds of multicellular plants, a **sporophyte** and a **gametophyte.** The sporophyte results from sexual reproduction and is diploid (2N). By meiosis it produces haploid spores that develop singly into haploid (N) gametophyte plants, which, without meiosis, produce gametes to

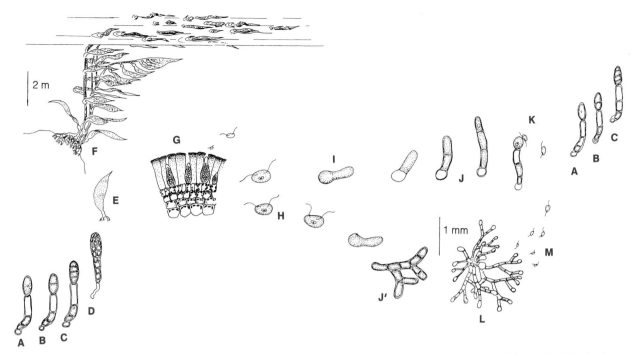

Figure 15-20 Life history of the kelp, *Macrocystis*. **A–E.** Cell division in the zygote and development of the first leaf-like lamina. **F.** Mature diploid **sporophyte**. **G.** Production of haploid **zoospores** by meiosis in specialized aeas of the lamina. **H.** Zoospores. **I.** Germinating zoospores. **J–K.** Development of female **gametophyte** with egg release. **J'.** Young male gametophyte. **L.** Mature male gametophyte releasing sperm. **K.** Fusion of egg and sperm to form zygote.

complete the cycle at fertilization, giving rise to sporophytes. Sporophyte and gametophyte may be virtually identical in algal life cycles, or one or the other may be dominant. In higher plants an important evolutionary theme is reduction and modification of the gametophyte in accomodating the reproductive cycle to dry terrestrial conditions unfavorable to the free-swimming gametes characteristic of algal gametophytic reproduction.

LAND PLANTS

Once plants similar to the brown algae became established along the shoreline, presumably there was a great tendency to move ashore. Lighting conditions would of course be even better there than in the shallow border waters and, initially, there would have been no competition for space. The ocean borders are an environment of limited dimensions filled with plant life, so it would appear inevitable that the comparably limitless terrestrial environments would have been invaded by plants. From the situation seen in the large brown algae, relatively minor adjustments would have had to be made in plant organization and function to accommodate the transition. Exterior surfaces required waterproofing and this, in turn, necessitated provision of access of respiratory gases to photosynthetic cells. Conducting systems would have required further development to carry water as well as nutrients, and holdfasts would have had to take on the additional role of water and mineral gathering organs.

Details of the transition of plants from sea to land are unknown. It is not even known whether it occurred more than once. Neither is it known whether the transition was directly from sea to land or if the intermediate staging area of estuaries and tidal marshes was used, as appears to have happened with animal life. In any case, two principal groups of higher plants emerged onto land. They are the **Bryophyta** and **Tracheophyta**. The former are the true mosses and liverworts, whereas the latter are the vascular

Figure 15-21 Life history of a moss. Habit sketch of a moss showing gametophytes **(A)** and the diploid sporophytes **(B). C.** Tip of gametophyte branch containing the male sex organs, the **antheridia (D)** that produce sperm **(E). F,** tip of gametophyte branch containing female sex organs, the **archegonia (G).** Sperm are produced mitotically and released when wetted. They are attracted chemically to the archegonia where they fertilize the egg. The resulting diploid zygote develops into the sporophyte **(B, H, I).** Haploid spores are produced within the sporophyte capsule. Released spores grow into haploid plants **(protonemata, J)** from which sprout the gametophytes **(K).**

plants: ferns, horsetails, and the dominant land plants, the seed plants.

Bryophytes. These small plants, principally confined to moist environments, have no true roots, stems, or leaves and lack a vascular system (Figure 15-21). Their reproductive cycle involves well-developed gametophyte and sporophyte generations, and the gametophyte produces motile male gametes that must swim to effect fertilization. Although these characteristics suggest that bryophytes might link the tracheophytes with marine plants, this does not appear to be so; they represent an independent and terminating lineage. The group is relatively small, including two principal types, liverworts and mosses (Figure 15-21).

Tracheophytes. By far the most successful land plants, the tracheophytes or vascular plants, solved all of the problems of terrestrial life that the bryophytes were incapable of solving. In addition, the tracheophytes evolved certain innovations that have allowed them to take advantage of all land environments from deserts to wet tropics, even cold regions where the living substance is frozen several months out of the year. The most successful of this group are the flowering plants with more than 250,000 known species. These plants range from delicate annual flowers to trees that may live 5000 years or that attain a height of 350 ft. Thus the tracheophytes include the longest lived and largest of organisms. This great host of vascular plants is the driving force of terrestrial life, capturing and converting into useful material the energy of the sun.

The reasons for the success of vascular plants on land are evident in (1) structural developments involving specialization of the plant into functional regions (roots, leaves, shoot) and the elaboration of vascular tissues interconnecting them, and (2) great improvements in reproductive mechanisms that reduce the requirement for water and improve the chances of wide distribution of reproductive products.

Primitive Vascular Plants

The most ancient of undisputed vascular plants are nearly entirely extinct psilopsids (Figure 15-22A). Al-

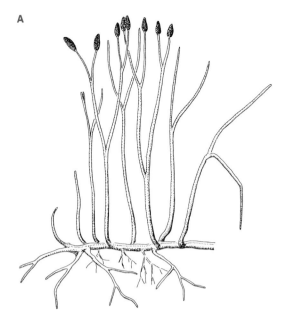

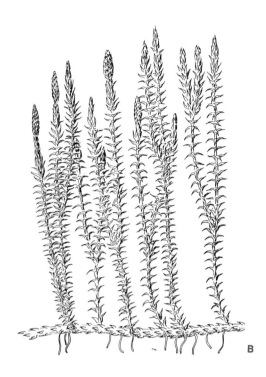

Figure 15-22 **Primitive vascular plants. A.** A member of the Psilophyta, which flourished more than 300 MY ago. Spore forming structures are seen at the tips of upright stems. **B.** A club moss, one of the Lycophyta. **C.** A horsetail, member of the Sphenophyta. The cone-like structures at the tips of some shoots are reproductive organs.

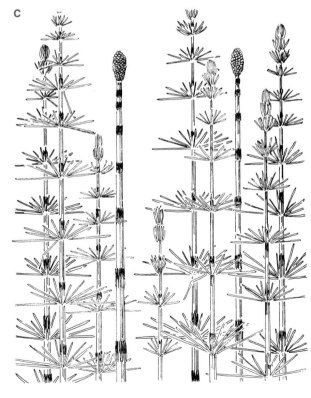

though known to have had xylem and phloem, they had no leaves. Photosynthesis was carried out by tissues in the periphery of stubby upright stems. The stems were attached to a system of rhizoids instead of roots. The life cycle is unknown, but comparison with the most closely related modern plants suggests that the dominant plant was a sporophyte whose spores produced a very small gametophyte that gave rise to motile sperm and eggs. The gametophyte probably developed under moist soil and was not photosynthetic, living as a **saprophyte,** consuming dead plant and animal substance.

The most primitive of living vascular plants are the club mosses (lycopods, Figure 15-22B) and horsetails (sphenophytes, Figure 15-22C). The club mosses are advanced beyond the psilopsids since they have leaves and roots. They were quite successful early in their evolutionary career, although they are now an insignificant element in terrestrial plant life. Once they formed great forests of large trees, which contributed massively to present-day coal deposits. The horsetails had a similar history; today they are represented by only about two dozen species, all confined to moist

THE VARIETY OF LIFE

environments. Both the club mosses and horsetails have similar life cycles, with a small gametophyte that produces motile sperm.

Ferns

The ferns, or filicinophytes, have a life cycle much like that of the preceding groups, with motile sperm, but they have overcome this handicap and have become very effective competitors of the seed plants in appropriately moist environments (Figure 15-23). There are more than 10,000 species of ferns living in many regions of the earth, although their center of dominance is in the tropics. There they form a major part of the vegetation in lowland forests, and even become treelike in the highlands. The reason for their margin of success seems to be the elaborate development of their leaf structure which enables them to remain competitive in dark environments such as the forest floor.

Seed Plants

Dominance of seed plants over other terrestrial plants is largely ascribed to the great improvements in reproductive biology characterized by pollen and seeds. In the seed plant, the gametophyte is parasitic on the dominant sporophyte so it does not lead a separate existence, as in lower plants.

Two kinds of spores are produced by the sporophyte but not released. Megaspores produce a female gametophyte, and this remains within the tissues of the sporophyte, eventually forming the egg (recall page 226). To allow outbreeding, a way to promote exchange of sperm from the gametophytes of different sporophytes was developed, namely, a tiny, heavily waterproofed and armoured male gametophyte, the pollen grain. Within the sporophyte, microspores produce the pollen grain, or gametophyte, which is designed to be borne from plant to plant by wind or other means. In only one seed plant, the ginko tree, the pollen grain produces a motile sperm; in all others the fertilization process is accomplished exclusively by growth of the pollen tube conveying the sperm nuclei to the egg. Thus, exchange of genetic material between plants is effected without need for a water-dependent sperm cell.

The seed is a remarkable adaptation for preservation of the embryo sporophyte against dessication and other hazards. It provides for wide distribution of embryos from the parental organism and carries a

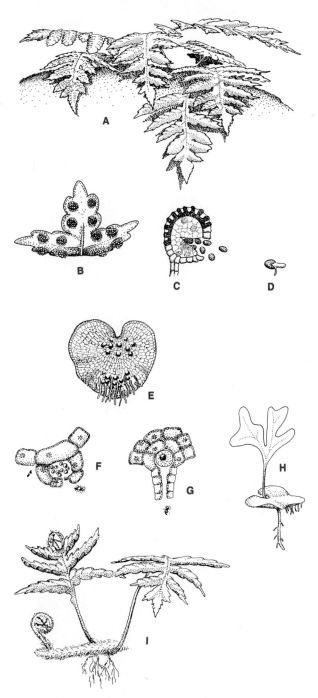

Figure 15-23 Life history of the fern. Mature diploid sporophytes **(A)** bear sori **(B)** on undersurface of fronds. Each sorus is composed of many sporangia **(C)** which produce spores by meiosis. The spores germinate **(D)** and develop into haploid gametophytes **(E)**. Each gametophyte bears sperm producing antheridia **(F)** and egg producing archegonia **(G)**. The young sporophyte **(H)** develops from the fertilized egg and spreads by means of underground stems (rhizomes, **I**).

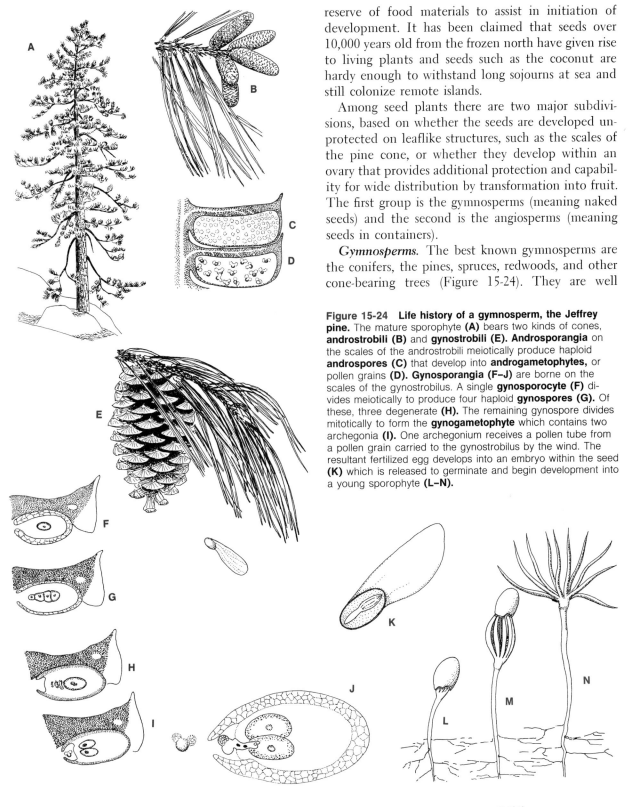

reserve of food materials to assist in initiation of development. It has been claimed that seeds over 10,000 years old from the frozen north have given rise to living plants and seeds such as the coconut are hardy enough to withstand long sojourns at sea and still colonize remote islands.

Among seed plants there are two major subdivisions, based on whether the seeds are developed unprotected on leaflike structures, such as the scales of the pine cone, or whether they develop within an ovary that provides additional protection and capability for wide distribution by transformation into fruit. The first group is the gymnosperms (meaning naked seeds) and the second is the angiosperms (meaning seeds in containers).

Gymnosperms. The best known gymnosperms are the conifers, the pines, spruces, redwoods, and other cone-bearing trees (Figure 15-24). They are well

Figure 15-24 Life history of a gymnosperm, the Jeffrey pine. The mature sporophyte **(A)** bears two kinds of cones, **androstrobili (B)** and **gynostrobili (E)**. **Androsporangia** on the scales of the androstrobili meiotically produce haploid **androspores (C)** that develop into **androgametophytes,** or pollen grains **(D)**. **Gynosporangia (F–J)** are borne on the scales of the gynostrobilus. A single **gynosporocyte (F)** divides meiotically to produce four haploid **gynospores (G)**. Of these, three degenerate **(H)**. The remaining gynospore divides mitotically to form the **gynogametophyte** which contains two archegonia **(I)**. One archegonium receives a pollen tube from a pollen grain carried to the gynostrobilus by the wind. The resultant fertilized egg develops into an embryo within the seed **(K)** which is released to germinate and begin development into a young sporophyte **(L–N)**.

THE VARIETY OF LIFE 255

adapted for cold and dry climates where they are so successful that they produce well over half of the world's lumber supply. Conifer leaves (needles) have reduced surface area and a dense cuticle that protects against dessication; they are usually evergreen. Pine cones are of two types, small pollen-bearing cones and large, seed-bearing cones. Depending on the species, a tree may produce both types or only one. Fertilization and embryo development are similar to the process in flowering plants except that double fertilization with the formation of an endosperm does not take place (See page 226). Reserve materials for the embryo are laid down by other means. This occurs with or without fertilization. The conifer is, to this extent, wasteful of its resources, in contrast to a flowering plant where the process of double fertilization insures that reserves are not laid down unless fertilization has taken place.

Angiosperms. Evidently quite soon after their appearance on earth the flowering plants became the dominant forms of plant life on land. The earliest of the angiosperms were flowering trees, and many such exist today, magnolias, maples and oaks, for example. However, much of the success of the flowering plants appears to have resulted from their giving rise to a markedly different plant form, the **annual plants**. Able to complete a life cycle from seed to resistant seed in one growing season, such plants are highly competitive, particularly in severe climates where there is only a short growing season due to cold or dryness. A second factor contributing to the prevalence of flowering plants has been, as we have already seen (page 227), their ability to entice insects to assist with pollen distribution. Finally, development of the fruit from the tissues surrounding the embryo has further facilitated distribution of flowering plants.

THE KINGDOM FUNGI

Although often called plants without chloroplasts, the **fungi** differ from all other organisms in so many respects that they are best segregated in their own kingdom. There are over 100,000 species of fungi, and each typically is adapted to a rather specific mode of life. A few are parasitic on animals (athlete's foot fungus), more on plants (potato blight), but the typical life habit is that of a saprophyte. Their great general importance to the world life system is that they break down dead material in the necessary recycling of the nutrients upon which life depends. In some instances fungi impinge upon our lives more directly. The yeasts used in baking and fermentation are fungi, and so are organisms like *Penicillium* whose poison helped spark the antibiotic revolution in medicine.

Fungi are thought to have arisen as motile, unicellular aquatic organisms that gave rise to multicellular forms, invading the land as well as retaining a strong representation in aquatic environments. The multicellular fungi form **hyphae**, filamentous strands of up to a few microns in diameter, which grow with great rapidity. The hyphae are feeding organs, secreting digestive enzymes and poisons and absorbing water and digested organic material.

The life cycles of fungi involve both sexual reproduction and spore formation, except in a few forms in which sexual reproduction is unknown. Water molds retain the evidently primitive characteristic of motile spores, whereas those adapted to terrestrial life employ airborn, dessication-resistant spores. In two large groups of terrestrial fungi, the sac fungi—organisms like *Neurospora* (page 115)—and the Basidiomycetes—the mushrooms, truffles, puffballs, and so on—the life cycle involves the joint action of many hyphae to produce a conspicuous fruiting body, a structure to facilitate spore formation and distribution.

The mushroom is a remarkable organism. Its life cycle begins with haploid hyphae growing underground (Figure 15-25). When opposite sexual strains of hyphae meet, they may fuse to form binucleate hyphae. In these, the nuclei do not fuse but remain side by side, dividing synchronously as the hyphae multiply to form what will be the familiar aboveground mushroom. This structure is fully formed before it commences to take on water, hence the great speed with which mushrooms and toadstools appear when it rains after a dry spell. Gills, on the underside of the mushroom cap, greatly increase the surface area. Here the nuclei of certain cells finally fuse. Each of these cells then undergoes meiosis, producing four haploid spores that continue the life cycle.

The slime molds (page 185) are included among the fungi although this placement is somewhat uncertain. They are of two sorts, one of which reproduces asexually and the other sexually. Both have several stages in their life histories, including a multinucleate stage and

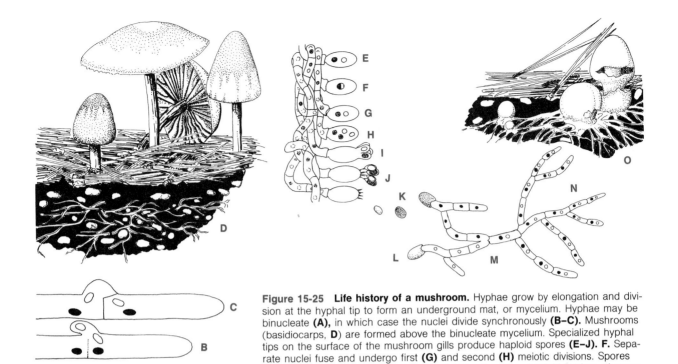

Figure 15-25 Life history of a mushroom. Hyphae grow by elongation and division at the hyphal tip to form an underground mat, or mycelium. Hyphae may be binucleate **(A)**, in which case the nuclei divide synchronously **(B–C)**. Mushrooms (basidiocarps, **D**) are formed above the binucleate mycelium. Specialized hyphal tips on the surface of the mushroom gills produce haploid spores **(E–J)**. **F.** Separate nuclei fuse and undergo first **(G)** and second **(H)** meiotic divisions. Spores form, **(I, J)** and germinate, forming haploid hyphae. Two haploid hyphae of appropriate mating type fuse, establishing the binucleate state. The mushroom begins as an underground "button" **(O)**.

an ameboid stage. At some stage in the life history of both types of slime molds, there is a phase of spore formation, which perhaps justifies their placement with the fungi (Figure 13-5).

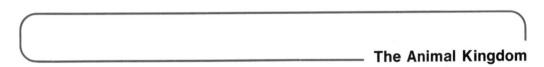

The Animal Kingdom

SUBKINGDOM PARAZOA: SPONGES

Sponges differ so much from other multicellular animal groups that they are isolated in a special subkingdom, the Parazoa, whereas the remainder of multicellular animals are placed in the Kingdom Eumetazoa, or true Metazoa. Sponges are predominantly marine, although there are a few fresh water forms. In size they vary from a few millimeters to several meters across. Usually they have no definite external shape; growth occurs by extensive branching. Although sponges possess distinct cell types that may be organized in functional regions, there is no tissue and organ formation as in higher organisms. Many cell types found in higher organisms are absent.

Figure 15-26 shows that a simple sponge is a filtering machine built around collar cells, **choanocytes,** which bear flagella emerging from a cytoplasmic collar. The beating of the flagella of large numbers of collar cells that line the interior compartments of the sponge induces a flow of water through the sponge. Nutrients are extracted from the water by collection on the choanocyte collars. Evolution of sponges from one particular group of flagellate protistans, the Choanoflagellata, appears probable because these protistans are morphologically identical to sponge choanocytes, and such cells occur only with great rarity in other organisms.

Sponge development also sets them off from the Metazoa. It begins with a curious fertilization process

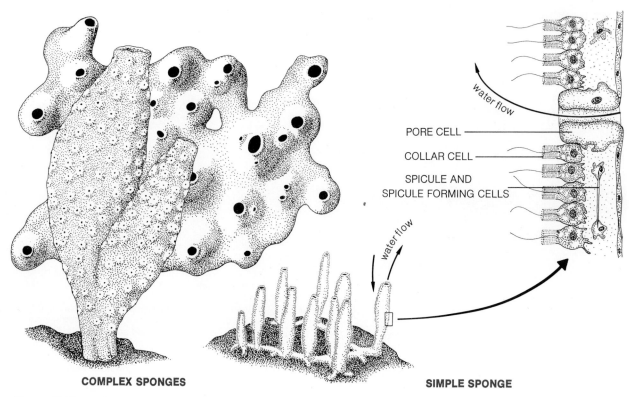

Figure 15-26 Sponges, a sidepath in evolution.

in which free-swimming sperm are taken up by choanocytes just as if they were food, particularly by choanocytes lying near a mature egg cell. The sperm is then transferred to the egg to effect fertilization and formation of the typical 2N animal zygote. A free-swimming larval sponge is formed by cell division and cell movements that are markedly different from the analogous process in higher animals.

During development and adult life of sponges, cells retain an unusual amount of developmental plasticity, which allows one cell type to convert into another, a rare event in higher organisms. To some extent this ability explains the remarkable regenerative abilities of sponges, first studied carefully many years ago by H. V. Wilson. He found that sponges could reassemble themselves into functional sponges after being broken up into single cells by squeezing through finemesh cloth. That this is not a wholly indiscriminate process was shown by making mixtures of cells from two distantly related species of sponges distinguishable by their colors. Each sponge that reaggregated from such mixtures was all of the same species; the sorting out did not result in **chiameras**, that is, animals with parts from more than one source.

All indications are that sponges represent a dead end in the evolution of multicellularity. They appear not to have given rise to any other forms of multicellular organisms. The most probable reason for this seems to be their almost exclusive choice of choanocytes as effector cells, rather than developing coordinated sets of muscle cells as have all other multicellular animals.

THE EUMETAZOA

The vast remainder of animal life, after exclusion of the sponges, is called Eumetozoa. From hydra to lobsters to starfish to man, this vast assemblage at first seems too diverse to have any logic of structure or organization. In actuality, the twelve phyla that make up the Eumetazoa do express a regular progression in complexity of organization. Sufficient broadly expressed common traits exist among phyla to allow construction of what is probably an accurate phyletic

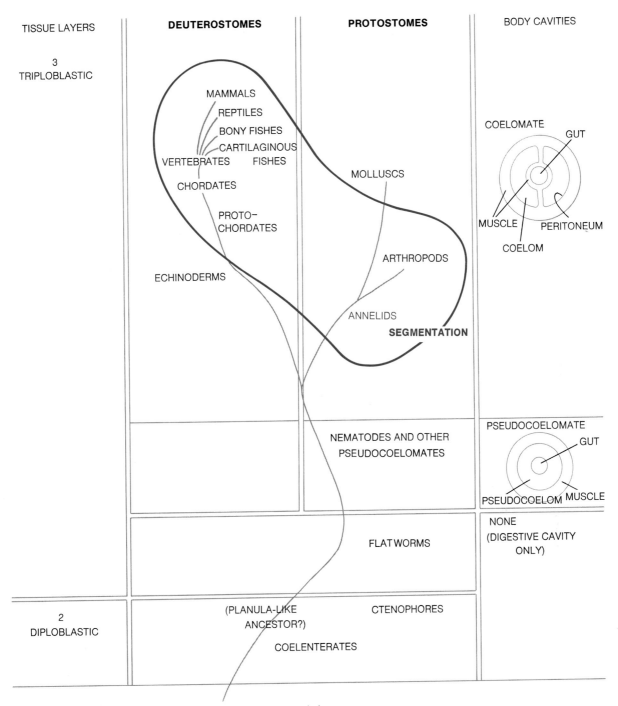

Figure 15-27 Fundamental characteristics of the eumetazoan phyla.

tree (Figure 15-27) showing the path of evolution of the modern animal phyla from their first multicellular ancestors.

Examination of Figure 15-27 shows that there are, at the root of the phyletic tree, organisms with only two layers of tissues (**diploblastic**) and that these give

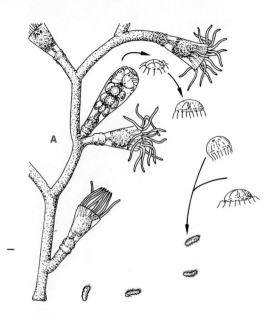
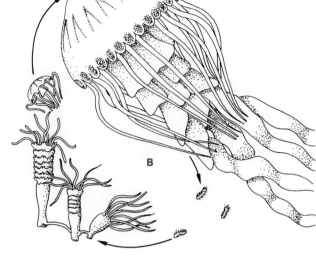
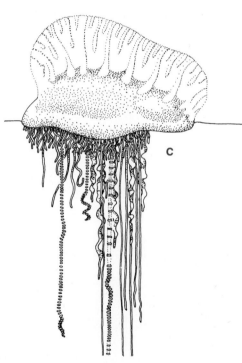

Figure 15-28 Representative coelenterates. A. Hydrozoan, illustrating alternation of generations. A branched colony spawns free-swimming medusae which continue the life cycle by sexually producing planula larvae which become the next polyp generation. **B.** A true jellyfish, in this instance shown reproducing by a polyp-like strobilus which produces small jellyfish by budding. **C.** A colonial hydrozoan, the Portugese man o'war. **D.** An anemone. **E.** Cross section of a generalized medusoid form to show its two-layered umbrella-like organization.

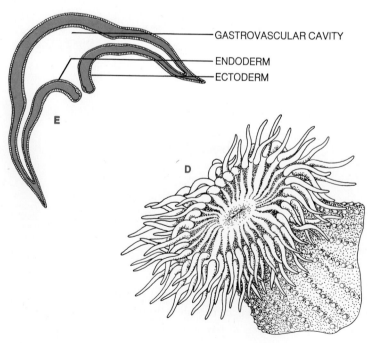

260　BIOLOGY

way to more complex forms with three tissue layers (**triploblastic**). Lower phyla on the tree have no true body cavity, or **coelom**, which is formed in one of several ways in higher organisms. Finally, in higher members of the phyletic tree, we see that the body is **segmented,** that is, composed of a series of closely linked modules containing elements of the major organ systems, whereas in lower members the body is nonsegmented. These variations in structure along with certain other characteristics, principally of an embryological nature, are the primary determinants of the basic phyletic relationships among animals.

Coelenterates

The basic design of a coelenterate is shown in Figure 15-28, from which it is evident that coelenterates are essentially double walled containers with a single opening to the exterior. The central chamber is the gastrovascular cavity, which, as its name implies, performs the multiple functions of digestion and distribution of materials. **Endoderm** forms the lining of this cavity, and in the simplest coelenterates it lies in more or less close contact with the second germ layer, the **ectoderm**, which forms the outer layer of the organism. The coelenterates are primarily **radially symmetrical**, which means that the body can be divided any number of ways into equal halves by cuts that pass through an axis running through the oral (mouth containing) end and the aboral end (the farthest away from the mouth). By contrast, most animals are **bilaterally symmetrical** and can only be divided into equal halves by a single cut in a specific plane. Nearly all animals have a well developed head end, but coelenterates do not show any sign of cephalization, the development of a head end, nor do most have left and right sides (Figure 15-29).

Alternation of generations occurs in coelenterates, although, unlike the situation in plants, all phases of the life cycle are diploid except for the gametes. The polyp and the medusa are the two body forms expressed in coelenterate life cycles. **Polyps** are attached to the substrate and move about slowly, if at all, while **medusae** are umbrella shaped forms that swim by contractions of the umbrella (Figure 15-28). The extent to which these two life forms are expressed in the coelenterate life cycle varies among the different types of coelenterates. The sea anemones are exclusively polyps and give rise to other polyps almost directly, with only a brief intervening free-swimming embry-

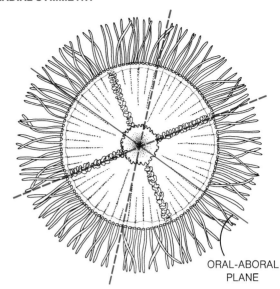

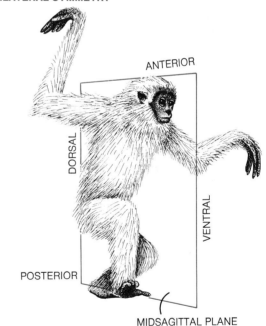

Figure 15-29 Radial and bilateral symmetry in animals.

onic stage, the **planula**. At the other extreme, jellyfish may complete their life cycles without a polyp stage or with a greatly reduced one. The hydrozoa are intermediate, with some members having well developed polyp and medusa stages and others, like *Hydra*, having no medusa stage.

Besides alternation of generations the coelenterates exhibit another type of structural variation known as **polymorphism,** in which a colonial organism is made up of individuals of strikingly different structure, which you recall was the case in the colonial *Renilla* (page 188). The value of these variations in body form appear to be to promote distribution of the species, in the case of the free-floating medusa, and, in the case of polymorphism, to permit functional specializations that would be otherwise difficult to attain with the simple tissue organization characteristic of the coelenterates.

Nematocysts. Nematocysts are unique to the coelenterates and represent what must be the ultimate in structurally complex secretions of cells. These offensive and defensive structures are actually formed as secretions within specialized cells called **nematoblasts,** and they appear to have no parallel anywhere else in the animal world. When completely synthesized they lie in the external surfaces of the coelenterate, especially in the tentacles of anemones and jellyfish. Each nematocyst possesses a trigger that, when tripped by an appropriate combination of mechanical and chemical stimuli, causes a small hatch to pop open and a tiny harpoon to dart out. In some nematocysts the harpoon and the hollow thread it carries are constructed so that they can saw their way into the tissues of the organism that caused their discharge. Nematocysts are able to immobilize small organisms. The nematocysts of some forms, such as the Portugese Man o'War (Figure 15-28) and certain small medusae, can be rapidly lethal even to man by injecting a toxin.

Comb Jellyfish

Closely allied to the coelenterates are the members of a small phylum of exclusively marine organisms, called the Ctenophora. Like the coelenterates they are radially symmetrical, but they differ from them in lacking nematocysts and in the possession of an unusual locomotor system consisting of eight rows of combs, actually short plates of cilia attached together and running meridionally from oral to aboral region (Figure 15-30). These rows of comb plates are connected by nerves to a simple balancing organ that caps the aboral end and which regulates their beating so as to control the movement of the organism. At certain times of the year ctenophores occur in such large numbers as to reduce significantly the survival of commercially valuable fish larvae since they are voracious predators on small organisms of the plankton.

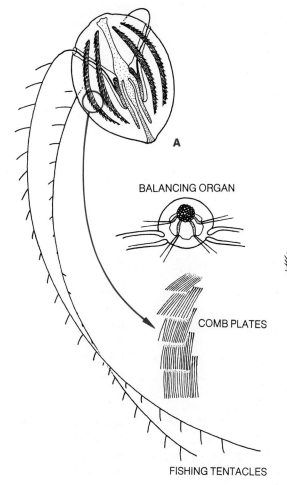

Figure 15-30 Ctenophores, or combjellies. A. A "sea-gooseberry," a common form, with fishing tentacles extended. **B.** Venus' girdle, a markedly flattened ctenophore which retains the basic locomotor and balancing structures shown in **A.**

From Radial to Bilateral Symmetry

The remaining animal phyla are bilaterally symmetrical, even the starfish and their relatives, which appear radially symmetrical as adults but which are plainly bilaterally symmetrical in their developmental stages. Given that the most primitive of the Eumetazoa are the radially symmetrical coelenterates and ctenophores, we have the problem of explaining how radial symmetry gave rise to bilateral symmetry if the radially symmetrical phyla are to be considered in the main stream of animal evolution.

Actually, it is difficult to see what the immediate advantage would be for an organism comfortably established in radial symmetry to begin to convert to bilateral symmetry. Although this might have happened, there is another evolutionary path that avoids this problem by suggesting that the ancestral bilateral animal was more like the larval form of coelenterates than like the adults. One of the best known "laws" of descriptive evolution is that the evolutionary history of an organism is mirrored in the stages of its development: "ontogeny recapitulates phylogeny" is the usual way it is phrased. This, of course, is a sensible law, since we have by now all kinds of evidence that the evolutionary process always builds upon existing structures. So, if the evolutionary route has been from A to B to C to D, we are not surprised to find that the development of D shows some signs of its antecedent structures, A, B, and C. And truly enough we find in the embryonic human a period in which gill slits are briefly present before being put to other uses, but present long enough to signal our ancestry. Such facts make it desirable to examine the planula larva of the coelenterate to see if it, presumably reflecting the nature of coelenterate ancestry, might tell us something of how the bilaterate phyla arose.

Planula Larvae and Flatworms. The planula larva consists of a solid core of endodermal cells covered with a ciliated ectoderm, the whole being shaped like a stubby cigar. One end has a greater supply of sensory cells than the other, and the larva always proceeds through the water with that end forward. The planula has a free-swimming existence of only a few hours before it settles and begins to transform into the adult. However, there seems to be no reason why such an organism could not survive indefinitely if it had a means of feeding. If it did, then a planula-like organism might well be a good model for the most primitive eumetazoans.

If we look among the simplest of the flatworms, relatives of the familiar *Planaria*, that are at the bottom of the bilaterate branch of the evolutionary tree, we find a group of small worms called the Acoela because they have no internal cavity of any kind. Although they do contain specialized cells that one would not expect to find in a coelenterate planula, if we discount these, we find that what is left is very much like a coelenterate planula. There are even muscle cells in the outer cell layer, which is highly characteristic of coelenterates. Yet these Acoela are adult organisms, not larvae, and they feed quite adequately with no gastral cavity. They have a simple mouth, but this leads only into an internal mass of nutritive cells. These cells engulf food just as a protozoan would and conduct digestion in a totally decentralized, cell by cell way. This remarkable similarity suggests that the evolutionary route to bilateral phyla was by such planuloid organisms. These are thought to have given rise to the coelenterates and the lower bilaterate phyla as well.

The planuloid scheme illustrates two important principles of evolution. The first is that specialization is usually an evolutionary dead end. Once a group of organisms has become highly specialized, it seems virtually impossible for the evolutionary process to convert them into something else. Thus the only relationships that the coelenterates have with the higher phyla appears to lie through the presumed planuloid ancestors of the flatworms, which assuredly would be less specialized than any adult coelenterate. The second point is an immediate derivative of this observation; the evolutionary process may avoid the blind alley of specialization by bypassing the specialized form and building upon its larval or juvenile forms, which are invariably less specialized. This process is called **neoteny,** meaning literally youth prolonged, and it essentially involves isolation of the adult from the evolutionary process by advancing the stage of sexual maturity into juvenile or larval life.

Advanced Flatworms

The acoele flatworms are the simplest members of the Phylum Platyhelminthes, in turn the simplest of the bilaterally symmetrical animals (Figure 15-31). Two of the three classes that make up the phylum are

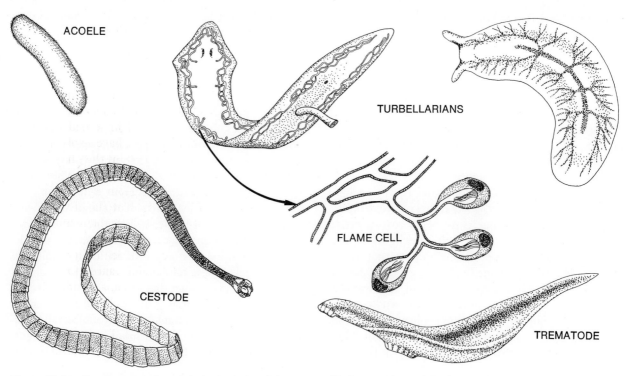

Figure 15-31 Flatworms may be relatively simple free living or parasitic forms with specializations such as suckers and attachment organs for maintaining station inside the host. Parasitic forms are also specialized for production of large numbers of offspring, as in the cestode (tapeworm) which produces immense numbers of reproductive segments (proglottids).

parasitic. These are the cestodes, intestinal parasites commonly known as tapeworms, and trematodes, which are principally internal parasites. They are major causes of human illness and inflict serious losses on domestic animals. The third group, the turbellarians, includes mostly free-living forms like the acoeles and planarians.

The planarian, a turbellarian, (Figures 15-31 and 13-12) shows the structural advances of the flatworms over the coelenterates. There is a well-developed front, or **anterior,** end possessing sensory structures—simple eyes and chemoreceptors—and a concentration of nerve cells, which principally serve the sensory cells and constitute the beginnings of a brain. Extending posteriorly from the brain are nerve cords, which are involved in sensory reception and muscle control. This **cephalization** of the nervous system represents more nearly what we find in higher animals than does the radial system of coelenterates.

The second major advance of the flatworms is that they have three fundamental tissues rather than the two present in coelenterates. This third tissue, the **mesoderm,** in flatworms solidly fills the space between the epidermis and the gastrodermis and is the source of many specialized cell types. In the flatworms the mesoderm forms elaborate crisscrossing sets of muscles, which give them incredible flexibility and mobility.

The digestive system is not markedly advanced over that found in coelenterates; there is still a single opening and no sign of the regional localization of various digestive processes seen in higher animals. An excretory-osmoregulatory system appears for the first time in the form of flame cell nephridia. The **flame cells** are cells with flagella that create a flow of fluid in tubes that ramify through the tissues and finally open to the exterior. Wastes are thought to filter into these tubes for conveyance to the outside. This kind of excretory system is found in many phyla that do not have high pressure circulatory systems, such as occur in the vertebrates. In the latter the flame cell nephridial system, has been supplanted by the renal system, which is able to capitalize upon blood pressure to filter wastes from the blood, as described on page 408. The flatworms are small enough and of sufficiently low metabolic activity that specialized circulatory and respiratory organs are unnecessary. They have elaborate reproductive systems, and often both sexes are represented in the same animal.

Worms and the Coelom

In the evolution of higher animals the appearance of a body cavity, or **coelom,** marks a signal advance. Although the coelom is formed in several different ways in various phyla and may have evolved independently at least twice, it performs the same important functions and is present in all major phyla. The most primitive form of the coelom is found in the roundworms of the Phylum Nematoda, of which the intestinal parasite *Ascaris* is perhaps the most familiar example. A cross section of *Ascaris* (Figure 15-32) shows the coelom as simply an open cavity in which the internal organs lie. A more advanced form of the coelom, as found in the earthworm, a member of the Phylum Annelida (Figure 15-33) is lined with a sheet of tissue called **peritoneum,** which forms mesenteries to support and anchor the internal organs.

In the nematodes, the lowest forms in which it first

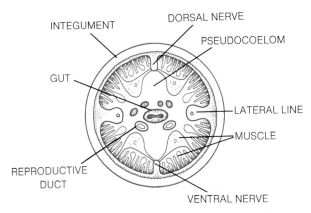

Figure 15-32 **Cross section of a nematode.** The body cavity has no lining. Muscle cells have long processes which extend to the nerve cord, taking the place of nerves running from the nerve cord to the muscles, as in other animals.

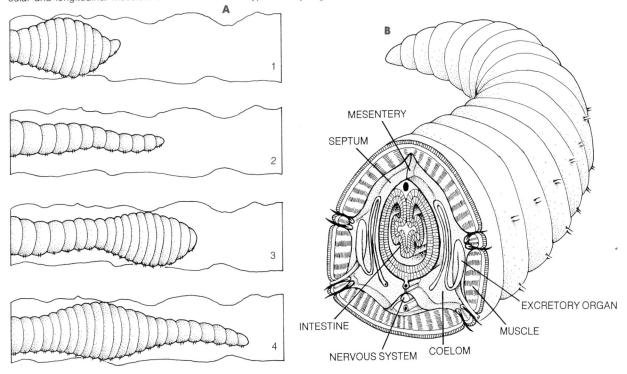

Figure 15-33 **Earthworm. A.** Locomotion involves shape changes in body segments brought about by reciprocal action of circular and longitudinal muscles. **B.** Cross section of typical body segment.

THE VARIETY OF LIFE 265

appears, the coelom acts as a substitute for a circulatory system. The coelomic fluid bathes directly or comes to within a few cell diameters of most cells, even in a large worm, and is stirred by body movements. The tissues of the body may exchange wastes and nutrients by way of this medium. Perhaps this was the first reason for the existence of the coelom, but its modifications in higher organisms have led to still other important roles. As internal organ systems became more complex, the presence of a body cavity permitted divergence from the simple tube-within-a-tube organization of the worms and permitted mechanical decoupling of the internal organs from the actions of the body muscle.

Segmentation

The final great modification of the coelom is first seen in the annelids, worms such as the earthworm, as division of the coelom by transverse partitions (**septa**) into a series of more or less identical chambers (Figure 15-33). Each body segment defined by these partitions can be thought of as a module containing representatives of each of the organ systems. Perhaps the immediate advantage accruing from segmentation was an improvement in locomotor mechanisms since the action of muscles on the hydrostatic skeleton could be more localized than is possible in a completely open hydrostatic skeleton.

Ultimately, segmentation had a more pervasive effect in establishing the segmental organization of the neuromuscular systems of higher organisms. In essence, segmentation led to a system of muscles in each body segment under the immediate control of a local collection of nerve cells (a **ganglion**) and with the local nerve cells controlled by a higher neural center, a brain. Whether this decentralized mode of control is the best way to run our bodies or not is a matter for speculation, but in lower organisms it was undoubtedly an advantage and this served to establish it permanently in our genetic heritage.

A way to illustrate the great advantage of segmentation to a lower organism is to consider what happens when an earthworm is cut in two. Because of the great degree of decentralization and the great capacity for regeneration that is characteristic of the annelids, no great harm is done. Each half seals itself and contains a sufficient representation of all organ systems to maintain life until the missing parts have been restored.

The Great Division in Animal Evolution

After reaching the level of the annelids, we come upon a great profusion of complex animal types, ranging from vertebrates to starfish, molluscs, and arthropods. Careful examination of all available evidence, particularly the evidence of patterns of early development, shows that these organisms fall into two major divisions, called the **Protostomia** and the **Deuterostomia** (Figure 15-27).

The basic differences between these two are of a developmental nature. In brief, the first opening into the embryonic gastral cavity becomes the adult mouth in the Protostomia, hence the name, which means first mouth. In the Deuterostomia the mouth forms by the appearance of a second opening. These and other differences show that some group of organisms ancestral to the modern annelids gave rise to the two great divisions, with the Platyhelminthes, Annelida, Mollusca, and Arthropoda forming the protostome branch and the Echinoderms and Chordates forming the deuterostome branch. The Echinoderms are the starfish and their relatives, whereas the Chordates are all of the organisms in our immediate ancestry from sea squirts through sharks to mammals.

Lessons from Arthropods and Molluscs

While it is only natural to regard our own phylum, the Chordata, as the culmination of the evolutionary process, this attitude should not divert us from appreciating the good points of other particularly successful phyla and perhaps, by comparing ourselves with them, learning a little more about why the chordates are now dominant. Actually, the contest with the **arthropods** for dominance has not yet been decided. One group within the phylum, the Class Insecta, is beyond doubt the most numerous class of animals and, as vectors of disease and consumers of our crops, its members are giving humanity a very good race. The other phylum, the Mollusca, while evolutionarily past its prime, has spawned the largest and most intelligent animals outside our own phylum, namely the cephalopods: squids and octopuses.

Arthropods

Representative arthropods, Figure 15-34, have in common an external, jointed exoskeleton, which sets them off from other phyla. Internally they are constructed very much like annelids, with a ventral nervous system and a dorsal blood vessel-heart. Segmentation is not always evident in all parts of the adult but is usually evident in the arrangement of muscles and the ventral nervous system in the posterior parts of the body; it is always evident in embryonic development.

Limitations Imposed By The Exoskeleton. The arthropod exoskeleton is an excellent example of an organ system that opened new horizons but, at the same time, placed restrictions upon its owners, cutting them off from other evolutionary pathways. The largest arthropods are marine. King crabs and lobsters, weighing from 15 to 40 lb, are the largest modern forms in the sea. The largest terrestrial arthropod is the coconut crab, which weighs only about 5 lb. The reasons for this restriction in size on land lie principally in the problem of the weight of an exoskeleton in proportion to body size and also on difficulties with molting and respiration that are associated with the exoskeleton in very large crustaceans.

On the other hand, the exoskeleton, dense and relatively impermeable to water, must have greatly eased the osmotic problems faced by arthropods in moving from the oceans to exploit new environments in fresh water and on land. This they did with dispatch, finally giving rise to the insects, which dominated the terrestrial environment for much of recent geological history and come close to doing so today.

Possession of the exoskeleton led to a definitely poor solution on the part of insects to the problem of ventilation. Gas exchange through the impermeant exoskeleton was provided for by development of the tracheal system, which conveys respiratory gases to and from the cells through a series of tubes that are air-filled over most of their length. Why a more "conventional" solution to the respiratory problem was not used we shall never know. Certainly even the annelids have circulating blood with good respiratory pigments that work in a way analogous to hemoglobin, so it is likely that the insects had that option as well. Probably the tracheal system won out because it is actually quite efficient in small air-breathing organisms. Indeed, the flight muscles of modern insects, which include the most metabolically active tissues known, are well supplied by tracheal respiration. However, the tracheal system places a definite upper limit on the size of the organism.

Consequently the insects have been cut off from the advantages of size, as detailed on page 181. Within their scale of being they have, however, been ex-

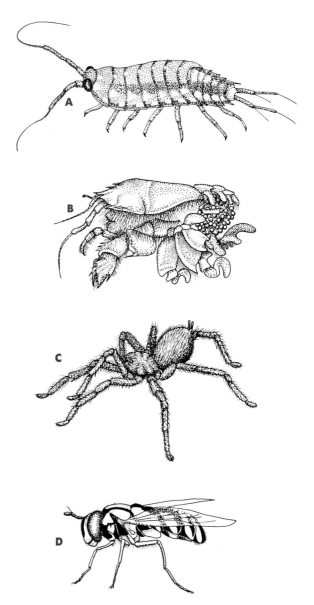

Figure 15-34 Representative arthropods. **A.** and **B.** crustaceans. **C.** spider. **D.** insect.

tremely successful. In fact, their small size has been an advantage in the invasion of many environments. With the development of flight, insects truly have major impact on all environments, aerial to fresh water, although oddly enough they are represented in mother ocean by few species, principally water skimmers.

Limitations Imposed by the Nervous System. Although insects are clearly well adapted to most environments, they are prevented from taking the final step to a higher level of environmental independence by their small size. This size limitation prevents the development of a nervous system capable of the elaborate forms of behavior that we see in mammals. Size alone does not make a brain smart, but there does appear to be a certain minimal size, which, if exceeded, seems to permit investment of some neuronal function in processes higher than the basic, vegetative control of body processes. Of course, within the brain we must consider the size of the neurons themselves. If they are very large, there cannot be very many of them per unit volume and this obviously would limit the abilities of the brain.

On this count the insects and other arthropods also lose. For some reason none of the phyla except the Chordata have the knack of increasing nerve impulse velocity by wrapping axons with layers of myelin (page 341). Since it is of great adaptive value to have neuronal conduction velocities as fast as possible, the arthropods, lacking myelinization, evolved another mechanism, namely, increasing the diameter of axons. Thus some arthropod axons are as much as 0.5 mm in diameter as compared with the $1-25\ \mu$ range in diameter found in vertebrate myelinated axons. Increasing axon diameter increases conduction velocity by improving the electrical conductivity along the inside of the axon. This is by no means as effective as myelination in increasing speed of the nerve impulse, and it has the great disadvantage of filling the nervous system up with a few very large cells.

Since the more neurons the nervous system has the smarter it is likely to be, organisms like the arthropods have made a compromise between rapid reactions and elaborate behavioral responses to changing environmental conditions. The result has been heavy reliance upon stereotyped or "instinctive" behavior. What this means, is that the nervous system, having few cells to invest in behavior, has arranged fixed systems of neural units to make appropriate responses to a limited range of particularly significant environmental or other cues. In other words, behavior in an insect essentially is modified only by the slow process of evolutionary selection rather than by the rapid learning from experience that is characteristic of the vertebrates, which have the advantage of large nervous systems with many small neurons. The stereotyped behavior system works well under the conditions for which it evolved, but it is largely incapable of producing useful adaptive responses if those conditions change, except as we have said, through the evolutionary process.

Molluscs

The molluscs are a remarkable group of diverse organisms that have placed great reliance on cilia in accomplishing their life functions (Figure 15-35). The form of the earliest mollusc is unknown, although it possibly was a segmented organism. There remains little sign of segmentation in modern molluscs except in the chitons, which adhere to rocks in the surf zone, and in the deep-sea living "fossil," *Neopilina*. In a general way all of the varied forms of modern molluscs can be derived from a snail-like theoretical form such as shown in Figure (15-35). This diagram shows well developed organ systems, and examination of the kinds of modern molluscs derived from this theoretical form shows how heavily reliance is placed on the use of sheets of ciliated epithelium in feeding and ventilation.

The age of molluscan ascendancy is well past and they are in only the most marginal way involved directly in our lives. The boring *Teredo*, the shipworm, damages wooden hulls and docks; some snails are intermediate hosts of important human and animal diseases; oysters, clams, and squids are a favorite part of the diet; giant squid have long preyed upon our imagination (Figure 15-36).

Squid and Octopus. The squid is a fiendishly well-designed predatory animal. Large squids, the "kraken" of ancient tales, give even sperm whales a bad time, and more ordinary-sized squids swim rapidly enough to capture fish almost at will. The secret of the squid's rapid locomotion is jet propulsion; water in the strongly muscular mantle cavity (Figure 15-37) can be almost explosively expelled through a tube that can be pointed to front or rear depending on the direction of motion required. A superb nervous system and sucker-bearing tentacles, which surround the mouth with its sharp beak-like jaws, complete the predatory equipment.

Although most molluscs are sedentary and do not maintain a high metabolic rate, the squid is continuously very active, and this activity is made possible by well-developed supportive, or vegetative, organ systems. The squids, alone among nonvertebrate animals, have a relatively high pressure circulatory system with capillaries rather than the open sinuses through which blood flows in the circulatory systems of most

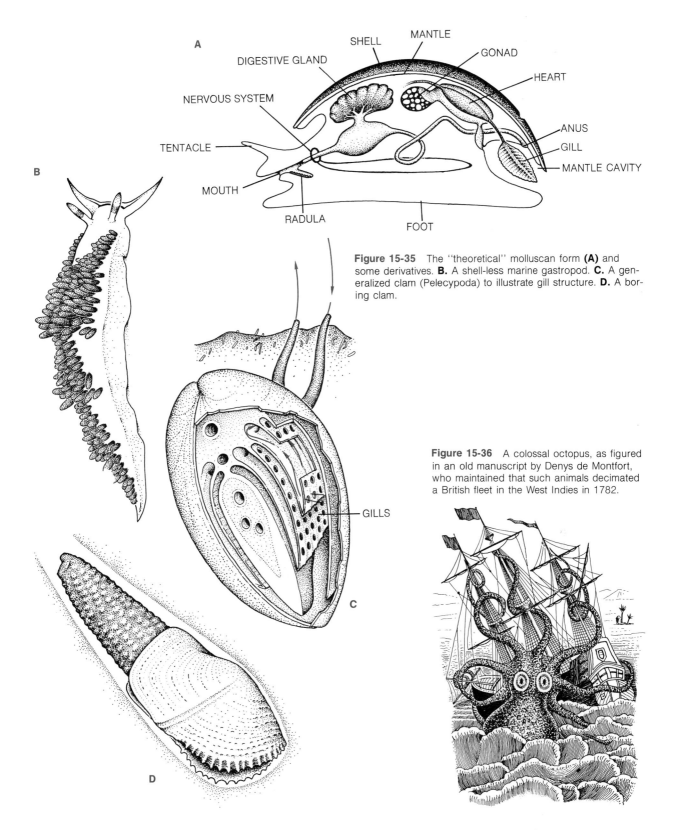

Figure 15-35 The "theoretical" molluscan form **(A)** and some derivatives. **B.** A shell-less marine gastropod. **C.** A generalized clam (Pelecypoda) to illustrate gill structure. **D.** A boring clam.

Figure 15-36 A colossal octopus, as figured in an old manuscript by Denys de Montfort, who maintained that such animals decimated a British fleet in the West Indies in 1782.

THE VARIETY OF LIFE 269

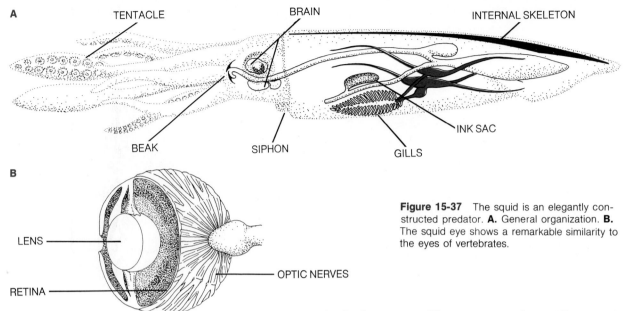

Figure 15-37 The squid is an elegantly constructed predator. **A.** General organization. **B.** The squid eye shows a remarkable similarity to the eyes of vertebrates.

other lower animals. Similarly, the respiratory system is well developed, with efficient gills through which blood flow is assisted by auxiliary hearts and which oxygenates a circulating respiratory carrier molecule, the copper containing protein **hemocyanin**. The octopus is similar to the squid in construction but has developed a maximally flexible body form adapted to its chosen life of predation by stealth from hiding places on the bottom.

None of these adaptations for a predatory life would be of much value without a well-developed nervous system, and assuredly we find that the development of the nervous systems of squids and octopuses is without parallel outside the vertebrates. The eyes of these animals represent a remarkable evolutionary convergence with the vertebrate eye (Figure 15-37), and they possess organs of equilibrium that are functionally analogous to the semicircular canals of the vertebrate inner ear. Although squids rely on giant nerve cells to improve reaction time, these neurons are principally in the peripheral parts of the nervous system. The brain, which represents the fusion of the several sets of separate ganglia of the primitive mollusc, is similar to ours in that it has regions composed of very large numbers of small neurons. As the arthropod example suggests, this type of nervous system ought to be capable of nonstereotyped behavior. This has been well demonstrated in experiments on learning in the octopus. The octopus can be rapidly trained to distinguish between several kinds of simple symbols by conditioning experiments in which rewards or punishments are presented in association with one or another symbol. The octopus has a good memory which is similar to ours in having both short term and long term memory mechanisms (See Chapter 20).

The Other Side of the Street: Deuterostomes

For further progress towards independence from the environment we must look to the deuterostome phyla. At first glance the Echinodermata, Protochordata, and Chordata seem to have little enough in common to consider them to be closely related. This is true of the adult forms but in their developmental stages the members of these phyla show their close relationship.

Something Completely Different: The Echinoderms

Although echinoderms start life with bilateral symmetry, they later become secondarily radially symmetrical. They include such beautiful organisms as crinoids, or sea lilies, and bottom dwelling or burrowing forms such as sea stars, sea urchins, and sea cucumbers the latter of which have returned to bilateral symmetry (Figure 15-38). It is a small phylum with limited impact on man. Sea cucumbers are eaten in some parts of the orient as *beche de mer*, sea stars raid clam and oyster beds and one, the crown-of-

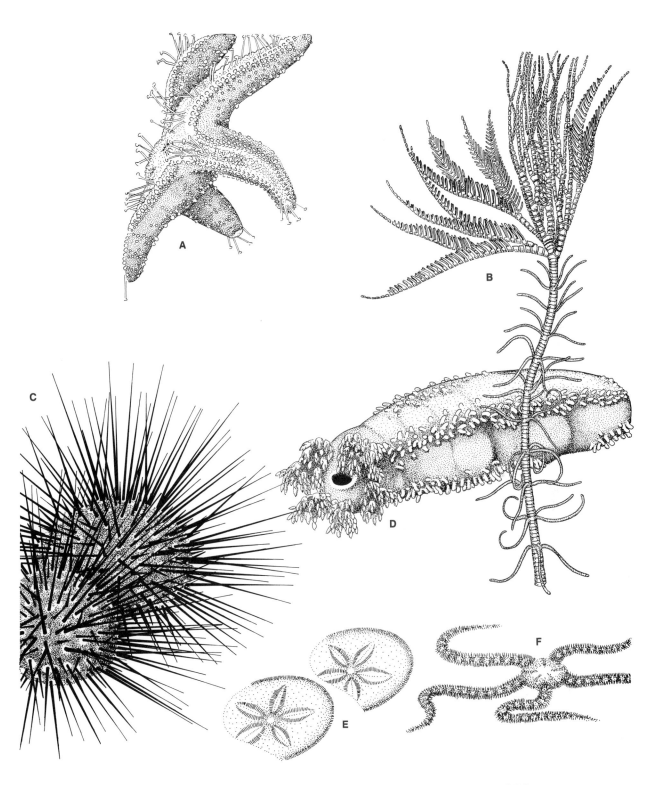

Figure 15-38 **Echinoderms.** **A.** starfish, **B.** crinoid, **C.** sea urchin, **D.** sea cucumber, **E.** sand dollar, **F.** brittle star.

THE VARIETY OF LIFE 271

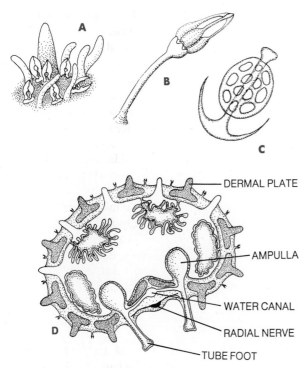

Figure 15-39 Echinoderm structure. A. Magnified view of surface of starfish showing respiratory papules and forceps-like pedicellaria. **B.** A three-jawed pedicellarium. Operated by muscles, pedicellaria protect the echinoderm from small organisms. **C.** Leathery skinned sea cucumbers have remnants of the skeletal system, elaborately shaped dermal ossicles. **D.** Cross section of a starfish arm to illustrate the locomotor system. Muscles in the ampulla force water into the tube foot to extend it while other muscles in the foot control the direction of stepping. Water is supplied to the ampulla by ciliary action in the water canal system which has one or more openings to the exterior.

The second phase of the experiment was to develop a stiff, armored exoskeleton (actually embedded in the exterior epithelium) and to rely for locomotion upon an entirely new mechanism, the water vascular system. The skeletal system is variably expressed in the phylum. In the sea cucumbers and particularly in the almost transparent, swimming deep-sea forms, the skeleton is virtually absent; however, in most genera it forms a stiffly flexible armour. In these, locomotion is achieved by hundreds of tube feet, which stick out through the armour and are moved by hydrostatic pressure generated in the water vascular system and by muscles (Figure 15-39). It is truly a wonderous system, and there is nothing else like it in the animal kingdom; but it leaves the echinoderms at an evolutionary blind end.

Protochordates

The unusual construction of the echinoderms appears to have prevented them from giving rise to other forms. However, the similarity of certain echinoderm larvae to the larvae of some protochordates suggests that there was a common ancestral form that gave rise to the two groups. As adults, the protochordates are ciliary feeders like the molluscs. The wormlike *Balanoglossus* carries on this activity while burrowing in the sea bottom. Sea squirts do the same while attached to objects in the sea, and their close relatives, the tadpole-like larvaceans and the salps, feed while floating free or swimming in the sea (Figure 15-40).

An adult sea squirt or tunicate has little about it to suggest affinities with the chordates. The epidermis secretes a protective tunic containing cellulose, a rarity in the animal kingdom. Much of the volume of the sea squirt is given over to a pharyngeal basket. Water passes by ciliary action into the pharyngeal basket through a buccal (mouth) siphon and out through an atrial siphon, which also discharges waste products and gametes. Mucous secreted in the pharyngeal basket collects food from the passing water and is taken into the digestive tract that lies at the base of the pharyngeal basket. Although this feeding mechanism suggests very little that relates to the vertebrate, it is a remarkable fact that a part of the pharyngeal basket called the **endostyle** is the forerunner of the thyroid gland in the vertebrates; even in the sea squirt it accumulates iodine, just as its vertebrate homolog accumulates iodine from circulating blood to manufacture the hormone, thyroxine. The nervous system of

thorns sea star, has recently been blamed for destruction of coral reefs by grazing on the coral polyps that lay down the reef structure.

The echinoderms are interesting to us as an example of an "experiment" in evolution that led into a blind alley. The first phase of the experiment, the reversion to radial symmetry, seems, as we would expect, to have either brought about great simplification of the nervous system or to have prevented its development. No elaborate sense organs are present, and there is no evidence of significant development of the central nervous system, as seen in bilaterally symmetrical animals with a definite head end; the greatest elaboration of the echinoderm nervous system is seen as a ring of neural tissue around the mouth.

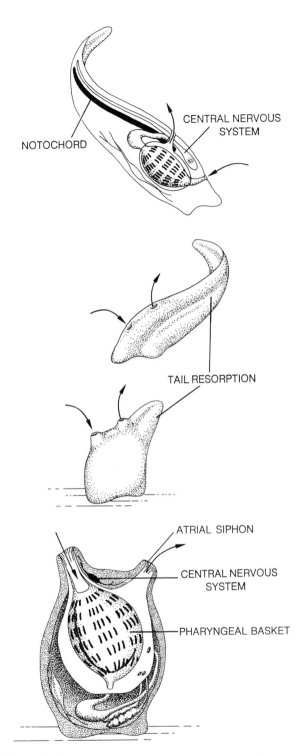

Figure 15-40 Protochordates. A free-swimming larval form is shown at top. Stages of metamorphosis into a non-motile sea squirt are shown below. Arrows indicate path of water flow through the pharyngeal apparatus.

the sea squirt is much reduced as would be expected of a nonmotile organism, and other organ systems are poorly developed except for the digestive tract and the gonads.

The adult sea squirt obviously requires a substantial reconstruction to convert it into a vertebrate, owing to its **sessile**, or nonmotile, life habit and its almost total commitment of its substance to the filter feeding apparatus. Its *larvae* are another matter. In fact, these larval stages are so like the vertebrates that they are called tadpole larvae (Figure 15-40). It is consequently thought that the link between the protochordates and chordates is through either a common ancestor, which was similar to the sea squirt tadpole larva, or through the process of neoteny, in which a larval form becomes precociously reproductive and, in effect, "discards" its adult form. The idea of neoteny is strongly supported by the fact that one group of protochordates, the Larvacea, looks like tadpole larvae as adults.

The Basic Chordate

Rather slight modification of the tadpole larva brings out the body form that is basic to all chordates, from shark to human. The primary chordate features (Figure 15-41) are (1) gill slits in the pharyngeal region at the anterior end of the digestive tract; (2) a dorsally situated, hollow nervous system; and (3) a dorsal, internal axial skeletal structure, which in the primitive chordates is called a **notochord**. The similarity of the basic chordate and the sea squirt larva is remarkable. Both possess a notochord and dorsal, tubular nervous system. Although the sea squirt larva lacks a mouth and gill slits structured as in the basic chordate, remember that these structures represent perhaps the most extreme specializations of the sea squirt adult, and even they are suggestive of the basic chordate condition.

Certainly within the animal kingdom it would be hard to find a closer match than the tadpole larva with the basic chordate. For example, zoologists for a time tried to show that the basic chordate was derived from an annelid or arthropod. Obviously this is a much greater stretch of the imagination because it would require turning the annelid or arthropod upside down to get its nervous system on top, and even then it would be solid rather than hollow. Other problems would remain: exoskeleton versus endoskeleton, and so on. It seems quite certain that the route to the chordate is through protochordate ancestral types.

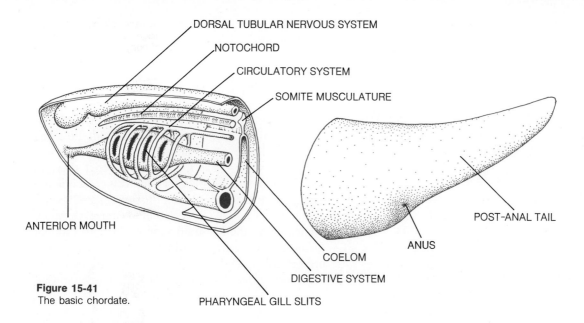

Figure 15-41
The basic chordate.

Figure 15-42 Lancelet. A. Structural features are characteristic of the chordates. **B.** Habit sketch. These small filter-feeders are found in shallow, clean sand along many warm coasts.

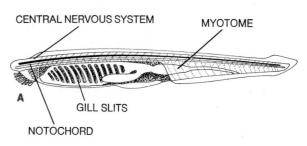

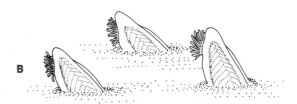

Vertebrates, the Dominant Chordates

Three groups of animals possess a notochord, the primary identifying structural characteristic of the chordate line. These include the sea squirts, the lancelets (Figure 15-42) and, by all odds the dominant group, the **vertebrates.** Lancelets are a small group with a superficially fishlike appearance that probably does not lie in the direct line of vertebrate evolution. The vertebrates are primarily distinguished from the other two groups of chordates by the possession of a vertebral skeleton of articulated (jointed) cartilagenous or bony **vertebrae,** which replaces the notochord as the axial skeleton. The vertebral skeleton is of great importance because it envelops and protects the central nervous system and provides attachment for muscles; it is stiff and supportive even of a large body, but is still sufficiently flexible to allow sculling motions of the body in swimming and the flexion necessary to terrestrial locomotion. Within the vertebrates there are two major groups, the aquatic, gill-breathing fishes and the terrestrial, lung-breathing tetrapods (four legged animals): the amphibians, reptiles, birds, and mammals.

Fishes

The earliest known vertebrate fossils are remnants of fishes that lacked jaws and are called agnathous fishes. A few somewhat similar fish exist today, the hagfish and lampreys, which are parasitic on other fish (Figure 15-43A). The first agnathous fish were probably detritus feeders that scooped up material from the bottom. They were eventually replaced by heavily armored fish provided with jaws, the placoderms (Figure 15-43B), now totally extinct. The reason for the armor of dense plates of bone lying just beneath the skin is obscure. According to one theory the armor was a defense against euryptids (Figure 13-20), which were gigantic predaceous arthropods, but it might as well have been simply for defense against each other because with their newly developed jaws they were able to set forth on a long evolutionary trail of predatory life. These ancient fishes gave rise to the two types of modern fishes, the cartilagenous fishes and the bony fishes.

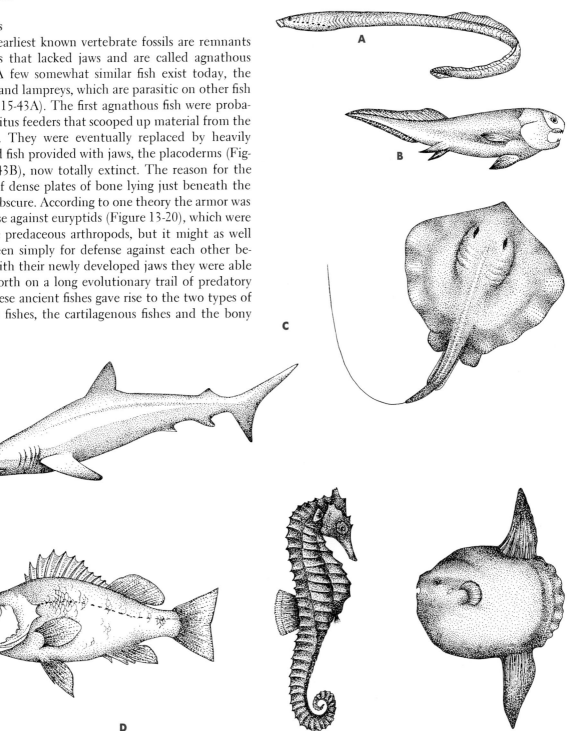

Figure 15-43 Fish. A. Lamprey, a jawless fish, B. Fossil placoderm, C. A shark and ray, cartilagenous fish, D. Three examples of the remarkable variety of bony fish.

THE VARIETY OF LIFE 275

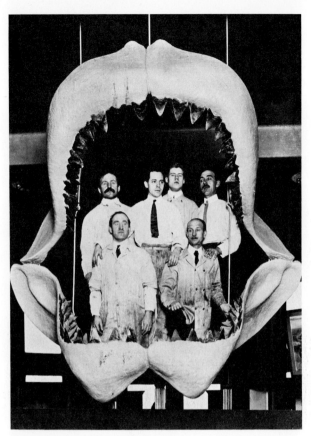

Figure 15-44 Ample documentation of the prominent status of cartilagenous fish in ancient seas, the jaws of a fossil shark. American Museum of Natural History.

Cartilagenous fishes are those fishes with noncalcified skeletons; that is, their skeletons are composed exclusively of cartilage. They have been most successful throughout much of recent geological time (Figure 15-44). The group includes the sharks, rays, and a much smaller group of fishes called the chiameras (Figure 15-43C). Although highly successful, the sharks and rays have been relatively conservative in body form and habit as compared with the bony fish. These fish, in which cartilage is replaced by bone during development, have invaded all aquatic environments from the depths of the deepest oceans to desert pools and have concomitantly evolved into a taxonomical nightmare of an estimated 41 orders (Figure 15-43D). By comparison there are only 29 orders of birds and 17 orders of livebearing or placental mammals.

The Transition to Land

The impetus for vertebrates to leave the sea must have been primarily the advantage of new environments and food sources not already heavily utilized by fishes. For such reasons fishes must have entered fresh water environments. The primary impediment to this transition seems to have been osmotic, and so it is thought likely that the transition was made gradually in wide, marshy river estuaries where mixing of fresh and sea water provides a gradient of salinity. This process of adaptation to fresh water involved modification of the functions of the kidney and other organs involved in water and salt regulation and also led to reduction of the osmotic inflow of water through the integument by increasing integumentary impermeability (recall page 198). This increase in waterproofing of the integument is, of course, also a requirement for emergence of vertebrates onto land where water would tend to flow in the opposite direction, from animal into the environment.

Invasion of fresh waters by ancestral fishes ultimately would have led to problems in respiration in stagnant waters. Fortunately, at least one group of fishes had developed lunglike outpocketings of the anterior digestive tract which are known as swim bladders. Their original and present use in most fish that have them is to adjust the specific gravity of the fish so that it may maintain a specified depth without muscular effort. Modern lungfishes with such organs use them for air breathing, gulping air through the mouth and filling the bladder, which is sufficiently supplied with blood vessels to effect a useful rate of oxygen and carbon dioxide exchange with the blood (Figure 15-45). It is presumed that this habit must have evolved early as fresh water fishes became trapped in stagnant pools, thus providing the beginnings of lungs even before the first truly terrestrial vertebrates appeared.

Some modern fishes are quite adept at slithering about on mud or wet rocks, as for example the mudskippers of tropical shores (Figure 15-46). Their method of locomotion is not truly that of the terrestrial vertebrate, which moves about supported on four legs or two. The transition from fins to legs is traced back to a group of fishes called the lobe fins. The living fossil *Latimeria* is an example (page 237). From this beginning came the tetrapod limb (Figure 15-47).

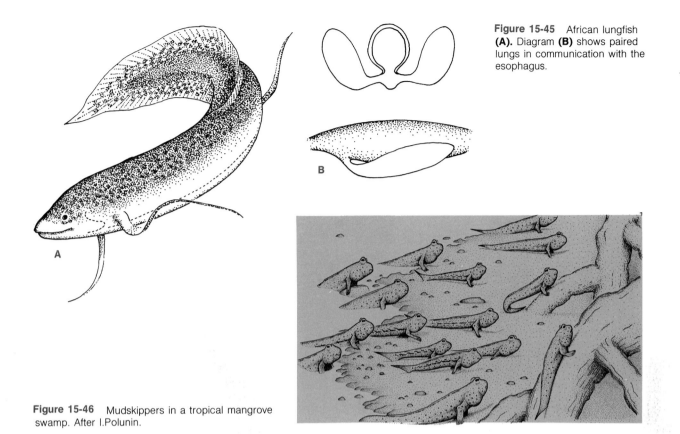

Figure 15-45 African lungfish **(A)**. Diagram **(B)** shows paired lungs in communication with the esophagus.

Figure 15-46 Mudskippers in a tropical mangrove swamp. After I. Polunin.

Figure 15-47 Evolution of the vertebrate limb, from fish fin **(A)** through reptile **(B)** to mammal **(C)**.

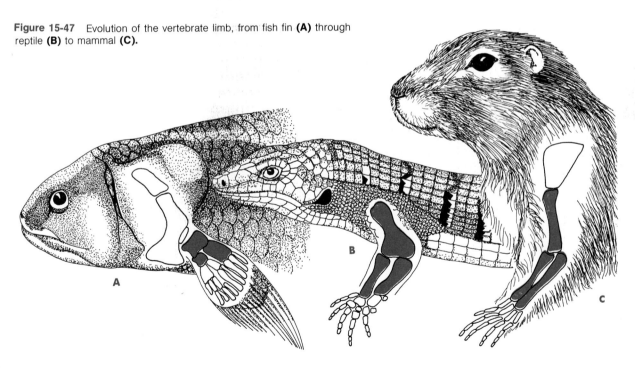

THE VARIETY OF LIFE 277

Figure 15-48 Amphibians. Upper left, salamander; lower left, tree frog; upper center, leopard frog; upper right, a semi-terrestrial toad. In center, nearly all terrestrial amphibians must return to water to mate. A tadpole is to the right of the mating pair of salamanders. At the lower right a frog expands its throat to form a sounding board when calling.

Amphibians

The earliest vertebrates to put these several characteristics together and step out onto land for extended periods were the amphibians, whose modern representatives are the moist-skinned frogs, newts, salamanders, and toads (Figure 15-48). The name amphibian implies that these animals are of the water and of the land, and this is quite true. Only a very few are sufficiently accomplished at water regulation to live very far from free water, and all need water to reproduce. Indeed, the final barrier to a completely free life on land for animals, as for plants, is the problem of achieving fertilization and protecting the developing embryo in the absence of free water. The amphibians simply spawn in water and leave their eggs there to hatch.

Reptiles

Amphibians gave rise to reptiles, truly terrestrial organisms that dominated the land for ages and even returned to the sea as fierce contenders with monstrously large early sharks (Figures 15-49; 15-11). Our modern turtles, crocodiles, snakes, and lizards are a weak echo of those giants. Yet their principal advancements over the amphibians were relatively trivial. There were improvements in the skeleton to allow the limbs to support the body better for rapid movement and there was development of internal fertilization and of a way to allow their young to develop out of water. This was the reptilian egg, which is virtually identical to the eggs of birds; both are designed to conserve water, to allow the embryo to ventilate, and to prevent poisoning from its own wastes during embryonic life (Figure 15-50).

Birds

The dramatically transitional fossil bird *Archaeoptryx* tells us that **birds** are flying reptiles with feathers (Figure 15-51). Although birds have many adaptations for flight, in particular a much lightened

Figure 15-49 Reptiles. Extinct dinosaurs range from the immense *Diplodocus* (A) to forms the size of modern lizards (B). Modern reptiles include the collared lizard (C), Galapagos tortoise (D) and vine snake (E).

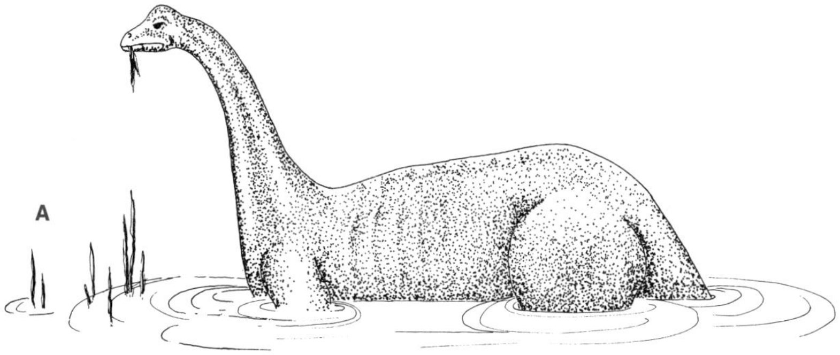

THE VARIETY OF LIFE 279

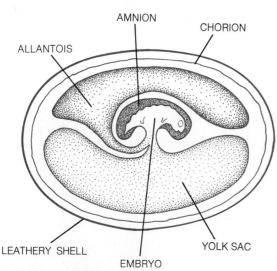

Figure 15-50 A reptilian embryo showing the embryonic membranes. In the reptile embryo the allantois stores nitrogenous wastes as crystals of uric acid. Fusion of allantois and chorion forms a respiratory organ. The yolk sac contains food reserves.

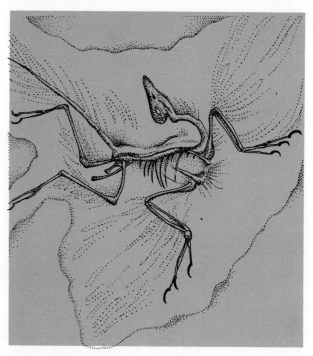

Figure 15-51 Perhaps the most famous of all nonhuman fossils, *Archaeoptryx*.

skeleton and a powerful respiratory system, perhaps their most significant advance over their reptilian ancestors is the fact that they are warm blooded, a property they share principally with the mammals, although some fish, reptiles, and insects are able to maintain elevated body temperatures under certain conditions.

Mammals

Early in their evolution the reptiles gave rise to the ancestors of the **mammals** and, indeed, the most primitive groups of mammals still living are the egg-laying platypuses and echidnas of Australasia (Figure 15-52). Mammals show many advances over the reptiles. Besides warm blood, perhaps the most significant advance is maternal care of offspring (Figure 15-53). The reptile lays eggs, usually burying them in a warm spot where their incubation is seen to by nature. The newly hatched young are plunged immediately, usually without parental help, into the fight for life. The slaughter is great, and to insure survival the reptile must invest a great deal of energy in egg production. Moreover there is no transmission of behavioral information from parent to young and usually no assistance provided to the young by the parent. By contrast, even the egg-laying mammals nest and care for the young and provide them with milk from mammary glands. The placental mammals do all this and go the egg layers one better by feeding and protecting the embryo inside the mother through a link with the maternal circulatory system called the **placenta**. At birth, the mammalian parent continues to provide protection, nutrition, and a greater or lesser amount of instruction in the ways of survival, a process which, of course, reaches its apex in the human.

There are no universally accepted reasons for the geologically sudden extinction of the giant reptiles and the roughly contemporaneous rise of the mammals. Some authorities believe the cause was climatic change, which either reduced the success of hatching of eggs left unattended in the typical reptilian fashion or was in some way deleterious to the adults. Some think that an age of increasing dryness might have reduced the extent of swamps that appear to have been necessary habitats to large plant eating forms. Whatever the cause, the disappearance of the giant reptiles produced a great environmental vacuum, unoccupied habitats into which the mammals expanded by great **adaptive radiation** (see page 296). Mammals

Figure 15-52 Monotremes, the spiny echidna and the aquatic platypus.

took to the air, returned to the seas, tunneled the earth, and dominated every nook and cranny of its surface (Figure 15-54). Out of this great cohort an insignificant group of shrewlike mammals gave rise to the earliest primates, whose dominant members are now the monkeys, apes, and humans, that are the subject of Chapter 17.

A CATALOG OF EXISTING LIFE AND ITS ORIGINS IN TIME

In this chapter the focus has been on the varieties of organisms with emphasis upon how their structure, function, and life histories have influenced the extent to which they now have free run of the earth. To make such a broad theme manageable it has been to a large extent two-dimensional. While evolutionary matters were to some extent unavoidable, there was no direct reference to them and no attempt to apply

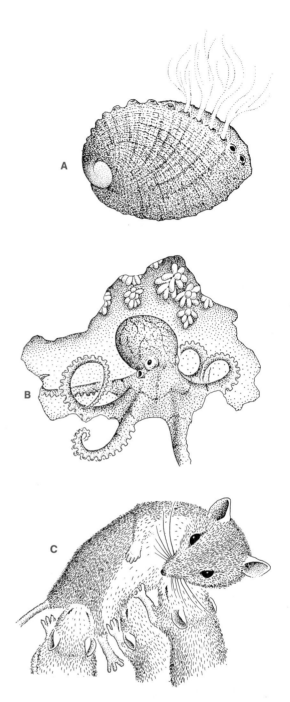

Figure 15-53 Maternal care of the young reaches its apex in the mammals. Many invertebrates, such as the abalone (top) releases eggs and sperm into the water and play no further role in the development of the young. The female octopus (middle) carefully protects her eggs until they hatch and then dies. Mammals, however, go far beyond this to actively support their young during juvenile life.

THE VARIETY OF LIFE

Figure 15-54 **Mammals. A.** macaque, **B.** antelope, **C.** bat, **D.** dolphin, **E.** lion, **F.** mole, **G.** kangaroo rat.

precise dates or to discuss rates of evolution. Undoubtedly many students will be interested in this material, and for them Appendix 2 provides an evolutionary time scale. It was also necessary totally to deemphasize formal taxonomy. For those who are interested, Appendix 3 presents a summary of, perhaps, the best accepted taxonomic system, the Five Kingdom System of R. H. Whittaker.

16 The Great Explanation

The unifying element of this text is evolution, the gradual and natural assembly from the nonliving of the infinitely branching chain of life, with each link giving rise to the next, on through time.

So far we have taken evolution almost for granted as we explored its ultimate basis in inherited chemical changes—mutations in genetic molecules. You have seen that there is no way for the forces of nature permanently to modify the living substance except through the survival or death of mutant forms that arise without reference to any specific environmental force. Nature does not specify the kinds of mutants that occur; it can only approve or disapprove of those that do arise by determining whether or not they survive. You are fortunate in having come upon the concept of evolution in this way, from the biochemical point of view. Knowing as you do from biochemistry that there is no way for nature to determine the *kinds* of mutations, you probably find it easy to accept the underlying randomness of the process.

But surely the last chapter, detailing the march of life towards forms of greater and greater complexity and independence of the environment, must have shaken your acceptance of randomness. It strains common sense to believe this vast assemblage could emerge from a background of random change. To help restore your belief in randomness you should remember that common sense does not ordinarily deal with billions of years or embrace such a great multiplicity of happenings. You should also remember that the story told in the last chapter, guided by hindsight, struck right down the center of a trail that had innumerable branchings left unmentioned. Beyond this, you should know that the study of evolution is far older than the revelations of molecular biology, which make the processes we describe a certainty, and that this classical study of evolution, without knowledge of the chemistry of the process, came to essentially the same conclusions. In this chapter we are going to cover part of the trail of the classical study of evolution because of its great historical importance and because evolution is too complex a subject to be wholly appreciated at the level of chemistry.

STABILITY OR CHANGE

If we go back to first principles, there are just two ways to account for the variety of life: either all forms of life have existed unchanged since some event of creation, or they have evolved with change from earlier forms.

In respect to the first hypothesis, the *hypothesis of stability*, there are two kinds of explanation. The first is that there is *no* explanation; the species have always existed and are immutable elements of nature. The second explanation is that speciation, separation into different species resulted from **special creation,** by actions of some entity beyond explanation by science. Infinite variations of this second explanation are possible since an entity beyond science, a supernatural force or being, may bring about any state of nature by any means it pleases. Thus, if variation is found in the species, it may be argued that this is not evolution in the sense that we have used it in this book but it is simply how the supernatural force creates; the variants are each special creations and not genetically related. Or adherents of a strict interpretation of the *Bible* might argue that the biblical statement that organisms were created and reproduce after their kind may allow for some evolution because "kind" in the biblical usage may represent a larger category than species and thus evolution could occur *within* kinds. It should, therefore, be clear that direct scientific evaluation of arguments for special creation can never be conclusive and is, in fact, an impossibility because science can only test that which is testable by its own precisely defined methods.

The scientific approach to the idea of special creation is to set up a scientifically testable alternate hypothesis. If that hypothesis is verified by scientific methods, it is considered proven and special creation by supernatural powers is defined as an *unnecessary process*. Science thus has not disproved special creation; it has only shown that there is an explanation compatible with scientific law that accounts for the variety of the species without invoking the supernatural. One is still free to choose supernatural explanations, but the choice must be on the basis of personal philosophy and studies lying outside the realm of science.

The second hypothesis, *that life evolved and that the species are genetically related,* has in one form or another had a long life and many believers. The obvious play of similarity and difference across the face of nature and the discovery of fossils similar, yet different, from living forms, were powerful arguments to early biologists. Although evolutionary change seemed likely to have occurred, there was, until the beginning of this century, much indecision as to its mechanism. In explanation there were two classes of hypotheses.

1. Species changed because certain members of the species acquired characteristics related in some specific way either to requirements imposed by the environment, or specified by either internal "drives" or external forces of a supernatural nature.
2. Species changed because organisms possessing randomly appearing changes survived or not according to the extent to which they could cope with the environment.

The best known theory of the first class is the **theory of inheritance of acquired characteristics** associated with the great French zoologist of the seventeenth century, Jean Baptiste Lamarck. He believed that demands created by the environment caused changes in organisms that are inherited. Everyone knows Lamarck's explanation of the length of the giraffe's neck. It was an amazingly persistent theory and still has a few adherents, although they do not include scientists working directly in the field of evolution. Interestingly, one of the rather few known examples of evident falsification of scientific data involves the apparent attempt by a scientist named Kammerer, or one of his assistants, to create evidence supporting the inheritance of acquired characteristics by injecting ink under the skins of toads. The whole story has been retold and analyzed in terms sympathetic to the theory of inheritance of acquired characteristics by Arthur Koestler in his book, *The Case of The Midwife Toad*. It makes fascinating reading and is a good test of one's evolutionary logic.

Theories involving internal drives or external, supernatural forces that herd organisms along evolutionary paths have been invoked with only the vaguest specification of the actual mechanisms by thinkers who seem unable to accept that the evolution of a species is a mindless process. They require a *goal* for the evolutionary process and some kind of *supervising power*. Such ideas lie outside the realm of science and again, as with special creation, are testable only to the extent that they may be shown unnecessary. If you are interested in such ideas you might examine the writings of H. Bergson, who defined a special evolutionary force as the *elan vital,* or the ideas of Teilhard de Chardin who espoused an inspiring but scientifically impervious philosophy involving collective evolutionary striving of all life towards perfection.

The hypothesis that evolution is brought about by survival of randomly appearing changes is primarily associated with the Englishman, Charles Darwin. His efforts, aided by those of his close associates—T. H. Huxley, A. R. Wallace, C. Lyell, J. S. Henslow, and others—are so centrally important to the development of biological thought that it is worthwhile to pause for a brief look at Darwin and his times.

DARWIN'S VOYAGE OF DISCOVERY

On December 27, 1831, a small British Admiralty survey ship, the 242 ton brig, H.M.S. Beagle, sailed from Plymouth, England, to circumnavigate the world (Figure 16-1). When she returned to England 5 years later, her young naturalist, Charles Darwin, went ashore with the basis for a scientific revolution incubating in his mind. His preparation for what was to be the greatest voyage of scientific enlightenment was modest. Darwin's family was financially secure and already known for medical and scientific skills. By their standards Charles did not get off to a particularly auspicious start in science. He gave up the study of medicine at Edinburgh and went to Cambridge, ostensibly to study for the ministry.

Although he was graduated, his studies did not lead him into the clergy. As a student his casual interest in zoology and botany continued to develop with no academic guidance except through the friendship that he established with John Stevens Henslow, who was Professor of Botany. Henslow was a major influence in Charles Darwin's development into a biologist, and it was Henslow who catalyzed all that followed by securing for Darwin the post of naturalist on the Beagle. The geological and biological observations that Darwin made on the voyage, especially in South America and the Galapagos Islands, triggered the patient collection of data and methodical development of ideas that finally came to the attention of the general public in a paper published in the *Journal of the Linnean Society* in 1858. This was essentially an abstract of a great work in progress, which was published in the next year with the title, *On The Origin of Species By Means of Natural Selection, or The Preservation of Favoured Races in The Struggle For Life.*

Darwin's Intellectual Baggage

When Darwin began his voyage his mind contained two systems of thought headed for collision. One was the then dominant explanation of biological diversity and of the significance of fossils attributed to the French zoologist and student of fossils, Georges Cuvier; the other was an evolutionary interpretation of the geology of the earth by the English geologist, Charles Lyell.

At that time Cuvier had thoroughly banished the Lamarckian idea of evolution in favor of his own system, explaining the stratification of the earth and the succession of fossils as resulting from a series of cataclysmic floods that had overwhelmed the earth. He thought the general extinction of animals and plants during each flood was compensated by a fresh creation, repopulating the world until the next innundation generated a new crop of fossils to be replaced by yet another act of creation. This concept of **catastrophism** fitted the literal interpretation of the *Bible*. The most recent of Cuvier's catastrophies was taken to be the biblical flood. Geological time was keyed on this event as either Antediluvian or Postdiluvian.

Lyell, on the other hand, supported the geological doctrine of **uniformitarianism**, which holds that the changes that have occurred in the earth in the past were brought about by forces identical to those presently at work. No universal floods or other remarkable disasters of a periodic nature were required. Although Lyell thus dispelled special creation from a role in explanation of the evolution of the earth, he was unable to do so as regards life. Darwin took the newly published first volume of Lyell's *Principles of Geology* with him, at the suggestion of Hooker, who said he should read but not believe it. However, Lyell's concept of uniformitarianism became the cornerstone of Darwin's developing philosophy of evolution.

Observations on the Voyage

During the 5 years of the voyage, Darwin collected and observed botanical, zoological, fossil, and geological materials on a wide scale. When he began he was obviously just keeping his eyes open for anything of interest and was not primarily oriented towards evolutionary considerations. Indeed, in the journals and letters which he wrote during the voyage there is only an occasional hint of what would ultimately be incubating in his mind.

On arrival in South America Darwin made several long trips inland while the Beagle continued its survey work. In Argentina he made large collections of animals and plants; and, near Bahia Blanca, he discovered a rich bed of "antediluvian" mammals and made important collections. He was taken by the similarity of many of these fossils to living forms. Thus a fossil armadillo was similar to ones still present in the region except for its giant size and, similarly, a fossil guanaco was twice the size of living ones.

Similarities of this nature began to erode the Cuvierian scheme and to replace it with thoughts of gradual descent of modern organisms from now ex-

tinct forms. Darwin's observations in the Andes convinced him completely of the correctness of Lyell's ideas of geological evolution, and the sight of fossil seashells high in the mountains drove home the immensity of the geological changes at work in the earth. Everywhere he was impressed with aspects of animal and plant distribution that seemed better explained by natural processes of migration and interaction with environmental variables than by special creation. Nowhere was this more evident than in the volcanic Galapagos Islands, 600 miles off the coast of Ecuador, where each of the principal islands of the tiny group has organisms distinct from those on the others but still bearing obvious relationships (Figure 16-1).

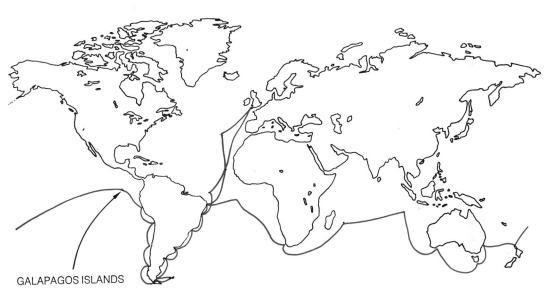

GALAPAGOS ISLANDS

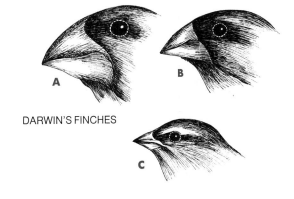

DARWIN'S FINCHES

Figure 16-1 **A sketch of Darwin in middle age, the route of the Beagle and heads of three of the finches of the Galapagos.** The beaks of closely related species of finches, **A** to **C**, modified for tasks from crushing hard seeds to probing for insects, evoked the following from Darwin, "Seeing this gradation and diversity of structure in one small, intimately related group of birds, one might really fancy that from an original paucity of birds in this archipelago, one species had been taken and modified for different ends." (*Journal of Researches*, 1845, J. Murray, London.)

The Incubation Period

Darwin returned to England, married, and finally settled in the country to rear his family and to continue his biological studies and writing. All the while he was a near invalid from an undiagnosed ailment, possibly Chagas disease, which some think he contracted in South America. During these placid years of country life, Darwin, having grasped, certainly before 1839, the essentials of the idea of evolution by means of natural selection, spent much of his time carefully amassing data and arguments for a great book on the subject.

Domestic Animals and Human Populations

Darwin gathered information from every conceivable source to throw light on the problem of the origin of species. For example, he recognized that the domestication of animals provided an excellent model for the evolutionary process. It was obvious that greyhound and bulldog looked as different as two species of wild canids, yet both must have come from a common ancestor since the time of domestication of dogs. The capstone of his arguments emerged when he read, for amusement as he said, the 1798 *Essay on the Principle of Population* by Thomas Robert Malthus. In this book Malthus recognized two opposing forces influencing the size of the human population. On the one hand, there was the tendency for infinite growth resulting from the great fertility of humans; on the other, there was the restraining influence of finite resources of the planet. Malthus looked at this interaction of factors negatively and saw the result as the limitation of population increase by disease, war, and starvation. Darwin, of course, saw it as natural selection at work, with only the best adapted of the population surviving to initiate the next generation.

Darwin had made friends with Lyell on returning to England. Eventually Lyell and Darwin's old friend, Hooker, became aware of the book that was taking form. They urged Darwin to complete it as soon as possible. Darwin delayed, continuing to amass data, quite possibly because he fully understood the upheaval that could be expected when he published. Perhaps he delayed both out of apprehension, for he had become very retiring, and out of the desire to present an airtight argument for evolution when he finally came out into the open.

Another Mind With the Same Idea

One day in 1858 Darwin received a letter and a manuscript from another Englishman, Alfred Russel Wallace (Figure 16-2), who was in the midst of a long, solitary collecting trip in the Malay Archipelago and what is now Indonesia. A man of minimal schooling, Wallace had cut his scientific teeth collecting animal and plant specimens in South America with the botanist Henry Walter Bates. On the basis of this good work, he had obtained transport in a British naval ship to Singapore and from there he set out on a long trip through the islands, supporting himself by shipping prize specimens of birds, insects, and other natural curiosities back home for sale to collectors. At one stage in his travels he wrote the manuscript that Darwin received. It was a theory of evolution nearly identical in evidence and argument to Darwin's own! Wallace, as a correspondent of Darwin's, knew that Darwin was interested in the problem of the changeability of the species. Thus it was logical that he send Darwin the manuscript, hoping for comments and asking that it be sent on to Lyell for publication if it had merit.

A lesser person than Darwin might well have rushed the publication of his own work, while dallying over Wallace's manuscript. Darwin was incapable of such a course. Moreover, he was at that time in no way ready to publish his great book. In this quandary he consulted his two friends, Lyell and Hooker. They generated a compromise in which Wallace's manuscript, together with two pieces of writing by Darwin on the subject, which Lyell and Hooker knew had been written some time earlier, were published together in the *Journal of the Linnean Society*. This was a reasonably fair solution to the problem of priority, and it gave Darwin time to polish up what he termed an *abstract* of the great work, the several hundred pages of the *Origin*.

The First Publications

In Appendix 4 parts of one of Darwin's two contributions to the joint publication in the *Journal of the Linnean Society* and the Wallace manuscript are reproduced. Darwin's contributions concisely state the now-familiar argument: overproduction of organisms and limitation of resources essential to life cause the resultant struggle for survival and consequently affect the composition of subsequent generations. He uses

Figure 16-2 A. R. Wallace

evidence from the breeding of domestic animals to support the argument and discusses sexual selection. This extract clearly represents the logical kernel of the *Origin*, even to a sentence that presages Darwin's troubles in finding a suitable mechanism of heredity (recall page 163), thus ". . . changes of external conditions would, from their acting on the reproductive system, probably cause the organization of those beings which were most affected to become . . . plastic." This is a way of saying inheritance of acquired characters. Darwin persisted in favoring the theory of inheritance of acquired characters and later wrote, "great weight must be attributed to the inheritance of the effects of use and disuse, with respect both to the body and mind." in the 1874 Preface to his book on human evolution (see Chapter 17).

The logical similarity of the two articles by Darwin and Wallace is remarkable but understandable since both had read Malthus and both had similar field experience, being thrown almost daily for several years into situations highlighting the endless variations of living organisms. It is unfair to attempt to judge the relative merits of the two pieces. Darwin's writing by no means represents his finished style since it was not intended for publication. Nonetheless, the power of Wallace's intellect is attested by his article because he wrote it, without aid of references or notes, while quite sick from "intermittent fever" in only 3 days from first to final draft!

RECEPTION OF DARWIN'S THEORY OF EVOLUTION

Thirty fellows of the Linnean Society were present on July 1, 1858, when the joint papers of Darwin and Wallace were read. Both authors were absent; Wallace was still in the tropics and Darwin was at home, ill and stricken over the death of his youngest son. There was no discussion and very little was said publicly about the revolutionary theory until after Darwin's *Origins* was published about a year later. The book sold well by the standards of those times, although it is interesting to note that, by 1876, when Darwinism had become relatively securely established, only 16,000 copies had been sold in England.

Soon, however, the reaction began. Darwin remained in seclusion, leaving the battle to his friends, especially to Thomas Henry Huxley, who called himself "Darwin's bulldog." Huxley's famous debate with Samuel Wilberforce, Bishop of Oxford, is among the best known of the early encounters between the pro- and anti-Darwin forces. Wilberforce entered the controversy early by publishing a major article, with coaching from anti-Darwin scientists, in which he ridiculed the theory by implying that the theory supports as close the most distant relationships and by asking to be shown some evolution taking place. Thus he wrote about "our unsuspected cousinship with the mushrooms" and asked if it is likely that turnips are turning into men, while noting that very close inspection has never revealed any tendency for algae to improve themselves by turning into coelenterates. Huxley was so incensed by this article that he could still get angry about it late in his life and so, most willingly, took on Wilberforce at the annual meeting of the British Association for the Advancement of Science in 1860.

Wilberforce spoke first, dealing only superficially with the science of the matter but making appeals to the heart such as asking the audience not to require that woman as well as man be descended from the beasts. He ended by turning to ask Huxley if he claimed descent from a monkey through his grandfather or his grandmother. Huxley immediately saw he had the Bishop with that. When it came time for him to speak he began with a careful and scientific defense of the theory, pointing out that Darwin did not hold that the ape is our ancestor but only that we both have descended through untold generations from some unknown common ancestor. In finishing, he neatly lifted the Bishop's scalp by remarking that a person has no reason to be ashamed to have an ape for his grandfather, but that:

if there were an ancestor whom I should feel shame in recalling, it would rather be a man, a man of restless and versatile intellect, who, not content with an equivocal success in his own sphere of activity, plunges into scientific questions with which he has no real acquaintance, only to obscure them by an aimless rhetoric, and distract the attention of his hearers from the real point at issue by eloquent digressions and skilled appeals to religious prejudice.

GREEN, J. R.: 1901.
Letters, p. 45.

Within the scientific community the theory, considering its huge impact on the structure of science, was accepted quite readily. There were virtually no holdouts among major scientists, although in the United States the father of American zoology, Louis Agassiz of Harvard, refused to come around. So it was when Darwin died in 1882 he and his theory were so acceptable that he found his rest in Westminster Abbey a few feet from the grave of Newton. Wallace lived on until 1913, maintaining as always that his own efforts were secondary to those of the great Darwin. Interestingly, he parted with Darwin when it came to human evolution (page 309).

EVOLUTION AND THE PUBLIC TODAY

Evolution soon found its way into legislative arenas, leading to ludicrous situations, as lawmakers, serene in their ignorance of science and secure in pious fundamentalism, sought to protect the minds of the young with legislation outlawing the teaching of evolution in the schools. The Tennessee "monkey trial" of a teacher, Scopes, for teaching evolution was a remarkable demonstration of the strange interactions that may occur when science "intrudes" upon everyday life. The law being contested, one prohibiting teaching of evolution in Tennessee, was actually shepherded through the legislatures of three southern states by fundamentalist William Jennings Bryan specifically to provide a publicity generating issue for his third try at the Presidency. The trial itself was pushed by a publicity-seeking school principal as a means of attracting industrial attention to his own small community. Actually, as attested by his own later public statement, Scopes never taught anything connected with evolution, about which he had no competent knowledge. He volunteered as a subject of prosecution, on the request of his principal, as a means of providing a service to the community by virtue of the attendant publicity.

Even today evolution remains conspicuously misunderstood by many well-intentioned people. We have a Creation Society dedicated to disproving evolution, and there have recently been strong and at least temporarily successful efforts to require equal time for competing theories in the public schools. Such activities illustrate the sometimes inflammatory nature of the discussion of evolution. Unlike any other major concept of science in modern times, in the minds of many it is impossible to treat evolution purely as science, which must necessarily lead us back to the absurdities of the Huxley-Wilberforce debate.

The Theory of Evolution Analyzed

In the remainder of this chapter we are going to analyze the theory of evolution by means of natural selection in a more formal way than before. You know some of the chemistry at the heart of the process; you have seen something of the great variety of life that the theory must explain and tie together; and you know a little of the historical development of the theory. Now let us go at it directly.

ANALYSIS OF THE THEORY OF EVOLUTION BY NATURAL SELECTION

To begin let us briefly state the fundamental Darwinian argument: *First* there is evidence from three general sources that evolution has taken place:

1. Fossils: Extinct species are found that are so similar to existing species that it seems likely that they are related.
2. Distribution of living species: Highly similar species are distributed over the earth in such a way that the most reasonable explanation of their similarity and distribution is origin from a common ancestor in a common geographical locale followed by speciation with migration away from that source.
3. Observations upon domesticated organisms: The many varieties of dogs, cattle, pigeons, garden and crop plants, and so on, resulting from selection by breeders serve as a model of, at least, the early stages of the process of speciation.

Second, having concluded from these three lines of evidence that evolution does take place, the process of evolution is taken to be *survival of the fittest* organisms emerging from a continual competition for limited resources. This statement, although commonly used as a slogan for the mechanism, is a tautology, an illogical statement resulting from the fact that the "fittest" organisms are by definition the survivors, so that what the phrase "survival of the fittest" actually means is "survival of the survivors." Thus, to grasp what Darwin and Wallace meant we must review their analysis based on the concepts of Malthus.

The Malthusian concept of population limitation is simple: *the reproductive potential of the species is greater than the ability of the environment to provide the necessities of life.* The inevitable result, according to Malthus, will be a scything down of the population by a variety of disasters. Darwin and Wallace took this idea of cycles of growth and disaster and applied it to all organisms, with the disasters converted to the ordinary day-to-day pressures and hard times of life: the competition for food and shelter, the avoidance of disease and predators, competition for mates, and so on. They reasoned that the growth potential of the population would insure that these pressures would remain high and that the result would be that not all organisms would survive long enough to reproduce. The survivors more often than not would be the strongest, or cleverest, or otherwise the best able to survive any of a myriad of stresses: *the fittest.*

Nature establishes barriers that must be surmounted. Each generation of organisms exhibits natural variations that can be inherited. Some of these pass the test, surmount the barriers, and contribute to the next generation, which is to that extent different from the last. Over the vast span of time these accumulated variations add up to speciation, to evolution.

THE THEORY IN MODERN TERMS

In modern terms the theory of evolution by means of natural selection can be divided into eight hypotheses (Table 16-1). For clarity, these eight hypotheses have been separated into three divisions: the *fundamental mechanism,* the *shaping process,* and *results.* If one accepts the hypotheses of a division as "proved" beyond reasonable doubt, then one must accept the process of that division as a part of the overall process of evolution. If one accepts all three divisions, then one must accept the Darwinian theory of evolution. We shall first explain these eight hypotheses simply and then go back over them in greater detail.

Clarifying the Hypotheses

***Hypotheses* A, B, and C.** Let us look briefly at each hypothesis in Table 16-1, not with regard to proof or disproof but only to clarify its meaning. Hypothesis A states what every animal breeder knows: if, for example, one wishes to increase the number of short-horned beef cattle in his herd, he merely prevents the long-horned cattle from breeding. The phenotypic composition of the herd eventually will change, to one of predominantly short-horned cattle.

Hypothesis B has been adequately described in Chapters 9 and 12, as was hypothesis C.

A Naturally Occurring Process. Hypothesis D requires more introductory explanation. It states that the kind of control of reproductive rate exercised by the animal breeder *will occur selectively in an uncon-*

Table 16-1 Hypotheses Basic to The Theory of Evolution By Means of Natural Selection

Division I—Fundamental Processes

Hypothesis A	Phenotypic composition of a population is alterable by selective control of reproductive efficiency.
Hypothesis B	If phenotypic change in a population persists over time it is the result of genetic change.
Hypothesis C	Mutations continually occur.
Hypothesis D	Selective effects on reproductive efficiency occur naturally

Division II—Shaping Processes

Hypothesis E	If alterations in phenotypic composition of a population brought about by the processes of Division I persist and become characteristic of the species, they are of positive adaptive value.
Hypothesis F	Useful adaptations are cumulative.

Division III—The Consequences

Hypothesis G	The processes of Divisions I and II result in speciation.
Hypothesis H	Speciation by these means evolved all life forms.

trolled state of nature. The inference, therefore, is that the phenotypic composition of natural populations does change. The mechanisms of and proof for this selective influence on reproductive rate we shall discuss in greater detail shortly.

Hypothesis E states that these changes of phenotype in a population, if they persist, are not just random changes, but rather reflect a significant *improvement* in the interaction of the members of that population with their environment.

Hypothesis F states that these changes usually occur as additions to other, previous changes. For example, man's vertical backbone represents a series of modifications of a backbone that was satisfactory for a four-legged animal.

Defining Speciation. Hypothesis G is self-explanatory, once **speciation** is defined as a process in which a sufficient number of phenotypic (and therefore genetic) changes occurs in a population so that members of that population are no longer able to interbreed successfully with individuals having the characteristics of the original unchanged population. Hypothesis H is also self-explanatory.

PROVING THE HYPOTHESES OF EVOLUTION

We are now ready, having set forth and explained the hypotheses that contribute to the theory of evolution, to treat each hypothesis in greater depth and with a critical regard to scientific verification.

Changing Phenotypes and Gene Frequency

A Model for Hypothesis A. It is necessary only to look around us for ample proof of hypothesis A: short-horned cattle that have been bred also for such other useful characteristics as short legs and beefy physique, highly specialized breeds of dogs and horses, the many vegetables and fruits that have had their phenotypes grossly altered for our benefit by selective breeding. Because the process of selective breeding employed by the animal and plant breeder is well known and simple, it is a good model to use in relating the genetic processes of Chapter 12 to the more general process of evolution.

What is the animal-breeder actually doing to the genetic constitution of his population when he produces short-horned cattle? First, let us agree that it is imperative that he deal with a *specific* population, usually his experimental herd. If his cattle mingle freely with his neighbor's, or if cattle buyers and sellers are randomly removing and adding cattle to his herd, his breeding program would not be successful. Evolution, therefore, *occurs to populations* even though the genetic changes that are occurring in that population can happen *only to individuals.* Only individuals have altered genes, but only populations evolve.

What Is Gene Frequency? Let us ask, specifically, what is happening to the genes in a population of long-horned cattle that are becoming short-horned? First, we know that horn length is determined by one or more allelic genes, as determined by Mendelian breeding experiments with the few short-horned cattle that appear spontaneously as mutations in any such population. The breeder's experimental population of cattle can be taken to represent, therefore, a **pool** of short-horn and long-horn genes—geneticists would call this a **gene pool**—with reference to those specific allelic genes. As most of the cattle are long-horns, with only a rare short-horn mutation appearing, we can say that the frequency of the long-horn gene is high in that gene pool and the frequency of the short-horn gene is low in the same gene pool. The geneticist would therefore refer to the relative gene frequency of these allelic genes in that gene pool, as a means of describing the most common phenotype that characterizes that population. A herd of short-horns would be described as having a very high gene frequency for the short-horn gene; a mixed population might have a gene frequency of about 50% for the short-horn and 50% for the long-horn gene.

A Tentative Definition of Evolution

Now we may explore what the animal breeder is doing to his population, but we shall use the geneticist's terms. In these terms, it is immediately apparent that the breeder, by selectively influencing the reproductive rate of individuals of his population, is, in fact, *changing the gene frequency of a specific gene pool*. He may reduce the reproductive rate of long-horn cattle first by **selecting** them and then by removing them from that population (separation), or by killing them preferentially for beef (selective survival), or by preventing the long-horn bulls from breeding. In any case, the net result *in that population* is that the gene frequency of the longhorn gene becomes smaller. As these gene frequencies change, the phenotypic composition of the population changes, until finally the population has evolved to 100% short-horn cattle that are no longer able to produce a long-horn phenotype except by mutation. Can we then define evolution as *a change in gene frequency of a gene pool, produced by some selective influence on reproductive rate*? According to the foregoing proofs of hypotheses A, B, and C, it appears that we can do so, but with the reservation that we shall subsequently consider the influences of hypothesis D and those of divisions II and III on such a definition.

Figure 16-3 Selection observed in nature. Two varieties of a moth, *Biston betularia* resting on a soot covered tree trunk. In the 100 yrs after 1850 the dark variant increased from less than 1 to approximately 90% of the total *Biston* population in soot polluted industrial parts of England. From the experiments of H. B. D. Kettlewell, University of Oxford.

Changing To A Natural Model

A New Moth Appears. To test the validity of hypothesis D, we must change models, for it is clear that the breeder's experiments do not take place in an uncontrolled state of nature. For our second model we turn to the rapidly industrializing regions of England around the turn of the century, where insect collectors began picking up what at first appeared to be a new species of dark-colored moth (Figure 16-3). It proved to be a dark variant of a well-known species. Evidently the coal based industrialization of the region showered enough soot on the surrounding woodlands to give the dark variant the edge on its light colored brethren in the avoidance of predation by birds.

Protective Coloration and Selection Pressure. The hypothesis was advanced that this darker subspecies in the industrial areas was the result of a changing gene frequency in the population of moths that occupied those areas, and that the increase in the frequency of the gene for darkness was due to some natural selective mechanism that reduced the reproductive efficiency of moths having the gene for light coloration. Since the dark moths blended more effectively with the soot-laden landscape of industrial areas (protective coloration), it seemed that the light moths might be easier prey: there was selective survival of the dark subspecies. Analysis of the stomach contents of birds in both industrial and nonindustrial areas, together with observation of feeding birds in these areas, confirmed the hypothesis. In the industrial areas, the birds dined preferentially on light moths, whereas in nonindustrial areas the birds selected against the gene for dark pigment by their feeding habits. Therefore, there is selection pressure for the dark gene in the industrial areas and selection pressure for the light phenotype in the nonindustrial areas. This shift in gene frequency in favor of the dark-gene moths, whose change in color is due to increased production of melanin pigment, is called **industrial melanism**.

Industrial Melanism Establishes Hypothesis D. Here is the case we need to establish the validity of hypothesis D—gene frequency change in an uncontrolled state of nature. Because this change occurs over a very short time span, we can say that selection pressure is very high, just as it is in the animal breeder's population. As we shall see, selection pressure as ordinarily occurring in nature tends to be extremely low, so low that changes in gene frequency large enough to be detected during a human lifetime are rare.

SOME MECHANISMS OF EVOLUTION

Low Selection Pressure Is the Rule

Reasoning from the preceding discussion, we might conclude that the major agencies of selection pressure that operate so slowly in nature are not primarily those caused by selective survival, that the natural balance of populations maintained in part by predators is not a major factor in shifts of gene frequency. Rather we might say that far more subtle, yet selective, influences on reproductive rates are responsible for most evolutionary change. "Nature, red in tooth and claw," as the early Darwinists put it, is the setting for very short-range, rapid evolution. "Survival of the fittest" is due principally to the influence on gene frequency of slightly different reproductive rates, brought about by low-pressure selective factors. Because these influences act so slowly and subtly in natural populations, we cannot test hypothesis D by witnessing this most common kind of phenotypic change in a natural population in the same way that we observed the change in the moth population or the animal breeder's population. Because natural evolution is so slow, we cannot readily test it experimentally by observing it happen. However, we can readily observe the *mechanisms* of evolution—the effects of selection pressures on reproductive rates—for they are evident about us most of the time.

Isolating Mechanisms

The Function of Isolation. Nevertheless, by a slight stretch of the imagination, we may continue to use the cattle and moth models to explore evolutionary mechanisms. Consider, for example, the mechanism of *isolation of a population*, necessary if gene frequencies are to change substantially and stay that way. We have already pointed out that the animal breeder would never achieve a 100% short-horn herd if he permitted long-horn gene flow into his population—it is only because reproductive isolation is achieved that gene frequencies can be moved toward 100%. And it is precisely due to *lack of isolation* that the moth dark gene frequencies in industrial areas will never come close to 100%, for there is constant light gene flow into these areas from surrounding nonindustrial areas.

The effect of isolation on these two model populations has its natural parallel in geological events. Continents separate (Figure 16-4); islands rise and fall; rivers, inland seas, mountain ranges, deserts perpetually form and change; all these events serve as potent isolating factors that affect evolving populations. A most remarkable example is the evolution in an isolated situation of an almost complete "matched set" of animals paralleling that of the placental mammals with which we are familiar—dog for dog, cat for cat, mouse for mouse, but each quite different in a major way from its placental counterpart (Figure 16-5). We refer, of course, to the marsupials of Australia, separated about 70 million years ago from all other mammals by the isolation of the Australian land mass.

The Subtlety of Selection Pressure

Make Short-horns by Pulling Teeth? Once the population is isolated, we are free to examine the more subtle actions of selection pressure on reproductive rate that may operate to change gene frequency. Consider, for the moment, a cattle breeder who has unlimited time at his disposal who wishes to produce short-horned animals *without* separation or prevention of reproduction. He could pull just one of the high-crowned, grass-grinding cheek teeth from the jaw of each long-horn calf, and by limiting the grass supply to a bare adequacy, ensure thereby that the long-horn animals would suffer a slight nutritional impairment. With only just enough grass to go around, the short-horns, with a full complement of cheek teeth, would get all the nutrition they require, leaving the table bare just short of the optimum requirements of the long-horns. Under these conditions, who can doubt that the short-horn cows would have more and healthier calves than their long-horn sisters—that more of their calves would survive to reproduce more effectively? This difference in reproductive rate could be only a fraction of a percent, but, given enough time, the gene frequencies in that gene pool would surely shift in favor of the short-horn gene.

Climatic Melanism in Moth and Man. This, however, is still an artificially induced change. What can we say about a low selection-pressure mechanism in the natural moth population? Here the analogy is harder to draw, unless we suppose some physiological consequence other than protective coloration from the marked pigmentation of the dark moth. Suppose, for example, it is a day-flying moth of the tropics, where we shall assume for our purposes that it is often exposed to bright sunlight and does not suffer predation by birds. Now, tropical radiation is physiologically damaging. Under these conditions, the darker moths might enjoy an advantage, as their heavy pigmentation could filter the sunlight. As before, the advantage need only be infinitesimally small to ensure a change in gene frequency in that isolated population so that, with the passage of say, 10 million years (the winking of an eye, as evolutionary processes go) only dark moths would be found in that region. This

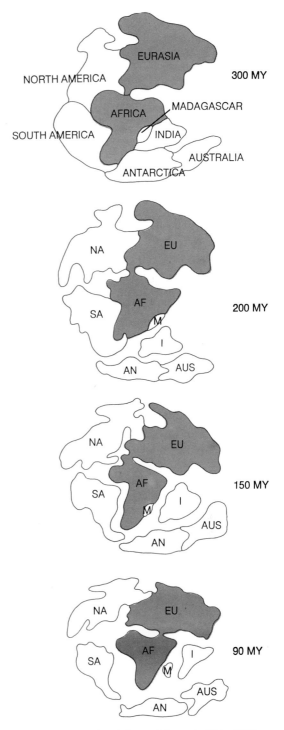

Figure 16-4 Vast slabs of the earth's crust, 60 to 120 km thick, move a few cm per year over the softer underlying mantle. Examination of the path of these movements through geological time explains why, for example, the fossils found in coal deposits in Europe and North America are similar.

THE GREAT EXPLANATION 295

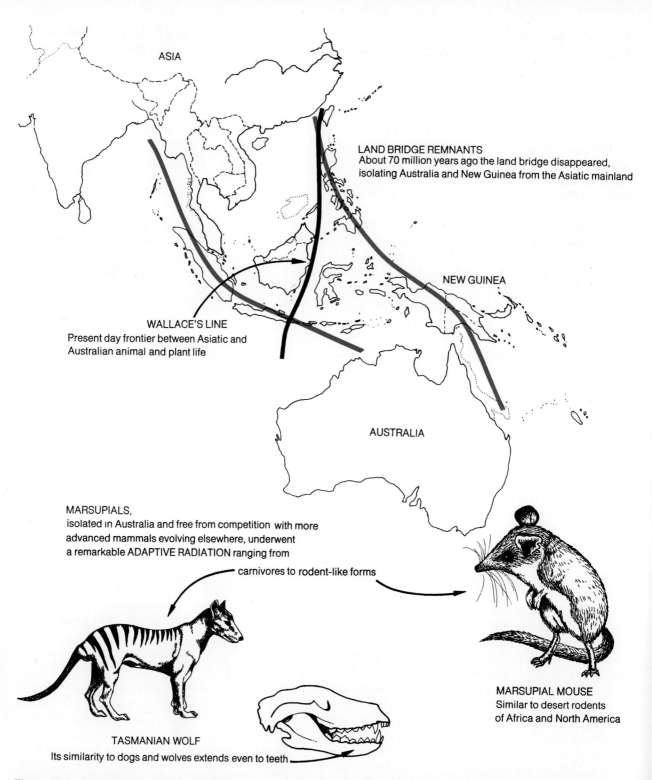

Figure 16-5 The adaptive radiation of marsupials in Australia and New Guinea. Absence of more advanced mammals allowed marsupials to fill the roles which otherwise would have been played by the placental mammals, as elsewhere in the world.

296 BIOLOGY

did not, as we well know, happen to the moths of England, but there is excellent reason to hypothesize that from this kind of selection pressure evolved the pigmented races of man that inhabit the tropical regions of Africa. Recently, studies of UV light destruction of the important vitamin folic acid in patients undergoing medical treatment with UV light add support to the idea that dark skin may provide a selective advantage to inhabitants of the tropics by reducing the destruction of important metabolites by solar radiation.

Sexual Selection and Territoriality

A Hereditary Style Preference. There is another even more subtle influence on reproductive rate that can act in combination with the foregoing mechanism to bring about changes in gene frequency. Suppose our cattle-breeder somehow could convince the females of the herd that long-horns are undesirable mates. Establishment of this pattern of behavior would ultimately have the same effect on the herd as castration of the long-horn bulls. Once again, given enough time, this preference need only be very slight indeed to cause a differential reproductive rate in favor of the short-horn gene. This kind of selection pressure is called **sexual selection**, and probably is operative to some extent in most higher animal populations. Because, like selective survival, sexual selection tends to reduce the reproductive capacity of certain phenotypes to zero, it can be a very potent factor in evolutionary change. The evidence, however, is that it usually has only a slight influence on reproductive rate because of the operation of other factors.

Territories Control Populations. It now seems for example, that many bird populations are limited automatically by a mechanism called **territoriality**. The male bird inherits an *instinctive* behavior, causing him to select and defend a rigidly defined territory in which he will reproduce. Bird songs often signal possession of a territory and the intention of the singer to defend it. In complementary fashion, the female instinctively selects a male that has a territory. Birds without a territory simply do not reproduce that year, and, as each bird lays a relatively constant number of eggs each year, the reproductive rate stays matched to the available territories in the environment of that species. Many other vertebrates have territorial breeding habits.

It is clear that selection pressure by means of sexual preferences could operate readily through this territorial mechanism to alter gene frequencies of any readily recognizable phenotype. But the mechanism would be complicated by genes governing territorial behavior, for the bird with the gene for aggressive partner acquisition (or some other mating facility) would tend to increase the gene frequency for *that behavioral characteristic*, without reference to sexual selection for some other characteristic.

The potency of sexual selection pressure may account for much of the variety and complexity of coloration common among birds and fish, as well as for the many complex instinctive behavior patterns observed in these animals. In certain cases, the characteristic selected for in mating-partner acquisition may not be obvious. For example, female fruitflies select outbred males in preference to inbred males. Presumably the reduced vigor, characteristic of the inbred male, is detected by the female, and causes the avoidance.

ADAPTATION AND THE CAPACITY TO EVOLVE

Establishing Hypothesis E

Adaptation Improves Reproductive Rate. Let us now critically consider hypothesis E—that these changes in the average genetically determined phenotype of a population, which are brought about by selection pressure, are **adaptive**. A close link is established immediately between selection and adaptation by the very definition of selection pressure and an analysis of its agents. If we define selection pressure as any tendency to increase or decrease the gene frequency of a specific gene in a specific gene pool by means of influence on the relative reproductive rate of individuals carrying that gene, then we have only to ask, "what are the causal agents of such a tendency?" in order to establish the link to adaptation. For it is surely true that whatever the agent may be—food supply, weather, predators—the change in phenotype will favor the reproductive capacity (and thus, generally, the overall well-being) of individuals carrying the gene, if that gene is increasing its gene frequency in that population. In other words, the changed gene frequency *must* result in an improved relationship with the environment on the part of the average members of the population, since only by an improved relationship will reproductive rate be maximized.

The Capacity to Evolve Is Universal. We can, if we wish, define adaptation in this way—the tendency of natural populations to change their average phenotype in such a way as to permit the maximum reproductive rate that the environment and physiology of the organism will support. This tendency is sometimes called **evolvability,** and it is a necesary characteristic for survival of all living systems that occupy a changing environment. Without adaptation to changing environment, life ceases. Because all environments change, genetic adaptation or evolvability is a necessary characteristic of life.

Examples of Adaptation from Nature
Origin of the High-Crowned Grinding Tooth. Consider the following example as an extension of the model in which the animal-breeder produced short-horn cattle by pulling one tooth from the jaw of each long-horn calf. Grasses, as dominant members of plant populations, first appeared about 20 million years ago. Grazing animals, the ancestors of our present-day horses, sheep, and cattle, until then fed on the succulent leaves of low trees and broad-leaved shrubs. With the dwindling of forests and the increasing domination of grasses, these animals were forced to feed increasingly on grasses, the new and successful members of the plant community.

It happens that grasses contain a much higher proportion of silicon than do the leaves of trees and shrubs. Silicon spicules are hard on the grinding teeth of plant-eating animals. Teeth wear away much more rapidly on a diet of grass than they do if softer leaves are available, especially if the grass is dried, as it often is in nature. It is certainly no coincidence, therefore, that high-crowned cheek teeth first appeared in the fossil skulls of grazing animals from geologic formations of about the same age as that of the appearance of the grasses. High crowns on the grinding teeth make the jaws a more long-lasting mill for grass eaters.

Animals in those early grazing populations that did not have the genetic prescription for high-crowned cheek teeth remind us of the long-horns that the animal-breeder deprived of one tooth—they would never do quite as well on a diet of grass as their cousins with high crowns. One has to suppose only a very slight difference in reproductive rate to ensure the ultimate disappearance of the shallow-crowned tooth. The agent of selection pressure in this case was a slight feeding difficulty produced by an environmental change: the corresponding genetic change was in the *adaptive* direction of high crowns, because only an adaptive change would serve to increase the gene frequency (by differential reproduction) and thus change the average phenotype of the population.

The Highly Adapted Grasses. The appearance of the grasses themselves was undoubtedly another example of this same kind of adaptive change, for the grasses, by virtue of their narrow leaves and silicaceous cuticle that prevent water loss, their rapid growth, and their thick, matted, water-seeking root systems, are much better able to survive and reproduce in a climate of limited and highly variable rainfall and temperature. In all such cases we must suppose that sexual reproduction, by permitting a rapid spread of desirable genes throughout a population by genetic mixing mechanisms, was the handmaiden of evolutionary change, always serving the primary function of adaptation.

Consequences of Failure to Adapt
Demise of the Reptile Rulers. One cannot doubt that the dominating populations of great reptiles, which ruled the earth for 150 million years, disappeared overnight (geologically speaking) because they were not able to change their gene frequencies to match environmental change. Perhaps the enormous size of the plant-eaters (which led to corresponding bulk in their reptile predators) was the ultimate extreme of that influence, presumably "the bigger, the better," that originally led to a surface-to-volume ratio problem (Figure 16-6). "Bigger" is not necessarily "better," if new environmental circumstances produce much sunlight and little shade, a reduced food supply, less water for floating enormous bodies, and newly evolved mammals to eat undefended reptile eggs. An enormous bulk requires enormous quantities of food, presents enormous heat-accumulation problems for an animal with poor body temperature regulating mechanisms, and is difficult to heave about out of the water for defense or for food seeking. One supposes that a genetic change in favor of smaller body size must have been too much of a strain on dinosaur genetic mechanisms in the time available. In any case the required adaptive change, whatever it was, could not be made in time. As a result, these eminently successful (as measured by the length of their dominance on earth)

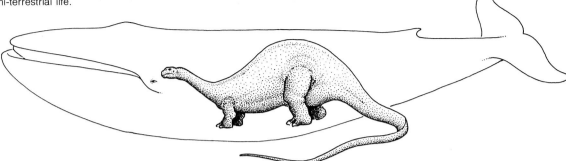

Figure 16-6 The immense dinosaur, *Brontosaurus,* is shown in comparison with the largest animal that has ever lived, the great blue whale (length about 30 m). Although much smaller than the whale, a dinosaur of this size must have been close to the physiological limits for even semi-terrestrial life.

giant animals dwindled away. Only a few large reptiles, such as the alligator, the Komodo Dragon, a few large turtles, and those eminently successful feathered dinosaurs, the birds, survive to remind us that worldly glory soon passes in the absence of adaptation. The lesson that we can draw from this is that although evolvability is a necessary characteristic of all life, *successful* evolution is not a universal characteristic of all species. The process of evolutionary change can result in adapted forms for whom a change in environment represents a dead-end from which further adaptive change cannot bring about retreat. We will discuss this question more thoroughly after treating hypotheses G and H.

Experimental Proof of the Adaptive Hypothesis

We conclude by observation of living plants and animals and their environmental interaction, by physiological considerations, and from fossil data that evolutionary changes are nearly always adaptive in the environment in which they occur. Yet to prove this by controlled experiment on natural populations is very difficult. Proof that the kind of selection pressure employed by animal breeders occurs in natural populations (as, for example, the English moth) would be difficult to establish experimentally because of the extreme slowness of the process. If that is true, then it is also true that the adaptive function of this selection pressure will be even more difficult to prove by controlled experiment because the observed process, yielding conclusive data, would be still slower. With an animal such as the fruit fly, which can produce a new generation every 10 days, one might in a few years demonstrate the positive result of gentle selection pressure on reproductive rate and thus on the average genotype. But to do this in a controlled environment, so that the population will change gene frequency in a clearly adaptive way, would take longer, possibly the 50 or 60 years that measured the increased success of the dark English moth. Yet no knowledgeable biologist doubts that it could be done.

Imperfection, Clutter, and Cumulativity

MARGINAL ADAPTATION AND VESTIGES

Back to the Drawing Board!

One of the difficulties of controlled experiments on adaptation is the potential imperfection of evolutionary changes; although almost always adaptive, they are not necessarily perfectly so. Commonly an adaptation is only marginally successful, persisting because it is at least workable. The human backbone fits this category for it is singularly undistinguished in its present usage by good engineering principles. Given the excellent physical properties of bone, connective tissues, and muscle, a competent engineer should be able to design a much better supporting structure for a bipedally erect mammal. Imperfection is not at all rare in evolved structures and processes, for as Theodosius Dobzhansky writes

> ... the fitness that is selected is the overall fitness of the organism to survive and reproduce, not the excellence of different organs, processes, and abilities taken separately. A consequence is that, especially in radical evolutionary constructions, the emerging product is an appalling mixture of excellence and weakness.
>
> THEODOSIUS DOBZHANSKY:
> *The Biology of Ultimate Concern.* 1967.
> New American Library, New York.

Evolutionary Baggage

Adding to this imperfection of evolutionary process is the clutter of evolutionary leftovers, persisting as vestigial structures and processes. The appendix in the human is often cited as a leftover from times when our ancestors had use for an extra intestinal pouch, probably to aid in the digestion of a bulky vegetable diet. We have other examples of vestigial organs and processes. Our toes have muscles that would serve a grasping, climbing (prehensile) function—if they still worked well. But they do not and so contribute little. There are thousands of hair follicles, distributed over most of our skin, but there is not enough hair remaining to serve any useful function. Yet each hair still has a tiny erector muscle.

Even the embryo, bridging a sensitive and chancy time of life, lengthens this costly and dangerous developmental process by forming structures similar to those that served a function only in our remote ancestors, a final twist to the mystery of evolutionary clutter.

The Clutter Establishes Cumulativity

A Wealth of Evidence. We can, however, gain some comfort from all the massive data that show the imperfection of evolutionary process and the attendant clutter. The comfort is that, in tracing the clutter, we have proved to the satisfaction of most critical scientists that evolution is indeed *cumulative* (hypothesis F). The many structures in plants and animals that parallel in inefficiency the backbone of man, the thousands of vestigial structures and processes, the repetition of earlier structures in the embryo—all these are convincing evidence that the evolutionary process consists of adaptive modification of countless adaptive modifications.

To these we must add equally massive data regarding the biochemical pathways forming the many complex molecules used in a multicellular organism. Even a glance at a chart of biochemical synthesis (Figure 16-7) reveals tortuous pathways involving precursors probably reminiscent of earlier cellular functions and having no present application. A biochemist could devise a much simpler way to make many of these molecules. The conclusion is inescapable that the steps of biochemical synthesis, like those of all evolutionary process, evolved one after the other in a cumulative sequence, with each step having adaptive value at the time it first evolved (see page 79).

A Lengthy Proof. Can we seriously require experimental verification of this process of accumulation? If so, we must be prepared to wait a very long time, for this would take even longer than controlled experiments on selection pressure and adaptation. To establish the cumulative character of evolutionary process, the experimenter would have to do all that he did to prove selection pressure and adaptation, and he would have to do it for at least *two* phenotypic characteristics, one after the other, with the second evolved as a modification of the first.

THE PROCESS OF SPECIATION

Observations of Living Populations

The Theory of Speciation. With the foregoing in mind, an easy transition to hypothesis G—the theory that these processes can and did produce speciation—is now possible. Having established the cumulative nature of evolution, we need only ask how to alter the genetic composition (genome) of members of an isolated population in such a way that they no longer successfully interbreed with members of the original population, and thus become a new species. One can imagine that speciation would occur fairly rapidly if several cumulative changes affected one phenotypic characteristic, for example the genital apparatus of sexually reproducing animals. Incompatibility in the process of fertilization should result in speciation in short order.

On the other hand, the accumulation of *slight* changes, involving many different genes, could require

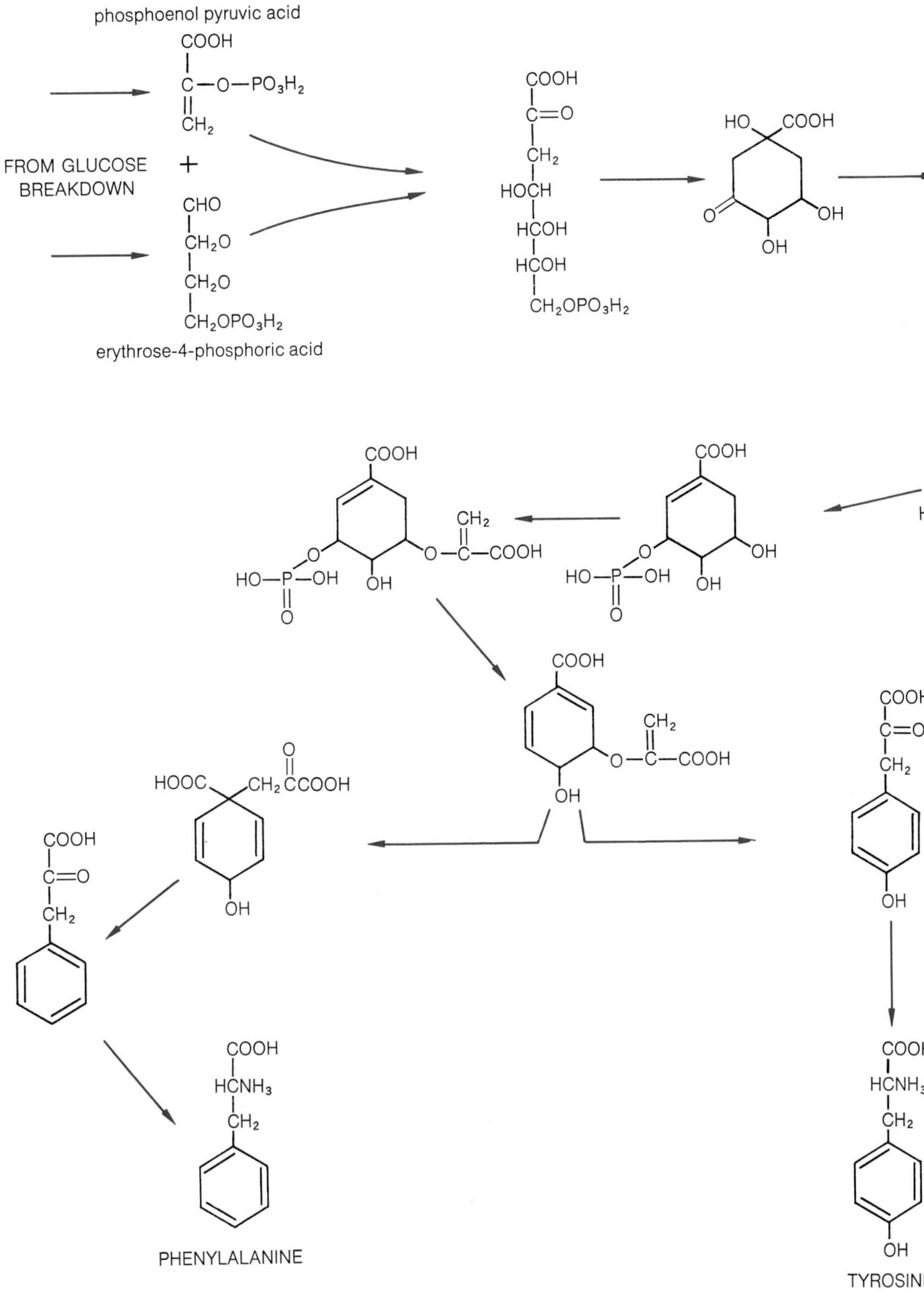

Figure 16-7 A typical biosynthetic pathway. Starting with straight chain precursors two amino acids containing aromatic rings are synthesized. While complex as written, the reactions shown actually involve directly 12 enzymes, ATP, NADPH, NAD⁺, and glutamic acid as an amino donor.

THE GREAT EXPLANATION

a very long time before the fertilization-fusion of the two genomes (one changed, one original) would result in genetic confusion and failure to produce progeny. In such a case, the evidence for speciation might not be complete reproductive failure. Hybrid progeny might be produced that would tend to have a reduced survival rate or some degree of sterility.

Racial Problem in the Frog Family. There is excellent evidence that speciation often follows the latter pathway. An illustration is the common leopard frog (*Rana pipiens*). It ranges from northern Canada to the Canal Zone, and over this entire area one frog looks about like another. One can breed a frog from Canada with one from Minnesota and have good success, or one from Minnesota with one from Oklahoma with equal success. The Oklahoma frog breeds fairly well with Mexican frogs also, but, when we attempt to breed a Canadian frog with a Mexican frog, the offspring are almost totally incapable of sustained existence; they die early in life or are sterile. Having made this discovery, one is now in a position to make other crosses and discover varying degrees of fertility, depending on how close together the parents occur geographically. Usually the closer, the better the fertility. Evidently the North American leopard frog is in process of speciation. One can, in fact, count some 23 races of this frog between Canada and Mexico.

Speciation in this case, no doubt, is instigated by selection pressure on the gene pool of these many populations, brought about by environmental differences, especially in temperature and water supply, over the vast geographical area that they occupy. The prevalence of certain genes governing body function is changing differently in the various populations of frogs. Speciation is not occurring rapidly in adjoining populations, because they are not sufficiently isolated. Gene flow occurs between neighboring gene pools and selection pressures are not very different. Sheer distance, however, increases genetic isolation in these nonmigratory animals and establishes effective isolation between the Canadian and Mexican leopard frog populations. Each is slowly and cumulatively adapting to the requirements of its own environment.

The Course of Evolutionary History

Though faced with the inability to provide long-term experimental confirmation of the components of the evolutionary process—selection pressure, adaptation, cumulativity, and speciation—it is difficult to justify serious reservations about the reality of evolution. Certainly each component of the process is strongly supported by data of the kind set forth in this chapter. The extent of this data, which we have here merely sampled, is so great as to defy logical organization *except* on the basis of evolutionary theory. To these hypotheses that support a theory of evolution as an ongoing process must be added the supportive evidence for hypothesis H—that the process of evolution did, in fact, produce all existing species. This evidence, even more mountainous than that which supports the other hypotheses, falls into three general categories:

1. The evidence from systematics (classification) of living organisms.
2. The evidence of the fossil record.
3. Physiological and anatomical continuities, past and present.

CLASSIFICATION AND EVOLUTION

Any array of organisms, evolved from a few common ancestors, necessarily reflects the course of evolution in the similarities of existing members of the array. The probability that a group of random, unrelated organisms, not produced as the result of evolutionary sequence, would organize readily into the branching pattern of relationship seen in typical phyletic "trees" is negligibly low—something like that of tossing a million pennies in the air and having them fall on the ground in the perfect, branching arborization pattern of a tree. This is another way of stating that the probability that our classification system actually reflects phylogenetic relationship is extremely high. It is not without significance that the father of taxonomy, Linnaeus, failed spectacularly to produce a workable system when he attempted to formulate a classification of rocks and minerals. Not until the science of geology made known the origins and chemical composition of rocks and minerals could a useful classification be developed.

Evolutionary Systematics

Modern systematists no longer accept the most obvious structural relationships as the basis for classification but actively seek out evolutionary (phyletic) significance as the ultimate criterion of the relationship of organisms that they study. In this way they have contributed hugely to the knowledge of evolutionary history and process. The fact that the present-day system of classification, based mainly on evolutionary criteria, adequately encompasses more than 1.5 million kinds of animals and 1 million kinds of plants can be taken without additional support as a proof that evolution did occur and, by inference, that it still continues.

FOSSILS

An Imposing Array

The evidence of the fossil record is remarkably extensive. Petrified tissue; buried bones; whole organisms preserved in amber, sediment, ice, sand, lava or tar; imprints in mud left by activity or by the encasement of an organ or organism long decayed away—all these and more carry a direct message from the past stating the reality of evolution. To make this fortress of support for evolution absolutely impregnable, we add the condition that these fossils must fit without conflict into two unrelated sources of data—on the one hand, they must be compatible with the evolutionary sequences worked out as a basis of taxonomic classification (fossils are classified just as are living organisms), and, on the other hand, the fossil sequences must be compatible with geologic information as to the age of the strata in which they were found. Each fossil must fit the classification scheme based on modern organisms, as well as fit appropriately into the time scale of geologic history.

We shall consider only one example of evidence from the fossil record from the many thousands available. It is a short history as evolutionary processes go—merely 50 million years.

Evolution of the Horse

Figure 16-8 sketches the fossil record of the evolution of the horse family over the past 55 million years. Fossils from suitably close intervals provide an almost unbroken continuity. The sequence is especially revealing with reference to the evolution of the single toe that we know as the horse's hoof. Equally as important as the continuity of this fossil sequence is its ability to satisfy both the systematist and the geologist. They discover that the evolutionary sequence depicted by the fossil line fits with what is known of the evolution of other single-hoofed animals, such as the donkey and the zebra, and with the origins of ancestral forms of other running, grass-eating ruminants, such as antelope and deer.

The historical **biogeographer** knows, from consideration of where the fossils in the sequence are found, that the sequence is not in violation of the known migratory routes and natural habitats of this evolving line. Very importantly, the geologist confirms that the fossils come from geological formations of the proper age. *Merychippus*, one of the ancestral horses, is found in formations about 25 million years old. If it were, for example, found also in formations 10 million or 50 million years old, the geologist would register his objections to the proposed evolutionary sequence. Actually, the correlation of a particular fossil species with a particular formation is usually so good that geologists often use fossil evidence to fix the age of an unknown formation when other data are lacking. Thus, such practical matters as the search for oil-bearing strata are materially assisted by a knowledge of **paleontology**.

Continuities—Past and Present

Fossils Are Not Always the Source. Consideration of horse fossils leads us naturally to the third line of evidence for the evolution of species, namely that of physiological and anatomical continuities. For the horse, we have discussed the establishment of a continuity based on evidence from the fossil record. However, similar continuities may often be derived without ever looking at a fossil.

The Vertebrate Heart and Skeleton. In some cases these continuities are ascertained from organs, or from animals that are not preserved well as fossils. For

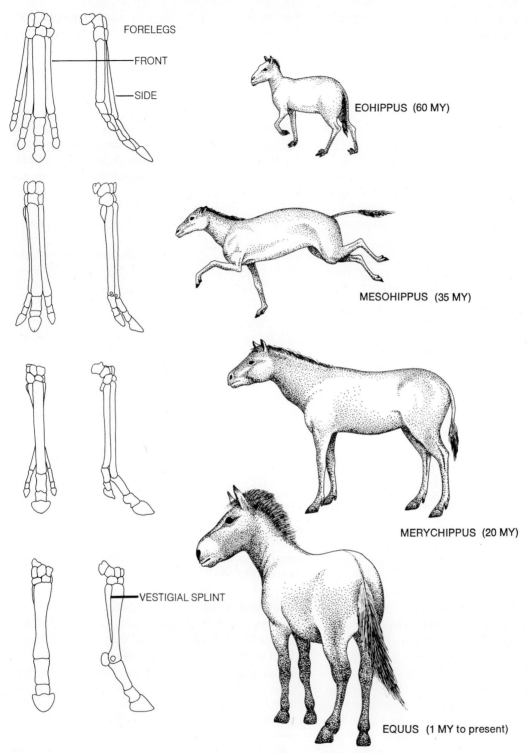

Figure 16-8 Representatives of one of the principal lines of horse evolution showing the tendency towards larger size and reduction in the number of toes. *Eohippus* (=*Hyracotherium*) was a browsing animal while *Merychippus* was a grazer.

example, it is well understood that the vertebrate heart evolved as a two-chambered organ in fish, a three-chambered organ in amphibia, became four-chambered in the reptiles, and finally reached its greatest four-chambered complexity in birds and mammals (Figure 16-9).

We do not find any fossil hearts, so how can we be sure of this continuity? First we must establish the evolutionary continuity of the sequence: fish, amphibia, reptiles, birds and mammals. This has been accomplished with excellent documentation of skeletal evidence from the fossil record. We can trace such transitions as gill arches to jaws and the smooth transition of arms and legs and the skeletal girdles to which they are attached, as early vertebrates evolved from water to land (Figure 15-47). Skeletal structures have many passages for nerves and blood vessels as well as many points of attachment for muscles; from these passages we can deduce much of the history of circulatory, muscular, and nervous systems. And, most importantly, all of these changes can be correlated with the geologic time scale, so that they fall into a regular time sequence of increased complexity of function and structure.

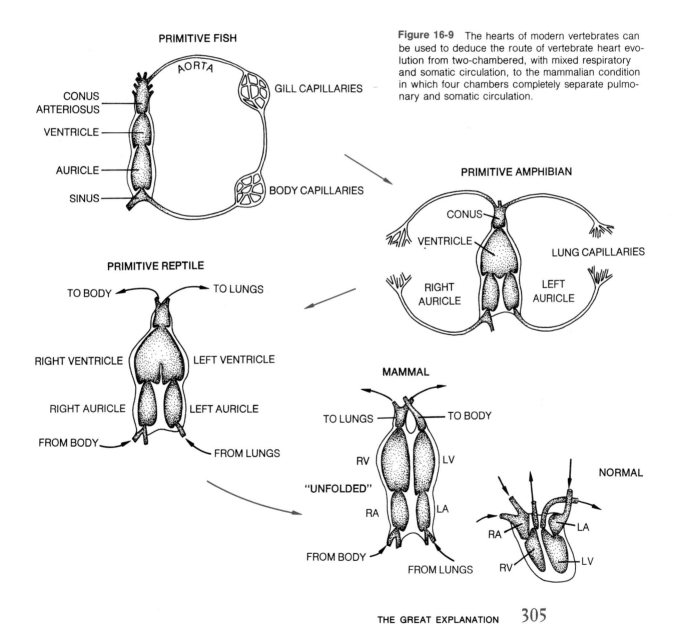

Figure 16-9 The hearts of modern vertebrates can be used to deduce the route of vertebrate heart evolution from two-chambered, with mixed respiratory and somatic circulation, to the mammalian condition in which four chambers completely separate pulmonary and somatic circulation.

THE GREAT EXPLANATION

Other Continuities. From such data, the evolution of the vertebrates brain is deduced, as in Figure 16-10. We can follow the evolution of the lung from the fish swim bladder. We can even speculate effectively on such purely physiological matters as the evolution of immunity in vertebrates from the variation in immunological mechanisms seen among living vertebrates.

These continuities among the vertebrates, with an equally impressive list among the invertebrates and in the plant kingdom, are so numerous that it would take hundreds of volumes to describe them all. Together, they are surely powerful evidence, when correlated with the taxonomic tree and with fossil evidence, of the validity of hypothesis H in Table 16-1.

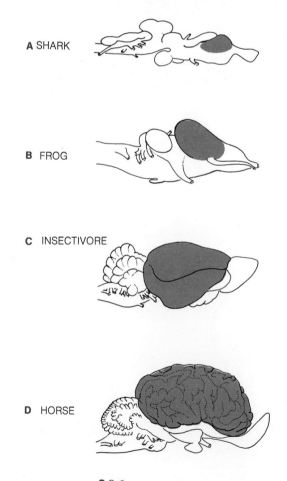

Figure 16-10 A series of brains of contemporary vertebrates illustrates the general route of evolution of the vertebrate brain. Increasing dominance of the cerebral cortex (color) is obvious.

A SHARK

B FROG

C INSECTIVORE

D HORSE

Gaps in the Fossil Record

A Sudden Leap Forward. Though the fossil record makes an enormously important contribution to evolutionary theory, this source of data poses questions that have proved to be a source of embarrassment to evolutionary theorists. For example, pre-Cambrian formations of 500–600 million years ago (see Appendix 2) contain only fossils of primitive, single-celled forms. Yet early in Cambrian times, at the beginning of the Paleozoic era, nearly all the phyla of multicellular animals as we know them today had appeared! Only the vertebrates are missing from the early Cambrian record, and even they had appeared by the beginning of the next era, a mere 30 or 40 million years later. It is as though life, "incubating" in single-celled forms for a billion years or more, suddenly evolved almost overnight into the great variety of complex multicellular animals.

One hypothesis in explanation supposes that these forms actually evolved over a much longer period of time, extending well back into the pre-Cambrian era, but that the record of this evolution was destroyed by violent geologic events at the beginning of the Paleozoic era. This must not be ruled out as a possibility, yet many paleobiologists are dissatisfied with the explanation, pointing out that the fossil records of the same kinds of animals survived subsequent geologic events of similar magnitude.

Is Sexual Reproduction the Answer? There is another possible explanation for this gap in the fossil record, based upon the observations and concepts presented in the preceding pages. Consider the problems that the genome must effectively solve in order for multicellularity to evolve—intercellular space, combined with cellular cohesiveness; a waterproof integument; the production of an internal environment, with circulating compartments and a pump; openings in the integument for ingestion and gas exchange; a vascular ventilating system; an alimentary system for digestion and absorption and the storage and distribution of foods; osmoregulatory and excretory systems; contractile tissue for movement and conductive tissue for integrated control—all hung on or in a suitable skeletal framework. It is not too great a strain on the imagination to suppose that mutation might make a significant step toward the establishment of any *one* of these characteristics. But one must suppose that the advent of functionally satisfactory multicellularity requires either an immense number of mutations in a single line of descent or else a *mixing of many mutations among successive generations of a single population.*

The first alternative is highly unlikely, and the mixing process of the second is not readily possible without sexual reproduction—there is no way that a mutation occurring to one member of a species can be combined with a different but useful mutation occurring to another member of that same species if reproduction is asexual (except by viral transfer). Only gametic fusion or exchange of genetic material permits this kind of cumulative mixing. It is highly improbable that the number and kind of complex requirements posed by the advent of multicellularity could be solved in the absence of sexual reproduction. So it seems likely that this gap in the fossil record may be explained by the advent of sexual reproduction. The reason that these many phyla appear suddenly in early Cambrian formations may well be that they did, in fact, evolve almost overnight, due to the vastly accelerated evolutionary process made possible by a mechanism of genetic mixing, sexual reproduction.

What is the Verdict?

Can we now regard as adequate the proof of the process of evolution? Hypotheses A through H of Table 16-1 have been presented, together with an outline of the evidence for each. The probability that the complexity of living forms was derived from some other, undetected process, unrelated to those of A through H, is so small as to be negligible. Nevertheless, it is clear also from our outline of evidence for hypotheses A through H that this evidence is partly circumstantial, even though mountainous in quantity. We cannot, for reasons stated, observe the complete process of evolution directly.

At this point you possibly feel that, whereas evolution in a general sense is well enough established by the evidence of this chapter, it is just that—too general, and without obvious application to the object of your primary interest in studying biology: the human. Perhaps *we* are somehow different with altogether different origins. You are in a position somewhat like that of a juror who has heard all about the actions of the accused's associates but nothing about the accused directly. The doings of the associates, although they might establish strong suspicions (it would be unlikely for *one* species, out of millions similarly evolved, to have had a different mode of origin), is not the best way to win a case in the halls of scientific justice. Fortunately we can do better. In Chapter 17 we shall show that the human has been subject to the same evolutionary processes that we have presented here.

17 Human Evolution

OUR PLACE IN THE COSMIC ZOO

Much of the uproar associated with the struggle for acceptance of the theory of evolution came from its implications regarding humanity. The debate between Bishop Wilberforce and Thomas Henry Huxley might have been considerably less acrimonious, and perhaps would not have occurred, if evolution were concerned only with lower organisms. The Bishop's effort to ridicule the theory by asking Huxley whether he was descended from apes on his mother's or his father's side was only one of an old family of complaints by humans at their gradual displacement from the center stage of Creation.

Assuredly, a persistent theme in history has been the steady displacement of humanity from the central place in the universal scheme of things, as envisaged by early thinkers, to a more and more peripheral role. We presume that the beginnings of this lie in unwritten prehistory and represent attempts by early humans to come to terms with the frightening unknown that enveloped them. Under such conditions it is undoubtedly fortifying to believe that the surrounding world is *for you* and manipulatable *by you*, rather than indifferent and unresponsive to the human presence. A relatively logical consequence of such an attitude was the situation in early historical times when the earth was considered the center of the universe, and humans were either in personal control of everything that mattered or were in favored communication with gods that resembled humans.

The wearing away of this attitude was hard work, which, in its beginnings, could readily earn the edifying reward of martyrdom. Astronomers struck many of the first hard blows, with ultimate success in at least putting into proper perspective the habitation of humanity, the earth. Our earth was demoted from the central axis of the universe to its proper role as a modest planet, attending a minor star of a smallish galaxy, in the suburbs of a universe of incomprehensible magnitude, whose pulse could hardly be expected to falter, whatever the fate of humanity, its planet, its star, or even its galaxy.

Insults as grave as this could have been tolerated if humanity, though relegated to the backwoods of the universe, still had a pedigree at least guaranteeing that its advent was unique and of such a nature as to ensure a privileged position in the universe. Thus, ideas of *special creation* were both welcome and hard to replace with scientific explanations even in the face of solid evidence. As is common in many intellectual pitched battles, much of the heat was generated out of the antagonist's ignorance about each other's position. By now, considerably better understanding has been attained and, for example, the principal religions of the world now largely accept evolution, many adopting the view that it is a vehicle for the actions of the Supreme Being.

Such attitudes, of course, are beyond the area of comment of biology texts, except for the observation that accommodations between religion and science are undoubtedly valuable in smoothing our hard path as we grope to understand ourselves and our significance. Consider the utility of the concept of soul in this restricted context. A scientist cannot scientifically say whether souls do exist or not, but might observe that the concept of soul is a device to forestall the agelong erosion of our uniqueness and to adjust our desire for primacy in the order of being to the facts of biological evolution. The concept of soul permits the believer to view human evolution with equanimity, and in conformity with the physical evidence, as an assembly line on which there developed a series of slack-jawed prehumans, endowed with sufficient mentality to understand the utility of stones in the arbitration of disputes. At one point on the assembly line the soul appeared, took control of the beast, and set its feet on the path to Significance in the Universe.

Even Alfred Russel Wallace, Darwin's steadfast admirer and companion in the study of evolution, took this attitude. He was, of course, in general a true believer in evolution by means of natural selection and readily conceded that natural selection could explain the lower attributes, the more animal qualities, of humans. But he argued that natural selection could not explain the attributes that defined the civilized human: artistic, musical, mathematical, and other skills. He wrote that such attributes

clearly point to the existence in man of something which he has not derived from his animal progenitors—something which we may best refer to as being of a spiritual essence or nature.... Thus we may perceive that the love of truth, the delight in beauty, the passion for justice, and the thrill of exultation with which we hear of any act of courageous self-sacrifice, are the workings within us of a higher nature which has not been developed by means of the struggle for material existence.

A. R. WALLACE: 1896.
Darwinism. Macmillan, London.

This defection by Wallace pained Darwin greatly. He hastened to write, even in the first edition of the *Origin*, that he did not attribute *all* changes in structure and mentality exclusively to natural selection. The concept of inheritance of acquired characters was evoked to explain such difficult points, thereby generating probably the most critical error in the great edifice of argument on evolution that he was erecting. The error is, of course, eminently excusable on the grounds of Darwin's lack of knowledge of the underlying mechanisms of inheritance.

Proving That We Evolved

Chapter 16 saves us a great deal of effort. We might assume, because we are organisms, that the mechanisms of evolution set forth in Chapter 16 apply to us. Therefore, we must have evolved. But we can be much more specific than this, and we shall show that we evolved just as surely as the organisms of the previous chapter by using precisely the same kinds of evidence as used there.

DARWIN'S VIEW OF HUMAN EVOLUTION

In 1871, 12 years after publication of the *Origins*, Darwin published *The Descent of Man, and Selection in Relation to Sex*. Even though fossil evidence of human origins, such as we now possess, was not then available, the *Descent* is still a very creditable introduction to human evolution. It sets forth clearly the major categories of essential evidence that scientists still work to complete.

Darwin looked for the telltale signs of natural selection in human populations. He demonstrated that humans vary and argued that the natural tendency for increase in numbers must cause a severe struggle for survival, with certain variants being more successful than others. In the search for information on human variation, Darwin continued his characteristic, painstaking accumulation of data from every conceivable source. Thus he wrote that, even in the natives of a restricted area such as the Hawaiian Islands, "An eminent dentist assures me that there is nearly as much diversity in the teeth as in the features." His search took him to the anatomical literature from which he noted the necessity of cataloging the many variations in the courses of the principal arteries for surgical purposes. Equally numerous variations in the organization of the human musculature were noted.

With great quantities of data on human *variation* in hand, Darwin turned to demonstrating the *struggle for existence*. He illustrated the breeding capacity of our species with an example from the United States. Observing that the United States population had doubled in 25 years, he wrote that, unchecked, "the present population of the United States [thirty millions], would in 675 years cover the whole terraqueous globe so thickly, that four men would have to stand on each square yard of surface." Clearly checks *are* applied, and, certainly in a primitive state of existence, the result would be evolution by means of natural selection.

Darwin was restricted largely to comparisons between the human and other recent animals in demonstrating that evolution had occurred. To relate us to lower forms he marshalled data of the following sorts:

Rudimentary organs. Presence in the human of rudiments of organs that are well developed in lower forms, and hence indicate a relationship, was documented. These include the **coccyx**, rudiment of a tail, and the **panniculus carnosus** muscles, which in lower forms are well developed to twitch the skin and ears and which persist to some degree in many of us.

Embryonic development. The resemblance of the human embryo to the embryo of lower forms was documented (Figure 17-1). He quotes one of the early giants of embryology, Karl Ernst von Baer, on the homologies of development, ". . . the feet of lizards and mammals, the wings and feet of birds, no less than the hands and feet of man, all arise from the same fundamental form."

Similarities of adults. According to Darwin, "It is notorious that man is constructed on the same general type or model as other mammals. All the bones in his skeleton can be compared with corresponding bones in a monkey, bat, or seal. . . . The brain follows the same law . . . every chief fissure and fold in the brain of man has its analogy in that of the orang. . . ."

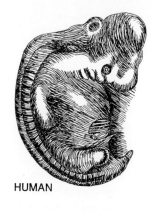
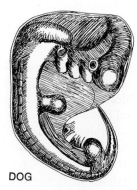

Figure 17-1 Darwin was impressed with the similarity in appearance of the embryos of the human (upper) and lower forms (dog, lower figure). (After C. Darwin, 1896. *The Descent of Man and Selection in Relation to Sex*. 2d. ed. D. Appleton, N. Y. Fig. 1.)

Characteristic of Darwin's wide-ranging interests was the inclusion of behavior in his study of homologies between the human and lower forms. Just a year after the appearance of the *Descent* he brought out a book wholly devoted to the subject, *The Expression of Emotions in Man and Animals*. Some of his collected data on the subject are slightly suspect since one is often prone to read more humanity into the behavior of animals than may actually be there.

In sum, Darwin's arguments marshalled around these three classes of data are quite strong, leading him to depart from his often diffident approach to conclusions and assert,

Thus we can understand how it has come to pass that man and all other vertebrate animals have been constructed on the same general model, why they pass through the same early stages of development, and why they retain certain rudiments in common. Consequently we ought frankly to admit their community of descent; to take any other view, is to admit that our own structure, and that of all the animals around us, is a mere snare laid to entrap our judgement. . . . But the time will before long come, when it will be thought wonderful that naturalists, who were well acquainted with the comparative structure and development of man, and other mammals, should have believed that each was the work of a separate act of creation.

C. DARWIN: 1896.
The Descent of Man and Selection in Relation to Sex.
2d ed., Appelton, New York.

A MODERN VIEW OF HUMAN EVOLUTION

There has been great progress since Darwin in the study of human evolution, but the work still lies within the framework that he provided. We now have the benefit of a rational theory of heredity which, of course, Darwin lacked almost totally. He also lacked information on fossils linking us with the primates, and this we now have in profusion. Beyond these two major additions, modern study of human evolution provides many satisfying details, which, however, do not truly represent significant conceptual departures from the outline he left us. For example, the homologies between human and primate have been carried from gross structure to the underlying apparatus that produces the structure. Modern studies of chromosome banding, protein sequencing, and the structure of DNA tell us what in a sense is obvious, *that like anatomy is made by like machinery*. It is, nonetheless, good to know these things, and in the following section we will examine some of these modern facts that bring Darwinism up to date.

WE SHARE COMMON FEATURES WITH MODERN PRIMATES

If we had separate origins from other animals, it ought to be difficult to fit us on any of the branches of the phyletic tree. In fact, we are well accommodated there as the Family Hominidae. Beside us are our nearest living relatives, the great apes—gorilla, orangutan, and chimpanzee—forming the Family Pongidae. The similarities between apes and humans are so marked that the two families are placed together in a

Table 17-1 The mammalian order Primates

	Common names	*Present distribution*	*Some distinguishing characteristics*
PRIMATES	primates	world-wide	Enlarged cerebral hemispheres; prehensile hands and feet with at least one digit opposable to the others; nails instead of claws; eyes directed forward; usually one pair of mammary glands; teeth include incisors, canines, pre-molars and molars.
SUBORDER PROSIMII	prosimians	tropical Africa and Asia	
Infraorder Lemuriformes	tree shrews	S.E. Asia	Squirrel-like and arboreal; claws instead of nails; least complex of primate brains; non-prehensile tail.
	lemurs	Africa, S.E. Asia	Nails on digits.
Infraorder Lorisiformes	lorises and bushbabies	Tropical Africa and Asia	No menstrual cycle.
Infraorder Tarsiiformes	tarsiers	S.E. Asia (1 genus)	Arboreal and nocturnal; extremely large eyes; hind foot and lower limb modified for leaping; monthly estrous cycle.
SUBORDER ANTHROPOIDEA			
Superfamily Ceboidea	New World monkeys	Central and South America	Many with grasping tails; thumbs not opposable to fingers.
Superfamily Cercopithecoidea	Old World monkeys	Europe, Africa, Asia	Thumbs opposable to fingers.
Superfamily Hominoidea			Largest of primates with strongly developed brachiating habit in pongids; digestive tract with appendix; teeth for omnivorous diet.
Family Pongidae	apes	Africa and S.E. Asia (orang-utan) world-wide	
Family Hominidae	humans		Largest relative brain size; no sexual difference in small canine teeth; prominent chin, reduced brow ridge, the least prognathous; least amount of body hair; big toe not opposable.

Superfamily, the Hominoidea. The Hominoidea, along with the two superfamilies of monkeys form the Suborder Anthropoidea. This Suborder includes all of the humanlike primates. The remaining primitive primates—tree shrews, lemurs, tarsiers—are placed in the Suborder Prosimii (Table 17-1).

Characteristics of the Hominoids

If we hominoids are figured together, hairless and in the same posture, our similarities are obvious (Figure 17-2). Let us examine these similarities a little more closely and at the same time inquire about the evolutionary forces that prompted their appearance.

Hands and Feet. Closer examination of hands and feet of anthropoids shows that they are all characterized by fingers that can grasp and thumbs and big toes that, at least to some extent, are opposable to the other fingers or toes of the same appendage (Figure 17-3). The figure also shows the hands and feet of a tree shrew. These small, squirrel-like animals (Figure 17-4) are considered to be either the most primitive primates or their closest relatives. They suggest what the earliest primates probably were like. Although they live in trees, their feet have much of the appearance of those of terrestrial animals, having claws and fingers that are poorly opposable. It is believed that from such tree-dwelling beginnings the remainder of the primates evolved, becoming more and more adapted to arboreal (tree) life. The hands and feet of chimpanzees and orangutans show this well in being constructed to facilitate grasping tree limbs. This was evidently not security enough for one of the superfamilies of the Anthropoidea, the New World Monkeys, the Ceboidea. They developed a "third hand," a prehensile tail.

Eyes. Arboreal life is believed to have led to several

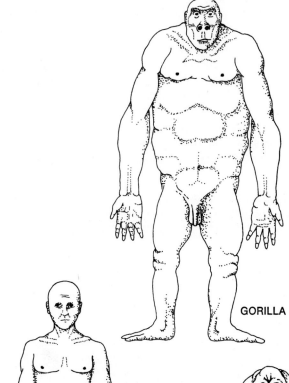
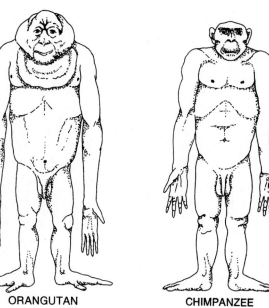

Figure 17-2 Our similarities with our primate relatives are clear. (After A. H. Schultz.)

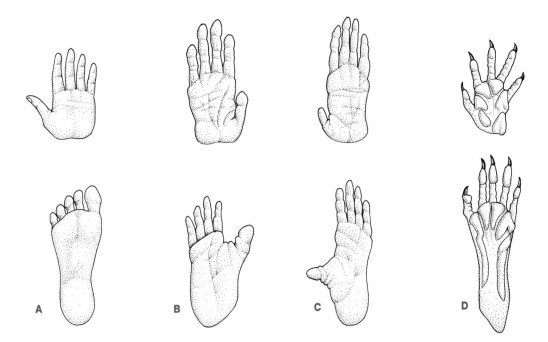

Figure 17-3 Hands and feet of some primates and a tree shrew. **A.** Human. **B.** Chimpanzee. **C.** Orangutan. **D.** Tree shrew. (After J. S. Weiner (1971) *The Natural History of Man*. Universe, New York.)

Figure 17-4 A Philippine tree shrew. (A. W. Ambler, National Audubon Society.)

other specializations that set the primates off from other mammals. For rapid motion among the branches of trees, good visual depth perception is important. This was achieved by migration of the eyes to the front, allowing the visual fields to overlap and providing the basis of binocular vision. This movement was facilitated by a switch in emphasis from the sense of smell to vision. The sense of smell is more useful to a terrestrial animal than to an arboreal one. As the importance of scent waned, the snout diminished, facilitating the forward migration of the eyes. The primate visual system was further improved by the development of color vision and the associated sharp day vision associated with the fovea (see page 352).

The Brain. It is generally believed that arboreal life must have been a bit more tricky than life on the ground. In a sense the complexity of life aloft is like moving from life in two dimensions into a three-dimensional world, although obviously this is not strictly true. At any rate it is believed that the presumed added complexity of life aloft not only resulted in the primate features just described but, most importantly, led to increased brain size relative to body weight. Actually, even the most primitive prosimians have rather large brains as compared with other mammals, and it is also true that the size of higher primate brains is not unique, except for man and his immedi-

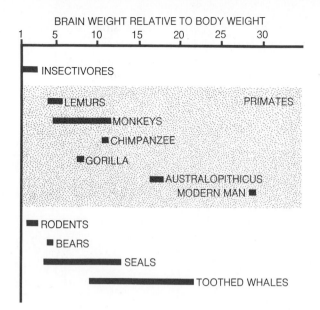

Figure 17-5 Relationship of brain to body weight for primates and some other mammals.

ate fossil ancestors. As Figure 17-5 shows, brain size in some seals and toothed whales even exceeds the brain size of the great apes. These, of course, also live aquatic "three-dimensional" lives, so perhaps there is something to the idea of the complexity of the arboreal world leading to increased brain size.

Posture, the Skeleton, and the Teeth. Life in the trees also resulted in widespread and distinctive modifications of the skeleton and posture. Habitual locomotion by swinging from branch to branch (**brachiation**) resulted in the typical erect trunk posture of primates and in the great flexibility of the limbs. Grasping hands evolved into remarkably effective organs for exploring and manipulating the environment.

Bipedal locomotion developed after the habit of brachiation was well established. Emphasis on bipedalism in the human has produced most of the distinguishing skeletal features that differentiate us from our ape relatives.

From the diet of small animals from the forest floor that was probably characteristic of the ancestral primates, the diet of the arboreal primates became more varied with the inclusion of fruits and other vegetable material. The development of the typical varied primate dentition which includes canines, incisors, and molars resulted (Figure 17-6).

Extended Juvenility. Primate reproductive behavior is also unusual in several respects. Perhaps because of the difficulty of caring for a large number of squirming offspring aloft, primates typically have only one infant at a time. Survival of the species consequently becomes dependent upon increasing the survival rate of the few offspring that are born. This has taken the direction of an increased level of postnatal and juvenile care, which is facilitated because primates typically maintain family groups and troops or clans of related adults, juveniles, and infants. While providing the infant with protection, the prolonged period of contact of the infant with parents, principally the mother and other females and infants, also provides opportunity for extensive training in the business of life. To a large degree the frequent sexual accessibility of the female primate, reflected by her monthly ovulatory cycle, is thought responsible for this basic element of primate social behavior.

Molecular and Genetic Similarities. Although the primate characteristics just described are sufficient to identify us firmly as members of that group, it is of great interest to examine molecular and genetic relationships. These are more closely related to the genetic blueprints of the members of the group, the DNA sequences of their genomes, by which their differences are ultimately determined. The goals of such studies are to identify genes held in common with other primates and to estimate the genic changes that have gone on as the primates evolved.

Antigenic relationships have been extensively measured for many living primates. The technique used requires making antibodies to a protein from one

Figure 17-6 Lower jaw of a chimpanzee. The varied dental armament reflects a diet which includes fruit and meat. After D. Pilbeam (1972) *The Ascent of Man*. Macmillan, New York.

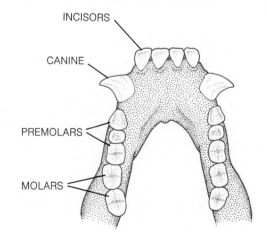

species and determining the extent to which these antibodies react with similar proteins from other primates (see page 500). The more intense the reaction, the more similar the compared proteins and the closer the two species involved are taken to be. Studies of this type confirm primate relationships established on the basis of more conventional taxonomic methods. Beyond this they are not particularly interesting because studies at this level give little or no indication of the difference in genetic constitution underlying antigenic differences. Antigenic reactions usually involve surface proteins of cells and it is not known how much genetic variation is responsible for a given degree of antigenic difference.

The comparatively recently developed technique of amino acid sequencing of proteins has the advantage of carrying the study of animal relationships closer to the genetic apparatus since, on the basis of differences in amino acid sequences, it is possible to arrive at estimates of the causal genetic variation.

Perhaps the best example of this type of investigation is found in the study of variation of amino acid sequences in the α and β chains of hemoglobin. These always consist, respectively, of 141 and 146 amino acids throughout the mammals, but there is much variation in the sequencing, or order, of amino acids in the two polypeptid chains, as shown in Table 17-2. As would be expected, the number of amino acid differences between human and other hemoglobins parallels the evidence from antigenic measurements and confirms the classification developed by classical methods. Thus, gorilla and chimpanzee hemoglobins are the most similar to ours, and the hemoglobin of the most distantly related mammals, the marsupials, show the largest number of differences in amino acid sequences.

Table 17-2 Variation in amino acid sequences for variation-prone positions in vertebrate β-hemoglobin illustrates the close relationships among the Anthropoidea and progressive divergence with more distant relationships. Amino acid variances from the human are indicated in red.

AMINO ACID HOMOLOGY IN VERTEBRATE β-HEMOGLOBIN

Group	Subgroup	Species	9	12	13	16	21	22	33	43	47	50	51	56	69	76	80	86	87	104	116
PRIMATE MAMMALS	ANTHROPOIDEA	Human	SER	THR	ALA	GLY	ASP	GLU	VAL	GLU	ASP	THR	PRO	GLY	GLY	ALA	ASN	ALA	THR	ARG	HIS
		Chimpanzee	SER	THR	ALA	GLY	ASP	GLU	VAL	GLU	ASP	THR	PRO	GLY	GLY	ALA	ASN	ALA	THR	ARG	HIS
		Gorilla	SER	THR	ALA	GLY	ASP	GLU	VAL	GLU	ASP	THR	PRO	GLY	GLY	ALA	ASN	ALA	THR	LYS	HIS
	INTERMEDIATE PRIMATES	Baboon	ASN	THR	THR	GLY	ASP	GLU	VAL	ASP	ASP	SER	PRO	GLY	GLY	ASN	ASN	ALA	GLN	LYS	HIS
		Macaque	ASN	THR	THR	GLY	ASP	GLU	LEU	GLU	ASP	SER	PRO	GLY	GLY	ASN	ASN	ALA	GLN	LYS	HIS
		Spider Monkey	ALA	THR	ALA	GLY	ASP	GLU	VAL	GLU	ASP	THR	PRO	SER	GLY	ALA	ASN	ALA	GLN	ARG	HIS
	LOWER PRIMATES	Slow Loris	SER	THR	ALA	GLY	ASP	ASN	VAL	GLU	ASP	SER	PRO	GLY	SER	ASN	ASN	ALA	LYS	ARG	HIS
		Lemur	ALA	THR	SER	GLY	GLU	LYS	VAL	GLU	ASP	SER	PRO	GLY	SER	HIS	ASN	ALA	GLN	LYS	LEU
MAMMALS	HIGHER MAMMALS	Rabbit	SER	THR	ALA	GLY	GLU	GLU	VAL	GLU	ASP	SER	ALA	ASN	ALA	SER	ASN	ALA	LYS	ARG	HIS
		Dog	SER	SER	GLY	GLY	ASP	GLU	ILU	ASP	ASP	THR	PRO	SER	ASN	LYS	ASN	ALA	LYS	LYS	HIS
	LOWER MAMMALS	Kangaroo	ASP	THR	SER	GLY	GLU	GLN	ILU	ASP	ASP	ASN	ALA	ALA	VAL	LYS	ASN	ALA	LYS	LYS	GLU
		Echidna	THR	THR	ASN	GLY	ASN	GLU	VAL	GLU	ASP	SER	ALA	GLY	THR	LYS	ASN	ALA	LYS	ASN	ARG
	OTHER VERTEBRATES	Chicken	GLN	THR	GLY	GLY	ALA	GLU	ILU	ALA	ASN	SER	PRO	GLU	THR	LYS	ASN	SER	GLN	ARG	ALA
		Frog	ASP	SER	GLY	GLY	LYS	HIS	VAL	THR	ASN	SER	ALA	HIS	ALA	LYS	ASN	ALA	LYS	ARG	ARG

Data from: Fasman, G. D., ed, Handbook of Biochemistry and Molecular Biology, vol 3, 3rd ed. 1976. CRC Press, Cleveland, Ohio.

The minimum number of mutations necessary to produce a given variation in amino acid sequence can be estimated. Viewed in the perspective of the geological time scale, these seem at first glance to be extraordinarily low. To have had no mutational difference arise between human and chimpanzee hemoglobin and only one between human and gorilla hemoglobin in the 14 million years since the ape and human stocks are thought to have diverged would suggest an astounding stability of the genetic apparatus. However, it must be remembered that hemoglobin is essential to life. Probably many mutational changes that occur in such a molecule are lethal. Thus, the true mutation rate for hemoglobin genes is probably far higher than the data of phyletic variation suggest.

Chromosome studies also confirm our close relationship with the great apes. The diploid number for all three of the great apes is 48 as compared with 46 for the human. Recent studies of chromosome banding by a fluorescent staining technique make it possible to identify homologous regions within primate chromosomes. Comparisons made in this way show greater affinity among human and gorilla or chimpanzee chromosomes than between human and orangutan chromosomes. It appears also that the reduction in human chromosome number did not result in any significant loss of genetic material since one particular human chromosome shows sufficient similarity to two chimpanzee chromosomes to suggest that it formed by the fusion of two chromosomes, thus reducing the genome to 46.

In summary, it is inescapable that our nearest living relatives are the great apes. Our anatomy is virtually that of a hairless ape, and anatomy is reinforced by identities in embryonic development and in physiology. If more evidence is needed, it is available in vestigial organs, the structure of chromosomes and even in the organization of DNA. But having admitted the kinship, we are still far from understanding the course of human evolution. We have clearly diverged greatly from the apes sometime in the past. The path of this divergence must be understood before we can know ourselves fully as primates.

Missing Links

If we are related to the apes, there must have been a common ancestor, a "missing link." To begin our search for the missing link (or links) and for the pathway that our species has followed from such an apelike ancestor, we can best start with a discussion of the second chapter in Darwin's *The Descent of Man*, entitled "On the Manner of Development of Man From Some Lower Form." It will be remembered that Darwin had no significant fossil evidence of early man to guide his thoughts. He had to rely almost totally on the evidence of anatomy and behavior of living primates. Even so, his picture of what the missing link must have been like and his analysis of the factors that set our ancestors off on the path to humanity was totally correct.

The Hand is the Trigger. Darwin, of course, saw in the overpowering intelligence of the human the essential difference between man and ape. He also realized that something was required to trigger the emphasis on brain development in the human line. The initiating factor was plainly evident to Darwin in perhaps the second biggest difference between ourselves and the apes, our bipedal stance. We walk erect, whereas the remainder of the primates walk on two feet awkwardly, if at all (Figure 17-7).

Darwin saw the great importance of bipedal locomotion. It frees the hands for activities other than locomotion. He wrote,

Man could not have attained his present dominant position in the world without the use of his hands, which are so admirably adapted to act in obedience to his will. . . . But the hands and arms could not have become perfect enough to have manufactured weapons, or to have hurled stones and spears with a true aim, as long as they were habitually used for locomotion. . . . If it be an advantage to man to stand firmly on his feet and to have his hands and arms free, of which, from his pre-eminent success in the battle of life, there can be no doubt, then I can see no reason why it should not have been advantageous to the progenitors of man to have become more and more erect or bipedal. They would thus have been better able to defend themselves with stones or clubs, to attack their prey, or otherwise to obtain food.

CHARLES DARWIN: 1896.
The Descent of Man and Selection in Relation to Sex.
2d ed. Appleton, New York.

Hand and Brain. Although the brain has many subtle ways to get information from the environment—the senses of smell, vision, touch, and hearing—it has only one means of completing the loop of interaction with the environment. It controls muscles operating various kinds of organs that do physical work. The more precise and adaptable this system of muscularly operated organs is, the more effectively the brain can act upon the information it receives. When the brain contacts the world through a crude paw or hoof, there is not much opportunity for sophistication and variety of interaction. When the contact is through a hand like ours, the horizons of interaction are tremendously widened. The hand, true enough, is superb as a device for throwing stones and making crude weapons, but that is only the beginning. The brain, equipped with this new probe, the hand, recently emancipated from purely locomotor duties, must have triggered the explosion of mental activity that led to the use and manufacture of simple tools, the mastery of fire, and ultimately to civilization.

Darwin's View of the First Human. According to Darwin the first humans would be expected to show evidence of recent descent from the trees in their posture and walking. Brain enlargement would be slight but noticeable, and, reciprocally with enlargement of the brain case, there would have been evident the beginnings of reduction of the massive, forward extending jaws and of the canine teeth. The earliest humans, he argued, would still have noticeable canines, but the growth in brain power would have provided better weapons than slashing teeth, so these hallmarks of primate ancestry would have started their decline to present dimensions. The fossil record confirms Darwin's logic, as we shall see.

OUR FOSSIL HISTORY

In Figure 17-8 we see artist's impressions of the fruits of searches that trace the ancestry of man into the Early Pleistocene Age, roughly 1 million years ago. By then the hominids who were clearly our ancestors existed. We must go back even farther, say to the beginning of the Miocene Age, 25 million years, to see the separation of the Hominidae and Pongidae.

The time scale of these happenings is as follows: Paleontologists tell us that the beginnings of the class Mammalia lie 200 million years ago in the Triassic Period as the age of the dinosaurs drew to a close. The most primitive primates did not appear until 70 million years ago as the mammals underwent their great diversification. It took another 20 million years for the first really apelike primates to appear. Then things began to speed up. The first indications of separation between hominids and pongids appeared in another 20 million years and the first humans appeared about 1 million years ago. This speeding up of evolution is highly characteristic of the hominids.

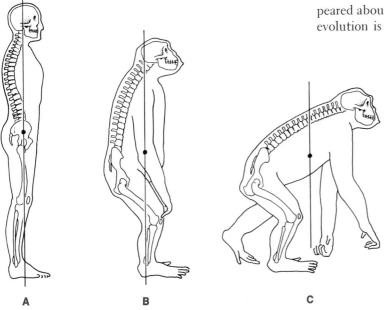

Figure 17-7 Primate bipedalism. The human **(A)** stands erect because the center of gravity allows the limbs to take an effort-conserving extended stance. To balance while erect the chimpanzee **(B)** has hip and knee joint displaced from the vertical line through the center of gravity and thus must expend relatively more energy to hold the leg joints in the bent position. Hence quadrupedal walking **(C)** is more efficient for animals with this limb configuration. (After Pilbeam, D. (1972) *The Ascent of Man, An Introduction to Human Evolution.*, p. 67. Macmillan, New York.)

African Beginnings

Scattered about central and southeastern Africa we find fossil evidence of the earliest hominids. The first species to be discovered was named *Australopithecus africanus,* meaning African southern ape (Figure 17-9A). We may refer generally to this and similar hominids as Australopithecines. The earliest of these were small, weighing at most probably 70 lb and standing 4 ft tall. From the time they appeared in the early Pleistocene until the mid-Pleistocene, the Australopi-

Figure 17-8 Artist's impression of our hominid ancestry. The sketches are to the same relative scale.

PROCONSUL

RAMAPITHECUS

AUSTRALOPITHECUS

PITHECANTHROPINE
(HOMO ERECTUS)

NEANDERTHAL
(HOMO SAPIENS)

CRO-MAGNON
(HOMO SAPIENS)

318 BIOLOGY

thecines grew in stature until they were essentially as big as modern humans. Indeed, there has been no important change in the size of hominids since then. The rapid increase in size of the Australopithecines may have been related to defensive-offensive matters, reflecting arrival at an optimum balance between being big enough not to be bothered by medium-sized predators and yet not too big to be expensive to feed.

Ape-man might be a good name to apply to *Australopithecus* because it has important characteristics in common with pongids. The brain was small; its volume of 500–600 cm^3 is about a third of ours. And *Australopithecus* had an ape's face. It protruded and bore a massive jaw, armed with large molar teeth. But the important characteristics that place it on the road to humanity were an erect stance and lack of large canine teeth.

How can we be sure the Australopithecines walked erect? After all, walking is behavior, and all that the paleontologist has to study of *Australopithecus* are bones and teeth, and not too many of them. Fortunately, enough skulls are available to show that the head was carried more erect on the spinal column than in pongids. The **foramen magnum**, the opening for the spinal cord that marks the balance point of the skull, projects downward instead of backward. Further, the structure of leg and pelvis also indicates erect posture.

Deducing Behavior from Bones

Admittedly, it is possible to figure out a relatively simple characteristic, such as posture, from bones. But can we go beyond this to learn anything else about the life of *Australopithecus*? Well, what about the reduced canine teeth? Lacking large canines, this animal had, in effect, checked its ape's defensive weapons at the entrance to a very dangerous saloon, namely, one without trees for leaps to safety. Remember that not every primate to take to the ground turned in its canines; the baboon is still all there in the tooth department. But note that baboons get around on all fours. They do not have their hands sufficiently free all the time to exploit them fully as manipulative organs. All of this suggests one thing: that by walking erect the Australopithecines had freed their hands for manipulative tasks and had found them, in fact, so effective in defense and offense that the large canines of their antecedents were unnecessary. In short, finding the naked fist of little utility, *Australopithecus* became a tool-user, certainly at least a rock-thrower, and probably was capable of making and using simple weapons.

Why Walk?

Why did our anthropoid ancestors leave the safety of the forest? Probably the answer lies in climate. At the time of which we speak the climate was tending toward cool and dry, with the consequence that the lush tropical forests that harbored early primate evolution were becoming more and more limited in distribution. As the forests shrank, it would have become advantageous to leave them and colonize the open ground. Once on the ground our ancestors found it advantageous to remain upright more continuously. Fortunately their previous mode of arboreal locomotion had accustomed them to an upright stance for short periods of time. Standing on the hind legs gives a better view of the surroundings and, as we have seen, frees the hands for more important work.

Before Australopithecus. Since the Australopithecines are distinctly in the human line of ascent, a search for a missing link between the ape and human families must focus farther back, into the Miocene and Pliocene ages, from 14 to 20 million years ago. Two extremely interesting fossils appear during this period. The oldest is *Dryopithecus* (*Proconsul* and others). On the order of five *Dryopithecus* species have been found in Africa, Europe, and Asia. These were small apes, weighing 40 lb or less; but they are important because various *Dryopithecus* species were probably ancestral to the chimpanzee and gorilla, and one of them is considered ancestral to the second Miocene fossil primate, *Ramapithecus*. This places the Dryopithecines in a most interesting position, just below the separation of the hominids and the apes, because *Ramapithecus* (Figure 17-9A) is taken to be ancestral to *Australopithecus*. Thus, these two genera give us a before and after view of the division point between hominids and apes.

Refinement and Spread of the African Innovation

In a way, *Australopithecus* seems to be the whole ball game. It walked; it used weapons. It was undoubtedly a most successful innovation because the next stage in our ascent, which we shall call the pithecanthropus stage (Figure 17-8), after the first fossil of this type to be named, was largely a perfection of the beginnings seen in *Australopithecus*. The Pithecanthropines were originally described as a species separate from ours, designated *Pithecanthropus*. Today its close affinity to our species is recognized by the name *Homo erectus*, although *Pithecanthropus* remains a useful term. *Homo erectus* must have been most successful. Instead of occurring only in the southern half of Africa, its fossils are found all over Africa and in Europe and Asia as well. In fact, the first specimens were discovered in Java (Java man) and some of the best-known fossils come from China (Peking man). Some of these are the fossils that were mysteriously lost during the Japanese invasion of China in World War II.

Homo erectus came on the scene about 1.5 million years ago and survived about a million years. It was a considerable improvement over the Australopithecines. Its teeth were almost modern and it was taller than its forerunners, but the face still protruded (Figure 17-9B). The critical difference was a matter of brains, which grew to 1200 cm^3. This is really not sufficiently different from the brain size of modern humans to worry about. Very intelligent persons may have brains as small as 1200 cm^3. Here we find what may be a secret of hominid evolution. Somehow the evolutionary process "concentrated" on improvement of the brain, and, as it attained its modern size so rapidly, it almost seems that the balance of human evolution has been largely a matter of learning how to use this organ!

Perhaps you miss the significance of this. Except in this instance, the course of evolution of the organs—limbs, eyes, and so forth—seems to have been nip and tuck, with minor improvements in a given organ being sufficient to provide just enough selective advantage to pull the species through. Evidently not so with regard to the hominid brain. At a volume of 1200 cm^3, we have a brain large enough for a nuclear physicist, yet it is first found in a being whose culture is not of the nuclear age, to say the least.

You may argue that just because a brain is so big does not mean that it contains all the machinery of high intelligence. This is a good point, but as far as we can tell, all primate brains are constructed internally along grossly identical patterns, and this means that brain volume is a good general indicator of intelligence, when body size is taken into account. In favorable human fossil skulls it is even possible to detect the major external features of the brain itself. These permit somewhat risky estimates of such behavioral abilities as speech. What we need to know is what went on in those extra 500 cm^3 of brains that *Homo erectus* possessed. Fortunately, with *Homo erectus* we have a little more evidence than just the fossils themselves in attempting to understand the lives of these early humans.

Hearth and Home in Mid-Pleistocene. *Australopithecus* evidently lived and died in the open. It is consequently very difficult to be certain that articles found with its remains were its possessions. For example, stones that might be interpreted to be tools are found in the same geological layers as the Australopithecines, but it is difficult to be positive that they shaped and used them.

Two related facts make things quite different as far as *Homo erectus* and later fossil humans are concerned. *Homo erectus* had a place to live, and was the Prometheus of our line. These humans lived in caves, and had fire, which was one of the reasons cave life was possible. Without fire, residence in a cave meant risking eviction or worse by the former occupants, which in Pleistocene times were animals to reckon with. We know that at least the northern *Homo erectus*, Peking man, had fire because hearths with the charred and broken bones of deer and other animals are found in their dwelling caves. Seeds are also found, indicating that Peking man was omnivorous. The presence of crude stone hand tools completes a picture of a more advanced cultural level than seen for the Australopithecines. It is perhaps worthwhile to speculate further and reason that, having a home to return to, Peking man did so. This implies a division of labor—females staying with the children in relative safety, while males hunted—and with that we see society beginning to form.

It's use of fire and caves allows us to understand the wide distribution of *Homo erectus*. The frigid north was no longer a serious problem. Humanity was thereby a long way along a new evolutionary route, figuring out new uses for an as yet largely unfathomed mental ability. By using the brain to master fire, our ancestors had become, without change in form, far more independent of the environment than any number of physiological changes might accomplish. *Cultural evolution was under way.*

Essentially Modern Man. The spark kindled by *Homo erectus* resulted, as early as 100,000 years ago, in the spread all over Europe, Africa, and Asia of beings who unmistakably deserve membership in our

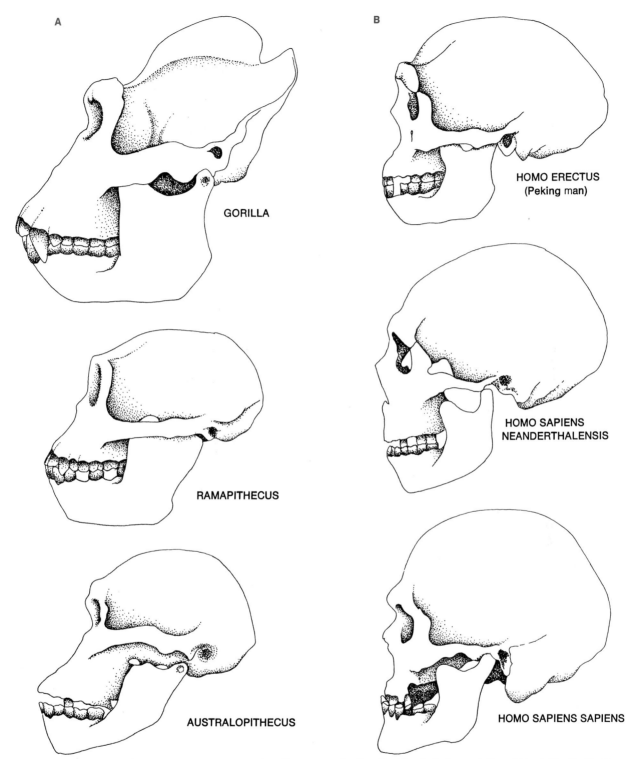

Figure 17-9 **A** and **B.** Skulls of modern human and gorilla compared with salient fossils in the evolution of Hominids. Skulls are all approximately to the same scale. Note trend towards reduction of brow ridge, "retreat" of the lower face (reduced prognathism), increasing prominence of the jaw in the human series. Compare Table 17-1.

Figure 17-10 Examples of ancient art from caves of the Dordogne and Hautes-Pyrenees, France. **A** and **E**, ibex. **B** and **C**, rhinoceros. **F**, mammoth. **C** and **D** are from the scene described in the text.

own species, *Homo sapiens*. From the 100,000-year point to as recently as 40,000 years ago, humans typical of the species *Homo sapiens neanderthalensis* lived over this entire vast region (Figure 17-8 and 17-9B). Neanderthal man had a fully modern brain case, but was of stocky build and the face had the protruding look and heavy teeth of *Homo erectus*. Culturally the Neanderthal race was far more advanced than *Homo erectus*. Their tools suggest that they used stone-tipped spears and made clothing from animal skins. The latter practice undoubtedly contributed in a major way to their ability to exist in the climatic shadow of the great Pleistocene glaciers that gave even southern Europe a subarctic climate. Most interestingly, they buried their dead with ceremony, suggesting that the concept of immortality may have taken root.

Collections of cave bear skulls in Neanderthal sites are considered to represent aspects of religion or magic. Discovery that some Neanderthal burials were purposefully decorated with flowers (identifiable by pollen remnants) evokes a sympathetic feeling towards those remote people.

Are We Modernized Neanderthalers?

Approximately 40,000 years ago Neanderthal culture was supplanted by a more modern one. Skulls and skeletal materials in some East European sites appear possibly to represent hybrids between Neanderthalers and these modern humans. Equally likely, across the wide expanse of Europe and Asia, there may have been continuous intergradation of types, with typical, heavy set Neanderthalers in the west and modern types in the east. Whether the two human types evolved from the same immediate stock or evolved separately and then hybridized, it seems certain that the Neanderthal people are represented in our ancestry.

Whatever their source, the next hominids to dominate the scene were of our own subspecies *Homo sapiens sapiens*. Cro-Magnon man, from South France, is a well known representative (Figure 17-9B).

These were hunters who took game as large as mammoths. They exhibited splendid workmanship in the construction of stone tools and in the shaping of spear throwers and other articles out of bone and antler. Their sculpture and cave painting are well known from localities in France and Spain. Examples of cave paintings (Figure 17-10) show in a compelling way both the acute observational powers of these people and their highly developed artistic talents. The economy of line with which the character of several animals in that figure is evoked is seldom approached by modern artists. The central part of the figure, showing a wounded bison, a naked man and a rhinoceros, as they appeared in a painted composition in the famous cave of Lascaux (Dordogne, France), was made between 15,000 and 20,000 years ago.

Evidently this scene represents something beyond the depiction of ordinary events. The man is shown with a bird's head, perhaps denoting membership in a particular clan, or possibly such an apparition was a part of mythology. The bison, intestines hanging, has been speared. Perhaps such figures were part of rituals designed to insure successful hunting. It is truly regrettable that the artists who produced these paintings left no realistic self-portraits. All drawings of themselves were usually the barest and conspicuously crude sketches. Since their skills would obviously have allowed accurate self portrayal, they must have had reasons not to do so. In some primitive societies within historical times accurate portrayal of specific individuals was considered damaging to the person depicted. This may be at the root of this, to us, unfortunate deficiency. Even so, we can only applaud the artists who inadvertently gave us this view of our remote past.

Whatever the purpose of their art, it was not casually done. Most of the known sites are in most difficultly accessible parts of the caves in which they are found. More accessible art may have been made only to disappear due to weathering, but the fact remains that these people were so intent on this activity that they went to great pains to do it.

Modern Man

PRIMITIVE MAN IN THE NEW WORLD

Primate evolution is centered in Europe, Africa, and Asia. However, in the Eocene many forms of animals, including early primates, were common to North America and Europe. With the opening of the Atlantic, the primates of the two continents were isolated and continued their evolution independently. In the Americas the platyrrhine monkeys were the result.

Invasion of North America by humans may have taken place as early as 40,000 years ago, judging by the age of stone tools found in Texas. This is a surprisingly early date because the only plausible route for large scale migration, *via* the Bering Straits, is thought to have been unfavorable for migration because of ice until 25,000 years ago. Although the land bridge between North and South America is thought to have formed only about 15,000 years ago, cultures several thousand years older than this are known from Chile, suggesting knowledge of water transport by these early Americans.

OUR ANCESTORS AND THE ENVIRONMENT

We tend to think of humans as having significant environmental effects only in recent times. As recently as 10,000 years ago there were, after all, only about 5 million people in the whole world, and these were unaided by the paraphernalia that we consider basic to a serious assault on the environment. However the delicacy of the balance of nature is illustrated by the fact that the rise of humanity did, in fact, seem to have a large effect on many kinds of animals. The advent of efficient human hunters in Africa, Europe, Asia, and later the New World, is correlated in time with the extinction of many large, formidable, and well established animals including the sabretooth tiger, dire wolf, and mastodon.

These extinctions are not believed to have been due so much to direct attack but more to indirect disturbances of the balance of nature. Thus the great predators are thought to have become extinct because more efficient human predation on their major sources of food starved them in lean winters. In other instances humans may actually have caused the increase in numbers of some animals that they hunted. They may have done this for bison in North America primarily by reducing the numbers of bison predators. It appears that other carnivores take relatively more immature, "easy," prey than a hunter provided with weapons enabling indiscriminate attack at all levels in the age-structure of the population. If so, the result of replacing other carnivores with human hunters is likely to be an increase in the prey population, unless the hunter is extremely efficient. Certainly several thousand years of bison subsistence by Indians left a bison population estimated at better than 60 million

Figure 17-11 The earliest pictoral record of the bow and arrow in a Spanish cave painting from the late Pleistocene. (After J. G. D. Clark.)

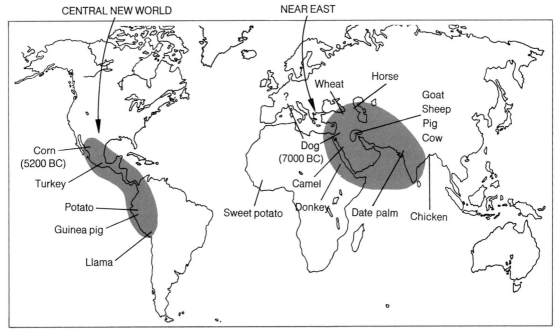

Figure 17-12 The spread of agriculture and domestication of animals.

at the beginning of modern exploration of the West. Extinction of the mastodon in North America may have been a consequence of ineffective competition with these huge numbers of bison.

Archaeological evidence of their kills and even cave paintings (Figure 17-11) demonstrate the skill of early *Homo sapiens* in hunting large game animals. This source of food was supplemented by fishing and by gathering of shellfish. There is clear evidence that these people took advantage of plant foods when they were available. Yet despite their effectiveness at using many food sources, people living at the hunting and gathering level were beset by uncertainty. They were unable to control the migrations of game animals or to insure that food plants reappeared from year to year. The result was that hunters and gatherers had to be continually on the move following the seasonal migrations of game and the ripening of plants. Uncontrollable variations in the natural supply of food meant instant hardship since the food reserves that migrant peoples could build up and carry about must have been small. Thus in the hunter-gatherer stage human population densities remained small.

HUNTERS TO FARMERS

The advent of farming, including both agriculture and domestication of animals, brought immediate disruption to the balance between human populations and nature. Human population estimates show this clearly. Although the human population of the world at the beginning of agriculture was about 5 million, after only 4000 years of agriculture the human population had grown to more than 85 million. Agriculture supported a 1700% population increase over 4000 years as compared with only a 160% growth in the last 20,000 years before agriculture. This rapid growth accelerated extinction of many kinds of game animals and the practice of agriculture began the still continuing process of changing the face of the earth through land clearing and attendant changes such as erosion.

How plant and animal culture began is unknown. The beginning must have required no special genius because it occurred independently in at least three major centers scattered over the earth (Figure 17-12). Humans long before agriculture had, of necessity,

HUMAN EVOLUTION   325

become accurate observers of nature to support their activities as hunters and gatherers. The cycles of animal and plant life must have been understood in detail. Hence only the smallest spark of innovation was probably required to initiate the agricultural revolution. It might have been discovery that seeds from wild harvested plants discarded at a camp site had conveniently matured when the camp was next occupied. Or perhaps suckling game animals, taken alive when their mothers were killed, might have been kept alive as a source of fresh meat. Discovery that they rapidly became tame and could be reared to a larger size before slaughter could have been the beginning of animal husbandry.

Obviously agriculture smoothed out the ups and downs in food supply experienced by hunter-gatherer economies. Large quantities of desirable plants could be grown in small areas, and supplies beyond immediate needs could be stored because they were available and because it was no longer necessary to wander far in search of food. Thus, agriculture not only made possible but also required establishment of village life. As agriculture improved, larger harvests from smaller areas of land allowed growth of large, concentrated human populations. These became the first civilizations. With a better and more reliable food supply than ever before, the members of these civilizations could devote a part of their time to pursuits other than the elemental search for food. It is not suprising, therefore, that in the same regions where agriculture began we see some of the earliest evidences of highly organized society, with division of labor and the development of written languages.

THE INHERITANCE OF CULTURE

The arrival of humanity at a cultural state characterized by written language is a good point to look back over hominid evolution and once again to consider the bothersome question of how we became so successful so fast. We have already seen that, except for brain *size*, man does not seem markedly different enough from his relatives to account for cultural differences. Physically, there is about as much difference between the modern human and the Australopithecines as between a horse and a burro. Yet consider the cultural differences: Incredible in the first instance, trivial in the second! So we are up against it; to understand modern man we must understand his cultural evolution. Yet Chapter 16, on which we have relied until now, seems able to help us very little, because it deals with the evolution of the vehicle, not of the concepts wrought by that vehicle.

Communication

Communication Is the Hereditary Transmitter of Cultural Evolution. Nonetheless, our knowledge of evolution possibly can help us, because it does suggest an analogy. Let us see if there was a rate change physical evolution, analogous to the rate change we see in the evolution of culture. There certainly was: Recall the sudden appearance of essentially all the major phyla long after the origin of life. Do you recall what accounted for this sudden upswing in evolutionary activity? It was, we think, the advent of sexual reproduction, allowing interchange of successful genetic material between individuals not of the same clone.

Just as the gene is the unit of biological evolution, we may think, a little less precisely, of the unit of cultural evolution as being the *concept*. This we will loosely define as the unit of understanding of the environment and of its manipulation. Now we are getting somewhere, because our analogy suggests to us that, just as biological evolution moved slowly before there was gene mixing (clonal versus sexual descent), cultural evolution will move slowly without concept mixing. And how do you mix concepts? By *communication*. Hence, to understand cultural evolution, we must look back over the evolutionary path and look for clues concerning the development of communication.

Communication Is Second Nature to Primates. As primates are predominantly group-oriented animals, there is in a sense a **preadaptation** for communication in our ancestry. Mammals that live in packs, clans, herds, or pods always have ways to communicate. These include facial expressions, body postures, physical contacts, odors, and noises, vocal or otherwise. Chimpanzees are so adept at their own modes of communication that there have been several successful attempts to teach them human symbolic speech.

The most successful of these efforts involved teaching chimpanzees a sign language based on hand signals or teaching them what amounts to a very simple form of writing. In this technique the chimpanzee is taught the meaning of a set of symbols which it becomes able to select and arrange grammatically to make statements and even to reply to messages conveyed by the same system (Figure 17-13). Earlier attempts to teach chimpanzees actually to speak evidently were failures only because of some physical inability to speak and not because of lack of mental ability to use simple language. This fact, of course, raises the question as to when human ancestors became able to speak since presumably factors that are measurable on fossils, such as brain size, would not necessarily indicate the presence of this ability. Some authorities have attempted to answer this question by reconstructing the vocal tract of Neanderthalers, using as their guide the structure of the base of the skull. Although these reconstructions suggest that Neanderthalers were at best capable of a very restricted range of vocalizations, they have been criticized and much more work will be required before a decision can be made, if at all.

Tools and Speech

Another way to estimate when our ancestors became capable of speech is to look at the archaeological evidence to see when human activities began that would have required a high level of communication between individuals. Such evidence would put the time of origin of speech much earlier because extensive tool using seems to be an example of a speech-requiring human activity. Perhaps the tools used by the Australopithecines, being simple cutting tools formed of convenient sized rocks sharpened by having a chip or two taken off, required no complex speech. They are really not conceptually all that different from the twigs that chimpanzees pick and "work" by stripping off leaves to make a termite collecting tool. But certainly speech must have become well established to support the tool and fire using culture of *Homo erectus*. In whatever way it happened and whenever it happened, this throwing open of the barriers between one mind and another brought cultural evolution into a phase of extremely rapid progress.

Speech alone was sufficient to carry many cultures to a high order of complexity, with the "genome" of culture—the history of its confrontations with the problems of living—being handed down only through oral tradition. Sooner or later, however, the accumulated burden of the cultural genome became too great for the inaccurate transcription methods of unaided human memory, and cultural advancement must have

Figure 17-13 A chimpanzee, Peony, responds to commands expressed by the characters affixed to the surface in the background. **A.** Peony—touch—Debbie (the trainer). **B.** Peony—give—banana—Debbie. (From the experiments of D. Premack, University of Pennsylvania.)

leveled off again. The advent of written language, undoubtedly from **pictographic** beginnings (Figure 17-10), provided the necessary accurate transcription and storage mechanism for cultural progress to modern times.

Curiously, at this moment our species seems faced with a new impediment to cultural evolution. Our cultural genome is becoming too large to handle, thanks to the printing press and the industry of many generations of scholars. No longer can one person learn the total of our cultural heritage. This is a serious problem because it tends to produce narrow specialists unable to communicate effectively with other members of society and because it becomes more and more difficult to identify useful information in the great volume that is poured out hourly by the printing presses of the world. Scientists are particularly afflicted by this problem. But there is some faint hope for them because the scientific part of our cultural heritage is sufficiently organizable to allow computerization. Several computerized systems of literature search and retrieval are well developed. However, they by no means completely solve the problem since ultimately humans must interpret and feed new information to the computers. Even this task is out of hand in some fast moving areas of science and technology.

THE VALUE OF LONGEVITY TO CULTURAL EVOLUTION

There is a clear tendency towards longer life in primate evolution. Of modern primates the primitive tree shrew lives only about 5 years, whereas the hominids live from 40 to 50 years. Careful estimates made on fossils show that, while *Dryopithecus* lived about as long as modern subhuman primates, there was a marked, progressive increase in the potential life span in the human evolutionary series, culminating in the Neanderthalers, whose potential life span was probably as long as ours, 95 years. The causes of this are obscure, but it is likely that they are related to the increase in brain size that was going on at the same time. For one thing, a big brain may make it possible to operate the body more effectively. It also seems that a long juvenile period is necessitated by the large brain, and, therefore, there would be natural selection for longevity to permit time for reproduction and rearing of young. The large brain necessitates a long juvenile period for the curious reason that the female birth canal is just large enough safely to let through a head containing a rather immature brain, which then must spend time maturing during the period of infancy.

Both of these aspects of the life cycle, the prolonged juvenile period and the great life expectancy, are useful to the transmission process in cultural evolution. The young are protected and taught a great deal at the parent's knee, preventing many bitter, self-taught learning experiences when they are on their own. Particularly when blessed with the ability to speak, the aged member of a human clan is able to bridge the cultural gap to a previous generation and thus make certain that facts learned are not lost by disrememberances of the young. Before writing, the memory was the only link between generations, and the longer an individual memory survived the stronger the link became.

THE VARIETIES OF MODERN *HOMO SAPIENS*

It is common knowledge that the modern world is populated by several races of humans, of which the men in Figure 17-14 are examples. The number of races ranges between 5 and 34, depending upon the authority consulted. The reason for this great variation in estimates lies in the fact that the definition of race is not based on clear, quantitative differences but rather on characteristics that are present in all races to some degree. You will recall that there is a precise, if not always useful, defining character at the species level: different species do not interbreed. Races, however, are populations that are capable of interbreeding.

Ordinarily, races do not interbreed because they are geographically or temporally isolated in such a way that they never have the opportunity. In the ordinary course of evolutionary events, races may evolve into species if their isolation from one another persists long enough for sufficient genetic variation to occur to make fertile interbreeding impossible. The human races are not approaching this stage of incipient speciation. All human racial hybrids are viable. Indeed, careful studies, as for example on racial hybrids in Hawaii, show no trace of any degree of hybrid disability. Human racial hybrids as far as we know are no more or no less viable than individuals bred within the several races.

The Races and Their Origins

Our discussion of the evolutionary history of *Homo sapiens* makes it evident that there was one major human type on earth during each of the later stages of evolution. Until as recently as 75,000 years ago these populations evolved and passed on their genetic heritage to the next stage in Africa, Europe, and Asia. Although this is a vast expanse geographically, it contains no absolute barriers to human migration and so we presume there was genetic exchange throughout

A B C

Figure 17-14 Three human racial types. Although they are of widely divergent origins and live in regions remote from each other it is clear that they differ from one another in only minor ways. **A.** Ainu, from Northern Japan. **B.** Australian Bushman. **C.** American Indian. (American Museum of Natural History and Museum of the American Indian.)

the entirety of the populated world. Thus, there is no reason to expect, and no data to suggest, that any of the human races can trace their ancestry exclusively to one or the other of the early human types. That is, Peking man did not give rise to the Orientals and the Neanderthalers did not give rise to Caucasians, or to Africans. Indeed, a simple, slightly foolish, calculation can show that all humans *must* be related through far more recent ancestors. Since each person has 2 parents, 4 grandparents, and 8 great grandparents we may calculate that we have 2^n ancestors n generations ago. If there is no significant amount of inbreeding (and this seems to have been an early taboo in human culture) and if there are 4 generations per 100 years, each of us would have had 2^{40} ancestors only 1000 years ago. Now that is 1,099,511,600,000 people, which is about 5500 times the 200 million population estimated for that time! The only way out of this absurdity is to admit that we are all rather closely related.

Origins of the Races

There is very little data and much speculation about the origins of the races. Generally speaking there were three major centers of racial development. These were, (1) Europe, Near East, and central Asia, (2) The Pacific and the Americas, and (3) Africa (Figure 17-15).

Caucasians were present in Europe by about 40,000 years ago as evidenced by skeletal measurements and even by a French cave painting of the period showing a white-skinned man with black hair. Their origins were mixed, representing intermixture between Neanderthalers and new arrivals from the East. Outward migration from Europe and Russia also occurred. Thus India was colonized by two separate migrations and even the Ainu, of the northernmost islands of Japan, are Caucasoids.

Mongoloids arose in Central Asia. Their characteristic facial features of forward placed cheek-bones with eyes and nose recessed into the face are thought an adaptation to cold. These peoples migrated into the Americas, establishing the North and South American Indian populations and, only about 3000 years ago, the Eskimo population. Mongoloids also moved south as far as Burma. The remainder of the Pacific peoples may also have had their origins on the Asiatic mainland. Their relationships to other groups are very

HUMAN EVOLUTION **329**

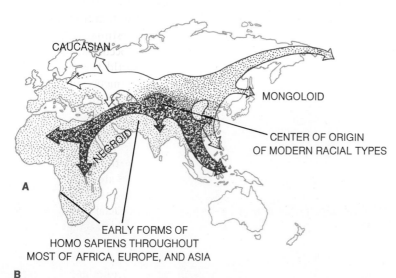

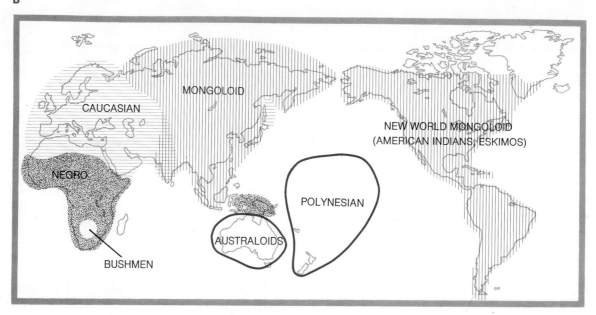

Figure 17-15 Human racial origins and distribution. A. One scheme of racial origins, the spreading of *Homo sapiens* from a central Asian center of origin. (After K. F. Dyer.) **B.** Racial distributions at the beginnings of recorded history.

obscure and, if they are related to the Mongoloids, it must have been prior to 50,000 years ago. These people include the Negroids of the Melanesian Islands, and the Negritos of the Philippines, Malaysia, and New Guinea Highlands. Although, as their names imply, these people resemble African Negroes, they are not directly related to them. Instead, it may be that the features that the Negritos, Negroids, and Negroes bear in common result from an ancient migration of peoples from Europe and southwest Asia into Africa and Australasia. The final group of Pacific people are the Australoids, who are found in the lowlands of New Guinea, Sri Lanka (Ceylon), and India. They also are probably of a very distant Mongoloid derivation from the Asiatic mainland.

The final great group of peoples are the Africans. Some of these are obviously mixed with Europeans, for example the Ethiopians. The true Negroes arose in West Africa, below the Sahara, and became highly diverse, varying from Pigmies to the tall Nilotic Negroes. Most of South Africa was initially occupied by Bushmen and Hottentots, whose origins are both very ancient and very obscure. Within the last several thousand years these people were displaced into marginal areas or assimilated by immigrant Bantu Negroes from West Africa.

Perhaps the best synthesis is to say that the modern races are the result of repeated mixings and remixings of three identifiable, ancient racial "families," namely African, Caucasoid, and Mongoloid. If the resultant array of modern races had remained as they were several thousand years ago and never developed significant means of long range transport, it is probably true that they would commence to speciate sometime in the far distant future. There is now, of course, little chance of this happening because migration and intermarriage have been tremendously increased in the past several hundred years as a consequence of the advance of civilization.

The Future of Racial Variation

Racial hybrids are rapidly becoming the norm in many parts of the world. Since Columbus's voyages of exploration, migration has been the rule. Europeans have migrated by the millions to the Americas, Australia, Africa, and South America. The Chinese, Japanese, and Indians have migrated all over the Pacific, and the Indians, particularly, have migrated to Africa. Perhaps a total of 15 million Africans were brought as slaves to the New World. These migrations and attendant activities of the migrators have, regrettably, caused the complete extinction of a few racial subgroups, for example the Tasmanians, who were actively hunted out. But by far the most common outcome has been hybridization. On a worldwide basis it is estimated that at least 5% of the world population is the result of hybridizations since that fateful year 1492. The principal part of these hybrids live in the Americas where 20% of the South American population are Indian-Caucasoid hybrids and where a like percentage of the North American population is composed of Negro-Caucasoid hybrids. Other areas of extensive hybridization are South-east Asia, Africa, and the Pacific islands, especially Hawaii.

What will happen in the future? It seems inescapable that the future will see a continuing increase in the number of hybrids. Social and legal barriers to hybridization are falling, and, of course, they were never particularly effective. Populations continue to move and set up new centers of hybridization, for example the recent United States military presence in Southeast Asia.

Will the final result be uniformity? In Hawaii, which is a much studied model system for worldwide racial hybridization, it has been estimated that complete racial amalgamation will occur in about two generations. Elsewhere in the world the process will be far slower because of the immense numbers of people involved and because of the still highly significant barriers to hybridization imposed by simple geography. If humanity persists long enough, racial uniformity may be approached but it is just as likely that new races will emerge from the amalgamations now in progress. Perhaps what we are witnessing is a return to a few racial types analogous to those out of which we arose, after a phase of racial profusion and incipient speciation that was nipped in the bud by the rise of civilization.

Is Human Racial Uniformity Desirable?

Here the biologist cuts close to the bone, for in asking this question he seeks to illuminate, with the dispassionate light of fact, matters that for generations have been passionately argued. Racial prejudice has caused excesses in both directions. Even practicing scientists have argued that racial mixing (leading to uniformity) is good, or bad, because the races are, or are not, significantly different. The well-meaning, seeing the tragedy of excessive racial prejudice, may try to avert it by arguing that the races are not significantly different. Others, sometimes confusing the culture of a particular race with its basic biology, argue just as sincerely that racial uniformity will pull the "advanced" races down to lower levels by diluting them with "inferior" genetic material.

This is particularly vehemently argued with regard to the intelligence of Negroes and Caucasians. However, to this very day no scientifically sound investigation of the differences in intelligence between whites and Negroes has been performed! Every study claiming to show inferior Negro intelligence can be criticized on the basis that it does not take differing cultural backgrounds into consideration. It is virtually impossible to devise an intelligence test that tests only intelligence. Rather, all such tests tend to evaluate what might be called the product of the interaction of intelligence with environment. Matters such as education, the richness of the environment, and even nutritional status, all are reflected in intelligence testing. Taking all this into consideration, the most that the biologist can say is that serious biological arguments against racial mixture have not been proved to exist and that, generally speaking, the larger the gene pool, the better our chances are going to be.

Unfortunately, too, the biologist suspects that psychological resistance to the attainment of racial uniformity may be an especially difficult bit of evolutionary clay to shake from our feet. Aside from straightforward predator-prey relationships, the tendency for strife among animals seems to be strongest among closely related forms, subspecies that in effect are in nearly direct competition with each other for the prerequisites of existence. In addition to this, there is the psychological fact that humans generalize. Since the races—or even nationalities—are usually by their form or behavior readily distinguished, there is a tendency to blame indignities suffered at the hands of one member of such an identifiable group on the entire group.

THE FUTURE OF THE SPECIES

It must be obvious that evolution is not finished with us. There is little certainty in our future, as we observed in the first chapter. Indeed, it is with some envy that we gaze on a cave painting from 20,000 years ago and wonder if we will survive as long as the painter's kind and whether our treasured artifacts will still be imaged in human eyes 20,000 years from now. Although much of the uncertainty in our future is politically generated, a significant part of it is generated by more straightforward biological problems.

Putting aside the significant possibility that we may make the world unlivable in the course of a few hours of nuclear "exchange," our culture is changing the world environment so rapidly that there is no time for natural selection to be of the slightest value in adapting us to the new conditions. That is one problem—our tinkering with the environment. The second one is that we tinker with ourselves. Some authorities argue that we have made our lives already so secure and have figured out so many medical ways to save the weak that natural selection is no longer at work.

This is a problem that is very difficult to evaluate. First of all, the idea of elimination of natural selection can now only apply to a small fraction of the world population, the highly advanced nations. And even within them there are places where life is still very hard and it is easy to see that only the strongest and cleverist survive. Secondly, if we divide the effects of natural selection into prenatal and postnatal effects, there is, even in the highly advanced nations, no evidence of very much change particularly in the early prenatal death rate. Thus we may assume that natural selection is still at work at this level.

The big change has been in postnatal death rates, which have been greatly reduced. Death by infectious disease was a very severe problem in early man, particularly as he began to congregate in villages. Now selection for disease resistance must be virtually completely arrested, but, as long as the medical art persists, this should cause no trouble. Selection against many types of hereditary diseases is counteracted by medical practices which permit many affected individuals to survive and reproduce. Many of these diseases are now detectable by genetic counseling and thus are preventable by abortion or other means. Thus the major hazard of modern times is probably not the diminishment of natural selection but in our rapid degradation of the environment.

Should We Control Our Evolution?

In asking this question we should distinguish between control and influence. It is already certain that we influence our evolution. Birth control and medical procedures together with environmental modification make this an absolute certainty. It is happening. It is also true that we have it in our power to influence our evolution in specific ways. First of all, this is possible because relatively few genetic changes are required to make large differences. This point is driven home by the remarkable fact that the great and swift progression in brain size from Australopithecine to Neanderthal size appears to have required modification of only a few hundred genes out of the estimated 300,000 genes (10% of the genome) that are expressed in the brain. Secondly, control of evolution is possible because the techniques are available to change gene frequencies and to modify the genetic structures themselves. Perhaps one of the most urgent and difficult tasks before us is to decide, first, whether we should actively try to shape the direction of our own evolution, and, second, if the decision is to do so, how should we proceed?

18 Human Physiology: Regulation

This far in our study of biology we have examined the origins and inner workings of cells and organisms, their variety, the mechanisms of their evolution, and finally we have focussed upon ourselves. Now, just as the last chapter showed that we evolved in common with other organisms, this chapter emphasizes our biological nature and builds up a description of our inner operation. We are embarked on the study of human biology, which will be our concern for the next several chapters.

We are guided in starting by the two most significant aspects of our physiology, namely

1. the immense development of our nervous system, and
2. the extremely effective integration of our organ systems under the control of the nervous and endocrine systems.

In what follows, organ systems, in the sense that they were described earlier, will not be considered in isolation, since you already know the principles of their operation. Instead, our interest will primarily be in how unified action of the organs is brought about by the neural and endocrine systems. These two controlling systems will be described in some detail, whereas other organ systems will be further described only as necessary.

THE CONCEPT OF HOMEOSTASIS

Homeostasis, coined from Greek words meaning "same state" by the American physiologist W. B. Cannon, is a term for the processes by which the organism utilizes its organ systems to maintain stability in life processes. Cannon was primarily interested in how one section of the mammalian nervous system, the sympathetic nervous system, acted to produce a number of complimentary responses of metabolism, of heart and blood vessels, and of other organs, all serving to protect the animal against damage. His work led him to the concept that the organism's internal environment could safely tolerate little change and that the tendency to change is opposed by the actions of the various organs. Cannon's ideas are rooted in the "milieu interieur," or interior environment, of Claude Bernard (page 197), who understood many years ago that stability of the internal environment of the organism is essential to a free living existence.

The central idea in homeostasis is that deviation from normal in any system of the organism is detected and corrective action taken by appropriate systems. Thus if we grow too hot, that fact is detected and we begin sweating. In 1947, the mathematician N. Wiener generalized such concepts into a branch of science that is called cybernetics, and whose principles are as applicable to machines as they are to organisms.

The cybernetic approach to understanding how self-regulating systems operate is to analyze them by seeking the answers to three questions: How does the system

1. detect change?
2. evaluate the change and determine what action should be taken?
3. effect restorative action?

A homeostatic mechanism must have a "memory" of all the conditions that are desirable. Whenever a change in the environment (either internal or external to the system) causes a deviation from these desirable conditions, there must be a receptor (internal or external) that detects this change. Once the undesirable change is detected by the receptor, a signal must be sent along some sensory pathway to a **central response selector,** which has both a memory of the most desirable condition and an ability to select the best method for returning the system to that desirable condition. The response selector sends out signals to other structures, **effectors,** that carry out the corrective response (Figure 18-1). As soon as corrective activity is undertaken by the effectors, information must be sent back to the receptors so that the response selector may evaluate the effectiveness of the corrective action.

Figure 18-1 The homeostatic relationship of organism with environment.

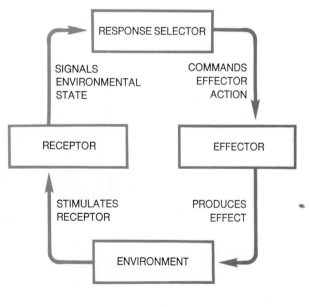

This relaying of information from the effector back to the receptor is appropriately called **feedback** (Figure 18-2).

POSITIVE AND NEGATIVE FEEDBACK

In some simple systems feedback may be directed back through the system to the effector so that the desirable condition is automatically maintained. For example, in a sunshine-watershed system (Figure 18-2) if solar radiation increases, more rain clouds may form, screening the ocean from the sun and reducing evaporation. With reduced evaporation, the clouds leave, the sun beats down, and evaporation rates again go up—only to go down as clouds reform. By this feedback control, evaporation rates from the ocean might be held within a reasonably narrow range. Such a feedback, one that tends to return the system from a swing in either direction back to the desired normal condition, is called **negative feedback.** In a suddenly accelerating automobile the driver's body slides back in the seat, and this may tend to pull his foot away from the accelerator pedal and reduce speed—another example of negative feedback.

On the other hand, feedback that causes the system to run away from the desired condition toward a complete loss of control would be called a **positive feedback.** Suppose if, in a rapidly moving automobile, the brakes are applied suddenly, throwing the body of the driver forward, with the result that the brake is pressed even harder. This decelerates the car more violently, the brake is pressed even harder, and the result may be loss of control. Positive feedback.

Negative feedback tends to return systems to some desirable or rest condition; positive feedback tends to cause systems to run violently away from the rest condition. In most biological systems the feedback control is not so automatic. The tendency is rather to feed back through the receptor-response selector portion of the control system so that there is less chance of an inappropriate or out-of-control response.

HUMAN PHYSIOLOGY AS HOMEOSTASIS

The three central questions of cybernetic analysis provide a useful way to dissect very complex biological

A RAINFALL FEEDBACK (NEGATIVE)

B BRAKING (NEGATIVE FEEDBACK)

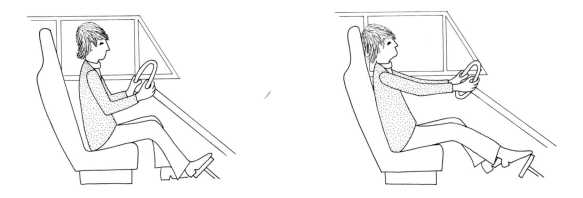

C ACCELERATION (POSITIVE FEEDBACK)

Figure 18-2 **Examples of feedback loops. A.** and **B.** negative feedback; **C.** positive feedback.

systems into manageable parts. To apply the concept to the human, we begin by observing that the human in broadest terms can be thought of as a homeostat that

1. maintains an appropriate internal environment, and
2. seeks an adequate external environment.

Viewed in this light it is immediately apparent that two organ systems are of paramount importance in directing organ system activities to attain homeostasis. They are the nervous and endocrine systems, which together serve as receptor, sensory pathway, and response selector elements of the homeostatic apparatus. Except for the tremendous development of the response selector elements, the brain and spinal cord, these systems in humans are altogether similar to those you have already considered. In general, the nervous system triggers short-term responses of the organ systems and serves as the receptor system for part of the endocrine system as well. The endocrine system typically triggers long-term responses in reaction to neural stimuli. Both systems are also controlled by effector response feedback (Figure 18-3).

COMPLEXITY AND ITS ORIGINS IN THE HUMAN MACHINE

Although the concept of the homeostatic human machine is simple enough, the actual homeostatic mechanisms involved in operating the organism are bewilderingly complex. Now everyone knows that the more complex a machine is the more likely it is to break down. Why, then, is the human, the most successful runner in the evolutionary race, so devoted to complexity? There are several reasons.

First, you understand from the study of evolution that the human machine was not built from the "ground up." It is the result of long ages of evolution. Our organs were not initially developed for our kind of life but are modifications of the organs of ancestors, and so on back through time. At no stage in evolution do these modifications involve a complete rebuilding along lines that might be ideal for that particular stage. They tend to be the minimum change necessary to allow survival. Consequently, we commonly find a solution to a physiological problem in the human reflecting the evolutionary history of the pertinent physiological systems as well as sound physiological

Figure 18-3 Control systems in complex organisms involve nervous and endocrine systems and effectors interacting in both externally initiated (black lines) and internally initiated (red lines) control loops.

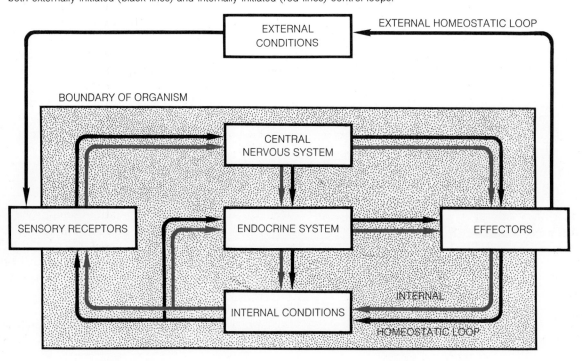

principles. Our erect stance illustrates this compromise with the past. A spinal column, evolved to serve an animal walking on four legs, is not always satisfactory for one walking on two.

The second requirement for complexity arises because the human, or any other mammal, requires an extremely stable internal environment. Simpler organisms often tolerate considerable variation in internal conditions and get along without elaborate control systems. Thus, a frog does not regulate its body temperature, but for this simplicity it pays a price. Its activity must vary with the temperature. Mammals, on the other hand, enjoy the advantages of precisely controlled, rather high body temperature. But these advantages have a price, too. The mammal has come to require—to be unable to live without—precisely regulated internal temperatures and must, therefore, invest in a control system for temperature regulation. That is, it has become more complex.

Third, because of the requirement for control, it becomes physiologically risky to place important aspects of the internal environment in the control of a single regulating system. Just as in ordinary engineering practice, the organism provides "back-up" control systems to maintain stability in the event of failure of its most important control systems. This idea has been called the *principle of multiple assurance*. We shall see it illustrated constantly in the following pages. For example, when we come to consider how proper oxygen concentrations are maintained in the tissues, we shall find two sets of oxygen and carbon dioxide sensors that help control breathing—one in the brain and one in the circulatory system. If one fails the other can carry on.

Finally, there is the obvious fact that humans must be complex just because they are large. As you know, size forces the development of communicating and conducting systems to insure adequate support for all cells and to insure that they all work properly together. The advantage of this investment in bulk is the opportunity for tissue and organ specialization and greater independence from the environment.

LEVELS OF CONTROL

It is already clear that control systems exist at all levels of complexity. We must not lose sight of cellular, subcellular, and molecular controls in the course of studying more obvious control systems involving neural and endocrine regulation of the major organs. These last are rooted in and dependent on control systems at lower levels of organization. Indeed, control systems at all levels of organization are so interdependent that it is almost meaningless to study control exclusively at one level.

An Example

The casual observer may see little connection between these facts:

1. One soon begins to pant if, for some reason, one's head is tightly enclosed in a paper bag.
2. Carbon dioxide slowly reacts with water to form an acid.

But there is actually much interconnecting these two facts, involving events at the highest and lowest levels of organization of the materials of life. This interconnection principally represents the *ventilation control system*, which adjusts breathing to the demands of metabolism. Ventilation will be discussed more fully later, but for the moment consider only the following:

Carbon dioxide, a waste product of cellular respiration, reacts rather slowly with the water of body fluids to form carbonic acid. This rapidly dissociates into hydrogen ions and bicarbonate ions (Figure 18-4). Acidity resulting from excess hydrogen ions in the circulation is damaging, and so part of the problem of oxygen and carbon dioxide transport is to prevent blood acidity from increasing as carbon dioxide moves from the tissues to the lungs for disposal.

As Figure 18-4 shows, the lowest level of control is represented by the reaction forming carbonic acid from carbon dioxide and water. Because the equilibrium in this reaction is such that a rapid buildup of carbonic acid does not occur, a good part of the carbon dioxide in the blood may remain harmlessly in solution. However, eventually, there would be an increase in acidity if no other controls were involved. Owing to the law of mass action, as the carbon dioxide concentration in the blood increased, the generation of carbonic acid would be favored. The remaining system of controls can be looked upon as a device to remedy this difficulty.

At the cellular level, the red blood cell influences carbon dioxide concentration in the blood by acting as a trap isolating carbon dioxide from the fluid part

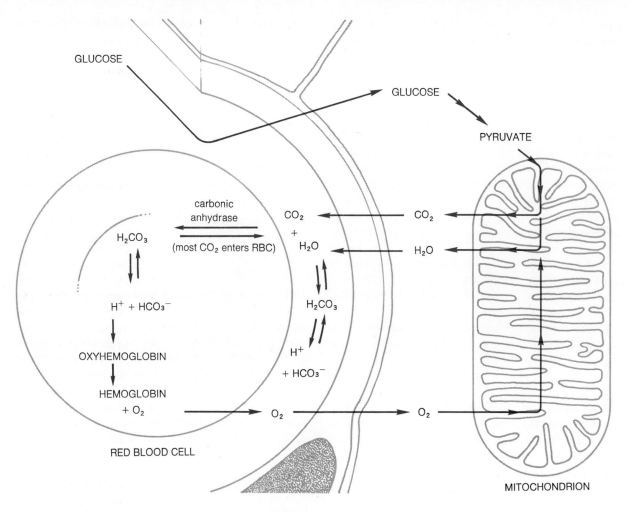

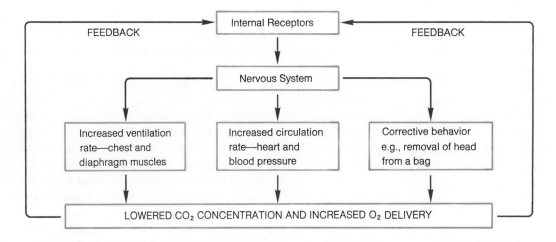

Figure 18-4 Levels of respiratory control extend from the molecular to the behavioral.

of the blood. When carbon dioxide diffuses into a red blood cell, the enzyme carbonic anhydrase promotes its conversion into carbonic acid at a much higher rate than occurs spontaneously in the blood. Since the dissociation products of carbonic acid are charged particles, they are trapped in the red blood cell because cell membranes, bearing charges themselves, do not readily pass ionized particles. Moreover, the red blood cell is heavily loaded with the respiratory protein **hemoglobin,** which acts as a weak base, taking up hydrogen ions as they are formed.

Hemoglobin, of course, is essential to gas transport because it is the oxygen carrier of the blood. Its ability to carry oxygen is *inversely* related to how many hydrogen ions it is carrying. This property of hemoglobin is most useful for the following reason: in tissues where metabolism is intense, there is a larger than ordinary demand for oxygen. In such places it is easier for hemoglobin to release oxygen because at these same sites it becomes more and more acidic, owing to the higher than ordinary levels of carbon dioxide resulting from the increased metabolic rate. This constitutes another kind of automatic control exerted at the molecular level: an end product of metabolism, carbon dioxide, acts upon hemoglobin to promote the release of a required metabolic reactant, oxygen.

The opposite sequence occurs in the lungs when the acid-loaded, oxygen-poor hemoglobin molecule is exposed to high oxygen concentration. As the hemoglobin takes up oxygen, hydrogen ions are released and this promotes the formation of carbon dioxide. In the lungs this is good because the carbon dioxide diffuses into the airways and is breathed out. The consequence is an elegant mechanism to promote gas exchange without serious effects on blood acidity: a mechanism in which blood, circulatory system, and lungs cooperate in a homeostasis that is ultimately a capitalization upon the well-known chemical law of mass action.

So far so good. We have seen what might be called the "steady state" of respiratory processes—how the system operates at the cellular and subcellular level under normal conditions. Now we come to why a sack over the head causes heavy breathing. Under such conditions, as one rebreathes stale air the oxygen concentration falls and the carbon dioxide concentration rises. Corresponding changes occurring in the blood are detected by two sets of sense organs, one in the large arteries near the heart and the other in the brain. Both signal to the part of the brain that operates the ventilating muscles, causing a quickening of the ventilating effort. This is an attempt to restore the concentration of oxygen and carbon dioxide to normal levels in the blood by promoting more effective gas exchange in the lungs. Additionally, these same stimuli cause an increase in heart activity, promoting more rapid circulation, which further improves gas exchange.

Finally, of course, even higher levels of control would be involved. Once it became obvious that increased respiratory activity did not solve the problem of breathing in a paper bag, the conscious parts of the brain would become aware of the situation and initiate some sensible remedy.

The Neural and Endocrine Control Systems

Now we are going to discuss the entire control apparatus that is formed by the neural and endocrine systems. Although for convenience we are going to divide the discussion so that it focuses first on the nervous system and second on the endocrine system, remember that both systems represent a continuum with components working together to regulate processes ranging from slow to fast and from widespread to specific in terms of the parts of the organism affected.

THE HUMAN NERVOUS SYSTEM

In one respect the nervous system is easy to describe because its basic component is one rather standard cell type, the nerve cell or **neuron.** Once we understand how the neuron works we are some distance towards understanding the entire nervous system. The entire route to complete understanding of the nervous system, however, is one that will take scientists a long time to traverse, if indeed the journey is ever completed. The reason is that the human nervous system contains over 200 billion neurons and, although they have similar basic structure and function, they do vary sufficiently among themselves and are organized into functional units of such complexity that we are very far from a detailed understanding of how the entire system works. Nonetheless, enough has been learned about the nervous system to give a general picture of its operation.

THE NEURON

At the cellular level the nervous system consists of the primary functional cells, the neurons, supported by various other types of cells that take no direct role in neural activity. Their role is to provide an optimum environment for the nerve cells. Most characteristic among these is a special kind of connective tissue called **glia** that fills in the spaces between nerve cells and acts as a mediator between the nerve cells and the blood supply. Some nerve cells are actually wrapped tightly in glia, which serves greatly to increase the speed with which they send signals from one to another.

Neuron Structure Optimizes Receipt and Transmission of Electrical Signals

The form of a typical neuron (Figure 18-5) is well adapted for handling information. The cell body

Figure 18-5 A diagram of a motor neuron which carries commands from the central nervous system to a specific group of muscle cells. Compare with Figure 13-22 which shows a neuron specialized for processing information within the brain.

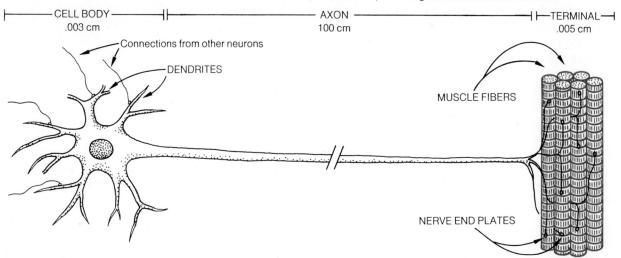

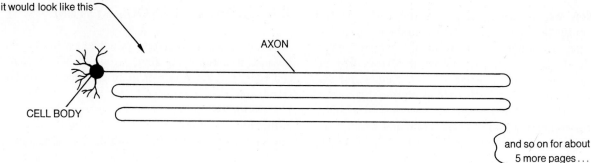

bears many-branched extensions called **dendrites**, which receive and process information, and a single **axon**, which may be hundreds of times the length of the dendrites. The axon carries information to other parts of the nervous system, or to organ systems, for example muscles, lying outside the nervous system. Axons terminate in **synapses**, structures that transmit information to other cells.

Language of the Neuron

This structure, obviously organized to convey information, immediately raises two questions: What information is carried? What is the language used? To answer in brief, the information that the nervous system uses in its operations includes the following:

1. Sensory data about the surrounding world.
2. Sensory data about the internal condition of the organism.
3. Commands going out from the nervous system to direct the actions of many organs.
4. The information involved in thinking, learning, remembering, and emotions.

As far as we know now, this immense amount of information is carried *within* nerve cells by a simple code of electrical pulses known as **nerve impulses.** However, information almost always passes *between* most nerve cells by a chemical transmitter link, which operates so that the quantity of the chemical transmitter substance moving from one nerve cell to another increases with increasing frequency of nerve impulses occurring in the transmitting nerve cell.

THE NERVE IMPULSE

The smallest information unit of the neural language, the nerve impulse, is an electrical pulse, as we have said. What is its source? All living cells are somewhat like batteries (Figure 18-6). Arrival of appropriate information at a nerve cell produces changes in the permeability of the nerve cell membrane, which short-circuits and discharges its cellular "battery" at that point. The great curiosity about nerve cells is that their batteries, in addition to short-circuiting, briefly go beyond discharge to shift electrical polarity in the opposite direction (the inside of the neuron actually goes from negative to positive). The electrical pulse thus produced occurs in only one small part of the neuron, but electrical currents that the pulse produces extend some distance along the nerve cell and have the ability to start this same process of short-circuiting and polarity change farther along the neuron. Thus the impulse can run along the neuron, reforming as it goes, at the expense of the electrical potential of the cell (Figure 18-6).

The nerve impulse is often compared with the burning of a fuse of gunpowder. The heat of the flame at one point ignites the powder at the next, just as the electrical currents from the nerve impulse start the changes in cell membrane permeability that result in a nerve impulse at the next point on the nerve. Of course, this comparison is limited in that a neuron, unlike a fuse, can carry hundreds of impulses per second. To accomplish this, the neuron must have some way to recharge the battery rapidly, a process that involves a poorly understood, energy-requiring metabolic "pump" that is part of the nerve cell membrane (Figure 18-6).

Speed of a Nerve Impulse

When you consider the lightning speed with which thoughts seem to occur or how fast your hand withdraws from something too hot for comfort, you might think nerve impulses travel extremely rapidly. Early students of the nervous system, who likened the nerve impulse to an electrical current in a wire, thought the nervous signals traveled just as fast as electricity in a wire. Actually nerve impulses in humans may travel as fast as 120 m/sec (269 mi/hr) or as slowly as a fraction of a meter per second. If a nerve cell is *myelinated*— that is, its axon is wrapped with an insulating sheath of myelin laid down by glial sheath cells (Figure 18-7A) —it conducts impulses very rapidly because the impulse actually does travel as an electrical current between the junctions between sheath cells, or **nodes of Ranvier,** in the covering of myelin (Figure 18-7B). The velocity with which nerve impulses travel along nerve cells depends on whether or not they are myelinated and on the diameter of their axons. The larger the diameter of an axon the more rapidly it conducts.

Starting a Nerve Impulse

Usually, not all parts of a neuron can produce a nerve impulse. Rarely does the impulse begin in the dendrites or the cell body. Usually the starting point is somewhere in the axon near the cell body, often at the first or second node of Ranvier in the case of myelinated neurons. Although the dendrites and cell body may not be able to sustain a nerve impulse, they can be short-circuited, or depolarized, by appropriate stimuli, and the electrical currents thus set

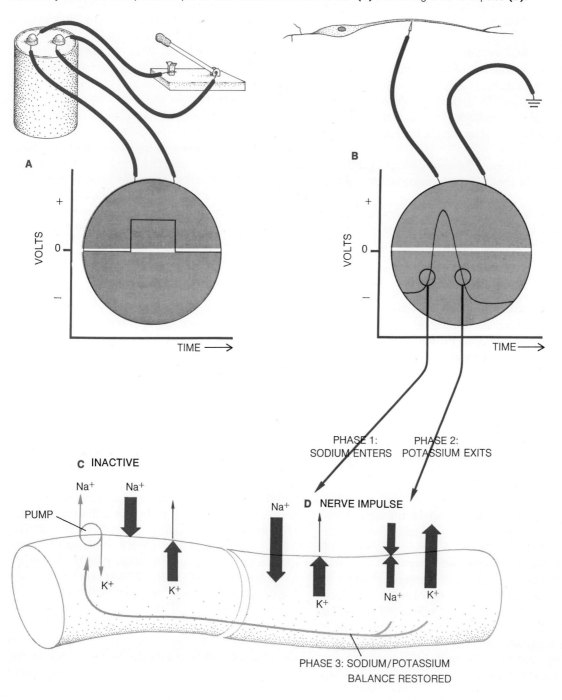

Figure 18-6 A nerve impulse is a biologically controlled and initiated electrical current. **A** and **B,** short circuiting of a battery and a nerve impulse compared. Ionic events in a nerve at rest **(C)** and during a nerve impulse **(D).**

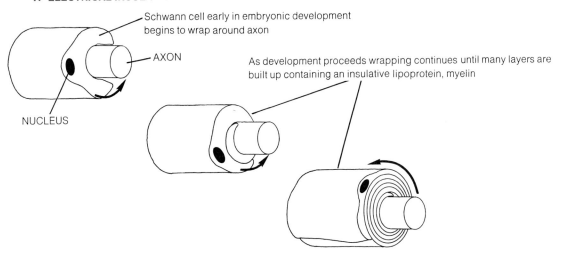

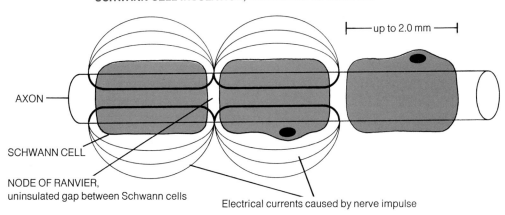

Figure 18-7 **Myelinization increases the speed of the nerve impulse. A.** The myelin sheath is laid down in concentric wrapping of the neuron by a Schwann cell from the neural crest of the embryo (see page 474). **B.** Between the gaps between cells of the myelin sheath, the Nodes of Ranvier, the nerve impulse travels as an electrical current, hence the high transmission speed of myelinated axons.

up flow through the cell and start an impulse in the axon.

Knowing that the depolarization of dendrites and cell body will start a nerve impulse, we then need to know about the kinds of events that cause this short-circuiting in the normal operations of the nervous system. Particularly, we need to know how a nerve impulse, running along in one nerve cell, can continue on to others with which it comes in contact.

Figure 18-8 shows a diagram of a **synapse**, that is, the point of closest approach of two nerve cells across which information flows. It is clearly apparent that the two cells are separated by a substantial space. Unlike its method of movement along the axon, the impulse does not get across this space by means of its electrical currents, except in certain rare types of synapses called **electrical synapses**. Instead, upon arriving at the synapse, the impulse causes the release of a **chemical transmitter**, often acetylcholine, which diffuses across to the next cell and starts a short circuit. Inside the nerve ending, acetylcholine is held in **synaptic vesicles** (see Figure 18-8) containing about

HUMAN PHYSIOLOGY: REGULATION   343

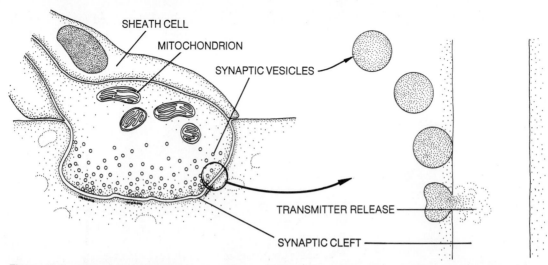

Figure 18-8 Diagram of a synapse. Arrival of the impulse at the nerve terminal causes synaptic vesicles containing acetylcholine to fuse with the cell membrane and release their contents into the synaptic space. Diffusion of acetylcholine across this narrow space to the adjacent cell sets up electrical activity in it.

70,000 acetylcholine molecules per vesicle, and is released in multiples of the contents of a single vesicle, depending upon how frequently impulses arrive at the synaptic ending. The more rapidly a nerve cell produces impulses, the more vesicles of transmitter will be released at its synaptic endings, thus producing more activity in the neurons its synapses contact.

Impulses are started in **sensory cells** by various physical or chemical stimuli, not by other nerves. In sensory cells the part of the cell analogous to the dendrite (input end) of an ordinary neuron is especially adapted to be short-circuited by some specific type of stimulus (light, heat, sound, chemicals, and so on), and no transmitter chemical is necessary for stimulation.

TRANSLATING THE NEURAL LANGUAGE

We have said that the neural language consists of nerve impulses. In any given neuron the impulses are of similar size; they occur in an all-or-none way. Thus the neural language seems to consist of only two words: "impulse" or "no impulse." But this is not really so because the *pattern* of occurrence of impulses (their frequency) carries information. The nervous system is therefore said to be a **frequency coding** system. A simple example is shown in Figure 18-9, where the response of a neuron that detects stretching of a muscle is shown relative to the amount it is stretched. The greater the stretch, the higher the frequency of impulses appearing in the neuron.

You might say, this is all very well, but how does the brain, which is receiving information from a particulary sensory neuron, know what kind of stimulus (chemical, light, and so on) is being applied to it? Because the brain has a very large number of sensory neurons to keep track of, you might easily wonder how it attaches *quality* to the *intensity* of the signals it receives from sensory neurons. What the brain does is assume that impulses coming from a particular type of sensory cell actually represent the intensity of the kind of stimulus that particular sensory cell is *designed* to receive. You can illustrate this by fooling your nervous system. Push on your eyeball with the eye shut and you will see colors. You have *mechanically* stimulated retinal nerve cells, but the brain is accustomed to interpret signals from them as light and does so.

So far we have implied that various degrees of excitation, ranging from none to maximal, were the only dimensions of the neural language. Actually some nerves may **inhibit**, that is, stop or reduce, the activity of the cells they innervate. Indeed, inhibition is of great importance in the normal operation of the

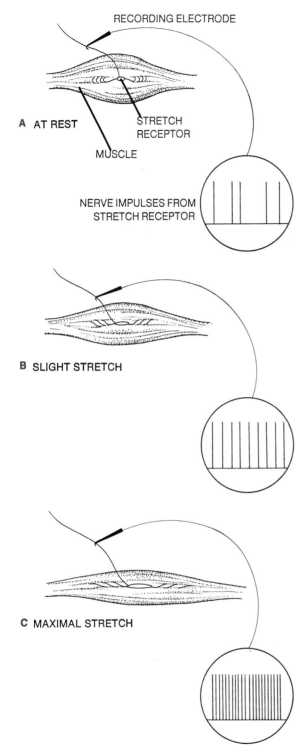

Figure 18-9 Recording from a stretch receptor in a muscle illustrates that the neural code expresses the intensity of a stimulus in terms of nerve impulse frequency.

nervous system. This is demonstrated by the terrible muscle spasms of strychnine poisoning. Strychnine prevents the normal inhibition of nerve impulses that occurs in the nervous system and uncontrollable muscular spasms result.

INTEGRATION IN THE NERVOUS SYSTEM

A nerve cell in a nervous system of any size at all is influenced by more than just one other neuron. In fact, a neuron may typically receive synapses from hundreds of other neurons. But despite the immense complexity of the *input* of information, there is usually only *one way out*, one axon. Thus we see that the highly complex input to a neuron must somehow be evaluated and reduced to a frequency-coded signal leaving by a single axon. Now perhaps you see a reason for the fact that nerve impulses do not begin in dendrites or cell bodies but farther down the axon. If impulses did commence in dendrites, the neuron would not be much more than a conductor of signals. Because the impulse actually begins at a distance from the dendrites and because all inputs impinging upon the dendrites influence, in a general way, the distant site of impulse generation in the cell bearing the dendrites, the neuron can be said to be **integrative**. That is, the output of the neuron represents in some way all elements of its input, excitatory and inhibitory.

HOW NERVES CONTROL MUSCLES

The relationship between a nerve axon and a muscle cell is similar to that between two synapsing nerve cells. There is a physical gap between the nerve endings and the muscle, and the arrival of the nerve impulse at the ending results in the opening of vesicles that release acetylcholine to diffuse across the space between the two cells and initiate an impulse in the muscle cell. This impulse is the same kind of impulse that occurs in nerves. As it sweeps along the length of the muscle cell, contraction is caused by tubular inpocketings of the cell surface that carry the impulse into the interior of the muscle cell and cause calcium ion release from vesicles (Figure 18-10).

In the majority of human skeletal muscle cells, the

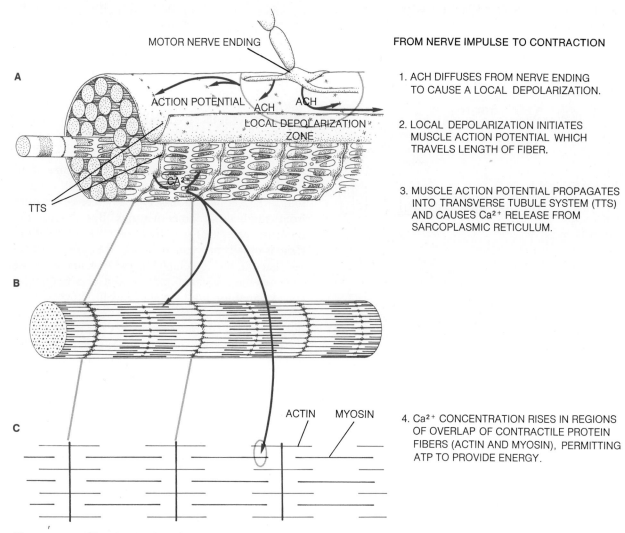

Figure 18-10 Neural activation of muscle. **A,** impulse spreads from synapse over muscle cell membrane and progresses into muscle fiber through a system of tubules with openings at the Z bands. **B** and **C,** impulses in tubular system activate contractile system by causing calcium release. Recall Figure 13-21.

result is a muscle contraction, that, like nerve impulses, is all or none; that is, only one strength of contraction occurs, if one occurs at all, and this constitutes the maximum capability of that muscle cell at the time. Obviously, we do not use our muscle *system* in that way; our movements are seldom the maximum that the muscles are capable of exerting. This seems to pose a contradiction between the all-or-none character of response of a single cell and the gradation of movement that we all know muscles, composed of many muscle cells, are able to achieve.

Two considerations resolve the contradiction. First, the duration of contraction of a single muscle cell is short in relation to the duration of any given movement of the whole muscle of which it is a part. Therefore, the amount of work that a muscle cell can perform during the movement of the muscle of which it is a part is some multiple of the work it does during a single contraction. If the muscle cell is stimulated at a high frequency, the cell may be **tetanized,** that is, stay contracted for a long period and effectively do work during the entire movement. Or the cell may be activated only a few times during the movement and do much less work. Thus the frequency with which muscle cells are stimulated is a way that the amount of work done by a muscle is regulated.

The second aspect of muscle control has to do with the anatomy of the relationship between nerve cells and muscle cells. When a muscle-controlling axon (**motor axon**) reaches a muscle, it branches into several endings that terminate on muscle cells. All muscle cells innervated by one axon form what is termed a **motor unit** because they must all work together. A nerve impulse in the main axon spreads into all of its branches and produces simultaneous contractions in all of the muscle cells in the motor unit. Because the central nervous system can control the number of motor axons to a given muscle that are activated, the number of active muscle cells in the muscle may thereby be controlled. Thus, in any given muscle, *the number of motor units activated and the frequency with which they are activated determines the strength and rate of contraction.*

Finally, in regard to motor units, there is a relationship between the delicacy of movements a muscle is ordinarily required to carry out and the average number of muscle cells in its motor units. Muscles that do not have a delicate role to play have very large motor units of several hundred muscle cells per controlling axon; as for example, in the large muscles of the legs. Muscles that must work with great delicacy and precision may have only a few muscle cells per motor unit, as in the muscles that move the eyeball, or even only one muscle cell per motor axon, as in our vocal cords. This is appropriate for the extremely precise operation of the vocal cords required in the production of sounds.

COMPONENTS OF THE HUMAN NERVOUS SYSTEM

The more than 200 billion neurons of the human nervous system are organized into two principal sections, the **central** and **peripheral nervous systems.** The central nervous system starts embryonic life as a hollow tube running the length of the body (see page 474). By growing more in some regions than in others, its two major components are formed. These are the **brain** and **spinal cord.** Closely associated with the spinal cord are two chains of nerve cells, the **sympathetic** and **parasympathetic** divisions of the nervous system.

In a general way we can say that the brain is concerned with (1) the control of many involuntary, "automatic" processes such as breathing as well as with the operation of voluntary muscles, those muscles that we use consciously, and (2) with all those processes that we call higher functions of the nervous system: reasoning, learning, memory. The spinal cord is a pathway connecting the brain with the peripheral nervous system, and it is also capable of certain simple actions independent of the cental nervous system. These are called **reflexes.** The sympathetic and parasympathetic chains are extensions of the spinal cord and they control elements of the body and activities that are not under conscious control, for example, operation of the musculature of the digestive system, regulating the adrenal glands, and so on. The peripheral nervous system is not as well defined as the foregoing; it includes (1) the neural connections between all of the structures of the body that are supplied with nerves and the central nervous system, and (2) many types of sense organs, which detect changes both in the environment and within the organism.

Brain

The human brain (Figure 18-11) is a large organ with most of its 1500 cm^3 average volume invested in the most anterior part, the **forebrain.** In the brains of less intelligent mammals or even the brains of our earliest prehistoric ancestors (as deduced from their fossil skulls), the forebrain is conspicuously smaller relative to the size of the remainder of the brain. You might therefore expect that the forebrain is the site of those higher functions of the nervous system (reasoning, learning, memory) that set man off from his less intelligent relatives. These functions involve reactions between the highly convoluted exterior rind of the forebrain, the **cerebral hemispheres,** and the deep-lying areas of the forebrain, which have many important specific functions. These areas include the **thalamus,** serving the cerebral hemispheres by relaying sensory information to them; and the **hypothalamus,** a center that controls or initiates many of the basic emotions (rage, fear, sex), body temperature, and even the desire to eat and drink. Finally, the hypothalamus controls many endocrine processes through its action on the pituitary gland.

The most posterior part of the brain, the connection between brain and spinal cord, is the **hindbrain.** One of its major parts, the **medulla oblongata,** is

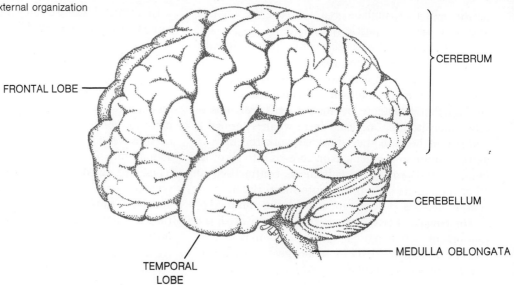

Figure 18-11 General external organization of the human brain.

absolutely essential to life because it automatically controls two vital processes, breathing and heart action. Other parts of the hindbrain, the **pons** and **cerebellum**, are essential to muscular coordination and the sense of balance.

Spinal Cord

We evolved from ancestors with segmentally constructed bodies, much as in the earthworm. In these ancestors the bulk of neural tissue was a trunk of nerves that ran the length of the body and gave off, in each segment, a set of nerves that served to gather information from that segment (sensory nerves) or to convey neural commands to the effector organs lodged there (motor nerves). Our spinal cord is still organized in this way, testifying to our beginnings.

The spinal cord runs in a protective bony channel formed by the segments of the backbone (**vertebrae**) from its origin in the base of the brain to the end of the spine. At regular intervals, pairs of spinal nerves are given off—paralleling the ancestral segmentation. These spinal nerves (including the 12 pairs of cranial nerves) are the sole avenues of communication between the central nervous system and the body (Figure 18-12).

A cross section of the spinal cord (Figure 18-12) shows that each spinal nerve has two pathways (**roots**) entering the cord. The **dorsal root** is sensory. The cell bodies of all nerves in the sensory root lie outside the spinal cord in the spinal sensory ganglion. All neural signals carried by the dorsal root are incoming and exclusively represent signals from sensory receptors. Thus the signal receiving, or dendritic part of a dorsal root cell may run, for example, all the way from the tip of a finger, where it might serve the sense of touch,

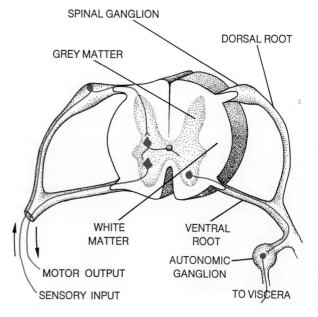

Figure 18-12 **Cross section of the spinal cord.** Cell bodies lie in the grey matter and axons, forming pathways running to and from the brain, are in the white matter. Central components of a simple reflex arc are shown, as is one of the pair of sympathetic ganglia present in each segment of the spinal cord.

to the spinal cord without an intervening synapse. The **ventral root** carries all outgoing signals—commands from the central nervous system to muscles and glands. Most motor nerves emerging from the ventral root proceed without intervening synapses all the way to the effector cells they innervate. Within the spinal cord both incoming sensory and outgoing motor neurons make synapses with neurons running longitudinally in the cord, forming connections with neurons at other levels within the cord and with the brain.

Autonomic Nervous System

The **viscera**—the collective name for the digestive, reproductive, excretory, circulatory, and respiratory systems—are served by the sympathetic and parasympathetic nervous systems, known together as the autonomic nervous system. Nearly all structures that one innervates, the other also innervates, producing opposite effects. For example, parasympathetic nerves slow the heart and sympathetic nerves speed it up (Figure 18-13). The two systems employ different synaptic transmitter chemicals—acetylcholine in the parasympathetic and adrenaline in the sympathetic. The connection of the autonomic system with the rest of the nervous system is by way of segmental nerves. Some of the cranial nerves (nerves coming directly from the brain) and a few from the lower part of the spinal cord lead to the parasympathetic system, whereas all the spinal nerves that supply the sympathetic nervous system issue from the middle region of the spinal cord (Figure 18-13).

These two systems, sympathetic and parasympathetic, are the "housekeeping" divisions of the nervous system, keeping the organs that they control operating properly by means of their mutually opposite effects. They may be influenced, of course, by the remainder of the nervous system, as, for example, by the psychological effects of fright causing increased heart action by way of the cardiovascular control centers in the brain, which increase the activity of sympathetic cardiac nerves or decrease the activity of the vagus, a parasympathetic nerve. Actually, one can get along reasonably well without the sympathetic system. In fact, its controlling chain of ganglia is sometimes surgically removed to relieve high blood pressure. But the price for this is the inability to adjust adequately to abnormal environmental conditions—high temperature and so forth.

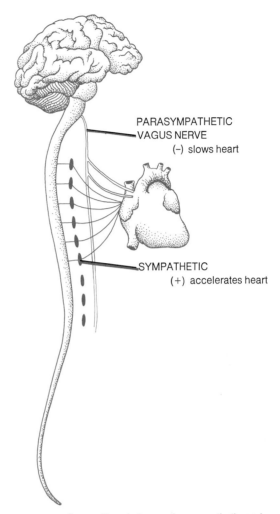

Figure 18-13 Connections between the sympathetic and parasympathetic nervous systems and the central nervous system consist of the vagus nerve and a series of sympathetic (autonomic) ganglia communicating with the spinal cord. The dual pattern of innervation characteristic of organs innervated by the autonomic system is illustrated for the heart.

Sense Organs

The parts of the nervous system that we have principally considered are mostly on the effector side of the feedback control systems that operate our internal machinery. Now we need to consider how information gets into this system.

The Kinds of Sense Organs. Considering all the aspects of internal functioning of man that must be measured and controlled, one might expect the variety of sense organs to be immense—even more so when one adds to the list of things to be measured the variety of stimuli of significance that we receive from

HUMAN PHYSIOLOGY: REGULATION

the outside world. Despite these considerations, receptors may be simply categorized as receptors of *light, temperature, chemical,* or *mechanical stimuli.* There are also pain receptors, but precisely what stimulus they respond to among those mentioned is not clear, and, in lower animals, electrical receptors.

The responsivity of a receptor for any one of these stimulus types is further specialized by the construc-

Figure 18-14 The properties of sensory receptor cells.

SENSE ORGANS CONTAIN SENSORY RECEPTOR CELLS THAT FUNCTION LIKE ORDINARY NEURONS **EXCEPT THAT**

A THEY ARE CAUSED TO GENERATE IMPULSES BY OUTSIDE INFLUENCES (STIMULI) OTHER THAN TRANSMITTER CHEMICALS

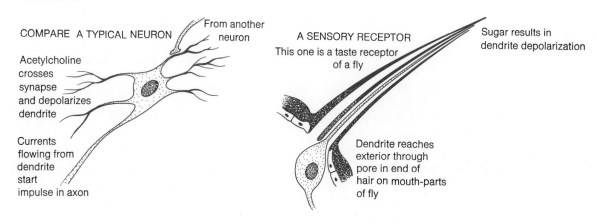

B THE SPECIFICITY OF THEIR RESPONSE MAY BE INCREASED BY ACCESSORY STRUCTURES WHICH FAVOR THE ACTION OF CERTAIN STIMULI AND REDUCE THE EFFECT OF OTHERS.

THUS SOME MECHANORECEPTORS ARE CELLS WITH CILIA

In the ear these cells are responsive to sound because sound vibrates a thin membrane which they touch

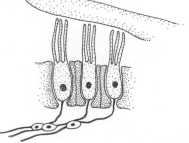

Thus the housing of the hair cell determines whether it is more sensitive to sound or to body movement and we have either an ear or an organ of equilibrium

In the semicircular canals similar cells are protected from sound by bone and are stimulated by the effect of turning of the body on fluid in a canal

If body suddenly turns this way fluid lags and cilia are bent

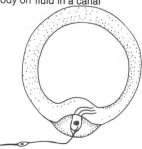

tion of the receptor. A receptor that basically responds to mechanical stimuli may because of its structure be specifically a hearing receptor, gravity receptor, blood pressure receptor, abdominal pressure receptor, and so on (Figure 18-14). That is, the receptor cell is activated, let us say, by movement, and this is true of all the mechanoreceptors that have just been listed. But the housing or position of each receptor determines what kinds of mechanical stimuli act upon it most effectively.

It is also useful to classify receptor organs in terms of the general scene that they view. **Exteroreceptors** monitor the external world; they are eyes and ears, the senses of smell and taste, and the mechanical and temperature sense of the skin. **Interoreceptors** watch over the internal physiological world; they include the many **chemoreceptors,** which monitor respiration and other internal chemical processes, **mechanoreceptors,** which measure phenomena ranging from the fullness of our urinary bladders to the position of our limbs or the adequacy of heart action, and **thermoreceptors,** which act in regulating body temperature.

How Sensory Receptors Work. All sensory receptors are **transducers.** This means that they use one form of energy to release another: a stimulus (one form of energy) impinges upon a sense organ, which then initiates nerve impulses (another form of energy) at a frequency governed by the kind and intensity of the stimulus (see Figure 18-9). In the sensory-transducer process the complete story of the molecular events of the transduction is unknown. In no case do we know all the details of how sensory cells change energy from one form to another. All that is understood is that an effective stimulus changes the ionic permeability of the sensory receptor cell in such a way that a resultant flow of electric current triggers nerve impulses. Thus awareness of our surroundings is ultimately dependent, as are so many other vital processes, upon the permeability of cell membranes.

Nerve impulses caused by stimulation of a receptor do not necessarily continue at a steady rate in response to a steady stimulus, because receptors have varying rates of **adaptation** to stimulation. Touch receptors adapt very rapidly, whereas receptors that measure blood pressure do not adapt at all. That is, when a touch receptor is steadily stimulated, it produces a brief flurry of nerve impulses and then falls silent until the stimulus changes in some way. A blood pressure receptor, on the other hand, produces impulses at a steady rate in response to blood pressure, even if the pressure does not vary for long periods.

THE PRINCIPAL HUMAN SENSE ORGANS: EYES AND EARS

The vast preponderance of information that flows into the brain of man from the outside world comes from the eyes and ears. Over 2 million neurons enter the brain from the eyes, a number nearly equal to the total number of sensory neurons that reach the brain from all other sources. The heavy reliance of man upon vision and hearing makes relatively unnecessary the third major external sense, the sense of smell; as a result it is poorly developed in man as compared with other mammals.

The Eye

The eye focuses light rays into an image upon a cup-shaped layer of **photoreceptor** cells that lie at the back of the eye in a thin sheet of neural tissue known as the **retina.** The rays are focused through a lens; muscles attached to the lens change its shape, thickening it for light rays from nearby objects and making it thinner for those from distant objects. The **iris,** a ring of pigmented, muscular tissue just in front of the lens, regulates the amount of light admitted to the retina by changing the diameter of its opening, the **pupil.** The iris and lens are under reflex control. These structures are shown in Figure 18-15.

Poor vision is often due to defects of the lens or other defects that prevent formation of a sharp image on the retina; these often may be corrected with glasses. Other defects in the optical system include cataracts, in which the lens becomes less transparent. Cataracts are common with aging or as a side effect of diabetes. Treatment involves removal of the defective lens, which may be replaced by an artificial lens or compensated for by corrective glasses.

How the Retina Works. The deepest layer of the retina contains two kinds of light receptor cells, the rods and cones, that are sensitive to light because of the pigments they contain (Figure 18-16A, B). Almost everyone knows two simple facts about vision: (1) that to see well at night one must have sufficient vitamin A in the diet, and (2) that if one is to detect an obscure

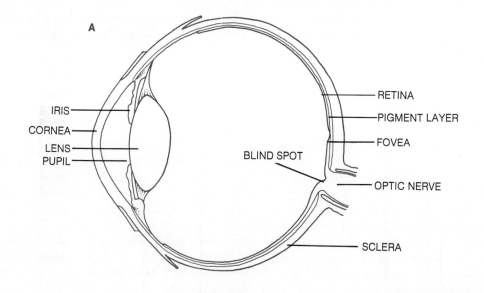

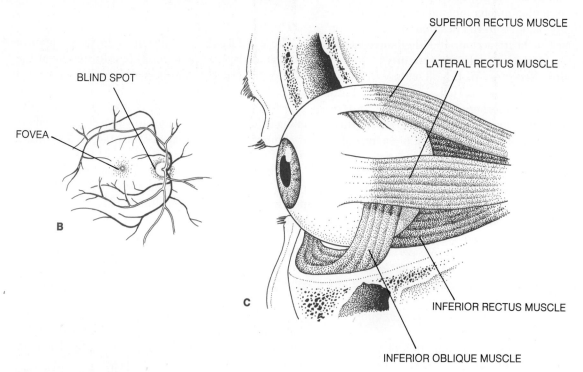

Figure 18-15 **The eye. A.** vertical section. **B.** surface view of fovea and blind spot. **C.** eye muscles.

object at night it is best to look at it out of the "corner," or periphery, of the eye. These facts tell us a lot about vision. Observation of the arrangement of **rods** and **cones** in the retina, shows that cones are more numerous in the center and that toward the edge there are only rods. Further, rods contain very large concentrations of a relative of vitamin A in association with the protein **opsin**; together the two

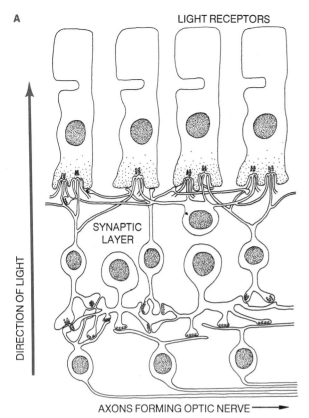 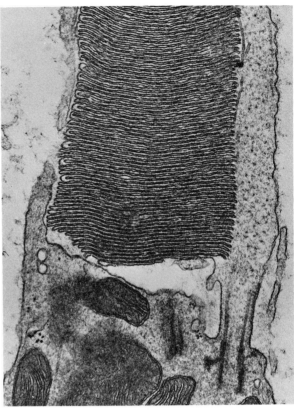

Figure 18-16 **The retina. A.** simplified diagram of microscopic organization of the retina in cross section. Light receptors, rods and cones, lie in the deepest part of the retina so that light entering the eye must traverse more superficial neural layers of the retina to reach them. Elaborate processing of receptor output takes place in the neural layers, as suggested by the elaborate synaptic interconnections shown. **B.** Electron micrograph of a mammalian photoreceptor rod showing the junction between inner and outer segments. The outer segment, consisting of tightly packed discs containing the visual pigment, is derived from a cilium whose characteristic microtubular structure is seen in the narrow link between inner and outer segment. (S. K. Fisher.)

are called **rhodopsin.** These observations lead to the concept that rods function in the type of vision known as night vision.

The highest concentration of the other type of photoreceptor cell, the cone, is towards the center of the eye in a small region known as the **fovea.** This region is not nearly as sensitive to dim illumination as the outer, rod-containing areas, but it is color sensitive. Colors of images on this region of the retina can be discriminated far more easily. Certainly you have noted the difficulty of making out colors at night; rods are not able to discern color. Although the chemical nature of their light-sensitive pigments is poorly known, cones are known to be of three types, either blue-, green-, or red-sensitive. Much evidence indicates that these three types of cones are the basis of color vision. For example, the various types of color blindness are usually attributable to specific cone defects.

Eye Movement. Six muscles move the eyes (see Figure 18-15), for the most part keeping the most sensitive part of the retina, the fovea, of both eyes aimed at whatever is being observed. This is done so well that one is not aware of the presence of a sizeable blind area in each eye, the **blind spot,** where the optic nerve exits from the retina.

Besides these aiming movements, the eye muscles impart a very fine tremor to the eye. Far from being a nuisance, this tremor appears to be essential to vision. If an image is completely stabilized on the retina, so that the same rods and cones are continuously exposed to an unvarying image, the image disappears! Evidently the photoreceptors require a changing stimulus to remain active, and change is ensured by the tremor. Whatever its mechanism, the inability to transmit a stable image is probably useful because there are structures within the eye that would otherwise cast blurring shadows upon the retinal cells; thus all the blood vessels and nerves that lie in the path of light to the rods and cones are not seen, because they move as the retina moves and form a stabilized image that the receptor cells evidently do not transmit.

The Ear

The paired organs of hearing and the closely associated structures that provide for the sense of equilibrium are basically mechanoreceptors. The organs of hearing are specifically sensitive to sound, usually airborne waves ranging in frequency from about 50 to 18,000 Hertz (Hz). Young humans may hear sounds at frequencies up to 20,000 Hz but the upper limit decreases with age.

The ear has three major parts (Figure 18-17). The **outer ear** focuses sound waves upon a thin, living membrane, the **eardrum**, which lies at the inner end of the ear canal. The **middle ear**, which the eardrum separates from the outer ear, contains three small bones that transfer sound-induced vibrations of the eardrum to the entrance of the actual organ of hearing, the **cochlea**. The middle ear is air-filled, and, to ensure equal air pressure on either side of the delicate eardrum, the middle ear is vented by way of the **eustachian tube**, which opens into the mouth cavity. Because the eardrum is larger in diameter than the opening into the fluid-filled cochlea and because of the lever action of the interconnecting bones, the middle ear apparatus is a mechanical sound amplifier, focusing the sound collected by the outer ear upon the oval window at the entrance to the cochlea.

The cochlea looks like a snail shell, with a partition bearing sensory cells dividing its canal lengthwise. Sounds arriving at the entrance to the cochlea set up waves in the cochlear fluid that affect different sensory cells according to the pitch (frequency) of the sound. This is a major part of the secret of how the ear manages to give the brain information concerning sound frequency. The sensory receptor cells (see Figure 18-17) are themselves not frequency-sensitive; they are made so by the fact that the structures of the cochlea that stimulate them *are* frequency-sensitive. The brain discerns a specific sound frequency by the activity of a specific set of cochlear receptor neurons

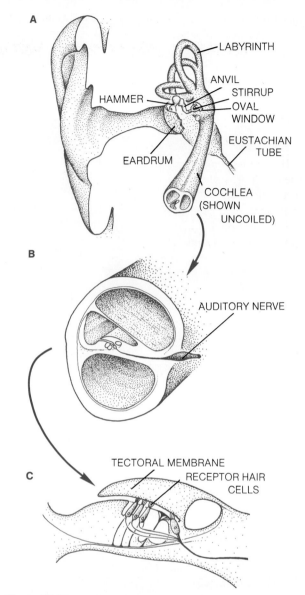

Figure 18-17 The ear. A. General view with structures in normal orientation except for cochlea, shown uncoiled and sectioned. **B.** and **C.** detail of sound detecting structures of the cochlea.

in a region unique for each frequency. The rate at which they make impulses serves as a measure of the intensity of sound.

Sound Localization. In addition to detecting frequency and intensity, the ears are able to sense the direction of origin of sounds. Our ears are far enough apart to provide an appreciable delay, on the order of 1 millisecond (msec), to sounds arriving at the two ears from a source on the axis running through the ears. Intensity differences also provide important directional cues. As you might expect, sounds originating at points equidistant between the ears cause difficulty; you cannot tell without head movements whether they come from ahead or behind, above or below.

The Sensitivity of Eyes and Ears

Both the eye and the ear function at very close to the maximum sensitivity theoretically attainable. Rod receptors are sensitive to 1 quantum of light. However, several rods must be activated before you can see; an insurance against spurious vision caused by random, spontaneous activation of rods by thermal energy or other means. The resolution of the eye is also remarkable. Lines so close together that they cast images only 1.0 nanometer (nm) apart on the fovea are distinguishable.

A few humans have a sense of absolute pitch, but anyone can tell the difference between two sound frequencies differing only by 20 Hz. The sensitivity of the ear is sufficient to allow it to hear sounds that oscillate the eardrum by only the distance represented by the diameter of a hydrogen atom.

THE ENDOCRINE SYSTEM

There are two ways to consider the functions of the endocrine system. Traditionally it has been considered quite different from the nervous system, sending its messages as molecules flowing through the circulatory system randomly to find their sites of action. At first glance this is quite a different picture from what we have seen for the nervous system, where electrical signals move rapidly through precisely defined nerve pathways to highly specific sites. With growth of our understanding of how both systems operate it has become apparent that, despite these obvious differences, the two systems operate in a fundamentally similar way. An endocrine gland cell releases a hormone that acts on another cell. This is exactly what happens at a synapse when one nerve cell releases a transmitter molecule to act on another cell, muscle, gland, or nerve.

The fundamental processes are identical, *secretion of information-carrying molecules.* The principal differences lie in the nature of the transmission link between cells, an unspecified pathway through the circulation in the case of the endocrine system. This contrasts with the rapid pathway through specialized cell extensions (axons) in the case of the nervous system. A consequence of this difference in the manner of signal delivery results in another difference between the two systems. Each endocrine gland must produce hormones uniquely tailored for the organs that they are intended to affect since they are essentially broadcast at random into the circulatory system. The nervous system, in contrast, uses a few different kinds of transmitter molecules to control a very wide range of organs because the application of the transmitter is directly onto the target cells. Thus the same transmitter, acetylcholine, is used for *all* voluntary muscle control in the vertebrate.

As we study these systems more similarities and points of cooperative interaction will be seen. For example, the same transmitter molecule, norepinephrine, is used by nerves of the autonomic system and is also released by an undeniable endocrine gland, the adrenal medulla. Indeed, as we have already seen, certain parts of the two systems are so inextricably cooperative that they are called a neuroendocrine system. In the final analysis, perhaps the term should be extended to the entirety of both systems.

COMPONENTS OF THE HUMAN ENDOCRINE SYSTEM

As shown in Figure 13-24, the human endocrine system consists of ten principal glands of definitely known function, plus others whose function is not completely known. Of these the pituitary is often called the master gland. This name is deserved for two reasons: (1) the pituitary is part of the principal route of communication between the nervous and endocrine systems, and (2) it controls, either directly or through other endocrine organs, many physiological processes. The remaining glands, either independently or by way of pituitary control, influence a bewildering array of processes. These glands include the hypothalamus, pineal, thyroid, parathyroid, pancreas, adrenal, and gonads. Further, the digestive system has its own endocrine system, concerned only with the coordination of the many processes involved with digestion (page 379).

The pineal and the thymus are poorly known

glands. The pineal has an endocrine role in lower vertebrates, and in some reptiles it functions as a third, upwards looking eye. In the human it may play a cooperative role with the hypothalamus in controlling pituitary function. In the adult human the thymus is poorly understood; in infancy and childhood it serves as a source of antibodies, part of the defense system against disease organisms (see page 500).

The Pituitary

During early embryonic life, the pituitary forms simultaneously from two sources—an inpocketing from the roof of the mouth cavity that meets an outpocketing from the floor of the embryonic brain. This dual origin is reflected in the anatomy and physiology of the adult pituitary (Figure 18-18), in which there persists an **anterior** and **intermediate lobe** derived from the embryonic mouth tissue and a **posterior lobe** derived from the embryonic brain. The intermediate lobe has relatively minor functions in the human, but both the anterior and posterior lobes are of immense significance.

Both lobes of the pituitary are functionally integrated with the hypothalamus, a part of the brain (Figure 18-18)—the anterior pituitary by way of a short system of blood vessels, and the posterior lobe by means of nerves extending into it from cell bodies in the hypothalamus. These nerves are of a type called **neurosecretory**; they release their transmitter molecules, actually not ordinary neural transmitters, into a portal system, that is a local bloodstream (see below).

The Pituitary Hormones. All pituitary hormones are either polypeptides or proteins. Six kinds are manufactured and released in the anterior lobe. Of these, the **growth hormone** has general effects on all the cells of the body, whereas the others influence only specific tissues or glands (adrenal cortex, thyroid, and so on), often causing them to synthesize and release hormones with more general effects. The two hormones of the posterior lobe, both peptides that have been synthesized, have highly specific effects. **Antidiuretic hormone (ADH)** helps regulate the osmotic properties (ion concentration, and so on) of the internal environment. **Oxytocin,** the other posterior lobe hormone, brings about release of milk from the breasts by causing contraction of the muscles lining the milk-producing ducts. It may also act during birth by promoting contraction of the muscles of the uterus.

Control of Pituitary Hormone Release. Release of all of the anterior pituitary hormones is regulated by a series of releasing factors, or "brain hormones," that are synthesized in the hypothalamus and conveyed into the anterior pituitary by blood vessels. These blood vessels make up what is called a **portal system** because they begin in capillaries and end in capillaries. Thus the portal system is able to gather a product with one set of capillaries and deliver it to a well defined spot with another. The two posterior lobe hormones are actually synthesized within the hypothalamus and enter the posterior lobe by transport in neurosecretory axons. Nerve impulses in these same axons evidently cause their release by mechanisms analogous to the release of neural transmitters at ordinary nerve terminals. The posterior lobe is simply a storage and release site for the neurosecretions of the hypothalamus.

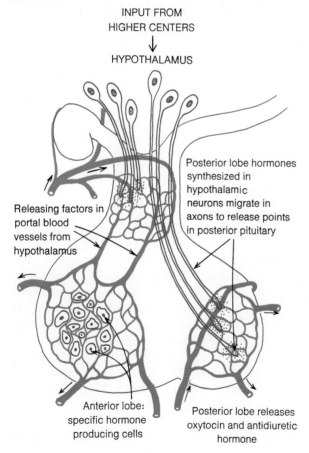

Figure 18-18 Relationship between hypothalamus and pituitary.

Regulation of the Concentrations of Pituitary Hormones. For most of the anterior pituitary hormones—perhaps for all of them—there are sensory cells in the hypothalamus that measure the circulating levels of the hormones from those subsidiary glands regulated by the pituitary. For example, the concentration of thyroid hormone in circulation is sensed by cells in the hypothalamus. These cells control the secretion of the **releasing factor,** which travels via the portal vessels to cause the pituitary to release **thyroid stimulating hormone (TSH).** You might wonder if this is an unnecessary step in the thyroid feedback control system. There might seem to be no need for releasing factors when control might be as satisfactory with (to stay with the thyroid example) thyroid hormone acting directly upon the TSH-releasing cells of the pituitary.

The explanation of this extra step is that the releasing factor mechanism provides a means for the *nervous system* to influence the pituitary, and through it most of the rest of the endocrine system. This link makes the neuroendocrine system a unified control system. Specifically, the releasing factor control loop provides an efficient way for the endocrine system to adjust to the environment (using sensory data gathered by the nervous system) and furthermore to adjust to the *psychological state* of the animal. There are many striking examples from the lower vertebrates. For example, some of these ovulate only upon the occurrence of an appropriate psychic stimulus. The female pigeon ovulates only when she sees another pigeon, as shown by the fact that she can be deluded into ovulating by the sight of herself in a mirror! Rabbits ovulate only upon copulation. Similarly, in mice olfactory stimuli may affect the reproductive cycle. In these situations, called **reflex ovulation,** the visual or other stimulus ultimately impinges upon the hypothalamus, causing the release of its factors controlling the gonadal hormones.

There is still other evidence to support the concept of feedback control based on circulating hormone levels. Removal of *one* of the pair of adrenal glands is followed by *compensatory hypertrophy,* an increase in size and hormone-producing activity of the remaining gland. The cause—diminished adrenal hormone levels following removal of one adrenal—calls forth increased secretion of adrenocorticotrophic hormone (ACTH), the normal hormone by which the pituitary activates the adrenals. Continuous exposure of the remaining gland to high ACTH levels causes the increase in size.

The two hormones of the posterior lobe of the pituitary are regulated also from within the hypothalamus and by stimuli acting upon the hypothalamus from elsewhere within the organism. Antidiuretic hormone release is governed by osmotic concentration receptors in the hypothalamus and also by detectors of blood volume that act upon the hypothalamus by way of sensory nerves. Release of oxytocin, the other posterior lobe hormone, is triggered by mechanical stimuli associated with the events of birth and suckling.

Thyroid Gland

One principal hormone, **thyroxin,** is produced by the thyroid gland (Figure 18-19). The hormone is manufactured in the gland from the amino acid tyrosine and iodine and stored in hollow nests of colloid-filled cells, or follicles. Thyroxin is essential to maintenance of normal metabolism, and in young animals it is essential to normal development. An uncorrected thyroid deficiency in a child causes it to become a cretin, a mentally subnormal dwarf. However, an adult can survive reasonably well with an inactive thyroid but will probably be overweight. Thyroid overactivity produces many abnormal phenomena—nervousness, weight loss, and so forth—all mostly attributable to elevated metabolic rate.

Excessive growth of the thyroid may result in **goiter.** Before the use of iodized salt became widespread, goiter was common in parts of the United States where there was a deficiency of iodine in the soil and hence in food. Lacking this essential ingredient of its hormone, the thyroid of an iodine-deficient individual would be forced into greater and greater activity by the elevated pituitary secretion caused by feedback from low blood concentrations of thyroxin.

Parathyroid Glands

Usually four of these very small glands (total volume in an adult is about 0.2 cc) lie embedded in the thyroid. Each gland secretes two hormones, **parathyroid hormone** and **calcitonin,** both polypeptides. They are essential to life because they regulate the concentration of calcium in the blood. When blood calcium falls too low, excitability of nerves and muscles increases until death results.

The hormones have several sites of action. They

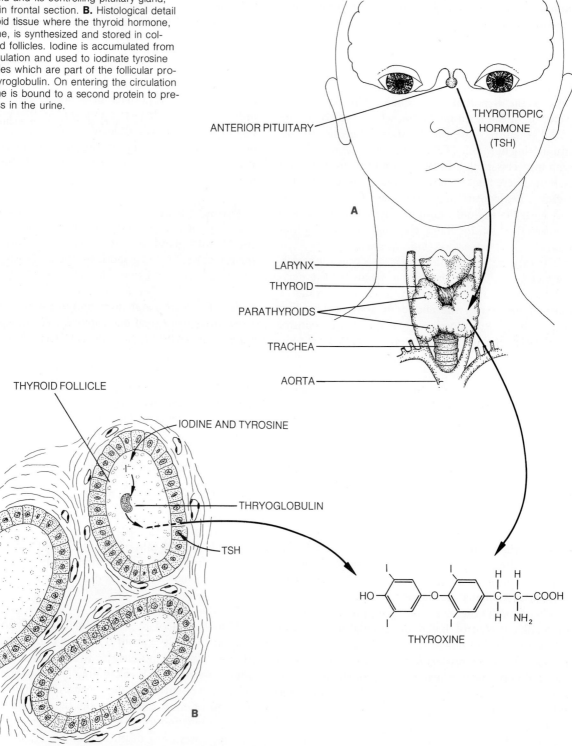

Figure 18-19 Thyroid system. A, location of thyroid and its controlling pituitary gland, shown in frontal section. **B.** Histological detail of thyroid tissue where the thyroid hormone, thyroxine, is synthesized and stored in colloid-filled follicles. Iodine is accumulated from the circulation and used to iodinate tyrosine molecules which are part of the follicular protein, thyroglobulin. On entering the circulation thyroxine is bound to a second protein to prevent loss in the urine.

influence the release of calcium from bone and may influence calcium reabsorption by the kidney and calcium uptake by the intestine.

Adrenal Glands

Each of the paired adrenal glands of man is formed from two completely independent endocrine organs,

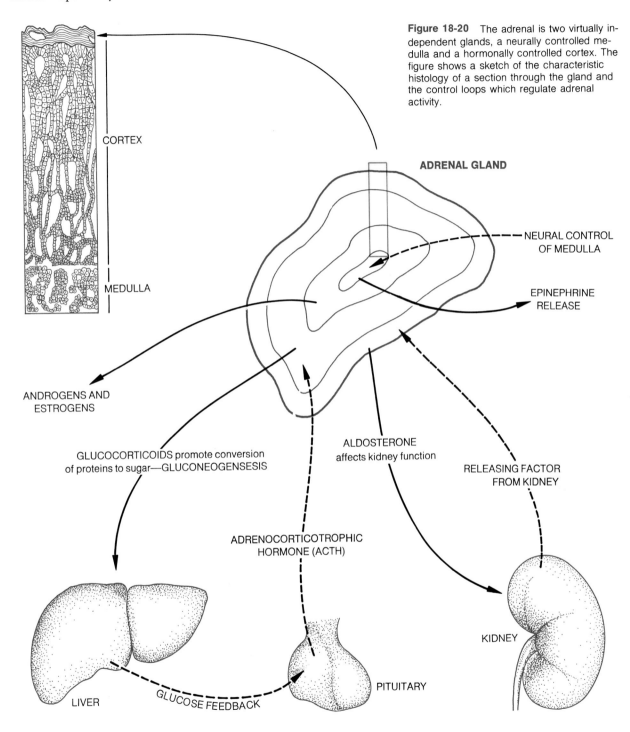

Figure 18-20 The adrenal is two virtually independent glands, a neurally controlled medulla and a hormonally controlled cortex. The figure shows a sketch of the characteristic histology of a section through the gland and the control loops which regulate adrenal activity.

HUMAN PHYSIOLOGY: REGULATION

the **adrenal cortex** and the **adrenal medulla** (Figure 18-20). In lower vertebrates these elements are physically separate. The medulla is primarily a neural tissue that secretes **epinephrine** (**adrenaline**) and smaller quantities of **noradrenaline**. The cortex is entirely different. In the adult human adrenal cortex there are three layers (Figure 18-20). Each produces its own set of hormones and is heavily infiltrated with blood vessels, facilitating distribution of the **adrenocortical hormones** to the body.

All of the many adrenocortical hormones are chemically related to the steroid cholesterol, and they may be divided into three groups on the basis of their actions. They include **sex hormones** (both male and female), **glucocorticoids** (which act to increase the level of glucose in the blood), and **mineralocorticoids** (which influence body salt and water balance by regulating the excretion of sodium). The most important role of the adrenal cortex is in the production of glucocorticoids and mineralocorticoids. Ordinarily, sex hormone production is so small that there is little or no effect on normal individuals. Abnormal conditions may increase the rate of production of these hormones, resulting in early onset of sexual maturity if they occur in the very young, or causing disturbances in the secondary sexual characteristics of adults.

The Pancreas

The pancreas is another multifunctional organ. Part of it is concerned with digestion and is not an endocrine organ at all. We shall hear more of it later. The endocrine part of the pancreas, for all of its great importance, consists of only a tiny fraction of the bulk of the organ—the islets of Langerhans, which are small groups of cells scattered all through the substance of the pancreas, making up only a small percentage of its total weight. Specialized endocrine cells occur in the islets. One type secretes a protein hormone, **insulin**, and another type secretes a polypeptide hormone, **glucagon**. Both hormones importantly affect metabolism: insulin acts to reduce the level of glucose in circulation, whereas glucagon acts to increase it. Control of release of the two oppositely acting hormones is maintained principally by direct sensing in the pancreas of the level of glucose in circulation. Insulin deficiency results in **diabetes**, in which the principal effects, usually fatal if uncorrected, are related to disturbed glucose metabolism. Glucose accumulates in the circulation and is unable to enter the cells, which require it as an energy source—causing starvation in the midst of plenty.

The Gonads

The gonads (Figure 13-24 and Chapter 21) perform two related reproductive functions. They produce the actual sexual products, eggs and sperm, and they produce sex hormones that play a variety of roles associated with reproduction, including development of male and female secondary sexual characters, insuring the psychological drive to reproduce, and controlling the environment in which the egg develops.

The ovaries are markedly cyclic in function. Largely inactive until the initiation of puberty, they become functional—producing eggs and hormones—during the period between puberty and menopause, a span covering the period between ages 11–16 and 40–50, respectively. At birth the ovary contains its lifetime supply of undeveloped egg cells (**oocytes**). When the ovary begins its mature life, these oocytes commence developing one at a time in **ovarian follicles**, according to a monthly cycle (Chapter 21), which is dependent on release by the anterior pituitary of two hormones, **follicle-stimulating hormone** (**FSH**) and **luteinizing hormone** (**LH**). Interestingly, the ovary itself establishes the cycling pattern of release of these hormones during embryonic life. The embryonic hypothalamus and pituitary are noncyclic and remain so during the remainder of life unless exposed to female sex hormone during development. The embryonic ovary produces sufficient female hormone to accomplish this and will do so even if grafted into a male embryo.

As the egg-containing follicle matures in the ovary under the action of FSH and LH, it produces the primary female sex hormones (**estrogens**), **progesterone**, and related steroids. These last are hormones that cause the uterus to prepare to nourish the embryo, if the egg in the follicle is fertilized (Chapter 21). In addition to these hormones, the ovary produces others. One is **relaxin**, a peptide. Together with estrogen, relaxin acts late in pregnancy to enlarge the pelvic birth canal, loosening the pelvic joints by converting cartilage to connective tissue.

The physiological actions of estrogen are widespread. Its developmental effect on the hypothalamus has already been noted. In adult life estrogens continue to act directly on the hypothalamus in the induction of sexual behavior. Estrogens promote growth in a wide variety of tissues, particularly those associated with reproduction. Maintenance and development of the vagina and uterus depend on estrogen, as does development of the breasts. Even the widely saluted female body form, wide hips and characteristically distributed fat, is estrogen-dependent, as is the characteristic distribution of facial and body hair of the female.

The testes produce spermatozoa and male sex hormones. **Spermatozoa** develop in **seminiferous tubules** and exit from these into the accessory ducts of the reproductive system when mature (Chapter 21). Masses of interstitial cells lying among the seminiferous tubules appear to be the major source of male hormones of testicular origin. (Remember that the adrenal cortex also produces these steroids.) Regulation of testicular function involves the same two pituitary hormones that control ovarian function—FSH and LH. There is little difference in hormone output between male and female pituitaries except in the *pattern* of their release. In the male, FSH and LH are released beginning with the onset of puberty and extending through most of the adult life. Both FSH and LH seem necessary for sperm production, whereas LH stimulates interstitial cell production of male hormones.

Male sex hormones are collectively known as **androgens**. The principal androgens of testicular origin are **testosterone** and **androsterone**. These hormones are essential for normal testicular function. Throughout the body they serve to increase protein synthesis, and they specifically control the development and maintenance of male secondary sexual characteristics such as heavy skeletal configuration, deep voice, hair patterns.

Pineal Gland

This gland, once thought to be the seat of the soul, is named from a fancied resemblance to a pine cone. It lies deep within the human brain and until recently was thought to be simply an evolutionary vestige of a "third eye" found in amphibians and reptiles. Present evidence is that it plays a significant role in the higher vertebrates, the birds and mammals, by secreting a hormone called **melatonin** (Figure 18-21), which it synthesizes from tryptophan. Melatonin lightens skin pigmentation by causing clumping of melanin (pigment) granules in pigment cells of the skin in amphibians, and, more importantly, it has a profound slowing effect on gonadal maturation.

In birds, production of melatonin is keyed to the

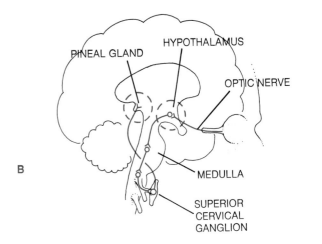

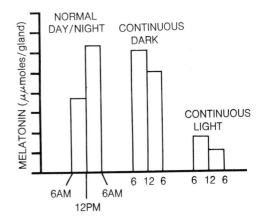

Figure 18-21 The human pineal. **A.** biosynthesis of the pineal hormone, melatonin. **B.** in the human, light affects the pineal by an indirect pathway. **C.** melatonin synthesis is heightened in the dark period of normal days and is higher in continuous dark than in continuous light, after Axelrod, J. et al, 1965, *J. Biol. Chem.* 240:949.

HUMAN PHYSIOLOGY: REGULATION

daily cycle, with the gland synthesizing more of the hormone in darkness than in light. The same is true for mammals except that the way the light signal is detected differs in the two groups of animals. In birds, the pineal retains enough of its primitive light sensitivity to produce the hormone in rhythm with the daily cycle by direct detection of light, even if the eyes are removed. Mammals have lost this direct pathway of light perception, and the gland is activated instead by a complex neural pathway involving the eyes and the sympathetic nervous system. Thus the pineal gland represents another example of a neuroendocrine mechanism that attunes the physiological state of the organism to the environment. In this instance the link is to the day-night cycle.

Other Glands and Hormones

Major discoveries about the endocrine system continue to be made. As we shall see later, much current investigation is centered on the precise molecular mechanisms of hormone action, but there are still new hormones being discovered. In the next chapter we shall consider digestive system hormones, and new candidate hormones of this type are frequently put forward for consideration. Also there are hormonal functions being proposed for the kidney and for the thymus gland, so we may expect many new developments from endocrinological research in the future.

One particular area of research holds great promise. This concerns an evidently large family of hormones produced in very small quantities by a large variety of cells. These hormones are called **prostaglandins** because one of the first known hormones of this group was thought to be produced by the prostate, a male accessory sex gland. The prostaglandins (Figure 18-22) are derivatives of fatty acids and are related to important constituents of the cell membrane, which may hint at their mode of action. A large number of these hormones have been isolated and undoubtedly many more remain to be discovered. They have a wide range of effects—on blood pressure, on the muscles involved in childbirth, on the secretion of digestive glands, and so on.

CELLULAR MECHANISMS OF HORMONE ACTION

Hormones are released into the blood and blood goes everywhere. So how does an endocrine gland produce a specific or limited effect? The endocrine system depends upon special properties of target tis-

Figure 18-22 Prostaglandins and their actions.

Prostaglandins C_{20} Fatty acid are precursors, forming two principal classes of hormones:	Characteristic Effects
(1) PGE series for example PGE_2 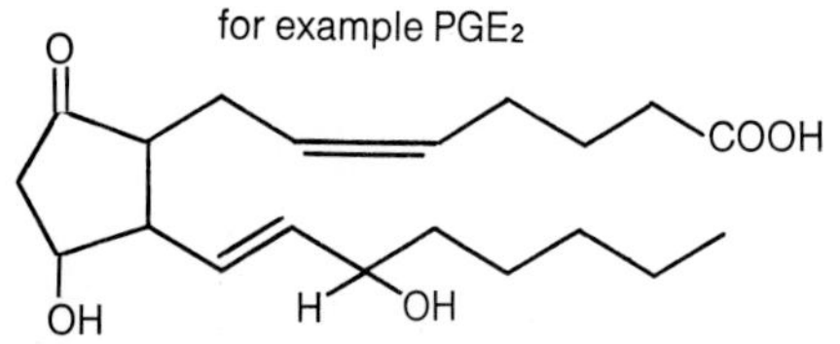	Increases motility of pregnent uterus. Decreases systemic blood pressure. Increases coronary arterial blood flow. Bronchial dilation
(2) PGF series for example $PGF_{2\alpha}$ 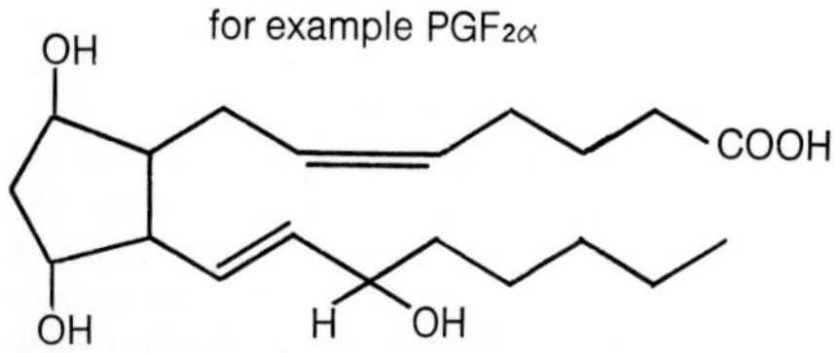	Induces labor, abortion (antifertility action). Increases systemic blood pressure. Bronchial constriction.

THYROID HORMONE ACTION
DEPENDENCE ON TARGET TISSUE RESPONSIVENESS

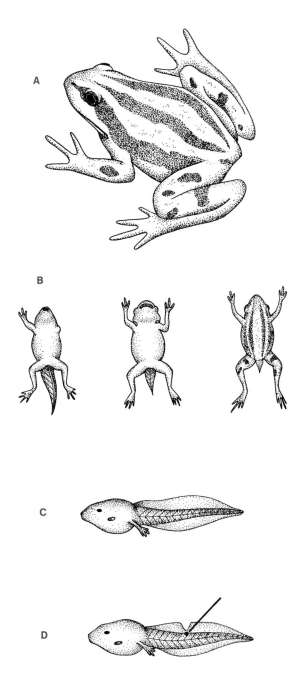

Figure 18-23 Target tissue responsiveness is a major factor in hormone effects. Thyroid hormone administered to adult frog **(A)** simply affects the metabolic rate. Applied generally or locally to tadpoles, profound structural changes are induced, including early metamorphosis **(B)** and local regression of tissues normally lost at metamorphosis **(C, D)**.

sues to respond selectively to hormones and thus to provide the specificity of effect. Hormones may reach nearly all cells in the body but only certain ones have properties allowing response. The degree of specificity of hormones may vary widely, however. For example, the thyroid hormone, in keeping with its controlling effect on body metabolic rate, acts on nearly all cells, whereas, in contrast, the pituitary hormone (TSH) that activates the thyroid gland acts only on the thyroid gland and is probably without direct effect on other tissues.

Further complexity is introduced by the age-dependent way in which cells respond to hormones. If thyroid hormone is given to a frog tadpole, profound changes in the form of the animal result. Thyroid hormone causes rapid metamorphosis into a young frog and this, of course, involves many morphogenetic changes (loss of tail, growth of legs) and even changes in behavior (Figure 18-23). In contrast, when thyroid hormone is given to an adult frog, there is only an elevation in the metabolic rate and no change in form.

The Cell Membrane Is the Target of Many Hormones

Many hormones such as epinephrine, thyroxine, and the polypeptide hormones of the anterior pituitary seem unable to enter cells and must achieve their effects on the cell membrane. The first step in the action of such hormones is combination of the hormone with a receptor molecule, which is a part of the cell membrane. The receptor molecule is tailored to accept only one kind of hormone, and this molecular specificity is the sole basis of the specific actions of the hormone. As shown in Figure 18-24 the next step in hormone action is for the enzyme adenyl cyclase to be activated as a consequence of the binding of the hormone to the receptor molecule. This enzyme converts ATP to 3′, 5′-cyclic adenylic acid, or cyclic AMP (c-AMP).

Cyclic AMP serves as what has come to be called the "second messenger" in hormone action, the hormone itself being the first messenger. Beginning with the pioneering work of Nobel laureate E. W. Sutherland in the 1950s, c-AMP has been implicated in more and more processes in which it acts as an intracellular agent for a hormone. The intracellular action of c-AMP is to lead to the release of enzymatic activity within the target cell. To do this enzymes in inactive

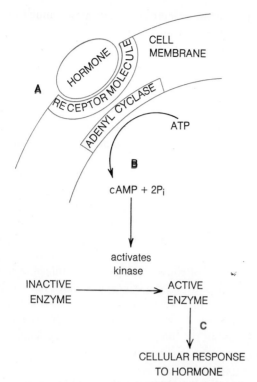

Figure 18-24 **Second messenger action of certain hormones. A.** hormone interaction with cell surface receptor causes c-AMP production within cell. **B.** second messenger, c-AMP, activates kinase which then converts enzyme responsible for cellular response to hormone from inactive to active state. **C.** active enzyme produces cellular response.

form must often be activated; AMP may turn on an activating enzyme (a kinase) that works to activate the remainder of the enzyme system. As a result the target cell becomes active in response to the enzyme. The role of c-AMP is a remarkable discovery and serves to rationalize using one basic mechanism that, until recently, was a bewildering assortment of hormonal processes.

Although the mechanisms are not nearly as well understood as second messenger processes, it also appears that some hormones act at the cell membrane to cause a specific change in permeability of the membrane, allowing entry or exit of some critical molecule.

Some Hormones Act at the Protein Transcription Level

In some ways the most remarkable discovery in endocrinology during recent years has been the finding that steroid hormones, which are able to pass through cell membranes, act at the level of protein synthesis. Hormones of this type are the sex hormones and adrenocortical hormones. These hormones enter the cell and combine with a special binding protein. Once combined, the hormone-protein complex enters the cell nucleus and brings about the formation of new RNA. Immediately, of course, one sees the possibility of a gene derepression mechanism (recall page 137) in which the hormone-protein molecule might combine with repressor protein and free a particular gene for action. At any rate, new RNA appears and soon thereafter new proteins.

Why Two Mechanisms of Hormone Action?

In a general way these two types of hormone action appear to be related to the *rates* of the processes that they control. The transcriptional level mode of action would tend to be slower starting and have longer persistence; it takes time to build proteins, and it takes time for them to wear out. The c-AMP system tends to be faster since it acts on enzyme systems that are fully formed and need only to be unbound; often such systems may become active in a few seconds.

FUNCTIONAL RELATIONSHIPS OF NEURAL AND ENDOCRINE UNITS

Now that we know something of the function of the basic parts of the neural and endocrine systems and of how they interact, it is appropriate to consider two actual examples of neural and endocrine function. We will begin with the simplest of all functional arrangements of nerve cells, the **reflex arc**. Then we will discuss a much more elaborate arrangement involving both the nervous and endocrine systems—the "fright" system controlling the adrenal medulla that secretes epinephrine (adrenaline), a hormone that functions to prepare the body to meet severe stress. Then, with insight from these examples, we can in the next chapters come to grips with just how these control systems supervise the incredibly complex machinery of the human animal.

The Reflex

The reflex arc is the simplest functional arrangement of cells in the nervous system (Figure 18-12). It may consist of as few as two neurons—one neuron conducting a signal directly to another neuron, which may then cause a response by an effector organ. The knee-jerk is a familiar example frequently used in medical examinations as a test of the state of health of the spinal cord. In this test the leg kicks involuntarily when a spot just below the kneecap is tapped lightly. Tapping there stretches a leg muscle very slightly and stimulates sensory receptors associated with it. They inform the nervous system that the muscle has stretched, and they do this by sending nerve impulses into the spinal cord according to the already familiar frequency code: the greater the stretch, the faster the rate of impulse generation. When impulses from muscle sense cells reach the spinal cord they set off activity in other kinds of neurons. Some of these send signals on into the conscious parts of the brain, so that you ultimately feel the tap.

However, at this point, we are more concerned with the reflex part of the knee jerk—that which causes the leg to kick forward before the conscious parts of the brain learns anything about it. As we have said, a message is already running up the spinal cord to the conscious parts of the brain, so that something has to happen fast if the leg is going to kick before anything consciously can be done about it. Remember that nerve impulses may travel as fast as 120 m/sec, giving the reflex only about 30 msec or so to act before the message goes from the leg to the brain and back. This speed is attained because the sense cell that initiated the message concerning stretch to the brain also synapses directly in the spinal cord with the motoneurons of the stretched muscle.

We see that there are two explanations of the speed of the stretch reflex: the impulse bypasses the brain, and it is monosynaptic. The signal must bridge only one gap between two nerves to activate the effector. Other reflex circuits may have a third neuron (an **internuncial neuron**) in the spinal cord, placed between the sensory input nerve and the motor output nerve; such paths are slower. The monosynaptic reflex circuit, potentially results in the most rapid possible response because not only is there no time lost sending impulses up the spinal cord to the brain and back (as must occur in conscious activity) but also because there is only the one synapse intervening between the sensory neuron and the neuron that produces the final action. The fewer synapses there are between neural input and output, the faster the circuit works, mainly because nerve impulses cross synapses much more slowly than they travel along nerves.

Because all of these events happen rapidly (the entire knee-jerk reflex requires about 20 msec from tap to kick), you might wonder just how it is known for sure that the knee-jerk reflex actually does not involve the brain. The simplest proof is that the reflex persists when the spinal cord has been sectioned in an accident, eliminating communication between spinal cord and brain.

Later, when we come to the matter of control of movement (page 413), it will be obvious that reflexes somewhat like this are important to normal operation of the muscles. But you will find, combined with these simple reflexes, others that inhibit the action of antagonistic muscles that might interfere with a desired movement. In addition, there are voluntary controls from the brain and there is even a special control system in the cerebellum, that compares what muscles do with what the brain has ordered them to do and initiates corrective action when necessary.

The Adrenal Medulla

A second more complex type of operating unit involves both the nervous system and the endocrine system. As we have seen the adrenal has an outer cortex and an inner medulla. In embryonic development the cells of the adrenal medulla come from a part of the embryonic nervous system and are thus closely related to nerve cells. But, instead of becoming nerve cells, they have become specialized to synthesize and release into the circulation two similar hormones, **epinepherine** and **norepinephrine,** members of a class of biologically active substances called **catechol amines.** For our purposes we will consider these hormones as one, with the name epinephrine. The relationship of the medullary cells with the nervous system is rendered even closer when we recall that one part of the nervous system, the sympathetic nervous system, releases epinephrine instead of acetylcholine at its synapses. Actually, one way to understand the adrenal medulla is to think of it as a tremendous mass of sympathetic nerves that release transmitter into the general circulation rather than directly onto effector cells in the usual highly specific way of nerves.

Now let us see what the adrenal medulla does. First of all, it helps to regulate the concentration of glucose in the blood. Although there are other mechanisms for blood glucose control (page 382), the adrenal medulla is quite important. Whenever the concentration of blood glucose falls much below 100 milligram percent (mg%, milligrams per 100 ml of blood), the lack of sugar directly causes the release of epinephrine. Epinephrine then promotes release of sugar into the blood from storage sites. This, then, is a purely endocrine function of the medulla.

To see what else the adrenal medulla does we must first say something about the functions of the sympathetic nervous system to which the adrenal medulla is so closely related. The sympathetic nervous system supplies nerves to many important organs—heart, blood vessels, muscles of the stomach and intestines, and the skin. Its actions include speeding up the heart, quieting the stomach and intestinal muscles, and shifting the flow of blood so that more runs through the skeletal muscles than through intestinal regions. Sympathetic responses are thus ideally suited for stressful situations, for switching the physiology of the animal from that of an animal contentedly loafing around digesting its food to that of one suddenly forced to fight for its life, faced with the possibility of massive loss of blood and with the need for every available ounce of strength.

Sympathetic impulses cause the heart rate to increase and cause blood to be shunted preferentially to the muscles that possibly will have to make a massive exertion. The sympathetic system may do all these things by sending neural signals individually to each organ. This is satisfactory in the ordinary course of events, but for the sympathetic system to prepare the organism for immediate bodily peril by releasing epinephrine individually in all the organs concerned is a little like calling up reserves platoon by platoon when the country is faced with a sudden military crisis. What the sympathetic system needs is a mechanism for a general physiological call to arms, and it has it in the adrenal medulla. Stimulation of the adrenal medulla by branches of the sympathetic system causes immediate, massive outpouring of epinephrine, which is carried throughout the body and rapidly assists in putting every responsive cell on a ready-for-action footing.

Complex as this appears to be, we still have not described the means of controlling this mechanism that prepares for trouble. Somewhere in the nervous system there has to be a device, or system of devices, that "knows" what trouble is and when it is time to ring the sympathetic alarm. Some of these devices, such as the pressure receptors that continually measure the blood pressure, are automatic. Any fall in blood pressure, as in excessive bleeding, triggers these receptors to activate the sympathetic system. The alarm sometimes may be triggered by psychic phenomena—the sight of impending danger. Alarms of this sort reach the sympathetic system from the conscious parts of the brain by way of the hypothalamus, that all-important link between neural and endocrine processes. Once again the point is made: the operation of any control system in the human organism is regulated at many levels of organization, from the biochemical to the highest neural levels.

19 Human Maintenance Physiology

The neuroendocrine control system has been displayed in broad outline in the last chapter. Now our interest is to see how it participates in physiological processes necessary to human life. In this chapter, we shall examine the steady-state activities of human physiology—what must be done just to stay alive: metabolism, respiration, circulation, temperature regulation, and excretion—all working to maintain a favorable internal environment. In the next chapter, we shall consider the operation of the human nervous system, made possible by these housekeeping physiological processes.

Feeding the Metabolic Furnace

All organisms obviously require continual input of energy and raw materials to maintain themselves. Therefore, it would seem appropriate to begin considering maintenance processes by asking: What are the requirements of human metabolism? How does the human recognize its metabolic needs and how does it satisfy them? These questions we shall answer insofar as is possible and, in doing so, illustrate the all-pervasive role of the neuroendocrine system.

METABOLIC REQUIREMENTS FOR LIFE

Metabolic requirements are described quantitatively in terms of the **basal metabolic rate (BMR)**. To support the normal metabolism of an average adult male the BMR must amount to around 2300 kilocalories per day (kcal/day). That is, food providing 2300 kcal of available chemical energy must be consumed daily simply to maintain the physiological status quo. This is a little more than the daily energy consumption of a 100 watt light bulb. During hard physical labor the metabolic rate rises to as high as 5000 kcal/day. Besides work load, other factors affect the metabolic rate. Energy requirements are higher in cold weather to compensate for loss of body heat. Age has an effect: children have higher BMRs relative to body weight than adults because, for one thing, they invest so much energy in growth. Females usually have lower BMRs than males. Table 19-1 indicates the energy content of some representative foods and the rate at which food energy is utilized in various activities.

Clearly, there is more to the problem than simply-

Table 19-1 Energy Content of Common Foods and Rate of Energy Utilization in Various Activities

A. Energy in Foods

Food	Measure	Weight (g)	Caloric Value (kcal/g)	Minutes to Utilize[1]
Avocado	1 fruit	280	1.30	120
Bacon	2 slices	15	6.00	30
Beer	12 fl oz	360	0.42	50
Beef, hamburger	3 oz	85	2.18	62
Bread, white	2 slices	15	2.70	14
Cola drink	12 fl oz	360	0.40	48
Candy, fudge	1 piece	28	4.11	38
Carrots, raw	1 carrot	50	0.40	7
Cheese, cheddar	1 slice	28	4.11	38
Chicken, breast	½ breast	94	1.65	52
Grapefruit, raw	½ fruit	241	0.19	15
Squash	1 cup	210	0.14	10
Tomato, raw	1 tomato	200	0.20	13
Ham	3 oz	85	2.88	81
Lettuce	1 head	454	0.13	20
Milk, whole	1 cup	244	0.66	54
Orange juice	1 cup	250	0.48	40
Yogurt, whole milk	1 cup	245	0.61	50

[1] For a 130 lb individual doing moderate work, 180 kcal/hr.
Caloric data: U.S. Dept. Agric., Handbook No. 8 (1963). Washington, D.C.

B. Energy Consumption During Activity

Activity	Kcal/Kg Body Wt/Hr	Kcal per 130-lb (59 kg) Adult/Hr
Sleeping	1.0	60
Sitting	1.1	65
Standing	1.3	74
Walking slowly	3.1	180
Ironing clothes	3.8	223
Gardening, light	4.1	244
Bicycle riding, moderate	4.4	260
Playing tennis	6.1	360
Swimming, slow crawl	7.7	454
Jogging (alternating with walking)	8.6	507
Running, cross country	9.8	578
Running, sprint	20.5[1]	1209[1]

[1] A level of effort sustainable for only a few minutes.

energy. The diet must also be properly balanced. Would you expect 2000 kcal/day of fat to ensure good health? Indeed not. Food intake must include proper representation of all the substances listed in Table 19-2, including water, protein, lipids, carbohydrate, certain minerals, and some 13 vitamins.

REGULATION OF FOOD INTAKE

A look at such a list of essential dietary elements is a little frightening. When food is required, the well-known hunger pangs start things rolling very nicely. But it is not obvious if those pangs can tell us anything about which of the numerous items shown in Table 19-2 are required. Hunger pangs seem only to command us to eat, without indicating *what* is to be eaten. However, it is true that somehow the organism tends to select the proper diet. For example, when a rat's adrenal glands are removed, difficulties with salt and water balance ensue. These may largely be corrected if the rat drinks salt water, which it does. Given the choice between pure and salt water, the adrenalectomized rat chooses salt water.

Something like this seems to work for us. Even small children are said to choose a balanced diet when fed cafeteria style. The mechanism of this remarkable phenomenon is unknown, but it must involve internal sensory mechanisms (recall Figure 18-3) that act on centers of the brain that control eating. It may well be true that, under natural conditions, what tastes good to us is what we need.

Actually, even if this diet-selecting mechanism did not exist, it would be rather difficult for any individual of reasonable income to suffer a serious dietary deficiency in the United States. Even eating more or less randomly from the immense bounty usually set before us, we usually have little to fear except growing fat. Tragically, severe malnutrition and starvation do exist in this wealthy country. Low income plus ignorance of how to formulate an adequate diet on a restricted budget combine in a significant element of the United States population to produce a state of malnutrition, making adults easy victims of disease and undermining their efforts to better their conditions. In children lifelong damage may result, even if in later life their diet is normal.

Social influences must be considered when studying a process with elements of voluntary control. Certainly the food habits of our associates and our own eating habits formed in the past have an enormous effect on food intake. These are not always good influences; much of the overeating that occurs in adult life can be traced to persistence of the gargantuan meals of adolescence without taking into account the lessened caloric requirements of the adult. Also the so-called "grandmother effect," indulgent overfeeding of the very young, appears to produce a lasting tendency to overeat.

Physiological Control of Food Intake

Food intake can readily be shown to be physiologically controlled, although frequently obscured by behavioral effects as just mentioned. First of all, there is a rough-and-ready control exerted by the stomach, whose muscular activity when empty generates hunger pangs that reach the brain by the sensory components of the vagus nerves. This is only part of the story, because cutting the vagus nerves does not eliminate hunger, even though hunger pangs are no longer felt. The remaining mechanism controlling feeding lies in the hypothalamus. There, two clumps of nerve cells (the lateral and medial nuclei) control feeding in a reciprocal way. The lateral nucleus initiates feeding. If it is electrically stimulated, the experimental animal eats at the first opportunity. Destruction of the lateral nucleus causes the experimental animal to starve to death, even in the presence of food. Quite oppositely, stimulation of the medial nucleus terminates feeding. Its destruction causes an abnormally large food intake that can lead to gross obesity.

This is well enough. We have two hypothalamic nuclei that regulate feeding. Now we must seek out the sensory part of the feeding control system. What information about nutritional requirements do these centers receive and how do they receive it? Reflection on the variety of dietary essentials (Table 19-2) indicates that it would be difficult to have sensory devices to transmit data concerning each of these to the hypothalamus. Instead it would be easier to monitor some single factor serving to reflect the general nutritional state.

Considering the overall metabolism of the animal, one compound stands out as universally essential. This is glucose. Appropriately, it is glucose in the blood that controls the hypothalamic feeding centers. As blood glucose diminishes, even though there are mechanisms to restore it from body reserves such as glycogen, the hypothalamic feeding center is triggered, inducing the desire to eat. Thus it is that a piece of candy or slice of bread eaten just before a meal will raise blood glucose and "spoil the appetite," a fact well known to dieters. Glucose control is critical: a drop of approximately 0.1% of the glucose in the blood will, unless corrected immediately, result in dizziness, convulsions, coma, and death. The nature of the receptors monitoring blood glucose is unknown, although they reside in the hypothalamus.

The Sense of Taste and Smell

Whereas the glucose receptor mechanism may certainly initiate feeding, what is eaten depends upon

Table 19-2 Essential Components of Human Diet

	Typical Adult Daily Requirements	*Sources*	*Examples of Action*
Water	2000 ml, roughly 1 ml/kcal food	All food and drink; oxidation of food	Universal solvent in biological systems; temperature regulation
Minerals			
Sodium	Adequate amount in ordinary diet	All foods	Osmotic balance; excitability of nerve and muscle
Potassium	Adequate amount in ordinary diet	All foods	Osmotic balance; excitability of nerve and muscle
Calcium	1200 mg	Milk, cheese	Bone formation; excitability of nerve and muscle.
Phosphorus	1200 mg	Milk, cheese, meat	Bone formation; energy metabolism
Magnesium	400 mg	Soybeans, leafy green vegetables, snails	Activator of enzymes; CNS excitability
Iodine	150 mg	Seafood, iodized salt	Metabolism control as part of thyroid hormone
Iron	18 mg	Meat, raisins, spinach	Oxygen transport and storage as part of hemoglobin and myoglobin; oxidation-reduction reactions as part of cytochrome
Zinc	15 mg	Beef, cheese	Cofactor of many enzymes such as carbonic anhydrase
Trace minerals (partial list)			
Manganese	5 mg (?)	Peanut butter, whole wheat bread	Enzyme cofactor
Copper	Adequate amount in ordinary diet	Beef liver, peanut butter	Enzyme cofactor; red blood cell formation
Cobalt	20 μg (?)	Meat, milk (as vitamin B-12)	Part of vitamin B-12 molecule
Fluorine	Trace	Addition to drinking water, about 1 part per million (ppm)	Tooth structure
Carbohydrates	About half the total caloric intake	Cereal grains, milk, fruit	Energy
Proteins	50–100 g	Animal protein more effective than most plant protein except soybean	Energy, essential amino acids for enzyme and structural protein formation

Based in part on *Recommended Daily Dietary Allowances*, 8th ed., National Academy of Sciences–National Research Council Publ. 2216, 1974, Washington, D.C.

the past experience and habits of the individual. These lead to choice of foods that pass the test of two sets of chemosensory guardians of the digestive system, the senses of taste and smell. Even in the adrenalectomized rat, the taste sense is the means by which the specific dietary requirement for salt is identified.

The sense of smell is usually the first chemosensory arbiter of food quality because it detects very low concentrations of chemicals borne in the air. Smell receptors lie in a patch of mucous membrane inside the nasal cavity (Figure 19-1). In the human, several million olfactory (smell) receptor cells lie in this patch. The olfactory receptor cell is a neuron that extends cilia (hairlike processes) into the overlying layer of mucus and an axon back directly into the brain. In lower animals it has been possible to record nerve impulses from single olfactory cells, and these recordings show that an olfactory cell can respond to a large number of different chemicals and that small

Table 19-2 Components of Human Diet (Continued)

	Typical Adult Daily Requirements	Sources	Examples of Action
Lipids			
Fatty acids (linoleic essential in diet)	About 40 g in the aggregate	Vegetable oils, animal fat, pecans	Richest energy source, 9 kcal/g; essential fatty acids
Fats		Meat, egg yolks	Cell membrane structure
Phospholipids			Precursors of vitamin D, cell membrane structure, bile formation
Sterols			
Vitamins			
Water soluble			
C (ascorbic acid)	45 mg	Citrus fruits, tomato	Prevents scurvy; connective tissue formation, adrenal function
niacin (nicotinic acid, nicotinamide)	20 mg	Meat, brewer's yeast	Prevents pellagra; component of coenzymes
riboflavin	2 mg	Brewer's yeast, liver, cheese	Part of coenzymes acting in cellular oxidations
thiamin	1.5 mg	Brewer's yeast, wheat germ, pecans	Part of many coenzymes
B-6 (pyridoxine, pyridoxal, pyridoxamine)	2 mg	Beef, banana, liver	Coenzymes in carbohydrate and protein metabolism
folacin (folic acid)	400 μg	Oranges, banana, spinach	Coenzymes in protein and nucleic acid metabolism
B-12 (cyanocobalamin, hydroxycobalamin)	3 μg	Beef liver, egg	Coenzymes in nucleic acid metabolism
pantothenic acid	5–10 mg (?)	Beef liver, broccoli	Part of Coenzyme A
biotin	Adequate amount in ordinary diet	Meat, most vegetables	Coenzymes in fatty acid synthesis, tricarboxylic acid cycle
Fat soluble			
A (retinol); provitamin is beta-carotene	2 mg	Whole milk, butter, yellow vegetables (for carotine)	Visual pigment; growth; maintenance of epithelia
D	Adequate amount in adult diet	Egg yolk, liver, milk	Prevents rickets in children, calcium metabolism
E (tocopherol)	15 mg	Leafy green vegetables, whole cereal grains	Antioxidant
K (phytylmenaquinone, menadione)	Adequate amount in adult diet	Cabbage, spinach	Blood clotting

variations among cells in relative sensitivity to various substances make the olfactory system a superb identifier of odors.

Taste receptors make the final analysis of the food as it is taken into the mouth and chewed. Rather than being neurons, like the smell receptors, these are special cells of the epithelium (skin) of the tongue and mouth lining, which are innervated by taste nerves. Several taste cells lie together in a structure called a **taste bud** (Figure 19-1). Each taste cell extends part of its surface into a taste pore, through which substances in the mouth contact the taste cells. After a short life span, each taste cell is replaced by another, which grows from the taste bud epithelium. All the taste cells in a given taste bud respond to the same stimulants, and it is possible to classify taste buds into those that are particularly responsive to sweet, salty, sour, bitter, and one or two other categories. Taste buds responsive to a specific taste category are localized in the same region of the tongue (Figure 19-1).

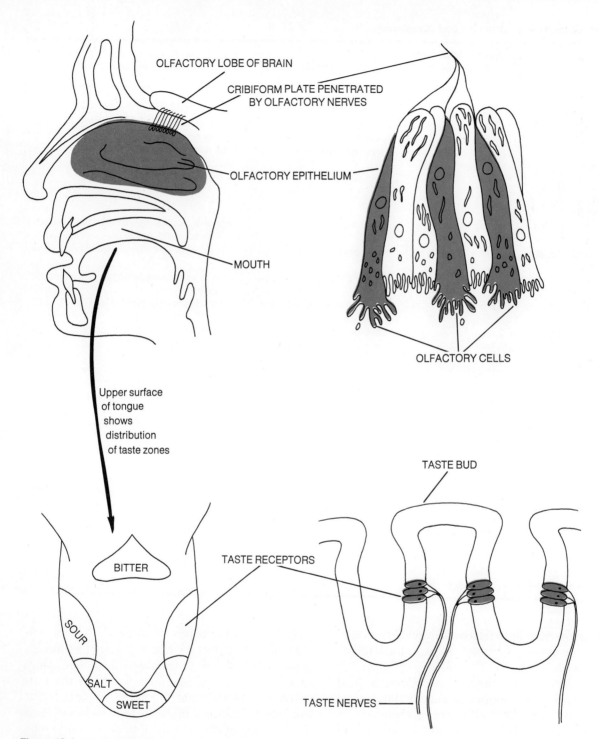

Figure 19-1 Human olfactory and taste receptors.

The Food Processing System

The digestive tract (Figure 19-2) is a long tube with regions modified to

1. perform a sequence of mechanical and chemical processes rendering food absorbable
2. absorb the resulting products into the circulation
3. eliminate wastes

These processes, together with the work of associated organs contributing to digestive tract function, are controlled by neural (voluntary and autonomic) and endocrine mechanisms.

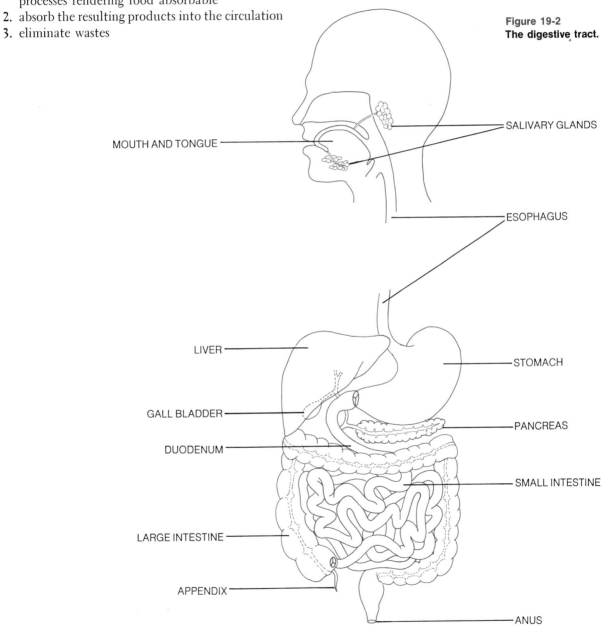

Figure 19-2
The digestive tract.

HUMAN MAINTENANCE PHYSIOLOGY

GENERAL NATURE OF DIGESTION

Chewing and Swallowing

Important steps preliminary to digestion take place in the mouth. The food is checked by taste receptors and is broken up into finer particles by chewing, facilitating access by digestive enzymes. **Salivary glands**, whose ducts open into the mouth, bathe the food in saliva, lubricating it to ease transport and contributing an enzyme, salivary amylase, that initiates starch breakdown to sugar (Figure 19-3). Salivation is controlled reflexly by sympathetic nerves. Food in the

Figure 19-3 Starch digestion.

STARCH DIGESTION

Two forms of starch:

α-AMYLOSE

250-300 1,4-α linked glucose molecules

AMYLOPECTIN

chains of about 25 glucose units with 1,4-α linkage, linked to each other by 1,6-α linkages.

Digestive process:

(1) α-amylase (salivary and pancreatic) acts on 1,4α linkages between glucose molecules.

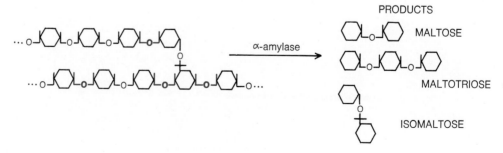

PRODUCTS
MALTOSE
MALTOTRIOSE
ISOMALTOSE

(2) Disaccharides of small intestine complete breakdown to glucose.
(3) Glucose is transported across intestinal wall.

mouth or even the sight, smell, or thought of food may trigger saliva production.

Swallowing starts voluntarily, but once food enters the esophagus, **peristalsis** automatically carries it to the stomach. In peristalsis a moving ring of muscular contraction forms behind the food mass and propels it along in a "milking" movement. Because the opening into the lungs, the **larynx**, also lies in the rear of the mouth, it is reflexly closed and breathing momentarily stops while swallowing takes place, insuring that the food reaches the right destination.

Action of the Stomach

The stomach is a large muscular bag (Figure 19-4), in which food is subjected to powerful churning action, further subdividing and exposing it to diges-

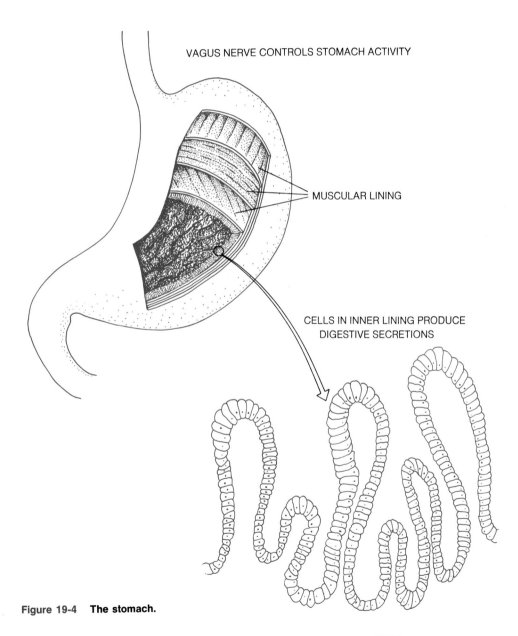

Figure 19-4 The stomach.

tive secretions produced by the stomach. These are produced in gastric glands (Figure 19-4) and include mucus, hydrochloric acid, and **pepsin** (gastric protease), an enzyme that begins the breakdown of proteins to amino acids by hydrolysing proteins into polypeptides.

Hydrochloric acid and pepsin secretion illustrate the ingenuity of our physiological machinery. Consider what the digestive system has to do: tear down foodstuffs into digestible subunits. The problem in doing this is that foodstuffs are composed of the same types of materials as the machinery doing the digesting. To proteolytic enzymes such as pepsin, protein is protein, whether in a steak or in your stomach lining. This creates a problem somewhat like trying to keep a fire localized in a wooden fireplace. The solution that the body employs is to synthesize such potentially destructive enzymes in an inactive form. Although pepsin is inside the cells of the gastric pits, it is in an inactive form called **pepsinogen**. Released into the stomach cavity, it is converted into active pepsin by hydrochloric acid, which the stomach also secretes. Mucus also helps prevent self-digestion by coating the stomach wall, protecting it from both pepsin and hydrochloric acid. Neural and hormonal controls insure that these chemicals are secreted only when food is present. Ulcers seem to be caused by some failure of these protective mechanisms allowing the stomach to digest itself.

Enzymes typically have pH optima, that is, particular hydrogen ion concentrations at which they work most effectively. Since pH varies markedly from region to region in the digestive tract, this serves also to turn on and off the activity of enzymes that have been released into the cavity of the gut. Because of its hydrochloric acid, the stomach contents have a very low pH, that is, they are extremely acidic. This condition stops the action of salivary amylase, and further carbohydrate digestion does not occur until the small intestine is reached. However, pepsin has a very low pH optimum and works very well in the stomach. The region of the gut beyond the stomach is alkaline (high pH), so the action of pepsin is largely confined to the stomach. Stomach hydrochloric acid also kills many kinds of bacteria in food.

The Small Intestine: Serving Both Digestion and Absorption

After a period in the stomach food is carried by peristalsis into the small intestine. This region of the digestive tube is well adapted to absorb products of digestion because of its very large surface area. The small intestine of an adult human is about 20 ft long. Even if the intestine were only a smooth tube, this would be a large surface, but the absorptive surface is further greatly enlarged by as much as 100 times the total *external* body surface by folds in the absorptive surface and by the presence of **microvilli** (Figure 19-5) on the cells lining the absorptive surface. Absorption is further assisted by an extensive bed of capillaries just below the absorptive surface and by lymph, or lacteal, vessels (see page 406), which are important in the absorption of fat.

Numerous enzymes and other factors aiding their action enter the cavity of the small intestine. They are produced by the small intestine itself, by the liver, and by the pancreas. These complete the work of digestion. Carbohydrates are hydrolyzed to monosaccharides, which are then absorbed (Figure 19-3). Proteins are broken down into peptides and amino acids, and these are absorbed by the small intestine (Figure 19-6). Here also fats are converted to fatty acids and glycerol. Both fat and its breakdown products are absorbable by the small intestine.

Absorption from the Small Intestine. Movement of the end products of digestion across the intestinal wall may only be a matter of simple diffusion, but commonly the process of **facilitated diffusion** is involved. For example, among the sugars, glucose and fructose move across the intestinal wall up to five times as fast as other sugars by combining reversibly with carrier molecules in the intestinal wall.

Of the vitamins, water-soluble ones diffuse freely through the intestinal wall. Fat-soluble vitamins (A, D, E, K) are absorbed poorly, unless there is sufficent fat in the diet to dissolve them. Vitamin B_{12} has a curious transport mechanism. A glycoprotein known as the intrinsic factor, secreted by the stomach, combines with the vitamin. The protein-B_{12} complex then is taken up, probably by pinocytosis. Persons with stomachs surgically removed get along reasonably well except that they can no longer get B_{12} from the diet. Pernicious anemia, in which red blood cells are greatly reduced in number, is the result, unless treated by B_{12} injections.

For a while after birth, an infant can absorb undigested protein, probably by pinocytosis. This permits passive immunization against infection by uptake of

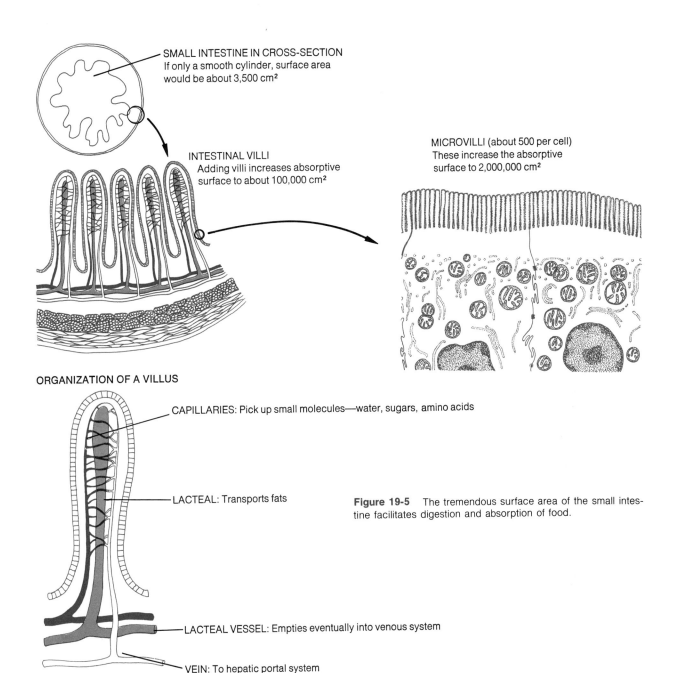

Figure 19-5 The tremendous surface area of the small intestine facilitates digestion and absorption of food.

protective antibodies (protein) present in mother's milk.

The Large Intestine

All digestion is essentially complete once the intestinal contents enter the approximately 4 ft of large intestine. However, the large intestine serves importantly to remove water from the intestinal contents. There is also excretion into the large intestine of ions such as calcium, magnesium, and iron. Finally, the large and generally harmless bacterial population of the large intestine acts on the intestinal contents. The products of intestinal bacterial action are ordinarily of no useful significance to man, although some vitamin B of bacterial origin does seem to be absorbed from the large intestine. Certainly, experimentally

PROTEIN DIGESTION

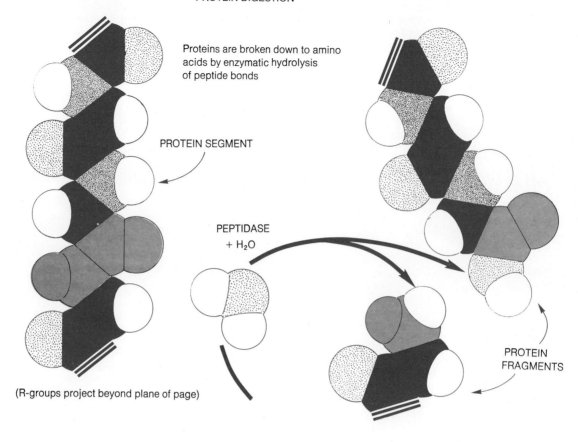

The digestive process in stomach and small intestine involves a series of peptidases with differing specificities:

(1) IN STOMACH, pepsin cleaves interior peptide bonds to break the protein down to peptides.

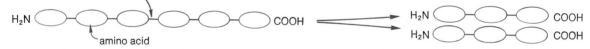

(2) IN SMALL INTESTINE, a family of enzymes secreted by the pancreas acts on specific peptide bonds, breaking peptides down to their constitutive amino acids or very small peptides.

trypsin, chymotrypsin ——————— split interior peptide bonds
carboxypeptidase-A ——————— acts on carboxyl terminal peptide bonds

amino peptidase ——————— acts on N-terminal peptide bonds

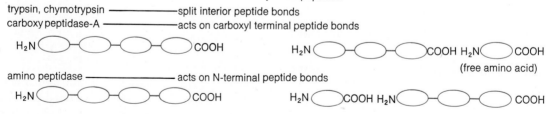

(3) Free amino acids and small peptides (2-6 amino acids) are taken up by cells lining small intestine
    di- and tri-peptidases—break down absorbed peptides to amino acids

Figure 19-6 Protein digestion requires the action of a family of peptidases acting on peptide bonds in specific locations in the protein molecule.

produced germ-free animals seem to suffer no harm by the absence of such bacteria. Every few hours the muscular action of the intestine forces its contents into its terminal segment, the **colon.** Here voluntary control resumes, and the colon is emptied by **defecation.**

COORDINATION OF DIGESTIVE PROCESSES

Even this hasty view of intestinal affairs verifies the complexity of digestion. With enzymes turning on and off, muscles churning, and glands secreting, digestion assumes the complexity of an assembly line running backward. Therefore, one suspects that extensive control systems are essential to smooth operation.

Actually, except for eating and defecation, the entire system operates without conscious control. Two control elements effect regulation: the **digestive nerve plexus** and the **digestive hormone system.**

The Digestive Nerve Plexus

Food moves through the digestive tube by the action of several layers of muscles operated by a semi-independent nervous system, a network of nerve cells called the digestive nerve plexus.

Although the plexus sees to the detail of food movement, the general level, or intensity, of its activity is influenced by sympathetic and parasympathetic nerves. These same autonomic nerves operate the **sphincters,** muscular valves between stomach and small intestine, with the anal sphincters guarding the terminus of the digestive tract, under both autonomic and voluntary control. Eating causes autonomic system activation of salivary gland secretions and, by way of the vagus nerves, release of bile from the gall bladder.

Digestive Hormone System

Digestive tract nerves have little to do with insuring timely secretion of digestive enzymes and related factors. These secretions are triggered by the food itself and by hormones secreted by the digestive system for the express purpose of initiating secretory activity (Table 19-3).

Entry of food into the stomach triggers release of the polypeptide hormone **gastrin.** Gastrin has two functions: it causes stomach glands to release hydro-

Table 19-3 The Principal Digestive Hormones

Source	Name	Stimulus to Secretion	Action
Stomach	Gastrin	(1) Food: chemical and mechanical stimulation of stomach wall (2) Psychic stimuli via vagus nerve	(1) Causes secretion of pepsin and HCl by stomach (2) Stimulates bile release (3) Stimulates pancreatic enzyme and alkali secretion
Small intestine (all from mucosa from upper part of small intestine)	Secretin	(1) Acid, peptides, fats (2) Vagus nerve	(1) Stimulates pancreatic secretion with low enzyme content, high alkalinity (2) Inhibits stomach activity
	CCK-PZ[1]	(1) Partly digested protein, fats (2) Vagus nerve	(1) Stimulates pancreatic secretion of trypsin, chymotrypsin, amylase, lipase (2) Causes gall bladder to empty bile into intestine (3) Inhibits stomach
	Enterogastrone	Fats	Inhibits stomach motility and secretion
	Vilikinin	Food in small intestine	Increases movement of small intestinal villi
	Duocrinin	Food in small intestine	Stimulates small intestinal secretions (from Brunner's glands)—a viscous mucus
	Enterocrinin	Food in small intestine	Controls production of small intestinal secretions from crypts of Lieberkühn—a clear, alkaline fluid

[1] See page 380.

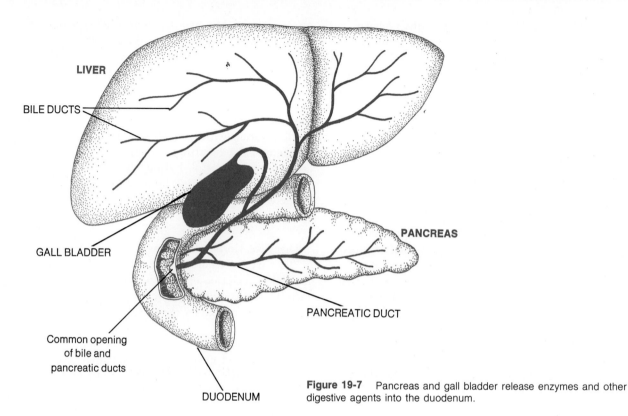

Figure 19-7 Pancreas and gall bladder release enzymes and other digestive agents into the duodenum.

chloric acid and pepsinogen; and it is carried through the blood stream to the pancreas, causing part of that gland to release digestive enzyme containing **pancreatic juice** (Figure 19-7). Pancreatic juice flows into the small intestine through the pancreatic duct in time to act on the food that originally triggered gastrin release. Later, as food enters the small intestine, it causes release of two hormones from the intestinal lining: **secretin** and **cholecystokinin-pancreozymin** (CCK-PZ). These have effects similar to gastrin on the pancreas and also cause release of bile salts from the liver into the gall bladder. CCK-PZ was until recently thought to be two hormones, hence the hyphenated name. There may be as many as 11 digestive hormones secreted by the lining of the small intestine.

Secretin is of historical interest as the first hormone to be discovered. In 1902 William Bayliss and E. H. Starling demonstrated its presence by causing secretion of pancreatic juice by injecting an extract of the wall of the upper intestine into the blood stream.

What Happens to the Products of Digestion

Although the past few pages describe a well-integrated system of digestive processes, we have still to consider another obstacle to the maintenance of the constancy of the internal environment. This is how the organism handles the flood of nutrient molecules provided by digestion. Serious problems might

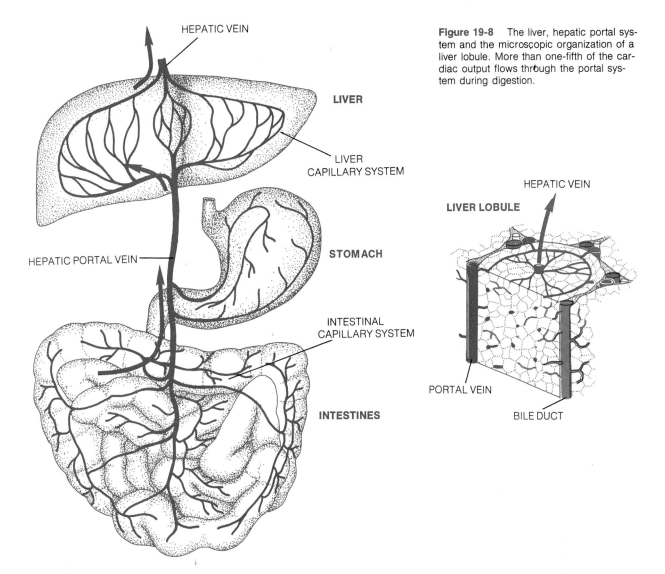

Figure 19-8 The liver, hepatic portal system and the microscopic organization of a liver lobule. More than one-fifth of the cardiac output flows through the portal system during digestion.

arise if this bounty were released directly into the general circulation.

THE WORK OF THE LIVER

Direct release does not occur because blood leaving the intestine is restricted to a portal system rather than entering the general circulation. This **hepatic portal system** channels all nutrients to the capillaries of a large organ, the liver (Figure 19-8). The liver may be considered a buffer zone between digestive system chemical input and the rest of the organism. It is of profound importance in the metabolism of the three major foodstuffs—carbohydrate, protein, and fat—in addition to being the site of many other important metabolic processes.

Organization of the Liver

Despite the numerous and complex functions of the liver, its cells are remarkably alike. These are organized into a structural unit, the **lobule** (Figure 19-8), in which the cells are surrounded by large **sinu-**

soids—irregular blood-filled spaces formed from capillaries. Blood enters the sinusoids from two sources, the hepatic artery (carrying oxygen-rich blood from the heart) and the hepatic portal vein (carrying food-laden blood from the digestive tract). After bathing the cells of the lobule, blood from these two sources enters the hepatic vein for eventual circulation to the rest of the body. Exposure to hepatic portal blood gives liver cells the opportunity to take up and modify the products of digestion.

However, not all metabolic products of the liver leave by the hepatic vein; some enter thin-walled collecting tubules, which join to form the **bile duct.** The collected fluid is **bile,** which the bile duct carries to the **gall bladder,** a storage organ from which the bile enters the small intestine. Bile is a mixture of waste products and of chemicals important to digestion. Bile wastes come from the blood oxygen-carrying pigment, hemoglobin, released from worn-out red blood cells. Bile salts are the constituents of bile that function in digestion. They make fats more readily digestible either by combining with them chemically to render them soluble in water or by **emulsifying** them. Without bile salts, nearly a quarter of the fats eaten would not be absorbed.

Like most organs of the digestive system, the liver receives nerves from both sympathetic and parasympathetic elements of the autonomic nervous system. These control the flow of blood through the organ, increasing flow during digestion and decreasing it at other times, by controlling the diameter of the blood vessels supplying the liver.

REGULATION OF THE GLUCOSE SUPPLY

The principal carbohydrates of the human diet are starch and two sugars, sucrose and lactose. Upon digestion these reach the liver via the portal vein as a mixture of three monosaccharides—glucose, fructose, and galactose. In the liver these last two are mostly converted to glucose, which is the key source of energy for metabolism. Because glucose is so important, one might expect a substantial amount of it in the blood to ensure a continual supply of energy, particularly during the periods of fasting between meals when the intestines and liver are not pouring sugar into the blood stream. Actually, there are only about 20 g of glucose in the body fluids between meals, and this is only enough to supply one's energy requirements for a little over 1 hr. That is, without resupply, the blood glucose level would begin to fall. This is a dangerous matter, mainly because the brain, which must be continually functional to sustain the minute-to-minute requirements of the organism, has no significant glucose reserve. Completely dependent upon blood glucose, its function becomes grossly disturbed as soon as blood glucose levels fall markedly. Quite obviously, the whole story of glucose cannot be before us. Somewhere in the body there must be a storage-and-release mechanism, smoothing out what would otherwise be feast or famine, excess glucose immediately after meals and not enough a few hours later.

Glucose Control

As one might expect, if only from the complexity of human physiological processes already evident in these pages, the control of glucose involves considerably more than just the liver. To understand its control we must also take into account virtually all tissues, but particularly fat, muscles, and three endocrine organs—the pancreas (islet of Langerhans tissue), adrenal medulla, and anterior pituitary. Basically, what all these do is to help control glucose levels by promoting glycogen and fat formation (Figure 19-9) from excess glucose and converting glycogen to glucose during times of low glucose levels in the blood. This is the storage and release mechanism that seemed necessary.

The Glycogen Bank

Liver and muscle are the two principal places to store glycogen (Figure 19-10). Although muscle can certainly assist in glucose control by making glycogen, it is unable to use the glycogen that it produces for any but its own purposes. The liver, in contrast, can make glycogen for its own use as well as supply glucose to the blood (and hence to the entire body). There are about 100 g of glycogen in the liver of an adult, enough to maintain normal blood glucose levels for about 6 hr. Muscles contain about 200 g. Besides this substantial reserve, the liver can make new glucose from two principal substrates. One is lactic acid, resulting from incomplete oxidation of glucose by muscle. This happens when the muscle is working so hard that it cannot obtain enough oxygen for complete

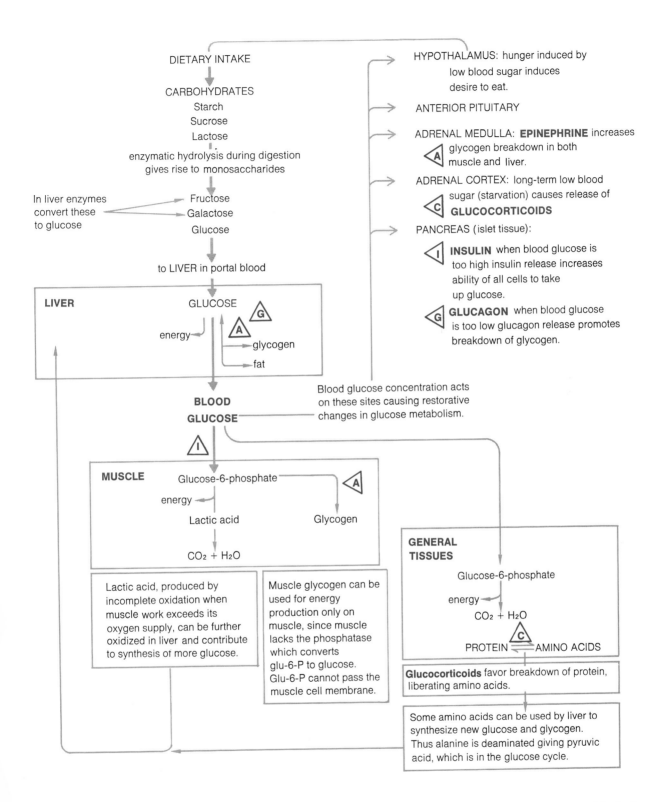

Figure 19-9 Glucose control involves a complex web of hormone mediated actions on virtually all tissues of the body.

HUMAN MAINTENANCE PHYSIOLOGY

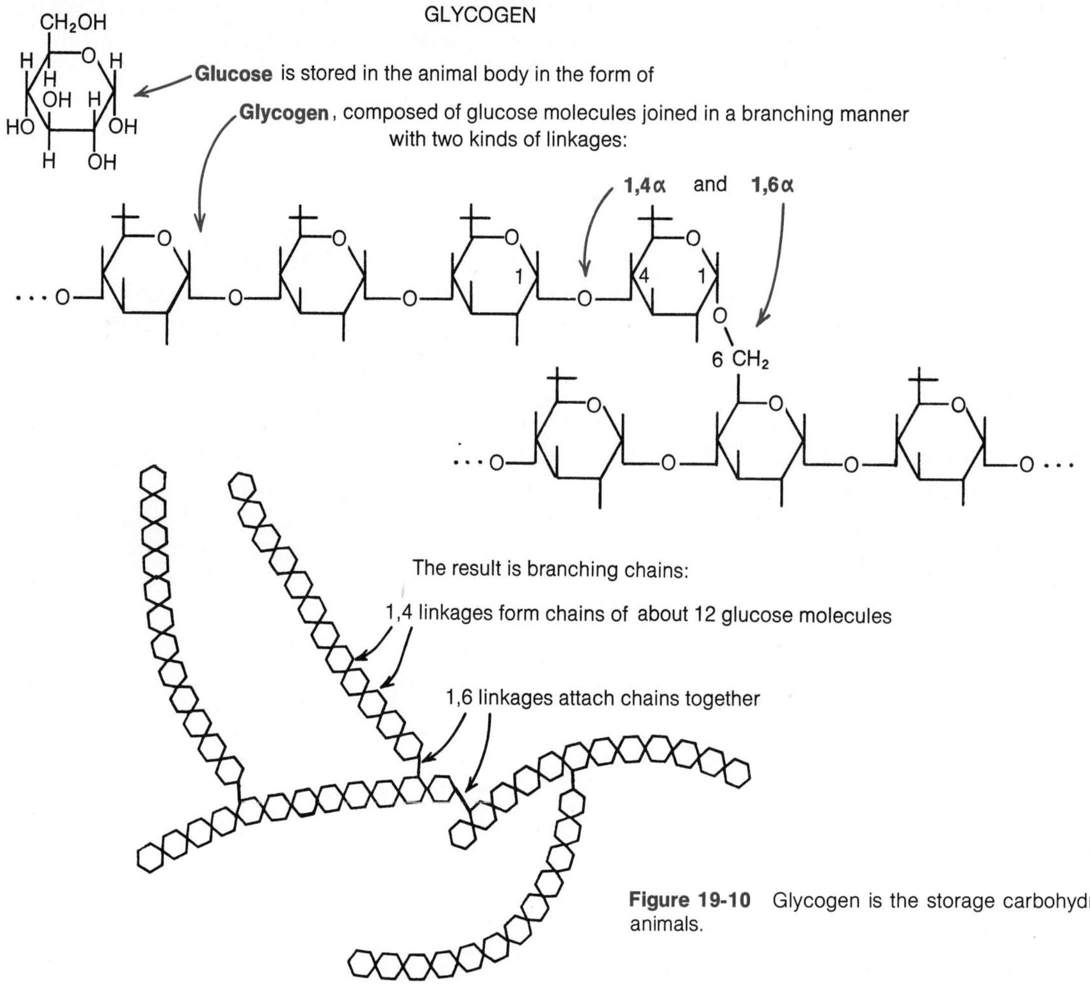

GLYCOGEN

Glucose is stored in the animal body in the form of **Glycogen**, composed of glucose molecules joined in a branching manner with two kinds of linkages: **1,4α** and **1,6α**

The result is branching chains:
1,4 linkages form chains of about 12 glucose molecules
1,6 linkages attach chains together

Figure 19-10 Glycogen is the storage carbohydrate of animals.

respiration (refer to page 72). The other principal substrates are amino acids, derived from the breakdown of protein. This is an emergency procedure. It is never carried very far, except in starvation, as it involves wastefully tearing down the fabric of the body to supply energy.

Because the reactions interconverting glucose and glycogen are reversible, the formation and breakdown of glycogen constitutes an automatic regulatory mechanism for glucose at the cellular level (Figure 19-9). Endocrine, and even indirect neural, control are superimposed upon this fundamental mechanism.

Controlling the Glycogen Reserve

The glycogen reserve control system has two major elements: one, involving the pancreas, adrenal cortex, and anterior pituitary, is concerned with relatively stable conditions; the other, involving the adrenal medulla, is active in emergencies.

Pancreatic islet cells make two hormones with opposite effects on blood glucose. **Insulin**, a protein, reduces blood glucose concentration; **glucagon**, a large polypeptide, raises it. Glucagon promotes glycogen breakdown to glucose in the liver by activating liver phosphorylase by a c-AMP second-messenger mechanism (see page 139). Insulin combines with a receptor protein on cell surfaces. The result is increased uptake of glucose and activation of the enzyme glycogen synthetase, which promotes conversion of glucose to glycogen, thus lowering blood levels of glucose. In fat cells, insulin promotes conversion of glycogen to fat by increasing permeability of the cell to glucose and inhibiting the enzyme responsible for fat breakdown, leaving active the enzymatic mechanisms for converting glucose to fat. Thus the two hormones of the pancreatic islets, insulin and glucagon, have effectively opposite actions: glucagon directly influences the glycolytic enzyme machinery of the liver cell to make glucose and insulin promotes the removal of glucose from the blood (Figure 19-9).

Release of these hormones is directly controlled by the concentration of glucose in the blood. Adrenal cortex and anterior pituitary become involved, if, as during starvation, blood glucose levels tend to remain low for long periods of time. The anterior pituitary then increases secretion of **adrenocorticotrophin (ACTH)**, causing the adrenal cortex to release steroid hormones known as **glucocorticoids**. These promote glucose formation from protein by processes that include increased breakdown of protein to amino acids and more effective uptake of amino acids, as well as increased glucose synthesis from amino acids by the liver. Besides ACTH, the anterior pituitary appears also to increase blood glucose levels by releasing another of its protein hormones, **STH (somatotrophin)** the growth hormone, which promotes liver glucose synthesis and reduces glucose uptake by some tissues.

Epinephrine, secreted by the adrenal medulla, effectively increases the amount of available glucose by increasing the rate of glycogen breakdown in both liver and muscle (remember, glucagon does this only in the liver). Ordinarily, epinephrine secretion is controlled directly by blood glucose, just as are insulin and glucagon. But during stress, massive epinephrine release is caused by psychic stimulation (fright, excitement), acting by way of the hypothalamus and vagus nerves. By this mechanism blood glucose concentrations may rise by as much as 20 mg per 100 cm^3 of blood per minute. This may occur in advance of anticipated violent use of muscles, a valuable preparation for stress.

Waste Heat from the Metabolic Furnace

The heat produced in metabolism is put to no practical use in most animals. In contrast, warm-blooded animals, mammals and birds primarily, conserve this heat, using it to maintain a constant body temperature. Because the chemistry of the body is markedly temperature-sensitive, this is as important to attaining homeostasis as the regulation of virtually any other aspect of the internal environment. Generally speaking, enzyme reaction rates are doubled for each 10° C increase in temperature. But variation in body temperature would do more than slow or speed the overall rate of living. Temperature variation also disrupts the efficient intermeshing of chemical reaction in cells, especially when the temperature sensitivities of enzymes are not all identical.

MOST ANIMALS DO NOT HAVE CENTRAL HEATING

In spite of the obvious influence of temperature on the operation of the chemistry of the body, most animals cannot control body temperature. There are few kinds of warm-blooded, or **endothermic**, animals, but these few are, partially because of their warm blood, the dominant animals in the world today. Most animals, instead, are **ectothermic**, which means that they have no effective way to conserve their metabolic heat in cold weather or to accelerate getting rid of it in hot weather. Such animals remain at temperatures very close to the average environmental temperature, unless they have behavioral traits, such as basking in the sun or seeking shade, that give them some measure of temperature regulation.

The disadvantages of cold-bloodedness are obvious. Besides the difficulties of operating the chemical systems of the body in the face of temperature variation, the temperature extremes that the nonregulating animal can tolerate are more limited than those of a temperature regulator. Brief exposure to hot desert sunlight may be lethal to an ectothermic lizard, whereas a dog, under similar conditions, might survive indefinitely by increasing its heat loss in ways that we shall consider shortly. Similarly, nonregulators become torpid at low temperatures that hardly inconvenience birds or mammals.

THE TEMPERATURE-REGULATING MECHANISM

We regulate body temperature analogously to the way one regulates the temperature of a building—by varying the amount of heat produced by the furnace and by varying the rate at which heat is allowed to leave the building.

The furnace in the example is obviously the total body metabolism, particularly that of the muscles. Our metabolic furnace produces about 70 kcal/hr at rest at a reasonable environmental temperature. The furance works harder and harder as the temperature falls, until at an environmental temperature of 0°C it is giving off approximately 250 kcal/hr.

How is this variation with external temperature brought about? Neuroendocrine controls are important. During brief exposure to cold, epinephrine released from the adrenal medulla increases the metabolic rate. If exposure to cold lasts a long time, thyroid hormone secretion is increased, producing a long-term elevation in metabolic rate. Further, the nervous system may become directly involved, causing involuntary **shivering**, a fine twitching of the muscles that generates heat. Also, voluntary muscular activity (stamping the feet, running) increases heat production.

The second aspect of temperature control, the regulation of heat loss, involves the surface of the body. Heat may be lost across the body surface by radiation, by contact with colder objects, or by evaporation of water. Our skin is constructed in ways that permit control of heat loss.

Skin (Figure 19-11) is formed in two layers. The **epidermis** is the protective, waterproofed outer layer. Its outermost cells are hardened (cornified) and dead or dying. As they wear away, they are replaced from below by new cells. The **dermis** lies beneath the epidermis. It contains a large amount of tough connective tissue, giving the skin pliable strength. Beneath

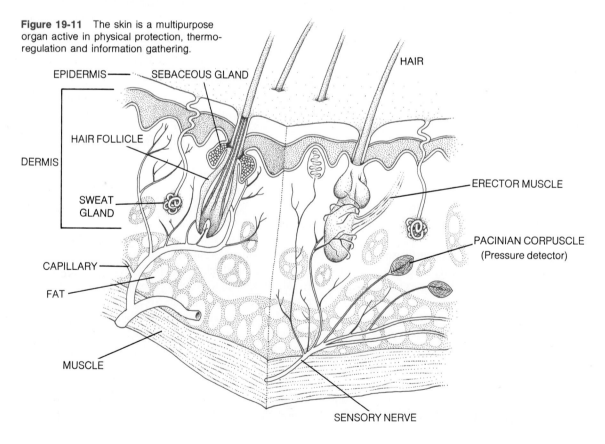

Figure 19-11 The skin is a multipurpose organ active in physical protection, thermoregulation and information gathering.

the dermis there are often insulating deposits of fat that assist in heat conservation.

Three structures found in skin influence heat loss. Hair impedes radiative heat loss by forming an insulating layer of still air just above the skin surface in animals that have a significant coat. The effectiveness of this insulating air coat can be varied by raising or lowering the air-entrapping hairs by means of tiny muscles attached to the base of each hair. Birds accomplish the same thing by ruffling their feathers (Figure 19-12). We retain a vestige of this mechanism. "Goose bumps" are dimplings of the skin caused by muscles attempting to erect our few remaining hairs in ineffectual response to cold.

Blood vessels lying in the skin are the second element of its heat-controlling equipment. Expanded to their greatest diameter, they carry much more blood near to the surface than if they are contracted. Expanded, they exert a cooling influence because the large quantity of warm blood that flows through the skin loses heat rapidly by radiation. When these vessels are constricted, the effect is to prevent heat loss.

Skin blood vessels of endothermic animals that live in exceedingly cold situations have been remarkably modified to control heat loss. In an animal such as a seal, living in cold, arctic water, there would be serious heat loss through the limbs. However, we find that arteries carrying warm blood from the body to the limb come into very close contact with veins carrying chilled blood from limb to body. What happens? Heat is short-circuited directly from outgoing arteries to the incoming veins, and thus is carried back into the body instead of being lost. Of course this means that the limb is cooler than the body, but that is no serious matter. Mechanisms of this sort are called **countercurrent exchange systems.** Even some large fish are able to maintain a high body core temperature by countercurrent heat exchange. Other important physiological systems employ the same principles to achieve local high concentrations of chemicals.

The third element of skin that serves temperature regulation is its complement of glands. Sebaceous glands secrete a fatty material onto the base of hairs and play no role in temperature regulation. However, sweat glands are very effective cooling devices. When our sweat glands are operating maximally, they secrete as much as 10 to 15 liters of sweat per day—roughly half the water in the body! This, of course, is most cooling because thermal energy is used to evaporate

Figure 19-12 Behavioral thermoregulation. An English sparrow ruffles its feathers to increase the insulative dead air space next to the skin. (National Audubon Society.)

all that water from the skin. At the usual temperature of the body surface, about 580 kcal are used up to evaporate a liter of sweat. Sweating, particularly in dry air, is thus an extremely effective cooling mechanism.

Humans have sweat glands distributed everywhere in the skin. In some animals, such as dogs and cats, they are mostly absent, but by panting these animals are also able to take advantage of the cooling to be derived from evaporating water. Panting, rapid shallow breathing, accelerates the evaporation of water from the mouth and lungs.

We Make Our Own Climate

Our lack of body hair is an evolutionary puzzle, especially when one notes the handsome pelts worn by the inmates of the monkey house of any zoo. Some students of human ancestry believe that human nakedness results from an early developed habit of wearing clothing made of the skins of animals. No longer needed, the self-grown finery is assumed to have been lost, along with lice and other vermin that cause disease, or at least disturb the long hours of meditation essential to an animal just learning to think.

Whether cause or result, hairlessness produces no thermoregulatory problem because all varieties of men, except the most primitive, have developed clothing excellently adapted to the environment. Until quite recently, the finest protective clothing a

Figure 19-13 Clever utilization of clothing allows human survival in otherwise unendurable climates; **A.**, polar regions, **B.** Sahara Desert. (University Museum, Philadelphia and United Nations.)

polar explorer could wear was made by Eskimos. In the bitterest cold, the skin clothing of the Eskimo maintains the body surface at semitropical temperatures. At the other thermal extreme, the clothing of natives of the North African desert is ideal, consisting of a complete covering of loose-fitting cloth that protects from the searing rays of the sun without greatly impeding the circulation of air (Figure 19-13A, B).

Wearing clothing is typical of the behavioral traits that greatly lighten the load on our physiological temperature-control system, enabling us to exist almost anywhere on earth.

The Body Thermostat

Since both endocrine and neural mechanisms are concerned with body heat production and its conservation or dissipation, it is logical to look again to the hypothalamus, the link between nervous and endocrine systems, to find the temperature-controlling system, the human thermostat. The hypothalamus contains two temperature controlling centers, one that initiates warming and another that initiates cooling. Each center seems to have temperature-sensitive neurons that measure the temperature of the blood coursing through that area of the brain. They are sensitive enough to trigger temperature-correcting responses upon changes in blood temperature of only $0.01°C$.

The two oppositely acting centers exert control through the sympathetic and parasympathetic nervous systems and through the endocrines. These control pathways are shown in Figure 19-14, which em-

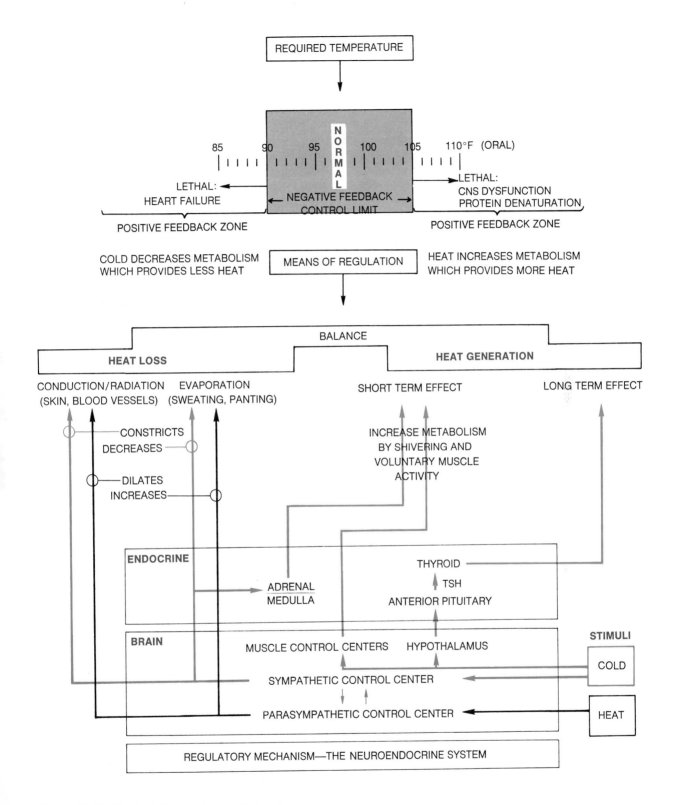

Figure 19-14 **The body temperature control system.**

phasizes the balancing nature of control action on the alternatives between heat generation and conservation. Thus, the sympathetic system, activated by the warming center, reduces heat dissipation by reducing blood flow through the skin and decreasing the production of sweat. The warming center also increases heat production through adrenaline release and by activating the thyroid gland during long cold exposure. Essentially opposite effects are produced by the cooling center, acting largely through the parasympathetic nervous system.

Shivering and voluntary muscular activity, which strongly increase heat production, are controlled by other parts of the brain that respond in large extent to signals generated by temperature receptors in the skin rather than in the brain.

This thermostat, with two sets of temperature receptors and with regulatory control, over sympathetic and parasympathetic innervation, adrenal and thyroid glands, and muscle, is able to regulate body temperature accurately as long as the temperature is not displaced beyond the limits of about 90° to 106°F (oral thermometer) by disease or some other agency. Below 90°F body heat generation is insufficient and man becomes ectothermic until he is warmed to within the regulatory range by applied heat. Low body temperature is not necessarily damaging, but, if the temperature is allowed to fall as low as 70°F, death results from heart failure. In the higher ranges of temperature, above 106°F, control fails because of the positive feedback effect of temperature on metabolism and because of disorders of the nervous system. Essential proteins may be denatured and death results. Simple procedures, such as cold water or alcohol sponge baths, can effectively keep body temperature out of this critical range during fever-inducing illness.

Fever

Substances produced by many disease organisms cause the hypothalamic thermostat to be reset to higher than normal values. The result is a fever, which is maintained by heat-retaining activities such as reducing the circulation to the skin, shivering, and voluntary activities. The agents that induce fever include living or dead bacteria and viruses. They act by causing release of **pyrogen,** a protein, from phagocytic blood cells. Drugs such as aspirin are able to block these fever inducing effects.

Whether fever is of value in fighting disease is not certain, although it is easy to see how it might be. It is also difficult to see why the phenomenon would be so widespread in cold blooded as well as warm blooded vertebrates if it were not useful. Experiments with nonhuman vertebrates generally show that fever does reduce the lethality of disease as long as the temperature does not exceed some limit characteristic of the experimental subject. Thus, the time honored method of wrapping up and going to bed to "sweat it out" may eventually receive the respect of scientific verification, and the practice of using fever combatting drugs (**antipyretics**) such as aspirin may be ill advised, except to control serious rises in temperature.

The Ventilating System

Although by now we have had a quick look at how the fuels of metabolism are handled, two other molecules critical to metabolism have not yet been discussed. These are carbon dioxide and oxygen. In this section we shall examine the means by which they are exchanged with the environment (**ventilation**), and, in the next we shall take up the circulation, the means by which these two gases and most of the other essentials of life are transported within the internal environment of the organism.

The fact that discussion of ventilation has been deferred until now does not belie its significance. Although it is obviously foolish to try to decide which organ system is the **most** essential to life, it is true that failure of the ventilative-circulatory apparatus does cause death very quickly. A person may survive, for weeks without food or days without water, but will die within minutes if the oxygen supply is interrupted. Because the supply of this vital molecule is so critical, the following pages will show that the ventilative-circulatory system is under direct control of the nervous system and is provided with elaborate safeguards to ensure continual optimal function.

MEETING PLACE OF BLOOD AND AIR: THE VENTILATIVE TRACT

As blood courses through the body, it supplies oxygen to and receives carbon dioxide from the energy-consuming tissues. The dilemma that faces the vertebrate is how to aerate stale blood, to rid it of carbon dioxide and replenish its oxygen. Earlier chapters of this text described the general nature of the dilemma: if the general body surface is made sufficiently permeable to allow aeration across its surface, then an unacceptable amount of body water is lost across the same surface. The problem is even worse for an animal as large as a human because, even if the entire body surface were permeable enough for gas exchange between blood and air, the surface-to-volume ratio would be so unfavorable that sufficient aeration to sustain life could not be maintained.

Lungs Solve the Problem

To increase the aerating surface of a larger terrestrial organism and still limit water loss, an invagination of the body surface is required, formed so as to result in a very large surface of contact between air and blood. This requirement is elegantly met by the lungs (Figure 19-15). Although the total volume occupied by the lungs is not terribly large, their gas-exchange surface is immense, because nearly all of the space within the lung is given over to thousands of tiny chambers, **alveoli**, in which air is separated from blood by very thin cells. From air to blood in an alveolus is a distance of only 0.5 nm, short enough to allow rapid diffusion between blood and air. The total gas-exchanging surface of the alveoli amounts to more than 70 square meters! At any one instant all of this large surface is engaged in aerating only about 60 g of blood, the total volume in the alveolar capillaries. To get an idea of just how thin this "film" of blood is, try painting the floor of a large room with 60 g of paint!.

The remarkable alveolar structure of the lung is there simply to facilitate diffusion of carbon dioxide and oxygen between blood and air. As blood moves through the alveolar capillaries, both of these molecules diffuse in opposite directions along their concentration gradients, clearing the blood of carbon dioxide and loading it with oxygen. Although these two gases, as we have seen (page 337), are carried in a special way once they are in the blood, it is important to remember that the final step in gas exchange is diffusion, just as in the lowest organism. We shall return to the matter of gas transport in the blood shortly.

WHY BREATHE?

By looking at Figure 19-15, you will observe an air path from the alveoli to the outside, by way of a system of **bronchioles** that join two main **bronchi**. The bronchi meet in a single tube, the **trachea**, which communicates with the exterior through the mouth cavity and nose. This distance between alveoli and outside air is too far for gases to diffuse in and out and yet maintain adequately steep concentration gradients of oxygen and carbon dioxide at the alveolar surface. The best possible solution would be to have fresh air in the alveoli at all times, because air contains the highest possible oxygen concentration and the lowest possible carbon dioxide concentration. Although this ideal is unattainable, a satisfactory compromise is achieved by sucking air in and out of the air passages of the lung in the rhythmic process of breathing. About 6 liters of air are breathed in and out per minute during normal breathing. This volume may rise to 60 to 70 liter/min during heavy exercise, a range in ventilation rate sufficient to allow adequate aeration of the blood under any reasonable circumstance. You should note that under resting conditions the bronchial passages contain about 500 ml of air, which is identical to the amount of air inspired per breath (tidal volume). In other words no air is pumped in or out of alveoli; air moves by diffusion between alveoli and bronchioles.

The Mechanism of Breathing

The lungs themselves do virtually no work in breathing. They hang in the **thorax**, an airtight cavity, with the **trachea** connecting them to the outside. When this cavity increases in volume, the resultant negative pressure causes air to rush into the lungs through the trachea. The opposite occurs when the volume of the airtight cavity is diminished, as illus-

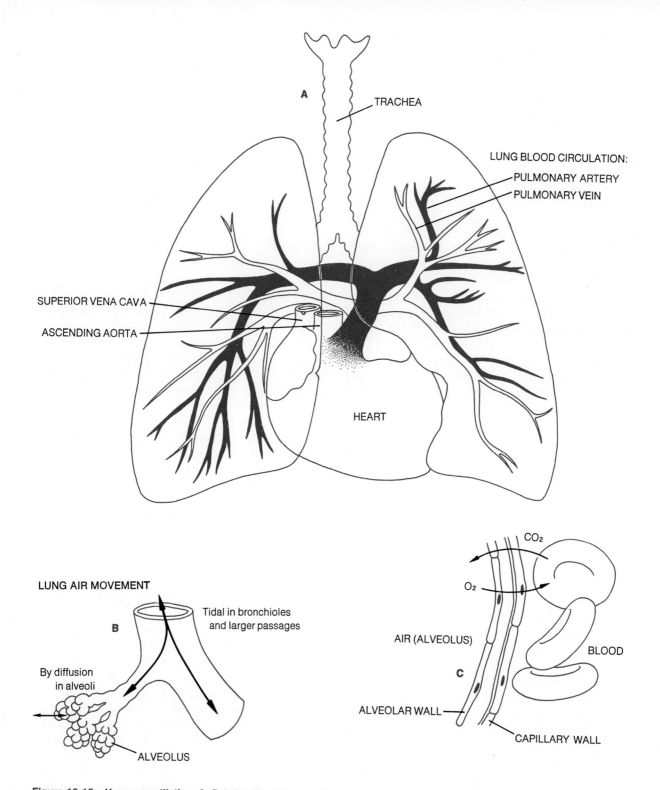

Figure 19-15 Human ventilation. A. Relationship of lungs to heart and major blood vessels. **B.** Alveolus, the site of gas exchange. **C.** Detail of alveolar wall showing that red blood cells are separated from alveolar air by less than their own diameter.

trated in the model of Figure 19-16. The thorax is bounded by the muscular rib cage and is separated from the abdominal cavity by a muscular **diaphragm.**

When the lungs are empty of air, the ribs hang downward from their attachments to the backbone and the diaphragm domes upwards into the thoracic cavity. Thoracic volume is then at its smallest. When air is taken into the lungs, rib muscles along with muscles of the shoulder and neck, pull the ribs into a more horizontal position, thus increasing the back-to-chest diameter of the thoracic cavity, while the diaphragm pulls down, increasing its length. The result of this increase in volume is, as shown in the model, an inrush of air through the only opening to the exterior, the trachea, that fills the lungs with air. At expiration, muscles operating the ribs and diaphragm relax and air is expelled, aided by the elasticity of the lung tissue, which tends to return, spring-like, to its rest position. Muscles of the wall of the abdomen also help in more violent expiration. This ventilation cycle is shown in Figure 19-16.

Protecting the Lungs

In discussing the skin, its hardened surface was said to provide protection against attack by disease organisms. This protection might be thought futile in the light of what we have just learned about the lungs: that they have a much larger air-exposed surface than the skin and that this surface is thin and highly permeable. Clearly we must look again at the lungs, because they must have some property that protects them from airborne contamination. We find this property in the lining of all air passages between nose and alveoli. This lining is somewhat like mechanized flypaper. Viscous mucus is secreted all through the air passages. It serves as the flypaper adhesive and traps inhaled particles, including bacteria. Mechanization is provided by the cilia present on most cells lining the respiratory passages. These cilia all beat towards the exterior so the mucus, with its load of trapped particles, is carried to the trachea, where it may be removed by coughing. Similar protection is also afforded by the nose, whose complicated mucus-secreting, ciliated passages filter, warm, and moisten inhaled air even before it reaches the lungs (Figure 19-1).

Speech Is a Byproduct of Ventilation

The expanded upper region of the trachea is the **larynx** (Adam's apple). Within the larynx two mova-

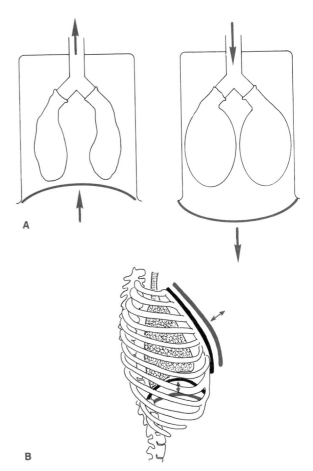

Figure 19-16 The mechanism of breathing. A, bell jar and balloons model illustrates negative pressure breathing. The diaphragm is represented by the flexible covering of the bell jar base. **B,** the rib cage and lungs in side view show that negative pressure breathing is effected by movements of the diaphragm and, in heavier breathing, upward and outward movements of the ribs.

ble folds of tissue partially block the airway. These are the **vocal cords.** Speech is produced by expelling air from the lungs, causing the vocal cords to vibrate, producing sound. Relaxed, the vocal cords vibrate slowly and low-pitched sounds are produced; under tension from associated muscles, they vibrate at high frequency to make high-pitched sounds. Sounds thus generated in the larynx are further modified by the tongue, by the lips and by the general shape of the mouth and nasal passages.

Artificial Respiration

As is evident from the model in Figure 19-16, the lungs also work satisfactorily if they are filled by posi-

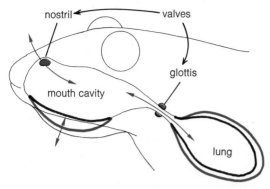

INSPIRATION (two steps of positive pressure):

1. a. Nostril open, glottis closed
   b. Floor or mouth descends
   c. Air enters mouth

2. a. Nostril closes, glottis opens
   b. Floor of mouth ascends
   c. Air is forced into lungs

EXPIRATION

   a. Nostril open, glottis open
   b. Muscles of body wall force air out of lungs

Figure 19-17 Positive pressure breathing in the frog.

Figure 19-18 Mouth-to-mouth resuscitation. (American Red Cross)

394 BIOLOGY

tive pressure, that is, by being blown up. This is how a frog actually breathes (Figure 19-17), and the fact that positive pressure will work for man forms the basis of a very effective method of artificial respiration, the mouth-to-mouth, or positive pressure, method, illustrated in Figure 19-18. In this simple method, the lungs of an unconscious person who has ceased breathing can be aerated enough to maintain life simply by blowing them up with the breath of the person administering first aid. Although exhaled air contains less oxygen than pure air, there is enough to support life.

If the heart has stopped, artificial respiration alone will not suffice. It must be supplemented by periodic firm pressure applied to the rib cage over the heart, which will promote some movement of blood. Properly used, these two techniques are extremely important life savers. They are simple to do correctly, and we have little excuse for not taking the time to learn them.

CONTROL AND OPERATION OF THE VENTILATIVE MACHINERY

Breathing is a reflex. Without having to think about it, the rate and depth of breathing are automatically adjusted to your oxygen requirements. Quite clearly, then, somewhere in the nervous system there must be a mechanism serving to control the ventilating muscles and to measure the minute-to-minute requirements for oxygen and the buildup of carbon dioxide.

Ventilation is directed from the medulla by neurons that sense the oxygen and carbon dioxide levels in the blood and control the operation of the ventilating muscles appropriately. This medullary ventilation control center can support ventilation by itself, but the type of ventilation produced by the center alone is not normal, consisting of strong gasps with long intervening pauses. Evidently there is more to normal ventilation control than the medullary center.

The remainder of the control system consists of two kinds of sensory receptors that supply information to the medullary respiratory center. One type consists of **stretch receptors** associated with the lungs. These inform the respiratory center about how the lung has responded to the neural commands sent out by the center to the ventilative muscles. That is, these receptors form the sensory part of a feedback loop. They stop the process of inspiration when the lungs are fully inflated and stop expiration when they are deflated. The second kind of receptor assisting the medullary respiratory center is **chemosensory**. Although the medullary center can itself monitor oxygen and carbon dioxide, there are other receptors that do this. They are strategically located to sample the blood in two extremely important arteries, the carotid (which supplies blood to the brain) and the aorta (the major artery of the remainder of the body). These receptors, the **carotid** and **aortic bodies**, do most of the controlling of respiration. Control falls exclusively to the medullary chemoreceptors only in abnormal conditions.

Thus we see that the ventilation control system (Figure 19-19) consists of (1) a center in the medulla, which directs the respiratory muscles and has associated chemoreceptors that can, in a pinch, maintain respiration, and (2) two kinds of sensory input from outside the brain: stretch receptors, monitoring the state of expansion of the lungs, and chemoreceptors, monitoring oxygen and carbon dioxide in the blood stream. The presence of two independent sets of oxygen and carbon dioxide receptors as part of this system is another example of the principle of multiple assurance, by which many physiological systems whose function is critical to life have multiple pathways, or backup systems, so that if one fails the others suffice.

THE RED BLOOD CELL: LINK BETWEEN TISSUES AND LUNGS

The ventilating system is simply a device promoting steep oxygen and carbon dioxide concentration gradients between blood and air. At the other end of the gas transport system the smallest branches of the circulatory system (the capillaries) serve similar ends, promoting diffusion between blood and metabolizing cells. But trouble arises with this scheme if one determines how much oxygen and carbon dioxide can be carried *in solution* in the blood. It develops, for example, that 100 ml of blood can carry in solution only about 0.3 ml of oxygen. This is not nearly enough. About 5 ml of oxygen per 100 ml of blood are required to support life in man. Thus we see the development of a new requirement for the circulatory

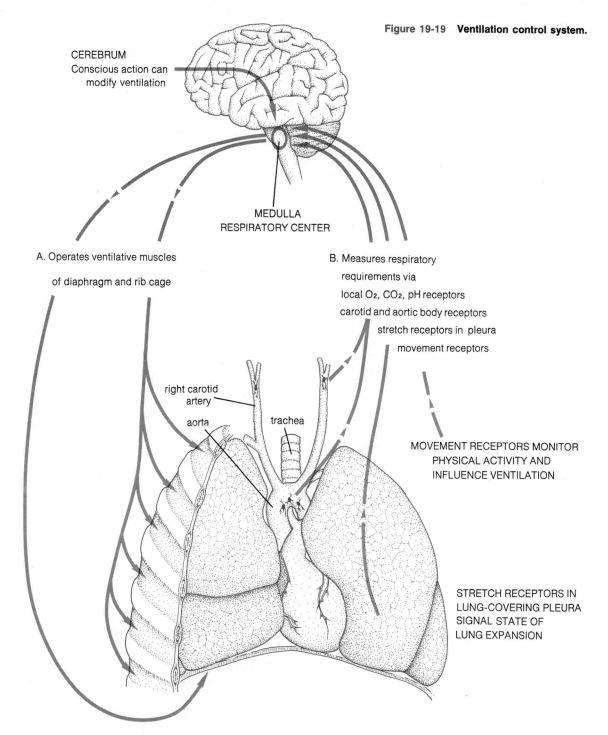

Figure 19-19 **Ventilation control system.**

system. In addition to providing a flow of blood that reaches out to all cells, the blood must possess a means of carrying oxygen more effectively than simply in solution. In vertebrates this means is furnished by the red blood cell (**erythrocyte**) which contains the oxygen and carbon dioxide carrying protein, **hemoglobin**. The circulating erythrocyte has lost its nucleus and assumed the shape of a biconcave disc, which admirably

release its load of oxygen; and second, the tendency of hemoglobin to pick up or release oxygen is related to the concentration of oxygen in solution around itself. This means that hemoglobin gives up nearly all of its oxygen in the tissues, where it is needed, and becomes completely loaded with oxygen during its journey through the lung capillaries.

Hemoglobin constitutes so much of the mass of red blood cells (95% of their dry weight) that it might seem simpler for the hemoglobin to be in solution in the blood. There are several reasons why this is not so. For one thing, if hemoglobin were free, the blood osmotic pressure would be enormous, and the blood would be so viscous that it would be difficult to pump through the blood vessels. It is also possible that a moderate-sized protein, such as hemoglobin, would experience difficulty remaining in circulation. Its usefulness in ventilation would, of course, be diminished if it were to leave the blood channels to any extent, interrupting its shuttling back and forth between lungs and metabolizing tissues.

Another important reason for confining hemoglobin to red blood cells has to do with the part it plays in carbon dioxide transport. As described in Chapter 18, one of the ways carbon dioxide is transported involves its reaction with water, catalyzed by carbonic anhydrase, and subsequent dissociation into hydrogen ions and bicarbonate ions, with the hydrogen ions combining with hemoglobin. If this reaction were to occur freely in the blood, there would be opportunity for hydrogen ions to become involved in other reactions, which might be harmful. Also, the bicarbonate ions would be free to disturb further the acid-base relationships of the body. With the hemoglobin in red cells along with the enzyme carbonic anhydrase, all of these undesirable possibilities are kept under control.

ADAPTATION TO VENTILATIVE STRESS

The nature of short-term adaptation to ventilative stress is obvious. Low oxygen or high carbon dioxide concentrations in the blood result in increased ventilation and circulation, improving gas transport. In addition to this, there may be changes in the number of red blood cells in circulation, especially in response to prolonged lack of oxygen such as might occur in

Figure 19-20 **Cross section of a red blood cell (diameter 7.5 μ) in a capillary.** K. Linberg.

serves the function of gas exchange by shortening the gas diffusion path (Figure 19-20).

How Hemoglobin Works

Hemoglobin (Figure 19-21) is an effective oxygen and carbon dioxide carrier because it combines reversibly with these gases and because the hemoglobin molecule has two important properties ensuring that it picks up and releases these molecules at the proper places. First, hemoglobin is built so that the more carbon dioxide it reacts with the more likely it is to

HEMOGLOBIN

Human hemoglobin, a protein of 67,000 molecular weight, is formed of four nearly identical protein chains. Each bears an iron atom residing in a porphyrin molecule (colored disks).

One of the four identical porphyrin molecules. One molecule of oxygen is carried by the iron atom, thus four oxygen molecules are carried per hemoglobin molecule.

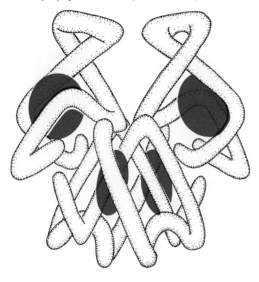

THE TRANSPORT OF OXYGEN BY HEMOGLOBIN

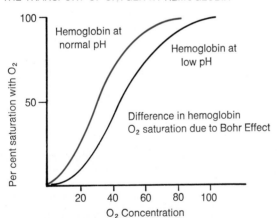

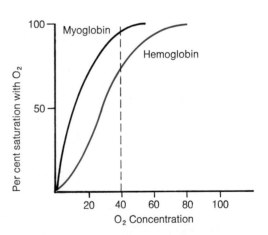

In the lungs, where O_2 concentration is high, hemoglobin is nearly saturated with O_2. As O_2 carrying hemoglobin reaches the tissues it gives up O_2 as the surrounding O_2 concentration falls. Black curve shows the effect of increased acidity on hemoglobin (Bohr Effect). This facilitates O_2 release at sites requiring O_2, since pH falls due to the CO_2 production.

Muscle cells may contain myoglobin, which is half a hemoglobin molecule with 2 O_2-combining iron atoms instead of 4. Its oxygen combining curve (black) is compared with that of hemoglobin. At any O_2 concentration myoglobin has a greater affinity for O_2 than has hemoglobin (dotted line) and can take up O_2 from it. The circulating hemoglobin of the human fetus also behaves like myoglobin.

Figure 19-21 Structure and function of hemoglobin. Structural representation after A. L. Lehninger, 1975, *Biochemistry,* 2d ed., Worth Publishers, Inc., New York.

mountain climbing. The average human male has about 5 million red blood cells per cubic millimeter of blood. These are produced mainly in bone marrow and released at a rate sufficient to keep the circulating level of red blood cells constant, in the face of the wearing-out rate of 3 million red blood cells per minute.

If one goes from sea level to the rarefied air of the high mountains, there is initially a sudden increase in the number of red blood cells in circulation. These cells come from a storage site, the spleen, that releases them in response to low oxygen conditions. If the stay at high altitudes is prolonged, the rate of production of red blood cells by bone marrow is accelerated by the action of the hormone **erythropoietin,** whose release seems to be triggered by the effect of oxygen lack on the kidney. This increased rate of production can be most significant. The best example is found among the natives of the Peruvian Andes, who live and work at the 18,000-ft level (persons normally living at sea level would fall unconscious if suddenly transported to an altitude of about 20,000 feet). Their red blood cell counts run about 7 million per cubic millimeter of blood.

The Circulatory System

Without sharing the results of their individual activities, the various organ systems cannot function. What good is it to ventilate a lung if the oxygen it receives can go no farther? This necessary form of communication—or, more accurately, *transport of materials* in contrast with the transport of *commands* carried on by the neuroendocrine system—is the function of the circulatory system. We have already discussed one element of the circulatory system, the red blood cell, because of its immediate significance to the functioning of the ventilating system. From this example the effective way in which the circulatory system mediates between the organs is clear. Now we shall turn our attention to the rest of the circulatory system, to the blood (of which the red blood cell is only one element), to the system of vessels (arteries, veins, capillaries, and lymph vessels that carry the circulatory fluid), to the heart, and to the neuroendocrine controls that adjust the work of the circulatory system to the demands of the body.

THE NATURE OF BLOOD

An average adult male contains about 5.0 liters of blood, a female somewhat less. Of this volume about half is fluid, or plasma, and the remainder is various kinds of blood cells. **Plasma,** the fluid of the internal environment, is extremely complex, containing an immense number of chemical substances essential to life—from simple inorganic ions to complex proteins—together with metabolic wastes of all sorts. Besides transporting all of these, the plasma serves importantly, because of the proteins it contains, as a regulator of the acid-base balance of the internal environment. Many of the plasma proteins are protective; they include antibodies, which attack many kinds of disease organisms, and other proteins essential to stopping bleeding. Finally, because of its heat-carrying capacity, plasma serves importantly in temperature regulation.

The most numerous cellular constituent of blood is the red blood cell. The remaining types, white cells (**leucocytes**) and **platelets,** are markedly fewer in number, there being some 5000 to 10,000 white cells and 300,000 platelets per cubic millimeter of blood. Blood cells are produced either in bone marrow, as are the red cells, or elsewhere.

White cells and platelets are strictly protective. White cells protect against invading microorganisms, whereas platelets serve exclusively in stopping blood loss after damage to a blood vessel. Although there are not very many white cells in circulation, there are many more wandering outside the blood stream, particularly in the lungs and intestines. There, where bacterial invasion is always a potential menace, these cells are immediately ready to go into action. Additional storage sites can release still more white cells in

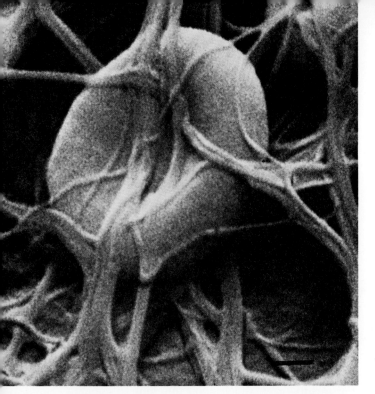

Figure 19-22 Fibrin strands in a blood clot enmesh an erythrocyte. Scanning electron micrograph by E. Bernstein. (Copyright 1971, American Association for the Advancement of Science.)

Figure 19-23 Blood clotting mechanism.

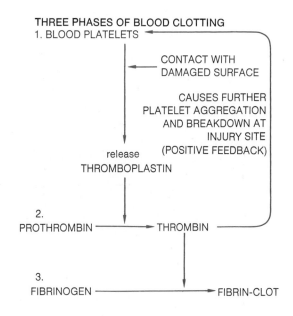

response to disease organisms, and all of these cells seem to be attracted to bacteria and other foreign bodies. White cells surround and actually consume these, thus preventing or slowing the spread of attacking organisms.

Blood platelets, formed as nonnucleated fragments of large bone marrow cells, stick to damaged regions of blood vessels. Although their mass alone might be of some value by physically plugging a tear in a very small vessel, the platelets exert much more useful effects. These result from the breakdown of the adhering platelets, with the consequent release of two substances of great importance in limiting blood loss. One of these causes the blood vessel to decrease in diameter, making it less likely that a clot formed to plug an injury site would be swept out by the flow of blood. The other substance is instrumental in forming the clot itself.

Blood clots are formed of dense meshworks of long protein fibrils entangling blood cells and debris from broken up platelets. The fibrils are formed from a soluble protein, **fibrinogen,** which is carried in the plasma. If a few of the terminal peptides of fibrinogen are enzymatically split off, the remaining molecule is able to polymerize with other such molecules, forming the long strands of **fibrin** (Figure 19-22). The processes that lead to fibrin production are triggered by an enzyme called **thromboplastin,** which is released from the blood platelets. This sequence of reactions is shown in Figure 19-23.

Sometimes, without an obvious injury, a blood clot may form in a blood vessel. Such a clot is called a **thrombus.** The medical condition is called **thrombosis** and can have serious effects should the thrombus block blood flow in a critical place. Coronary thrombosis, for example, is a serious condition in which a clot blocks circulation in one of the coronary blood vessels that serves the heart muscles.

HEART AND BLOOD VESSELS

Now that medical science has laid low the great killers among infectious diseases, the attention of researchers turns more and more to failure of heart and circulatory vessels. "Heart attack" and "stroke" are the common names applied to failure of parts of the circulatory system, and their current prevalence as causes of death emphasizes the critical role played by the circulatory system.

The Role of the Heart

Figure 19-24A, B, showing the route of blood through the heart and major blood vessels, reveals that this powerful muscular organ is really two pumps, one sending blood through the lungs, and the other supplying blood to the remainder of the body.

Each side of the heart—each pump—contains two pumping chambers, connected by valves that allow blood to flow in only one direction when the chambers are contracted by their muscular walls. The upper pair of chambers are the **atria**. The left atrium (on the animal's left) receives oxygenated blood from the lungs by way of the pulmonary veins. The right atrium

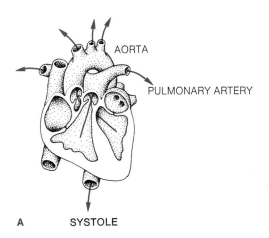
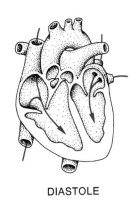

Figure 19-24 Route of blood through heart **(A)** and systemic circulation **(B)**.

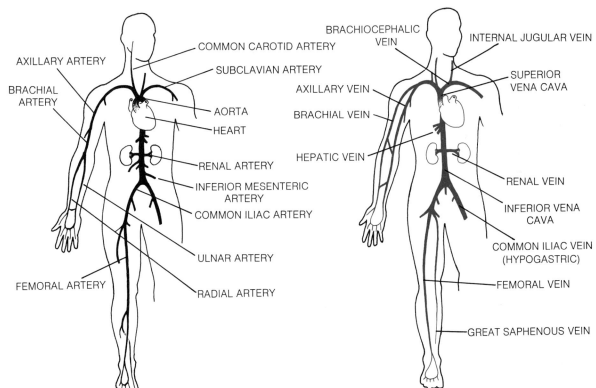

HUMAN MAINTENANCE PHYSIOLOGY

receives venous blood from the remainder of the body. The atria are not as muscular as the two lower chambers into which they empty, the **ventricles.** Atria do not have to be strongly muscular since they fill more or less passively as venous blood flows in. When filled, the atria have only to propel their contents into the ventricles. No appreciable amount of blood re-enters the veins when the atria contract because the one-way valves found throughout the venous system prevent backflow.

In the cycle of the heart action, following nearly simultaneous atrial contractions, the filled ventricles contract. Because they must force blood into extensive systems of blood vessels that build up considerable resistance, the ventricles are very muscular. Once ventricular contraction has built up sufficient pressure, the valves that guard the openings into the arteries leading from the ventricles snap open, and blood is forcefully ejected into the arteries, producing the pulse. Blood from the right ventricle enters the **pulmonary artery,** which supplies the lungs, and blood from the left ventricle proceeds into the **aorta,** which distributes blood to the body (systemic) circulation (Figure 19-24). After completing this two-step cycle, the heart rests briefly before beginning another.

The heart contraction cycle rate varies in the adult from about 70 beats per minute to about 200 per minute during heavy exercise. From this, it follows that the heart must beat nearly 2.6 billion times in the average human lifetime. This would be a remarkable performance record for any pump, not to mention one constructed of living tissue.

The tissue of which the heart is constructed does, however, have certain peculiarities that set it off from other muscular organs. Heart muscle cells resemble ordinary striated muscle (the voluntary muscle of arms, leg, and so forth) except that they branch and connect with each other (Figure 19-25). Heart muscle also differs from ordinary striated skeletal muscle in several other respects, the most important of which is that the signal to contract can be transmitted from muscle cell to muscle cell, whereas each cell in ordinary muscle must be independently stimulated by a nerve. In fact, certain of the heart muscle cells become quite nervelike in function, serving to transmit the signal to contract rather than joining in the work. These cells are called **Purkinje fibers,** after their discoverer. Their distribution and action in the human heart are shown in Figure 19-25.

Regulation of the Pump

The heart contraction system also differs from ordinary muscle in that it can, by itself, contract rhythmically. Early in embryonic life, the developing heart begins to beat before nerves extend to it from the central nervous system. Long before birth, however, the heart is reached by nerves that, throughout life, modify the rate and vigor of its beat.

The medulla is the source in the brain of neural control of the heart. There, a heart-inhibiting center continually exerts a slowing influence on the heart by way of parasympathetic vagus nerves, which act on the areas of the atria that initiate heartbeat. If these nerves are cut, the heart almost always speeds up. Sympathetic nerves that act principally on the ventricles accelerate and strengthen the heartbeat. These also are controlled from the medulla.

What Influences Heart Activity?

Two varieties of stimuli influence the action of the heart. One is the "fight-or-flight" response to stress, resulting in general sympathetic discharge. The heart, of course, is included in the series of adrenaline-initiated events that prepare the organism for danger. Its contribution is to accelerate and strengthen its beat.

The other stimulus influencing the heart is pressure. This makes sense because the heart is a pump, and one way to measure the effectiveness of a pump is to measure the pressure it maintains in the pipes it serves. The heart has pressure receptors that measure blood pressure both entering and leaving the heart. Measurement on both sides of the blood stream is necessary because the pressure does not remain the same all the way around the circulatory system. On the output side of the heart, pressure is measured by receptors located in the wall of the left ventricle, in the arch of the aorta, and in the carotid artery. Thus the major outgoing branches of the circulation are monitored. When these receptors register excessive pressure, the heart-inhibiting center in the brain is signalled to slow the heart. If the measured pressure is too low, the sympathetic nerves are called into action to strengthen the heart beat.

Excessive pressure on the input side acts on receptors in the right atrium, which cause the heart to accelerate. This receptor mechanism of the right atrium acts in an interesting way to increase the blood supply to the body muscles when their requirements are increased by exercise. Return of blood to the heart

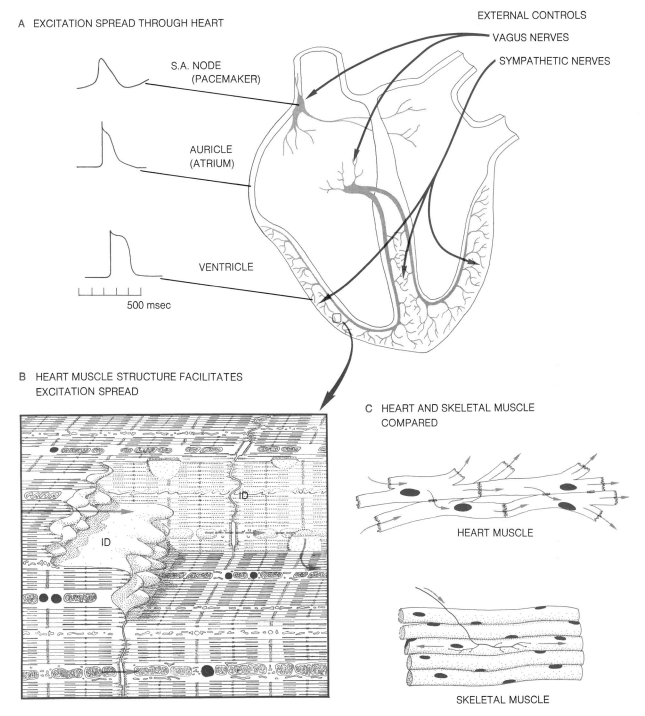

Figure 19-25 Control of the heart. A. Electrical recordings from the heart show that the signal to contract begins in the sinoatrial node and spreads to the right auricle and then to the left. There is a brief delay in the atrioventricular node and then the signal spreads through the Purkinje fibers to the ventricular muscle. **B.** A diagram of the structure of heart muscle shows sites (ID, intercalated disks) where excitation transmission from muscle cell to muscle cell by electrical current flow occurs. After Sjöstrand, F. S., et al, 1958, *J. Ultrastructure Research* 1:271. **C.** In heart muscle excitation flows from muscle cell to muscle cell while in skeletal muscle each cell must be individually excited by a nerve.

Note 19-1. *William Harvey and the Discovery of the Circulation of the Blood*

Although it is true enough that we take the circulation of the blood to be obvious, it was not an easy matter to demonstrate before the time of the microscope. The proof was accomplished sometime before 1653 by William Harvey, an English Court Physician. His work on the circulation, *An Anatomical Disputation Concerning the Movement of the Heart and Blood in Living Creatures,* still stands as a remarkable example of the scientific method.

To understand properly what Harvey accomplished, we must remember he worked when the function of the blood was unknown; one common theory of the times was that the blood was simply a cooling agent for the organs. Furthermore, the capillaries connecting arteries and veins were unknown. Yet Harvey, by performing simple experiments on man and a variety of lower animals, showed that the blood must circulate, that is, go around a circuit

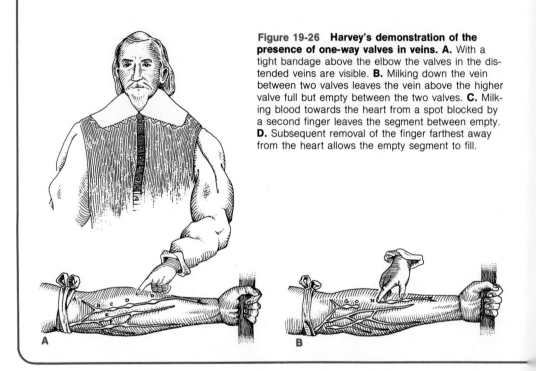

Figure 19-26 **Harvey's demonstration of the presence of one-way valves in veins. A.** With a tight bandage above the elbow the valves in the distended veins are visible. **B.** Milking down the vein between two valves leaves the vein above the higher valve full but empty between the two valves. **C.** Milking blood towards the heart from a spot blocked by a second finger leaves the segment between empty. **D.** Subsequent removal of the finger farthest away from the heart allows the empty segment to fill.

depends a great deal upon the general muscular activity of the body, which milks the returning blood back to the heart through one-way valves in the veins. Consequently, when muscles become active, this very activity increases the rate of return of blood to the right atrium, stretching it and initiating, via its receptors, a call for increased heart action.

The Blood Vessels

You undoubtedly already know that blood begins its journey from the heart to the tissues in **arteries** and returns in **veins** (Note 19-1). Because arteries carry blood under pressure from the heart, they are heavily wrapped with elastic connective tissue, enabling them to stretch under the hammer blows of ventricular contractions. As the arteries divide into smaller branches, a wrapping of smooth muscle becomes more prominent. Its function is to control the diameter of the smaller arteries and thereby aid in regulation of blood pressure and, by contracting more in one region than another, control the distribution of the blood to the various parts of the body.

from arteries to veins and back again, and he determined a great deal about how this was accomplished.

Harvey did two important things. First, he determined quantitatively that it was impossible for the body to produce in a given period of time the total amount of blood that could be estimated to leave the heart during that period. Therefore, the blood that exited the heart could not be all new; it had to circulate, going through the heart many times in the course of Harvey's measurements. He demonstrated this quite simply, by measuring how much blood the ventricle could hold and then estimating how much blood it pumped out at each heartbeat. This quantity could then be multiplied by the number of beats for a given amount of time to arrive at the total volume of blood ejected during the same period. Harvey estimated that the ventricle might eject as little as half an ounce of blood per beat and that the heart might beat about 2000 times per hour, which would mean 1000 oz of blood pumped. The absurdity of the idea that the blood did not circulate is immediately evident by determining the amount of blood pumped over a day: 24,000 oz or 1500 lb.

Having thus proved that the blood had to be reused, Harvey turned to the problem of what route it followed from the heart and back again. He worked out the entire story except for the capillary linkage between arteries and veins. This was a discovery that had to await the development of good microscopes. Harvey figured out the roles of the various chambers of the heart by direct observation and experiment on living hearts of many kinds of animals and by examination of the hearts of human cadavers. He then demonstrated that the blood entered a limb through the arteries and left it through the veins by very simple experiments. Putting a tourniquet around an arm showed that arteries became distended above the block and deflated below. A finger on the deflated part could feel the blood rush in when the tourniquet was loosened. In contrast, veins became distended below the tourniquet. He even demonstrated that the valves found in veins, which had been discovered by one of his teachers, were so constructed that they could pass blood only towards the heart. At dissection he observed that a probe run through a vein could be passed only in the direction of the heart. A probe passed away from the heart was blocked by the one-way valves. Then he turned to a living human arm and did the experiment shown in Figure 19-26.

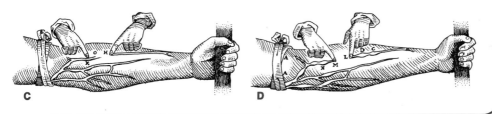

As the arteries continue repeatedly to divide, they become **arterioles** whose walls are relatively thinner because they now have only to contain a placidly flowing stream, rather than the pulsing torrent confined by the major arteries. This taming of the blood stream is the result of the increased resistance to blood flow caused by the increasing area of blood vessel walls, as arteries grow smaller and increase in number.

Ultimately all connective tissue and muscle are lost as the arterioles become **capillaries,** which are simply tubes constructed of a single cell layer of endothelium, the smooth lining of the entire circulatory system. Here, where flow is the slowest and vessel surface in contact with tissues is the largest in the circulatory path, the essential role of the blood is played: capillaries are the only site of exchange between blood and tissues. Both water and dissolved substances filter through the capillary endothelium to the tissues, propelled by blood pressure and often along osmotic and diffusion gradients. This blood filtrate is **lymph.** It also contains white blood cells, leucocytes, which wander in and out of the blood stream, passing through the junctions between endothelial cells in their continual scavaging for disease organisms and tissue wastes.

Lymph percolates among the cells and the vital exchange of raw materials for wastes takes place. To complete the transport cycle, waste-carrying lymph returns to the blood stream for cleansing. Most of it directly reenters the capillaries further "downstream," but some enters another system of capillarylike lymph vessels, which finally return the lymph to the circulation by way of a single duct emptying into the right subclavian vein. This same lymph-return system also receives lymph from the fat-transporting lymph vessels (lacteals) of the small intestine.

CONTROL OF BLOOD PRESSURE AND BLOOD DISTRIBUTION

Practical familiarity with household plumbing leaves one with the knowledge that maintenance of adequate water pressure in the mains does not necessarily mean that water is properly distributed through the house; the basement may flood while the second floor shower manages only a trickle. The human circulatory system has similar problems. Consequently, although we have already seen how pressure is maintained in the mains (heart and aorta), we have now to determine how circulation is controlled in the periphery. The mechanisms involved are more complicated than in domestic plumbing; besides an automatic leak-repairer (the clotting mechanism), peripheral blood pressure and even the regional distribution of blood is regulated by neuroendocrine mechanisms acting to control blood vessel diameter.

The total circulatory control apparatus is most effective; it shunts blood to organs in response to demand (thus, blood supply to the leg muscles may rise from 500 to 5000 ml/min during exercise), and it enables us to survive blood losses up to four times as great as we otherwise could.

The essential control element in peripheral circulation is already known to you in the form of the system of smooth muscles in the blood vessel walls. The extent to which this musculature is contracted or relaxed determines where the greatest volume of blood flows. By muscular control of a system of **arteriovenous shunts**—blood vessels that bypass blood directly from arterioles into venules—small areas of the capillary exchange system may even be completely bypassed. Thus, circulation control depends on regulation of heart action and on control of blood distribution.

To show how blood distribution is controlled, let us examine what happens in various circumstances. First, take the example of bleeding. If bleeding is severe enough, blood pressure falls. In response to this, the muscular linings of the blood vessels constrict, decreasing the volume of the circulatory system and thus restoring sufficient pressure to ensure continued adequate circulation to the brain and heart. By this it must be clear that blood vessel constriction is not generalized but can be brought about in specific regions of the body.

Although the response to bleeding is dramatic enough, exactly similar although less pronounced reactions of the circulatory apparatus occur continually. For example, during body movements continual changes of blood vessel diameter are made to compensate for gravitational effects. The momentary giddiness that you may feel upon standing suddenly after lying down for a while is due both to slowness of the circulation-adjusting reflexes to compensate for the increased difficulty of pumping blood upward to the brain and to the tendency of blood to accumulate by gravitation in the lower parts of the body.

Consider next how the circulation to specific regions of the body may be controlled. In some instances, an organ may have pressure receptors associated with its arterial supply. These may reflexly control arterial diameter (and hence blood flow) by way of the medulla. In addition to these neurally mediated controls, there is local chemical control of blood vessel diameter. Although we know that this occurs, its mechanism is not understood. It appears that a chemical factor, or perhaps several, is released from tissues that lack oxygen, which directly relaxes the muscles of nearby blood vessels, increasing blood flow into the area requiring oxygen.

In summary, the circulatory system is regulated by a family of control mechanisms that act either on the entire circulatory system or on specific regions, ensuring a stable overall blood supply as well as serving the particular requirements of specific regions and organs.

The Excretory System

An adequate volume of body fluids is critical to life, as we learned in studying the circulation. Reconsideration of Chaper 13 should make it clear that more than mere volume is involved. An injection of distilled water might restore body fluid volume after severe bleeding, but the results would be bad. Certainly this circulating medium must be stable in terms of its *composition* as well as its volume.

Now, as the blood circulates, its composition tends to change unfavorably for two reasons: blood accumulates metabolic wastes and becomes depleted of substances essential to life. Many organs are concerned with restoring blood to its normal state, either by eliminating or adding substances. The lungs remove carbon dioxide and supply oxygen; the liver, among other things, supplies sugar to the blood and removes broken-down hemoglobin, which is excreted by way of the bile duct. Thus, the excretory system in a broad sense is a rather hard system to define, as this bartering with the blood stream, which is its essential nature, goes on in many places. However, one organ, the kidney, is principally concerned with maintaining the purity of the blood (and hence of the entire organism) and so we must now see how it operates and how it is controlled.

ORGANIZATION OF THE KIDNEY

The anatomy of the kidneys and their associated plumbing, together known as the **urinary system**, is shown in Figure 19-27. Kidney function is not particularly hard to understand. Large blood vessels enter and leave the kidneys. The entering arteries finally break up into capillaries. The essence of the structure of the kidney is that these capillaries are in contact with a system of tubes that finally drain into the bladder. Blood in the capillaries is still under pressure; consequently water and solutes (up to molecules of about 70,000 molecular weight) filter through the capillary walls from the blood into the kidney tubules, aided by osmotic pressure and diffusion gradients. This is the first phase of kidney action—**filtration** of the blood. Figure 19-28 shows that the region of the kidney where filtration takes place is composed of a regularly repeated series of tiny structures known as **nephrons**. There are about 1,200,000 of these in each kidney of man, a number fixed at birth. If the number of nephrons is reduced by disease or by loss of a kidney, the remainder may increase in size but new ones cannot develop.

In the first phase, filtration, you would be correct to assume that about as much damage as good may be done, as filtration is nondiscriminatory and does not differentiate between molecules of wastes and molecules of essential substances. If the molecule is small enough, it usually goes through into the nephric tubule. But this really shows the "genius" of kidney design. Rather than having mechanisms to scour the blood of wastes or of ingested harmful substances (which might vary from time to time and thus require more and more cleansing mechanisms), the kidney simply throws almost everything out and then retrieves the essentials. The essentials are well defined and relatively unchanging, and so the kidney has the much easier task of retrieving what the body requires rather than the ever-changing job of eliminating what is not needed.

Looking again at Figure 19-28 you will see that the nephric tubule on its way toward the ureter is wrapped about by the venule issuing from the capillary tuft where filtration occurred. Here resides the mechanism of the second phase of kidney function—**selective retrieval**. Essential substances either diffuse or are carried back into the venule by active transport; there may also be secretion of certain molecules from the venule into the tubule.

As an example of how the nephron works consider how it treats glucose. Under normal conditions essentially no glucose appears in the urine although the concentration in the blood plasma is about 100 mg/100 ml. Essentially this entire amount is filtered from the blood into the tubule, but it is returned to the capillary blood almost totally by active transport. If the active transport mechanism is damaged or if its capacity is exceeded, as by large amounts

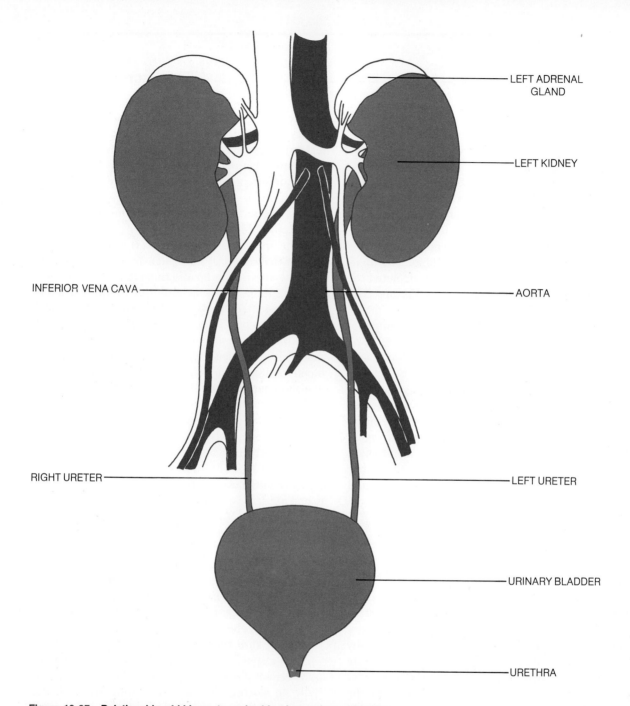

Figure 19-27 Relationship of kidneys to major blood vessels and bladder.

of glucose in the blood, glucose appears in the urine. The kidney behaves in this way towards other important molecules, such as amino acids, having special transport mechanisms for all of them.

THE KIDNEY AND OSMOREGULATION

It must be obvious to you what sorts of metabolic wastes the kidney eliminates. Primary among these is

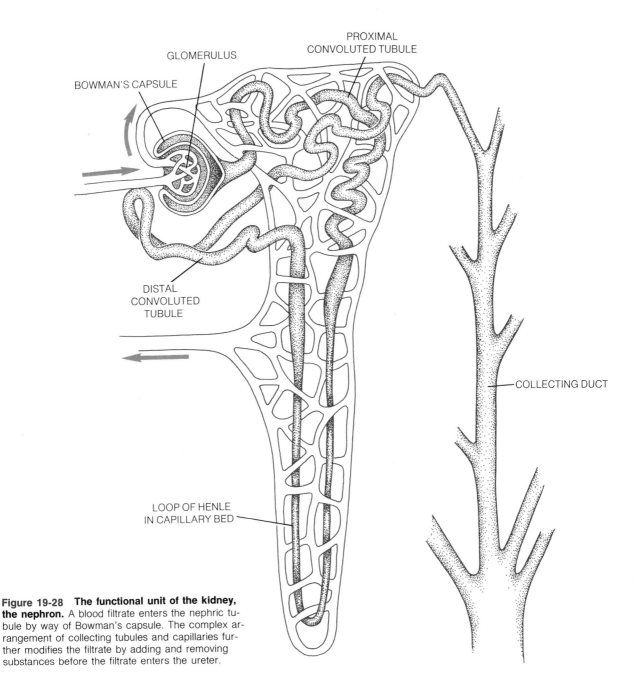

Figure 19-28 The functional unit of the kidney, the nephron. A blood filtrate enters the nephric tubule by way of Bowman's capsule. The complex arrangement of collecting tubules and capillaries further modifies the filtrate by adding and removing substances before the filtrate enters the ureter.

the nitrogenous waste urea. But it is not as obvious that the kidney plays another extremely important role as the site of regulation of water and electrolytes in the body.

The kidney must have a great effect on the amount of water in the body because it processes about 180 liters of body water per day. Ordinarily this results in 1 to 2 liter/day of water leaving the body as urine. The concentration of osmotically active molecules, or **osmolarity,** and the volume of body water are measured by osmoreceptors in the hypothalamus and by pressure receptors in the larger blood vessels. Together these control the release of **antidiuretic hormone (ADH)** from the posterior pituitary. ADH causes the

HUMAN MAINTENANCE PHYSIOLOGY   **409**

nephric tubule to reabsorb water. This means that more of the 180 liter/day of water that filter into the nephric tubules can filter back into the blood again, resulting in less urine. This increases the volume of fluid in the body and also decreases its osmolarity, as about the same amount of salts continue to be lost by the kidney irrespective of the action of ADH. If the animal needs to increase its fluid volume, ADH secretion increases and the nephric tubules return more water to the blood. Under these conditions the same stimuli also trigger a thirst center in the hypothalamus, leading to the sensation of thirst and the increase of fluid volume by drinking.

If ADH secretion does not occur, there ensues a condition known as diabetes insipidus, which is characterized by drinking large volumes of water and the production of large volumes of urine.

As you know, the organism must carefully control the concentrations of several electrolytes. Two of these in particular, sodium and potassium, are regulated by the kidney, acting under neuroendocrine controls. For example, sodium is filtered and then reabsorbed from the nephric tubule. The extent of reabsorption is dependent on adrenal cortical hormones. Release of these hormones is triggered by the anterior pituitary acting under the influence of osmoreceptors in the hypothalamus.

20 The Brain

WHAT MAKES US DIFFERENT?

Homo sapiens, thinking man, the name we modestly have picked for ourselves, salutes that aspect of our biology that decisively sets us off from other animals. Indeed, the remarkable fact about the human physiology that we have already discussed is our physiological similarity to other animals. Even the human activity, which we shall loosely call thinking for the moment, is in its basic nature represented in other animals. But, although qualitative differences are hard to identify, quantitatively there is no comparison between us and other animals when it comes to higher functions of the nervous system.

To understand the difference the brain makes we shall build on what you have already learned about the nervous system. We begin with what you know about the workings of nerve cells individually and as parts of both simple and relatively complex functional systems of nerve cells—reflexes and homeostatic controls. In fact, we shall begin with reflex control of muscle at the level of the spinal cord and then track muscular controls into the brain. There is good reason for doing this: first of all, it must be clear that it is only through control of muscles that the brain achieves external effects; and, second, in the way the brain controls muscles we see a relatively simple example of how the brain functions. But even in this regard you must be warned that we are still not fully informed about even these relatively simple matters. We know only enough to sketch them in general terms. And further, when we proceed to more complex levels of brain function, to the realms of thinking, memory, and emotions, you will see that our ignorance is truly impressive.

The effects of this ignorance are serious. Understanding the human brain represents as important a frontier of scientific knowledge as there can be. This is obvious because the brain is the tool with which all knowledge is accumulated. If we do not understand it, how can we trust what it tells us? Another demonstration of the importance of understanding the brain is to recall that it is the extent of development of the brain that most significantly differentiates us from all other animals. The brain has not only led us to our present precarious glory, it has caused most of the misery of our species. One brain may generate a *Mass in B Minor,* a *Hamlet,* or a theory of relativity; another may lead a nation to racial murder or generate other dreadful acts. Obviously such immense powers must be understood. One important way to promote this is for scientists to seek out knowledge of the physiological basis of brain function. You will see in this chapter that, although the most exciting beginnings have been made, we are just embarked on this great quest.

WAYS TO STUDY THE BRAIN

Although our approach to the study of the brain is rooted in physiology—after all this is a biology text-

Note 20-1. *Loss of the Link Between Short and Long Term Memory; The Case of H. M.*

Perhaps owing to a boyhood head injury, H. M. began experiencing seizures at age 10. These increased to as many as six an hour and he could no longer lead a normal life. Since relief from epilepsy is frequently obtained by removing part of the affected temporal lobe of the brain, this treatment was undertaken. The damaged area proved to be diffuse and could not be located, so part of both temporal lobes and the hippocampus were removed (Figure 20-1). There was a dramatic reduction in the frequency of seizures. Unfortunately, there was also a virtually complete loss of the link between short and long term memory. Both memory stores continued to function, but only the most meagre additions could be made to long term memory. H. M. read newspapers and forgot them in 15 min. His intelligence was unaffected and actually increased slightly in certain categories. Thus he could still work difficult crossword puzzles, but he soon forgot them. He was lost in the present, saying of himself, "Every day is alone in itself, whatever enjoyment I've had, and whatever sorrow I've had."

Some years later a follow up study showed essentially no change. A detail of one of the tests performed then dramatically illustrated his difficulty. This was a test of short term memory in which the subject listens to a series of numbers and must state whether a number subsequently presented was part of the series. A person with poor short term memory will do well only if the two numbers are close together in time of presentation. H. M. performed normally on this test but *only if he had a card before him at all times carrying a message telling him what he was to do!*

H. M.'s preoperative long term memory remained good. He could recall events from before the operation and did not forget persons from the preoperative period. He could recall events from that time normally. He simply could not make further additions.

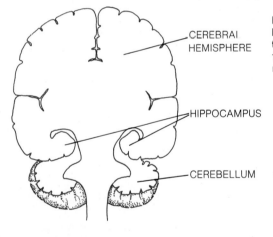

Figure 20-1 A vertical section of the brain as viewed from the rear to show the hippocampus. Bilateral destruction of this region prevents establishment of new memories.

book—it is important to know that there are other very important ways to study the function of the brain. While ultimately we would hope to do so, it is not necessary to understand each interaction of every nerve cell in the brain to learn a great deal about how it works. If physiology may be termed the study of the brain from the bottom up, **psychology** and **psychiatry** consider it from the top down. Both approaches supply useful answers to questions about brain function, and both approaches assist each other.

To illustrate how top-down procedures work, a psychologist might study learning by observing learning behavior in whole animals under a variety of experimental conditions. Thus, it has been discovered that, if a rat is given a massive electrical shock to the brain a few minutes after learning something, it will forget, whereas if the shock is applied several hours later, the learning is retained. Experiments of this type gave rise to the idea of at least two kinds of memory stores in the brain. One is *short term memory,* and it differs from *long term memory* by storing its information in a more easily disturbed form. This explains how a newly learned task can be eradicated from memory by electrical shock and provides a useful conceptual basis for other studies of learning.

However, a whole animal study does not tell us *where* in the brain memories are located and does not tell us *how* they are maintained and called forth when we remember. The psychiatrist or **neurologist,** both medical practitioners devoted to the study of mind and nervous system, provide further information on at least the first of these questions. They find that certain kinds of damage to specific parts of the human brain produce specific defects in the memory process, thus indicating that those parts of the brain are either the location of part of the memory system or otherwise important to memory, for example as pathways to or from memory storage. A celebrated instance, described in Note 20-1, shows that damage to a region of the brain called the hippocampus may result in inability to transfer information from short to long term memory without effect on either memory system.

Finally, physiologists investigating the physicochemical basis of memory understand that all information going into the nervous system must, at least initially, travel as electrical signals, nerve impulses. Some physiologists theorize that short term memory is easily disturbed because it might consist simply of patterns of nerve impulses that keep the memory alive only so long as the pattern is preserved. Thus an electrical shock, which would be bound to disturb such electrical patterns, must have a profound effect on short term memory. Similarly, since long term memory is not easily disturbed, it follows that this process must be based on some mechanism that is more stable than patterns of nerve impulses. One theory is that the long term memory stores represent some kind of permanent chemical change in nerve cells. Consequently, the physiologist teams with the biochemist to attempt to figure out how nerve impulses can produce permanent chemical changes in neurons of a sort that may be used to key memories into the brain.

Muscle Control: The Everyday Business of the Brain

The physiological study of muscle function is well understood and serves as an excellent introduction to the way the central nervous system works.

Skeletal muscle is the immediate vehicle of expression of the actions of the nervous system. Without control over skeletal muscle the nervous system cannot long survive, because such vegetative functions as ventilation, which are essential to life, require continual participation of skeletal muscle controlled by the central nervous system. Without appropriate control over skeletal muscle there would be no way for one human nervous system to share with others the knowledge it has gleaned, as it would be incapable of speech or writing; thus the growth of human civilization—or, indeed, of the meanest animal pack behavior—could not have occurred without brain control over muscles of "expression."

The fundamental basis of skeletal muscle control is well known to you already from Chapter 18. Each muscle cell is caused to contract by impulses reaching it over a motor nerve; this nerve controls the amount of work done by varying the frequency with which impulses arrive. Further control of the work done by the entire muscle, of which the cell is a part, is achieved by the extent to which the population of cells within the muscle is activated. It is, of course, a long way from such simple muscle actions to the coordinated interplay of many muscles that occurs in the simplest movements of limbs and body or in the generation of speech. Let us examine what is known of such coordination processes.

REFLEX CONTROL OF VOLUNTARY MUSCLE

All voluntary muscles have within them sensory nerve endings that contact special muscle fibers and

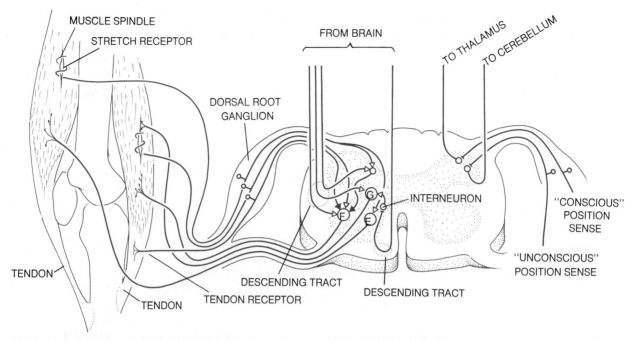

Figure 20-2 Diagram of the motor control system of a skeletal muscle and its antagonist. E and F are motor neurons to muscle fibers of an extensor-flexor pair of muscles. Dark arrows are inhibitory and light arrows are excitatory synapses. The diagram represents the "minimal" neural circuitry present. See text for details.

form with them structures called **muscle spindles** (Figure 20-2). When the spindles are stretched, the muscle is reflexly caused to contract because the spindle sensory neurons activate motor neurons in the spinal cord that operate the majority of the muscle cells. This arc actually constitutes a negative feedback system, a system that resists change (see Chapter 18). Thus, when the muscle is stretched, sensory impulses activate the muscle motor neurons and continue to do so until the muscle shortens enough to reduce the sensory input to the motor neurons to the prestretch, or rest, level.

This serves to hold muscles in the same position. Whenever some external force stretches them, they reflexly contract to the rest condition. But how, then, does one move a muscle usefully? Surprisingly, the same negative feedback system is often used. Rather than sending directions from the conscious parts of the brain directly to ordinary muscle cells, nerve impulses are made to flow to the special muscle cells that are part of the sensory muscle spindle (Figure 20-2). If the flow of these impulses *increases*, the spindle muscle *shortens*. This relieves the strain on the spindle receptors, causing them in turn to diminish the level of excitation of the muscle motor neurons. The result is that the muscle relaxes until the spindle muscle fibers are sufficiently stretched to restore the activity of the spindle receptors to normal. If it is required to *shorten* the muscle, the brain allows the spindle muscles to *extend*, by diminishing the flow of nerve impulses to them and the resultant excitation of muscle motor neurons induces shortening of the entire muscle.

Muscles typically operate in pairs, one, the **agonist**, working against the other, the **antagonist**. If such a pair is to move effectively the structure that they power, one muscle must relax while the other contracts. This cooperation is ensured by a further ramification of the reflex muscle control system just described. The spindle sensory neuron from the agonist, in addition to having excitory synaptic endings on the motor neurons of its own muscle, has inhibitory synaptic endings on the motor neurons of the antagonist, and vice versa.

This control system, which tends to hold muscles in the same position, is called the **gamma control system** and is shown in Figure 20-2. In addition, voluntary muscles may be controlled by a more direct control system called the **alpha system.** In the alpha system, control signals proceed relatively directly from the conscious levels of the brain to the muscles, causing them to contract and overriding the gamma control system. Why two systems? The gamma system is particularly useful in controlling muscles that must perform either a steady, long term task or perform a task requiring great precision of motion.

Consequently, we see that the gamma system is used in control of the postural muscles, which must act steadily over long periods of time to hold the body in whatever stance it has assumed. By "setting" the muscle receptors by signals from the brain, the system can subsequently run spontaneously without the continual attention that would be required if the muscles were being operated directly under brain control. When muscles must perform relatively slow and precise movements, the gamma system is also most useful because its sensitivity can be greatly increased by causing partial contraction of the receptor muscle fibers of both members of agonist-antagonist pairs, making the system very resistant to perturbations that are not directly commanded by the nervous system. On the other hand, when the muscular system is called upon to make very rapid and less precise motions, as for example when throwing something, the alpha direct command system is used. Since alpha control would be interfered with by the gamma system, were the two systems to be functioning independently, the gamma system is activated in parallel so that both the muscle fibers to which the muscle receptors are attached and the main muscle fibers are contracted simultaneously, insuring that no interfering control signal is generated by the muscle receptors.

HIGHER CONTROL OF MUSCLE ACTION AND BODY MOVEMENT

Let us reflect on what happens when we put our muscles to some simple task, as, for example, tearing the telephone directory in half. We do not consciously say to ourselves, "Ah, ha! Here we have the Manhattan Telephone Directory. Tearing it in half will require x g of force applied thus and then do so." What actually happens is that we begin to apply force rather sparingly and then judge how much additional force is required by observing the results so far achieved. Observation goes on at both the conscious and subconscious levels of the brain. Muscle receptors signal to subconscious levels while consciously we feel, see, and hear the feat in progress. Somewhere in the brain all these signals come together; somewhere in the brain their reports of action achieved are compared with the action initially desired, and corrective signals are sent to the musculature.

Comparing Intention With Result

The central place where the signals come together is the cerebellum, which lies attached to the brain stem beneath the overhanging cerebral hemispheres (Figure 20-3). A person with damage to this structure walks in a "drunken" fashion. Voluntary movements occur sloppily, with great awkwardness upon starting, stopping, or changing direction. Even a simple task, such as moving the arm to touch something with the finger, may involve large oscillations in the path followed by the finger rather than the smooth, flowing movement made by a normal person (Figure 20-4). Such observations have long suggested that the cere-

Figure 20-3 The motor cortex controls voluntary movements indirectly by way of the basal ganglia and the cerebellum.

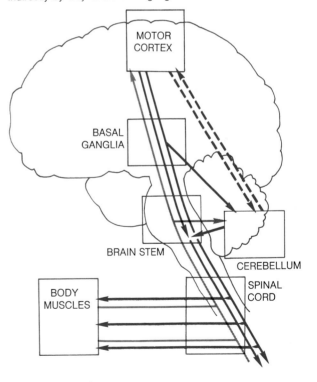

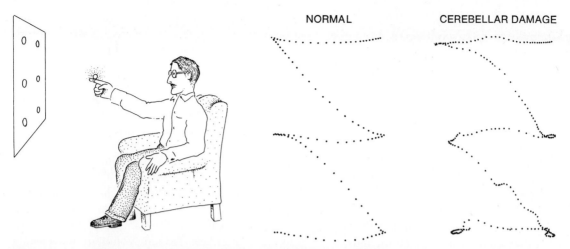

Figure 20-4 Damage to the cerebellum illustrates its importance to movement control. A patient with damage to the right cerebellum attempts to move his right or left index finger smoothly between six points. A flashing light is attached to the finger so that its movements may be recorded in a photographic time exposure in a darkened room. The finger controlled by the undamaged side of the cerebellum performs smoothly while the other shows irregularity in rate and direction of movement. (After Holmes, G., 1939, Brain *62:* 1.)

bellum must play a mediating role between the brain proper and the body in controlling voluntary movement.

This is not a "chain of command" mediation. The cerebral cortex, the center of initiation of conscious activity, does *not* issue commands to the cerebellum, which the cerebellum then conveys to the muscles. The cerebellum, in fact, has no direct way to control muscles. Its only outgoing (efferent) nerve impulses go into motor centers in the brain stem and to the cerebral cortex, from which the cerebellum also receives impulses. These facts, together with its rich sensory input, suggest how the cerebellum may function. "Copies" of commands to muscles from the cerebral cortex may pass to the cerebellum and there be compared with sensory impulses (from virtually all types of receptors, including the eyes and ears) that monitor the actions in progress. Presumably the result would be that signals would be sent back into the brain stem and cortex indicating what further actions should be taken to achieve the desired result. Another crucial function of the cerebellum, maintaining balance of the body, is made possible by input from the organs of balance, the semicircular canals and otolith organs.

The Way In and the Way Out

The lower parts of the brain contain many centers where incoming and outgoing signals are passed from one neuron to another. It is important to understand how these centers function in a general way if we are to see how the cerebrum exerts its effects upon the remainder of the brain. The **thalamus** is the center that receives sensory information destined for the cerebrum from all sources except the sense of smell. If we examine Figure 20-5 we see that neurons receiving sensory information from mechanoreceptors (touch, pressure, and position sense) or pain and temperature receptors ascend through the spinal cord along specific tracts.

In the spinal cord or medulla many of these **decussate,** or cross to the opposite side of the nervous system, and then continue into the thalamus. Thus stimuli from the right side of the body enter the left thalamus and go from there to the left side of the cerebral cortex and vice versa. Signals from the cranial nerves, including those subserving hearing and vision, also enter the thalamus. All of these pathways are highly organized and continue to the cerebral cortex along paths that preserve the spatial relations of the input. Damage to the thalamus not only produces sensory defects, as might be expected, but also commonly produces chronic and severe pain, which the subject refers to some part of the body. Consequently, it is believed that the thalamus may be the principal site of origin of the sensation of pain. It is also involved in generation of certain kinds of generalized emotional responses.

Subconscious inputs to the brain also exist. The

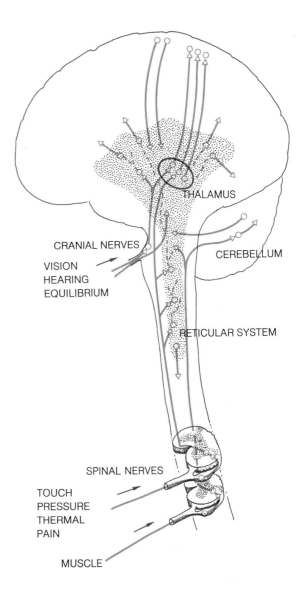

Figure 20-5 Thalamus and reticular system. Each half cerebral cortex receives sensory input from one of a pair of thalami, which are capable of crude sensory sensation but pass sensory input on to the cortex for finer analysis. The reticular system (shaded in the figure) lies in the midbrain and medulla and performs many functions, regulating sensory input and consciousness as well as such housekeeping functions as respiration and heart rate.

cerebellum receives information independently of the thalamus by spinal tracts which supply signals from muscle receptors directly to it and also to the **reticular formation,** which forms the central core of the midbrain-medulla region. The reticular formation consists of many groups of nerve cells elaborately interconnected by a network of fibers, hence the term reticular. The system has ascending (ingoing) and descending (outgoing) components. The output to higher levels from the ascending system determines the level of consciousness of the individual, as we shall see a little later.

There are two direct routes for both conscious and subconscious commands to proceed from the cerebral cortex to the periphery. One goes to the motor neurons of the cranial nerves and serves as the control route for speech, eye movements, and operation of the facial and chewing muscles. The second route descends through the spinal cord and controls motor neurons controlling the neck, thoracic, pelvic, and limb muscles. Neurons in this system terminate directly on spinal motorneurons controlling the digits, or otherwise terminate on interneurons, thus morphologically emphasizing the very great importance of precise hand and foot movements to the well-being of the organism.

The Command Center

In the lower vertebrates the cerebral cortex is not of great importance. The entire structure may be removed from a frog and it still can see and generally do what frogs do. A similar operation on a human would completely extirpate its humanity. It would be blind, deaf, and unable to communicate or move, except reflexly. In vertebrate evolution, the most marked developments in the brain occur in the forebrain. The earliest development of the forebrain is the hypothalamus, which throughout the vertebrates retains an important constellation of functions. It is also the forerunner of the first part of the cerebral cortex to evolve, the olfactory cortex. In fishes, the hypothalamus and olfactory cortex are by far the dominant

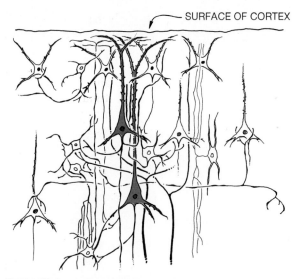

Figure 20-6 Neurons with large synaptic areas are complexly interconnected in the cerebral cortex.

forebrain structures. In amphibians, the dorsal thalamus appears and it is the beginning of further cortical structures that culminate in the neocortex of mammals. Out of the neocortex emerges the huge cerebral cortex of man and the primates.

The human cerebral cortex is estimated to contain 100 billion nerve cells. These are complexly organized into several identifiable layers (Figure 20-6), which occupy only the outer 2.5 mm of the cortex. Due to the extensive folding of the cortex the surface occupied by the cells of the cortex is quite large, approximately 2300 cm^2.

THE WORK OF THE CEREBRAL CORTEX

To try to understand how the cerebral cortex works we might logically proceed from the example of the spinal cord. After all the cortex has been shown to have specific paths in and out. Perhaps then it is just a very fancy reflex maker. Perhaps if you have a mechanistic outlook on life you would hasten to agree, since this would eliminate the necessity of looking for such physiologically embarrassing entities in the cortex as personality, will, morality, conscience, or even the soul. But if we wish to proceed on this basis we must concede immediately that there is no comparison between a spinal reflex and the simplest activities of the cerebral cortex in terms of richness of input, of possible pathways between input and output, and of output. The cortex receives input from essentially all the sense organs, from the dominating flood of information from eyes and ears down to the most obscure tactile receptor.

This, however, is only the beginning, because the cortex is not a creature of the moment, responding only to current stimuli in their immediate context. Since it is the residence of memory, it has the capability of evaluating present stimuli in the light of previous experience. Since it is the locus of reasoning ability, it can formulate its response to present stimuli in the light of remembered responses to previous stimuli. Indeed, these considerations make it ludicrous to speak of the cerebral cortex as a reflex maker in any strict sense of the word. Nonetheless the concept is useful as a device to make the point that the cortex is not the residence of a supernatural family of processes that are unassailable by the physiologist. The physiologist assumes, until it is proven otherwise, that the cortex functions according to the principles that govern the activity of any neuron or synapse.

Whether the physiologist will ever be able to understand the cortex completely on these terms is an open question. But if the brain evades this type of analysis, we must assume, until proven otherwise, that the evasion is based solely on the sheer number of neurons and the richness of their interconnections and not upon principles that lie outside natural laws. When one considers not only the sheer number of neurons in the cortex but the fact that each cortical neuron probably has between 100 and 1000 synapses with other neurons, it seems highly probable that the richness of the behavior generated by the cortex can be explained by this great network.

LOCALIZATION OF CORTICAL FUNCTIONS

The long path towards understanding how the cerebral cortex operates is traced to the late 1800s when physiologists began to study the effects of localized electrical stimulation and removal of small areas of cortex. Regions were found in which minimal electrical stimuli produced movements of specific parts of the body, and when those regions were removed paralysis of those same parts resulted. All such regions together are called the **motor areas** of the cortex, and they are arranged in a similar way in all mammals. As shown in Figure 20-7 the amount of cortical area devoted to control of a given part of the body is related to the complexity of the movements that part performs. Thus the area devoted to the trunk is small as compared to tongue or finger areas.

Are there sensory areas corresponding to motor areas? There are, but their study had to await the development of surgical techniques and sufficient medical reasons for operations on the human brain. Electrical stimulation of the brains of conscious patients is possible because the brain itself does not react with pain signals to direct stimulation and because other pain associated with surgery can be eliminated by local anaesthesia. Under such conditions, by stimulating a particular spot and then asking the

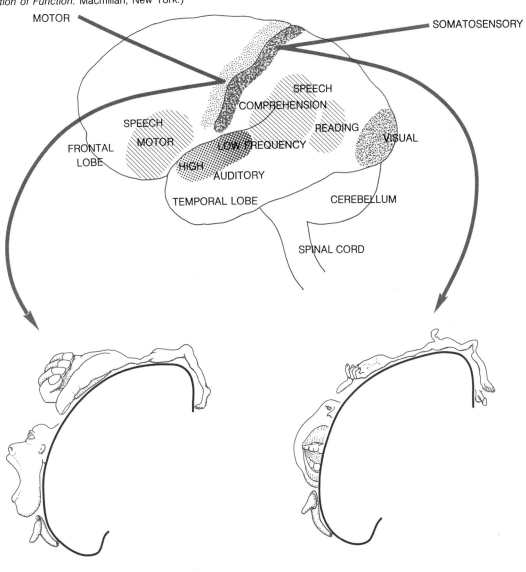

Figure 20-7 Map of functional areas of the cerebral cortex. The lower figures are sections of the motor and somatic sensory regions with the extent to which they are committed to various parts of the body indicated by the relative size of the parts of the homunculi. (Sensory and motor homunculi after Penfield, W. G. and T. B. Rassmussen (1950) *The Cerebral Cortex of Man: A Clinical Study of Localisation of Function.* Macmillan, New York.)

THE BRAIN

patient what is perceived, a sensory map of the cortex can be constructed.

Three sensory areas are found (Figure 20-7). One parallels the motor area and is called the **somatosensory** area. It receives stimuli from the body and is represented by a map in which the magnitude of the sensory imput is reflected in the area devoted to a particular body region. In addition, there are specific areas for hearing and vision. In keeping with our human emphasis on vision, the visual area is larger than the entire somatosensory area. Analogously to the somatosensory map, the auditory cortex is laid out like a functional map of the organ of hearing, the cochlea. As stimuli are applied from one end of the cortical auditory area to the other, the frequency of apparent sound perceived varies in a regular way. The organization of the visual cortex is extremely complex and is far from being fully understood. Nonetheless, it is clear that the visual cortex is spatially organized on the basis of a complex but highly modified "map" of the retina. Stimulation of the visual cortex produces sensations of unstructured flashes of light, which appear to come from various parts of the visual field.

Called **visual phosphenes**, these nonvisually induced flashes of light are being used by biomedical engineers as a way to salvage some use of the visual system in persons who are blind from defects occurring peripherially to the visual cortex. If a grid of several stimulating electrodes is implanted so that each electrode activates a different part of the cortex, the patient can learn to associate activity of each electrode with the particular phosphene that it generates. Then the electrode array can be connected with a system of light detectors so that certain patterns of phosphenes are generated by certain patterns of light. Eventually it may be possible to use such a system to construct a reading device capable of translating printed letters into identifiable phosphenes.

Secondary Sensory Areas

The sensory areas just described are known as the primary sensory areas. Secondary sensory areas or **sensory association areas** surround them. These become active whenever the related primary sensory area responds to sensory input. The association areas are responsible for a more sophisticated level of analysis of sensory input. For example, damage to the secondary somatic sensory area causes loss of spacial perception of particular parts of the body, even though perception of stimuli from those parts persists. Or, damage to the secondary visual area may obliterate the ability to interpret complex visual input, although it does not cause blindness. Thus, written words may not be recognized even though perfectly understood when spoken.

HIGHER FUNCTIONS OF THE CEREBRAL CORTEX

The areas of the cortex so far described are devoted to fractionated views of the total sensory experience. In the **temporal lobe** (Figure 20-7) we find evidence of higher and more integrated levels of analysis. A person with damage to this area may be able to read perfectly well, but still be unable to define the thought that the words present. Electrical stimulation of specific sites in this area may evoke complex and detailed memories in contrast to the simple sensory manifestations resulting from stimulation of the primary sensory areas. Extensive damage produces a wide range of high level defects, although the patient is able to function reasonably well in terms of sensory perception and other processes. A particularly fascinating— and tragic—example is detailed in Note 20-2.

The **prefrontal lobes** of the cortex (Figure 20-7) have always been suspected as being the primary residence of the mental properties that differentiate the human from other primates, since they are substantially larger in the human than in other forms. In addition, they mature more slowly than other parts of the cortex, not achieving full growth until the end of childhood. Beginning with the first documented case of severe damage to the prefrontal lobes, a medically famous accident in 1848 in which a railroad worker had most of his frontal lobes carried away by an explosion in which a tamping iron entered his skull below the eye and exited in the anterior midline (Figure 20-8), there have been many instances recorded in which such accidents produced marked personality changes.

Considering the magnitude of the loss of tissue these changes must be considered rather slight and might even go unnoticed by observers who knew the victim only slightly, particularly if the observer was unacquainted with the victim's behavior before the

Note 20-2. *Sublieutenant Zasetsky: Autobiographical Notes After a Severe Brain Injury*

During the Second World War a well educated young Russian, Sublieutenant Zasetsky, suffered irreversible damage to his left cerebral hemisphere. He survived and was studied by the eminent Russian neurologist, A. R. Luria. The following quotations are from Zasetsky's own diary, composed with unbelievable effort due to the effects of his wound, on higher cerebral functions, and published by Luria, with his own comments on the case as *The Man With A Shattered World; the History of A Brain Wound* (1972, Basic Books, New York.)

Just before being wounded Zasetsky recalls:

I made the rounds again, talked with each of my men . . . I looked to the west, to the opposite bank of the Vorya where the Germans were situated . . . we had to get through somehow . . . everyone stepped up his pace and moved on across the icy river. The Germans waited silently . . . Then all at once there was a burst of fire from their side machine guns . . . bullets whistled over my head, I dropped down for cover. But I couldn't lie there waiting . . , I jumped up . . .

Then he was hit and later wrote these recollections of his first period of consciousness:

Somewhere not far from our furthest position on the front lines, in a tent blazing with light, I finally came to again . . . At first I couldn't even recognize myself, or what had happened to me, and for a long time didn't even know where I'd been hit . . . I seemed to be some newborn creature that just looked, listened, observed, repeated, but still had no mind of its own . . . By the end of the second month I recalled who Lenin was, understood words like sun, moon, cloud, rain, and remembered my first and last name . . .

His recollections continue in the rehabilitation hospital:

I still have to read syllable by syllable like a child . . . I've had a hard time understanding and identifying things in my environment . . . when I see or imagine things in my mind . . . I still can't think of the words for these right away.

His therapist notes their first conversation:

What town was he from?
 "At home . . . there's . . . I want to write, just can't."
Any relatives?
 "There's . . . my mother . . . and also—what do you call them?"
Read this page.
 "What's this? No, I don't know, don't understand, what is this?"

He writes later about his condition:

Ever since I was wounded I haven't been able to see a single object as a whole . . . When I look at a spoon, at the tip, I'm amazed . . . I only see the tip and not the whole spoon . . . sometimes I'd actually get frightened when the spoon got lost in my soup.

Often, I even forget where my forearm or buttocks are and have to think of what these two words refer to . . . When the doctor says: Hands on your hips! I stand there wondering what this means . . . During the night I suddenly woke up and felt a kind of pressure in my stomach. . . . but it wasn't that I had to urinate—it was something else. But what? I just couldn't figure it out. Meanwhile the pressure in my stomach was getting stronger every minute. Suddenly I realized I had to go to the john but couldn't figure out how. I knew what organ got rid of urine, but this pressure was on a different orifice, except that I had forgotten what it was for.

Zasetsky continued his remarkable diary for many years, but, sadly, it reflected no improvement.

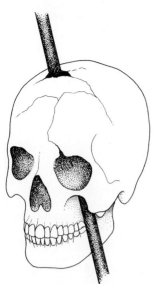

Figure 20-8 The skull of Phineas Gage showing the route of the tamping bar through his brain.

accident. In general, damage to the prefrontal lobes produces two categories of effect. The victim shows great lack of inhibition in social behavior and is prone to rapid changes of mood. Secondly, the victim shows great distractability, finding it very hard to pursue a line of thought. One might say the person with prefrontal damage lives in the present, reacting immediately to stimuli without consideration of alternate plans of action and showing a generalized inability to plan a course of action into the future.

Since it is a relatively simple surgical procedure to isolate the prefrontal lobe from the remainder of the brain by separating its connections, **leucotomy (lobotomy)** was performed in the 1940s and 1950s on a number of persons suffering from profound mental disturbances characterized by overwhelming anxiety. The operation was successful to the extent that it produced individuals with no concerns outside of the immediate present, but the overall effects were so damaging to the individual's behavior that the operation is no longer used for such purposes.

Right and Left Cerebral Hemispheres

In constructing a functional map of the cerebral cortex we have not yet considered some remarkable consequences of the fact that the cerebral cortex is a bilateral structure, with a right and left half. For most bilaterally displayed functions, such as the control of muscle masses in the left or right half of the body, the right side of the cerebral cortex controls the left side of the body and the left controls the right. This follows quite unremarkably from the crossing over of nerve tracts as they proceed into or out of the brain from the spinal cord. But what about speech? This process can be said to have a singular output. We do not talk on the left side or on the right side, we just talk. Therefore, it is worth inquiring as to which side of the brain controls the phenomenon of speech, or whether both sides participate.

The facts are that in most persons the left cerebral hemisphere controls not only generation of speech but also its comprehension. A few left-handers may be the other way around, or even have both hemispheres participate in the process. There is no obvious reason for this. It just happens, but, quite interestingly, the right side of the cerebral cortex can control speech if it is stimulated to do so early enough in life. If a very young child has its normal left side speech center damaged, the right side swiftly picks up the task and there may be little or no retardation in learning to speak. A similar defect in an older child results in a period of inability to speak, but then the process switches to the right side and the child goes through the language learning process again. Later in life the transfer can rarely be made and the victim of a defect in the left hemisphere speech area has permanent speech problems.

The Split Brain

The fact that the normal individual has a speech center in the right cerebral hemisphere that seems to contribute nothing to speech raises fascinating questions. Does the right cerebral hemisphere have other seemingly redundant areas? This would at first glance seem likely because we have seen nothing that the right side of the cerebral cortex does except control movement on the right side of the body and receive sensory stimuli. What about all of the other regions of the right cerebral hemisphere outside the motor and sensory areas? Have their functions been totally supplanted by the dominant left cerebral hemisphere? If the example of the transfer of speech from left to right hemisphere early in life is any indication, it seems likely that the two halves of the cerebrum at least start life with equal potentialities, so it would seem to be a great waste of evolutionarily hard-won brain tissue to, in effect, throw half of it away in adulthood.

The problem, of course, is to find out what is in the

right hemisphere, since the left hemisphere is so dominant. Only if the techniques are available to, as it were, question the right side of the brain and get its answers independently of the left side can we answer the questions that the bilaterality of cerebrum raises.

Fortunately the brain is constructed so that the left and right cerebral hemispheres can be questioned independently. The left and right hemispheres are connected by relatively restricted pathways, principally the **corpus callosum,** about 5 mm thick, containing millions of nerve fibers running between the two hemispheres. These connections have been severed experimentally in monkeys, but the most interesting data on separated cerebral hemispheres comes from a few humans who underwent section of the corpus callosum in the course of treatment of epilepsy. These patients have been exhaustively studied by R. W. Sperry and his associates. After recovering from the immediate effects of the operation, the split-brain patients appeared normal to casual observation. However, under appropriate experimental conditions they behaved as if they had two separate higher nervous systems, each with radically different properties.

Examination of Figure 20-9 will assist in explaining how this comes about and how the two halves of the cerebrum are tested. The figure includes a diagram of a split brain from above. Note that, although the optic nerves supply information from each eye to each half of the cerebrum, the division of the optic nerves at the optic chiasma is such that the right half of the visual field of each eye goes exclusively to the left visual cortex and the left half to the right visual cortex. Consequently, if something is seen exclusively in the right visual field it will be transmitted exclusively to the left hemisphere, and vice versa. In Sperry's experiments it was possible to do just this. A characteristic result is shown in an experiment in which the word "hatband" was displayed so that the right visual field saw "band" and the left saw "hat". When asked what had been seen the patient responded "band." Since the right visual field conveyed information to the left brain, which controls the speech center, the left brain conveyed only what it had seen, the half word "band."

This outcome is not too surprising, and the experiment leaves hanging the very important question about what was going on in the right side. Perhaps in the adult the right side is unable to perceive anything without the help of the left hemisphere. We have no

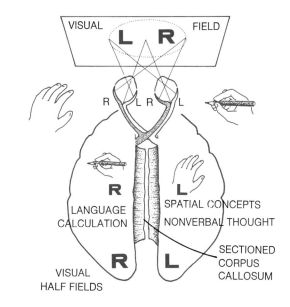

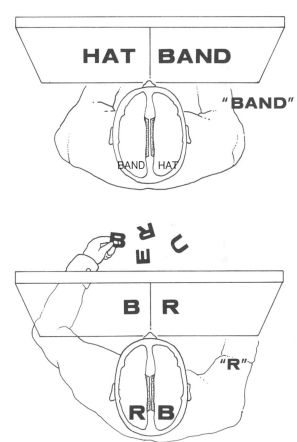

Figure 20-9 Illustrations of the method used by R. W. Sperry to test independently each half of the brain of a split brain patient.

Note 20-3. *Electroencephalography*

In 1929, Hans Berger showed that very small electrical signals (about 100 μV) could be recorded from the scalp. He believed, and it was subsequently proven that these signals represented the massed electrical activity of the brain. In the years since, the electroencephalogram (EEG) has proven to be most valuable in the study of the function of the brain and in identifying abnormalities. For example, signals indicating initiation of motor activity may be seen in the EEG, as shown in Figure 20-10A which shows activity (readiness potential) preceeding voluntary movement. A rising signal is initiated less than a second before the movement begins, as indicated by the vertical line. Figure 20-10B shows EEG records for a normal adult. The first of each pair of records is taken from the front of the head and the second is from the occipital (rear) part of the head. While awake and not performing active thought the brain produces the EEG shown in the two records. Alpha waves (8 to 18 per second) are characteristic of this state, particularly in occipital records. The alpha waves disappear when the subject engages in active thought. They also disappear with drowsiness and the onset of sleep, to be replaced by large amplitude, slower waves, as shown in the remaining three sets of records. A pathological state is shown in Figure 20-10C, which shows the large, regular waves characteristic of an attack of *petit mal*, a disorder characterized by brief episodes of convulsions. The record shows the beginning of the attack and, after a deletion of part of the record, its termination. Time between the arrows is 15 seconds.

Figure 20-10 **The electroencephalogram. A.** Electrical activity preceeding a voluntary movement, whose time of occurrence is indicated by the vertical bar. **B.** Examples of the normal EEG; see Note 20-3. **C.** Record of an attack of *petit mal*. **A** after Deecke, L., et al. (1969) *Experimental Brain Res.* 7: 158. **B** and **C** after Kooi, K. A. (1971) *Fundamentals of Electroencephalography*, Harper and Row, New York.

---

way of knowing as long as we confine ourselves to systems of questioning that require verbal answers since the right side has no control over speech. We know, of course, that, working normally with the left side, the right side must be able to register information since after all we see out of both sides of our eyes and the images are whole and well assembled across the binocular visual field.

The second experiment shown in Figure 20-9 tests a response system that is available to each half cerebrum, namely the motor and tactile sense system. Two letters are projected, one to each hemisphere and then the subject searches, without using vision, with each hand, seeking to find a carved out letter corresponding to the letter each respective controlling hemisphere saw. The task is readily completed by each hand-hemisphere. Sperry noted that frequently one hand would come upon the object sought by the other hand. In such cases there would be no sign of recognition and the hand-hemisphere would go on searching until the object registered on its own visual field was located. The two half brains, therefore, separately are able to carry on a complex task of perception but cannot communicate their perceptions to each other without some external link, such as speech. Further, since the brain is still cross-connected below the cerebrum, it follows that all conscious brain activity that normally does flow from one cerebral cortex to the other does not find routes through these lower parts of the brain.

These operations to divide the corpus callosum were done in midlife. Considering the plasticity shown in the left to right transfer of the speech center in the very young, it is of interest to inquire if the young brain is able to correct such a defect and be able to speak with control from both hemispheres. This seems to be the case in an instance of normal lack of development of the corpus callosum. This person, studied by Sperry, showed none of the deficits exhibited by operated patients. For example, when asked what her right brain had seen she was able to answer correctly, and the same was true for the left

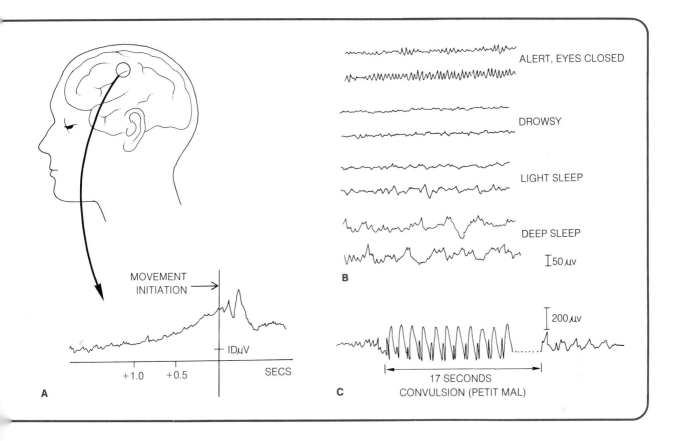

side. The two halves of the cortex seemed to be working in parallel.

Laborious accumulation of information from patients with damage to either left or right hemisphere as well as studies on split brain patients indicate that the dominant and nondominant hemispheres play different roles in higher thought processes. In general, the dominant, usually left, hemisphere seems to be specialized in handling quantitative, and, of course, verbally organized material. The nondominant hemisphere handles nonverbalized thought and seems to excell over the other hemisphere in what may be called spatiotemporal material. This separation in function is supported by several kinds of observations. For example, a Russian composer is reported to have continued excellently to compose music after severe damage to his left hemisphere left him unable to write down the notes of his compositions. Nevertheless, he could compose, remember, and play them. In further illustration, the left brain usually has a harder time drawing three-dimensional subjects than does the right hemisphere in split brain patients.

Recently, investigators have taken advantage of the fact that it is possible to determine electrically whether one side of the brain is active or not by measuring brain waves, or the **electroencephalogram (EEG)** (Note 20-3 and Figure 20-10. This technique makes it possible to get an indication if the normal person uses one side of the brain for one type of mental task and the other for another. The results indicate that the two sides of the brain are specialists. When a person is writing, or even just thinking about writing, the left side of the brain shows an EEG indicates of attention and activity whereas the EEG of the right side indicates that it is loafing along. When the subject is doing a musical or artistic task, it is the right side that seems to be at work. Thus it appears that the nondominant side of the cerebrum contributes importantly to higher thought mechanisms. In keeping with this, it is generally observed that persons who congenitally lack connections between the hemispheres are seldom of more than average intelligence.

Learning and Memory

At this point we know a little about the locations of various motor and sensory activities in the cerebral cortex, and it is clear that the cerebral hemispheres are the locus of the higher processes that we call thinking, learning, and memory. These are the brain activities that raise us above the rest of the animal kingdom, yet it is also true that these processes differ only qualitatively between humans and the lower animals. Because of this, it is useful to discuss the higher functions of the nervous system in a phylogenetic way.

PLAIN AND FANCY LEARNING

One way to relate human mental processes to the activities of simpler brains is to say that the kinds of rewards or goals that "satisfy" brains are a function of their sophistication. Those rewards that satisfy the most simple animals are the basic necessities of life, and the activity of the brain in attaining them is describable in terms of straightforward homeostatic loops. Brains of greater complexity are able to comprehend less direct routes to the attainment of the necessities, and consequently their activities—goal seeking—are more complex.

In the human, this evolution of sophistication has gone so far that the connection between homeostasis and brain activity sometimes becomes virtually impossible to unravel. Beyond this, the functional capacity of the human brain has become so large that it seems not to need to devote its entire energies to the immediate necessities of survival. Thus, alone among animals, humans are able to look beyond their requirements for survival and follow pursuits that seem to have no relationship whatsoever to survival. The goals have progressed from the basic necessities of living to those we traditionally associate with the "intellectual life," the search for knowledge for the sake of knowing, the appreciation of the arts, and the pursuit of pleasure. Yet who is to say that even these are totally dissociated from the necessities of life? Knowledge in its purest form has a way of eventually producing results that are useful for survival. We must admit that the brain as an organ of evolutionary progress is infinitely more varied in its adaptive mechanisms than any other organ that might contribute to survival.

To some degree or another, all organisms rely on instinctive behavior, which we may call innate "programming". At some time in the evolutionary past, each piece of behavior that we now call instinctive became fixed genetically in the nervous system so that it comes into play when triggered by the appropriate stimulus without the necessity of relearning in each generation. The advantages and disadvantages are obvious. The organism does not have to undertake the possibly risky relearning process in each generation and does not need to devote a large fraction of its nervous system to the machinery necessary to reestablish this particular bit of behavior. Thus we see that instinctive behavior tends to dominate in organisms with small nervous systems, with the insects serving as the ultimate example.

The disadvantage associated with instinctive behavior is equally obvious. Since the instinctive behavior is fixed, the possessor cannot readily modify its behavior to adapt to changing conditions that might render the behavior useless or actually detrimental. A classical illustration is the mud-dauber wasp, which builds and stocks its nest with food for its larvae according to a series of instinctual commands. Once the nest is complete, instinct commands the wasp to hunt prey and fill the nest to a certain level before laying its egg on the food supply thus provided. If the experimenter carefully removes the bottom of the nest so that as food is placed in the nest it falls out, the wasp is locked in an instinctual trap. It cannot comprehend the difficulty and perform adaptive behavior such as repairing the damage to the nest. Instead, it simply continues indefinitely to bring new food supplies and drop them into the bottomless pit. In this pattern of instinctive behavior, each operation must be completed before the next is started and, if there is a failure of the type described, the organism is unable to proceed to the next step.

However, even the simplest organisms are not totally locked within a limited repertoire of instinctive behavior because they have some ways to adapt their neural responses to changing conditions in the environment. Of course, you may ask if such changes in behavior are legitimately classified as learning. Certainly in their simplest form they are remote from learning in the sense of the process that you apply to solving chemistry or math problems. However, if we consider learning across the entirety of the animal kingdom, we find that it can be generally defined in terms that include both the highest and lowest forms of the process. We may say that **learning** is simply persistent modification of behavior because of experience. The modifications that we include in the category of learning do not include *fatigue* (change of a response simply because of exhaustion of the organism or part of its nervous system), *variations in attention* (the overall level of alertness of the nervous system), or *effects that directly modify the reactivity of peripheral parts of the nervous system,* as for example sense organs. Thus a bright flash of light in the eyes followed by a change in behavior with reference to dimly seen objects would hardly be called learning.

Habituation

Habituation is the simplest behavior that we know to fall within the limits of our definition of learning. In habituation, the response of the organism to a regularly repeated sequence of nearly identical stimuli is modified. The shadow reflex of a marine worm is a good example. When a shadow is cast over its head end, the worm instantly withdraws into the protection of its tube. When this is repeated at regular intervals each time the worm reappears, the withdrawal response rapidly disappears. A simple model of habituation is diagrammed in Figure 20-11. A stimulus, the shadow (S), initiates a reflex, which results in a response (R). This would be exactly similar to the knee-jerk reflex, except that possibly more neurons are involved.

But how does habituation occur? That the disappearance of the response is not due to fatigue is shown by varying the character of shadow stimulation—perhaps by changing its timing, or the darkness of the shadow—whereupon R reappears in its maximal form. Therefore, the organism has some way to evaluate events subsequent to S and to determine whether R is appropriate. If S is not followed by some related consequent stimulus, S_R, for example, jarring of the worm tube, as by a predator that might originally have cast the shadow, then S is probably harmless and does not warrant R. We therefore add to the diagram an additional loop that can turn off R if S_R indicates R is not the appropriate response to S. Thus we have a simple learning "circuit."

Sensitization

Behaviorally somewhat more complex is the process of sensitization, in which a response initially brought about by a particular stimulus (S) *transfers* to other stimuli (S_1, S_2, and so on). This is not like conditioning, discussed in the next section, in which S and S_1 always occur in pairs. With sensitization, it is as though the sensitizing stimulus, S, has closed neural switches that channel the response to many other kinds of stimuli so as to produce one particular R (Figure 20-11). For example, a fish swimming quietly in an aquarium might pay no particular attention to minor stimuli, such as a hand waving outside the aquarium, tapping on the glass, and so on. But if the fish is given a stimulus bothersome enough to produce agitated swimming, perhaps an electrical shock, then for a time afterwards it is very apt to exhibit the same escape response to previously ignored minor stimuli. Behavior of this type is probably advantageous. If the fish of our example had been sensitized by a predator instead of by an electrical shock, then subsequent stimuli might be due to the predator lurking about, and an escape response might be useful.

Conditioning

In conditioning an already-learned stimulus-response relationship is transferred to another specific stimulus. Consider a very common example: many a household cat has learned that if it dashes into the kitchen upon hearing the sound of its dish being set on the floor, it will be rewarded with food. Now, if the cat's owner buys an electric can opener, the cat will hear a new stimulus, the whir of the can opener, just before the familiar dish noise. After hearing this sequence a few times, the cat, more than likely, will begin showing up in the kitchen before the dish noises occur. The cat has learned the relationship of the two noises to the reward and now needs to hear only the first to respond. From such observations we derive the concept of conditioning (Figure 20-11): an animal having a definite response (R) to some stimulus (S) may learn to make the same R to another stimulus (S_1) if the two stimuli are presented together prior to the reward. The animal is said to be conditioned to

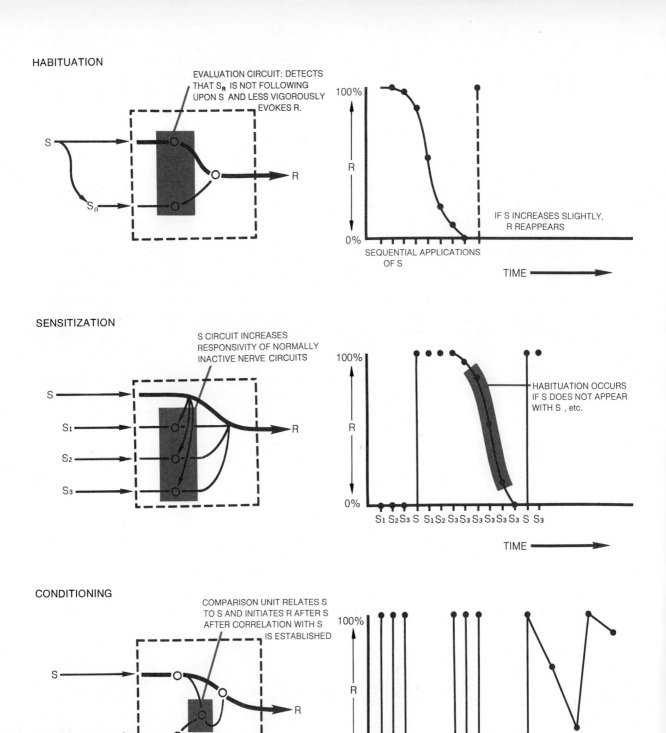

Figure 20-11 **Models of simple learning processes.**

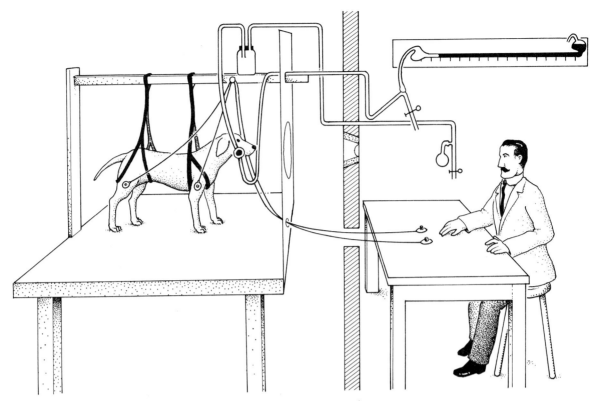

Figure 20-12 Arrangement of one of Pavlov's conditioning experiments. The experimenter observes dog from another room and is able to apply stimuli and measure saliva production.

respond to S_1 by pairing S_1 with S, which already produces a response.

Early experiments by the Russian Pavlov on conditioning actually involved a part of the nervous system that one might think was beyond higher nervous control. For example, dogs, like many mammals, begin salivating upon sight or smell of food. This is a response involving the sympathetic nervous system. In an experimental situation (Figure 20-12), after a bell was sounded a few times along with presentation of food, Pavlov discovered that salivation occurred even when the bell was sounded without the presentation of food. Thus a conditioned response to the bell was generated, and this obviously involved interaction of the brain with the autonomic nervous system.

Once established, a conditioned response is not necessarily permanent. If the conditioned stimulus (S_1) is repeatedly presented without the associated reward, the response becomes sporadic and finally disappears. Some trace of the conditioned state persists, for, if after the response has disappeared, S_1 is not presented for a time and then is presented once more, the response may reappear for a while. Or, if the response has been completely abolished, retraining by the original process of associating S, S_1, and R often is easier.

BIOFEEDBACK

Pavlov's experiment in which a dog is conditioned to salivate on hearing a bell clearly demonstrated that conditioning could involve the autonomic nervous system. Although the ability to condition the autonomic system clearly might have some practical applications, as for example in controlling blood pressure, the technique was not extensively applied until after Neal Miller and his associates demonstrated the remarkable extent to which autonomic control could be attained with the proper type of conditioning. Prior to their work, it had been generally thought that reward-type conditioning would not work for autonomic processes. Whatever success had been achieved was suspected to involve conditioning of voluntary muscular processes, which could then have affected autonomic behavior, thus giving the spurious effect of direct conditioning of the autonomic system.

As shown in Figure 20-13 this criticism was gotten around by conditioning a rat with all of its voluntary musculature completely paralysed by an injection of curare, which blocks voluntary neuromuscular junctions. Such a preparation requires artificial respiration and, of course, it is not possible to reward a paralysed animal in the usual way by giving it a bit of food. Instead, an electrode was implanted in the pleasure center of its brain (see page 438). Normal rats with such implantations will work incessantly at pushing a lever to administer shocks via electrodes to the pleasure center. The heart rate was monitored by electrical circuits, which could trigger reward stimuli through the pleasure center electrode whenever the heart rate changed, according to the experimenter's requirement, either speeding or slowing. As the graph in Figure 20-13 shows, the experimental feedback system is extremely effective at changing the heart rate and this, of course, must be a direct effect via the innervation of the heart since all voluntary muscles are inactive.

Today biofeedback is an extremely popular phenomenon. For medical purposes it has been clearly of value in certain limited areas. For example, if a pattern of innervation is changed because of surgical reconstruction of the face, biofeedback may assist the patient to regain control of facial muscles far sooner than otherwise. In this instance biofeedback training is conducted with the conditioning stimulus consisting of the electrical signals produced by muscles over which it is desirable to attain control. Biofeedback has also been used to gain control of certain types of headache caused by muscular activity that interferes with blood flow in the head.

Biofeedback involving the alpha rhythm of the electroencephalogram (EEG) is a controversial subject. The **alpha rhythm** (Note 20-3) appears when the brain is not paying attention to the environment or thinking intensely. Just closing the eyes is enough to increase its frequency of appearance. In the popular mind, the alpha rhythm quickly came to be associated with meditative states, with creativity and relaxation. Once this idea had emerged it took little time to develop biofeedback devices that use the alpha rhythm as the conditioning stimulus. Alpha biofeedback machines are widely sold and often advertised as devices assisting in attaining a more tranquil state, achieving more, stopping smoking, or just generally getting one's self together. Initial experiments suggested that the alpha rhythm could be made more predominant by using a conditioning procedure in which lights indicated when the rhythm was evident. It now seems that these experiments are suspect. The principal difficulty with alpha rhythm training is that

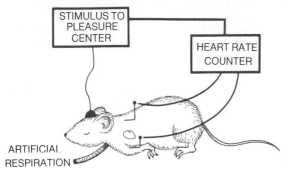

Figure 20-13 In an experiment showing conditioning of the autonomic nervous system an experimental animal with voluntary muscles paralyzed can be trained to slow or speed its heart in response to electrical stimuli applied to the pleasure center of the brain. (From the experiments of Jay A. Trowill.)

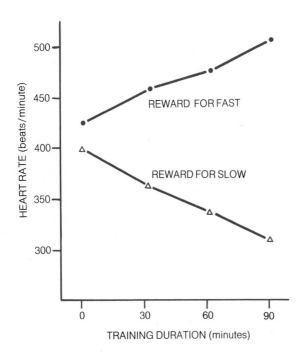

the rhythm appears spontaneously when the subject becomes bored and ceases to pay attention to the indicator lights or other conditioning stimulus. Thus the tranquil state, with increased alpha, appears spontaneously, but probably no more than if the subject became bored in a more conventional manner.

PERCEPTUAL LEARNING

In this, the most complex form of learning, we see the greatest possible flexibility in the relationship between stimulus and response. In the form of perceptual learning known as trial-and-error learning, the animal generates actions more or less randomly. If one of these produces a desired result, the animal may learn to associate action with result. Thus, one might wish to open a locked door and attempt to do so by testing a series of keys randomly. If one works, it is immediately identified and thereafter used to open that particular door.

Finally, in the highest form of perceptual learning, no overt trial-and-error procedure is required at all. This is **insightful learning**. In terms of the locked door situation something like this might happen: the series of keys would be examined and certain ones eliminated from consideration without testing because they are obviously not door keys. From among the remainder, one might note that the manufacturer's name on one of the keys is the same as the one on the lock. Then the insightful act occurs. As the name on both lock and key are the same, it is more likely that that key will open the door than any remaining in the set, and so it is the one selected.

From this it is evident that, at one level, insightful learning is simply trial-and-error learning conducted entirely within the mind. The first round of elimination, for example, simply might have involved comparing key size with keyhole size by looking rather than by testing. The second stage is a different matter, since the characteristic that is used to relate key to lock is not immediately apparent in the learning situation. The insightful learning sequence requires appreciating that, if a particular lock and key are produced by the same manufacturer, they are more likely to fit than if they come from different factories.

How widespread is perceptual learning among lower animals? Early experiments on primates suggested that they lack higher forms of learning. In one illustrative example a chimpanzee was taught to get a cup of water from a particular tap and use it to put out a small flame that blocked access to a reward. It did this very well, but when the chimpanzee was placed in a situation in which water was visible, readily available and nearer than the accustomed tap, it persisted in using water from the tap to solve the problem.

This outcome was taken as evidence that the chimpanzee is unable to generalize, the implication being that its learning is simply a form of conditioning. However, one must note in criticism that the experimenter got exactly what the experiment was designed to demonstrate. The chimpanzee was, in fact, conditioned in the strict sense of the word, and no possibility of reward for a generalized solution of the problem was built into the training situation. Modern behaviorists, observing chimpanzees in nature and particularly investigators teaching chimpanzees language, come to different conclusions. With respect to studies on language acquisition, many examples of perceptual learning have emerged and, indeed, the behavior of chimpanzees such as Sarah (page 327) or Lana, a chimp that converses using a specially constructed computer input, leads to the consensus that perceptual learning ability is a prerequisite to attain language skill at the level these animals demonstrate.

In David Premack's investigations of Sarah many examples of high level learning emerge when the experimental situation is designed to detect them. For example, after Sarah had learned a symbol for chocolate she was then introduced to a symbol meaning brown, which color was not then present, by the array of symbols meaning "Brown color of chocolate." Subsequently she was able to pick a brown colored disc from among several colored discs. We would have to agree that this is an example of perceptual learning.

This admission, of course, does not answer our original question about how widespread this phenomenon may be among animals, but it does appear that we should be most careful in denying this capacity to animals unless we are satisfied that they have been fairly tested. Not only must the test be designed to "reveal" perceptual learning, but it must be constructed to eliminate opportunity for the "Clever Hans" phenomenon to occur. Clever Hans was a trained horse purported to be able to count by stamping a hoof when the trainer commanded it to count to a certain number. It developed, on more sophisticated testing, that Clever Hans had, instead of a finely honed linguistic and mathematical ability, the capacity of detecting minor and probably unconsciously generated cues from the trainer. Regretably, this phenomenon is at the bottom of the ability of our household pets to understand our commands.

MEMORY

Memory, consciously accessible storage of experience in the brain, is an essential adjunct to any significant degree of learning. It is a phenomenon about which we have very little data. Little with certainty is known about how information is stored within the memory system; neither do we know much about the equally difficult problem of how memories are recalled.

Easily the most fascinating series of studies on memory were performed by Wilder Penfield and his associates on a series of over 1100 patients undergoing surgical treatment for epilepsy. In a large fraction of these, treatment involved electrical stimulation of the temporal cortex while recording the responses of the conscious patient in a search for the site of initiation of the epileptic attack. The remarkable conclusion from these studies was that stimulation at specific sites in the temporal region evoked fully structured memories, as illustrated in a typical conversation between surgeon and patient in Figure 20-14.

The completeness of these memories is surprising, considering the general haziness of most of our own memory traces, particularly when they are from long ago and seldom recalled, as was sometimes the case with the memories that Penfield evoked electrically in his patients. Penfield was so impressed by the completeness of recall in the course of his studies that he came to think that the brain normally remembered everything to which it attended. If true, this obviously makes the problem of recall even more interesting since the implication is that much of forgetting involves loss of contact with still-stored memories.

Short and Long Term Memory

As we saw in the beginning of this chapter there is evidence that memories are established in two steps, the first being more easily disrupted than the second. The first phase, short term memory, as we have seen, is easily disrupted by events that would be expected to disturb patterns of electrical activity in the brain—massive electrical shocks or a blow to the head. In the human, if a blow is severe enough, amnesia is produced, and commonly this affects recent memories more severely than older ones, again suggesting that recent memories are less firmly established than older ones. Indeed, although the amnesia victim may have virtually complete recovery of memory, events in the interval just preceding the accident are often never remembered. Long term memory seems not to depend on such transitory processes as neural activity in the brain, since they are not lost by exposure to low temperature, as in hibernation, when the cortex becomes quite inactive.

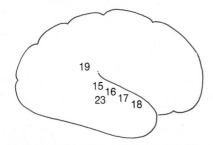

Figure 20-14 Stimulation of specific locations on the cerebral cortex evokes specific memories. From the studies of W. Penfield, 1958, *The Excitable Cortex in Conscious Man*, C. C. Thomas, Springfield, Ill.

MAP OF ELECTRICAL STIMULATION SITES OF RIGHT CEREBRAL HEMISPHERE OF ONE OF PENFIELD'S PATIENTS.

NOTES MADE DURING OPERATION (LOCAL ANAESTHESIA)

Stimulus point	Observations
15.	"I hear singing."
15.	Repeated. "Yes, it is White Christmas." When asked if anyone was singing, she said, "Yes, a choir." When asked if she remembered it being sung with a choir, she said she thought so.
16.	"That is different, a voice—talking—a man."
17.	"Yes, I have heard it before. A man's voice—talking."
17.	Repeated without warning. "Yes, about the same."
18.	"There is the sound again—like a radio program—a man talking." She said it was like a play, the same voice as before.
19.	"The play again!" Then she began to hum. When asked what she was humming, she said she did not know, it was what she heard.
19.	Repeated. Patient began to hum. She continued at the ordinary pace of a song. "I know it but I don't know the name—I have heard it before. I hear it, it is an instrument—just one." She thought it was a violin.
15.	Repeated (26 minutes after last stimulation at 15). "White Christmas," she said it was the orchestra playing.
17.	Repeated (24 minutes after last stimulation at 17). "Yes, the play again."
18.	Repeated (21 minutes after last stimulation at 18). "White Christmas."
23.	"The play—they are talking." When asked who, she said, "The men are talking." When asked who they were, she said, "I don't know."

It has already been mentioned that long term memory may possibly be chemically coded. Evidence for this hypothesis is that, if chemicals that block protein synthesis are administered to animals so as to stop synthesis of protein during the brief period when the memory of a training task is being established in long-term memory, usually the memory is lost. This implied mechanism of memory coding in terms of protein synthesis is supported by the rather common failure of short-term memory in old age, when it appears that many of the characteristic senile changes are due to failing protein-synthetic mechanisms (see Chapter 23).

Because protein synthesis necessarily involves RNA, some investigators suggest that neural activity directly codes memories in RNA-base sequences, in a system analogous to the genetic code. This would, of course, fit with the protein-inhibition experiments, because RNA controls living systems through the specific proteins that it synthesizes. In support of the RNA theory, some experimenters have demonstrated that increases in nerve cell RNA concentration occur upon learning and have even reported transfer of learning from one animal to another upon injection of RNA derived from the brain of a trained animal into that of an untrained one.

Evidence supporting chemical coding is appealing because, if chemical coding does occur, the storage capacity of the brain is tremendously increased over what it would be if storage occurred only in nerve circuits. However, we have a long way to go before chemical storage is proven. The experiments done so far are not satisfyingly clear-cut. Blocking protein synthesis, for example, might erase fragile short-term memories nonspecifically, because such treatment must be generally deleterious; beyond this, the chemical-transfer experiments have been extremely difficult to repeat. Besides the difficulties with experiments, there is the problem that we know of no way for neuronal activity to have such a specific control over cellular metabolism as to allow the specification of base sequences in RNA synthesis. Finally, getting the memory *into* a chemical code is only half the problem. How do you get it out? How, in the split second that is usually required for memory recall, does the brain search through a "file" of chemically coded memories and translate the proper ones back into nerve impulses?

Anywhere and Everywhere. The experiment in which electrical stimulation of one particular spot on the cortex brought forth a particular set of memories suggests that memories, in whatever form they are stored, must be stored in specific regions. Actually, there is much evidence that this is *not* so, and it may be that the localization associated with the electrical-stimulation experiments comes from some other cause, possibly from stimulation of one of several neural access routes into the memory stores. The reason for believing this comes from experiments in which laboratory animals suffered no loss of training even though the cerebral cortex was covered with a network of shallow incisions, which ought to have caused disturbances in recall if memories were stored in specific locations. Indeed, large areas of the cerebral cortex can be removed without serious disturbance of memory or of other mental processes; this is confirmed in humans who have suffered brain damage. Such observations have suggested that a large part of the activity of the cortex depends on the total mass of neural tissue present and not on any specific small part of it: memories and thought may reside "anywhere and everywhere" in the cortex.

INTELLIGENCE

Intelligence—Product of Learning Ability and Memory

Since man depends so much on his brain, there have been many attempts to arrive at a quantitative measure of mental ability—intelligence. This we define as related to learning ability and memory. If intelligence could be measured accurately, such information would be very useful, indeed. For example, it might be possible to place students more accurately in the proper vocational training programs. Or again, it might greatly assist in evaluating the effectiveness of new teaching methods. Unfortunately, measuring intelligence is a very difficult matter. What we might call the **potential intelligence** of any person—the maximal intelligence that might be exhibited if all potentialities were developed to the fullest—is almost never equal to the **apparent intelligence**—the intelligence that can actually be measured.

This is due to two general types of factors. One is the conditions of life, which to a greater or lesser extent prevent the development of intelligence to its maximum. These conditions may be either biological

Figure 20-15 Experiments with young rats show that environmental quality influences brain development and intelligence. Rats living in stimulating environments, as shown, are more intelligent and have increased brain size and nerve cell complexity when compared with rats kept in sterile environments. (E. L. Bennett.)

or cultural. Simple examples of the former are birth complications and childhood malnutrition. In the course of a difficult birth, the fetus might suffer a lack of oxygen. This might irreversibly damage brain cells and thus significantly diminish the capacities of the brain. Malnutrition, especially the protein lack that is so characteristic of poverty, can impede the maturation of the brain during childhood and, again, significantly diminish the capacities of the brain. Even if the biological maturation of the brain is ideal, cultural factors may impede attainment of full capability. The richness or variety of the environmental stimuli that impinge upon the developing mind seems to have a great influence. It seems that the more active is the developing mind and the greater variety of the things that it does, the more nearly its abilities develop to the maximum. This phenomenon is seen even in laboratory rats reared under "culturally" sterile or rich conditions. Those reared in a rich environment—one with many objects to explore, for example—were shown to be better learners when tested later, and there is evidence that the structure of their brains is favorably affected by the rich environment (Figure 20-15).

The second factor affecting apparent intelligence is also culturally influenced, in that it is nearly impossible to eliminate from *intelligence tests* biases due to the type of culture in which the tested individual was reared and is living. The first generally acceptable intelligence tests were devised to measure the intelligence of school children and were evaluated in terms of how the score that an individual made related to subsequent success in school. Obviously there must have been bias in these tests in favor of intelligence factors that fitted the child best for a school environment. Subsequently, intelligence testing has undergone many modifications to rid test procedures of cultural bias, but we are still by no means certain that all such bias has been eliminated. Certainly, cultural bias becomes more difficult to eliminate from testing the older the subject is. For this reason intelligence tests are most accurate for children.

In the attempt to differentiate between the effects of heredity and the environment, great reliance has been made on studies of identical human twins in which intelligence is compared between pairs reared together and apart. Since the twins are genetically identical, it is argued that differences seen between twins reared separately must be of environmental origin. Unfortunately, the scientific community appears to have been deluded in the principal study of this type, one conducted by Sir Cyril Burt, according to recent reinvestigations of his massive study of 53 pairs of identical twins reared separately. Burt reported a very high correlation of intelligence between members of such pairs, although not as high as the correlation between members of pairs reared together. Although this is not surprising and may well be true, the unfortunate fact is that Burt's data shows signs of forgery. For example, some of the statistical evaluations are too uniform to be plausible. Further investigation has revealed more difficulties with the studies, including the possibilities that Burt's coinvestigators in a number of critical studies may never have existed. Regretably, Burt died in 1972, before these suspicions were raised, and his records have been dispersed or are too inadequately organized to support a detailed reinvestigation.

Intelligence Quotient

The results of intelligence testing procedures are usually presented in terms of the **IQ**, or **intelligence quotient**, which is the tested mental age of the individual divided by chronological age, times 100. If a child of 5 years of age performs on a mental test precisely as the average 5-year-old would, then the IQ is 100. If the child performs as well as a 7-year-old, then the IQ is 140. The mean IQ is 100. Individuals ranking as low as 70 are able to lead normal lives. Those with lower IQs are classed as mental defectives. Approximately 0.1% of the population has IQs above 140. Anyone with an IQ of above 130 is classed as "gifted."

Does IQ Differ Among the Races?

This is a sensitive question because claims of racial IQ differences are frequently used as justification for racial discrimination. Moreover it is an extremely difficult question to settle because the markedly different cultural backgrounds of the races make it difficult to judge the reliability of the testing procedures. Since the human races are genetically different, a useful way to consider the question is to inquire first into the genetic basis of intelligence.

There are certainly numerous examples of clear effects upon intelligence of specific genetic differences (Table 20-1), but these are deleterious effects. We have yet to identify a specific gene that produces high intelligence. Instead, the inheritance of intelligence appears to be polygenic. Thus we see the wide, smooth spread of levels of intelligence throughout the population, instead of the choppy—some high, some low, few in-between—distribution that would be expected if only a few genes were determinants of intelligence. This being so, and because the races do have the majority of their genes in common, we would not expect a preponderance of the many genes determining intelligence to occur in one race or another. This is indeed what seems to be the case.

When the IQs of American whites and blacks are compared, the blacks average only about 11 points below the whites. Although small, even this difference is in question because of aforementioned doubts about what IQ tests really measure. Traditional forms of the IQ test tended to favor individuals from white, middle-class society with orientation towards going to school. Racial differences tend to diminish as such biases are rooted out. However, there can probably never be an intelligence test that compares two different ethnic groups without bias. Thus, IQ data provides no basis for the discrimination in schooling that has sometimes been suggested—isolation of whites from blacks to prevent each from impeding the other's scholastic advancement. Since the best available estimates tell us that 85% of human diversity—including intelligence—is due to differences among individuals within races and only about 6% is due to differences between races, the best course of action is to tailor education to fit the individual, rather than the race.

Table 20-1 Genetic Disorders Affecting Brain Function

Condition	Disorder of Metabolism
A. Many gross malfunctions of the brain are directly traceable to metabolic disorders	
Cretinism: cretins are dwarfed, mentally defective individuals	Thyroid inactivity—inability to form thyroxine or due to maternal iodine lack during pregnancy; partially curable with early thyroid hormone and/or iodine
Palsy: also known as Parkinson's disease; affected individual shakes continually, has difficulty making coordinated movements	Central nervous system deficient in a neural transmitter chemical, dopamine; can be alleviated by feeding L-DOPA, a metabolic precursor of dopamine
Galactosemia: mental retardation appearing in infancy	Caused by a recessive mutant that makes the infant unable to convert the sugar galactose (from milk sugar—lactose) to glucose; readily diagnosed by testing urine for galactose and almost wholly alleviated by removing milk from diet
Tay-Sach's disease: fatal cerebral degenerative condition of infancy	Caused by a recessive mutant that causes accumulation of a specific ganglioside, due to lack of an enzyme required for its breakdown; gangliosides are lipids important in cell membrane function.
B. Others are clearly of genetic origin but the nature of the metabolic disorder is not yet known	
Mongolism: an extreme mental defective	Invariably associated with the presence of 3 (trisomic) chromosomes 21 instead of the usual 2.
Klinefelter's disease: mentally defective male with malformed testes	Associated with the presence of an extra X chromosome; the male is XXY instead of XY.

Can Intelligence Be Improved?

Genetically determined potential intelligence cannot be modified until that remotely distant and improbable day when it may be possible to perform "genetic surgery," to change the genetic makeup of the individual. However, since potential intelligence probably rarely develops to the genetically determined limit, there is much room for beneficial modification of phenotypic, or expressed, intelligence. We have already seen evidence of how this might be accomplished. Careful attention to child health is critical. The mother must have adequate nutrition to ensure a healthy fetus, and the delivery must be expertly performed to limit the chances of birth damage. Adequate nutrition during childhood is essential. In addition to these purely biological considerations, the evidence just cited of the culturally enriched rats strongly confirms the commonplace observation that a varied and intellectually stimulating environment during infancy and childhood improves apparent intelligence.

Modification of Brain Function

In addition to the long-term modifications of intelligence that we have been discussing, there are many other possible modifications of brain function and many ways to achieve these modifications. To illustrate these, we shall discuss two normal types of function-modifiers, sleep and the emotions.

SLEEP AND KEEPING THE BRAIN TURNED ON

Sleep is a regularly recurring necessity. The normal person requires from 6 to 8 hr of sleep per day. Although there are verified records of evidently normal people requiring only 3 hr of sleep per day, it is generally recognized that prolonged sleeplessness is deleterious. Yet, at least some persons, usually with assistance, are able to stay awake for very long periods without harmful effects. For example, in 1965, a high school student, Randy Gardner, as a science fair project stayed awake for 264 hr and 11 min, whereupon he slept for only about 15 hr and then voluntarily stayed awake for 24 hr before returning to a normal sleep cycle. Typically, however, sleeplessness produces deterioration in sensory acuity, memorizing ability, quickness of reaction, and other measures of the quality of mental activity. Prolonged deprivation of sleep may produce hallucinations and, of course, sleep deprivation is a well known means of "brain-washing." Certainly, we feel an overpowering urge to sleep at intervals and cannot avoid going to sleep at least once a day without heroic measures. Thus, despite the well documented examples of persons able to do without it for long periods of time, sleep does appear to be a necessity. Certainly it does not appear that sleep could possibly persist through evolutionary time if it were not essential, for, as one student of the phenomenon has written,

If sleep does not serve an absolutely vital function, then it is the biggest mistake the evolutionary process ever made. Sleep precludes hunting for and consuming food. It is incompatible with procreation. It produces vulnerability to attack from enemies.

A. RECHSCHAFFEN: 1971.
The Control of Sleep,
in W. A. Hunt, ed.: *Human Behavior and its Control.*

We find further evidence of the importance of sleep in that it occurs according to a natural rhythm that persists in the absence of all obvious environmental cues such as day and night. When isolated from environmental cues, as in a deep cave or a heavily insulated underground bunker, humans still sleep periodically according to a circadian rhythm (page 222). In Figure 20-16, records of human daily cycles are shown. These illustrate that, when the subject is in isolation from environmental cues, the sleep cycle continues, although it advances regularly through the day. This is because the human activity cycle has an approximately 25-hr periodicity, which is normally reset every day by recurrent environmental cues. Without the cues the cycle is said to free-run with a 25-hr periodicity and thus progresses through the 24-hr solar day until the subject returns to a normal environment, whereupon the cycle becomes entrained once more with the solar day.

Experiments in which the EEG is recorded during sleep show that there are two principal sleep phases. In **slow wave sleep (SWS)** the EEG shows slow, synchronized waves as compared with the waking state. In this stage heart and respiratory activity are stable and the muscular system is inactive. At intervals SWS changes to **desynchronized sleep (DS)** in which the brain waves return to the irregular pattern of wakefulness. In DS sleep respiration and heart activity are irregular, voluntary muscles may be active, and **rapid eye movements (REMs)** take place behind closed eyelids. These are so characteristic that this phase is often called REM sleep. By waking subjects from these two stages of sleep it has been determined that dreaming occurs during REM sleep.

The significance of the two stages of sleep remains obscure. Subjects deprived of REM sleep, by being awakened every time they enter it, compensate by getting more than the usual amount of REM sleep when finally allowed to sleep undisturbed. Further indication that REM sleep is of importance comes from the fact that the amount of REM sleep varies with age. Infants spend nearly half of their sleep time in REM sleep, and this falls to less than a quarter of the total sleep time in maturity. One theory of the physiological role of sleep holds that REM sleep may be an activity essential to maturation of the brain and to the maintenance of synaptic relationships.

Regretably, systematic capitalization on the sleeping activity of the brain seems to escape us. Sleep-teaching techniques in which material is "taught" by repetitive aural presentation during sleep appears to be totally ineffective if allowance is made for learning that occurs during moments of wakefulness.

Dreaming is proof of organized activity in the cerebral cortex during sleep. Usually dreams make little sense to us and they are, in fact, not ordinarily remembered unless one awakens immediately after they have occurred. However, there are enough examples on record to verify that the sleeping brain often performs constructive mental tasks. In Chapter 1, when we were discussing the origins of scientific ideas, we

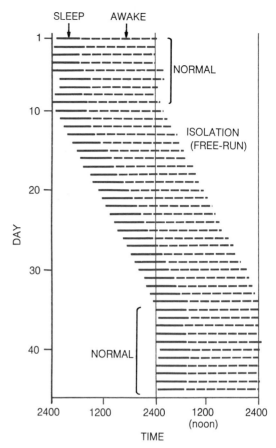

Figure 20-16 Sleep and wakefulness in the human. Solid lines indicate sleep, dotted lines wakefulness. In isolation, when the subject is active or not without restriction and without clock or environmental time cues, a free-running cycle of about 25 hours is followed. (After Dement, W. C., 1972.)

noted the example of Otto Lowei who obtained the idea for a crucial experiment from a recurrent dream. Moreover, creativity during dreaming is not confined to scientific endeavor. Robert Louis Stevenson attributed the plot of *Doctor Jekyll and Mister Hyde* to a dream; and Coleridge claims to have composed the poem "Kubla Khan" in its entirety in a dream, but was able to write down only part of it upon awakening, before an interruption dispelled the rest forever.

Although it is obviously extremely difficult to perform scientific studies of such matters, it does appear quite likely that the brain does perform such remarkable feats while asleep. In most instances such creative efforts are primed by a great deal of immediately previous conscious effort devoted to the same problem. Coleridge had just read some lines about Kubla

Khan, and Lowei had, of course, been fretting over his scientific problem for a very long time. It is intriguing to think that problem solving during dreaming is due to release from suppression of the nondominant side of the brain, a suggestion that is supported by the often nonanalytical, often allegorical, nature of dream generated solutions, which seems appropriate to what we know of the manner of operation of the nondominant side of the brain.

The **ascending reticular system** (Figure 20-5) has primary responsibility for the state of wakefulness. The nonspecific signals that it sends to the cortex represent a synthesis of the total sensory input and cause a general arousal of the cortex. This parallels the specific sensory inputs that make the cortex aware of details of the state of the body and of the surrounding environment. Lesions in the ascending reticular system that interfere with this flow of generalized sensory input produce a state of permanent sleep. Electrical stimulation of regions of the brain stem that control the flow of arousal signals produce what appears to be normal sleep in a reversible way. There is also evidence that sleep inducing chemicals are produced either as a cause or consequence of sleep. Transfusion of blood leaving the brain of a sleeping animal will induce sleep in one which is awake.

EMOTIONS

Expression of the emotions—fear, rage, and so forth—involves both the cortex and the lower, more primitive, parts of the brain as well as the autonomic nervous system. In general, the cortex acts to keep the emotions under control. Thus, an experimental animal with much of its cortex removed may fly into a rage at the slightest touch. There appear to be centers within the brain that specifically control various types of emotion. Stimulation of specific areas in the hypothalamus can produce fear or rage, for example. There are even pleasure centers. Experimental animals can readily be trained to stimulate these centers by means of implanted electrodes connected to a pedal that they must press in order to deliver a weak shock to the pleasure center. Once the association between working the pedal and stimulation of the pleasure center is established, the experimental animals often work the pedal almost continuously, in preference even to eating.

The Tendency Toward Violence

The ease with which rage is induced in decorticate animals illustrates one of the major dilemmas of modern man. His cerebral cortex sits on an emotional powder keg. Much of human behavior, especially the tendency toward violence, seems to represent manifestations of more primitive brain functions that have come down to us from earlier times, when, perhaps, violent behavior was essential for survival. Latent in the lower brain, these tendencies seem only poorly to be kept in check by the cerebral cortex, the "rational" brain. The peculiar thing about human violence is that it is directed against other humans. Violent behavior is common enough among lower mammals, but this is almost invariably addressed toward individuals of other species that threaten existence or serve as food. Much conflict among members of the same species is ritualized to such an extent that conflict is terminated by some signal of submission by the defeated before serious damage is inflicted.

When we look for the source of man's own peculiarly lethal form of intraspecific violence in his evolutionary history, we see no clear explanation. This form of violence seems to have been with us for a long time, for in the bones of prehistoric man we see evidence of cannibalism. But when we consider living apes—chimpanzees and gorillas—as examples of what the behavior of human apelike ancestors might have been like, we discover the unsettling fact that these animals are remarkably peaceable. The temper tantrums of a male gorilla, indeed, are among the most impressive of animal threat displays, but they *never* result in violence to his own kind, and rarely to any other, for that matter. Commonly, an episode of roaring and chest thumping terminates with a few flat-handed swats at the ground or the destruction of a bush, and nothing more. As in lower animals, this display is just that—a behavioral mechanism intended to communicate certain facts about rights to property, mates, or the desire for privacy. It does not lead to limb-rending fights *among* gorillas. As a matter of fact, recent field studies demonstrate that our ape relatives lead remarkably tranquil lives, with clans intervisiting without trouble and with none of the difficulties over homestead rights that occupy so much of the time of lower animals.

Thus we are led to the disturbing possibility that human violence, although it may be precipitated by lower centers of the brain and although it usually may be controlled by the cortex, when it does break free, receives its lethal turn *from* the cortex. Man, uniquely among animals, recalls the past with precision and sees future events as a consequence of the present and past. Every confrontation he has is more than an affair of the moment. It is the result of a past history of similar confrontations and an indicator of future ones. Certainly, since man knows the meaning of death, it would seem a logical step for him to reward a submissive gesture from an adversary with a deathblow, permanently ridding his world of at least one problem.

Thus, human violence may be a natural phase in the evolution of insightful intelligence. Our particular tragedy seems to be that any possible evolution beyond this brutal stage toward one favoring more peaceable resolution of human conflicts seems to be outstripped by the fertility of our minds at devising means of wholesale killing. Our tendency toward lethal violence is bad enough person to person. Translated into the collective behavior of societies, it constitutes a terrible danger to human existence.

PAIN

Like the emotions, pain may take command of the individual's behavior, diverting all activity to its demands. Ordinarily pain is highly adaptive since it compels the organism to remedy the problem for which it is the danger signal.

Signals from pain receptors enter the nervous system by two routes, which generate two types of pain sensations. Pain signals entering the nervous system through large-diameter neurons initiate sensations that allow identification of their source. More slowly arriving pain impulses are initiated by small-diameter neurons and enter the higher neural centers through the diffuse pathways of the reticular system. These signals are believed to produce the emotional, or suffering, aspects of pain.

Commonplace experience has probably already told you about these two types of pain sensations and their different rates of travel to the levels of consciousness. For example, if you step into a bath that is too hot, as your foot enters the water you perceive a sensation not much more uncomfortable than touch and which produces the feeling that something is wrong. This sensation does not enable you to determine whether the foot is too hot or too cold. An instant or so later, signals arrive over slower pathways that elicit the sensation of heat and then pain. The delay in slow pain pathways is compensated for by the fact that severe pain initiates reflex withdrawal responses within the spinal cord. Thus when you burn a finger the arm is already withdrawing by the time the brain becomes aware of the pain.

The extent to which pain is felt appears to be influenced by two levels of control, one at the entry points of pain nerves into the spinal cord and the other in the lower centers of the brain. Within the spinal cord large- and small-diameter nerve fibers serving touch and pain sensations synapse on neurons in the dorsal horns of the grey matter. From there pain axons cross the cord and ascend to the brain in a tract of nerves called the **ventrolateral pathway.** The importance of this pathway is evidenced in the human by the fact that it can be sectioned to relieve pain at lower levels on the opposite side of the body. Before pain signals enter this tract, they are subject to modification in the dorsal horn synaptic region. For example, it has long been known that stimulation of the large diameter nerve fibers can produce a degree of *analgesia,* or relief of pain in the region of the body served by those fibers. Evidence of this type has given rise to the **gating theory of pain,** which holds that, within the dorsal horn region, certain types of neural inputs are able to depress the activity of pain cells and, therefore, diminish the perception of pain.

There is also evidence that pain sensations may be modified once they reach the lower center of the brain. Experiments on laboratory animals show that electrical stimulation in certain regions of the midbrain produces localized regions of analgesia in the body. Damage or stimulation of other sites may produce continual pain. Finally, it is well known that the perception of pain is greatly influenced by the overall mental state. Persons who are highly excited or who have their attention wholly occupied by immediate events, for example, accident victims or soldiers in combat, may briefly tolerate dreadful wounds without noticing them.

Acupuncture

Acupuncture is an ancient part of Chinese medicine, having been fully systematized in writing by the eighth century B.C. On at least two occasions it has been introduced into western civilization prior to the present rush of interest kindled by the use of acu-

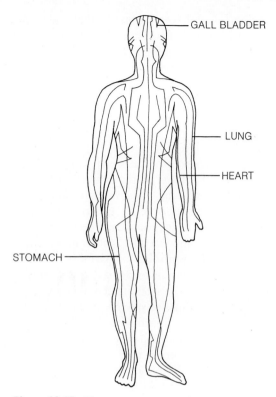

Figure 20-17 **The acupuncture meridians.**

puncture by Chinese surgeons for anaesthesia during major operations. According to traditional theory, the body contains a source of vital energy that runs through channels, or meridians (Figure 20-17) to the organs. Disease is thought to interfere with this flow. Diagnosis of illness is accomplished by detection of abnormal energy flow by observations on skin, eyes, tongue, and breath, by pulse measurements and, in modern times, by measurements of skin electrical resistance. Treatment of disease or induction of anaesthesia is accomplished by placing needles in certain of several hundred spots in the 12 meridians running the length of the body and rotating, vibrating, or passing small electrical currents through them. Acupuncture points designated according to acupuncture theory may be remote from the structures they are said to influence and often bear no physiological or anatomical relationship to them. For example, parts of the external ear may influence organs throughout the body owing to the outline resemblance of the ear to the shape of a fetus. However, recent observers note that Chinese surgeons actually pay little attention to the meridian theory and seem to use any location and as few as one needle to induce anaesthesia.

In seeking to explain the mechanism of acupuncture, the gating theory of pain limitation comes to mind. If stimulation produced by the acupuncture needles somehow manages predominately to excite the large neurons that are theorized to close the pain gates in the spinal cord, then acupuncture would at least fall within the limits of a respectable physiological theory. Unfortunately, the concept does not fit the simplest form of the gating theory, which would require that the needles be placed in tissues whose sensory innervation enters the spinal cord at the same level traversed by incoming pain stimuli. Since the ear lobes are common sites for needle placement during thoracic or abdominal surgery, we would need to postulate a more centrally placed gate or gates to make the gating theory applicable. This is perhaps not too far fetched an idea, since our scientific folklore abounds with evidence of what has been called the counter-stimulant concept of pain control in which, for example, loud sounds—music or even random noise—can make an otherwise exciting session with the dentist bearable.

The efficacy of acupuncture may have more mundane explanations based on psychological suggestion. Observers of the Chinese report that patients are thoroughly indoctrinated with the idea that acupuncture works and, beyond this, are, from childhood, conditioned to regard surgery as uniformly good and no big thing to undergo. Thus one visitor to a children's hospital saw smiling five-year-olds cuing up for tonsillectomies which they underwent without complaining, fortified with only a quick anaesthetic throat spray and *no* acupuncture. To this we may add the observation that surgical pain may be somewhat overestimated and that much of the discomfort of surgical procedures may be precipitated by fear.

It is true, in fact, that, although the skin is quite sensitive and cutting of muscles produces some pain, most of the internal organs generate little or no pain upon being cut, although they do when stretched or when pressure is applied. Many of these observations were made in the early 1900s when it was rather common to perform a variety of major operations using only local anaesthetics on the skin. Thus it may be that much of the success of acupuncture as practiced in China is due to the strong belief of the patients in the value of the procedure and to the relative insensitivity of many of the organs to surgical procedures. Finally, it should be noted that small amounts of local anaesthetics and sedatives are typically used in support of acupuncture in Chinese medical practice.

DRUGS AND THE CHEMICAL CODING OF BEHAVIOR

Our view of the brain so far has emphasized localization of function in terms of structure. However, it must be obvious that brain function is dependent on the generalized metabolic processes characteristic of any living tissue as well as upon the many specialized physiological and biochemical processes associated with the unique properties of nerve cells, namely, generation of nerve impulses and transmission of signals between nerve cells by means of chemical transmitters. Since early times it has been obvious that drugs like ethanol, opium, and caffeine affect brain function. Slowly this knowledge of drug effects on the brain has matured and become firmly based on scientific concepts. As the science of **neuropharmacology**, it has contributed immensely to the understanding of brain function.

In particular, the science of neuropharmacology tells us that there is a chemical map of the brain, just as there is a structural map. Understanding this chemical map allows subtle intervention with brain function in relatively specific ways by administration of drugs to enhance or suppress neurochemical processes in various brain regions. This capability provides a major tool for unravelling the mystery of brain function. Furthermore, it is not necessary to emphasize the immense practical significance of drugs that affect brain function. Their misuse represents a major public health problem, whereas their proper use is basic to many of the major advances in medicine (Figure 20-18).

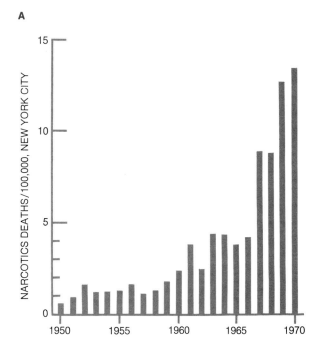

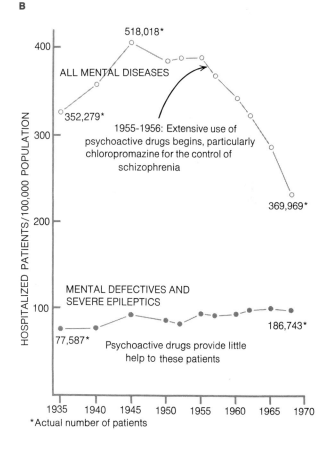

Figure 20-18 Two faces of the drug use coin. A. Narcotics death in New York City increased 1300% in the 20 years from 1950 to 1970. Data from Borden, M. M. (1971) Legal Medical Annual, Appleton-Century Crofts, New York. **B.** Use of psychoactive drugs to reduce the necessity of hospitalization of mental patients. Treatment costs are vastly reduced and the chances of recovery are improved by return of patients to a more normal environment. (U.S. Public Health Service)

Neural Transmitters in the Brain

You already know that there are two kinds of neural transmitters in the peripheral nervous system, acetylcholine and noradrenaline. These two transmitters also function within the brain. However, in the brain noradrenaline is only one of several related transmitters called **biogenic amines** or catecholamines (Figure 20-19). Thus we find within the brain **cholinergic nerve terminals,** which release acetylcholine (ACH), and several types of terminals that release biogenic amines: **adrenergic terminals,** which release adrenaline or noradrenaline (NA); **dopaminergic terminals,** which release dopamine; and **serotonin releasing terminals.** There are other central nervous system transmitter molecules. Some are relatively simple, such as the amino acids glycine and glutamic acid, whereas others are structurally complex. Among the latter are a recently discovered family of peptides, the enkephalins and endorphins, which are thought to be endogenously produced transmitters with properties similar to morphine. Recent experiments suggest that the release of these may help to explain the placebo effect (page 9) and acupuncture.

Transmitter Geography

An extremely important point about the distribution of some of these transmitters in the brain is that their pattern of distribution is such that one kind or another of transmitter is predominately involved in transmission within specific regions of the brain, or they perform different functions when present within the same region. For example, ACH injected into a particular site in the hypothalamus causes drinking, whereas an injection of NA at that site causes eating. This pattern is so striking that early workers in the field theorized that there are two systems in the brain serving to control lower brain functions such as the emotions and behavioral arousal. These systems were called the **ergotropic system,** which brings about behavioral arousal and uses noradrenaline and adrenaline as the transmitters, and the **trophotropic system,** which controls restorative actions of the nervous system and uses serotonin as the transmitter. Although later studies have shown that this dichotomy of systems is not as clear cut as originally thought, the concept of two systems balanced against each other and operating with chemically different transmitter processes is still quite useful to understanding the action of drugs on the brain.

Cellular Sites of Drug Action

The mechanism of action of many drugs that affect the central nervous system is known to involve interference with or facilitation of synaptic transmission in one or both of the two major synaptic classes, cholinergic and adrenergic. In order to understand how these drugs act, we must consider some of the details of synaptic action in the two classes of synapses. Figure 20-19 shows the essential details of the synaptic process in cholinergic and adrenergic synapses.

The transmitter cycle in a cholinergic synapse (recall page 343) involves (1) synthesis of acetylcholine and its packaging in synaptic vesicles within the nerve terminal, (2) release of acetylcholine from vesicles by the action potential, (3) reaction with a receptor molecule in the postsynaptic cell, (4) breakdown of acetylcholine by the enzyme acetyl cholinesterase (ACHase), and finally (5) uptake of the breakdown product choline by the nerve terminal for use in synthesis of more acetylcholine. Transmission of the neural signal from nerve terminal to postsynaptic cell is accomplished during the time that acetylcholine is attached to the receptor molecule. Any drug that prolongs or increases the reaction of transmitter with receptor molecule causes the synapse to be more active than ordinary, whereas any drug that shortens or diminishes transmitter action diminishes the action of the synapse.

Among the drugs that increase synaptic action in cholinergic synapses are substances that inhibit ACHase. When ACHase is inhibited, any given ACH molecule remains active longer and the synapse thereby is more active. This fact is illustrated by the tactics of treatment of the disease myasthenia gravis. This progressive disease involves loss of voluntary muscle control because nerve terminals become more and more unable to release sufficient amounts of ACH to control muscles. Temporary relief is provided by administration of chemicals such as prostigmine or DFP (Di-isopropylfluorophosphate) which, given in proper dosage, sufficiently inhibits ACHase to make up for the disease caused reduction in ACH. A similar dose of DFP supplied to a normal person might produce excessive synaptic activity and result in spasms and loss of muscle control. DFP is, in fact, a chemical relative of the military "nerve" gases, which achieve their lethal action in exactly this way.

Nicotine and curare are examples of drugs that influence cholinergic synapses by directly influencing

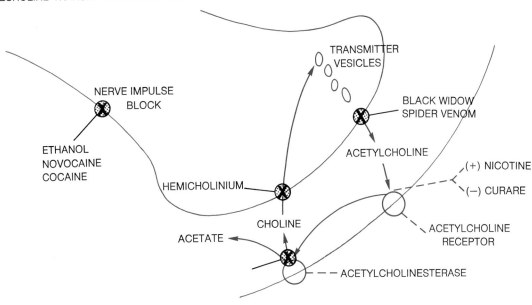

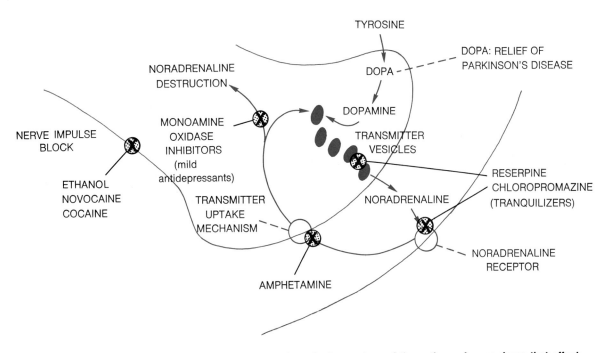

Figure 20-19 Two major types of brain synapses, cholinergic and adrenergic, and the actions of some drugs that affect them. In general, drugs that affect synaptic function may affect neural activation of the synapse (ethanol), transmitter synthesis or uptake by transmitter vesicles (reserpine), transmitter release (black widow spider venom), transmitter action (anticholinesterase), and transmitter re-uptake by nerve ending after activity (amphetamine).

the receptor molecule. Curare produces neuromuscular paralysis because it reacts with the receptor and prevents ACH from acting upon it. Nicotine behaves as though it is ACH in reacting with the receptor and thus increases the effect of any ACH that is normally released. This is the principal physiological basis of the sense of well being produced by smoking, namely a widespread general augmentation, or facilitation, of cholinergic synapses throughout the nervous system.

Curare causes neuromuscular paralysis by blocking transmitter action on the postsynaptic cell, and a highly similar form of paralysis may be caused by botulinum toxin. This toxin, produced by anaerobic bacteria that frequently grow in improperly sterilized canned food, is one of the most toxic compounds known. It combines with the membrane of the nerve ending and blocks release of ACH vesicles. The effect is irreversible and the poisoned nerve ending remains out of action until sufficient new synaptic membrane structure is synthesized. Black widow spider venom also acts on the nerve ending. It causes destructive changes that cause rapid dumping of ACH followed by further destructive changes that render the synapse nonfunctional for a long period of time.

Finally, in terms of illustrating the types of action that drugs may have on cholinergic synapses, we come to a family of compounds, of which hemicholinium is an example. These interfere with synthesis of the transmitter. In the cholinergic synapse ACHase breaks ACH down into acetate and choline extracellularly. Acetate is plentiful in all cells, but choline is relatively rare; thus, the nerve ending has a mechanism in its membrane that transports choline back into the ending for reuse. Hemicholinium acts by blocking this uptake and thus serves to reduce the amount of ACH available and causes paralysis of the synapse.

In the adrenergic synapse (Figure 20-19) we see a family of mechanisms similar to those just described for the cholinergic synapse. Synthesis of the transmitter begins with the amino acid tyrosine and proceeds through several steps to noradrenaline (NA), which is packaged in vesicles that are larger and denser than ACH vesicles. Upon neural release, NA combines with a postsynaptic receptor molecule to achieve its transmitter action. It may then be broken down by an enzyme in the postsynaptic membrane (catecholamine-O-methyl transferase, COMT) or, as is more likely, it may be taken up intact by a transport system in the nerve ending. Once returned to the nerve ending, NA may be repackaged in vesicles or broken down by a second enzyme, monoamine oxidase (MAO), which serves to scavenge any excess free NA that is not packaged.

Within this biochemical arena lie sites of action of many drugs. Tranquilizers act in various ways to diminish the amount of NA available for synaptic transmission. For example, chloropromazine blocks the receptor molecule so that it is inaccessible to NA, reserpine disrupts for a long period of time the production of transmitter vesicles, whereas disulphiram inhibits the synthesis of NA. At the other end of the spectrum, antidepressants, or mood-elevators, block NA uptake (this being the adrenergic analog of anticholinesterase action at a cholinergic synapse) or inhibit its breakdown by MAO. The amphetamines are the best known example of blockers of uptake. However, their action is highly complex. While blocking NA uptake by the nerve ending, amphetamine is taken into the ending where it interferes both with MAO and the normal process of accumulating NA in transmitter vesicles. The result is continual leakage of NA from the ending. All these effects are excitatory. Subsequently, however, amphetamine interferes with further release of NA and has a depressive effect. Chronic use leads to accentuation of depressant relative to stimulatory effects. In low concentrations cocaine acts as a mood elevator by blocking NA uptake. Several mild antidepressants act by inhibiting MAO, and probably produce their effect by the consequent slow leakage of NA from the terminal independently of the normal synaptic vesicle release mechanism.

Dopaminergic terminals are a special class of nerve endings in which dopamine is the transmitter. The best known pathological condition involving such terminals is the disease called parkinsonism. In this condition it is thought that a cluster of dopaminergic terminals involved in motor control is defective, suffering atrophy or otherwise becoming less effective. As a result the patient suffers severe and continual muscular tremors. One means of alleviating this disease is to administer very large doses of DOPA, a precursor of the transmitter dopamine. The affected cells are thus able to increase their total output of dopamine and the tremors are reduced. This is not a totally successful treatment because the large doses of DOPA that are required, 8 to 12 g/day, produce unfortunate side effects.

Some drugs influence *both* adrenergic and cholinergic synapses by blocking the transmission of action potentials into them. In high doses cocaine, in addition to its action to prevent NA uptake, does this. Its close relative novocaine is a commonly used nerve blocker that lacks the NA synaptic effect. Ethanol achieves a substantial fraction of its depressant effect on the nervous system by dissolving in membrane lipids and thereby interfering with impulse transmission. The hallucinogens, LSD and mescaline, may achieve their effects by blocking impulse generation in specific groups of neurons in the brain.

To understand the action of drugs like caffeine, we must reexamine a transmitter system that we first discussed in connection with hormone action (page 363). You will recall that some hormones act on their target cells by regulating the level of cyclic adenosine monophosphate (c-AMP) in the target cell. Within the target cell c-AMP then influences the action of critical enzymes that generate the response of the cell to the hormone. To this scheme we may now add the information that some types of cells in the nervous system are activated in an analogous way. In this instance, c-AMP reacts with a membrane molecule to produce a prolonged change in cell permeability to an ion critical to impulse generation. For example, permeability to sodium ions might be slightly increased, making the cell more excitable. Now in such systems an intracellular enzyme, phosphodiesterase, is responsible for turning off the c-AMP effect by breaking it down into AMP (Figure 18-24). Caffeine and related compounds, such as theobromine, act at this point to inhibit phosphodiesterase and thus facilitate the action of whatever transmitter causes the initial production of c-AMP. Thus caffeine acts as a stimulant.

Self-produced Opiates

The brain produces no substances such as morphine. Therefore, investigators recently were surprised to find that certain neurons in the brain contain receptor molecules for morphinelike molecules. Although the morphine receptor molecule might react to morphine much as the ACH-receptor reacts to nicotine, the mystery was that there was no known natural transmitter molecule to react with the receptor and thus to give it a role in brain function. Since it is unlikely that a receptor would exist without a corresponding natural transmitter molecule, a search was commenced for a natural, or endogenous, transmitter for the opiate receptor. Such a transmitter would be expected to act much like morphine, to be, in other words, a natural sedative.

Soon a group of five peptides were isolated from brain and pituitary and found to have opiate effects. These molecules are known as **enkephalins** and **endorphins.** Three of them were found to be identical with amino acid sequences in β-lipotropin, a poorly understood polypeptide isolated from the pituitary. Since these fragments of β-lipotropin have somewhat diverse effects on homeostatic processes, such as the regulation of body temperature in addition to acting as opiates, the idea has emerged that they may comprise an entirely new system of transmitters involved in homeostasis.

Investigators working with these newly discovered molecules find that they produce remarkable effects such as **catalepsy** in addition to rendering experimental animals insensitive to standard tests of pain sensitivity. It is possible that malfunctioning of this proposed new transmitter system may account for certain mental disorders that so far have escaped physiological or biochemical definition. For example, the auditory hallucinations of two schizophrenic patients were halted by administration of a drug known to antagonize binding of opiates to opiate receptor molecules.

Why So Many Different Types of Transmitters in the Brain?

This brief discussion of transmitters naturally evokes the question of why there are so many of them since, after all, the great profusion of neural pathways in the brain would seem to provide an ample degree of specificity of signal transmission. Could not the brain function adequately merely on the basis of the organization of signal transmission achieved by specific nerve tracts without the necessity of the additional coding provided by a variety of transmitters? One might reply by noting that there are at least two possible reasons that make many different transmitters useful.

First, many nerve cells are sensitive to more than one transmitter and respond differently to each. For example, ACH delivered to a synapse may be excitatory whereas glutamic acid delivered by another nerve

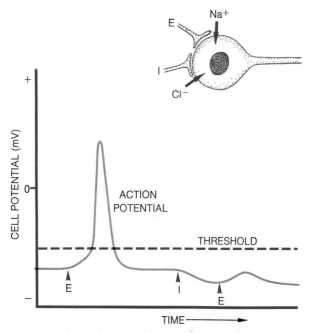

Figure 20-20 Excitatory synaptic action (E) involves increased sodium entry with movement of the neuronal resting potential towards the firing threshold. Inhibitory synapses (I) commonly act by increasing permeability to chloride, thus moving the rest potential away from the firing threshold. The figure shows that the two types of synaptic action are additive.

cell to another synapse on the same nerve cell may be inhibitory, reversing the action of ACH (Figure 20-20). So a given neuron may either increase or decrease its activity in response to the total input that it receives from surrounding nerve cells.

Second, some transmitters may actually travel some distance through the brain and produce widespread effects on those cells that are specifically sensitive to them. It might be theorized that changes in behavior might require small adjustments in the sensitivity of many widely dispersed nerve cells. Although the adjustments required might be achieved by a complex neural network, they might also be achieved by release into the brain environment of transmitter molecules from one or a few sources, in a way analogous to how NA, released from the adrenal, spreads through the body to affect many different organs. This idea is supported by the fact that the brain is remarkably well protected from chemical influences arising outside of it by a living cellular screen, the blood-brain barrier. The presence of this barrier, interposed as it is between each brain cell and the nearest circulatory element, suggests that brain neurons must be exquisitely sensitive to variations in their chemical environment.

21 Human Reproduction

Now our attention turns from the operation of the complete organism to the means by which the organism is constructed. The process is remarkable. It begins with the fusion of two single cells, the fertilization of an egg by a sperm. The resulting **zygote**, contains genetic information (DNA) from both parents. In the course of many subsequent cell divisions, this information directs the differentiation of cells into tissues and organs and ultimately into a mature organism.

In gross terms, the process of development is astonishing enough. In roughly 9 months of development the human embryo increases in weight from a fraction of a milligram to about 4 kg in the course of multiplying from one cell to approximately 50,000 billion cells. Although the necessary energy and raw materials come from supplies continually provided by the mother, the *direction* of the entire process originates within the zygote. This directive process is so remarkable, so important, and at the same time so poorly understood, that embryonic development is still one of the major problems of biology.

The Human Reproductive System

Early in the life of the human embryo, cells of the germplasm, destined eventually to form either eggs or sperm, migrate through the tissues from a site actually outside the embryo proper into the developing male or female gonad. Under both genetic and hormonal influences, the embryonic gonad becomes either a testis or ovary associated with other tissues and organs essential to its function in the adult.

MALE REPRODUCTIVE SYSTEM

The male reproductive system (Figure 21-1) consists of sources of sperm and the necessary apparatus to convey them safely into the female reproductive tract.

Sperm form within the two testes, which also have the function of producing male sex hormones, principally testosterone. **Spermatogenesis** (sperm produc-

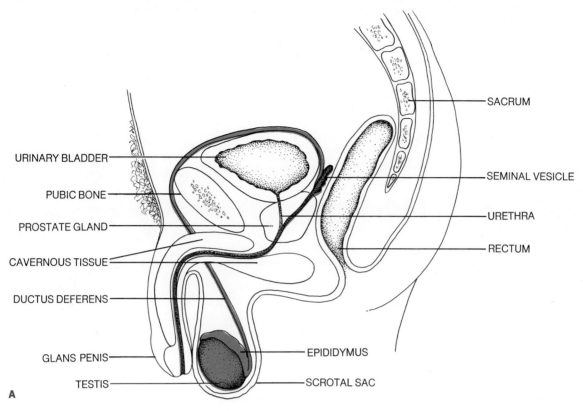
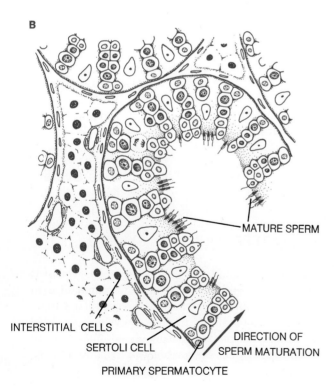

Figure 21-1 The male reproductive system. A. General mid-line view. **B.** Histological detail of sperm formation. Cross section of seminiferous tubule showing sperm maturation, beginning with primary spermatocyte at periphery of tubule. Mature sperm line the lumen of the tubule and are sustained by Sertoli cells. Interstitial cells lying between tubules are source of testosterone.

tion) takes place in **seminiferous tubules,** which occupy most of the volume of the testis. These tubules come together to form a single duct, the **epididymus,** which exits from each testis. It joins with the **vas deferens,** which leaves the scrotal sac and terminates within the **prostate gland** in an **ejaculatory duct** which opens into the **urethra.** Accessory sexual glands include, besides the prostate, the **seminal vesicle** and the **bulbourethral glands.** At sexual climax the ejaculate, consisting of spermatozoa and secretions of the accessory glands, reaches the exterior through the urethra.

Sperm Formation

The seminiferous tubules are solid until puberty. At that time, under hormonal influence, they develop central cavities and their germinal cells commence cell division, which continues throughout the reproductive life of the individual. Some of the resulting cells become primary **spermatocytes** and undergo meiosis. Each spermatocyte that enters meiosis produces four **spermatids** with the haploid chromosome number. The spermatids become sperm by undergoing structural changes fitting them for movement and egg penetration.

The spermatid grows a long flagellum, and the remainder of its substance is rearranged into an **acrosomal cap** covering a head piece that contains the nucleus and a middle piece containing mitochondria (Figure 21-2). Some cytoplasm is discarded and the sperm becomes so highly specialized as a device to transport genetic material that it seems unable to survive for long without assistance. This is provided by the **Sertoli cell,** another cell characteristic of the seminiferous tubule. Differentiating sperm are found in intimate connection with Sertoli cells, which are believed to provide the sperm with essential nutrients.

As sperm are formed they are transported from the seminiferous tubules and stored principally in the epididymus but also the vas deferens. The rate of use and production of sperm influences the reproductive efficiency of the male. A sperm count in excess of 1,000,000 per ejaculate is necessary to insure a high probability of fertilization. The sperm count falls below this number if sexual activity is less frequent than once every five days or more frequent than every 24 hr. Sperm are gradually and continually voided with the urine in the absence of ejaculation.

Sperm are relatively inactive while still within the male. Even as they move into the epididymus they are still undergoing maturational changes. The acrosome is incomplete and cytoplasm is still being lost. Only after passing through the epididymus are sperm able to accomplish fertilization and, in some animals including man, they require a further period of time in the female tract before they can fertilize an egg. This is called the **capacitation period.** At ejaculation, sperm from the epididymus and the vas deferens are forced into the urethra by muscular contractions of the duct walls. By this time they are mixed with the other secretions that make up the **semen.** This is a very complex mixture containing sperm nutrients and substances that activate the sperm and stimulate the female reproductive tract to render it chemically more favorable for survival of the sperm. The average volume of the ejaculate is about 3 ml.

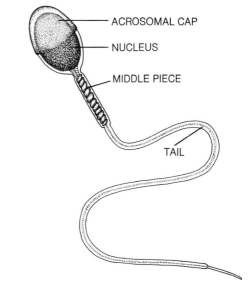

Figure 21-2 **Human spermatozoan.**

FEMALE REPRODUCTIVE SYSTEM

The female reproductive system (Figure 21-3), in addition to producing gametes and sex hormones, has the added roles of providing a favorable environment for embryonic development and of effecting the critical transition of the embryo from an essentially parasitic life in the uterus to a free-living existence. Within the female reproductive system, egg cells are produced in limited numbers by the paired ovaries. At birth, the female human possesses about 2 million germ cells in her ovaries. From this maximum they swiftly decline to less than 400,000 by the seventh year and to perhaps 10,000 by age 40. Most are lost by atrophy. In the lifetime of the female less than 500 become mature egg cells. Virtually none are left by menopause.

After puberty, maturation of germ cells occurs as the fundamental process in the approximately 28-day **menstrual cycle.** The ovary contains large numbers of primary germ cells distributed around its outer surface, or cortex. Certain of these become surrounded by a layer of cortical cells to become a primary **ovarian follicle.** As the follicle matures, the germ cell grows until it is about 0.1 mm in diameter. The containing follicle also grows with proliferation of the follicle cells and the appearance of a liquid filled cavity (Figure 21-4). Finally, the follicle ruptures and releases the

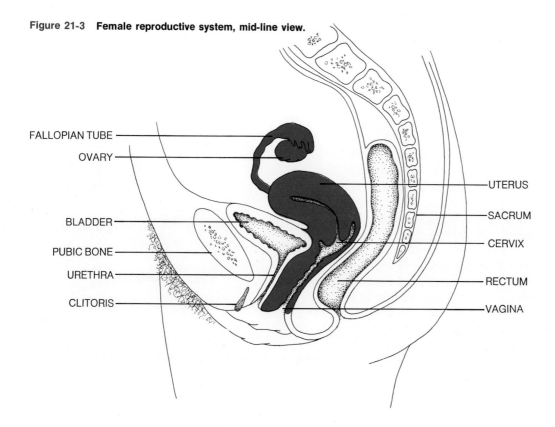

Figure 21-3 Female reproductive system, mid-line view.

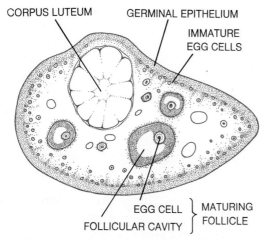

Figure 21-4 Human ovary showing development of ovarian follicles.

Meiosis in the female mammal differs from the process in the male in that each germ cell undergoing meiosis gives rise to only one mature ovum (Figure 21-5). The reason for this is that the two cell divisions of meiosis in the female are unequal, resulting in almost all of the cytoplasm at each division going into one daughter cell. The other cell, consisting mostly of nuclear material, is called a **polar body** and does not survive. This peculiarity insures that at least one cell out of the four derived from each germ cell entering meiosis will have enough cytoplasmic reserves to see it through the early stages of development when external food sources are unavailable.

When the ovum is released it is technically loose in the body cavity, but it is immediately caught in the mouth of the **fallopian tube** and conveyed by cilia and the action of smooth muscle along the fallopian tube towards the uterus. If sperm are available, somewhere in the fallopian tube fertilization occurs and the zygote initiates development. The fallopian tubes open into the **uterus** which is an organ fitted to receive, protect, and feed the embryo. The lowermost part of the uterus is the **cervix,** which opens into the **vagina.**

ovum, still surrounded by a layer of follicular cells. Prior to this time the ovum has undergone the first of the two cell divisions of meiosis and will undergo the second when it is fertilized.

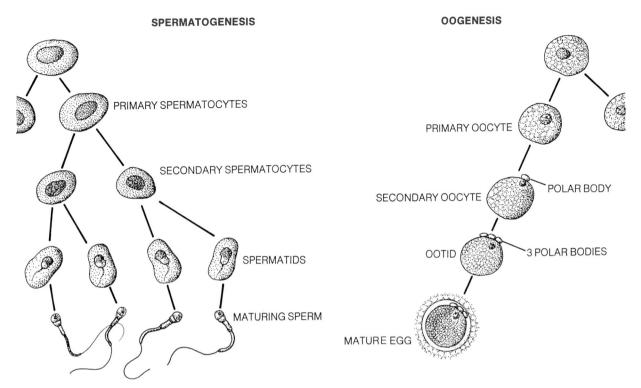

Figure 21-5 Meiosis in the female mammal results in one ovum and three, non-viable polar bodies per germ cell entering meiosis, while the result in the male is four spermatozoa.

This muscular canal extends between the uterus and the external **genitalia** and receives the **penis** during copulation. Both the vagina and uterus are capable of remarkable distention to allow passage of the fetus at birth. The external opening of the vagina may be covered partially by a membranous structure, the **hymen**, which sometimes is an obstacle to first intercourse. The external genitalia lie at the opening of the vagina. They are a pair of fleshy folds, the **labia majora** and **labia minora**, and the **clitoris**. The clitoris is the homolog of the penis. It contains erectile tissue, such as is found in the penis, and is densely supplied with nerves whose stimulation contributes to sexual arousal.

HORMONAL CONTROL OF HUMAN REPRODUCTION

Although the importance of hormonal control to physiological processes has been explained in earlier chapters, its importance to the human reproductive system requires emphasis. Every step in the reproductive cycle is hormonally regulated, and even the development and maintenance of the reproductive organs is controlled by hormones. Despite their importance, in no case do we have full understanding of how reproductive hormones act.

The Onset of Puberty

As we saw in Chapter 18, the gonads are under control of the master endocrine gland, the anterior pituitary. Except for a small amount of activity immediately following birth, the pituitary does not trigger the activity of the sex glands until the organism is physically mature. In males this delay may be as much as 14 years, but it is often substantially shorter for females. The process of sexual maturation that results is called **puberty**.

We do not know what tells the pituitary, after 12 to 14 years, that it is time to trigger production of gametes and production of sex hormones. We know, at least, that once reproductive cycles are initiated it is the hypothalamus (page 356) that regulates pituitary

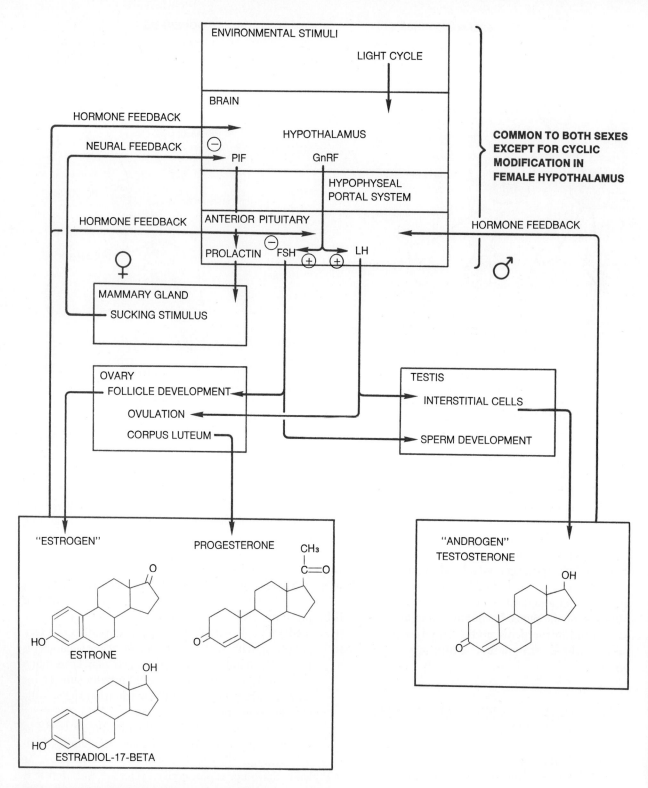

Figure 21-6 **Regulatory processes in human reproductive biology.**

hormone release. The hypothalamus produces polypeptide releasing factors for pituitary hormones. These reach the anterior pituitary by way of a portal blood system that joins the capillaries in the hypothalamus with capillaries in the anterior pituitary. A single decapeptide releasing factor (GnRF) controls both LH and FSH whereas an inhibitory factor (PIF) suppresses prolactin release.

In vertebrates with seasonally regulated reproductive cycles, the hypothalamus is sensitive to day length, possibly through the pineal (page 361), and thus can adjust the cycle to the season. What initiates the first reproductive cycles in animals like the human is unknown.

FSH Is the Trigger

Whatever the timing mechanism, at the appropriate stage of development the pituitary releases FSH, the **follicle stimulating hormone**. This protein hormone induces maturation of the ovaries and testes, making them capable of producing eggs and sperm. A second hormone, the **luteinizing hormone** (LH), causes ovulation and development of the corpus luteum in females. The identical hormone in the male causes the interstitial cells of the testis to produce male sex hormones, principally the steroid, **testosterone**.

Some of the interactions of these hormones and releasing factors are diagrammed in Figure 21-6.

DEVELOPMENT OF THE HUMAN MALE

The Role of Testosterone

With the surge of testosterone in the male at puberty, most of his growth systems respond with a profound increase in activity. The boy increases his food intake, the sex organs mature rapidly, the rate of growth of certain bones accelerates to make him taller

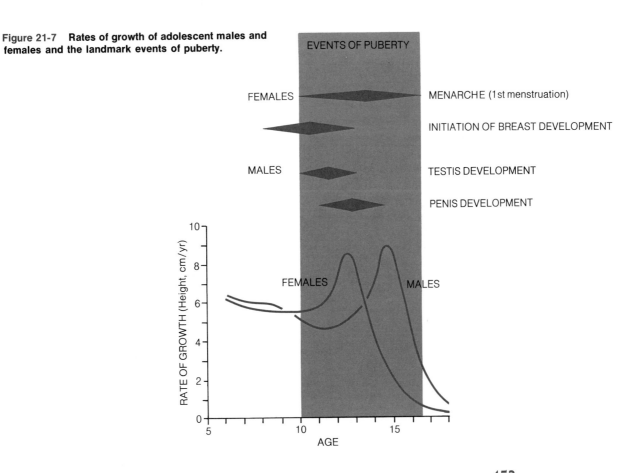

Figure 21-7 Rates of growth of adolescent males and females and the landmark events of puberty.

(Figure 21-7) and his shoulders broader, the skin grows hair in the characteristic male pattern, and the facial skin particularly may display an array of annoying blemishes. His muscles become stronger than a girl's muscles will ever become due to a difference in muscle fiber, not just to development from exercise. (Recall the use of male sex hormones ("anabolic steroids") in muscle building in athletes.) The larynx enlarges, causing the voice at first to crack and then to deepen in tone. Additional profound effects of testosterone appear as behavioral changes in the young male. He often alternates from restless, strenuous physical activity to a virtually comatose condition. At his very best he may be mentally groggy. A grumpy disposition may change suddenly to violent, aggressive behavior. His interaction with girls is often confused and contradictory.

Although the behavior of the male at puberty is influenced by the kind of culture in which he lives, we do know that the aggressive behavior is principally related to the production of testosterone. Clearly, testosterone produces profound systemic changes in virtually all tissues of the pubescent male.

A Glandular Source of Aggression

The aggression that appears in the human male under the influence of testosterone is not unique. Males of many species are unmanageably aggressive during the mating season when testosterone levels are at their highest. The continuous reproductive behavior of the human, as contrasted with the cyclic behavior of most other mammals, together with the customs imposed by civilization tend to make this form of aggression relatively mild in man. Perhaps this is not an altogether bad physiological inheritance from our evolutionary past. Certainly, this form of aggression has good influences in that it is often directed toward protection of mate and family.

In the male at puberty, the time course of hormonal changes and behavioral confusion is about 2 years (see Figure 21-7). The boy gradually matures and becomes adjusted to the continually released testosterone. Testosterone levels stabilize under control of a negative feedback loop regulating pituitary release of gonad-stimulating hormones. Through such mechanisms of homeostasis, concentrations of male hormones remain reasonably constant so that over the years the male is able to stabilize his physiological and behavioral reactions to their powerful effects.

REPRODUCTIVE CYCLE OF THE HUMAN FEMALE

The human female is faced with a different situation in that all during her reproductive life she experiences the effects of cyclically varying hormonal levels. Her ovaries became active at puberty, just as did the male gonads, under the influence of FSH and LH from the pituitary. However, in the female, puberty occurs almost 2 years earlier, at an average of 11 years of age (Figure 21-7). FSH causes the differentiation of female germ cells into gametes. The ovaries begin to manufacture eggs. FSH also causes glandular cells in the ovaries to produce the female hormone, the steroid estrogen. Estrogen induces in girls an analogous process of sexual maturation to that which testosterone induces in boys.

In the female the vagina and the uterus grow and mature rapidly at this time. The breasts also begin to grow and form tissue able to manufacture milk after childbirth. The bones of the pelvic girdle (hip bones) change their growth rates so that the hips widen to provide an opening in the pelvic arch of sufficient size to provide passage for the fetus at birth (see Figure 21-12). Androgens are manufactured in the adrenal glands. They are responsible for the development of pubic hair and the overall growth spurt of the whole body.

Sex Specific Behavior Starts at Birth

It is generally accepted that the behavioral differences between girl and boy babies are produced by differing concentrations of these same adrenal androgens, which are in higher concentration in the male. Thus, boy babies tend to be significantly more vigorous and willful than are girl babies. However, the more fundamental differences between male and female are induced only by the more powerful testosterone which greatly increases in concentration in the male at puberty.

It is probably true that the female is never exposed to significant amounts of an androgen as powerful behavioral actions as testosterone, and it may well be that the less difficult puberty of the female is the result of a hormonal balance between estrogens produced by the ovaries and adrenal-gonadal androgens. The few estrogens that appear in minor concentrations in the male at puberty appear to be inadequate to offset the effects of testosterone. Offsetting her more placid experience during puberty, the mature female experiences continuing rounds of hormonal shock because in the normal reproductive cycle her system has little time to become accustomed to one kind of hormone before being exposed to another (Figure 21-8).

Female Cycle: Preovulatory Phase

Subsequent to the **menarche,** or first menstrual period, the normal female maintains 28-day menstrual cycles into her 50s, except for periods of pregnancy and lactation. To describe the cycle, its first day is arbitrarily designated as the day vaginal bleeding begins. As shown in Figure 21-8 the subsequent days of the cycle are marked by complex changes in hormone output and structural and functional changes in both ovary and uterus. In the preovulatory phase several ovarian follicles commence growth under the stimulus of FSH. One of these dominates and matures enough to ovulate, typically on the fourteenth day of the cycle. As the follicle matures ovarian output of estrogen increases, accompanied by small amounts of progesterone. Reaching a peak about 4 days before ovulation, estrogen levels in circulation have a feedback effect on the brain-hypothalamus-pituitary control system, causing a surge of LH release. This is the major trigger of ovulation a day or so later.

During the preovulatory period, the uterus is in its proliferative stage in which, under the stimulus of estrogen, the **endometrium** undergoes changes in preparation for receiving the fertilized egg. Vascularization of the endometrium is greatly increased and the uterine glands undergo development.

Female Cycle: Postovulatory Phase

Upon ovulation, the LH surge promotes development of a new hormone source in the ovulated follicle. This is the **corpus luteum,** which produces yet another steroid, **progesterone.** Under its influence the uterus enters the secretory phase in which glandular and circulatory preparations to support embryonic development are completed. This process is also supported by continued production of ovarian estrogen.

The postovulatory phase can, of course, have one of two outcomes. Fertilization and initiation of development may or may not occur. In the absence of fertilization, the uterus looses its endocrine support, menstruation commences, and the cycle repeats. With the initiation of embryonic development, a hormone produced by a fetal tissue, the **chorion,** prevents the hormonal changes that would lead to breakdown of the endometrium.

The Initial Events of Pregnancy

The corpus luteum matures and causes progesterone and estrogen levels to rise as the egg after release moves through the fallopian tube towards the uterus. The result is that, by the time the egg reaches the uterus, the heavy vascularization and secretory activity of this organ have prepared it to receive and nourish the early embryo, should fertilization have occurred. In this case, the embryo in effect digests its way into the endometrium (see Chapter 22 for details) and jointly with uterine tissues begins formation of the placenta, which is essential to supply the needs of the embryo throughout the developmental process.

These early developmental events involve a critical transfer of an endocrine function from the mother to the embryonic chorion. In the event that a pregnancy is not established the corpus luteum soon begins to reduce its output of progesterone and to atrophy. However, with the development of the embryonic chorion, this structure produces **chorionic gonadotrophin,** which supports the action of the declining amounts of pituitary-produced LH. This prevents atrophy of the corpus luteum, and it continues to produce progesterone for the support of the endometrium. Ultimately the corpus luteum becomes nonfunctional after 6 to 8 weeks of pregnancy, but by that time the placenta is producing enough estrogen and progesterone.

Fertilization Failure and Menstrual Flow

If fertilization does not occur, levels of estrogen and progesterone fall. The endometrial arteries are particularly sensitive to estrogen and, as a result, undergo contractions that diminish blood flow. Consequently, in this **ischemic** phase, the superficial parts of the endometrium degenerate and are lost in the ensuing menstrual flow. Normally several days are required for this process. During or following the menstrual period, FSH levels begin to rise again and a new cycle of follicular development begins. What determines which follicle will mature is unknown, but, since alternate ovaries usually undergo follicular development, it is possible that a maturing follicle has an inhibitory effect on those nearby in the same ovary.

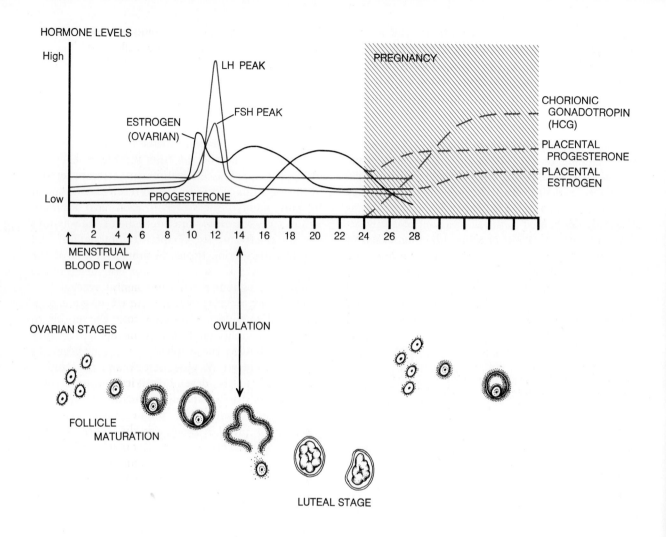
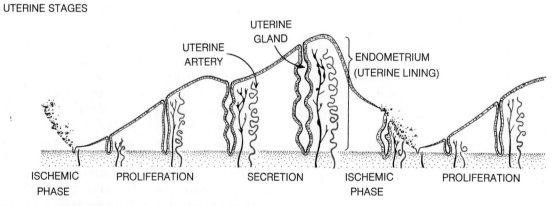

Figure 21-8 The human female reproductive cycle. Relationships of hormonal levels with ovarian and uterine events are shown. Dotted lines in graph refer to hormone levels in pregnancy.

What Makes the Female Reproductively Cyclic?

Since, as Figure 21-6 shows, the same set of releasing factors and gonadotrophic hormones are produced in males and females, it is of interest to inquire how the system becomes cyclic in the one sex and not in the other. Although there is no direct experimental evidence for the human, experiments on rats strongly suggest that the principal cause of the difference between the male and the female is the action of male sex hormone on the brain-hypothalamus at an early critical stage of life. Thus, if a newborn male rat is castrated, it may later be shown, by observing an implanted ovary, that it is secreting pituitary hormones in a sequence capable of producing a normal female ovulatory cycle. Or, if a newborn female receives an injection of testosterone, she becomes nonovulatory (see freemartin, page 481).

That the site of action is the brain-hypothalamus and not the pituitary is shown by pituitary transplants. If a normal rat has her pituitary replaced by the pituitary from a nonovulatory, testosterone treated female, she continues to exhibit normal sexual cycles. The reciprocal experiment, replacement of the pituitary of the testosterone treated female by one from a normal female does not relieve the condition.

Estrogen Affects Emotions

There can be little doubt that the hormones active in the female sexual cycle have effects on behavior. In rats, for example, the female at the time of sexual receptivity (coinciding with ovulation) leaves her burrow and lays scent trails over a wide territory to attract males. Similarly, in the human female, since the egg passing through the fallopian tube may be fertilizable for only a few hours, it seems probable that behavioral changes might take place to favor copulation. The demonstration of such effects in the human is complicated by social influences, which also influence the self-reporting of other symptoms that occur during the menstrual cycle. Thus a recent study in which women were led to believe they were premenstrual when they were not, showed a high instance of reporting of premenstrual symptoms. When allowance is made for such effects, it still appears that depression is common premenstrually, when estrogen and progesterone levels are lowest, and that there is a feeling of well being close to ovulation, when estrogen levels are high. That these fluctuations in behavior can reach serious levels is attested by data from a suicide prevention center showing a very high proportion of calls from women in premenstrual and early menstrual stages. One peculiar effect of the sexual cycle on female behavior is increased sensitivity to a variety of odors around the time of ovulation.

THE SEXUALLY EXCITED MALE

The male has little more to do reproductively than to continue to manufacture and release sperm so as to ensure fertilization. In a mature male, there is normally a continuous supply of sperm available in the testes and associated ducts (Figure 21-1), where it remains until the nervous system causes discharge. These neural commands are normally initiated only when certain sensory receptors are effectively stimulated by interaction with the female. These include visual, touch, and pressure receptors of the skin and olfactory receptors.

Sexual Pheromones. In many lower animals it is readily shown that special chemicals called **pheromones** are important in communication between individuals. The pheromone released by the sexually receptive female moth is a well-studied example. Only a few molecules per cubic milliliter of air is enough to prompt the male to fly upwind to find her. Or, among mammals, the marking of a wolf's territory by urination is another example, with which urgent business all dog owners are familiar.

Among our primate relatives, pheromones are involved in sexual behavior. Odors produced by the female monkey are of great importance in capturing the interest of the male. These odors are produced by the normal bacterial flora of the vagina under the influence of estrogen. These odorants, collectively called **copulins**, are volatile fatty acids. Hormonal regulation of their production is demonstrated by the fact that their rate of production rises during the egg maturation phase of the menstrual cycle and by the fact that their production undergoes no monthly rise in females taking birth control pills (Figure 21-9). To demonstrate that copulins make the female monkey attractive to the male, male monkeys are trained to press levers repeatedly to attain access to a female in an adjacent cage. A female with copulins removed by vaginal douches excites no lever pressing, whereas a

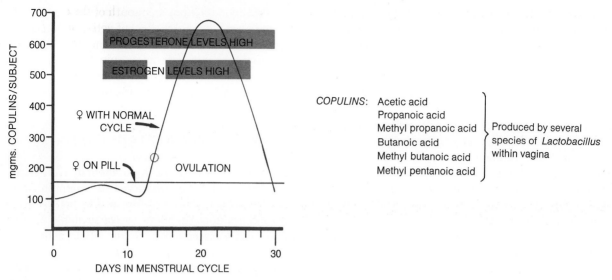

Figure 21-9 Copulins, volatile fatty acids produced in the vagina under the influence of estrogens.

female on the pill, ordinarily sexually uninteresting, excites lever pressing if she has been painted with copulins from an artificial source.

In the visually oriented human, the sense of smell has been assumed to play an insignificant role in behavior. Nonetheless the human female produces copulins.

Erection and Ejaculation

Psychic stimulation and the stimulation from sexual foreplay induces erection through parasympathetic control of the arteries and veins supplying the cavernous tissue, causing these structures to become engorged with blood (Figures 21-1 and 21-2). In normal individuals in the waking state, erection is followed by ejaculation only after a period of adequate stimulus of the glans penis. This stimulation eventually reaches the response threshold of an ejaculatory center in the spinal cord, and ejaculation occurs. There is a rapid contraction of the smooth muscle lining the seminal vesicles, with resultant ejection of seminal fluid into the urethra. This results in reflex contraction of associated striated muscles, which completes ejaculation. As ejaculation proceeds, further reflex muscular actions prevent backup of the ejaculate into the bladder and add essential components to the seminal fluid from the seminal vesicles, prostate, and bulbourethral glands. Substances supplied by these glands cause the sperm to commence swimming.

ORGASM AND HUMAN SEXUALITY

In the male, ejaculation is accompanied by local and generalized pleasurable feelings generally termed **orgasm**. This response in the male is virtually inevitable. In contrast, the sexual response in the female is more complex and is by no means always completely experienced in each copulatory act. Orgasm in the female is to a substantial degree a learned response, profoundly influenced by her general emotional state and environmental conditions. Sexual arousal in the female is relatively prolonged and is characterized by vaginal secretions and distention together with secondary processes such as enlargement of the nipples. Vascular distention analogous to that occuring in the male enlarges the clitoris and distends the superficial part of the vagina. The only physiologically obvious sign usually associated with orgasm in the female is a series of muscular contractions of the superficial part of the vagina. Orgasm is not followed by a refractory period, as in the male, and multiple orgasms can occur.

Human sexual behavior is not particularly amenable to rigorous scientific study. This, together with the societal restrictions built up around human sexual behavior, has resulted in much detailed nonsense being written on this highly complex activity. Actually, not much more is physiologically known about orgasm than is divulged in the preceeding brief paragraph. In recent years the subject has enjoyed more rigorous scientific attention, and some useful information has been forthcoming. Thus, Masters and Johnson[1] conducted a valuable study, but one that has been criticised because its observations of intercourse in a laboratory environment may have deprived the observed behavior of its higher emotional content. Other, necessarily more limited studies, have attempted to include this element, for example the Foxes[2], who privately analyzed their own sexual activities. Despite these notable attempts, the scientific examination of human sexual behavior is still in a primitive state perhaps never to be outgrown, owing to the delicacy of the psychophysiological processes under study.

FERTILIZATION

Sperm Cells Are Extremely Fragile

Human sperm cells have a brief lifetime, approximately 24 to 36 hr. To achieve even this short life expectancy, a number of constituents of the seminal fluid are necessary, including energy sources, acidity regulators, and other chemicals of as yet unclear function. At best, out of the millions of sperm released at orgasm only a few thousand will reach the vicinity of the egg. Of these most contribute enzymes that disrupt the still-adherent coating of ovarian follicle cells that invest the egg, allowing only one of their number to accomplish fertilization. Penetration of the egg by more than one sperm, which would cause genetic disaster, is prevented by a reaction of the egg surface, which occurs within a few seconds after sperm entry and prevents further penetrations by sperm.

---

[1] Masters, W. H., and V. E. Johnson: 1966. *Human Sexual Response*. Churchill, London.

[2] Fox, C. A., and C. B. Fox: 1971. A comparative study of coital physiology with special reference to sexual climax. *Journal of Reproduction and Fertility* 24: 319–36. Or see I. Singer: 1973. *The Goals of Human Sexuality*. Wildwood House, London.

Sperm are deposited at the mouth of the cervix and reach the egg by a combination of active and passive factors. Although sperm may swim at up to $100\ \mu/\text{sec}$ in reproductive tract mucus, in animal experiments inactivated sperm were transported as fast as normal sperm. Thus muscular movements of the uterus and fallopian tubes and ciliary action must be of great importance.

Penetration and Fertilization

Arriving near the egg in the fallopian tube, the sperm is possibly directly attracted to it by chemical messengers released by the egg itself. The DNA-filled head of the sperm contacts the egg membrane and adheres. Egg membranes reach out and enclose the sperm, giving the impression that the egg has consumed it, much as any wandering cell might ingest food. Once engulfed, the sperm does not survive as a cell; only its nucleus persists to join the egg nucleus and complete the transfer of information, which is called fertilization.

As the sperm penetrates, the egg is caused to complete the second meiotic division. The egg chromosomes become surrounded with a nuclear membrane, forming the female **pronucleus.** Within 12 hr the male nucleus meets and fuses with it. At about 24 hr after fertilization, the first cell division of the zygote occurs.

Multiple Fertilization and Twinning

More than one egg may be fertilized if more than one is present. This is typical in animals such as dogs and cats in which several follicles mature at once. Among humans it is not highly unusual for both ovaries to release eggs during the same cycle, providing an opportunity for twin fertilization. The twins resulting from these two eggs are **fraternal** twins. They are not identical, being no different from ordinary brothers and sisters, except that they develop simultaneously. Some females seem to have inherited the tendency for frequent, multiple ovulation, for it is a fact that twinning of this type is associated with certain families.

Identical twins, on the other hand, arise from a single fertilized egg (Figure 21-10). For some reason not understood, after the first cleavage the two cells separate completely. Cleavage then begins again in both cells and continues through embryogenesis and

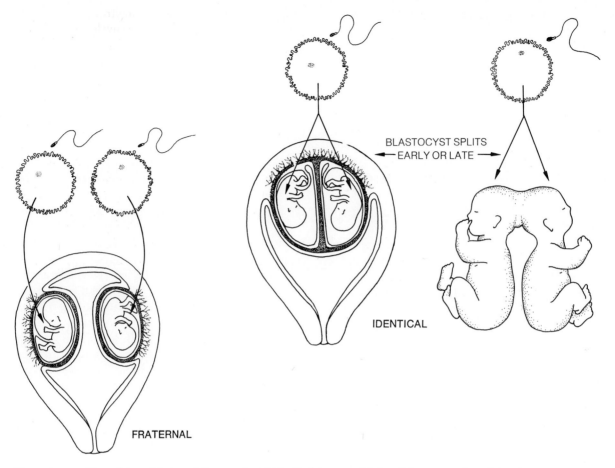

Figure 21-10 The origins of fraternal (two egg) and identical (one egg) twins. When separation of the cleaving cell mass into two occurs later in development Siamese twins result. Note that identical twins share the same placenta.

development of two complete organisms. Since the twins arise from the same fertilized egg, their cells carry identical hereditary information. The result is that the twins are identical in appearance and, of course, are always of the same sex. The frequency of twins in humans is about 1.2% with about 70% of twins being fraternal. Twins grow more slowly during late fetal life and often are born prematurely.

THE EVENTS OF PREGNANCY

Implantation

After fertilization, the egg completes the process of cleavage, with cells dividing approximately every 22 hr, during the 3 to 5 days required to move down the fallopian tube to the uterus. On reaching the uterus the embryo is in the **blastocyst** stage (Figure 21-11A). The inner mass of blastocyst cells will become the embryo proper, whereas the outer cells, or **trophoblast**, will form embryonic membranes and placenta. Implantation takes place rapidly, with trophoblast cells actively invading the endometrium, destroying and engulfing maternal cells. Soon this reaction gives way to cooperative formation of the placenta out of maternal and embryonic cells. **Ectopic** (out of place) implantation occurs rarely. Since viable sperm may reach the abdominal cavity, it is thought that occasional failure of the egg to enter the fallopian tube results in fertilization and implantation outside the uterus. Advanced stages of development are sometimes attained and require surgical intervention.

Coincident with implantation, the mother undergoes other hormonal changes in addition to those causing the corpus luteum to continue producing progesterone. Associated events include the enlargement of the mother's breasts and the occurrence of morning sickness. A desire for unusual foods may appear as part of the behavior pattern of pregnancy, perhaps reflecting changing physiological needs of mother and embryo. These symptoms are usually temporary, and the greater part of pregnancy is characterized in most women by excellent physical and mental health.

Miscarriage

Hormones produced by the embryo signal the progress of development. Should there be serious misdirection of early embryogenesis, a reduction of progesterone output causes breakdown of the endometrium, and a miscarriage (loss of embryo) may occur with an abnormal or heavy menstrual flow. The embryo is much more likely to miscarry during the first month of embryogenesis. Following this critical period, the embryo becomes increasingly resistive to miscarriage. It seems likely that the number of miscarriages during the first month of pregnancy is much larger than reported, because it is difficult to be certain of the occurrence of pregnancy and miscarriage at this early stage.

Formation of the Placenta

Normally the embryo very soon surrounds itself with a membranous sac, the **amnion**, filled with the **amniotic fluid**, which is of appropriate composition to form a near-perfect external environment for the newly forming organism. The chorion forms around the inner amniotic sac (Figure 21-11A). Through the amniotic sac and chorion, a tubular system develops that might be likened to a plant root system. These rootlike vessels, which pass through the amnion as the **umbilical cord**, divide into millions of tiny rootlets that are rich in blood vessels and grow thickly down into the heavy vascular tissue of the maternal endometrium. The mat of rootlike vascular tissue thus formed, bound to the outside of the chorion, eventually surrounds the embryo outside of the amnion. This combination of maternal endometrium and embryonic chorion, forming a complete envelope around the amnion, becomes the **placenta** (Figure 21-11B).

Nutrient and Waste Pathways

The tubular system that extends through the umbilical cord from the embryo to the placenta contains blood vessels through which the blood of the embryo is circulated by the developing heart. It is a closed circulation which relies on diffusion mechanisms together with complicated controls to ensure that nutrient molecules diffuse inward from the maternal blood and that waste molecules diffuse outward (Figure 21-11A, B). Embryo blood does not normally mix with maternal blood. There is always at least one membrane separating maternal and embryonic environments.

In summary, then, the embryo is bathed gently in amniotic fluid and enclosed by the membranous amnion, which is itself enclosed in the placenta. Embryonic blood is pumped via the umbilical cord to the placenta and returned. This stable condition prevails until birth.

The details of human embryonic life are set forth in the next chapter.

BIRTH

After an average of 266 days (9 lunar months) labor is initiated. The initiating causes are not clear, but one appears to be steroid hormones released by the fetal adrenal cortex. This idea is supported by the fact that the longest human pregnancies result when the fetus has a defective adrenal cortex and by experiments on sheep involving fetal hypophysectomy and ACTH (adrenocorticotrophic hormone) replacement. In the sheep, labor is not initiated if the fetus has been hypophysectomized but can be started if the hypophysectomized fetus is injected with ACTH.

As birth approaches, the uterus becomes more and more sensitive to the pituitary intermediate lobe hormone **oxytocin**, which causes it to contract. Maternal hormones also cause a relaxation of the joints between the bones of the pelvic girdle. The consequent spreading of the pelvic arch coincides with powerful, rhythmic muscular contractions of the uterus, which

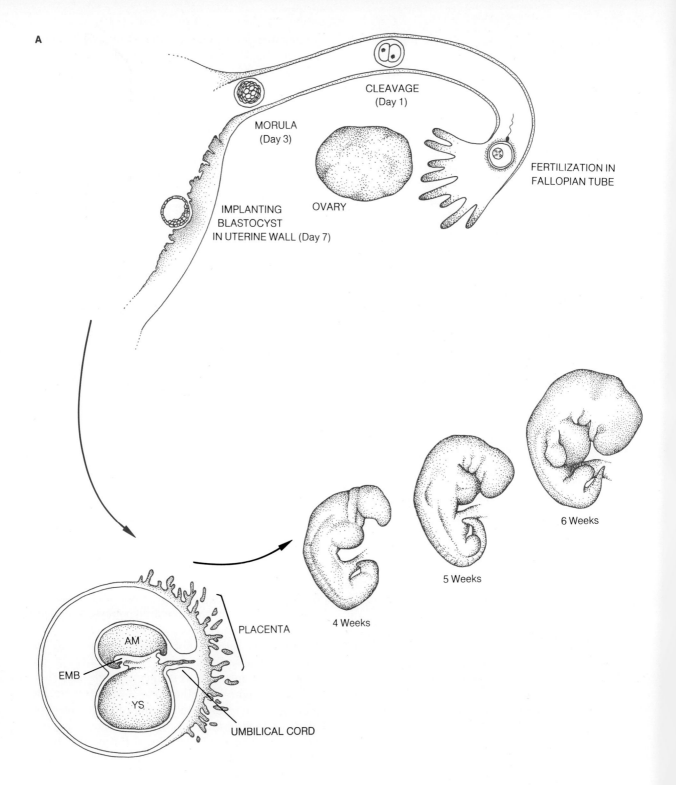

Figure 21-11 The stages of human development. A. fertilization through the sixth week. **B.** from 7 weeks to term. Stages are not drawn to the same scale. AM, amnion; YS, yolk sac.

462 BIOLOGY

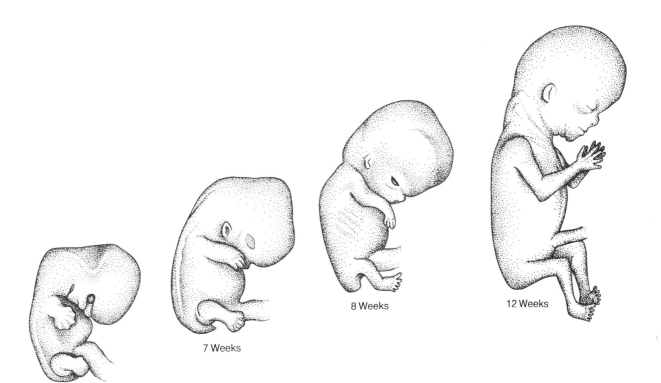

6½ Weeks
7 Weeks
8 Weeks
12 Weeks

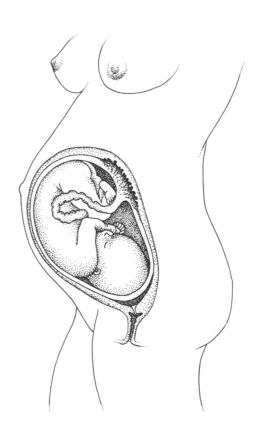

REPRODUCTION

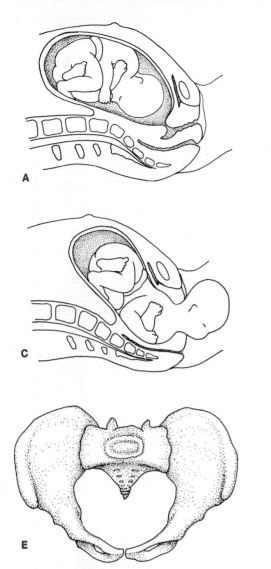

Figure 21-12 The events of birth. A. initiation. B. cervix dilating, C. head presentation. D. birth of infant complete, with expulsion of afterbirth in progress. E. view of birth canal from above in a skeletal preparation.

move the fetus through the vaginal canal. Typically the process works exceedingly well. In from a few minutes to a few hours the baby is born (Figure 21-12).

There is no period in life when so many adaptations have to be made in so short a time. Remember that for the first 9 months of life the fetus has really been a parasite on its mother, protected in a temperature-regulated, aqueous environment. The placenta has served its every need, from respiration to excretion.

Then, at birth, the fetus is suddenly expelled and its connection with the placenta severed. In this short interval, its organ systems must take up their lifetime tasks. Probably the most immediately critical problem at birth is establishment of pulmonary ventilation and establishment of adult blood circulatory pattern. The lungs must inflate with air for the first time and take over the process of gas exchange that an instant before was performed by the placenta. The circulation must shift, through the closing of the *foramen ovale* between the auricles of the heart (see page 492) and other changes, from an oxygenation circuit through the placenta to the adult circuit with oxygenation occurring in the lungs.

Hormonal changes in the mother have not yet run their complete course. Messengers are diffusing through the internal environment, signaling the change from pregnancy to motherhood and lactation. Glandular tissue in the breasts initiates milk production. Behavior at this time is highly oriented to care of the new infant. The mind-influencing consequences of hormonal changes may continue to be a source of concern, for there often occurs a condition called "postnatal depression" during which the new mother is morose and inconsolable. Little is known about the causes or function of this phenomenon, but it is, usually of short duration.

Birth Control

Although there is usually a prolonged menstruation following birth, eventually the pituitary hormones FSH and LH begin once more their rhythmic pattern of secretion. Ovulation may occur less than a month after childbirth unless lactation and suckling continues. The neural stimulus of suckling impinging on the hypothalamus promotes continued release of prolactin, which is necessary to continued milk production. At the same time it seems that prolactin may have a direct effect on the ovary to cause ovulatory failure. Evidence of this effect is seen rather clearly in human societies that reproduce at nearly the maximum rate. For example, the Hutterites (a Protestant sect) in the United States breast feed and use no means of contraception. They have families in excess of 10 children with an average birth interval of 22 months. If an infant dies soon after birth, the interval falls to 17 months, suggesting that breast feeding has about a 5 month inhibitory effect on ovulation. Until recently, probably more births were prevented worldwide through lactation than by all forms of contraception.

On a worldwide basis, the average number of viable children per female is five. Without consideration of world population problems (Chapter 25) this is usually more children than a family can cope with for financial and social reasons. Thus if only for reasons of improving the conditions of family life, birth control methods have long been with us and grow more accepted as time passes.

The most direct approach to birth control is elimination of sexual contact. Whereas complete elimination of sexual activity (celibacy) may be possible for unmarried and unattached individuals, it is not often that it has emerged as a workable means of regulating reproduction by married couples.

Humans maintain a very high level of sexuality, prompted by the frequent ovulations of the female and the constant reproductive competence of the male. Sometimes religious vows, together with isolation, suffice to eliminate sexual activity, but more often the only certain antidote for man's sexual motivation has been castration, complete removal of the gonads. Castration is, of course, of little value as a voluntary birth control technique because of the drastic effects it may have on male sexual behavior, owing to the loss of the testes, source of testosterone. This difficulty is gotten around by the remarkable operation said to have been practiced, for the purposes of birth limitation and improvement of the race, by certain clans of Australian aborigines. As shown in Figure 21-13, the procedure involves "surgical" diversion of the urethra in the male, again a practice not destined for widespread acceptance.

Today infertility without influence on sexual motivation is attained surgically by separation and ligature of the vas deferens or fallopian tubes. Sterilization of the male by section of the vas deferens is a relatively minor procedure increasingly practiced in this country and in countries with population problems, such as

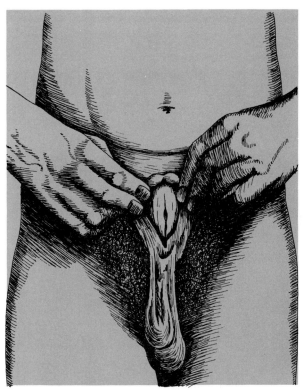

Figure 21-13 A method of birth limitation once practiced by native Australians. After pounding the penis into insensitivity an opening was made in the urethra just above the scrotum. Semen was diverted through this opening during copulation.

REPRODUCTION   465

India (see page 574). The procedure is not to be undertaken lightly, since the operation to restore fertility in the male by rejoining the vas deferens is very expensive and not highly successful.

Sterilization of the female is done by removal of a segment of each fallopian tube or by otherwise blocking them by tying or applying a clip or band. Menstrual and sexual functions are unaffected, since the ovary and uterus are left intact. Until recently the most commonly used female sterilization method took advantage of the fact that for 48 hr after giving birth the fallopian tubes are still near enough to the abdominal wall to be reached through a small and cosmetically insignificant incision. When the female is in the normal, nonpregnant state, tubal sterilization can be accomplished through an incision in the vaginal wall. Both of these procedures are relatively complicated medically and require the use of surgical facilities and general anaesthesia.

Recently sterilization methods involving **endoscopy** have come into wide use because they may be conducted on an outpatient basis and often with only local anaesthesia. One of these procedures is shown in Figure 21-14. The abdomen is penetrated by the endoscope, which is a device allowing viewing of the fallopian tubes. Then the device for interrupting the fallopian tubes is guided into place under visual control of the physician. Such methods have much to recommend them. The risk of infection is lower than with the previously described procedures and the operation is rapidly and inexpensively done. In one large study complications occurred in only 1.7% of the cases.

When the family is considered complete, surgical sterilization by one of these methods is a popular birth control method. Pills, which might have deleterious effects, do not have to be remembered and procedures do not have to be followed that might interfere with sexual behavior. Nearly 1.3 million surgical sterilizations were performed on males and females in the United States in 1975, making this the most extensively used birth control method for couples in which the female is over 30.

REVERSIBLE TECHNIQUES OF BIRTH CONTROL

Coitus Interruptus

And Judah said unto Onan, Go in unto thy brother's wife and marry her, and raise up seed to thy brother. And Onan knew that the seed should not be his; and it came to pass, when he went in unto his brother's wife, that he spilled it on the ground, lest that he should give seed to his brother.

As we see from these lines from *Genesis*, coitus interruptus may be the most ancient birth control technique still in use. Indeed, on a worldwide basis it may still be among the most frequently used methods. It is a risky technique because it requires the mental discipline to withdraw before ejaculation and because in a significant fraction of males ejaculation commences somewhat imperceptibly, depositing small numbers of sperm before the main ejaculate is emitted. Coitus interruptus, in that it does not allow a natural orgasm, is likely to result eventually in dissatisfaction for one or both partners.

Sperm-blocking Devices

There are many devices intended to permit copulation but to block the entry of the sperm into the uterus. These include a rubber or animal membrane sheath, the **condom**, worn over the penis; a rubber

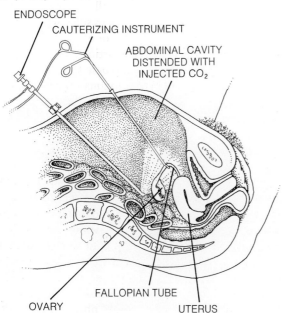

Figure 21-14 Sterilization by endoscopy. The abdomen is temporarily expanded with carbon dioxide.

Table 21-1 Effectiveness of Methods of Contraception

Method	Effectiveness (pregnancies per 100 woman-years)
No contraception	88
Most effective methods	
Sterilization	
male (vasectomy)	0.15
female (fallopian tube section)	0.06
Oral contraceptives	0.1–0.7*
Highly effective methods	
IUD	3
Diaphragm and cream or jelly	12
Condom	12
Less effective methods	
Rhythm methods	14–20†
Chemicals (foams, jellies, etc. without diaphragm)	20
Douche after copulation	35

* Higher rates include effect of missed pills; combined pills (progestin + estrogen) more effective than sequentials.
† Lower rate represents a closely supervised group.
DATA FROM VARIOUS SOURCES. See M. E. Calderone ed.: 1970. *Manual of Family Planning and Contraceptive Practice*. Williams and Wilkins, Baltimore; Speidel and Ravenholt; in R. G. Wheeler, et al (eds.): 1974. *Intrauterine Devices*, Academic Press, New York.

diaphragm in the vagina to block the cervical opening to the uterus; and **spermicidal jelly** or cream, deposited in the vagina to kill sperm cells before they enter the uterus. Such techniques are often used in combination with one another to reduce further the probability of fertilization (Table 21-1).

Limited Sexual Activity

In view of the observation that ovulation usually occurs about midway between menstrual periods, the **calendar rhythm** system calls for sexual abstinence during the 5 days preceding and the 5 days following this midpoint (Figure 21-8). Careful application of this plan reduces but does not eliminate the probability of fertilization since many females have irregular cycles. In women who have irregular cycle lengths, therefore, this method is unacceptably poor for either family planning or birth control.

Establishing the time of ovulation by the associated rise in body temperature is thought by many to be a more effective method than the calendar rhythm technique for many women. If the basal body temperature is taken each morning before arising, a marked rise in temperature often is observed at one point during the cycle. By following the temperature cycle for several months it is possible to chart the days unsafe for intercourse by counting as the day of ovulation the last day before the rise in temperature. As in the rhythm method, to avoid fertilization one should abstain from intercourse for 5 days before and 5 days after ovulation. Of course, in establishing the characteristic day of ovulation no data should be taken when one has a fever from illness. During the months when temperature is taken to pinpoint the probable day of ovulation, alternate contraception should be used to avoid unwanted fertilization. Perhaps the safest way to use the temperature method is to limit intercourse to the period after a 2-day upward shift in temperature.

In addition to the temperature method a timing method based on observable changes in cervical mucus has been proposed as an improvement on the straightforward calendar method. In this technique the amount and quality of cervical mucus is evaluated as an indication of onset of ovulation. During the early phase of the menstrual cycle, under the influence of rising levels of estrogen, mucus increases from virtually none to a substantial, cloudy, sticky yellowish or whitish discharge. Immediately before ovulation the discharge becomes clear and lubricative. After about three days, the discharge diminishes and again becomes cloudy and sticky.

It must be emphasized that all of these methods involving abstinence based on signs such as the history of previous cycles, temperature, or cervical mucus, are risky and are not based on particularly good data. Moreover, these methods are based on a series of assumptions that are only approximately true, namely that one menstrual cycle will be like the one preceeding, or that the egg is fertilizable for only 24 hr and that sperm are effective for only 48 hr.

Prevention of Implantation

According to legend, camel drivers in antiquity discovered that by placing silver rings or even stones in the uteri of their camels they could avoid the complications of pregnancies in them while making long desert crossings. In recent years an analogous technique has become much used with the development of various types of intrauterine devices (IUDs). Within the uterus the IUD probably functions by blocking implantation or causing phagocytic destruction of the early embryo. When the IUD is removed normal fertility can be expected.

Presence of an IUD can cause difficulty if pregnancy does occur. Use of a particular IUD, the Dalkon Shield, has been correlated with a number of uterine infections resulting in spontaneous abortion and sometimes death of the mother. One possible cause may have been that the particular plastic fiber "tail" initially used with the Dalkon Shield served as a wick to introduce bacteria from the vaginal canal into the uterus. This experience has considerably tempered hopes that the IUD represented a safe form of contraception which did not involve hormonal or other intervention. Women using IUDs are recommended to keep careful watch for possible pregnancy and seek the earliest possible medical consultation in the event that pregnancy is suspected.

The Pill

Probably the most generally successful female birth control technique is the birth control pill. In essence the pill induces those conditions which, during pregnancy, inhibit ovulation. Additional fertilizations do not occur during pregnancy because during that time no eggs are formed and released. It has already been mentioned that high progesterone levels occurring after ovulation induce a rich vascular endometrium in the uterus and also inhibit further ovulation. Figure 21-8 shows that both progesterone and estrogen levels remain high during pregnancy, and it has been found that any phase of the cycle can be induced by applying the estrogen-progesterone ratio characteristic of that phase. So it is that the birth control pills contain both progesterone and estrogen, or analogs of these molecules, to induce the ratio characteristic of pregnancy, during which no further ovulation takes place. To be exact, the induced condition is not that of pregnancy, for the placental hormones produced after implantation are missing. Rather, the induced condition is a great prolongation of that short period between ovulation and implantation when estrogen-progesterone levels are higher.

Whatever the mechanism, there is no question but that, if properly used, the pill inhibits ovulation with a very high degree of effectiveness. There are, however, some precautionary considerations. For example, once started, the pills must be taken faithfully; when stopped, ovulation is enhanced and chances of a pregnancy are considerably increased. Many physicians use these pills, not for birth control, but to increase fertility in previously childless couples. A 3-month course of pills, followed by total discontinuance, results in pregnancy in a large number of cases. On the other hand, when used for birth control, the pills must be discontinued immediately if, despite their influence, pregnancy does occur. The reason is that the normally high progesterone levels of pregnancy, augmented by the progesterone of the pills, produces an abnormally high level of male hormone, which can have an undesirable influence on a female embryo. If continued for any substantial length of time during pregnancy, the girl baby may show evidence of heavy musculature, excessive hair, and may be affected in her later sexual development.

A more serious difficulty, although fortunately of rare occurrence, is a poorly understood interference by the pill with calcium metabolism. It is well established that calcium ions figure importantly in the blood-clotting mechanism. In a very limited number of cases, the hormones of the pill disturb calcium balance and this, in turn, produces a tendency toward thrombosis. In this condition blood clots form inside the blood vessels, blocking circulation. Thrombosis in a vital organ is usually fatal, especially in the brain, where impaired circulation causes a stroke. There has been a sufficient number of cases of thrombosis of liver, kidneys, heart, and brain correlated with the use of certain kinds of birth control pills to lead some physicians to recommend regular prothrombin (clotting mechanism) tests, both when the pills are started and occasionally thereafter. These effects appear to be greater in the 35–44 year age group than in younger females. In general, it is conceded that the pill offers the best combination of ease of use and reliability now available. Nearly 12 million women in the United States and 50 million worldwide are believed to be using some form of oral contraceptive.

ABORTION

Failure or lack of use of contraceptive or implantation prevention techniques results in large numbers of unwanted pregnancies. Changing social attitudes and liberalization of legal restrictions (Figure 21-15) have led to large increases in the number of legal abortions.

> "We recognize the right of the individual, married or single, to be free from unwarranted governmental intrusion into matters so fundamentally affecting a person as the decision whether to bear or beget a child. That right necessarily includes the right of a woman to decide whether or not to terminate her pregnancy."

Figure 21-15 The U.S. Supreme Court on abortion, January 22, 1973.

Thus, in California alone there were only about 5000 legal abortions in 1968, but there were 145,000 in 1976. There were slightly over 1 million legal abortions performed in the entire United States in 1977.

Methods

Once pregnancy occurs there are several means of termination. In the early weeks of pregnancy abortion is simply achieved by uterine aspiration. This procedure involves removing the conceptus and much of the endometrium associated with it by vacuum applied to a tube inserted into the uterus. Although this is a fairly complex medical procedure it usually requires no hospitalization. Recently this technique has been modified as the minisuction technique and appears to be destined for widespread use, particularly in parts of the world where elaborate clinical facilities are not available. Using a modified 50 ml syringe to provide suction, paramedical personnel can provide uterine aspiration in the several weeks following a missed period, usually without the need for anaesthesia or dilation of the cervex.

Prostaglandins hold much promise for terminating pregnancies once techniques of application have been worked out. Certain prostaglandins (page 362) are extremely effective at inducing uterine contractions. Experiments still being evaluated show that incorporating them in vaginal suppositories produce highly effective emptying of the uterus up to 7 weeks after the last menstruation.

The Problem

Abortion has been with us since the beginnings of civilization; even Hippocrates supplied a formula. Yet in the civilized parts of the world, until quite recently, if a woman wished to have an abortion she had to submit to illegal practitioners or worse. As women have gained sufficient legal control over their bodies to make abortion legal in many countries over the past few years, an often acrimonious confrontation between two well-meaning groups has resulted. On the one hand right-to-life groups emphasize the defenselessness and humanity of the embryo (Figure 21-16) and maintain that abortion is murder, a sin. Pro-abortion groups hold that the pregnant woman is the sole, rightful arbiter of her person, which, they maintain, includes the unborn fetus.

The arguments between pro-life and pro-abortion factions boil down to traditional religious and legal definitions of life versus emerging views of the importance of the individual's rights to control his or her person. The arguments on both sides are known to most of us in detail because the conflict is now peaking and, indeed, this is no place to summarize them.

Figure 21-16 A human fetus. (R. Rugh.)

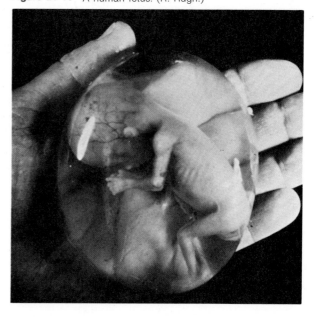

However, these arguments do hinge in part on a rather clear biological question about which a lot of nonsense tends to be generated. This is the question of when the conceptus becomes a life in the sense that it deserves the rights of any human.

To many pro-life adherents this is a nonquestion. They argue that the new person, fully endowed with rights, comes into existence with fertilization. It cannot be denied that this is the first instant at which an egg or sperm can have any future. It also cannot be denied that the idea of full rights upon fertilization has the dual advantages of being a clear-cut event and of relieving the authorities of the difficulty of timing conceptions to determine if any particular abortion breaks the law.

The zygote does have an element of completeness about it. All genetic information necessary to produce an undoubted human is present. The architect, plans in hand as it were, is awaiting action of the builder with her materials. However, this view of the situation does raise some question about completeness. Just as the idea is not legally the act, the plans are not the house, and the zygote has no future without the mother. Is it better then to at least talk about the early stages of development as a partnership? Assuredly, the senior partner has obligations to the junior partner. It would not be embarking on the path to humanity without her actions, after all. But one might view that same obligation as easily as a gift since, without those same maternal actions, the conceptus would not exist. Then, too, we must remember that the senior partner has her rights. Pregnancy is a risk and perhaps one has a right to avoid a risk even if it costs a potential life.

Obviously this is not a wholly scientific line of discussion. One gets this far in the discussion of abortion on the basis of personal philosophy, religion and attitude towards the law. If we return to a strictly biological context we see two guideposts of a reasonably definite sort which might provide a basis for deciding whether or not an abortion is the eradication of an unusual bit of tissue or of a life in the larger sense.

For one of these guideposts we might use the criterion of ability to survive if separated from the mother and the other guidepost might be the criterion of consciousness. Simply in the naming of the stages of development we sense that the embryologist perceives guideposts of this type, that some later stage of development is qualitatively different from preceeding stages. Embryo is the name applied to the human conceptus through perhaps the first nine weeks of development. By the end of that time it clearly looks human and is beginning to work like one. It is for the remainder of development called a **fetus.** However, independent survival is out of the question in the ninth week. The absolute earliest time when survival is possible is after about 25 weeks, and this with much medical assistance. By then the fetus may weigh a little under 1 kg and its lungs, the organ system most immediately essential to survival, may be sufficiently mature to support respiration. The outlook at this early stage is bleak, and the very few survivors are likely to be abnormal. Nonetheless, we mark 25 weeks as a turning point in fetal life signalling the beginning of relaxation of total dependence on the mother. Since this condition is characteristic of full-fledged humanity, then one might, reasoning biologically, be more hesitant about aborting after 25 weeks than before.

What about the other guidepost, the matter of the establishment of consciousness? This is even a less precise matter than the criterion of independent survival. By 25 weeks the fetal nervous system has many functions. It conducts regulatory processes, operates the heart, and controls voluntary muscles. At birth such an early fetus will react as if with displeasure to the new environment. To this extent consciousness exists. Perhaps, again speaking biologically, this criterion further supports the twenty-fifth week as the pivotal time. If one demands more of consciousness to label it human, then this guidepost disappears into the vagueness of definitions of human consciousness. To go to the logical limit, one might argue that abortion does no harm unless its victim understands the difference between life and death and is aware of its own personality. Obviously this is no acceptable criterion for it is not met until childhood is well advanced.

Thus we are left with the twenty-fifth week as a biologically logical decision point. It is fairly close to the decision time as is usually legally described for legal abortion, and this means, since laws tend to express the common sense of society, that it is a decision point that is morally comfortable to most people.

22 Human Development and Fetal Health

In the last chapter we saw how human life is initiated and how provision is made by the uterus for nurture and protection of the embryo as its self-assembly proceeds during the 9 months of pregnancy. In this chapter we will describe some of the salient events of human embryonic development and point out particularly important health problems that face mother and fetus during this critical period.

LIFE OF THE EMBRYO

We have already seen how fertilization occurs in the fallopian tube and cleavage proceeds as the embryo is moved towards the uterus. By the time it enters the uterus the embryo is a thin sphere of trophoblast cells surrounding an inner cell mass destined to become the embryo proper. The whole process of development which this simple appearing structure is about to undergo may be roughly divided into these stages: (1) *cleavage*, by this time already well advanced, (2) *establishment of the basic structure* of the embryo, and (3) *maturation of function* of embryonic organ systems. The major events of these stages are sketched in the following account.

Cleavage
When cleavage begins the zygote has a large amount of cytoplasm relative to the amount of nuclear material present. As cleavage progresses there is no growth in mass, since the embryo is still existing on its own food reserves. In the human, cleavage seems to occur quite irregularly with a jumbled, solid sphere of cells resulting. Since nuclear material increases at each division while the remaining cytoplasm is parcelled out among more and more cells, the nuclear-cytoplasmic ratio rapidly changes to more typical levels. However, the first clear sign that assembly of the embryo is beginning is seen when this solid sphere of cells forms a cavity known as the **blastocoel**. Most of the cells forming the wall of the blastocoel are **trophoblast** cells destined to participate in formation of structures such as the placenta and embryonic membranes outside the embryo proper. At one side of the hollow sphere of cells there is a small collection of a second type of cells, the **inner cell mass,** which becomes the embryo.

Since each cell of the embryo has an identical set of chromosomes and receives cytoplasm from a common source, the cytoplasm of the zygote, it is difficult to understand what determines the fate of cells at this early stage of development. Being alike, why should they ever become different? Why should one cell become part of the trophoblast and another part of the inner cell mass? Similar questions are applicable to innumerable decision points in the life of the embryo.

Causes of Early Differentiation
Embryologists have long known that differentiation, the process of specification of form and function,

begins at various times in development in the several types of organisms that have been studied. You are already familiar with the experiment that Driesh performed on the sea urchin embryo (page 15) to show that any cell in the very early embryo could form a small but normal embryo. However, there are embryos of other organisms in which quite the opposite is true. In some of these, even certain regions of the cytoplasm of the uncleaved egg have their fate precisely defined. Cells that receive this material during cleavage perform precise and unique roles in construction of the organism. A well-known example of cytoplasmic influence is found in the ctenophores (page 262). A particular region of the zygote cytoplasm has the ability to determine exclusively the development of the comb plates, the locomotor organs of the adult. If some of this cytoplasm is removed, a specific defect in the comb plates results and it cannot be repaired by other cells or other cytoplasm.

Study of many examples of development ranging between the two extremes of the sea urchin and ctenophore types of early differentiation has led to the conclusion that the developmental process is a continuum. In some organisms the fate of parts of the embryo is fixed early and irrevocably, sometimes even in the uncleaved egg. In others this stage occurs late in development. When determination of fate occurs before cleavage the determining factor is often of maternal origin since the tissues of the mother control egg formation. On the other hand, in the frog it is the point at which the sperm enters the egg that determines the laying down of the fundamental axis along which the embryo forms. Often environmental factors can provide the initial directions. A very clear example occurs in the egg of the marine alga *Fucus*, the brown seaweed found attached to rocks in the tidal zone. Before the cleavage of the *Fucus* zygote the cell pushes out a process that is to become the rootlike holdfast organ of the plant. External factors, such as the direction of incident light, the presence of acidity or carbon dioxide, and the presence of other developing *Fucus* embryos determine this early step in the development of the plant (Figure 22-1).

Experiments conducted primarily on mouse embryos, which can be cultivated artificially through the cleavage period, show that the mammalian embryo does not undergo early fixation of cell fates and that determination of which cells become trophoblast and which become the inner cell mass depends on environmental conditions. Lack of fate determination is shown by experiments in which *one* cell of the two-cell mouse embryo can be put into a foster-mother and develop normally, eventually becoming a fertile adult mouse. Even more remarkably, it has been possible to fuse together two early mouse embryos at the eight-cell stage with a normal adult the result. Such adults can be shown to be *chimaeras* (mixed animals) if the fused embryos are from distinguishable genetic strains. For example, mice from fused embryos may have patches of varied coat color due to contributions from cells from one embryo or the other, showing that the mice have *four* parents instead of *two*.

It is possible that human chimaeras occur. Instances are on record of persons having two chromosomally distinct cell populations, for example with normal female cells (two X chromosomes) and normal male cells (one X and one Y).

Quite simply the environmental factor that determines which cells form the inner cell mass and which contribute to the trophoblast is outside versus inside. In the blastula stage, before the gastrocoel begins to form, some cells must obviously be on the inside and some on the outside. The outer ones become trophoblast cells. The determining factor is their *position* and not that they are *particular cells*. This is shown by

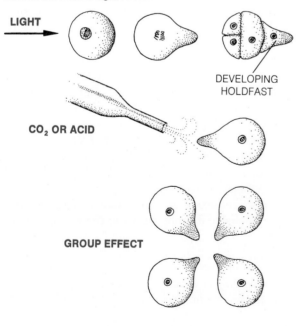

Figure 22-1 External factors determine the axis of development in the marine alga, *Fucus*.

experiments in which a giant embryo is formed by fusing several embryos around one central one. When this is done, nearly all of the cells of the centrally placed embryo appear in the inner cell mass rather than contributing both to inner cell mass and trophoblast as would normally be expected.

Establishment of the Basic Structure of the Embryo

During the second and third weeks of development, the human embryo proceeds from a simple mass of cells to a flat plate in which the basic tissue rudiments for further development are established. During the second week, the inner cell mass initiates this process by forming two flattened layers of cells, **endoderm** and **ectoderm**, lying in contact with each other and forming the **germinal disc** (Figure 22-2). In the third week, a longitudinal furrow appears in the ectodermal surface of the germinal disc. This furrow is called the **primitive streak.** Cells of the ectoderm migrate to and then proceed through the primitive streak downwards into the space between endoderm and ectoderm. Some migrate laterally to form the third germ layer, the **mesoderm** (Figure 22-3). Others migrate towards the future head end of the disc to form the first indication of the axial skeleton, the **notochord.**

This process of cellular migration is called **gastrulation.** It results in establishment of the basic form of the embryo and the appearance of the three fundamental tissue layers out of which the organs of the fetus will be formed. The *ectoderm* will form the nervous system, the skin, the enamel of the teeth and certain glands such as the anterior pituitary and mammary glands. The *endoderm* forms the inner lining of

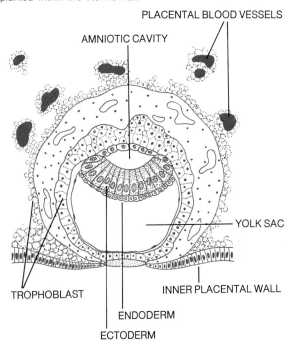

Figure 22-2 Cross section of an early human embryo implanted within the uterine wall.

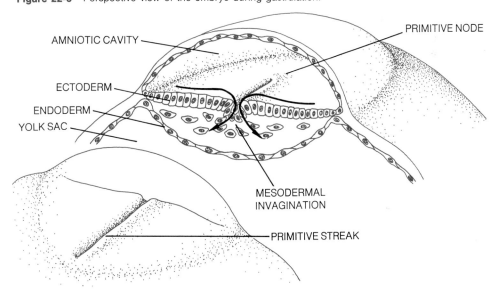

Figure 22-3 Perspective view of the embryo during gastrulation.

HUMAN DEVELOPMENT AND FETAL HEALTH

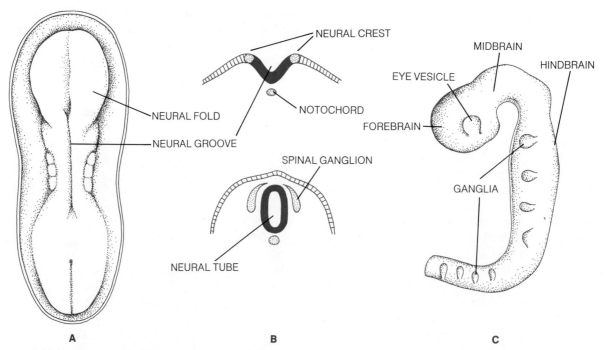

Figure 22-4 Early development of the nervous system. **A,** dorsal view. **B,** cross-sections showing folding of the neural groove to form the neural tube. **C,** lateral view of a later stage in development with the major regions of the nervous system defined by differential growth and folding of the neural tube.

the digestive tract, the lungs, the inner lining of the bladder and urethra, and glands such as liver, pancreas, thymus, thyroid, and parathyroids. *Mesoderm* forms muscle and skeleton, the kidneys, gonads, spleen, adrenal cortex, and the membranes that line the internal cavities of the body.

Organ Formation

With the three fundamental tissue layers established, the embryo swiftly proceeds to form major organs. By the eighth week of development the embryo is fully sketched out. It has the obvious appearance of a human although it weighs only about 5 g.

Nervous System. The earliest sign of the developing nervous system is a thickening of the ectoderm overlying the notochord. This thickening is the **neural plate.** The edges of this plate rise and the central part sinks down forming a longitudinal **neural groove** (Figure 22-4). At about where the future neck will be the groove closes to form a hollow **neural tube.** As development proceeds the tube continues to form in both directions, forming the hollow, dorsally situated nervous system characteristic of all vertebrates. The rear part of the tube becomes the spinal cord, while the head end of the tube expands into a series of **brain vesicles** that eventually grow and fold upon themselves to form the brain (Figure 22-4).

The **neural crest** is formed as the neural tube closes. Neural crest cells emerge from the edges of the folding neural groove and come to lie between the completed neural tube and the overlying ectoderm. These cells have unusually varied fates. Some form the sensory ganglia of the spinal cord. Others migrate along nerves and wrap around them as electrically insulating Schwann cells. Neural crest cells also form the sympathetic nervous system, become pigment cells, tooth-forming cells, the covering tissues of the brain (meninges) and even cartilage that contributes ultimately to the bony structure of the jaws and face.

Embryonic Induction. This varied spectrum of roles served by such a restricted group of cells raises again the question of fate determination during embryonic development. We have already seen this process at work in the very early stages of development, and the story of the neural crest cells makes it clear that similar questions arise at every turning in the complex path of development. What do we know about these processes in later development?

Mesoderm and the Nervous System. As we have emphasized, probably every cell in the embryo carries in its DNA a complete set of instructions for the entire organism. Thus we cannot explain how the nervous system and its parts form by saying that genes are parcelled out during development so that, for example, some cells only get DNA that forms nervous system whereas others get DNA specific for other parts of the embryo. Actually, when you think of it, even if DNA was sorted out, system by system, during the early cell divisions of development we still would not have a final explanation of the mechanism of differentiation. Obviously the answer would lie in the unknown process that sorted out the DNA. We would still have a mystery.

If every cell in the embryo and adult organism does have an identical genome, then under normal conditions every cell must have a very large number of inactive genes. Certainly all cells of an organism must have a group of genes in common that are always active, maintaining the basic processes of cellular life. In a very simple organism like a bacterium this would include nearly the whole genome. But, as organisms became more complex, the group of genes controlling processes unique to different *regions* of the organism or unique to different *periods* in life must have grown immensely. Surely if all of this latter group were to be simultaneously active, the results would be fatal—if only because of the burden of such a level of synthetic activity.

Viewed in this way the process of embryonic induction is seen as ultimately a problem of turning on and off of genes in proper sequence at proper times and places. Molecular biologists provide good models of how such processes might work by showing how products and other factors control the activity of bacterial genes (Chapter 9). However, the problem in developmental systems is clearly more complex because it must involve influences that arise from myriads of sources from outside the cell or even from outside the organism, as in the very simple case of *Fucus*.

Embryologists have long known that embryonic tissues have mutual influences on their differentiation. The nervous system is one of the classic arenas for the demonstration of such effects. For example, in the embryo of the frog it has been shown that the nervous system forms where it does, not because of some special quality of the ectoderm cells at that place but because there they are in contact with underlying mesoderm cells. Thus, if mesoderm is transplanted under ectoderm destined to form flank or belly skin, the result is the formation at the transplant site of a new nervous system (Figure 22-5).

The nervous system in turn controls the develop-

Figure 22-5 Embryonic induction. Mesoderm destined to control neural development is transplanted from one frog embryo (**A**) to another (**B**). The result is development of a secondary partial body axis (**C,D**) under the influence of the transplanted mesoderm.

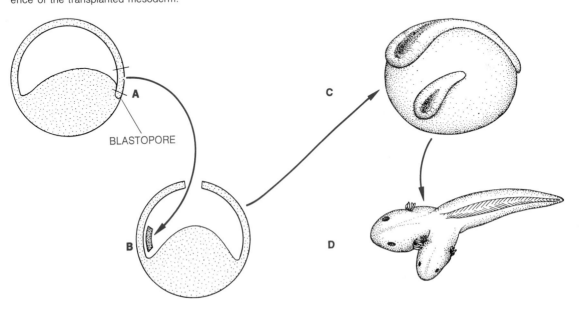

ment of other tissues. The eye is a good example of this hierarchy of control. The eye forms as an outpocketing of the forebrain commencing at about day 22. The outpocketing forms the neural part of the eye, but the lens is formed from the overlying ectoderm of the head under the influence of the underlying eye tissue. This is shown experimentally in the frog embryo either by grafting a developing eye beneath ectoderm that does not normally form a lens or by grafting ectoderm from another site in place of the normal lens-forming ectoderm. In both instances a normal lens forms.

These examples show that the embryo takes form because of a complex web of tissue interactions. Precisely how these interactions take place remains a mystery. However, certain of them have been shown to be caused by chemicals that diffuse from one cell type to another. This can be shown by experiments in which tissues that ordinarily interact in development are grown in tissue culture separated by an artificial membrane with pores so small that only the smallest of particles can get through. Such barriers permit the same type of interaction to occur that would occur in the normal course of development. The inference is that molecules diffusing from one cell to another initiate specific patterns of gene action in the recipient cell.

Heart and Circulatory System

The first sign of the developing circulatory system is seen at about 25 days of development when mesodermal blood forming cells (**angioblasts**) appear as isolated clumps in the mesoderm of the anterior end of the embryo. These clumps join together to form blood vessels filled with primitive blood cells. As development proceeds, blood cells are formed at other sites such as the liver, spleen, and thymus and, finally, in bone marrow, the blood forming site of the adult. While blood vessels are forming within the embryo, other vessels form in a similar fashion outside the embryo. Eventually they join, forming the connection between the embryonic circulatory system and the placenta.

The red blood cells of the embryo contain hemoglobin and function just as do the red blood cells of the adult, except that their gas exchange site is the placenta rather than the lungs. However, the red blood cells of the embryo have hemoglobin that is slightly different from adult hemoglobin. It contains the amino acid **isoleucine,** which adult hemoglobin lacks. Gradually during development fetal hemoglobin is replaced by synthesis of adult hemoglobin. At birth, fetal hemoglobin represents about 80% of the total hemoglobin in the fetus and drops to zero usually during the second year of life. It may play a special role in embryonic life because it has a higher affinity for oxygen than maternal hemoglobin (Figure 19-21).

The first major blood vessels of the embryo are a pair of **dorsal aortas** (Figure 22-6). In the anterior end of the embryo these become continuous with a pair of **heart tubes,** which initially lie separated by the forming gut. As growth continues, these heart tubes move to the midline and fuse together to construct the rudiment of the heart. The heart rudiment grows more rapidly than the surrounding tisues, and thus is forced to fold upon itself. Septa form within the folded, single heart tube and the adult four-chambered heart results (Figure 22-7).

The dorsal aortas continuing anteriorly from the heart clearly proclaim our ancestry. During development they faithfully form a series of aortic arches, which, in a primitive vertebrate, would have supplied blood to the gills (Figures 22-6E, 22-8). Although the six aortic arches of primitive vertebrates never appear at the same time in the human embryo, they do appear serially and help form the circulatory system of the adult. A parallel system of veins develops in synchrony with the arterial system. Both the venous and arterial systems make connection with the placental circulation through the large umbilical vein and artery that open into the embryonic vena cava and aorta respectively.

The system of extraembryonic blood vessels provides us with still another telltale indication of our ancestry. Because the embryos of lower vertebrates gain nutrition during development from yolk, there is in such forms, a well-developed system of blood vessels going to and from the yolk. If one examines a human embryo (Figure 22-6) these same vessels are seen going to and from a rudimentary yolk sac. Although these yolk sac, or **vitelline,** vessels would thus appear to be useless vestiges of the past, this is by no means so. They contribute in a major way to the formation of the complex circulation of the liver and to formation of a major vein, the hepatic portal vein.

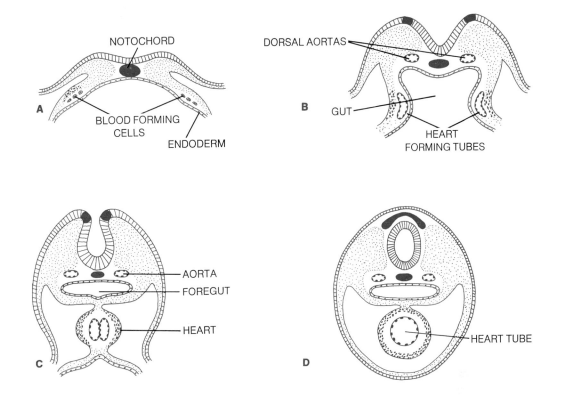
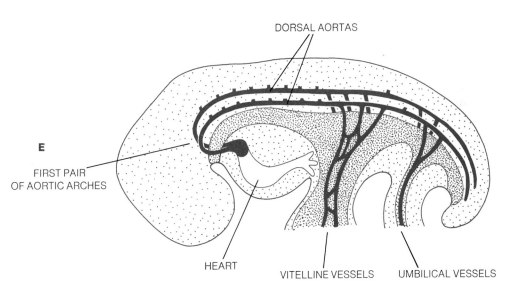

Figure 22-6 **Development of heart and major blood vessels. A–D,** stages in fusion of paired dorsal aortas to form heart tube, shown in cross sections of embryo. **E,** lateral view of the embryonic circulation and its connections with the extraembryonic membranes.

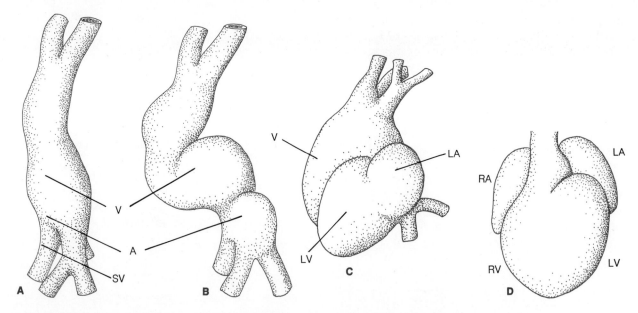

Figure 22-7 Stages in development of the mammalian four-chambered heart from a fused pair of primitive aortas. RA and LA, right and left aortas; RV and LV, right and left ventricles; V, ventricle; A, auricle; SV, sinus veinosus.

Madeover Gills

Thus the embryo teaches us a lesson in evolution, using as a text leftover blood vessels from our past. There is much more to that text and it is instructive to pursue it a little further. If we return to the matter of aortic arches we can readily see that the evolutionary vestiges of the gills persist in more than blood vessels. Examination of the head of a 5-week old human embryo reveals a series of four pouches on each side of the embryonic pharynx (Figure 22-8). These reflect the structure of the gills in our ancestry. It is interesting to see what becomes of them in the course of development. Like the aortic arches, they are not wasted. The first two pairs have skeletal elements that ultimately form the three tiny bones of the middle ear. The first pouch becomes the middle ear cavity and eustachian tube. Tissues developing in the first two pouches become the external ear canal

Figure 22-8 Fates of the aortic arches. **A,** aortic arches and pouches in a five week embryo. **B,** fates of the arches somewhat later in development. Parts of the jaw, ear and pharyngeal structures are seen laid down in cartilage derived from the arches. **C,** section through the embryonic pharyngeal region of the neck, indicating the contribution of the pharyngeal pouches to the adult.

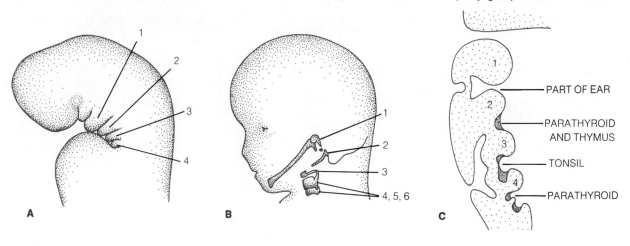

and part of the eardrum. Finally, the remaining three pouches form tonsils, thymus, and parathyroid glands.

Processes of this sort, in which ancient structures are made over into new organs, might not seem to be the most efficient way to build a mammal. This would perhaps be true if nature had the ability to start from scratch. Brief reflection indicates that this is an impossibility. In nature it is not possible to switch models, say from fish to mammals, as one might switch the manufacture of cars from one generation of models to another. This can be done for cars because there are builders external to things being built. But with living systems the external builder does not exist; the builder is the thing being built. Consequently, in life, the only way to achieve a "model" change is by modification of previous models so gradually that there is no interruption in the course of life. There is no way to effect such an immense renovation of genetic substance as would be required to abolish the ancient developmental ritual of gill formation and conversion into other structures in favor of direct development of those structures.

Muscles, Bones, and Kidneys

Besides forming the circulatory system, mesoderm forms muscle, bone, connective tissue, and the excretory units of the kidney. Early in the third week of development, the initiation of mesoderm differentiation is seen in the formation of **somites**, paired blocks of cells on each side of the neural tube (Figure 22-9). Each somite forms a set of muscles for its segment of the body, and these become innervated by spinal nerves of that segment. In the lower vertebrates this segmental organization of the musculature is evident in the adult, but in the human the modifications of the trunk musculature attendant on the development of the musculature of the limbs so modifies segmental organization that it is not readily apparent. Even so, the segmental origin of the trunk muscles of the adult is evident from their pattern of innervation, which is always traceable back to the segmental level of origin in the spinal cord. The somite also gives rise to the vertebral column and supplies the subcutaneous layer of the skin, the **dermis**.

Outside the somite, mesoderm that does not become involved in formation of muscle, skin, and vertebral column forms the excretory units of the urinary system. In the course of vertebrate evolution three types of excretory organs are found and the develop-

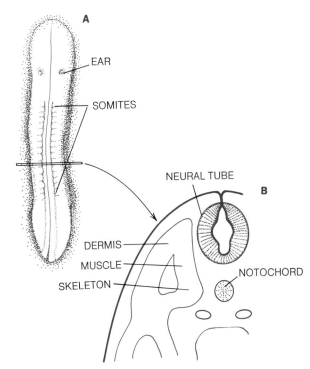

Figure 22-9 Somite formation. A. Dorsal view of an early embryo showing the appearance of paired blocks of somites along the neural tube. **B.** Cross section of embryo showing the placement of tissues within the somite and their ultimate fates.

mental history of the excretory system parallels this evolutionary history. In the most primitive type, the excretory unit opens directly into the **coelom** (body cavity) and in lower vertebrates the excretory organs are distributed along the length of the trunk.

The embryo goes through these stages relatively faithfully. At about day 25 the human embryo has excretory units opening into the coelom and at least vestigial excretory units running the length of the body. These structures have not been shown to function in the human embryo. By the sixth week of development they are replaced by a permanent kidney of adult type (Figure 22-10). During the second half of development, this permanent kidney becomes functional. Urine is produced and released into the amniotic cavity. Some is swallowed by the fetus, enters the fetal blood stream from the intestinal tract and is eliminated by exchange with the mother's blood in the placenta. This urine is principally water because the placenta does the work of excretion at this stage.

Figure 22-10 Three excretory organs appear sequentially in development. A. Cross section of an early embryo showing the most primitive excretory organs of the series. These open directly or through a duct into the coelom. **B.** Lateral view of the embryo showing the relationship of the segmentally arranged primitive kidney structures to the definitive adult kidney. **C.** A cross section shows that the primitive kidney has the same fundamental structure, a Bowman's capsule and glomerulus-collecting duct, that is present in the adult kidney.

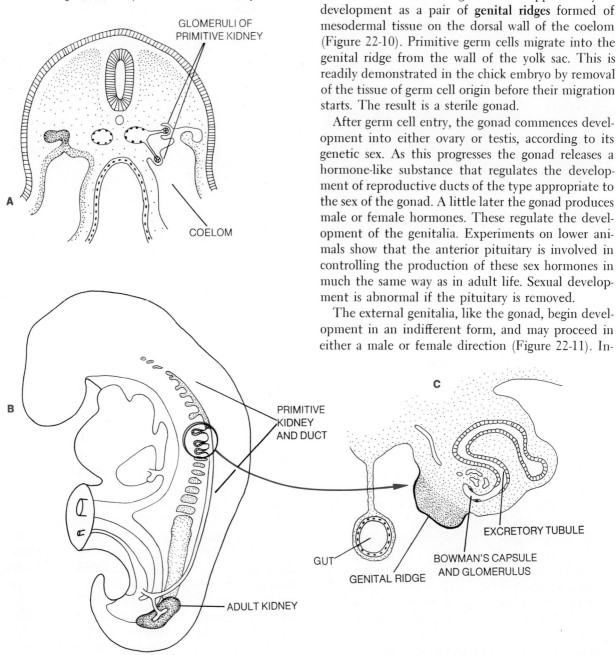

Sexual Development

The genetic sex of the embryo is determined at fertilization. However, early in development the gonads and external and internal genitalia pass through an indifferent stage and their final form—male, female, or intersexual—can be influenced by a variety of factors. The indifferent gonads first appear early in development as a pair of **genital ridges** formed of mesodermal tissue on the dorsal wall of the coelom (Figure 22-10). Primitive germ cells migrate into the genital ridge from the wall of the yolk sac. This is readily demonstrated in the chick embryo by removal of the tissue of germ cell origin before their migration starts. The result is a sterile gonad.

After germ cell entry, the gonad commences development into either ovary or testis, according to its genetic sex. As this progresses the gonad releases a hormone-like substance that regulates the development of reproductive ducts of the type appropriate to the sex of the gonad. A little later the gonad produces male or female hormones. These regulate the development of the genitalia. Experiments on lower animals show that the anterior pituitary is involved in controlling the production of these sex hormones in much the same way as in adult life. Sexual development is abnormal if the pituitary is removed.

The external genitalia, like the gonad, begin development in an indifferent form, and may proceed in either a male or female direction (Figure 22-11). In-

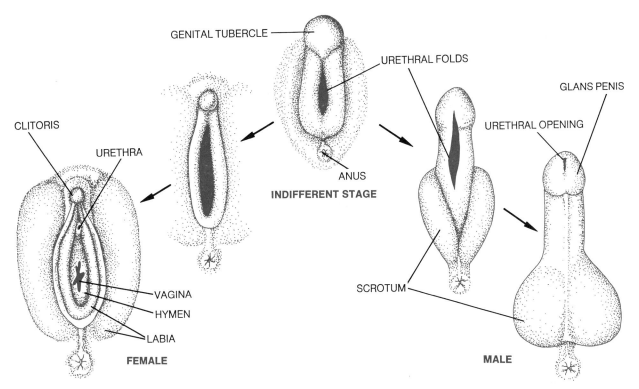

Figure 22-11 Differentiation of male and female external genitalia from the indifferent primordial stage.

tersexual forms are produced by imbalance of sex hormones during the course of this process. Sometimes imbalance is caused by malfunction of the adrenal cortex, which can produce male sex hormones. A female fetus suffering from this adrenogenital syndrome will have ovaries, but the male hormones produced by the adrenal cortex will produce various degrees of masculinization, ranging from simple enlargement of the clitoris to genitalia that look almost male. There are many other possible causes including, in mammals such as the cow, the influence of a male on a female twin. The placental circulations of two such twins are sufficiently connected to allow sex hormones from the male twin to influence sexual development of its female sibling, making it a sterile freemartin.

The Endoderm in Development

Early in development the endoderm forms a spacious sac that constitutes the ventral surface of the embryo and encloses the yolk sac. As the embryo grows, the endoderm is dragged along forming an anterior and posterior tube with a central opening to the yolk sac (Figure 22-12). The anterior region is the foregut; the central part is the midgut, and the posterior region is the hind gut. At the end of the first month of development the foregut meets and opens into the oral cavity. This has formed as an inpocketing of ectoderm beneath the forebrain. In this region the gill pouches form. In addition to the organs formed from these structures, the oral part of the gut is also the source of the thyroid gland. Like the gonad, the thyroid soon comes under the influence of the anterior pituitary, itself formed as an inpocketing of the dorsal region of the oral cavity. The respiratory system is the final major derivative of the anterior gut. The bronchial tree buds off the ventral surface of the foregut and begins the many subdivisions into bronchi characteristic of the mature lung. Alveoli appear late, in the seventh month.

The remainder of the foregut forms the esophagus, stomach, and duodenum. The liver is initiated as a diverticulum of the foregut, which connection persists as the bile duct of the adult. The pancreas has a

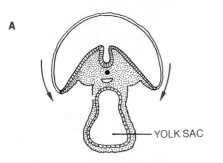

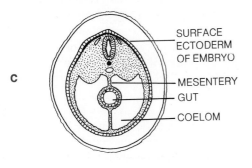
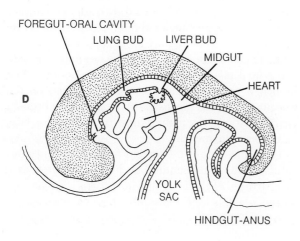

similar origin from the foregut. The midgut forms the small intestines and some of the colon, and the hindgut gives rise to the remainder of the colon and the inner part of the anus.

HEALTH OF THE FETUS

The overall mortality rate of embryo and fetus during development is difficult to determine because many spontaneous abortions occur during the first two weeks of pregnancy before the mother realizes that she is pregnant. Subsequently, about 15% of pregnancies result in spontaneous abortion by the twenty-fifth week. Mortality continues at a low rate throughout pregnancy, concluding with a mortality of about 1% at birth. These figures are, of course, significantly influenced by the socioeconomic state of the mother and the quality of medical care that she receives during pregnancy and at delivery.

It is believed that over 50% of spontaneous abortions are due to genetic defects. The incidence of genetic defects increases with age of both parents and is reflected in the fact that neonatal (newborn) mortality doubles for the 40–45 year maternal age group as compared with mothers between 18 and 30 years of age. Most of these genetic defects cannot be remedied. However, careful attention to maternal health during pregnancy and the provision of good medical care during pregnancy and delivery can reduce fetal mortality and, perhaps more importantly, help insure that those infants who are born have improved chances of healthy lives.

Maternal Nutrition

Undernutrition, which means inadequate intake of protein and carbohydrate as well as a badly balanced diet, such as occurs in genuine famine, fortunately is rare in the United States. When it does occur there is an increased incidence of premature births and those fetuses that go to term suffer retarded growth. Various

Figure 22-12 Development of the gut and related structures. A–C, cross sections showing folding of the body wall to enclose the gut with formation of mesenteries. **D,** longitudinal view showing the persistent connection of the gut with the yolk sac in a late stage of the enclosure of the ventral body wall. Lung and liver buds are forming and the fore- and hindgut are about to break through to form oral cavity and anus.

estimates indicate that between 30 and 60% of the world population suffers sufficient malnutrition to be in this condition. The hard-bitten view might be that this is a beneficial feedback serving to limit population increase in overpopulated regions of the world. But it must be remembered that the effects of undernutrition on the fetus are felt long after the births of those who survive. These grow into enfeebled adults whose probably inefficient efforts in the business of surviving further lower the general standard of living.

An important reason for this persistent effect of fetal undernutrition is easily seen in the fact that the growth of that most critical of organs, the brain, is particularly rapid in the latter half of pregnancy and in the first 6 months of infant life. At 6 months of age the infant's brain has reached half its adult size. Its body won't catch up until age 10. Stuffing the baby with good food after this period of rapid brain growth does little or no good. The brain by the time of birth has made almost all of the nerve cells it is ever going to make. If many of these are absent or stunted by undernutrition, the stigma of brain damage is a lifelong burden.

When the mother is on a normal balanced diet of about 2000 to 2200 kcal/day, which includes about 55 g of protein, very little change is necessary during pregnancy. In the United States the only vitamin deficiency that appears with any frequency involves folacin (folic acid). The normal intake of folacin is recommended to be doubled to about 400 μg/day during pregnancy. Since this vitamin is involved in nucleic acid metabolism, it is not surprising that a deficiency results in poor fetal growth and miscarriages.

Drugs

The construction of the placenta, providing intimate contact between maternal and fetal blood, allows entry of virtually all the contents of maternal blood, except very large molecules and cells, into the fetus. Consequently, excessive use of both alcohol and tobacco produce smaller than normal and weakened infants. Many kinds of medication must be strictly avoided. The most disastrous of these in recent history has been the drug **thalidomide,** which produces various degrees of abnormal limb formation (Figure 1-6). Just one adult dose (100 mg) between the days 35 and 45 of pregnancy almost certainly produces malformations. This drug acts at the critical stage of limb formation and is almost uniquely toxic to the human. Since large doses of thalidomide administered to experimental animals show little or no effect on embryonic development, many other similarly tested and presumed safe drugs may also have toxic effects on the fetus. It is thus a wise precaution for the expectant mother to avoid nonprescribed medication. In like vein, she should reduce her exposure to such environmental contaminants as insecticides, automobile exhausts, and hairsprays, remembering that the fetus she is carrying is just as exposed to such agents as she is and is undoubtedly far more sensitive.

The difficulty of protecting the fetus from harmful chemicals is compounded by the fact that some chemicals seem not to produce damage until long after birth. An example is **diethylstilbesterol (DES)**, a steroid related to sex hormones, which was used about 20 years ago to help prevent spontaneous abortion. Unfortunately, DES was later found to affect reproductive tract development so that daughters of DES treated women have a high incidence of reproductive tract malignant tumors, beginning at puberty. Sons of DES treated mothers appear to have a high incidence of cysts developing in association with the testes and to have sperm abnormalities that may affect reproductive ability. Although the effects on males have not yet been shown to involve malignancy, researchers fear malignancy will appear when these males enter the period of life when they are most prone to reproductive system cancer, from 50 to 60 years of age.

Maternal Disease

The fetus is susceptible to several viral diseases that cross the placenta and become established in the embryo, which has little resistance to them. The most serious of these is **rubella,** a trivial, measles-like disease of the adult. Yet if a woman contracts rubella within the first 3 weeks after conception, there is a 50% chance of fetal malformations involving eyes, ears, and heart. Mental retardation is a common result. The virus has its greatest effect during the period of organ formation. By the third month of pregnancy, rubella produces malformations in only about 10% of cases.

Public health authorities estimate that 10,000 defective children were born in the United States during the rubella epidemic of 1964-1965. Immunological tests are available to determine whether women are immune, and results show that about 85% of the adult population has immunity. A rubella virus vaccine has been in use since 1969 and has been recommended to nonimmune women contemplating pregnancy, but should not be taken within 3 months before becoming pregnant.

It has recently been estimated that **cytomegalovirus** (CVM) is an even more serious cause of mental retardation than rubella. Perhaps 3000 infants per year in the United States are seriously retarded by this cause and probably many more have significant but less obvious effects. CVM is common. Antibodies indicating previous or presently active disease are found in up to half of some United States populations. There is no vaccine available for this disease at present. Some scientists even question the advisability of widespread use of such a vaccine because CVM is a herpesvirus. Members of this class of viruses have been implicated as possible causal agents of certain types of cancer (see page 509).

Many bacterial diseases of the mother affect the fetus. Syphilis is one of the most severe of these. In the course of maternal syphilis there may be several bouts of spirochaetemia, or general distribution of the syphilis spirochaetes in the maternal blood. These organisms readily pass through the placenta and infect the fetus, which has virtually no resistance to the disease. Hence the severity of the infection is usually greater in the fetus than in the mother. Syphilis is extremely infective. Nearly 80% of untreated mothers in the early stages of syphilis infect their fetuses. Fetal mortality is high, with 25% dying before term and another 25% dying in early infancy. Prevention of fetal infection is relatively easy, since infection of the fetus rarely occurs before the fourth month of pregnancy. This allows time for the mother to be treated with penicillin, which will also cure a fetal infection if it has occurred.

Immunological Interactions between Mother and Fetus

It is obvious that the fetus must produce a large variety of proteins different from the mother's since mother and fetus are genetically different. If any of these proteins enter the maternal circulation, her immune system will treat them like any other foreign protein (antigen) and make antibodies to them (see page 501). This has a potential for disaster for the fetus because maternal antibodies are able to cross the placenta. Should this happen, various harmful effects may result. The best known example of this phenomenon is hemolytic disease of the fetus, or **erythroblastosis fetalis**. In this disorder the mother makes antibodies to an antigen present on the membranes of fetal red blood cells that enter her circulation due to some accident. These antibodies enter the fetus and react with fetal red blood cells. This increases the rate of fetal red blood cell destruction and ultimately results in fetal hypoxia (lack of oxygen) and jaundice (accumulation of metabolic byproducts of red cell breakdown). When untreated there is a 50% mortality from this condition with survivors often suffering from effects such as cerebral palsy and deafness.

The conditions producing this disease are now well known. About 15% of whites and about 1% of the nonwhites carry genes that make it impossible to form a certain surface protein on red blood cells. All other persons carry alleles of the gene that produce this surface protein. Because the first studies of this factor involved tests with the rhesus monkey, the genes are called the **rhesus factors.** The dominant alleles are designated as **Rh**. There are several, but we will consider them as one in this discussion. Individuals either homozygous (Rh/Rh) or heterozygous (Rh/rh) for the rhesus factor have the rhesus protein on their red blood cells. Individuals homozygous for the recessive alleles (rh/rh) lack this protein and are called Rh negatives.

If an Rh negative mother has an Rh positive fetus (Figure 22-13), her immune system will treat the rhesus protein on any fetal blood cells that enter her circulation as foreign protein and produce antibodies to it. These enter the fetus, causing erythroblastosis. Barring some accident, fetal blood cells rarely enter the maternal blood stream until the end of pregnancy, since a rupture of fetal placental blood vessels is required to effect transfer. The consequence is that often the first Rh positive child of an Rh negative mother is born normally. Successive pregnancies increase the likelihood of the symptoms developing.

There are several methods of treatment that reduce the incidence and severity of the disease. Mothers at risk are easily identified by blood typing procedures conducted on both parents. Development of the disease can be detected by detection of Rh antibodies in the mother's blood. Progress of the disease in the fetus can be followed by a test for the breakdown products of red blood cells conducted on samples of amniotic fluid. When the disease is severe, the fetus can often be saved by blood transfusion in the uterus. In this

RHESUS FACTOR

1. **Rh POSITIVE** individuals have Rh-antigen on their blood cells.

2. **Rh NEGATIVE** individuals have no Rh-antigens
   BUT they produce antibodies to Rh-antigens if
   exposed to them by: a. blood transfusion
                       b. fetal-maternal blood exchange

3. **Three genes** determine the Rh trait.

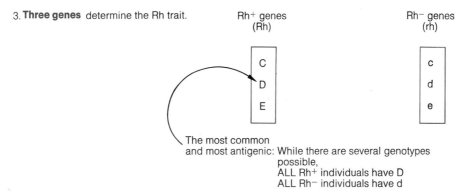

The most common and most antigenic: While there are several genotypes possible,
ALL Rh+ individuals have D
ALL Rh− individuals have d

4. **Maternal-fetal effect**

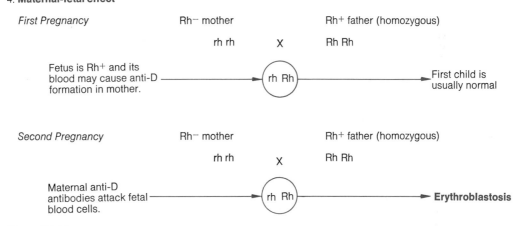

Figure 22-13 **The cause of erythroblastosis.**

procedure, Rh negative blood is injected into the fetal abdominal cavity. Enough is taken up from there by the fetal circulation to make up significantly for the destruction of fetal blood cells. At birth, a complete blood exchange can be performed, giving the newborn enough time to make a new supply of its own type of red blood cells in the absence of the destructive maternal antibodies, which are washed out by the transfusion.

An important preventive of this condition is to make certain that no woman who is Rh negative receives an Rh positive blood transfusion. This would immediately induce the formation of antibodies that would act on her next Rh positive fetus. It is also possible to prevent the formation of Rh antibodies in a mother carrying an Rh positive fetus by giving her antibodies just after birth to inactivate fetal blood cells that have entered her circulation before her own immune system can react to them. This is particularly useful at the end of a first pregnancy when fetal blood cells may enter the maternal circulation for the first time during childbirth.

Maternal-fetal incompatibility with respect to the ABO blood group (Figure 22-14) may also produce erythroblastosis. Some authorities argue that as many as 6% of very early embryos die because of ABO

ABO BLOOD GROUPS

A ANTIGEN/ANTIBODY DISTRIBUTION

GENOTYPE	PHENOTYPE	BLOOD CELL ANTIGENS	BLOOD PLASMA ANTIBODIES
OO	O	NONE	anti-A anti-B
AA AO	A	A	anti-B
BB BO	B	B	anti-A
AB	AB	A,B	NONE

TRANSFUSION POSSIBILITIES
RESULTS OF MIXING CELLS (C) OR PLASMA (P)
+ = no reaction, ± = slight reaction, − = agglutination

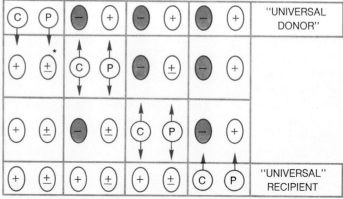

*Plasma transfusion effect is minimal because transfused antibodies are so diluted by recipient cells that no significant clumping occurs.

B POSSIBLE MATERNAL-FETAL ABO INCOMPATIBILITY
(Considering the effect of maternal antibodies on fetal blood cells)

Normal ↓   Possible reaction ↘

	FATHER					
MOTHER	OO	AA	AO	BB	BO	AB
OO anti-A anti-B	OO	**AO**	**AO** OO	**BO**	**BO** OO	**AO BO**
AA anti-B	AO	AA	AA AO	**AB**	**AB** AO	AA **AB**
AO anti-B	AO OO	AA AO	AA AO OO	**AB BO**	**AB** AO **BO** OO	AA **AB** **BO**
BB anti-A	BO	**AB**	**AB** BO	BB	BB BO	**AB** BB
BO anti-A	BO OO	**AB AO**	**AB** BO **AO** OO	BB BO	BB BO OO	**AB** BB **AO BO**
AB none	AO BO	AA AB	AA AO AB BO	AB BB	AB BB AO BO	AA AB BB

Figure 22-14 The ABO blood group in blood transfusion and maternal-fetal incompatability.

incompatibility as, for example, between a type O mother and her fetus by an AB father.

MONITORING FETAL HEALTH

Until recently the state of fetal health was determined only poorly and with difficulty by heart sounds, measurements of fetal growth obtained by palpation of the mother's abdomen, and x rays with their attendant dangers. Two new techniques have greatly increased the ability of the physician to monitor fetal health. Beams of ultrasound can be directed at the fetus so as to create an image constructed from the "echoes" of the sound beam from fetal tissues (Figure 22-15). This technique provides much of the information that x rays provide, seemingly without damage. However, the most powerful new diagnostic technique recently developed is **amniocentesis.**

Amniocentesis, from the Greek words meaning membrane pricking, is a relatively simple procedure in which a hypodermic needle is inserted through the mother's abdominal wall and into the amniotic cavity of the fetus. Amniotic fluid is withdrawn for diagnostic procedures (Figure 22-16). Chemical analysis of the fluid reveals certain abnormalities, for example, the pigmented materials indicative of erythroblastosis. More importantly, the amniotic fluid contains cells derived from the fetal bladder, skin, trachea, and the

Figure 22-15 Ultrasound image of an eight month fetus. Head is to the lower right and spinal cord towards the top.

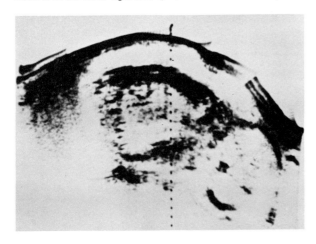

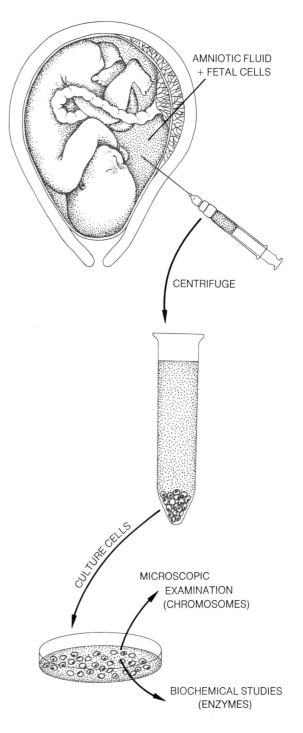

Figure 22-16 The technique of amniocentesis. Amniotic fluid is withdrawn and centrifuged. Biochemical tests may be performed on the fluid and the cells cultured for microscopic examination of chromosomes or for other biochemical tests.

amnion. These cells may be studied in various ways to detect a great number of fetal disorders. Sex of the fetus is quickly determined by a microscopic examination of the cells to reveal the presence of characteristic sex bodies, densely staining regions on the nuclear membrane. Knowing the sex of the fetus is important when the presence of a sex-linked disease such as hemophilia is suspected.

A more prolonged procedure in which the fetal cells are grown in tissue culture is needed for other tests. After enough cells have been grown in culture, some of them may be stopped in metaphase of cell division by the drug colchicine. This permits examination of the chromosomes to determine if the chromosomal abnormalities characteristic of certain inherited diseases are present. For example, this technique will show the presence of Down's syndrome (page 177). Cultured cells may also be used in enzyme studies, which are able to reveal the presence of nearly 70 different metabolic disorders. **Tay-Sachs disease** is detected in this way by the absence of an important enzyme, hexosaminidase A, which is necessary for normal metabolism in the central nervous system. Since this disorder is invariably fatal in early infancy, amniocentesis is now routinely performed when the prospective parents have been diagnosed as carriers of the disease. For the same reason all older pregnant women are recommended to have amniocentesis since they run a heightened risk of Down's syndrome.

Amniocentesis is by no means as widely utilized as it might be. Although as many as 400,000 women are appropriate candidates for amniocentesis every year in the United States, it was performed on only 3000 in 1975. It is, of course, no complete solution to the problem of birth defects. It offers no cures, only the possibility of diagnosis of a hundred or so defects. Regretably, the only response to nearly all of these conditions is either abortion or the saddening decision to continue a pregnancy virtually certain to result in a defective child.

23 Problems in Human Life: Disease and Aging

INTRODUCTION

Eventually we meet our end from accident, disease, or the inevitable process of aging and the diseases to which aging allows free rein. Though inevitable, death can be postponed, statistically speaking, and, most importantly, we can ensure sufficient freedom from disease and physical disability to make any lengthening of our span more than just a burden to be borne infirmly to the grave.

This is a large topic for a single chapter, and we must proceed with great selectivity. Furthermore, as you read, you must always be aware that we are dealing with principles. No matter how specific the examples of this chapter may seem, you must remember that the answers to your own health problems, beyond principles, are to be sought from your physician, not from textbooks!

THE IMMEDIATE CAUSE OF DEATH

From your previous study of the integration of organ systems in the operation of complex animals such as man, you could readily predict the *immediate* cause of death almost irrespective of the disease, accident, or condition that brought it about. The immediate cause of death is usually failure of the oxygen supply to the brain. Only a few minutes of **anoxia** are sufficient to kill brain cells essential both to conscious activities and the operation of basic control systems. Lack of oxygen generally brings about death more rapidly than any other lethal effect; therefore, it is the most urgent problem to be faced in dealing with medical emergencies (recall page 390). Because the oxygen transport chain between outside air and brain cells is long and complex, there are many points where it may be fatally interrupted. Suffocation, as by drowning, mechanically blocking the trachea, or choking on food, prevents gas exchange with the environment. Certain diseases and poisons act similarly, by stopping the action of muscles responsible for ventilation. Interruption of gas transport by inactivation of hemoglobin by carbon monoxide, by blood loss, or by interruption of circulation (as by blockage of critical blood vessels or by heart failure) are all rapidly fatal. And finally, certain agents act fatally by preventing the utilization of oxygen by the cells.

Whereas the immediate cause of death usually is thus clearly seen to be anoxia, the contributory causes are as varied as one might imagine. Some of these we shall explore in this chapter.

Poisons

A poison, according to the dictionary, is an agent that, when introduced into an organism in small amounts, produces an injurious or deadly effect. Clearly, this is an extremely broad category but, quite roughly, poisons may usefully be grouped into three classes: Those that

1. destroy living tissues by direct chemical action.
2. interrupt respiratory or metabolic processes.
3. interfere with the function of the nervous system.

The first class includes many poisons that small children find in medicine cabinets, around the kitchen sink, or in the garage—strong acids, such as hydrochloric (muriatic) acid, sulfuric acid (oil of vitriol), oxalic acid (a common constituent of some drain cleaners); strong bases, such as lye (sodium hydroxide or caustic soda); organic solvents, such as carbon tetrachloride (in some dry cleaners). Most of these produce such severe tissue damage to the alimentary tract (organic solvents may erode liver or kidneys) that even immediate treatment may not be successful. By far the best tactic in regard to these substances is *prevention*—careful storage of dangerous chemicals, particularly so that they are positively out of investigatory range. Children die in appallingly large numbers each year as the result of failure to observe this precaution.

Of the respiratory poisons, the most infamous is cyanide (HCN, a common fumigating agent), which has an affinity for the ferrous iron of the cytochromes. With the iron tied up by cyanide, these electron-transport enzymes cannot function, and the cells soon die from lack of energy in the form of ATP. Carbon monoxide (CO), as from car exhausts and tobacco smoke, is also a respiratory poison. It combines with hemoglobin in the red corpuscles in lung capillaries, with the result that these critical oxygen-transporting molecules are unable to carry sufficient oxygen to sustain life. Exposure to carbon monoxide is an example of positive feedback. As hemoglobin is inactivated by carbon monoxide, the resultant lack of oxygen at cellular levels causes increased ventilating effort, just as it does at high altitudes. Increased ventilation promotes more inactivation of hemoglobin by carbon monoxide, which increases ventilation still more.

Poisons acting on the nervous system are a large, varied, and poorly understood group of compounds. They all cause death by disrupting the neural control systems that regulate respiration and circulation. In some instances the cellular site of action is well known and, indeed, such substances are useful to the experimenter in the study of neurophysiology.

One of these, a protein produced by the bacterium *Clostridium botulinum*, is probably the most extremely poisonous substance known—20 million molecules will kill a mouse. There being considerably more than 20 million cells in the mouse nervous system, it is clear that botulinus toxin must act at very critical sites. These are the neuromuscular endings, where only a few molecules of the toxin prevent the release of acetylcholine from nerve terminals. This effectively cuts off the nervous system from control of striated muscle, with obviously lethal consequences. *C. botulinum* is of practical interest because it can occur in inadequately sterilized canned or plastic-packed foods that provide the anaerobic growth conditions it requires.

Another mode of disruption of neural function is illustrated by the toxin produced by a related bacterium, *Clostridium tetani*. These organisms can multiply in wounds that are anaerobic, such as the puncture produced by a nail, and cause the nearly always fatal disease **tetanus.** The toxin acts on the central nervous system to prevent inhibition, which is as essential to normal neural function as excitation. The resultant spasms prevent normal respiration and other muscle-operated processes. Strychnine, an **alkaloid** derived from a plant, the nightshade (*Strychnos*), acts in a similar way, causing excitation by suppression of inhibition in the central nervous system.

Many insecticides in common use are nerve poisons. DDT acts to prolong the ordinarily brief permeability of the nerve cell membrane to K^+ during the nerve impulse, with the result that many nerve impulses are generated under conditions that ordinarily produce only a few. Thus originate the tremors characteristic of DDT poisoning, "the DDTs." Similar results are attained by a different mechanism in the case of parathion, an insecticide that disrupts the functioning of the nervous system by inactivating the enzyme acetylcholinesterase, which normally breaks down the synaptic transmitter acetylcholine after it has acted. With this enzyme out of action, acetylcholine persists far longer than normal at the synapse, and the result is that more than the normal number of impulses occur in the "downstream" cell in response to stimulation. The result is ultimately the same with both insecticides—death due to disruptive hyperactivity of the nervous system. The unfortunate thing is that these substances are very nearly as toxic to us and to other vertebrates as they are to their intended targets. The serious consequences of indiscriminate use of such compounds are discussed in Chapter 26.

Finally, in regard to poisons having their primary effect on the nervous system, you should review Chapter 20, where the influence of drugs on higher neural functions is discussed. Many such drugs must assuredly be classed as poisons, even though it may be only rarely that they directly disrupt physiological function in an obvious way. Ordinarily they affect mental function sufficiently to prevent the user from living a normal life and all too frequently cause the user to suffer permanent physical damage. For example, suicide may result from the acute behavior changes caused by LSD and, of course, we are even more familiar with risks of personal destruction from alcoholic drivers.

Disease

Human disease is usefully described in four categories: *organic failure, pathogenic invasion, nutritional deficiency,* and *cancer*. Organic failure involves physiological malfunctions that are of internal origin and that may involve two types of problems: failure of a control system, as in diabetes, or structural malfunction, as in the rupture of a blood vessel. In pathogenic invasion, externally originating disease organisms, which range in size and complexity from viruses to large parasitic worms, enter the body and act either by destroying tissues directly or by producing poisons. Nutritional deficiencies range from the deleterious effects of specific vitamin or mineral deficiencies (for example, the **goiter,** stemming from iodine lack) to the more generalized effects of protein or carbohydrate starvation. Cancer cannot be as precisely defined because its causes remain obscure. In some known instances, viruses induce cancer, and this action may be facilitated by other agents such as radiation and certain chemicals, which also may act independently. Cancerous cells multiply and invade other organs, often with fatal effects.

ORGANIC FAILURE

Organic failures, of course, include many dysfunctions that have already been described as originating from external causes. Thus many poisons, such as strychnine, produce neural control system failure. Similarly, a disease organism might bring about structural malfunction by, to name one example, invading the heart valves and damaging them sufficiently to prevent the heart from doing its normal work. But here we are concerned with organic failures that do not have such immediately obvious external origins.

Congenital Malfunctions

Considering the complexity of development, it is indeed a tribute to natural selection that so many of us are born essentially normal, but *mistakes in development* do occur and these constitute a significant fraction of the organic failures that beset man. In addition, we include here *inherited disorders,* which are attributable to our genetic makeup rather than to developmental accident.

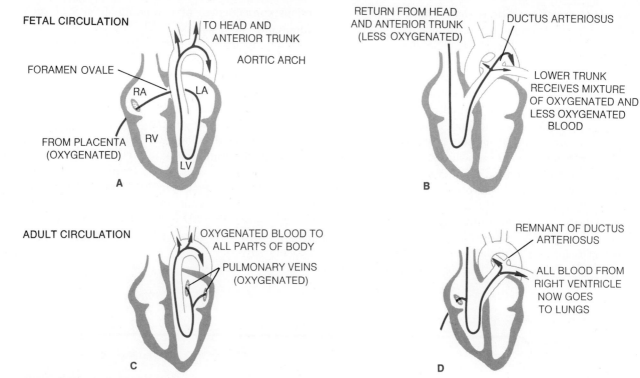

Figure 23-1 Congenital defects. To accommodate to change from fetal to free life the heart rapidly shifts from a circulatory pattern by which fetal blood is oxygenated by the placenta to one in which gas exchange is accomplished in the lungs of the newborn. The situation in fetal life is shown in **A** and **B** where oxygenated blood from the placenta is seen to enter the right auricle and cross directly to the left auricle through the foramen ovale, and thence to the systemic circulation, avoiding a useless circuit of the lungs. Blood from the right ventricle is largely diverted from the pulmonary circuit by way of the ductus arteriosus. The adult configuration, **C** and **D**, normally is attained by immediate closure of the foramen ovale, owing to blood pressure changes at birth, and this is followed by a gradual growing shut of the ductus arteriosus. Failure of either of these processes results in congenital heart disease.

One of the best illustrations of developmental mistakes and of the possibility of correcting many of them is the case of birth defects of the heart (Figure 23-1). If an infant afflicted with such a defect survives long enough, the heart often can be restored to normalcy by surgical repair.

Similarly, many inherited defects are sufficiently correctable by medical means to permit a normal life span (Table 23-1; recall Table 20-1). This is an instance in which medical progress may be a mixed blessing. Untreated individuals possessing many such defects would probably die before reaching reproductive age, and this weeding out would reduce the prevalence of the defect-causing genes. The tendency is now for such genes to be preserved, through medical support of their carriers to reproductive age.

Organic Failure in Later Life

To illustrate the broad spectrum of diseases in this category we mention two types of malfunction: diabetes, which can be considered a metabolic control-system failure, and critical circulatory-system failures, which represent structural malfunctions.

Diabetes. Diabetes results from malfunction of one portion of the blood sugar control system—that portion found in the pancreas which secretes insulin and glucagon (page 383). The pancreas of a diabetic person, often because of an inherited tendency, secretes insufficient insulin. It will be recalled that one of the effects of insulin is to influence the membranes of certain cells so that they can transport glucose from the blood into those cells as a nutrient. Without insulin, even though blood sugar remains high, the

Table 23-1 Examples of Correctable and Noncorrectable Genetic Disorders

Condition	Primary Effect	Treatment or Outcome
A. Correctable		
Diabetes	Sugar metabolism and circulatory system	Insulin and diet control; circulatory effects often become severe in later life
Hemophilia	Clotting defect	Temporary alleviation by injection of missing clotting factors; usually fatal before middle age
Galactosemia	Inability to convert galactose (milk sugar) into glucose	Removal of galactose from diet prevents liver damage and mental impairment in infants
Phenylketonuria (PKU)	Inability to convert phenylalanine to tyrosine	Reduction of phenylalanine in infant's diet prevents severe mental retardation
Hare lip and cleft palate	Embryonic development of palate	Surgery on young, favorable cases prevents impaired speech and facial disfigurement
B. Noncorrectable		
Sickle-cell anemia	Abnormal hemoglobin	Fatal when homozygous
Glycogen storage diseases	Enzyme deficiencies in carbohydrate metabolism	Often fatal
Albinism	Defect in tyrosine metabolism	Normal except for extreme light sensitivity
Thalassemia	Anemia with underproduction of hemoglobin	Fatal when homozygous
Cystic fibrosis	Defect in glycoprotein metabolism	Fatal in midlife

cells die from lack of nutrition. Before adequate treatment was devised, nearly every case of diabetes was fatal. The treatment that was found to be effective consisted of restoring the ailing control system by the injection of insulin, currently obtained from the pancreatic tissue of cattle. Insulin therapy does correct the sugar metabolism problems of the diabetic, but unfortunately there are almost always circulatory disorders associated with the disease, and these are not correctable. A frequent result is loss of vision and other severe circulation problems that often force amputation of limbs.

Diabetes, Diet, and Health. There is a form of diabetes called **marginal diabetes,** that can be controlled by diet. All the characteristics of severe diabetes are present but in milder form due to a smaller insulin deficit. This condition is not rare in our population. Probably there are in addition a great many more borderline cases of insulin shortage that never attract medical attention and diagnosis. This is especially true of persons over 30 years of age, for something about the aging process seems to aggravate marginal diabetes, bringing more acute symptoms during the peak of maturity or later. Most importantly, from the successful treatment method for marginal diabetes we can see an example of the way nutritional needs affect the metabolic mechanism of the human body in health, sickness, and during the aging process.

Consider first the effect of overeating of carbohydrates on a marginal diabetic. Because such an individual's cells can accept only a limited amount of glucose, due to insulin lack, blood sugar rises to very high levels—**hyperglycemia.** The kidneys attempt to remedy the situation by excreting a copious sugar-laden urine. If water intake is insufficient, the consequence is dehydration. Even with sufficient water, the increased kidney load causes mineral loss and poor osmoregulation. In either case, the result is an inadequate internal environment that causes the individual to lose weight—a unique case of losing weight by overeating. As can readily be appreciated, the diet of the marginal diabetic should strictly limit the amount of nutrient intake, especially carbohydrates.

On the other hand, if the dieter cuts back too much on food intake, unpleasant results follow. First, because blood sugar is too low for insulin-starved cellular membranes to transport readily, the cells of the body become starved for sugar. All the glycogen reserves in liver and muscles are converted by the liver to maintain blood sugar. When these are gone, the fat reserves are attacked and converted to sugar. Finally, when all other reserves are exhausted, proteins are broken down to amino acids, deaminated, and converted to sugar. Tissue starvation is followed by weight loss, degeneration, and ultimate organ failure and death.

Circulatory-System Failures. As medical practice becomes more effective at controlling disease and prolonging life, we become relatively more susceptible to circulatory-system failure, commonly known as CVA. Certainly, heart and circulatory disease is now the major killer in the United States (Figure 23-2).

Although **CVA** stands for two different causes of death (**cerebrovascular accident** and **cardiovascular accident**) they are really the same kind of event occurring in different places. To understand CVA, we must recall that both brain and heart cells metabolize at a high rate, yet have almost no reserves of nutrients and oxygen. They are totally dependent upon a constant and adequate blood supply. Without it, either organ will sustain permanent damage in a matter of minutes. Further, we recall that the ultimate exchange between the fluid of the internal environment and the cells of a tissue occurs in the capillary bed of that tissue. Any serious interference with the blood supply in a brain or heart capillary bed will therefore result in almost immediate damage. Both kinds of CVA reflect just such an interference—rupture or blocking of blood vessels, with the result that tissues normally served by those vessels soon cease to function. If the circulatory interruption occurs in the blood supply of the brain (in which case the CVA is called a "stroke"), the severity of damage will depend on the location. As long as the major automatic motor-control centers remain undamaged, the patient may survive, but rarely without some serious impairment of function. Damage to major control centers, of course, results in death. If the CVA occurs in heart capillaries, the damaged tissue causes the heart to cease function slowly, with resultant lowered blood pressure, oxygen starvation of the brain, unconsciousness, and death.

Causes of CVA. Because heart attacks kill more United States citizens than any other disease, they warrant a closer look. The CVA of the heart, just

Figure 23-2 Progress in the reduction of the death rate.

CAUSES OF DEATH IN USA: 1900 ☐ COMPARED WITH 1974 ■

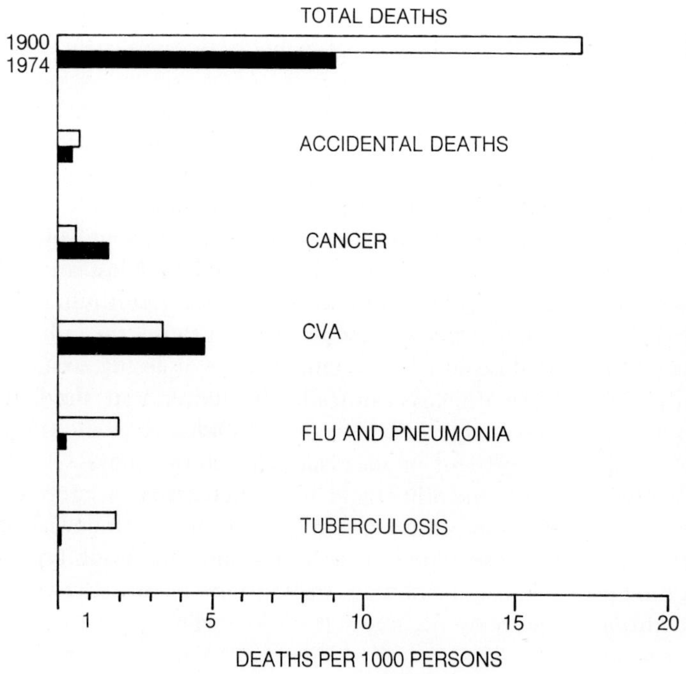

Category	Deaths as % of all deaths	Explanation of changes between 1900 and 1974
ACCIDENTAL DEATHS	4.2 / 5.5	LITTLE CHANGE —offsetting effects of improved health care are new accident causes, e.g., automobile
CANCER	3.7 / 18.6	INCREASE—(1) reduction in other death causes (2) increase in some types of cancer, e.g. lung.
CVA	20.1 / 52.3	INCREASE—(1) reduction in other death causes (2) increase in some CVA types probably related to diet, smoking, environmental causes.
FLU AND PNEUMONIA	11.7 / 2.8	DECREASE—(1) antibiotics (2) improvements in sanitation, preventive medicine, general medical care.
TUBERCULOSIS	11.3 / 0.2	

DEATHS PER 1000 PERSONS

described, impairs its function by causing death by oxygen starvation of part of the heart muscle. Very commonly, CVA is brought about by **arteriosclerosis**, a condition that nearly all United States adults have to some extent. In this disorder, the ordinarily smooth endothelial lining of blood vessels becomes roughened and their diameter decreased by an accumulation of fatty deposits, accompanied by rapid multiplication of endothelial cells to form obstructing *plaques* of cells. Although the restriction in blood-vessel diameter is serious enough, the arteriosclerotic condition is also dangerous because the roughened blood-vessel walls are apt to cause formation of blood clots (see page 400). These may further impair circulation by breaking loose and lodging in smaller downstream vessels.

The **coronary** arteries, which supply the heart musculature, are particularly susceptible to such interference. Fortunately, the anatomy of the coronary arteries provides some protection to the heart. Although three major branches of the coronary arteries supply roughly equivalent thirds of the heart muscle, even complete obstruction of flow in one does not kill one third of the heart muscle. The actual damage would be much smaller, because the branches of the coronary arteries interconnect, providing bypass channels that can take over some of the blocked circulation. Without these *collateral arterial channels*, blockage of one major coronary artery branch invariably would prove fatal.

Beyond this, the heart itself is susceptible to three other classes of disease. The *rhythm of contraction* may be impaired by damage to the electrical conducting system of the heart. *Damage to heart valves*, often produced by chronic disease such as rheumatic fever, may cause inefficient pumping. And finally **hypertension**, high blood pressure, may increase the work load of the heart so much as to cause failure.

Prevention and Treatment of Circulatory Disease. The major factors in reducing the incidence of circulatory disease are preventive measures involving diet control, exercise, and abstinence from smoking. Exercise appears to be particularly valuable in delaying or preventing the onset of circulatory disease, both by its direct effect on the vitality of the circulatory system and by aiding in prevention of obesity.

Diet control is still subject to controversy. It goes without saying that being substantially overweight is an almost certain invitation to circulatory disease because of the added work load of the heart. Beyond this, there is confusion regarding the role of specific elements of the diet in bringing on arteriosclerosis. Because the material that clogs the blood vessels in arteriosclerosis is primarily fatty, suspicion naturally focuses on the effects of fat in the diet. Although one cannot perform experiments on humans, there is evidence that high-fat diets fed to laboratory animals increase the incidence and severity of arteriosclerosis in them. Furthermore, there are certain "natural" experiments that suggest that the same is true of man. These indicate that heart disease is far more common in Western Europe and in the United States, in "affluent" societies, than among African natives. Dietary fat is implicated in this situation, as fats represent only 17% of the total caloric intake of the Africans and between 35 and 40% of the Europeans and Americans. Of course, as in most "natural experiments," many other uncontrolled factors may influence the outcome. For example, it is probable that the African gets more exercise than the European or American and that this undoubtedly would be a factor influencing the incidence of heart disease in the two groups.

Once circulatory disease has developed, several types of drugs are of value. If blood clots are the difficulty, there has been some success in dissolving them with injections of proteolytic enzymes, and the possibility of formation of more clots may be reduced by anticoagulant drugs, such as coumarin. Over short periods of time, the vigor of the heart action can be increased by drugs, such as nitroglycerine and digitalis. Nitroglycerine dilates the coronary arteries, improving heart muscle oxygenation, and may also reduce the work of the heart, lessening the demand for oxygen. Digitalis strengthens the action of the heart by influencing the flow of ions involved in conduction and contraction of cardiac conducting and muscle cells.

In favorable circumstances arteriosclerotic deposits are sufficiently localized to allow surgical repair. The blocked region may be removed and replaced by a piece of synthetic tubing; or, *endoarterioectomy* may be performed, a process in which the defective artery is opened and the blocking material dissected away.

If heart disease originates from difficulty with the conducting system within the heart, certain types of conducting-system failure can be corrected by surgical implantation of a *cardiac pacemaker*, a battery-powered device that electrically triggers the heartbeat.

Organ Transplants and Substitutes

Until recently, failure of a vital organ meant death. Now, the possibility of artificial organ substitutes and organ transplants offers hope of prolonging life after organ failure. Primarily, such techniques are applicable to the heart and kidney. Both are relatively simple to transplant, as the required connections to the recipient body are principally large blood vessels, or, in the case of the kidney, the ureter. These are easy for a competent surgeon to suture. Moreover, the main functions of these two organs—pumping and cleansing blood—are quite simple to duplicate by machinery. There seems to be no hope for the successful transplantation of brain or spinal cord. Although these organs might well be kept alive in a new host, the regrettable fact is that the central nervous system has virtually no regenerative ability and thus could never send out nerves to establish control over the organs and muscles of the host.

The use of organ transplants and substitutes is fraught with difficulties, although these seem to be yielding to the onslaught of medical research. The primary difficulty with organ transplantation lies in the rejection of the transplant by the **immune system** of the host. Kidney grafts are commonly made between siblings or between parents and children, as their genetic similarity reduces the likelihood of rejection. Grafts between identical twins are, of course, ideal. Even when grafts must be made between genetically dissimilar individuals, the intensity of the rejection phenomenon can be reduced by immunosuppressive treatment, which renders the host immune system incapable of reacting to the graft during the period of its accommodation to the host. The most promising of these treatments involves the use of antibodies to lymphocytes, the blood cells that are principally involved in rejecting foreign proteins, which of course includes those of both grafts and disease organisms. This makes such treatments a risk, because they leave the patient defenseless for a time against almost any infection.

Two considerations make it clearly advisable to seek the development of organ substitutes. First is the problem of finding immunologically compatible organs for transplantation and the associated dangers of suppressing rejection responses. Even if this problem were solved, we come to the second, the difficulty of providing sufficient donor organs. Because donor organs must be in good condition, they ordinarily must come from individuals who die from accidental causes near enough to major hospitals to ensure adequate care of the organs in the interval between death of the donor and their transplantation.

It is already quite feasible to substitute artificial devices for the heart and the kidney on a short-term basis (Figure 23-3). Immediately after a heart attack, a short interval of assistance from an extracorporeal heart pump provides an invaluable period of reduced labor that can greatly assist the recovery of a damaged heart. In the event of kidney failure, periodic removal of wastes from the blood can be accomplished by a blood **dialysis** apparatus, a technique that is so successful that there are now many centers in the United States providing dialysis for nearly 40,000 patients. The most difficult problems now to be faced are the construction of devices so small and effective that they are suitable for continual use without limitation of the patient's activities and to reduce costs from the present staggering $25,000 per year per patient.

The Morality of Organ Transplantation and Substitutes

Before a vital organ is transplanted, two moral and legal problems lying outside the realm of pure science must be faced. It must first of all be established that the donor is in fact dead, and this raises the difficult question of what constitutes death. Because medical techniques are now so effective in substituting for, or sustaining over short periods, such vital functions as respiration and circulation, the age-old "signs" of death are now meaningless. Perhaps the best criterion lies in the electrical activity of the brain, detected by electroencephalography (page 424). If brain waves are undetectable for something over 24 hours, it is impossible for conscious activity ever to reappear; the patient can be considered medically dead and viable organs made available for transplantation, assuming family consent. Medical and legal authorities are currently at work refining this definition of death and working out procedures that will both protect donors from premature certification as dead and that also will ensure that death is certified with sufficient timeliness to make available organs that are in the best possible condition for transplantation.

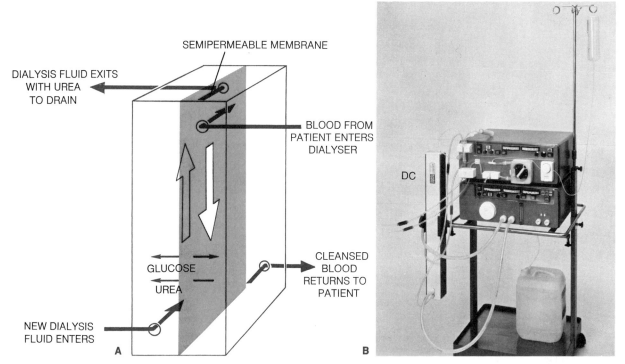

Figure 23-3 **Blood dialysis. A.** Principle of the technique is movement of blood wastes from blood across a semipermeable membrane to a dialysis fluid whose composition insures that only wastes are lost. For example, presence of glucose in dialysis fluid eliminates the concentration gradient that otherwise would cause its loss from blood. **B.** A dialysis unit suitable for home use. DC is the dialysis chamber. The remainder for the apparatus consists of pumps and devices to monitor and control blood and dialysis fluid pressure, to detect the presence of bubbles or leaks, and to add anticoagulant to blood. **B** supplied by Gambro, Inc., Barrington, Ill.

The second problem arising in this area applies both to organ transplantation and to the use of organ substitutes. This is the question of who shall benefit. There are just not enough organs available for transplantation and, to mention one pressing example, not enough blood dialysis machines available to serve all who need them. The result is that the medical profession is called upon to make decisions involving problems that are more than strictly medical. The physician can readily decide who, among several possible patients, is the most likely to benefit *physically* from transplantation or the use of an artificial organ, but how is it to be decided which of two equally good medical risks is otherwise the more deserving of treatment? At this juncture, such medical intangibles as the responsibilities borne by the patient or his value to society must be considered, and society as a whole, not merely the medical specialist, must make the decision.

PATHOGENIC INVASION

The conquest of the diseases caused by pathogenic invasion can justly be called an *accomplishment* of medical science. Whereas the reduction of death rates and the alleviation of the suffering due to cardiovascular disease or, as we shall see, to cancer or to the finality of aging are expectations somewhat indistinctly seen in the future of medicine, the pathogenic invasive diseases are largely subdued. We know what causes nearly all of them; we know how to protect the individual against nearly all of them, and we have the means to cure ourselves of most. The steps leading to these fortunate circumstances were many and difficult, and they were often dangerous to the medical scientists who made them. Many died accidentally or as willing experimental subjects. In Figure 23-4 these steps are outlined. From it you can see that the following critical stages occurred:

Figure 23-4 Four essential steps to rational disease control.

THE CONQUEST OF MICROBIAL DISEASE

Until *observation* of the course of disease and the circumstances of infection provided accurate data, no adequate *concept* of the nature of disease could be formulated. No rational methods of prevention and care could be developed.

Therefore, the critical first steps in disease conquest were:

1. RECOGNITION THAT DISEASES HAVE SPECIFIC CAUSES

This began with development of the concept of *contagion*—that disease spread from one sick person to another, rather than occurring spontaneously, or as an action of the nonliving environment, for example "bad air."

2. IDENTIFY WHAT PASSES FROM ONE SICK PERSON TO ANOTHER

This critical step required:

(1) The invention of the microscope, permitting microbes to be seen.
(2) The development of experimental techniques allowing:
   (a) *isolation of a suspect microorganism* from a sick individual, and
   (b) *growing a pure colony* of that organism, and then
   (c) *causing the disease* by administering some of the colony to a healthy animal.

3. THE CONCEPT OF IMMUNITY

Even before the causes of diseases were known it was clear that individuals once having had a certain disease never again contracted it and that having certain mild diseases might confer immunity to some related, more severe disease. Thus infection with cowpox was first used for vaccination against smallpox. With disease organisms known, they could then be used in various ways to confer immunity. They might, for example, be killed and then injected to produce long-lasting immunity.

4. CHEMICAL ATTACK ON DISEASE

Since bacteria were readily killed outside the body by antiseptics like alcohol or carbolic acid, it was logical to seek chemicals which could attack bacteria inside the body. Obviously, substances more specific in action than alcohol or carbolic acid are required. Among the earliest were arsenic compounds which were to some extent a "magic bullet" against syphilis. But chemotherapy, the chemical attack on disease, was actually never very successful until the advent of the sulfa drugs in the early 1940s. The blossoming of the Age of Chemotherapy occurred with the development of penicillin.

---

1. The demonstration that disease did not occur without a causing organism.
2. The demonstration that many diseases conferred immunity on individuals once infected.
3. The demonstration that many diseases could be quelled by specific chemical treatment.

In what follows, we shall consider the implications of these major advances toward freedom from pathogenic invasion.

To Each Ailment Its Own Bug

You may have read of the frenzied reactions in the Middle Ages to the "Black Death," a plague that we now know is caused by a microorganism. Even though people had no way of knowing the cause of the disease, they soon came to understand by trial and error that the disease struck more frequently in large concentrations of people. Thus we read that the wealthy would retire to the country during the plague season, while those unfortunate enough to have to remain in town would isolate themselves as best they could. This was wise, for, as we now know, one form of plague, *pneumonic plague,* is transmitted readily from person to person. But the relationship between dying rats, fleas, and human plague mortality went unnoticed until well into the nineteenth century (Figure 23-5). Having no scientific basis to build on, early plague fighters relied on the most outlandish disease-control methods. Bonfires were burned and cannons fired to cleanse the air. "Bad" air was an early favorite as a disease agent. Consider malaria, mosquitoes, and swamps and you will see why! For good measure,

INSECTS AND DISEASE
Insects and other ARTHROPODS such as mites, ticks and water fleas are important disease transmitters.

The may act as:

1. MECHANICAL CARRIERS — Flies, cockroaches and other filth-contacting insects may *physically* transport disease organisms to uncontaminated food or directly to uninfected persons. The insect is *not* essential to transmission. There are alternate routes of infection and the disease organism does not undergo any critical stage in its life cycle in the insect.

2. ESSENTIAL TRANSMITTERS — Many diseases *require an insect* for the usual mode of *transmission* and may even require an insect for *completion* of certain *stages of the disease organism's life cycle*.

Consider as examples these two diseases:

PLAGUE — Primarily a bacterial disease of rodents, plague is transmitted between rodents by their fleas. Humans become involved when so many rodents are killed that their fleas seek alternate food sources which include man. In the human, involvement of the lungs makes direct spreading between persons possible (pneumonic plague). Plague exists in both city rats and "country rats," principally ground squirrels in the U.S.A. Thus these cycles are possible:

This part of the cycle occurs only when the natural host of the city flea, the city rat, is reduced in numbers by plague. Thus a rat epidemic often signals an oncoming human epidemic.

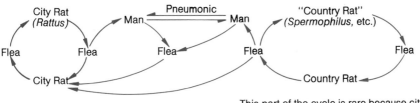

This part of the cycle is rare because city and country rats and their fleas don't often mix, and because people rarely come in contact with infected country rats.

MALARIA — This protozoan requires a mosquito to carry it from one human to another but also must carry out an essential part of its life cycle in the mosquito.

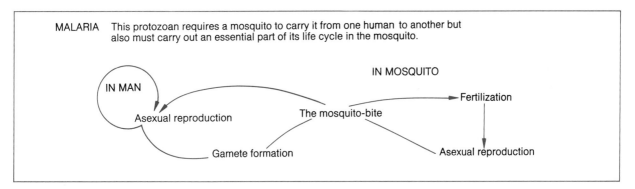

Figure 23-5 Insects transmit disease as passive carriers or as essential elements of the life cycle of the disease organism.

witches and other suspect types were rounded up and put to death, often with medieval ingenuity. Certain diets were prescribed and so, of course, was bleeding, which might be accomplished with leeches. Obviously all this was useless, and no progress could be made in the conquest of such diseases until their true causes were ferreted out.

Almost simultaneously with the invention of the microscope, the first microorganisms were seen, including the bacteria, among which are numbered the

PROBLEMS IN HUMAN LIFE: DISEASE AND AGING   499

causes of many diseases. It then remained to isolate a specific kind of bacterium from an animal infected with a disease and to show that it really caused that same disease when administered to a healthy animal. From this beginning we have come by now to understand that other living forms besides bacteria are also guilty of pathogenic invasion. They include viruses, protozoa, and lower plants (in addition to bacteria) among microorganisms, and a large number of parasitic multicellular animals.

Defense Against Disease

Natural Mechanisms. Certain of the defenses of animals against pathogenic invasion are obvious, such as the mechanical barrier against entry provided by the skin. Others are more subtle, and we were a long time in coming to understand them. Even so, they are in a sense obvious from the simple fact that we often recover from disease without specific treatment. If we examine the course of such events in the light of modern knowledge, these subtle defenses become clear.

The stages of a bacterial disease may roughly be described as *onset, acute stage,* and *recovery. Onset* includes the initial infection, usually through nose or mouth via air, food, or drink; an incubation period of several days to several weeks, during which the bacteria build up to a massive reproductive effort; and the beginning of that effort, when the first symptoms of the disease, often fatigue, headache, high temperature, and malaise appear.

In the *acute stage* that follows, the defense mechanisms of the body gather their forces to offset the massive reproductive effort of the bacteria. At first, the bacteria multiply very rapidly, whereas the defense mechanisms come into action more slowly. During this time the *symptoms* become more pronounced and more characteristic of that particular disease (pain in specific locations, rash, and so forth). The patient may not feel quite as badly at this point as he will later, at the height of the pitched battle, when bacterial toxins, bacterial and body-cell breakdown products, and wastes are at higher levels in the blood.

Gradually, as more defense mechanisms come into play, the bacteria begin to die more rapidly than they can reproduce. Once this point is reached, the decline in bacterial population occurs very rapidly, and the *recovery* stage begins. During this last stage, all vestiges in local pockets of infection are eliminated and tissue repair is accomplished. Toxins and wastes must be cleansed from the blood, and control systems affected by bacterial toxins must be brought back into balance. Too short a recovery period, with insufficient rest, may allow a resurgence of the bacterial population, causing a *relapse.*

The course of a disease thus reflects the balance of power between a rapidly reproducing bacterial population and the resistive processes. The bacteria pit their sheer reproductive and invasive powers against the population-reducing mechanisms of the defender. It is simply a balance of two forces. Anything that affects either of these forces influences the course of the disease.

As elements of natural resistance, two basic internal lines of defense are important, one of which is found throughout the animal kingdom and the other only among vertebrates, reaching its maximum efficiency in mammals, including man. The first of these lines of defense consists of cells that attack any and all chemicals or cells that are not normally found in the organism. Most animals possess *phagocytic scavenger cells* that continuously consume foreign materials, including any disease organisms they encounter.

The second line of defense takes somewhat longer to bring into operation. This is because it must first "recognize" that an invasion has occurred, and, in response, manufacture a population of special cells. These cells, in turn, sense the nature of the invader and manufacture specific molecules that will inactivate *that particular* infecting organism. This line of defense is the *immune system,* and the protection it confers is termed *immunity.*

The parent cells of the immune mechanism are blood cells, free-moving lymphocytes of two types: B lymphocytes, produced in bone marrow, and T lymphocytes, produced in the thymus. Lymphocytes are responsible for two types of immune response, namely, the production of circulating **antibodies** that react in a highly specific manner with foreign molecules, which are called **antigens,** and for the destruction of foreign cells.

Virtually any macromolecule can serve as an antigen to evoke the production of an antibody capable of reacting exclusively with it, even when the antigen is an artificial one that the organism could never previously have experienced. For a long time it was not understood how this remarkable feat of biological

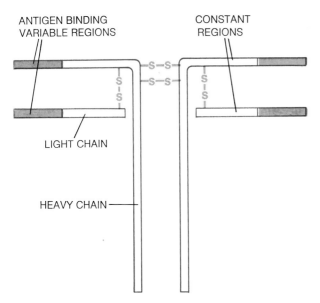

5 BASIC TYPES OF Ig APPEAR IN BLOOD AND VARIOUS SECRETIONS SUCH AS TEARS, SALIVA, AND MILK. MOLECULAR WEIGHTS RANGE FROM 160,000-900,000.

Figure 23-6 Diagram of immunoglobulin (Ig) molecule.

chemistry is accomplished. Determination of the structure of a typical antibody molecule in 1968 by Gerald Edelman and his associates at Rockefeller University contributed definitively towards the eventual complete solution of this important problem. They found that the antibody molecule, which was already known to be a protein of the globulin type, **immunoglobulin**, is constructed of four chains linked together in a Y shape by disulfide bonds (Figure 23-6) and has a total molecular weight of 150,000. Each molecule has two antigen binding sites, and these are in a region of the molecule that is structurally variable in comparison with the remainder of the molecule, which always has the same amino acid sequence. Different genes are believed to control these two parts of each chain; that is, there are two genes for each chain.

The following hypothesis is suggested to account for the huge number of different kinds of antibodies that can be formed. Each lymphocyte contains the genes for all types of antibody configurations that the organism is capable of forming, but the lymphocytes exist as clones. These differ from each other in having different sets of immunoglobulin genes active, with the remainder permanently repressed. Thus, each clone makes a different basic kind of antibody. This *clonal selection theory* has been well supported by studies on single cells.

To produce an immune response, an antigen reacts with a surface protein on one of the few circulating B lymphocytes of the particular clone that manufactures the antibody that is the most reactive with it. The antigen also reacts similarly with T lymphocytes, causing them to release a *cooperation factor*. This factor activates any B cells that have combined with antigen and causes them to convert to *plasma cells*. These

Figure 23-7 Antibody formation according to the clonal selection theory.

IMMUNE SYSTEM
Antigens evoke development of clones of lymphocytes that combat them

(1) Clones of T and B cells develop, each with a specific surface antigen-detecting and combining protein (immunoglobulin, Ig).

(2) Self recognition phase
During embryonic life proteins (potential antigens) of the embryo suppress T and B cells that bear Ig's that are reactive with them.

prevents reaction with self

(3) Functional Phase
(a) Detection. Foreign antigen reacts with surface Ig of T or B cell.

foreign antigen

(b) Immunity appears
B-cell clone multiplies, increasing the number of cells capable of reacting with that particular antigen. "Memory" cells form. These have long-lasting ability to recognize antigen, and produce rapid reaction to subsequent exposures to it.

T-cell clone multiplies. Resultant cells aid B-cells and attack foreign cells directly.

undergo rapid cell division and tremendously increase their synthesis of antibody. During this process it is believed that further structural perfection of the antibody to react with the inciting antigen takes place, possibly by small translocations within the variable region genes producing corresponding changes in the variable region structure of the finished antibody. This scheme is diagrammed in Figure 23-7.

Disease Susceptibility and Hereditary Variations in the Immune System

In recent years it has become evident that hereditary variations in the immune system can have highly specific effects on disease susceptibility. The predominant example is an extremely complex system of cell-surface proteins coded by perhaps as many as 1000 closely linked genes on human chromosome VI. The proteins coded by these genes are dimers consisting of a low molecular weight globulin and a glycoprotein. They form part of what is called the **histocompatibility system** because they are importantly concerned with cell recognition. The system of genes on chromosome VI is called the HLA system (human leucocyte antigen) since they may be detected most readily by tests on leucocytes. Because this large group of genes is closely linked, there is a high probability that they will be inherited together within families. However, since there is some crossing over among loci, over the entire human population there are millions of different combinations.

The first association of specific combinations of HLA alleles with disease was with the cancer, Hodgkin's disease. Subsequently the system was linked to many other diseases. In the case of an inflammatory arthritis, ankylosing spondylitis, the connection with certain HLA alleles is so strong that genetic evaluation is actually an aid to diagnosis of the disease, which is difficult to identify in its early stages. Other diseases strongly associated with the HLA system are the cancer retinoblastoma, other types of rheumatoid diseases, one of the types of diabetes, multiple sclerosis, and various serious "inflammatory" diseases such as psoriasis. A remarkable characteristic of these associations between HLA loci and disease is that the disease in many instances is triggered by a specific infection. For example, two of the HLA-associated forms of arthritis are correlated with previous infections by dysentery causing organisms.

There is no generally accepted theory as to the

Note 23-1 *The Eradication of Smallpox*

Figure 23-8 An early portrayal of a smallpox victim at the peak of pustule development.

underlying mechanisms of these associations. One possibility is that the HLA cell-surface proteins may be used by viruses or bacteria as sites of recognition

Smallpox, an extremely infectious virus disease, is of ancient origin; its characteristic facial scars are found on Egyptian mummies. The disease received its name from the depressed rounded scars remaining after the pustules (Figure 23-8) had healed. In the seventh century, smallpox was carried into Europe, and from there it spread about the world with the advance of European exploration. The disease was highly fatal even in Europeans, and its progress among newly exposed populations, such as the New World Indians, was explosive and caused immense mortality.

In early times ways of circumventing the worst effects of the disease were known. Since everyone caught the disease sooner or later, it was reasoned that one might as well try to contract a mild case. Even a mild case conferred permanent immunity. Thus, the Chinese had a method in which smallpox was contracted by inhaling a dried powder made from the pustules of a recovering patient, and in Europe the process of *variolation* was widespread. Variolation involved direct innoculation of infective matter. These methods were actually of value, since the death rate, from such deliberately contracted infections was only on the order of one tenth the mortality from spontaneous infections. Even more safe and effective was an innoculation of cowpox, a closely related disease. Cowpox was hardly ever directly fatal and conferred immunity to smallpox.

The English physician William Jenner knew these facts and conducted a study that confirmed them. He published his results in 1798, and thereafter the practice of vaccination became widely accepted. There was some resistance, however. For a time critics argued that vaccination could have unsuspected, untoward effects, as depicted in Figure 23-9.

One of the problems in achieving widespread vaccination was in maintaining the virus in good condition without the cumbersome necessity of keeping an infected cow on hand. Often the infective matter was kept dried on a thread, but the virus was short lived under such conditions. Thus it took two tries to get a viable sample across the Atlantic to Thomas Jefferson. Eventually the development of a potent dried vaccine made a worldwide attack on the disease possible.

In 1967 the World Health Organization mounted such a program aimed at total eradication of the disease. At that time there were large regions where smallpox was endemic—Brazil, Africa, and large regions of Asia—and there were perhaps as many as 15 million cases per year. On the order of $1 billion per year was being spent by the non-endemic countries on quarantine and vaccination just to keep the disease away. By 1975 smallpox was eliminated in Brazil and Asia and only Ethiopia and neighboring Somalia remained infected in Africa. Ethiopia was reported clear in 1976. However, 800 cases were found in Somalia during the early part of 1977. Although the political and military situation in that part of the world makes control efforts difficult, it appears that the last cases have been located and the disease is completely eradicated.

It is believed that the total costs of the eradication program have been less than $270 million, of which only about $90 million have been provided by international assistance funds—something of a bargain.

Figure 23-9 A cartoon that appeared in Jenner's time making clear one view of the dangers of vaccination with cowpox.

during invasion. Thus, a particular configuration of HLA alleles might render the carrier susceptible to a particular disease causing organism. If true, then to a significant degree it would appear that much of our susceptibility to many types of diseases may not be as chance dependent as formerly thought. We may actually be born with specific susceptibilities. Far from being a discouraging thought, this theory raises the possibility of eventually being able to determine the specific susceptibilities of the newborn and to take protective action *before* exposure to the disease triggering organisms or conditions.

Assisting Natural Defense Mechanisms Against Disease. Since our single-handed battles against disease are not always successful, it is fortunate that it is possible to assist the natural defense mechanisms. There are two important ways to do this, namely, (1) by assisting the action of the immune system and (2) by directly attacking the invading organisms by chemical means.

The immune system may be assisted by either **active** or **passive immunization**. In the former, killed or weakened disease organisms are administered, and these, while producing only minor symptoms, cause the immune system to build up antibodies and to be prepared to produce more very quickly, should an infection of that organism subsequently occur. This kind of protection is effective against many diseases, including polio, smallpox, and diptheria, which consequently are all virtually eradicated now in the medically advanced regions of the world. Some diseases, like mumps or measles, themselves automatically confer immunity by similar mechanisms, so that we ordinarily have them only once. (Note 23-1).

Passive immunization involves the injection of antibodies that are specific to disease organisms or to other antigenic substances (even snake venoms). These are protective for only short periods and confer no lasting immunity. Consequently they are administered only after probable disease exposure. A good example is the treatment of rabies, which requires such a long and painful series of antibody injections that it is initiated only if absolutely necessary.

The antibodies used in passive immunization cannot be purified and are administered along with sizable amounts of other antigenic material from the animal in which the antibodies were formed (for example, horse blood serum or chick embryo tissues). This can lead to dangerous consequences if the immune system reacts to these foreign antigenic substances in a certain way. The recipient may become sensitized to these substances, or more properly, **allergic**. Then, upon a subsequent administration of the same antigen, an allergic reaction may occur. At its worst this may result in immediate death, owing to various causes such as heart and respiratory failure, brought on by violent constriction of the smooth muscles that line the air passages of the lungs and the blood vessels of the heart. **Serum sickness** is a more mild reaction, involving fever and swelling at the site of injection.

Chemical attack on disease-causing bacteria has become highly effective since the development of *antibiotics* (Figure 23-10). Although much is yet to be learned about these miraculous molecules, it is known that at least some of them do their job by interfering with the bacterial protein-making machinery. In Chapter 8 we saw a summary of the flow of information in the living cell: from DNA, to mRNA, to ribosomes, to enzymes, to cellular processes. Clearly, any agent that will block this chain of command prevents essential genetic information (without which the cell cannot survive) from influencing cellular operations through the synthesis of appropriate enzymes. It is believed that nearly all antibiotics kill bacteria by doing just that, though the precise details of this process remain a target for continuing research. In one case, at least, it appears that the antibiotic molecule is similar enough to tRNA in geometry and behavior to act as a substitute in the protein-making process. However, once locked into place on the ribosome, it does not link up an amino acid as it is supposed to do. This failure breaks the chain and, of course, produces a faulty protein. Why these antibiotics, which originally came from plants can confuse protein synthesis in bacterial cells and not do so in the cells of the human body, which have very similar mechanisms, is not yet known.

Probably slight changes in the bacterial protein-synthesis machinery prevents action of the antibiotic, for it has been observed that, when a strain of bacteria is continually exposed to a certain antibiotic, it often results in the selection of a mutant strain that is unaffected (recall page 155). Alternatively, antibiotic resistance may stem from development of the ability to break down the antibiotic molecule. The existence of antibiotic-resistant bacterial strains is well known in hospitals where, obviously, extended exposure of bacteria to antibiotics occurs. Particularly notorious in this regard are staphylococcus bacteria. Approximately 90 percent of "staph" strains isolated in hospitals are antibiotic-resistant, as compared with about 10 percent from other environments. "Staph" infections range from acne-like skin disorders to rapidly lethal

ANTIBIOTICS act by selectively attacking phases of microbial metabolism that differ sufficiently from host metabolism to avoid deleterious side effects

CATEGORY	TYPICAL MODE OF ACTION	EFFECT	MEDICAL USE
ANTIMETABOLITES **Sulfanilamide** 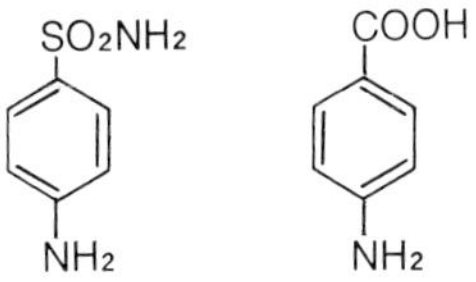	Structural analog of para-amino benzoic acid (PABA) that inhibits synthesis of folic acid, which contains PABA. Animals require folic acid in diet but bacteria must manufacture their own.	Folic acid is an essential coenzyme in DNA synthesis. Cell growth stops when folic acid is absent.	These were the first of the modern antibiotics. Current use has diminished with the development of more effective antibiotics. They are, however, still useful for urinary tract infections and when allergy to penicillins exists.
CELL WALL DISRUPTORS **Penicillin G** 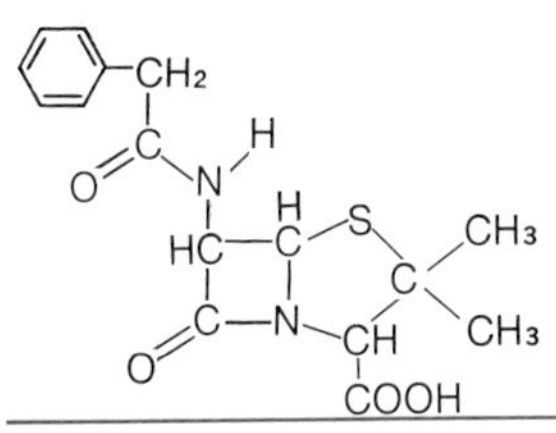	Prevents formation of cross-links between building block molecules that form bacterial cell wall.	Under normal osmotic conditions bacterial cytoplasm is hypertonic to the medium. When cell wall strength is diminished by penicillin action, the bacterium swells and ruptures.	The most important antibiotics. They have virtually no side effects, except for allergic manifestations, and act on a large spectrum of bacteria including syphilis, gonorrhea, anthrax, pneumonia, and streptococcus organisms.
PERMEABILITY AFFECTORS **Valinomycin** 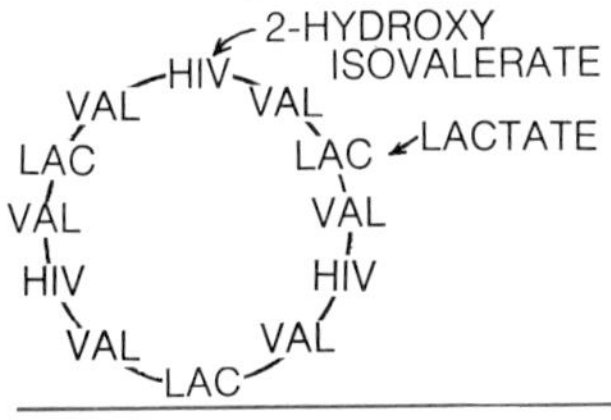	Lodges in cell membrane and facilitates passage of certain ions, potassium in this instance.	Loss of potassium kills cell by indirect biochemical actions, such as disruption of glycolysis, and by osmotic effects.	Valinomycin is too toxic for medical use since it acts on animal as well as bacterial cells. Similar compounds are useful as antifungal agents in treatment of skin infections.
PROTEIN SYNTHESIS INHIBITORS **Streptomycin** 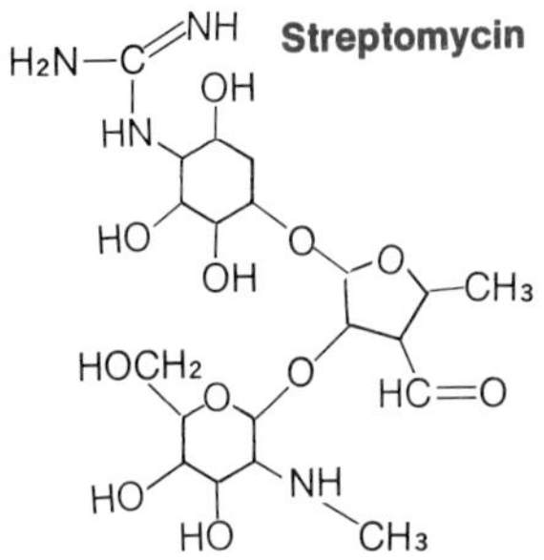	Binds to ribosomes and causes certain specific "misreadings" of genetic code.	Kills bacteria by inhibition of protein synthesis and production of defective proteins.	The first effective anti-tuberculosis drug.
Tetracycline 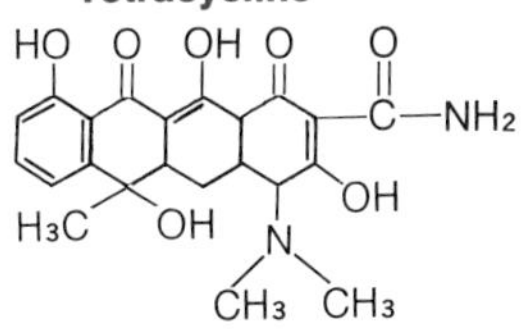	Prevents binding of animoacyl-tRNA to ribosome (refer to Figure 9-5)	Kills or has bacteriostatic action, by blocking protein synthesis.	Tetracyclines act on a very large variety of microbes, including cholera, typhus, and the rickettsial agent of Rocky Mountain Spotted Fever.

Figure 23-10 A few examples of the mechanisms of antibiotic action.

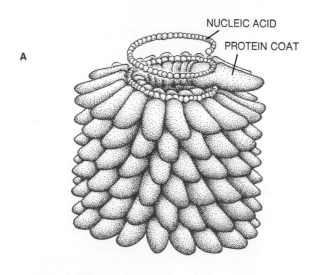

A. VIRUS STRUCTURE AND ACTION

This diagram of the virus causing mosaic disease of tobacco shows the essential components of viruses. They are:

(1) A set of DNA or RNA coded instructions for making more virus.

(2) An encasing protein coat which protects the nucleic acid core and determines the host specificity of the virus—which cells of which organism it will penetrate.

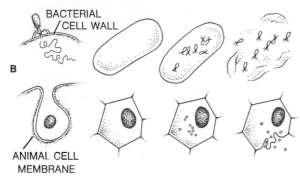

B. LIFE HISTORY OF A VIRUS

(1) The virus enters a cell. Methods vary. Some viruses of bacteria leave the protein coat outside with the nucleic acid entering through an enzymatically digested hole in the bacterial cell wall. Animal viruses are usually taken up by phagocytosis and then their protective coating is digested away.

(2) The virus multiplies. The nucleic acid of the virus takes control of the host cell and directs replication of virus protein and nucleic acid. During this period the virus cannot be recovered from the cell.

(3) New viruses assemble. Protein and nucleic acid combine finally to form hundreds of infective virus particles.

(4) Virus is released. As briefly as 20 minutes after initial infection, the cell may rupture releasing infective virus. Or, virus particles may be released sequentially over a long period of time.

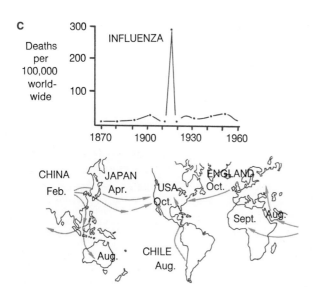

C. THE PROBLEMS OF VIRUS DISEASE CONTROL

Influenza is a world wide disease which changes continually. Before 1890 it was not a serious problem. Subsequently it grew rapidly more serious, culminating in the most lethal world-wide epidemic ever experienced—a total mortality of nearly 20 million people in 1918. Probably the epidemic was spread by the massive movements of troops between the U.S. and Europe.

Commonly the epidemics of influenza which sweep around the world begin in the Orient, as shown at the left for the "Asian Influenza" epidemic of 1957.

Unfortunately there are several types of influenza and this prevents an effective vaccination program. If the outbreak can be identified in time, vaccines can be prepared by growth of virus on chick embryos and supplied in sufficient quantities to protect the most endangered—the elderly and those with chronic respiratory disease.

Figure 23-11 The natural history of viruses.

blood infections, and so it is particularly distressing that they give rise so readily to antibiotic-resistant strains. Because of this tendency of bacteria, it is folly to use antibiotics indiscriminantly. They should only be taken under medical supervision, in sufficient dosage to give rapid eradication of the disease, and never for trivial causes. Even with these precautions, the antibiotic attack on disease organisms requires the continual development of new antibiotics as resistant strains multiply.

Virus Control. Although antibiotics have bacterial infection at least precariously under control, the story is still grim where viruses are concerned. Viruses are very difficult to attack chemically. The reason for this becomes clear when we examine their peculiar structure (Figure 23-11). They are not, in fact, living things at all. They do not share with the living cell any of those characteristics of life set forth in earlier chapters except the capacity to reproduce and evolve and these they cannot do without the cooperation of a living cell. A virus is little more than a string of DNA, tightly packed into a sausage skin of protein. When a virus attacks a cell, a specific portion of its exterior, presumably by chance or by some chemical affinity, is applied, to the cell membrane. Enzymes open both the protein coat of the virus and the membrane of the target cell at that point, whereupon the viral DNA slides into the cell, leaving its jacket outside. Once inside, viral DNA utilizes the metabolic and synthetic machinery of the cell, so that thereafter the cell devotes itself to production of more viral DNA in protein jackets as much as to its normal functions. When the cell is full of new viral particles, which it has manufactured because of the influence of the infecting viral DNA, the cell is ruptured to release these particles, which are then free to infect other cells. The cell that was the target of the original attack usually dies. Clearly, only an antibiotic that would also disrupt the protein-making machinery of human cells could confound the reproductive process of infecting viruses—and such an antibiotic would kill uninfected cells as well.

In a few specific instances viruses may be chemically treated in stages of their life cycles that have sufficient differences from host cell biochemistry to allow a selective attack. Thus certain influenza viruses may be prevented from entering host cells by prior administration of the drug amantidine. In other instances drugs are known that interrupt the assembly phase of the viral life cycle. While promising, these approaches to treatment of viral disease are still a feeble shadow of the chemical armament available for treatment of bacterially caused diseases.

Interferon. In 1957, A. Isaacs and J. Lindemann, of the National Institute of Medical Research in London, made a notable discovery that may eventually prove to be of practical value in combating viruses. They found that the body has a special defense mechanism against viruses. What happens is that, whenever a virus attacks a cell, the cell produces a glycoprotein which contains only about 200 amino acids. This protein makes other cells resistant to any viral infection. This substance, which has been named **interferon,** evidently has the remarkable property of protecting against *any* virus although it is unfortunately host specific; human interferon works only for humans.

Interferon works in very low amounts, with perhaps only a few molecules required to protect a cell. It has an interesting indirect action. It binds to the cell surface and this, in some unknown way, initiates within the cell synthesis of a new protein, which acts to inhibit replication of the attacking virus. The virus can still get in, but once inside it cannot replicate. Interferon can be isolated from human white blood cells and it has been used in experiments to prevent experimentally produced common colds. Quite remarkably, preliminary tests by a Swedish group show that interferon may improve survival of patients with a particularly malignant type of cancer.

Preventing the Spread of Disease

To wait until one is attacked by disease to resist is stupid. One has no assurance of winning, for one thing, and for another, the cure itself is apt to have deleterious effects upon health. Consequently, it is extremely important to emphasize prevention. The tactics of prevention are quite straightforward. They involve (1) improving the disease resistance of the individual, and (2) preventing exposure of the individual to disease organisms.

Disease resistance is optimized by maintaining a high general level of health by adequate nutrition, shelter, and so forth, by personal cleanliness, and by immunization when feasible. Exposure prevention is, of course, facilitated by mass immunization by isolation of disease carriers, and by environmental sanitation. The last includes the elimination of physical sources of disease, such as drinking water contaminated with human excreta, and elimination of animal carriers of disease, such as the mosquitoes that carry malaria (Figure 23-5).

Cancer

According to the National Cancer Institute, the cancer death rate in the United States has increased at a rate of 1 percent per year during the more than 40 years of accurate record keeping. This rate of rise is probably due to a number of factors in addition to an actual increase in cancer incidence. These include improvement in diagnosis of causes of death, improved record keeping, and most importantly, reduction of other causes of *morbidity*, coupled with the fact that the chances of developing cancer increase with age. It is thought, however, that some fraction of this increase is due to the annually greater presence in the environment of substances believed to promote induction of cancer. In consideration of this it is perhaps ominous to note that in three of the last 23 years the national cancer rate took sharp upturns from the annual 1% increment. In 1952 the rate rose to 1.9%. In 1972 it rose to 3.35%, and in the first half of 1975 it reached 5.2%. What these percentages mean is that 665,000 persons in the United States had cancer detected in them in 1975, a rate of 176.3 per 100,000 population. This compares with 105.9 per 100,000 in 1933, so the cancer rate had almost doubled by 1975. The tragedy of these numbers is incalculable: of those 665,000 victims of cancer in 1975, about 380,000 will have died of cancers within 5 years and the aggregate cost of treating them will come to better than 3 billion dollars. Remember that these statistics are *only* for the United States.

CANCER IS UNREGULATED GROWTH

For an organism to maintain form and function, certain kinds of cells (such as blood forming, epidermal, and those of the intestinal lining) must grow continually at a carefully regulated rate so as to achieve exact replacement, whereas others may grow cyclically (as in the uterine lining) or grow not at all (as in the brain). The cancerous state brings about two changes in this integration of growth and replacement of cells. Primarily, it involves release of otherwise normal cells from growth limitation and permits an unending cycle of DNA replication and cell division. Secondarily, it brings about a tendency for the cancerous cell to lose its affinity for its normal neighbors and to be carried into or otherwise invade other tissues and organs.

There are potentially as many different kinds of cancer as there are types of cells in the body. If they arise from epithelial tissues, they are called **carcinomas;** they are called **sarcomas** when they originate from nonepithelial tissues. Thus, a cancer involving bone making cells (osseous tissue) is called an osteogenic sarcoma, or a cancer of the epidermal cells of the lip is called an epidermoid carcinoma. Cancers that involve blood cells and spread through the circulatory system are known as **leukemia,** after the white blood cells which are primarily affected. When a cancer spreads from the primary site of origin, it is said to have **metastisized.** Some types of cancerous growths are not fatal, or malignant, and may grow slowly for years as benign tumors. Such tumors can reach remarkable size if they do not mechanically interfere with body processes, as is attested by a uterine tumor weighing 89 lb which was removed from a woman who weighed 8.5 lb less than the tumor, and survived. Malignant cancers have a variety of fatal outcomes. They may interfere with essential body functions, as for example in lung cancer or cancer of the liver, or they may cause death in nonspecific ways, as by invading and rupturing the walls of critical blood vessels. Some of the most serious types of cancers are listed in Table 23-2.

Table 23-2 Incidence and Signs of Some Major Types of Cancer

Type	Mortality 1976, U.S.A.		Warning	Characteristics
	Females	Males		
Breast	33,119 (30.1)*	284 (0.3)	Lump in breast	The major cancer killer of women.
Respiratory System	21,514 (19.5)	69,617 (66.6)	Long-lasting cough	The major cancer killer of men. Substantially preventable and curable by elimination of smoking and regular chest x-rays.
Digestive System	48,000 (43.6)	53,729 (51.4)	Bleeding, change in voiding habits	Relatively easily cured by timely surgery.
Genital Organs	23,129 (21.0)	21,332 (20.4)	Unusual uterine bleeding, urinary difficulty	"Pap test" is a simple and effective monitor for uterine malignancy.
Mouth and Pharynx	2,383 (2.2)	5,713 (5.5)	Hoarseness, persistent sore throat	Higher incidence in males in part due to smoking habits.
Leukemia	6,500 (5.9)	8,556 (8.2)	Generally poor health, anemia in children	Surgery impossible but long remissions are common with chemotherapy.
All Types	171,906 (156.0)	205,406 (196.6)		

* The first figure is the total number of deaths and the figure in parentheses is the number of deaths per 100,000 individuals, male or female. Data from Dept. Health, Education and Welfare (1978) Monthly Vital Statistics Report: Final Mortality Statistics, 1976.

THE CAUSES OF CANCER

There have been many theories as to the causes of cancer. The fact that certain tumors called **teratomas** tend to form tangled masses of identifiable tissues and organs led to the idea that cancers might be caused by sudden resumption of growth by embryonic cells which had for some reason stopped development early in life. Other theories have emphasized external factors, and today it is true that perhaps more than 80 percent of cancers can be shown to be highly correlated with environmental factors.

Specific chemical substances have long been implicated as **carcinogens**, or initiators of cancer. In 1775 scrotal cancer, then an occupational disease of chimney sweeps, was thought to be related to soot and by now literally hundreds of carcinogens are known from direct animal experimentation or by strong implication from human associations with specific chemicals. One of the most striking such instances is the 30% of cancer deaths in United States males that is attributable to cigarette smoking (Figure 23-12).

There are well substantiated instances of cancer related to ionizing radiation. In the 1920s painters of radium dials on watches received huge exposures, and virtually all died of cancer. Studies of the use of diagnostic x-rays on the uterus during pregnancy show a clearly increased cancer incidence in children from such pregnancies. Long term studies of the Japanese nuclear bomb victims show equally clear effects, particularly with regard to leukemia.

Many other less well specified environmental conditions appear to be correlated with certain types of cancer. Thus, among the Japanese the stomach cancer rate is 7 times the rate in United States inhabitants, but in Japanese Americans the rate is substantially the same as for others in the United States; clearly something in the Japanese national diet is implicated.

Viruses as Cancer Agents. Over 50 years ago Peyton Rous, of the Rockefeller Institute, obtained the first evidence that a virus could induce cancer by transferring a sarcoma from one chicken to another by injection of material passed through filters so fine that only virus particles could pass. Many other examples in lower animals soon followed, for example the viral transmission between rabbits of Shope's papilloma. Demonstration of viral agents of human cancer is currently a matter of immense interest, but progress is

TOBACCO AND HEALTH

There are two ways to study the effects of tobacco on health:

1. Study the effects of tobacco and substances in tobacco on experimental animals.
2. Study the medical records of people who smoke and compare them with non-smokers. The data below is from several studies of this sort compiled and analyzed in "Smoking and Health", a book published by the Public Health Service.

Mortality Ratios—The data is presented in this form:

$$(MR) \quad \frac{\text{The number of deaths in a group under study}}{\text{The number of deaths in a similar but non-smoking group (control)}} = \text{MORTALITY RATIO}$$

Thus, if 500 people died during the study period in the study group, and 100 died in the same period in the control group, then

Mortality Ratio = 500/100 = 5,

We conclude that whatever was under study is quite lethal. A mortality ratio of 1.0 means, of course, that the phenomenon under study has no relationship to mortality.

A DEATHS FROM ALL CAUSES RELATED TO SMOKING:

1. What and How Much Smoked

Cigarettes	MR
less than 10/day	1.45
10-20/day	1.75
21-39/day	1.90
more than 39/day	2.20
Cigar 1-4/day	0.93
more than 5/day	1.20
Pipe	
1-9/day	0.92
more than 9/day	1.05

CONCLUSION: Cigarettes are a dangerous risk, cigars and pipes are not too dangerous

2. How You Smoke

	MR
no inhaling	1.45
slight inhaling	1.65
moderate inhaling	1.83
deep inhaling	2.00

CONCLUSION: Risk is increased with greater inhaling

3. How Long You Have Smoked

	MR
less than 25 years	1.46
26-34 years	1.74
more than 34 years	1.78

CONCLUSION: Risk is greater the longer you have smoked

WHAT DOES THIS TELL US? All that can be said so far is that there is danger to life experienced by people who smoke. But, we can't tell WHAT the danger is or if it might be that people who smoke represent a class of individuals more likely to die from causes not related to smoking.

B CIGARETTE SMOKER MRs RELATED TO CAUSE OF DEATH:

If the relationship between smoking and death is really a direct one, then it should show up in the manner of death of smokers as compared with non-smokers

1. Deaths likely to be related to smoking

Emphysema	14.1
Larynx cancer	13.1
Lung cancer	10.0
Esophagus cancer	6.6
Stomach ulcers	5.0
Bronchitis	4.5

2. Deaths not likely to be related to smoking

Accidents, violent death	1.1
Cancer of prostate gland	1.6
Kidney disease	0.9

CONCLUSION: The increased risk of death characteristic of smokers is directly related to smoking

Figure 23-12   The relationship between smoking and certain diseases.

slow. Certain isolation of a virus from a tumor is difficult, and the mere fact that it is accomplished does not necessarily mean that the virus is the causal agent of the cancer. Direct proof would require a transmission experiment in the human, and that is ethically unacceptable. Moreover, cancer appears often to have a long incubation period, which would make such an experiment awkward since it would have to continue for many years.

Thus far the best example of implication of a virus in human cancer is Burkitt's lymphoma, a disease limited to the natives of Africa and New Guinea. Most remarkably, the virus isolated from Burkitt's lymphoma is the Epstein-Barr virus (EBV), which is almost certainly the causal agent of infectious mononucleosis. EBV is very common in humans, usually being contracted in early childhood without symptoms and transferred by way of saliva. Mononucleosis occurs when young adults pick up the virus after having somehow avoided it during childhood. Why then does it produce cancer only in Africa and New Guinea, if indeed it is the causal agent? There is now no answer although one possible clue lies in the fact that most individuals who develop the cancer also have malaria. Yet malaria is common in other parts of the world where Burkitt's lymphoma does not occur, South East Asia for example.

Steadily improving knowledge of the biology of viruses is leading to answers to these difficult questions, answers that may clarify the role of viruses in cancer and provide a rational explanation of the actions of the many cancer inducing agents. An important clue comes from the fact that it is now known that all human cell lines that are able to multiply indefinitely in tissue culture carry viruses. Many normal cell strains appear to be able to divide only about 50 times before dying out (see page 515), whereas those which have no such limit are always shown to harbor viruses. EBV is found, for example, in all human lymphoid cell cultures capable of indefinite growth in tissue culture. This "model cancer" strongly supports the studies of viral transmission of cancer in animals and requires a closer examination of the relationship of viruses to their host cells.

Viruses reproduce in two ways in host cells. In what may be called overt viral infections, such as in polio or measles, the virus undergoes straightforward and rapid reproduction within the host cell, often destroying it and releasing large numbers of virus particles. Viruses may also accomplish more subtle infection by incorporating their own DNA into the chromosomal DNA of the host so that they are replicated with the host DNA at every cell division. In the evolutionary sense this is an advantageous tactic for the virus since it insures viral multiplication with little or no damage to the host. In this state the virus is said to be in the *provirus* form. Subsequently it may also be reproductively advantageous for the virus to do two further things: first, to induce the host cell to undergo more than the normal number of cell divisions for which it is fated, and, secondly, to transform from the provirus form to the overt form so as to produce a large number of infective virus particles. These processes might favor transmission of the virus to new hosts and would give the transformed virus a selective advantage over other kinds of viruses in the same host, which, having not transformed, would die with the host without opportunity to induce new infections.

Investigators of the properties of viruses associated with cancers believe that somehow cancer causing viruses are able to accomplish these changes by taking control of cell processes that regulate growth and host cell affinity for other cells. It is further theorized that viruses of this type exist as proviruses in almost all individuals and, as such, do no harm until some accident releases them from restraint and allows them to trigger cancerous behavior in the host cell. This theory offers an explanation of the mechanism of action of many cancer inducing agents such as radiation and many types of chemicals. Radiation might act by breaking DNA strands and thus either directly releasing provirus or causing its release during DNA repair. Since many of the known cancer inducing chemicals are also known to cause mutation, it seems likely that they may act similarly by bringing about the release of provirus. If the theory is true, our cancers are simply the consequence of the reproductive tactics of a family of viruses that have been with us for so long as to be nearly perfectly adapted as human parasites.

TREATMENT AND PREVENTION OF CANCER

The body has relatively poor defenses once cancer develops; the immune system is of some value in younger persons and that is about all. The best hope for a favorable outcome is early identification of the cancer and its complete removal by surgery before it has had an opportunity to metastasize. For this reason perhaps the most effective protective action that the individual can take, besides avoiding contact with environmental carcinogens, is to become familiar with the early signs of cancer (Table 23-2), to conduct self-examination regularly for early signs of cancer, and to have regular medical examinations.

Although early surgery is the most certain therapeutic measure, oftentimes the location of the primary tumor or its spreading makes surgery impractical. Then other techniques, which selectively kill rapidly growing cells, may be used. These include ionizing radiation and drugs that interfere with cell division. Such agents obviously are generally damaging to the body, since many normal types of cells are in continual division, and must be used with the greatest care. In a few instances, most notably in the case of juvenile leukemia, such drugs have achieved great prolongation of life.

Environmentally Related Cancer

Avoidance of environmental pollutants that may be carcinogenic is difficult because there are so many of them and because it is at present difficult to measure what constitutes dangerous doses. There are some obviously serious examples. Asbestos workers, in addition to suffering from a chronic lung disorder, run very high risks of lung and gastrointestinal cancer. Since asbestos is widely used in construction work, it is a serious problem but it must be remembered that the asbestos workers have far higher exposures than the general population. However, any optimism that this observation might engender is offset by the possibility that low doses might induce cancer after very long periods of time and in the long run be as serious as the effects of short term large doses.

By all odds the most clear cut example of environmentally caused cancer is concerned with tobacco smoking. Lung cancer has progressed from a relatively rare disease in the early years of this century to become at present the most serious form of cancer. Mortality in the United States for 1975 is estimated at about 88,000. The source of this increase has been extensively sought and makes a most instructive story, illustrating not only the difficulty of this type of medical research but also how much more difficult health problems become when they involve human habits and economics.

The story begins with the observation by physicians that victims of lung cancer were very commonly smokers. This observation correlated with the rising incidence of lung cancer, because cigarette use had also increased, in the interval between 1910 and 1962, from 138 to 3,958 cigarettes per person per year. Although this fact alone was insufficient to indict tobacco, evidence from many sources, including medical studies of smokers and nonsmokers, and animal studies of the effects of the chemicals in tobacco smoke, indicated that cigarette smoking is a major cause of lung cancer. (Figure 23-12). In fact, as detailed in a U.S. Public Health Service book, *Smoking and Health*, the lung cancer death rate for male smokers is nearly 1,000 percent higher than for nonsmokers.

What was the result of these findings? Virtually nothing happened! Cigarettes continued to be advertised, except on television, and sold as before. Smoking, through habit, imitation, and intensive advertising, has become, it would seem, an almost essential part of our lives (Figure 23-13). Thus, the total consumption of cigarettes in the United States continued to rise at least through 1975. In 1978 the total number of adult tobacco users in the United States was estimated at 48 million, roughly 22% of the population. The tobacco consumer, of course, does not have lung cancer and is secure in the feeling that if it has not gotten him yet it never will; besides, he finds smoking such a comfort. The tobacco industry evidently does not believe that the implications of *Smoking and Health* are either sufficiently well founded or serious enough to warrant more than minor changes in their product (filter-tips and "low tar and nicotine"), let alone to withdraw them from the market. So there arises an astonishing situation in which, even after the strong arguments of *Smoking and Health*, there is essentially no change in tobacco production or use. And further, owing to the belief that the small tobacco farmer could not survive economically without growing tobacco with federal subsidy, the final irony in this curious situation is that one arm of government pays to have tobacco grown while another campaigns against its use.

Unproven Cancer Cures

Cancer therapy has long been an area for unproven cures, as we mentioned in Chapter 1. The fact that the disease is often greatly prolonged, even though the outcome is certain, has made its victims easy

Figure 23-13 A glance at any public ash tray demonstrates the *sang froid* of the U.S. populace.

marks for almost any treatment proposed. At the same time the occasional spontaneous remissions that occur in the progress of the disease often provide false testimony as to the efficacy of an unproven cure undertaken at the time of the remission. This plus the well-known tendency of people to remember when things work and forget when they don't often builds up strong lay followings for unproven treatments. A few years ago a substance called Krebiozen was touted as a cancer cure, and this has recently been replaced in popular fancy by Laetrile, a substance isolated from apricot pits and largely consisting of the biomolecule amygdalin. Laetrile enjoys widespread public approval; 50,000 cancer victims in the United States are believed to have used it in 1977. Legalized in 14 states (1978), it may soon enjoy nationwide legalization if a U.S. District Court order forbidding the U.S. Food and Drug Administration from interfering with interstate commerce of Laetril is sustained.

Believers in Laetrile theorize that amygdalin is specifically taken up by cancer cells, which, they believe, contain a specific enzyme that metabolizes amygdalin, causing release of toxicants destructive of the cancerous cell. Regretably there is no evidence to support the theory. Amygdalin in moderate doses is probably harmless, although the National Cancer Institute has warned that samples tested in its laboratories were "... microbially contaminated and unfit as pharmaceutical products for human use ..." The tragedy of the controversy over its use is that a strong polarization has developed in which the supporters of Laetrile depict orthodox medical treatment of cancer as butchery or poisoning and thus tend to lead cancer victims away from their only proven chance of help (Figure 23-14).

Figure 23-14 A cartoon from *The Choice* illustrates the attitude of proponents of laetrile towards conventional treatment of cancer. By permission, *The Choice,* Los Altos, California.

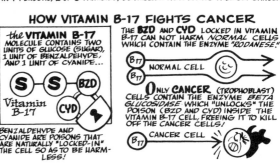

Additional controversy arises from consideration of the rights of incurable cancer patients. Should they not be allowed *any* treatment that gives them hope or at least comfort? Why not let them avoid facing death as long as possible? No one with normal sensibilities can argue against this position, but it must be noted that it raises a logical dilemma. It would seem that it would be necessary either to pronounce the patient as incurable before permitting use of Laetrile, or any other unorthodox treatment, or let *any* patient in *any* stage of the disease do as he or she pleases. Obviously being pronounced incurable would markedly dilute any solace to be derived from unorthodox treatment while the alternative of providing access to unorthodox treatment at any stage of the disease could have terrible consequences.

Aging

Assuming avoidance of premature ensnarement by virus, bacteria, CVA, badly steered cars, and the infernal weed, we arrive finally at the unavoidable ensnarement, aging, whose only cure is death. All forms of death that we have seen earlier in this chapter have at least a logic to them. They are really almost all accidents, either of the development or of the experience of the organism, and as accidents they are to some extent both understandable and avoidable. But aging seems different; when it comes there is no avoidance, and, superficially, its necessity is incomprehensible. Physiologically, we know that organisms are open systems and ought not to wear out, as their fabric is continually renewed. Further we sense the loss our species suffers when we see disappear, in the death of an old person, a lifetime's accumulated wisdom. Certainly these thoughts require that we preface the discussion of aging with the question of why organisms should die at all.

FOR THE GOOD OF THE SPECIES

If we consider organisms in general, aging, terminating in natural death, is undoubtedly beneficial to the perpetuation of the species. If there is no protective or instructive relationship between parents and offspring, then as soon as parents have reproduced they are largely superfluous. Indeed, they might actually hinder the coming generation by competing with it, and so, from the point of view of the species, the only advantage of life continued after the first reproductive period would be to further ensure the next generation by additional reproductive periods. There are many instructive examples in lower animals: the males of spiders are frequently killed and eaten by their mates immediately after insemination, and in innumerable species both sexes die immediately after reproduction.

These arguments are sound, however, only so long as the contribution of the parents to the next generation is understood to be strictly confined to the biological aspects of producing young. When the parents protect the young, instruct, or otherwise assist their survival, a more prolonged life cycle becomes advantageous. The extreme of this condition exists in the human, whose period of physiological development lasts for several years after birth, and who continues to accumulate *information* useful to survival long after the purely biological reasons for life have disappeared. Recalling Chapter 17, we may say that to the extent that human evolution has become *cultural*, so has aging and death become a disadvantage.

MECHANISMS OF AGING

There are three plausible theories of the basic mechanism of aging.

1. The post-reproductive deleterious mutation theory.
2. The error catastrophy theory.
3. The programmed death theory.

According to the **post-reproductive mutation** theory there is vigorous evolutionary selection in favor of reproduction early in life and against deleterious mutations that are expressed during the period of life prior to reproduction. However, subsequent to reproduction, there is little selective effect. The individual's genes are already established in the next generation, after which event the parent's fate is largely immaterial. It is thought that a consequence of this fact would be accumulation of deleterious mutations that act late in life. These would accumulate, since they occur after and do not impede reproduction. Thus, the aging process is seen as a consequence of the accumulation in the genome of such late acting mutations. The Russian geneticist Z. Medvedev has suggested that this might be an explanation of the "excess" DNA, great numbers of repeated DNA sequences, found in the eukaryote genome. This material may represent multiple copies of essential genes to serve as spares to replace genes put out of action by mutation. There is probably no doubt that such mutations do occur and do contribute to the aging process, but it also seems quite likely that processes such as those postulated in the error castastrophy theory also contribute.

The **error catastrophy theory** holds that cells in all parts of the organism accumulate mutations as a function of time. These **somatic** mutations have a variety of causes such as natural radiation, mutation inducing chemicals, and probably many other unknown factors. If the cell has no way to repair these mutations, it is only a matter of time until finally sufficient damage is done to the genetic apparatus to render it unable to produce correct copies of the protein molecules necessary for life, and the cell dies. The accumulation of such damage must ultimately weaken the organism until it can no longer survive.

At least in some instances there must be more to the aging process than the ideas put forth in the first two theories because it is clear that there can be definite genetic determinants of aging—aging can be **genetically programmed.** Some of the most remarkable examples occur during embryonic development, when certain cells die early as a normal part of development. Thus it is found that within the spinal cord more prospective motoneurons than are necessary initially form, and those that are not needed quickly disappear. Even more striking is the well-studied example of development of the vertebrate limb in which the embryologist John Saunders showed that important phases of the shaping of the limb depend upon selective cell death at specific sites. Even if removed to another position in the embryo, these cells die on schedule.

It has been argued that most normal cells from the higher vertebrates have a specified term of life measured as the number of cell divisions they may undergo before death. This term is known as the **Hayflick limit,** after the biologist who discovered that normal fibroblasts in tissue culture are able only to divide a certain number of times before death. If the cells tested are from human embryos, they are able to divide about 50 times. The older the cells which are tested in this way, the fewer divisions they are able to undergo in tissue culture. Similarly, if cells are taken from humans suffering from progeria, a form of genetically determined premature aging, they have a correspondingly lower division potential. However, the fifty division limit may not be universally applicable since epithelial germinative cells of tongue and skin divide many hundreds of times and cells from organs with strong regenerative powers, such as the liver, also exceed the Hayflick limit.

The Hayflick limit can be shown to be independent of some peculiarity of the tissue culture procedures employed because the limit can be demonstrated by serial passage of identifiable cells through living animals. For example, a strain of mammary gland cells may be passed through a series of host animals which have had their own mammary tissue removed and the approximate number of cell divisions determined before the grafted mammary tissues fail to grow. The evidence is that the cells are capable only of a limited number of divisions and that these are relatively independent of the total duration of time in which they occur. This fact, which suggests that the death limit is approached as a function of how many times the cell division process takes place rather than simply as a function of time, argues that the error catastrophy theory cannot explain the Hayflick limit. Some authorities think there is a built-in device in the DNA replicative machinery that progressively degrades the replicative process with each cycle of DNA replication.

Further support for the idea that an important aspect of aging is some built-in process of this type comes from the remarkable fact that cells ignore the Hayflick limit if they are cancerous. The cancerous cell seems to be potentially immortal. Since there seems to be no way that random mutations causing aging could be suppressed by the cancerous state it is more likely that the effect of cancer is to act on some rather limited control process. We are left with a strange paradox: cancer, one of our principal killers, may harbor a secret of longevity.

Another possibility is that there is an age-dependent failure of DNA repair enzymes (page 128). Evidence supporting this concept comes from experiments on tissue cultures of persons with progeria in which it is shown that there is virtually no repair of x-ray damage to DNA. Such repair proceeds swiftly in cultures of normal cells.

THE SPAN OF LIFE

Persons born in the United States in 1975 have a life expectancy of 70 years, a marked increase over the 47 year life expectancy of 1900. In underdeveloped parts of the world, life expectancy is usually far lower owing to many external causes such as malnutrition, poor sanitation, and limited health care. There are, however, reports of exceptional human longevity in three restricted parts of the world: the high valleys of the Ecuadorean Andes, the Hunza valley of the Himalayas, and the Caucasus Mountains of Russia. While birth registration records are such a recent innovation that it is difficult to document reports of persons living more than 150 years, it is probable that a larger than typical number of healthy individuals live to nearly 100 in these three regions.

The Russian population has been extensively studied with the conclusion that a group of environmental conditions are the principal factors favoring longevity. These people are relatively isolated from the outside world and hence are protected to some degree from infectious diseases. There is very little social stress on the oldsters; almost all feel that they still contribute effectively to the community. They eat a low calorie diet consisting mainly of vegetables and all engage in rather hard physical work. Such is the prescription that any physician would recommend for longevity in our own society. Evidence for genetic factors favoring longevity is not well developed although some authorities consider that the low disposition to cancer and heart disease in the elderly Russian group is indicative of such factors. Obviously it will be difficult to determine with precision the relative contributions of genetic and environmental factors in this situation.

We have already seen evidence of genetic control of aging in at least one human pathological condition, and if we examine the whole of the living world we find such a wide range of life spans that it would seem necessary for there to be strong elements of genetic control. For example, among the coelenterates some hydroid polyps live only about a week whereas certain sea anemones are known to live at least 70 years. Similarly, parrots commonly live for more than 50 years, whereas most other birds are old at 5 years. Finally, among plants there are the remarkable bristlecone pines of which there are a few living specimens exceeding 3500 years of age. Since there is in most instances no evolutionary advantage for survival beyond the first effective period of reproduction, it is not surprising that most organisms live only a very few years at most.

HUMAN AGING

In physiological terms the aging process is seen in all organ systems. Bone mass decreases and cartilage and ligaments calcify. Joints lose their elasticity and very often undergo arthritic degeneration. Muscles decline in bulk and become weaker. Skin and hair, which have been actively undergoing mitosis throughout life, undergo degenerative changes, particularly in the instance of individuals who have had extra strain placed on the skin by excessive exposure to sunlight. The thymus is completely atrophied and with it is lost much of the ability of the body to react effectively to new antigens. Sexual behaviour may persist into advancing age, long after reproductive function is lost. Arteriosclerosis commonly appears and with it greatly increased chance of stroke or heart failure. The brain is an extremely interesting example of the aging process since it clearly ages without neuronal mitosis, this having ceased shortly after birth. Nonetheless, senile dementia, accompanied by widespread degeneration of brain cells, even in the presence of good circulation, is not uncommon and loss of short-term memory is one of the well-known early signs of aging.

When the aged finally die it is often difficult to define precisely the cause of death. Rather slight perturbations in health are often enough to overcome the enfeebled resistive powers of the body. Exposure of the population to a new disease such as a new flu strain very often carries off a high percentage of the aged due to the lack of responsivity of their immune systems. Unusual environmental stress also selects against the aged. During the 1952 Killer Smog in London, mortality for persons aged 15-44 years increased only 1.6 times, whereas it increased 2.8 times for those aged 65-74 years. Since in developed societies the incidence of cancer rises with age, many of the elderly die of this cause, and stroke and heart disease continue to claim an ever increasing toll.

Care of the Aged

In the United States a major health care crisis has been created by the great increase in life expectancy coupled with the fact that the final period of relative helplessness tends to become longer the older a person is at its onset. Although we now have about 300 percent more nursing homes for the elderly than there were even as recently as 1960, these still are able to accomodate only about 5 percent of the aged population, and the quality of care in many of them is suspect. The nursing home itself, however effective its medical care, exemplifies the great sociological problem of aging. Our society places a premium on youth that is so great that many old persons feel forced to withdraw from general society and develop a sense of uselessness that debases the quality of their lives and may hasten death.

To a large extent this problem is exacerbated by the great mobility of our society, which breaks up family groups, and by the general tendency to move away from the extended family of earlier times in which several generations formed a self-fulfilling community. The result has been for those oldsters who can afford it to concentrate among their peers in retirement homes and villages where their physiological requirements may be well enough served but where the sense of purpose in life is lost or artificially supported by what are basically kindergarten activities for the aged. The prospect for the low income oldster is markedly bleaker. Frequently isolated from the family, such persons often live in virtually total isolation and either die quickly of poor nutrition or of the consequences of poor general maintenance, or they are collected by public agencies and filed away to await death in hospitals or homes of various types.

HEALTH CARE DELIVERY AND MEDICAL RESEARCH

Beyond what we can do for ourselves by personally heeding the already known requirements for good health, as citizens we may help ourselves by supporting local and national efforts to improve health care delivery and to facilitate medical research.

The quality of health care received depends on where treatment begins. If one is stricken with a heart attack within range of a modern hospital and provided the best possible supportive treatment by a paramedic team on the way to that hospital, the chances of survival and return to useful life are markedly greater than if the attack occurred in the hills of Tennessee, or in innumerable other places in this country where one is hours away from adequate care. Already the medical community knows enough to effect a tremendous reduction in mortality if optimal medical care could be supplied early enough to all citizens. To achieve this goal is an almost insurmountable problem today owing to lack of medical personnel and the suboptimal geographical distribution of those who are available, to the rigid traditional structuring of medical services which prevents greater use of paramedical personnel, and to the costs of facilities. The establishment of regional hospitals in which optimal facilities are concentrated is well advanced in some parts of the country. Continuation of this development along with the provision of a widespread network of paramedical facilities capable of effective early treatment and rapid transport of patients to such centers appears to be the only achievable goal for the immediate future on the long route to adequate health care delivery.

If we shift our view to a broader perspective, it becomes obvious that perhaps the most wisely spent dollar in the health field is the dollar spent on medical research. The reason for this is simply that with adequate knowledge it is usually cheaper to prevent a disease or accident than to treat it once it has occurred. The treatment of polio is a remarkable illustration of this principle. Poliomyelitis is a virus disease of the nervous system that commonly produces lasting severe paralysis or kills its victims. It has been nearly completely eradicated in the developed nations by an immunization procedure that is almost totally safe and whose aggregate cost per individual is only a few cents. In the United States in 1952, just before widespread immunization of the population was attained, there were 52,000 paralytic cases. Treatment then was wholly symptomatic. Since nothing could be done directly to get at the cause of the disease, all efforts were directed towards largely ineffectual efforts to reduce muscle wastage due to the paralysis, mechanically to assist the breathing of polio victims with respiratory paralysis, and generally to make the patients comfortable. The costs were immense.

Although humanitarian reasons require that we make such expenditures, if only to reduce suffering, it seems obvious that the most expensive disease is the one for which there is only symptomatic care. Only a fraction of the total expenditures for one year on polio care, when devoted to development of the immunization technique, was all that was required to eliminate the disease as a significant problem. In connection with our earlier discussion of pure and applied science (page 11) it is interesting to note that some of these expenditures were made in the realm of pure research and without reference to the polio problem. For example, the basic work of John F. Enders and his colleagues on the development of tissue culture techniques for mammalian cells were essential to the development of the first immunization technique by Jonas Salk.

A recently completed program to wipe out measles further illustrates the economies of prevention over cure even in relatively harmless diseases. Between 1961 and 1972, 60 million doses of live measles vaccine distributed at a cost of $3 per treatment prevented 24 million cases of measles, 2,400 deaths, and 7,900 cases of mental retardation in the United States. Savings of $1.3 billion are estimated. Many other examples of the importance and economy of basic medical research are evident in the recent medical literature. They question spending billions as we do on what is largely symptomatic treatment of many diseases, such as cancer and circulatory disorders, without making adequate expenditures on basic research, which may lead to their eventual eradication.

24 The Environment: Organization of the Ecosphere

ASSAULT ON THE EARTH

About 10,000 years ago our ancestors were only one of many competing species. Numbering only a few million worldwide, their chances for survival then were not much better than those of their competitors. From this modest beginning humans quickly reached a level of dominance over the earth that has never been equaled. Humans perfected hunting and gathering, then developed agriculture, and finally accomplished the industrial revolution. Each step in this progression allowed growth in human populations through increased availability of food and through improved protection from the elements and from disease. Consequently, the earth today supports over 4 billion people. The increase is not slowing. By 2077 there should be more than ten billion of us.

If numbers count, we seem to be doing all right. Numbers do count, but now they count *against* us because we are overgrowing the planet. There are so many of us, and we are so technically adept and at the same time so callous of our earthly heritage, that we are seriously disturbing the age-old balance among organisms and with the physical world. Vital renewable resources are used more rapidly than they can be replaced and nonrenewable resources are expended with no thought of the future generations that must live without them. Many scientists and public figures believe that this plunder and destruction has gone too far to be stopped by voluntary means. They believe that the growth of human populations can now only be arrested by any of a number of kinds of unavoidable calamities. They may be correct. If we are to avoid this dire outcome, corrective measures must immediately be taken, not merely within the present generation but within the next few years.

What is the problem? It is simply this: Normally all organisms on earth are regulated by a system of feedback controls that does not allow a single kind of organism to become so dominant that it can damage the planetary life system. Recently in earth history humanity has escaped these feedback controls and has so efficiently exploited nature that humans are potentially able to increase in numbers up to limits defined only by the total resources of the planet. As we approach these limits, which cannot be exceeded for obvious reasons, we generate disaster for ourselves. And, through the damage we do to the worldwide environment, we threaten the existence of nearly all forms of life.

This tragedy, a runaway species destroying itself through overgrowth exceeding the resources of the planet, is not unexpected. Recall that Malthus, who influenced Darwin and Wallace, recognized that a species tended to increase "geometrically" until checked by some external force. Malthus saw starvation, disease, and war as the great checking forces on human populations.

Just as the tragedy is not unexpected, its cause and cure are clearly evident. We have only to stop increasing in numbers and to make some relatively simple adjustments in the way we live and exploit the resources of the planet in order to restore stability to the thin shell of life that covers the earth. In short, there are too many of us and we are too intelligent any longer to be controlled by the natural system of checks and balances that work for most organisms. Therefore, we must achieve this control by voluntary means. *On a worldwide basis we must regulate our numbers, apportion our resources, and limit our intrusions on nature.*

Many wise observers are aware of this and their comments have been widespread. Book stores are crammed with books detailing the probable coming disaster, its causes, and its cure. But even in countries where this information is widely distributed, little is done to meet the crisis; and in many third-world countries people are too much concerned with the immediate problems of survival to react to the looming ultimate disaster.

Politics and Life

The concensus among authorities who have thought carefully about the future of humanity is aptly stated by Sir MacFarlane Burnett:

There are three imperatives: to reduce war to a minimum; to stabilize human populations; and to prevent the progressive destruction of the earth's irreplaceable resources.

Biology and the Appreciation of Life, 1966, p. 29.

Thus stated, the great problems of human survival involve a most difficult mixture of politics, economics, and science. Even as science clearly shows a path to survival, political decisions may divert us. The obvious example is war. Because future wars probably will involve nuclear weapons and because it has been amply demonstrated that these have long-term detrimental genetic effects (page 597) in addition to their immense explosive effects, presumably no nation would consider using them. Quite the opposite seems to be true. More and more nations are acquiring nuclear weapons and it seems only a matter of time before some nation, setting immediate local gains above the general good of humanity, puts them to use.

Similarly, prevailing business practices typically interfere with optimal use of resources. Nonreplaceable resources, such as oil and natural gas, are exploited at the most profitable rate for the producer rather than at rates that conserve them for the optimal use of all humanity. These matters are, of course, not appropriate to courses in biology, but they must be remembered while the limits beyond which humanity may not proceed without causing disaster are defined in courses in biology and the other sciences. It must be remembered particularly that political and economic decisions cannot budge these limits on a worldwide basis. The essence of our tragedy is that there is no general understanding by the populace that these limits exist and that the unchanging laws of nature are the rules of the game that we are playing. Unless the laws of nations and worldwide economic policies soon become compatible with the natural order, humanity will in all probability experience the final embarrassment of a fate that it knows how to avoid.

In the following pages we will explore the scientific basis of these predictions. We shall proceed by first describing some of the major concepts of the science of **ecology**, which deals with the interactions of organisms and the environment. Then we shall apply these concepts to the study of the influences of humanity on the world environment.

Ecology, Study of the Ultimate Complexity of Life

Those branches of biology that we have already studied each focus on some relatively limited aspect of life. Thus, genetics is concerned only with the mechanisms of inheritance, and even systematics, although it does include all organisms, is concerned only with how they are related to one another. In contrast, the study of ecology involves all aspects of the interactions of organisms with each other and with the environment.

Ecology was originally defined in 1869 by Ernst Haeckel as the total relationship of organisms with the living and nonliving environment. Today **ecology** (from the Greek, *oikos* = the home) is defined in a more restrictive way as the *study of the interactions that determine the distribution and abundance of organisms*. Even with this limitation, it is obvious that ecology must cross many of the traditional boundaries between the sciences. To understand why a certain number of organisms of certain kinds exist in a particular place the ecologist must employ data and concepts from geology, paleontology, geochemistry, and climatology, as well as from a broad spectrum of the biological sciences such as biogeography, physiology, biochemistry, systematics, and parasitology, to name but a few. The attitude of the ecologist is one of synthesis, the bringing together of many branches of science to explain a central problem in biology.

LEVELS OF ECOLOGICAL ORGANIZATION

In biology we find that the processes of life can be arranged as an ascending series of levels of organizational complexity. Each higher level is more complex than its predecessor because it builds on all the preceding ones. So far in our studies we have been principally concerned with the lower levels of this series, the chemical, cellular, organ, and individual organism levels. Ecology begins at this last level and proceeds to levels of greater complexity, which are characterized by interactions among organisms and ultimately by the interactions among all organisms on the planet and with the planetary physicochemical environment.

The Black Box Concept

It is evident in ascending the series of organizational levels to the domain of ecology that we are proceeding into areas about which less and less is known. This must be obvious, since our study of the biology of just one organism, the human, has shown us how much is still unknown about its workings. Since the human is the best known of higher organisms, it must follow, when we undertake to study the interactions of humans with each other and with the environment (human ecology), that we build on an insecure foundation. The same comment is even more true of any other complex organism that the ecologist might wish to study. Is it then possible to study ecology in a useful way, or is its scientific basis too insecure?

Although it would be ideal if biology at all less complex levels of organization could be completely known before undertaking studies at the ecological level, this goal will probably never be attained. Yet ecological research does proceed and does produce useful answers. This can be done in the face of ignorance of many of the functional details of the organisms under study by use of what has been called the black box technique.

According to the black box technique, it is possible to evaluate the role of any component of a system in the functioning of that system by understanding only that component's *transfer function*. It is not necessary to understand all of the details of what goes on in that component, as long as the transfer function is understood. Transfer function is simply another way of saying "input-output relationship," or "what a component does when something is done to it."

Vending Machine Ecology

Consider vending machines as an example. To understand how vending machines fit into the "ecology" of, let us say, a school cafeteria, it is not necessary to understand the complex details of their inner workings. All one has to know are the transfer functions of the various types of machines found in the cafeteria. For one machine the transfer function might be, "Twenty cents in—candy bar out." Continued observation might reveal that others have different transfer functions, such as "Fifteen cents in—coffee out" or "Fifty cents in—sandwich out."

This information renders it possible to understand how the vending machines enter into the lives of the users of the cafeteria and even makes possible certain kinds of predictions about the behavior of the whole system. For example, the machine with a transfer function involving coffee would probably be more active in the early morning and late at night than the sandwich vending machine. Such "ecological" data would be valid irrespective of what one believed actually went on inside the vending machine, about which one needs to know no more than one knows about the contents of a "black box."

However, the black box concept must be used judiciously, since it can lead to error. For example, a chemical analysis of the input and output of the several vending machines might lead to the conclusions that they convert silver and nickel into complex biomolecules. Obviously the transfer function as stated, although it satisfactorily accounts for the interactions of customers and vending machines, cannot represent the true internal process of the slot machines because what appears to happen violates the laws of nature.

This observation might prompt further study, which would reveal that the vending machines actually have another transfer function for one type of human, the vending machine owner. For him, the transfer function is "food in—money out." Thus, still without understanding how the insides of the machine works, the careful observation of transfer functions shows clearly its role in the ecology of the cafeteria as an exchange point at which one class of humans exchanges money for food provided by another. Exactly similar methodology allows the ecologist to understand the interactions between the members of very complex systems up to and including all life on earth without the necessity of understanding many of the details of the life processes of all types of members of these systems.

As an example, consider a farm yard containing pigs, chickens, and cows. It is a simple matter to determine the transfer functions of these three types of organisms and then use them to analyze the economics of the farm community. The data show clearly which of the three types of animals have transfer functions that are the most favorable to the human members of the community. Thus, the pig is the most efficient converter of relatively low-grade input (principally garbage) into protein and fat, whereas the cow is less efficient, requiring input of expensive grains for maximal meat or milk production.

Autecology—Species Ecology

The least complex level of ecological analysis involves the study of a species population. Species ecology, or **autecology**, is devoted to understanding how all environmental factors, both biological and physicochemical, bear upon the survival of a particular species population. *Populations* are localized groups of a single species in which all members are not prevented from interbreeding by physical barriers. If we wished to study the autecology of the pig in the farmyard example, the horizons of investigation would shift from the analysis of input-output to all factors bearing on the survival of pigs in this particular farmyard. Besides the actions of the farmer, these would include environmental conditions such as tem-

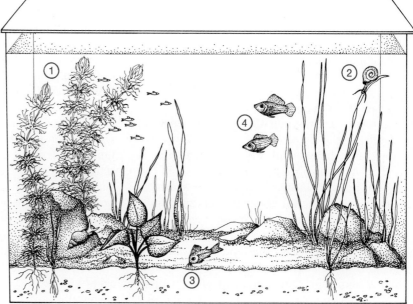

Figure 24-1 A balanced aquarium is a readily comprehensible model of an ecosystem containing photosynthesizers (1), grazers (2), scavengers (3) and predators (4) in addition to a varied microflora and fauna, all in isolation save for sunlight.

Figure 24-2 The earth photographed from space. The prevalence of water—in clouds, oceans, and icecaps—is the hallmark of our planet. NASA.

perature and the influences of other organisms, for example, disease-causing bacteria and parasitic worms.

Community Ecology

When the focus shifts from a single species to study of the relationships of all populations, plant, animal, and microbial, in a particular area, the endeavor is called **community ecology.** A community is defined in terms of the numbers and principal varieties of organisms of which it is composed—for example, a grassland community or a coral reef community. The interrelations of the component organisms of a community are described on the basis of transfer functions. This can become an immense task because even the simplest natural communities involve hundreds of kinds of organisms. Thus, a complete community level analysis of the artificial barnyard community would involve, besides the organisms already mentioned, a long list including soil bacteria and fungi, insects, small mammals such as rats and mice, birds, and vegetation.

Ecosystems Ecology

A more complete analysis of community function is obtained when the relevant physicochemical environment is considered in addition to all the organisms present. An **ecosystem** consists of a community plus its nonliving environment. Except for energy input in the form of sunlight, an ecosystem is the smallest component of the natural world that is completely self-sufficient. A well known artificial example is a "balanced aquarium." It is completely sealed off from the world and yet survives indefinitely with an input only of sunlight. Examination of Figure 24-1 shows how complex such a system is, with animals, plants, and microbes providing essential materials for each other in a never-ending series of cycles of energy and material, all ultimately driven by the sun. In nature it is impossible to find ecosystems that are so clearly isolated from the remainder of the world as this artificial example. However, it is possible to find ecosystems with well enough defined boundaries to allow analysis of their function in ways that are analogous, although much more difficult to carry out, to how one would study the energy and material cycles—the "metabolism"—of a balanced aquarium.

The Ecosphere

Finally, we come to the *ecosphere*. This is the grand total of all ecosystems on the planet. It includes all life on the planet together with all physicochemical systems that affect life. At this ultimate level the concept of the ecosystem as a "balanced aquarium" emerges again in a dramatically undeniable form (Figure 24-2). Just as the balanced aquarium is set off from the rest of the world by its glass walls, the ecosphere is isolated from other planets by the inhospitable vastness of space. From outside, the ecosphere at present receives little material input, only space "dust" and meteorites.

The ecosphere must survive on its own material resources. If some critical component is consumed totally, the ecosphere must somehow do without or die, wholly or in part. The ecosphere has, in fact, only one important link with the remainder of the universe. Like the balanced aquarium that dies if placed in the shade, the ecosphere depends on the sun for the vast majority of its energy input. The ultimately possible life span of the ecosphere is directly linked to the life span of the sun, which has enough fuel to burn at least another several billion years. Then, having consumed most of its nuclear fuel, the sun will undergo a final outburst, reducing the earth to a lifeless cinder, and then subside.

A DELICATE BALANCE

The overwhelming mass of the earth generates a sense of security in those of us who are confined to its surface. Earth implies stability, resistance to change—Mother Earth. How could such an immensity be altered? How could it be made inhospitable to life except by cosmic forces? Regretably, such questions do not reflect the true perspective. Viewed in more accurate perspective, as the astronauts see it, the earth's susceptibility to change, and consequently ours, is evident. The perspective from a few thousand miles out in space is of a small planet covered with a vanishingly thin, livable rind, the ecosphere.

From space the dominant visual impact of the ecosphere is caused by water, the cobalt blue oceans and the overlying, changing layers of white clouds (Figure 24-2). Earlier chapters have detailed the importance of water to life. It is the essential environment in which life arose and in which many organisms are still confined. Although water is only one of the requirements of life, a brief consideration of the state of this essential factor in the ecosphere convinces us of the tenuousness of our residence on the planet, of the delicacy with which the life-preserving balance is maintained.

Water and Global Temperature

The presence and physical state of water on the earth depends in the first analysis on our distance from the sun. If the mean distance from earth to sun were only about 5 percent less than actual the oceans would not have condensed and earth would have become a thickly clouded planet much like Venus. If the distance were a little over one percent greater the planet would probably be locked in a perpetual ice age. At the existing orbital distance the average surface temperature of the earth is 15°C. This allows the presence of both ice and free water. However, geologists tell us that the records of the past show that this balance between ice and water is readily disturbed without change in our distance from the sun (Figure 24-3).

The ice-water balance can change because of small temperature changes occurring for several reasons. There is some evidence of small, relatively short term variations of solar energy output and there may also be variations in the transparency of the earth atmosphere to incoming radiation. Prominent, however, among possible explanations is the phenomenon of *precession*. This is an extremely slow cyclic change in the angle of the earth's axis with reference to the plane of its orbit around the sun, caused by gravitational interaction with moon and sun. Precession influences the amount of sunlight that strikes the northern and southern parts of the globe and thus favors or inhibits ice formation in the polar regions.

The best evidence indicates that these temperature-affective processes continue to act and that we consequently may expect extensive glaciation to reappear in the northern hemisphere over the next few thousand years. This reappearance of the great ice sheets of the past will probably not be rapid or severe enough to cause any general extinction of life, but in the past "ice ages" have had immense effects on the course of evolution. At the very least, should we survive to witness it, the next great episode of glaciation will have profound cultural effects. Large parts of the northern hemisphere will become uninhabitable due to the overburden of ice, and coastlines will be radi-

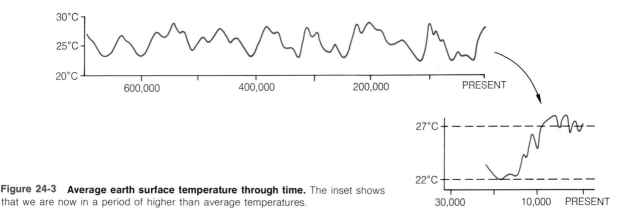

Figure 24-3 Average earth surface temperature through time. The inset shows that we are now in a period of higher than average temperatures.

cally altered by the retreat of the seas caused by entrapment of water in the glaciation.

Carbon Dioxide

Remarkable testimony to the delicacy of balance characteristic of the ecosphere comes from evidence that human activities may, inadvertently or intentionally, hasten or delay such global phenomena as the advance of the glaciers. Pollutants emitted into the atmosphere may, depending on their nature, either reflect sunlight back into space, and thus diminish the temperature of the earth, or trap heat that might otherwise be radiated back into space, thereby raising temperatures. One of these pollutants is carbon dioxide.

Normally carbon dioxide would not accumulate to detrimental levels in the atmosphere because its release, primarily by the respiration and decay of organisms, is balanced by photosynthesis and solution in the oceans. However, in the 20 years that atmospheric carbon dioxide has been systematically measured, there has been a steady increase. The current value of 332 parts per million (ppm) in the atmosphere will probably double within the next 50 years. The cause of this increase is the burning of fossil fuels (gas, oil, and coal), and it is believed that the situation will worsen as many nations accelerate burning of their large coal reserves to replace fast disappearing gas and oil. One of the effects of carbon dioxide in the atmosphere is to trap heat radiated from the earth. This so-called **greenhouse effect** (page 616) might be powerful enough, if carbon dioxide concentrations in the atmosphere continue to rise, to cause a net increase of as much as 2°C in the surface temperature of the earth. This does not seem large but actually it would almost certainly have serious widespread effects, one of which would be to cause sufficient melting of the polar ice caps to raise the sea level by as much as 80 m.

These examples, both natural and human-induced, of how global temperatures are affected are only a few among many known or suspected influences on the surface temperature of the earth. The ramifications of these effects are extremely complex because all biological and chemical processes on the face of the earth are so interrelated that change in one usually sets off a long chain of action and reaction through the system until balance is restored. Consider, for example, just one major process important to determining carbon dioxide levels in the atmosphere—photosynthesis. What factors affect photosynthesis? Is the deforestation under way in many parts of the earth sufficient to have an effect? Will pollution of the seas have a significant effect through the poisoning of phytoplankton which are responsible for better than half of the earth's photosynthesis? Or what if the planet does get warmer because of the carbon dioxide added to the atmosphere by fossil fuel burning? Won't the warming speed up photosynthesis and thus reduce the amount of carbon dioxide again? Or will the warming cause an increase in cloud cover, reflecting more sunlight back into space, reducing photosynthesis, and causing another round of carbon dioxide increase and warming? On and on go the actions and reactions. How are we to predict the outcome? Which perturbations will have significant effects and which will be counteracted by other processes and so be without effect?

Matter and Energy Flow in the Ecosphere

In order sufficiently to understand how the ecosphere functions, so that we may begin to answer questions like those in the previous section as well as innumerable others of a similar nature bearing directly on the quality of life on the planet, we must treat the totality of life on the planet—the biosphere—somewhat like a single organism. Viewed in this light, the basic questions are more easily identified. Obviously, if the world organism, or whatever we want to call it, is to live, it has two requirements, like any lesser organism. It requires energy and it requires specific kinds of matter. Thus it would seem that a route to understanding the workings of the ecosphere would be to trace through it the flow of energy and matter. Once these pathways are established, it should be relatively simple to determine how the world organism responds to environmental perturbations of the type just mentioned.

THE DRIVING FORCE

The ecosphere receives energy from three sources. Gravitation and internal forces within the earth, such as radioactivity and the movements of the earth's crust and volcanism, have significant effects, but by far the most important source of energy is solar radiation. Only about one two-billionth of the total radiation emitted by the sun arrives at the upper level of the atmosphere. Since the energy output of the sun is probably relatively constant over extremely long periods of time and since the maximum variation of the distance of the earth from the sun is only about 3%, the amount of energy reaching the upper atmosphere is uniform, amounting to 1.94 cal/cm² min, which is termed the **solar constant**. As detailed in the following paragraphs, the fraction of this energy that actually reaches the surface of the earth is small, but it is still so large in terms of our everyday experience as to be almost incomprehensible. Thus it has been estimated that the daily solar output arriving at the surface of the earth is enough to melt a 35-ft thick ice layer over the whole earth.

Energy Equilibrium

Since the temperature of the earth is not rising nearly enough to account for this tremendous input, the earth must also be losing energy at a rate nearly equal to the rate of input. How is it lost? Most solar energy reaches the earth in the form of radiation in the ultraviolet and visible wavelengths that lose energy in reactions with matter. There are several results. Some energy is re-radiated into space as less energetic, infrared radiation. A small fraction is captured in photosynthesis, serving as the total energy source for the biosphere. Larger amounts are involved in heating the earth, air, and water. These processes ultimately power the winds of the atmosphere, cause evaporation of water (thus producing our weather), and generate ocean currents. The weather and ocean current cycles can be looked on as the circulatory mechanisms of the ecosystem since these processes promote the exchange of energy and materials between the three great material reservoirs—land, water, and air.

Ultimately, however, energy *in* must equal energy *out*. This fundamental requirement is rapidly met, for most of the energy arriving at the surface of the earth is used to heat components of the ecosphere and is re-radiated into space, either directly or as a consequence of work done by the heated components, for example, in the movements of masses of air or water. The retention time of some energy that enters the biosphere can, however, be quite long—even millions of years—as in the case of organic matter stored in fossil fuels.

Solar Energy at the Earth's Surface

Two factors determine how much of the solar energy that arrives at the upper atmosphere actually reaches the surface of land and water. The first is reflection and absorption in the atmosphere, and the second is the angle of incidence of sunlight, which is related to the latitude, that is, to distance from the equator. The effect of angle of incidence of sunlight can be visualized from Figure 24-4. If we imagine a unit surface at right angles and precisely aligned with a beam of sunlight of the same unit cross-sectional

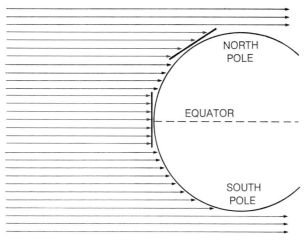

Figure 24-4 Influence of latitude on solar heating of the earth's surface.

area, then the surface will obviously intercept all of the sunlight in the beam. It will receive the maximum possible amount of energy from the sunlight. Let us say that this is the situation at the Equator. Now, since the earth is spherical, as we move the unit surface either north or south from the Equator, the angle between the unit surface and the incident beam of sunlight will steadily diverge from a right angle until, at the poles, the unit surface will actually be parallel to the incident beam. Thus, although the area and energy content of the beam as well as the area of the unit surface are invariant, as the angle of incident sunlight changes the fraction of sunlight from the beam that strikes the unit surface becomes smaller. Thus the polar regions are always cold and the equatorial regions are always hot.

The fate of solar radiation entering the atmosphere is as follows: Virtually all high energy ultraviolet radiation is absorbed in the ozone layer and ultimately reradiated into space at longer wavelengths. Of the remainder, about 35% is immediately reflected back into space from clouds, the gases of the atmosphere, and the surface of the earth itself. About 14% is absorbed in the atmosphere and thus contributes to air movement. Sunlight directly reaching the surface amounts to about 30% of the total, with about 4% of this being immediately reflected. Another 25% of the solar input also reaches the surface as infrared radiation from the atmosphere and as indirectly reflected visible light. Of the 30% that is directly incident on the surface, only an average of about 1% is utilized in photosynthesis (Figure 24-5).

Solar Energy, Wind and Water Currents

Circulation of materials between the three reservoirs of the ecosphere—land, water, and air—depends on the fluidity of air and water. Winds carry water vapor over land where it falls as rain or snow, leaching materials out of the earth and carrying them to the sea. Within the immense reservoir of the seas—which are 70% of the area of the earth and about 270 times the mass of the atmosphere—circulation of materials is brought about by currents that cause both vertical and horizontal mixing. The primary driving force of these circulatory mechanisms is solar heating.

Important to understanding how solar energy causes atmospheric circulation are measurements showing that atmospheric temperatures are highest at low altitudes and decrease steadily up to 12 km, where the temperature begins to increase again until reaching 0°C at 50 km due to heating caused by ultraviolet absorption in the ozone layer. These facts, together with the observation that the atmosphere is quite transparent to visible light, show that most atmospheric heating is brought about by reradiation of solar energy from the earth. Incoming sunlight interacts with surface matter with the result that infrared is radiated back into the atmosphere where it causes heating at low altitudes as it is absorbed by water vapor and carbon dioxide in the atmosphere. This is the basic cause of wind and weather as illustrated in Figure 24-6.

Although the process is almost infinitely complicated, as attested by the notoriously unreliable weatherman, the fundamental logic is straightforward. As air at low altitudes is warmed, it rises and cools by expansion in the rarified upper atmosphere. Rising warm air is replaced by horizontally moving cooler air, producing winds. The figure shows three primary zones of winds caused in this way: the trade winds, just north and south of the Equator; the westerlies, centered half-way between the Equator and poles; and a polar circulation. The equatorially centered doldrums are characterized by little wind and a very slow rise of air so that this zone is a barrier to the exchange of air-borne materials between the northern and southern hemispheres. Since the atmosphere is not rigidly coupled to the earth, it lags behind as the earth

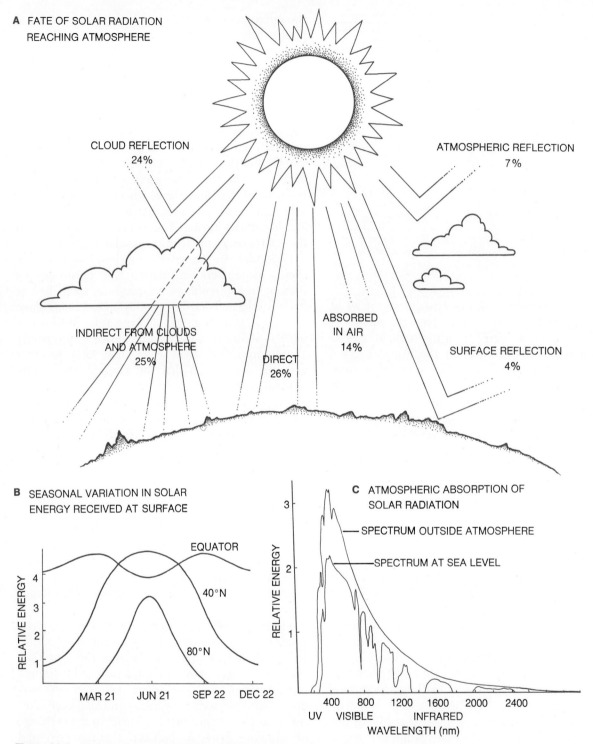

Figure 24-5 Solar energy at the earth's surface. A. Fate of solar radiation as it penetrates the atmosphere. **B.** Latitudinal effect on solar heating at the earth's surface through the year. **C.** Absorption of specific wavelengths of solar radiation by the atmosphere. **B.** after Rumney, G. R., 1968, *Climatology and the World's Climate.* Macmillan, New York. **C** after U.S. Air Force Cambridge Laboratories, 1965, Handbook of Geophysics and Space Environments. U.S. Air Force.

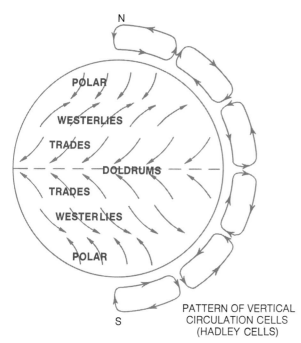
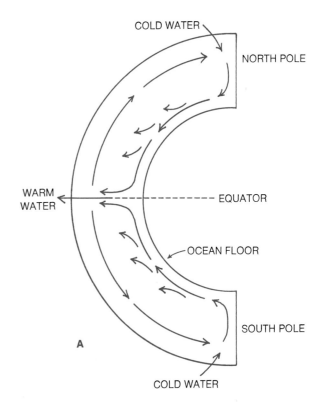

Figure 24-6 (Above) Atmospheric circulation is driven by solar energy to form the primary zones of wind direction. Where these zones interact zones of vertical circulation, Hadley cells, occur.

Figure 24-7 Water flow in the oceans. A. Solar heating affects vertical circulation. **B.** The principal surface currents in Atlantic and Pacific Oceans. N.E.C., North Equatorial Current; E.C.C., Equatorial Counter Current; S.E.C., South Equatorial Current.

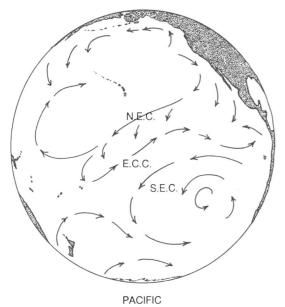

PACIFIC

ATLANTIC

THE ENVIRONMENT: ORGANIZATION OF THE ECOSPHERE

rotates beneath it. The result is that in the northern hemisphere winds move counterclockwise around a low pressure area whereas the opposite is true in the southern hemisphere.

Ocean currents are driven both directly and indirectly by solar heating. Vertical movements are caused by rising of warm water, which is less dense than cold water, at the equator and sinking of cold water at the poles. The result is two great vertical loops of circulation between the poles and equator, one in the northern and one in the southern hemisphere (Figure 24-7A). The fundamental cause of horizontal currents in the upper waters is coupling of air movements with the water. Thus, the steady trade winds produce, in both the Atlantic and Pacific, a pair of westward equatorial currents (Figure 24-7B). Between the westward equatorial currents there is an eastward equatorial countercurrent, which serves to return some of the water moved by the westward equatorial currents.

Although primarily caused by coupling with the winds, surface currents are profoundly and complexly affected by the land masses that they encounter. Thus, in the Pacific when the westward equatorial current encounters the Asiatic land mass, it splits, forming a southern component that contributes to the equatorial countercurrent and a northern running component that forms the Japan Current. This current recrosses the North Pacific and, on encountering the North American land mass, divides to run north and south as the Alaska and California Currents. In an exactly similar way in the Atlantic, the westward equatorial current produces the Gulf Stream, which flows north and recrosses the Atlantic. In doing so it spawns a south running Labrador current, which keeps Maine coastal waters continually cold (Figure 24-7B). Surface oceanic currents have effects on weather, for example the tempering effect of the warm Gulf Stream on coastal North America and Europe, as well as serving to mix materials that enter the sea. Note however that the Equator is as much a barrier to water mixing as it is to air mixing between the northern and southern hemispheres.

Chemical Cycles in the Ecosphere

Long before reaching the mouth of a great river mariners sense its nearness by the marked discoloration of the transparent blue water of the ocean by chocolate colored sediments carried out to sea by the river's flow (Figure 24-8). On a world-wide basis this flow of the land into the sea is estimated to amount to about 1.2 cubic mile ($2 km^3$) of solids per year. Given the millions of years that such wearing away of the land has been in progress, it must be obvious that stream deposits in the ocean must be part of a cycle. Somehow there must be a return of material to the land to match this outpouring. Otherwise the ecosphere should have long since reached a point of uniformity, perhaps as a vast, shallow sea dotted with small points of low lying rock.

MAJOR GEOLOGICAL CYCLES

The cycle of material that these observations demand is but a small fraction of an immense cycle of exchange of material between the crust of the planet and the underlying, hot, semifluid mantle. Volcanoes bring mantle material to the surface in brief, locally violent eruptions. Their contribution of material is difficult to estimate because, besides the volcanoes on land, it is estimated that nearly 10,000 more lie beneath the surface of the oceans. More significant over the vast duration of geological time are convection currents in the mantle which move the overlying crust. Since the crust is broken up into a number of independent plates, the major continental masses, as they move bear onto each and may cause one to submerge into the mantle while the other rises (see Figure 16-4). At other points on the crustal surface there is a compensatory movement of mantle to the surface. The time scale of such renewal of the planetary surface is immensely long as compared with the cycles of materials in the shallow surface layers of the crust.

The deposition and uplift of sedimentary materials takes place on a time scale measured in millions of

Figure 24-8 Discharge of soil from the Colorado River into the Gulf of California is clearly visible from an altitude of 270 mi. (NASA Skylab.)

years. These sediments are deposited out of water, and they now cover about 80% of the land mass. They are formed in three ways:

1. From inorganic chemical reactions in sea water that result in deposition of insoluble materials, such as manganese nodules.
2. From particles of rocks or volcanic ash transported by rivers or winds and ocean currents.
3. From the activity of living organisms, for example, the calcium carbonate deposits produced by the skeleton-building activities of corals and single celled foraminifera, or the siliceous deposits manufactured by diatoms and radiolarians.

BIOGEOCHEMICAL CYCLES

Thus the essentials of the circulatory processes in the ecosphere begin to emerge. Underlying the entire system is an extremely slow exchange of material between the surface crust and the earth's mantle. Materials rapidly are moved about the crust, between land and water and air, by the agency of water and air flow. Material that moves from the land into the seas returns to the land by a variety of geological and chemical processes that may include the activities of living organisms—*biogeochemical cycles*.

Now we narrow our focus to living systems and examine how they participate in these grand cycles of materials, insuring satisfaction of their own material requirements and, in many instances, themselves playing essential roles in the geological cycling of the materials of the earth's crust.

Residence Times—the Black Box in Action

Although it seems obvious that the chemicals essential to life must cycle through the reservoirs of the ecosphere, it might also seem that the system is too large to permit useful study. True enough, precise measurements, for example, of the chemical contents of all the streams entering the oceans cannot be obtained and neither can the rate of uplifting of sediments be determined exactly. However, it is possible to make sufficiently good estimates of the contents of the reservoirs, and of rates of flow into and out of them, to obtain estimates of what is going on that are useful for many purposes. The procedure is an example of black box analysis, mentioned earlier as the savior of the quantitative ecological researcher. In this instance the procedure is the measurement of residence (or turnover) time.

Suppose that we have a 100-liter reservoir of water with a flow in and out of 0.01 liter/min. We divide the volume of the reservoir by the rate of addition or removal and get as an answer 1000 min. This is the **residence time** of water in this particular reservoir. What does the answer mean? It simply indicates the time required either to add or to remove an amount of water equal to the total in the reservoir. This would not necessarily mean that in 1000 min all of the old water would be replaced by new water, since to do this the incoming water could not mix at all with the old water. The time required completely to replace the old water in the reservoir is called the **renewal time** and, in nature, this is always longer than the residence time.

Calculation of a simple relationship such as residence time can usefully clarify many questions regarding movements of materials in the ecosphere. To illustrate, let us return to the fact that many organisms in the ocean deposit calcium carbonate in marine sediments. Now even if you did not know that this happened, a calculation of the residence time of calcium in the oceans would tell you that there had to be some process at work keeping the concentration of calcium low in ocean waters. For calculation of the calcium residence time the total calcium in the oceans is estimated to be 6×10^{20} g, and the input is estimated at 5×10^{14} g/year. Dividing input into reservoir size yields a residence time of 1.2×10^6 years. What does this tell us? It is immediately obvious that 1 million years is a short residence time in terms of the total time during which similar rates of calcium addition have probably been maintained, perhaps between 1 and 2 billion years.

This must mean, therefore, that the oceans are not a *permanent* dumping ground for calcium that is washed out of the land. If this were so, the amount of calcium presently in the oceans ought to be many times the present value. Thus the residence time measurement tells us to look for a mechanism that removes calcium at a substantial rate. And, of course, we find it in the biological process that forms insoluble calcium carbonate from calcium ions in solution and in the geological events that return the calcium carbonate to the land in uplifted sediments.

The conclusion that the ocean is not simply a dead-end dumping ground for continental run-off is reinforced by residence time measurements of the other major inorganic constituents of sea water. These reveal no value larger than about 100 million years, which is still small compared with the probable age of the oceans. We conclude that the ocean is indeed part of a circulatory system for inorganic ions. In Chapter 26 we will see other applications of residence time measurements to the problem of pollution.

THE WATER CYCLE

The blood of the ecosphere is water (Figure 24-9). Driven by the energy of the sun, it carries, either suspended or in solution, all the elements essential to life. Most of the earth's water is in the oceans, which hold about 80% of the total. About 1.2% is in ice, fresh water bodies, and the air. Most water not in the oceans is held by the land in the soil, some 18.8%. Water that falls on the land as rain finds its way back to the sea by three routes with radically different residence times. About half of the rain falling on land evaporates and may fall again elsewhere, on land or sea. Roughly a quarter runs directly into streams emptying into the sea with a residence time measured in weeks. The remaining quarter soaks down through soil and rocks to enter the *water table*, a relatively shallow zone in which the earth is saturated with water. How long water remains there without being pumped out in wells or flowing into the ocean is unknown, but it is probably thousands of years.

The energy expended in driving the water cycle is a large fraction of the total entering the lower regions of the atmosphere. About 23% of arriving solar energy is expended in evaporating water, which eventually falls as rain. In the ocean this great expenditure of energy serves annually to evaporate water equivalent to a depth of 1 m over the entire surface of the ocean. In equatorial regions this is large enough to increase the salinity of surface oceanic waters. In addition to evaporation, water also enters the atmosphere as small droplets called aerosols, caused by interaction of wind and waves. This also results in some circulation of entrapped dissolved substances by the atmosphere.

The geographic pattern of rainfall greatly influences the distribution of life on land. The rainfall pattern is markedly nonuniform, ranging from essentially none

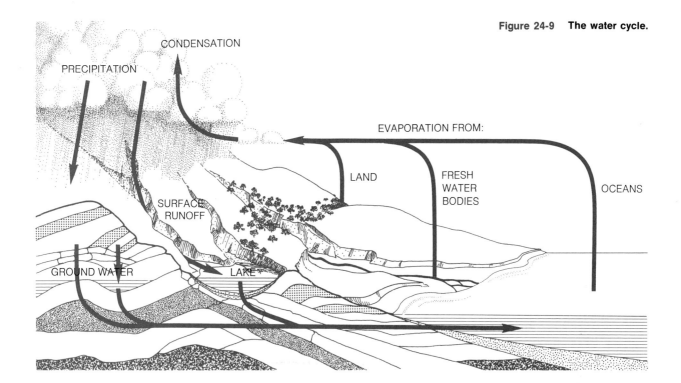

Figure 24-9 The water cycle.

to as much as 1000 in./year. There are two principal causes of this nonuniformity: wind patterns and terrain. Coastal western North America illustrates both factors. The coast from Washington to Alaska receives large amounts of rain because it is in the zone of the westerlies. These winds receive a heavy load of water vapor in warm southern regions and become cool as they move northward. They thus become progressively less able to retain their burden of water, which then falls as rain. Farther south, coastal regions nearer the equator are dry because they are in the zone of the trade winds. These easterly winds proceed towards the equator from cool to warm regions so they have an increased carrying capacity for water and little tendency to produce rain.

As winds move inland from the coast, they encounter terrain features that influence rainfall. Outstanding among these are the great ranges of mountains paralleling the coast. Moist air moving inland from the ocean is forced to higher altitudes by the mountains. Once cooled to a critical temperature, they tend to drop much of their load of water on the windward sides of these ranges. Reciprocally, as the now relatively dry winds descend on the inland sides of these mountains, they are warmed and thus are not likely to deposit rain.

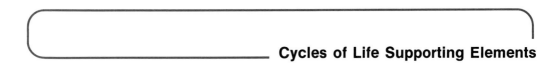

Cycles of Life Supporting Elements

Having reviewed some of the principal mechanisms responsible for circulation of materials within the ecosphere, we now examine the details of circulation of three life-essential elements: carbon, nitrogen, and phosphorous. The cycles of these three elements are of interest because of the variation in principal reser-

THE ENVIRONMENT: ORGANIZATION OF THE ECOSPHERE

voirs that they demonstrate, and because of the variety of biological systems that are incorporated in their cycles. Thus the principal reservoirs in the carbon cycle are sedimentary rock, the oceans, and the atmosphere, whereas the principal reservoir for phosphorous is only sedimentary rock. Biological processes are minimally concerned in the phosphorous cycle, whereas in the nitrogen cycle biological mechanisms, principally bacterial, are active at virtually every important point. The phosphorous cycle is also of practical interest because phosphorous is the most common limiting factor in plant growth.

THE CARBON CYCLE

The carbon cycle (Figure 24-10) is of pivotal importance because it is the route by which solar energy usefully enters the biosphere by way of photosynthesis. This subject, the energy cycle, is discussed on page 540.

In the basic cycle, carbon circulates among a complex system of reservoirs, primarily *sedimentary rocks* containing carbonate; the *oceans*, and the *atmosphere*; *photosynthetic organisms*; *consumer organisms*; and *decomposer organisms*. Carbon dioxide is the key intermediary molecule in the cycle because this is the form in which carbon enters the biological phases of the cycle, in photosynthesis. *Photosynthetic organisms* (producers) utilize solar energy to generate oxygen, energy-rich organic molecules, and low grade heat. *Consumers*, principally animals, use these organic molecules in their life processes, generating heat and returning some carbon dioxide to the reservoirs. Ultimately, various *decomposer organisms*, principally bacteria, oxidize plant and animal remains with the production of more heat and return of carbon dioxide to the reservoir.

Examination of the structure of the carbon cycle reservoirs shows that the basic cycle has many important complications. First of all, one may divide the reservoirs into short and long term types; second, there is a complex set of interactions between components of the short term reservoir. The short term reservoir has two parts, the oceans and the air. The carbon of the oceans is principally bicarbonate ion (HCO_3^-)

Figure 24-10 The carbon cycle.

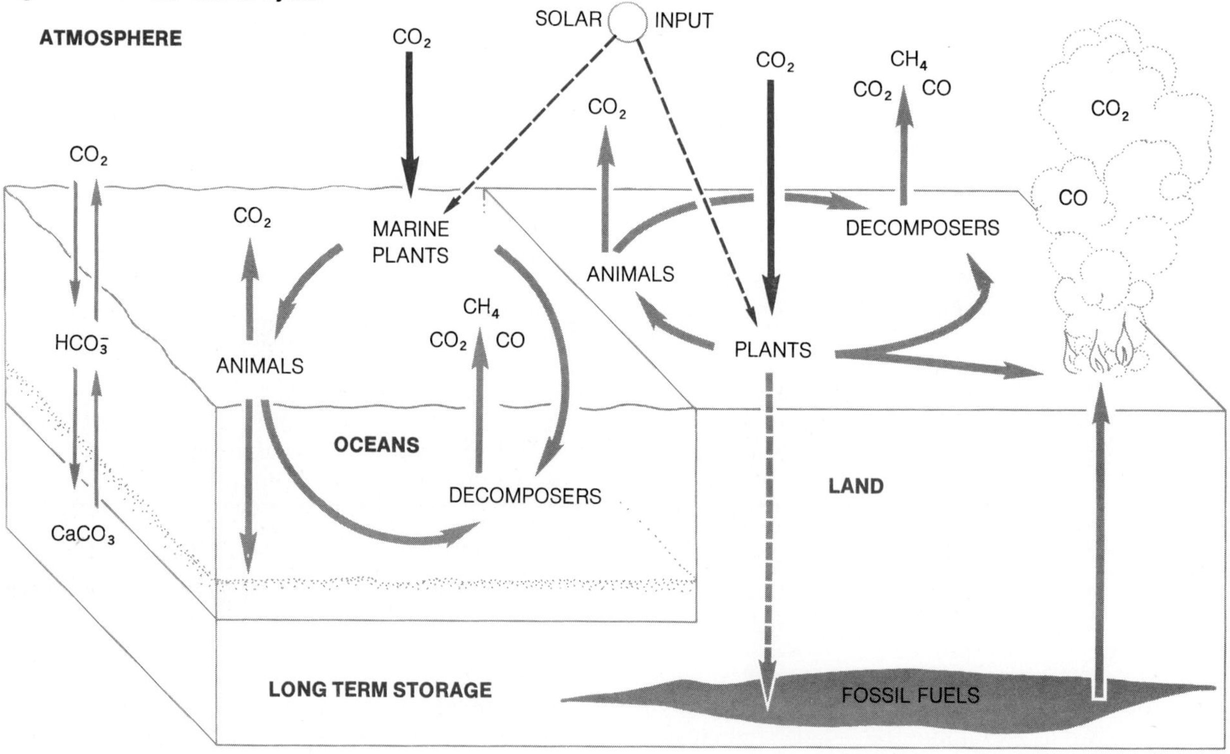

while in the air carbon dioxide (CO_2) is dominant. The oceans contain about 60 times as much carbon as does the atmosphere, and there is free exchange between the two. When CO_2 dissolves in water the following reversible reactions take place:

$$CO_2 + H_2O \rightleftharpoons \underset{\text{carbonic acid}}{H_2CO_3} \rightleftharpoons H^+ + \underset{\text{bicarbonate}}{HCO_3^-} \rightleftharpoons H^+ + \underset{\text{carbonate}}{CO_3^-}$$

These reactions are readily reversible and pH dependent. Under acidic conditions, the reactions are driven to the left with the result that CO_2 tends to be evolved into the atmosphere because it is poorly soluble in water. Under alkaline conditions the reactions are driven to the right, increasing the capacity of the solution to take up CO_2.

The atmospheric part of the short term reservoir contains other forms of carbon besides CO_2. Carbon monoxide (CO) and methane (CH_4) are present in very small amounts in the ratio of 1 CO : 25 CH_4 : 2800 CO_2. Under natural conditions both CH_4 and CO are products of decomposition. CO is rapidly converted to CO_2 in the atmosphere and has a residence time of only about a month. CH_4 is ultimately converted to CO_2, but with a longer residence time of 4 years, approximately equal to that of CO_2. CH_4 production is not high enough to allow significant levels to accumulate in the atmosphere except in very localized conditions, as, for example, during anaerobic decomposition in swamps when enough may be evolved sometimes to be ignited, producing the will-o'-the-wisp. The short residence time of CO suggests that it could never reach significant concentrations. This is no longer true because of combustion products from automobile engines. Under stagnant air conditions in large cities, automobile emissions can raise CO levels to about 1000 times normal in heavy traffic areas.

Long term storage of carbon takes place in two forms. Sedimentary rocks store carbon as carbonate in immense quantities estimated at 1700 times the total amount of carbon in the sea, air, and organisms. This huge store has a residence time of something more than 300 million years. Approximately one fifth as much carbon as is in the sedimentary carbonate stores is in another long term storage form, fossil fuels (coal, gas, oil). Except for coal, accessible fossil fuels can be considered to be totally expended relative to the time scale of this discussion.

Regulation in the Carbon Cycle

The carbon cycle elegantly illustrates the interaction between living and nonliving systems characteristic of biogeochemical cycles. First of all, remember that the composition of the atmosphere is largely attributable to photosynthesis. Present day oxygen and carbon dioxide concentrations in the atmosphere are determined by the slight excess of photosynthetic oxygen release over CO_2 release by metabolism. From the biological point of view, it is important to note that this atmospheric balance hinges on an infinitesimally small amount of one form of carbon as compared with the total reservoir of carbon. This form is, of course, CO_2, which amounts to only 1.5% of the total carbon in short term stores (all categories except carbonate rocks and fossil fuels). When all stores are included, CO_2 is only about 0.0009% of the total.

Maintenance of atmospheric CO_2 at present levels in the presence of such immense amounts of other forms of carbon is accomplished by three stages of control. The first stage is conversion of carbonate in solution into insoluble forms, such as calcium carbonate ($CaCO_3$), which can be considered to be out of circulation over periods far longer than the residence time of CO_2 in the atmosphere. The second stage of control involves the reversible solution of atmospheric CO_2 in bodies of water, thus allowing the waters of the earth to serve as a fast acting buffer for atmospheric CO_2. The final level of control is at the biological level and is based on the fact that rates of photosynthesis are sensitive both to CO_2 and O_2 concentrations. An *increase* in CO_2 favors an *increase* in photosynthesis. This serves to *diminish* CO_2 concentrations, since CO_2 is a substrate of photosynthesis. Greater than normal concentrations of O_2 act oppositely and reduce photosynthesis. Thus both of these molecules act in a negative feedback loop controlling rates of photosynthesis and thereby stabilizing the concentrations of both CO_2 and O_2 in the atmosphere.

When we look farther afield from the carbon cycle, evidences of its interactions with other systems are readily apparent, showing again that no element of the ecosystem can function alone, as perhaps the preceding discussion of the carbon cycle in isolation might suggest. Thus we find, for example, that the phosphorous cycle is directly linked with the carbon system in several ways, including one quite obvious way—since phosphorous is often the limiting nutrient in photosynthesizing organisms. More broadly, precise regulation of the carbon cycle atmospheric phase is important to worldwide average temperatures because atmospheric CO_2 is a major determinant of surface temperatures by absorbing heat, reradiated as infrared radiation from the earth's surface. And finally, since the carbon cycle is the route of entry for energy and organic molecules to the entire biosphere, the cycle obviously has a pervasive influence over all life-related materials cycles.

NITROGEN CYCLE

Organisms require nitrogen in reduced form for synthesis of essential molecules such as amino and nucleic acids. Since the principal nitrogen store is atmospheric molecular nitrogen (N_2), the critical process in the nitrogen cycle (Figure 24-11) is nitrogen fixation. This is the reduction of N_2 to ammonia (NH_3) and other biologically useful forms of nitrogen. Nitrogen fixation under natural conditions is accomplished in two ways, by inorganic chemical processes in the atmosphere—photochemical, electrical (lightning)—and by biological processes mediated by certain bacteria, fungi, and blue-green algae. Human activities, such as the use of artificial fertilizers, now provide an alarmingly high fraction of the total nitrogen movement between reservoirs, a situation that must be watched with care.

The atmospheric nitrogen reservoir consists almost totally of N_2, some 2.8×10^{20} moles, or 79% of the total nitrogen store within the ecosphere. There are, however, six other forms of nitrogen in the atmosphere in amounts that do not approach even 1/10,000 the amount of N_2 present. Some of them have important effects, which will be discussed later. There is extensive exchange of various forms of nitrogen between the atmosphere and land and sea, amounting to roughly an interchange of 15×10^{12} moles/year. The total nitrogen store on land and in the sea, consisting of organic nitrogen compounds in living or decaying systems and dissolved in water, is only 0.04% of the total ecosphere nitrogen store. Sedimentary rock contains about 20.2% of the total ecosystem nitrogen with most of it locked up as organic nitrogen

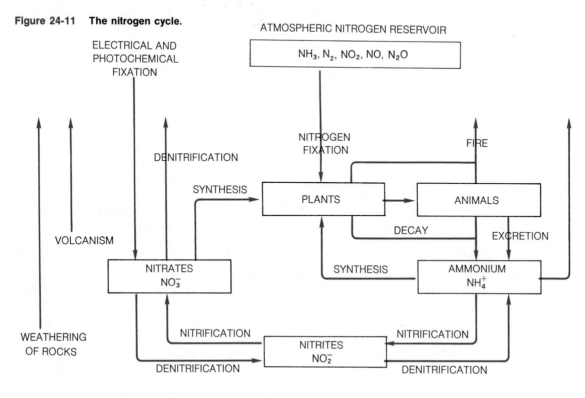

Figure 24-11 The nitrogen cycle.

in fossil fuels. At present the dominant exchange between sedimentary nitrogen and the rest of the ecosphere is due to withdrawal of fossil fuels. Relatively small amounts of nitrogen enter the sedimentary reservoir.

The Biological Nitrogen Cycle

The biological nitrogen cycle consists of three phases:

1. Nitrogen fixation.
2. Nitrogen utilization in higher organisms.
3. Nitrogen scavenging.

Nitrogen fixation accomplishes nitrogen input into organisms from atmospheric N_2. Nitrogen utilization involves transfer of organic nitrogen from the primary fixers through consumer organisms. Finally, nitrogen from the wastes and decaying tissues of organisms is largely converted again into biologically useful forms by a bacterial scavenging cycle.

Biological fixation of nitrogen is accomplished by symbiotic nitrogen fixation by a number of species of bacteria of the genus *Rhizobium* and by certain photosynthetic bacteria and blue-green algae. On land, the major nitrogen fixing activity is carried out by *Rhizobium* symbiotes and some free-living bacteria, whereas blue-green algae are the major nitrogen fixers in the oceans.

Symbiotic Nitrogen Fixation

Nitrogen fixing bacteria of the genus *Rhizobium* are widespread in soil. They can fix nitrogen only when they occupy nodules on the roots of legumes (clover, peas, beans) and a few other plants. The bacteria enter through root hairs and become established in sacs within nodules formed of root tissues. These nodules contain hemoglobin, a unique occurrence in the plant kingdom, one which appears to be essential to nitrogen fixation. Production of hemoglobin is caused by the presence of the nodule bacteria since it is not present when the plants are sterile.

Nitrogen fixation is biochemically complex; the overall process is as follows:

$$N_2 + 6 H^+ + 6 e^- + 12 ATP \rightarrow 2 NH_3 + 12 ADP + 12 P \text{ (inorg)}$$

The reaction requires a bacterial nitrogenase enzyme in which the active unit is formed from two protein molecules containing molybdenum and iron. The enzyme binds N_2 and reacts with a special electron carrier protein (**ferredoxin**), oxidizing the carrier and picking up $6 H^+$ to generate two molecules of NH_3. Vascular tissues of the plant are well developed in the vicinity of the nodules and carry the NH_3 to synthetically active tissues for incorporation into biomolecules. The predominant incorporation reaction is amination of α-ketoglutarate, forming glutamate. This is then available for transamination reactions (transfer of amino groups) with other α-keto acids to form the corresponding amino acids.

Nitrogen containing biomolecules circulate from primary fixers through consumer organisms. Some is lost through excretory processes (as ammonia, urea, and uric acid) and through decomposition with production of ammonia. Nearly all of this nitrogen is recovered by conversion of ammonia to nitrite (NO_2^-) and then to nitrate (NO_3^-), which can be assimilated by higher plants, completing the biological cycle.

Nitrogen Scavenging

The recovery of waste reduced nitrogen compounds from the biosphere is accomplished by highly specialized bacteria in a process called *nitrification*. Nitrification occurs in two steps. In the first the soil bacterium *Nitrosomonas* converts ammonia to nitrite. This remarkable organism is a chemoautotroph. It uses ammonia as its energy yielding fuel and obtains necessary carbon by direct fixation of carbon dioxide. Energy is obtained from ammonia by an ammonia dehydrogenase, which feeds electrons from ammonia into a respiratory chain coupled with oxidative phosphorylation. The second step in nitrification is a further oxidation of nitrite to nitrate by another chemoautotrophic bacteria, *Nitrobacter*. This bacterium obtains nearly all of its energy from the oxidation of nitrite and, like *Nitrosomonas*, obtains carbon from carbon dioxide.

Nitrate Denitrification and Assimilation

Some bacteria, such as *Pseudomonas*, and a few fungi use nitrate as an oxygen source in poorly aerated soil, and this results in loss of nitrate to the biological cycle as gaseous N_2. However, the majority of the available nitrate is taken up by green plants and assimilated into useful ammonia again in a two-step enzymatic reduction of nitrate to nitrite and of nitrite to ammonia.

The foregoing illustrates how life supporting nitrogen input from the atmospheric N_2 reservoir depends on bacterial activity. To a lesser extent this biological activity is supplemented by photochemical and lightning activated N_2 fixation.

Nitrogen Compounds in the Atmosphere

Although N_2 is by far the predominant form of nitrogen in the atmosphere, several other forms of nitrogen have significance far outweighing their low abundance. These include nitrous oxide (N_2O), nitric oxide (NO), and nitrogen dioxide (NO_2). These compounds are naturally produced, since large amounts of N_2O and ammonia (NH_3) enter the atmosphere as the result of decay. At high altitudes N_2O is photochemically converted to N_2 and NO. Then NO reacts with ozone (O_3) to form NO_2:

$$2N_2O + light \rightarrow 2NO + N_2$$

$$NO + O_3 \rightarrow NO_2 + O_2$$

NO is recovered in a coupled reaction between NO_2 and the monoatomic oxygen that forms in the natural breakdown of O_3 by ultraviolet light:

$$O_3 + UV \rightarrow O_2 + O$$

$$NO_2 + O \rightarrow NO + O_2$$

These reactions and others linked with them, have great practical significance because the introduction of nitrogen oxides into the upper atmosphere interferes with the normal reformation of O_3 from O and O_2. Human activities are currently estimated to inject nitrogen oxides to the extent of 3% of the total nitrogen that enters the atmosphere from all sources. A recent evaluation of N_2O entry consequent upon bacterial denitrification associated with the use of nitrogen fertilizers indicates that their continued use at present rates will probably deplete the ozone layer by 15% or more during the next 100 yr. With the addition of nitrogen oxides emitted by high altitude aircraft, there is the possibility of sufficient destruction of the ozone layer to produce a hazardous increase in the amount of ultraviolet light that reaches the earth's surface.

The introduction of nitrogen oxides into the lower atmosphere may also have serious consequences. NO_2 can combine either with water vapor, forming nitric acid (HNO_3), or with various cations, forming ammonium nitrate and so on. Although this process is part of the natural nitrogen cycle, human activities can amplify its effects. Perhaps a more serious consequence results when oxides of nitrogen combine with hydrocarbons, both of which are produced during combustion, forming *photochemical smog* when sunlight is plentiful and atmospheric conditions are right.

Ammonia is another important form of nitrogen that cycles from the surface into the atmosphere. It is converted to ammonium (NH_4^+) and hydroxyl ions (OH^-) in reaction with water vapor. NH_4^+ leaves the atmosphere in rain as ammonium sulfate, or it may be oxidized to N_2.

Fertilizers

Biologically available nitrogen is frequently a limiting factor in intensive agriculture. Consequently, modern farming has become so dependent upon heavy nitrogen fertilization that the amount of fixed nitrogen compounds produced industrially is equal to 80% of fixation accomplished biologically. In the past, natural deposits of sodium nitrate ($NaNO_3$) and potassium nitrate (KNO_3) were heavily exploited for fertilizer and other industrial uses. Today the most important source of fixed nitrogen for fertilizer production is the Claude-Haber process for ammonia production. Pure hydrogen and nitrogen gases are reacted at high temperature and pressure in the presence of iron (catalyst) to form ammonia. The principal impetus for perfecting the synthetic ammonia process was caused by Germany's isolation from natural sources of nitrates during World War I. For use as fertilizer, ammonia may be injected directly into the soil or applied in irrigation water, or it may be used to synthesize various solid fertilizers such as ammonium nitrate or ammonium sulfate.

PHOSPHOROUS CYCLE

Critical biomolecules such as nucleic acids and ATP, as well as the skeletal structures of vertebrates, contain phosphorous, making it essential to life. Phosphorous is made available by a cycle (Figure 24-12) that is immediately seen to differ from the carbon and nitrogen cycles by its lack of an atmospheric reservoir. Aside from the phosphorous circulating in the biosphere, the principal reservoir for phosphorous is in phosphate rocks, principally apatite, $Ca_3(PO_4)_2$. Rela-

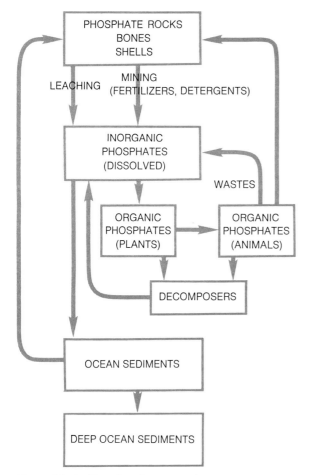

Figure 24-12 **The phosphorous cycle.**

factor in photosynthesis, excessive growth of plants ensues. This process, called **eutrophication,** can totally disrupt aquatic ecosystems with serious consequences (see page 621).

On a worldwide basis, large scale mining and conversion of phosphate reserves to fertilizer may lead to serious shortages of phosphate for agriculture. A large fraction of soluble phosphates used in agriculture is naturally converted to insoluble phosphates after application to the soil. These may, over very long periods of time, be converted to soluble form, for example, by nitric acid from the nitrogen cycle. However, the net effect is that phosphate is mined and used more rapidly than the limited supply is renewed. We thus face eventual phosphate shortages, which may force development of more energy expensive phosphate sources in order to support agriculture.

There is a total of approximately four times as much phosphorous in the terrestrial as in the marine biosphere. On land and in the sea, phosphorous cycles from photosynthesizers through consumers with recycling effected by decomposing organisms. The phosphorous cycle is leaky, or incomplete, in that there is on land some loss of phosphate into insoluble forms and there is a slow loss of phosphate from land into the oceans, from which there is only poor natural return, except over geologically long periods of time. In the bulk of the ocean, below the heavily populated surface layers, the availability of phosphate is regulated by the solubility of an extremely large amount of phosphate salts, principally calcium-phosphate. Solubility of calcium phosphate rises with increased acidity of the ocean, which could occur because of an increase of carbon dioxide concentrations. This might occur, for example, from a falling off of photosynthesis due to phosphate shortage. Linking of carbon and phosphate cycles in this way thus is potentially able to stabilize levels of photosynthesis in the seas.

Return of some of the phosphorous loss from the land to the sea is accomplished by fishing and, particularly in earlier years, the mining of guano deposits for fertilizer. Guano is the consolidated excreta and carcasses of marine birds that have accumulated for millenia in huge, minable deposits in roosting areas along the shore. Since these birds feed on fish, guano represents a sizeable fraction of phosphate from the marine biological system. It is, in fact, a rich fertilizer consisting of up to 20% phosphoric acid and 12% fixed nitrogen.

tively small amounts of phosphate weather out of this reservoir. This fact, coupled with the loss to biological systems as insoluble compounds of a substantial part of biologically available phosphate means that phosphate is often a limiting factor in plant life. In areas of intensive agriculture, this frequently occurs. India is a particularly tragic example because there an already bad agricultural situation is worsened by the necessity of using cow dung as fuel, an activity that accelerates the loss of phosphates by conversion to insoluble forms.

Human intervention is a significant factor in the phosphorous cycle. Large quantities of phosphates are mined and used as fertilizers and for other uses, such as in detergents. The result is that some fresh water streams and lakes have a great excess of biologically available phosphate from run-off and sewage. Since, in such bodies of water, phosphate is often a limiting

Energy Flow in the Biosphere

The energy necessary to power the biological parts of the cycles just described, besides several others not discussed, originates in the sun. Once captured by photosynthesis, this energy flows through the biosphere until it is finally consumed. There is, of course, no complete cycling of energy, since at each transfer from one organism to another there are large energy losses and the recipients themselves expend much of what they receive. Ultimately, as has been pointed out, energy in must equal energy out.

Energy flows through several clearly defined trophic levels, which are conceptually similar on land and in the seas (Figure 24-13). Strictly speaking, energy input to the biosphere is accomplished by autotrophs of two types, **chemoautotrophs** and **photoautotrophs**. The former, bacteria that utilize inorganic chemicals such as ammonia or sulfur as an energy source, have a trivial input as compared with photoautotrophs, that is photosynthetic organisms. Photosynthetic organisms constitute the first trophic level, the level of primary producers. On land these are primarily higher plants, whereas in the sea primary producers are algae. The second trophic level is composed of herbivores, which eat primary producers. Beyond this stage trophic levels become somewhat blurred by the complexity of routes of energy flow. Typically, however, herbivores are eaten by carnivores and they supply one or two further levels of carnivores. The analysis of energy flow is made complex because there are omnivores, animals that eat both plant and animals, and a large group of decomposers and scavengers which partially recycle energy.

THE ULTIMATE LIMIT: PRIMARY PRODUCTION

In order to determine the ultimate limits to life in terms of energy supply, it is necessary first to determine the total amount of energy captured and organic matter synthesized by photosynthesizers that can be made available to other organisms. This is called the **net primary production**, and it is obtained by subtracting the metabolic expenditures of the photosyn-

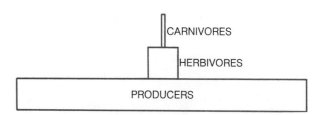

Figure 24-13 A typical trophic pyramid. Width of each level indicates relative numbers of organisms.

thesizers from the total energy that they produce.

How is productivity determined? There are many techniques. For phytoplankton in the open sea, a measured amount of organisms can be provided with radiocarbon labeled carbon dioxide ($^{14}CO_2$), and the amount of $^{14}CO_2$ converted into organic matter in a given time under standard light conditions is then determined. After correcting for respiratory losses, the net productivity is obtained. For terrestrial and shoreline measurements, the techniques are more difficult because the photosynthesizing organisms are not very regularly distributed in space and because conditions usually require use of cruder techniques than the labeled carbon dioxide method. For an annual, agricultural crop a harvesting method is used. When growth is completed, representative areas are completely harvested, roots and all, and their dry weight is determined. This figure, corrected for losses such as those due to premature harvesting by insects, provides a figure for net productivity in terms of weight of plant material. These values can be converted to energy equivalents since it is possible to determine energy content by **calorimetry**, the controlled combustion and measurement of heat produced. To measure the productivity of a tropical forest or a salt marsh obviously requires more guesswork, but the techniques are similar.

Table 24-1 shows values obtained by such techniques for primary production of the major environments on earth. Examination of the table shows immediately that net primary production varies markedly over the earth. What are the causes? Low productivity of categories such as ice, rock, and desert is easily explained by the environmental extremes of tempera-

Table 24-1 Net primary production of the biosphere.

Primary Human Food Producing Areas	Area ($\times 10^6$ km²)*	Net Primary Production ($\times 10^9$ DMT/yr)†	Biomass ($\times 10^9$ DMT)†
Agricultural land	14	9.1	14
Temperate grassland	9	5.4	14
Oceans (open sea, near-shore, estuaries)	361	55.0	3.3
Other ecosystems			
Tropical forests	20	47.0	900.
Other forests and woodlands	37	37.2	822.
Savanna	15	10.5	60.
Lakes, streams	2	1.0	.04
Swamps, marshes	2	4.0	24
Tundra, alpine	8	1.1	5
Deserts, rock, ice	42	1.37	13.5
Total land	149	117	1852
Total marine	361	55	3.3
Total earth	510	172	1855.3

SOURCE: Table 4-2, R. H. Whittaker: *Communities and Ecosystems*, Macmillan, New York, 1970; and Table 15-1, R. H. Whittaker: The biosphere and man. In H. Lieth and R. W. Whittaker: (eds.); *Primary Productivity of the Biosphere*, Springer-Verlag, New York, 1975.

*1 km² = 10 hectares (ha)
†Dry Metric Tons/year.

ture and dryness and by lack of adequate soil. But why is open ocean so much less productive than the continental shelf and near-shore regions of the seas, or why is agricultural land not as productive as swamps, marshes, and tropical forests?

Primary Productivity in the Sea

The low productivity of the open sea may be surprising since you might have been led to believe that the seas are a great and virtually untapped source of food for humanity. Actually, when one considers the entire world ocean, its production is substantial, about 55×10^9 dry ton/year. However, this is quite low on a unit area basis. The cause is based on two factors, the amount of light available to the entire volume of the ocean and limitations in critical nutrients. Light enough to support significant net production penetrates only a little way into the depths. The *compensation point*, the depth at which photosynthetic accumulation of energy and material is exactly offset by consumption (net production = zero) is at best about 120 m. Thus, the vast bulk of the ocean is inhabited by other than primary producers. Moreover, even in the photosynthetic zone, the concentration of living organisms is quite low as compared with many terrestrial ecosystems. Most oceanic photosynthetic organisms are single celled phytoplankton. For these to grow in sufficient density to give noticeable color to the water is a rarity over the whole ocean. The measure of this factor, the amount of living organisms present at a given time, is the **biomass**. For the open sea, biomass is only about 3 mg dry weight/m², whereas for agricultural land the value is about 1 kg dry weight/m². Thus, as far as biomass is concerned the open sea is really a desert.

The major factor limiting primary producer biomass in the open sea photosynthetic zone is nutrient shortage. Phosphorous is usually considered to be the primary limiting factor among the essential elements. Nutrients present in the photosynthetic zone are rapidly incorporated by living phytoplankton. Recycling in the surface layers is poor because there is continual loss when organisms and materials sink into the depths where photosynthesis is not possible. Mixing of deep with surface waters to renew nutrients at the surface is usually prevented by the presence of a **thermocline**, a temperature discontinuity that prevents mixing of warm surface water with cold deeper waters. Consequently, the only places in the open sea where phytoplankton grow in very high concentration are where *upwelling* of water from the depths brings fresh nutrients to the surface.

The principal sites of upwelling are around the Antarctic continent, where there is no thermocline, and along the western sides of continents, where the prevailing winds move surface waters offshore to be replaced by nutrient-rich deeper waters. These regions are the sites of the principal fisheries of the world. Tremendous plankton-feeding, baleen whales are supported in Arctic and Antarctic seas where they feed on zooplankton which, in turn, feed on the rich phytoplankton blooms that occur at the end of the long polar night. Upwelling coastal currents provide the basis of the food chains that support the tuna fishery based on our West Coast. Farther south, perhaps the world's richest fishery is the anchoveta fishery based on the Peruvian upwelling. The very high primary productivity of near-shore regions is due to improved nutrient conditions from upwelling and also from nutrient runoff from the land. Shallow waters also provide attachment for large algae such as the West Coast *Macrocystis*, which is a prodigious contributor to primary productivity.

Primary Productivity on Land

Terrestrial primary production is characterized by greater variety and complexity of organisms and greater variation in environments than marine productivity. Thus, whereas the dominant marine primary producers are free-floating, single celled algae, terrestrial primary producers range from single celled to the largest living organisms, the great trees of the redwood forests. In the ocean the predominant variable factors affecting life in the photosynthetic zone are limitations of certain nutrients, and to a lesser extent temperature. In terrestrial environments we add to these the availability of water and local soil conditions, which include physical form in addition to nutrient content. Finally, primary productivity on land is, under natural conditions, greatly influenced by *ecological succession*, a type of long term interaction among plants and the environment that determines the kinds of plants that are found in a particular environment.

Magnitude of Primary Production on Land. The land outproduces the sea both in terms of net primary productivity per unit area and in terms of total net primary production, even though there is much more ocean area than land (Table 24-1). However, the relatively high productivity of the land undergoes wide fluctuations. Deserts produce less on the average than the open ocean, whereas several terrestrial environments, such as rain forests and marshes, equal or outproduce the most productive marine environments, namely, near-shore and estuaries. Agriculture by no means necessarily increases the primary productivity of land over its productivity in a state of nature; rather it serves to focus productivity on generation of foodstuffs in a manner and with timing of production suitable for efficient harvesting. However, the productivity of some crops, such as rice or sugar cane, does rival the highest natural productivities.

Geography of Terrestrial Primary Productivity. The principal ecosystems of land and sea are classified into **biomes,** groups of ecosystems with similar characteristics. Ecosystems in a particular biome have similar types of producers and consumers although the actual species present are usually different. The biome classification system is particularly useful in analysis of terrestrial primary productivity because the characteristics and pattern of geographical distribution of terrestrial biomes very clearly demonstrate the roles of water and temperature in regulation of terrestrial primary productivity. As shown in Figure 24-14, the seven terrestrial biomes are generally distributed in parallel bands encircling the globe. For example, the principal desert biome circles the world at the same latitudes in both the Old and New Worlds. It includes the Sahara, Arabian, Gobi Desert complex, extending across Africa and Asia, and the North American deserts of the western United States and Mexico.

Adaptations of Primary Producers to Limiting Conditions

Although they contribute little to world productivity, the plants of the desert and semidesert biomes well illustrate the remarkable abilities of vegetation to adapt to limiting conditions. Consider, for example, water limitation. Plants typically require large amounts of water. It is necessary for successful germination and early growth of seedlings. Moreover, in photosynthesis there typically occur large transpirational water losses through the stomata, which must remain widely open so the plant may garner sufficient carbon dioxide from the air.

Many desert plants have been able to accomodate these biological functions to a dry environment by keying their periods of greatest growth and of seed germination to the rare but intense rains characteristic of deserts. Some plants produce seeds containing water-soluble germination inhibitors. Since these can

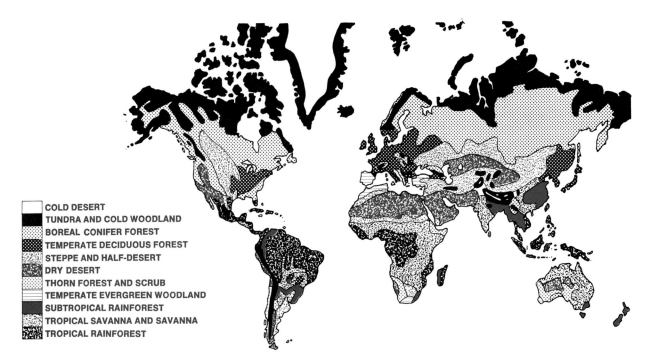

Figure 24-14 **The geography of terrestrial productivity.** After Clapham, W. B., 1973, *Natural Ecosystems.* Macmillan, New York.

only be washed off by heavy rain, it is insured that development occurs under the best possible conditions. Desert plants may also give off growth inhibitors that render the soil around them inhospitable to other plants, thus decreasing the competition for water.

Plants from hot and dry climates often have biochemical mechanisms to assure the lowest possible water loss relative to photosynthetic fixation of carbon dioxide. In our earlier discussion of photosynthesis (page 87) it was stated that the typical acceptor of carbon dioxide is ribulose-1,5-diphosphate. In hot and dry climate plants, the acceptor is often phosphoenolpyruvate. The advantage of this switch is that the phosphoenolpyruvate system is more effective than the standard system in extracting carbon dioxide from the air. The net effect is to improve the conservation of water while maintaining a very high level of carbon dioxide fixation. Succulent plants carry this one step further by taking up large quantities of carbon dioxide at night, when lower air temperatures reduce water loss through the stomata. The carbon dioxide is stored in large water vacuoles and it is removed during the day for participation in the light phase of photosynthesis.

Tropical Rain Forests

Lying at the opposite end of the primary production spectrum from deserts, tropical rain forests undoubtedly represent the greatest profusion of plant and animal life on land. They lie in three great regions: the Central American and Amazon forests, Africa centering on the Congo River drainage, and a vast arc from southern India through the Malay Peninsula to Indonesia and New Guinea. The high productivity of these regions is simply explained by calling them natural greenhouses. They are hot and wet the year around. Temperatures average 25°C through the year and annual rainfall is in the neighborhood of 200 in./year.

The highest known species diversities in both plants and animals occur in this biome. Although far northern environments might have only three to five dominant plant species and temperate zone forests might have perhaps 20 species of trees per hectare (0.01 km^2), in tropical rain forests there may be as many as 70 species of trees per hectare. This profusion of plant life insures that virtually no light reaches the forest floor without traversing a photosynthetic leaf, a condition that maintains throughout the year since most rain forest trees are continually in leaf.

Such ecosystems are thus understandably highly productive, but it is of great interest to a world ever in search of food that they do *not* convert easily to agriculture. Although the prairie biome has been essentially totally converted to agriculture with great success, similar attempts at converting tropical rain forests are unsuccessful for two principal reasons. First, as might be expected in such intensively active systems, most available plant nutrients are tied up in cycling between vegetation and rotting plant material (humus), with very little actually in the soil. Consequently, most of the nutrient is removed when land is cleared for agriculture. Secondly, the soil that typically underlies tropical rain forests is **lateritic**. Laterite soils (from the Latin word for brick) are rich in iron and are claylike. When exposed to intense sunlight, as when land clearing is in progress, laterite soils 'are converted to an extremely hard, bricklike consistency and are almost impossible to cultivate.

CONSUMERS

Net primary productivity supports the remainder of the biosphere. A large array of heterotrophic organisms transfer the photosynthetic harvest through a series of trophic levels until all biologically available energy is extracted. Major losses occur at each transfer of energy and material between organisms. The inevitable result is a smaller biomass of primary consumers (herbivores) feeding on primary producers. Biomass shrinkage continues in each successive stage of transfer. These drastic losses insure that such food chains rarely involve more than five trophic levels. Wastes and dead organisms from each tropic level enter a detritus food chain. This includes scavenging animals and finally bacteria and fungi. The detritus food chain completes the breakdown of biomolecules and insures return of maximal quantities of inorganic nutrients to the soil or water for another cycle of primary production.

Transfer Costs Between Trophic Levels

Analysis of losses between trophic levels shows that herbivores harvest only about 10% of net primary production. There are three principal reasons for these losses:

1. Herbivores cannot harvest *all* of the plant material in the ecosystem if subsequent generations are to be produced. Thus losses must occur in the form of plants that go unconsumed. Ultimately such material finds its way into the detritus chain.
2. Not all plant material can be usefully harvested by herbivores. The cellulose containing and often silicaceous cell walls and structural elements of plants are totally indigestible by virtually all animals. Exceptions are those that have cellulose digesting symbionts, such as the rumen bacteria of cattle and the intestinal protozoa of termites. Further losses occur because differences between animal and plant metabolic systems render some plant biomolecules useless to animals.
3. The herbivore must spend large amounts of harvested energy and material in its own growth and maintenance and in the harvesting process itself. These expenditures are always very large, especially in mammalian herbivores that use energy to maintain body temperature.

Thus losses at the first level of tropic exchange are extremely high. Losses continue at high levels in subsequent exchanges, but transfers between animal trophic levels tend to be somewhat more efficient due to the greater biochemical similarity between animals than between plants and animals. This advantage is offset by the greater energy that higher trophic level carnivores tend to expend in obtaining food. The result of these successive losses is a steep pyramid of productivity correlated with an equally steep *biomass pyramid*.

Transfer Strategies

The virtually infinite ways that organisms have evolved for eating primary producers and each other make strictly linear food chains unlikely. The linear route from primary producer to herbivore to two or three levels of carnivores and finally to the detritus chain is more a concept than a reality. Some of the detours that occur in the flow of energy and material are of the greatest practical significance. Moreover, they have inherent interest as examples of evolutionary ingenuity.

Detritus Chain. The detritus chain is a major short-circuiting of the linear herbivore to carnivore food chain. It receives input from all trophic levels from primary producer through the ultimate carnivore. In terrestrial ecosystems, the detritus chain is actually more important than animals for breaking down plant material.

Detritus organisms separate into two classes on the basis of their feeding mechanisms. Animals such as earthworms and soil insects feed on detritus in the usual animal way, ingesting and digesting food in an internal digestive system. In shallow marine environments, bottom dwelling organisms such as the lugworm, *Arenicola*, play a similar role. Frequently such organisms feed indiscriminantly, ingesting the substrate, extracting the nutrient dispersed within it and depositing the rest as feces. The total magnitude of such activity is staggering to comprehend and plays an important role by aerating and promoting circulation of materials through the substrate.

Detritus organisms may serve to return organic matter to higher trophic levels. For example, earthworms scavenge humus, converting it into food for birds and small mammals.

The final step in the detritus chain is accomplished by bacteria, yeasts, and fungi, collectively known as **saprobes.** They are distinguished from other detritus organisms in that they feed by absorption of organic molecules, often predigested by secretion of enzymes into the substrate. The final results of the work of saprobes depends on environmental conditions, particularly on the ratio of oxygen to organic material. If the ratio is high enough to support aerobic respiration, then detritus is broken down to its inorganic constituents, a small fraction of humic organic molecules, carbon dioxide, and water. When there is insufficient oxygen, saprobes rely on anaerobic respiration, and the breakdown of detritus is incomplete. Substances such as organic acids, alcohols, and methane are produced and may have deleterious effects on the ecosystem, often causing further cycles of incomplete oxidation of detritus.

Because of this, excessive enrichment of ecosystems by human activities such as discharge of sewage and other organic wastes into rivers and lakes is a major cause of environmental deterioration. Properly handled, of course, fermentation of detritus can be put to good use. One interesting example is commercially useful recovery of methane or methyl alcohol from sewage and plant wastes. Methane can actually be recovered by drilling shallow wells in old garbage fills and can be used for heating. Methyl alcohol produced by fermentation of wastes has many uses, including automobile fuel. For this purpose it has the advantage of being less polluting than petroleum fuels, since its combustion products are only carbon dioxide and water.

Food Chains and Food Webs. In the detritus chain we have seen two predominant feeding mechanisms, uptake of organic molecules across external cell membranes and more or less random engulfment of food dispersed among particles of substrate. Feeding mechanisms become much more varied at higher trophic levels. They include filter feeding, symbiosis, parasitism, and predation (Figure 24-15).

Marine food chains rely heavily on filter feeders to harvest unicellular algal primary producers. Because these phytoplankton are small, the herbivores that feed upon them are also small, consisting primarily of small zooplankton arthropods called copepods (Figure 24-15). Swimming activity of the appendages of the copepod sweeps water containing phytoplankton into a sieve formed by bristles on the mouthparts. Copepods form the first consumer level of long food chains involving successively larger predators, culminating in such top-level predators as squids, penguins and other sea birds, sharks, tuna, and the toothed whales. More direct incorporation of zooplankton into very large consumers is accomplished by the baleen, or whalebone, whales, which have their own filtering mechanisms for directly harvesting small crustaceans.

The fact that light effective for photosynthesis penetrates only a slight distance into the immense depths of the sea means that deep sea organisms have no direct access to primary producers. Food reaches the great depths in two general ways. It may simply fall out of the photosynthetic zone as living material or detritus or be carried there in the form of organisms that move back and forth between shallow and deep waters. These organisms undergo a daily cycle of vertical migration of up to several hundred meters, rising with onset of darkness and descending at dawn. While in the surface waters they feed, and upon descending to the depths fall prey to deeper living animals or otherwise eventually contribute their harvest to the deep water ecosystem.

Figure 24-15 Feeding mechanisms as seen in aquatic organisms. A. Saprophytic: a water mold grows on a dead fish. **B.** Predatory: an octopus at work. **C.** Detritus feeding: a lugworm, *Arenicola*, feeds by extracting nutriment from randomly engulfed sand and mud. **D.** Filter feeding: a copepod makes a feeding current with its swimming appendages. **E.** Parasitism: two parasitic copepods, *Pennella*, attached to a fish. **F.** Symbiosis: luminous bacteria, lodged harmlessly in an organ beneath the eye, light the way for a flashlight fish, *Photoblepharon*.

Terrestrial food chains are typified by large variation in size of primary consumers reflecting the large variation in size of primary producers. Indeed, terrestrial vegetation is substantial enough to support directly very large grazing animals such as cattle or elephants. By contrast, in the ocean, a food chain terminating in a similar sized animal, except for the baleen whales and plankton-feeding sharks, would undoubtedly involve many energy transfers, from copepods to small fish to one or two further levels of larger fish. This is an important difference in assessing the relative value of the seas and land for supporting the food requirements of humanity. Losses of primary productivity are far less between vegetation, grazing mammals, and man than they are in any marine food chain except perhaps the one terminating in baleen whales. Even if these whales are used directly for human food the overall harvesting efficiency is lower than for harvesting terrestrial grazing animals because of the large energy expenditures currently necessary for hunting down these vanishing animals.

Food Webs: Stability from Complexity. It is obvious that, if a consumer is such a dietary specialist that it can only eat one kind of food organism, it has closely linked its own survival to the survival of the organism that it eats. Thus if koala bears run out of eucalyptus leaves, they starve. As might be expected, such **monophagous** animals are rare. Instead, most animals increase their chances of survival by being able to feed on several varieties of food organisms.

Although there may be preferences for one component of the diet, animals typically are able to switch to other foods to accommodate seasonal or other changes in the abundance of various food sources. The result is an ecosystem characterized by complex food webs providing many alternate routes of energy flow through consumer trophic levels.

It seems clear that, as the complexity of a food web increases, there ought to be a tendency for increasing stability of the ecosystem. This is because in a complex food web the loss of a particular type of organism would be unlikely to produce large fluctuations in numbers of other organisms that use it for food since they would have alternate sources. There must, of course, be an upper limit to the complexity that an ecosystem can tolerate. This seems to be determined by the extent to which the carrying capacity of an ecosystem can be divided among individuals of different species and still support a sufficient number of each species to insure reproductive stability.

Population cycles in the far north illustrate population instability due to simplicity in food chains. In the north, where a predator may have only one predominant species of prey, effects on population stability are easily observed. The classical example is the Arctic Fox, *Alopex lagopus*, whose major winter prey over much of its range is a small rodent, the lemming, *Dicrostonyx*. Lemming populations have a natural cycle with peaks approximately every 4 years. The cause of this is obscure. Probably various factors such

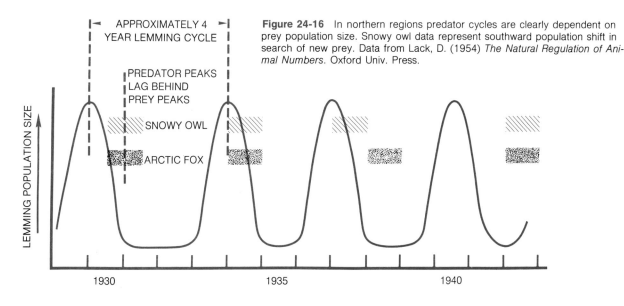

Figure 24-16 In northern regions predator cycles are clearly dependent on prey population size. Snowy owl data represent southward population shift in search of new prey. Data from Lack, D. (1954) *The Natural Regulation of Animal Numbers*. Oxford Univ. Press.

as starvation and disease within the crowded lemming population and, to some extent, predation bring about the rapid decline in numbers of lemmings. Whatever the cause, the decline in numbers of lemmings is soon felt by their natural predators, namely the Arctic Fox, which is seen in large numbers in starving condition, and the Snowy Owl (*Nyctea*), which attempts to avoid starvation by ranging south into the northern United States in search of other prey (Figure 24-16).

The importance of complexity to population stability is seen when we examine predator-prey cycles in more southern regions, where there is an abundance of animals in both categories. Any given predator usually has many other species upon which it can subsist should one preferred prey species diminish in numbers through the effects of predation or other causes. Thus, the foxes of the midwest have available, as food throughout the year, dozens of species of shrews, rodents, and rabbits, not to mention birds and insects. Under these conditions the foxes do not undergo an explosive population increase because of the secondary predators that prey upon them. These were formerly large mammals such as the mountain lion. Now that these have been, for all practical purposes, exterminated, the major secondary predator is man.

Regrettably, we do not often play this role of secondary predator with intelligence. We tend to eradicate completely those animals upon which we prey, and this simplification of the complex web of predator-prey relationships very commonly has detrimental effects upon ecosystem stability. For example, the extermination of all the foxes in a region, possibly for their valuable pelts or because they are taking an occasional chicken, is likely to cost far more than these immediate gains in terms of field- and stored-crop losses to the rodent populations suddenly released from the control of fox predation.

Our Role in the Biological Energy Cycle

This example of our effect on ecosystem stability leads to the question of what overall effect we have on matter and energy cycles. Our influences are many and pervasive. We have already observed how human population growth has led to conversion of virtually the entire worldwide prairie biome to agriculture. We have caused the extinction of any number of species. Our interference has arrested or changed the age-old interplay of environmental effects and interactions among organisms that are essential to ecosystem stability. In addition to these relatively direct and specific interventions, we are conducting a wholesale poisoning of the ecosphere by pollution. The matter of pollution is the subject of Chapter 26. Now let us focus our attention on some examples of more specific effects on biological energy flow caused by our agriculture, hunting, and ecosystem manipulation.

AGRICULTURE

Agriculture involves slightly more than 10% of the earth's land area and represents the primary effort to feed our swelling numbers. At the same time it constitutes a major perturbation of the ecosphere. Since the extent of this perturbation can only increase as efforts to feed ourselves are heightened, it is important to consider the impact of agriculture, first in terms of the extent to which it is capable of supporting humanity, and, second, in terms of its impact on the environment.

Let us begin by estimating where we stand. Can we feed ourselves now? If we use the crude yardstick of caloric intake—which does not consider specific dietary requirements—the U.S. Department of Agriculture estimates that about 56% of the world population now has an insufficient calorie intake. Serious famines occur nearly every year; 12,000 persons are estimated to die of starvation every day.

How Much Food is Currently Being Produced?

Such unpleasant facts suggest that either there may be insufficient food production even now, or that food distribution is faulty, or both. Although it is common knowledge that food is not evenly distributed, perhaps this is not the whole problem. Perhaps even

Table 24-2 Potentially available food and an estimate of the fraction actually used by humans.

1. **What is the maximum potentially available?**
   Total world net primary productivity = 172×10^9 DMT/yr*
   (from Table 24-1)
   Convert to calories by assuming
   3 kcal/gm: $\times (10^6 \text{ g/DMT} \times 3)$
   Potentially available food energy = 5.16×10^{17} kcal/yr

2. **How much of this potentially available amount is actually harvested for human use?**
   If we assume that all food produced is directly used by humans with none left over, then we may estimate the fraction of world net productivity that is converted to human food as follows:
   (a) **Low estimate**
      4019×10^6 people use 2000 kcal/day
      $(4019 \times 10^6)(2000 \times 365)$ = **2.93×10^{15} kcal/yr, or 0.57% of world net production**
   (b) **High estimate**
      4019×10^6 people use 3000 kcal/day
      $(4019 \times 10^6)(3000 \times 365)$ = **4.40×10^{15} kcal/yr, or 0.84% of world net production**

3. **We conclude that the human harvest of world net primary productivity amounts to less than 1 percent of the annual total.**

*Dry weight metric tons per year

today the world cannot produce enough food to feed us. One way to examine this possibility is to make some rough estimates of how much food the earth ought now to be able to produce with current techniques. Then we can determine if, in practice, these levels of production are achieved by comparison of our theoretical data with data on world agricultural productivity such as is gathered by the FAO, the Food and Agricultural Organization of the United Nations.

To begin, let us take the data on net primary production from Table 24-1. There we find that the total net primary productivity for the world is 172×10^9 dry metric tons/year. As shown in Table 24-2, this figure can be converted to its energy equivalent by assuming an average value of 3 kcal/g dry weight of plant material. This works out to 5.16×10^{17} kcal/year. By comparison, our total use of fossil fuel and nonphotosynthetic energy sources, such as water power and nuclear energy, currently is less than 10% of this stupendous amount.

Of course the critical question is how much of this great accumulation of energy can be converted into human food? One very rough but instructive way to get an approximate answer to this question is to assume what seems obvious, namely that, irrespective of food distribution problems, there isn't much more food produced than enough to feed the present world population of a little over 4 billion persons. The average daily caloric intake of humans varies widely from over 3000 kcal/day in some of the industrialized countries to less than 2000 kcal/day in India. If we take 2000 kcal/day as the worldwide level of intake, we find that the 2.93×10^{15} kcal/yr required to feed the human population is only 0.6% of the total net productivity. Even at 3000 kcal/day the total required is only 0.8% of total net productivity. It would seem there is a lot of potential food lying about. Let us see how efficiently we are able to harvest it.

To do this we first proceed from net primary productivity data to estimates of how much food is available from those regions of the earth that are major food sources. The calculations are shown in Table 24-3. We are concerned with three primary regions as indicated in that table. These are agricultural lands, grazing lands (temperate climate grasslands), and the oceans. If we look at the figures for agricultural lands we begin to see why so little of the net primary productivity ends up in human stomachs. About half of it is unfit for human consumption (by present standards!), being roots, stems, and other inedible parts. If this seems a lot, reflect on how little of the net productivity of an orange tree one can easily eat. Of the

Table 24-3 World food production estimated from net primary production.

Agricultural Land
Net primary production = 9.1×10^9 DMT*/yr
Reduce by the following:
  50% unusable for human food
  30% losses (pests, spoilage, seed)

  80% unavailable for human food, leaving 20%, or 1.82×10^9 DMT/yr
Feed 25% of human unusable fraction to domestic animals at
10:1 conversion rate, producing 1.14×10^8 DMT/yr
Total available from agricultural land 1.93×10^9 DMT/yr

Grazing land (temperate grassland)
Net primary production = 5.4×10^9 DMT/yr
Reduce by the following:
  10% land unused, leaving 4.86×10^9 DMT/yr
  25% amount actually grazed, leaving 1.22×10^9 DMT/yr
  10:1 conversion efficiency, leaving 1.22×10^8 DMT/yr
  10% losses, leaving 1.09×10^8 DMT/yr
Total available from grazing land 1.09×10^8 DMT/yr

Oceans
Net primary production = 55.0×10^9 DMT/yr
Reduce by the following
  92% transfer loss from phytoplankton to zooplankton, leaving 4.40×10^9 DMT/yr
  92% transfer loss to 1st. carnivore, leaving 3.52×10^8 DMT/yr
  92% transfer loss to 2d. carnivore, leaving 2.82×10^7 DMT/yr
    Human harvest is at 1st and 2d carnivore level with
    about 1/3 of the species non-harvestable, leaving
    available 2.53×10^8 DMT/yr
    If capture efficiency is 5% the yield is 1.26×10^7 DMT/yr
    Less 10% losses, leaving 1.14×10^7 DMT/yr
Total available from marine productivity 1.14×10^7 DMT/yr
Estimated total world food production based on productivity data 2.05×10^9 DMT/yr

* Dry weight metric tons per year

remaining half, there are tremendous losses to spoilage and to pests, especially in the agricultural systems of the developing countries. In many parts of the world such losses are greater than 30%. Finally, a small fraction must be reserved for next year's planting, thus leaving only about 20% available for food.

Some of this loss can be recovered by feeding material not used directly for human consumption to domestic animals, as indicated in the example. Particularly in the case of ruminants such as cattle, the fermentation of otherwise indigestible plant parts ultimately provides a rich source of calories. In the industrialized nations we have the bad habit of feeding cattle on grains that are actually fit for human consumption. This wastes much food energy because the ratio of conversion of grain to beef is about 10 to 1. However, for the purpose of our example we are assuming that this waste does not occur and that all domestic meat production is based either on agricultural wastes or on grazing on nonagricultural land.

It is difficult to estimate the fraction of temperate grasslands that are in use for grazing. We estimate the use factor at 90% and that 25% of the net productivity is eaten by grazing domestic animals. This must finally be reduced by a factor of 10 due to the trophic level loss mentioned previously.

Initially, marine net primary productivity looks huge, but it dwindles rapidly in passing through the complex food chains that finally convert it into organisms harvested by fishermen. An 8% trophic level transfer efficiency is commonly assumed by fisheries experts. Thus the initial 55.0×10^9 dry metric tons/year of marine primary productivity dwindles to 4.4×10^9 at the herbivorous zooplankton level. The fisheries harvest is taken about equally from the next two higher trophic levels. But of these it is estimated that about one third represents species that are not utilized in commercial fisheries and it is further estimated that the fisheries are so inefficient as to harvest only about 5% of the desirable species. Thus, as the table shows, the oceans are not likely to contribute massively to the nutrition of the world population, unless we develop ways to fish at lower trophic levels. Fish in the diet, of course, is of great importance in contributing needed animal protein and other food constituents.

In summary, the total caloric value for agriculture, domestic animal production, and the fisheries, as estimated for net primary production data amounts to 6.16×10^{15} kcal/year. To see how this figure compares with FAO estimates of actual food production turn to Table 24-4. Here the FAO harvest estimates for 1973 are tabulated alongside our estimate from primary production. As indicated in the table, certain corrections must be applied, namely, for use of high quality crop calories in livestock feeding in the industrial countries, for losses, and for the production of inedible crops—such as rubber and tobacco—which would otherwise inflate the data. The data from these two different estimates are in relatively good agreement. But this is not as important as the fact that these two estimates are rather close to the "high" estimate in Table 24-2.

These three estimates converge on about 5×10^{15} kcal/year as the current productivity of the planet for supporting the dietary requirements of our

Table 24-4 Comparison of World Food Production by Two Methods of Estimation (All Values are in Kilocalories, on Basis of 3 kcal/g, or by Actual Values Calculated for FAO Data)

	Estimate from Primary Production	Estimate from FAO Data for 1973*		
	World	Industrialized Nations†	Developing Nations	World
Agriculture	5.46×10^{15}	2.95×10^{15}‡	2.69×10^{15}	5.64×10^{15}
Meat, milk, eggs from agricultural wastes and grazing land estimates (actual figures for FAO).	7.80×10^{14}	5.44×10^{14}	2.85×10^{14}	8.29×10^{14}
Fisheries	3.42×10^{13}	2.41×10^{13}	3.98×10^{13}	6.39×10^{13}
Uncorrected Totals	6.27×10^{15}	3.51×10^{15}	3.01×10^{15}	6.53×10^{15}
Corrections				
1. For use of crop calories (edible by humans) in meat, milk, egg production: 25% in industrialized and 5% in developing countries. Correction not necessary for primary production estimate.	None	-9.52×10^{14}	-9.98×10^{13}	
2. Losses between harvest and consumption—pests, spoilage, waste. Already calculated for primary production estimate. Assume smaller values for FAO data since these obtained after harvest: 5% for industrialized and 10% for developing nations	Included	-1.75×10^{14}	-3.01×10^{14}	
3. For fraction of primary production estimate due to production of non-edible crops, from FAO data.	-5.75×10^{14}			
Corrected Total Productivity Estimates		2.39×10^{15}	5.69×10^{15}	
World Productivity Estimates Compared	5.69×10^{15}			4.99×10^{15}
Food available per capita		6400 kcal/day (population = 1012×10^6)	2300 kcal/day (population = 3007×10^6)	

* United Nations Food and Agricultural Organization, *The State of Food and Agriculture*, 1974.
† Principally Europe, U.S.S.R., North America.
‡ Includes wheat, barley, rye, oats, corn, rice, millet, sorghum, beans, sugar, potatoes, apples, citrus, bananas, sunflower seed, vegetable oils, soybeans.

4 billion people. We thus have to conclude that the cause of malnourishment today is inequitable distribution of food since the data from these several estimates suggests that there is on the order of 3400 kcal/day available worldwide. These figures cannot provide much hope of improving human nutrition because it does not appear likely that political realities and social attitudes will soon change sufficiently to produce a more equitable distribution of available food. Beyond this, there are probably enough transportation problems to prevent equitable food distribution even if the other impediments were overcome. Hence we must be pragmatic and assume that 5×10^{15} kcal worldwide production is no more and no less than what it takes to insure 2000 kcal/day to the members of hard-pressed populations in the developing nations.

Food in the Critical Years Ahead

We must, however, look to the future since the curve of population increase continues to trend upwards. We ask if the planetary food production system is already working at maximum capacity and, if it is not, how many more people can it feed? The answers suggested to these questions are all over the map. On one hand there are those who believe that we have already gone well past the carrying capacity of the planet and that the optimum population is about a quarter of the present world population. Optimum is, of course, an important word in such calculations, since it is obvious that optimum in terms of the quality of life is undoubtedly well below the maximum sustainable carrying capacity of the planet. If we ignore such considerations and also assume that there will be maximum cooperativity among nations in terms of food production and distribution, it has been estimated that the planet could support about 30 times the present population, or $120,000 \times 10^6$ people. Life under such conditions is difficult to imagine, so let us confine ourselves to a more comprehensible exercise by determining how many more people the planet can support with only relatively modest changes in the means of food production.

Improving World Food Production

As far as agricultural production is concerned, more food can be produced by putting more land into production and by improving crop yield per hectare. As for putting more land into use, it is unlikely that more than 15% additional agricultural land of reasonable quality can be activated. In the United States, for example, virtually all high quality agricultural land that was out of production in the days of grain surpluses is now back in production. Further additions to crop acreage will require the use of marginal land and land requiring irrigation. This raises the spectre of markedly escalating food costs occasioned by rising fertilizer requirements, the price of irrigation water, and possibly, ultimately, the need to use desalinated water.

The Green Revolution Concept

The most recent massive attempt to produce higher yielding crops has been conducted under the name **Green Revolution.** Initiated by the Rockefeller Foundation as a program to improve wheat strains for growth in Mexico, the effort has spread to other crops, particularly rice, an important staple of the developing world. Norman Borlag, who received the Nobel Prize for his work in this area, envisions the increment in productivity generated by Green Revolution crops as buying time to permit development of a lasting solution to the population problem.

Unfortunately, the Green Revolution appears to have bought less time than anticipated. To illustrate, consider what has happened to a certain new rice strain in the Philippines. Green Revolution crops are highly responsive to fertilizer and, when grown under carefully controlled conditions, they are spectacular producers. Traditional agricultural practices in the Philippines may produce only about 1.5 metric tons/hectare of rice. In contrast, one of the high yield strains developed at the Philippine Rice Institute typically yields from 8 to 10 metric tons/hectare under experiment station conditions. Regretably, yields in the hands of the farmer are only a little better than for the old strains.

The reasons include poor insect and weed control, inadequate watering regimes, and insufficient fertilizer. To some extent these problems can be corrected by further instruction of farmers in the techniques required for cultivating the new rice. However, it must be remembered that insect and weed control either require intensive hand labor or expensive chemicals, and that fertilizer is expensive and prices can be expected to rise in parallel with oil prices. The result is that the peasant farmer cannot attain maximum yield from the new rice strain without greatly increasing production costs. The crop must consequently be sold at a higher price, and this will have the effect of impeding the distribution of the increased yield to the very people who need it the most.

It thus appears that there is unlikely to be a "technological fix" that will greatly improve agricultural productivity. There is, however, one tactic that may be used significantly to increase our food resources and that is for us to eat lower in the trophic sequence. We see from Table 24-4 that the consumption of meat is much larger in the industrialized than in the developing nations. The estimate in the table indicates that perfectly edible grains were used to generate these meat calories to the extent of about 9.5×10^{14} kcal/yr. At 3000 kcal/day, that is enough to feed 868×10^6 people, or 22% of the present world population. And remember that this change in habits would not mean forswearing meat. There would still be fish and there would be meat grown on agricultural wastes and by grazing on uncultivated land without grain supplementation.

We might eventually be able to go much farther with this concept and harvest the sea at lower trophic levels, thus more effectively taking advantage of its great primary productivity. We have foolishly disposed of the whales that could do this for us (see page 559). It remains a very perplexing problem to invent machinery that can harvest the immense but thinly spread crop of oceanic plankton, as efficiently as they might still be doing.

FEEDING OURSELVES IN THE NOT TOO DISTANT FUTURE

With all these limitations in mind let us see how world food production might realistically be improved to accomodate the population increases that are likely to occur before, by design or by accident, world population growth levels off or begins to decrease.

Planning for 6 Billion in 2000 A.D.

There seems little doubt that we will add another 2 billion people to the world population within the next 25 years. Most of these new mouths to feed will appear in the developing nations, which are least able to provide food for them. As indicated in Table 24-4 the average caloric intake by people in the developing

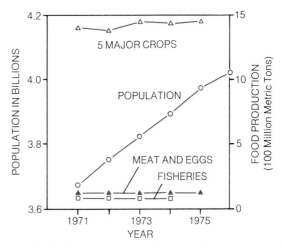

Figure 24-17 World food production is not keeping pace with human population growth. The five major crops plotted are wheat, barley, corn, rice, and potatoes. Data from United Nations Monthly Bulletin of Statistics, 1977.

nations is 2400 kcal/day, which confirms that a large number of people in these regions are now at starvation level, considering the inequalities of distribution. Since about two-thirds of the additional 2 billion people will be born in the developing nations, it is clear that their food problem, acute now, must rapidly worsen. In the industrialized nations there is still a food surplus, as shown by the fact that there are now potentially available 6400 kcal/day for each person. However, as indicated, the grain surpluses of the past are gone so that even the great grain producing nations, Canada and the United States, are at the whim of the weather. Thus, although the industrialized nations have enough food now to feed themselves and to help out the developing nations, this condition will not persist for long, even in the face of the relatively modest population growth that industrialized countries will experience between now and 2000. What can be done?

Based on recent experience, the hope of increasing food production to accomodate population growth may not be realized. As Figure 24-17 shows, there has been no increase in major food production indicators since 1971, and some experts believe there actually has been an overall decline of 2% in world food productivity in the intervening years. During the same period there has been a sixfold increase in fertilizer costs, due in large part to rising oil prices, and this in turn has contributed to at least a tripling of wheat and rice prices during the same period.

Table 24-5 The Implications of Feeding 6 Billion People in A.D. 2000

	Industrial Nations	Developing Nations
Present Status		
Population	1012×10^6	3007×10^6
Total food available	2.39×10^{15} kcal/yr	2.60×10^{15} kcal/yr
Food calories per person per day	6400	2300
Situation in A.D. 2000		
Population Estimate	1680×10^6	4340×10^6
(Assumes two thirds of growth occurs in developing nations)		
Food required to meet present standards	3.92×10^{15} kcal/yr	3.64×10^{15} kcal/yr
How the additional requirement might be met		
Agriculture		
1. Increase yield/average by 15% *and*	$+4.42 \times 10^{14}$ kcal/yr	$+4.03 \times 10^{14}$ kcal/yr
2. Eliminate grain fed livestock		
3. Cut spoilage and pest losses by 50%	$+8.75 \times 10^{13}$ kcal/yr	$+1.50 \times 10^{14}$ kcal/yr
Meat/Milk/Eggs	No change	No change
Fisheries		
Increase yield 5% by harvesting "trash" fish as fish meal protein supplement	$+1.20 \times 10^{12}$	$+1.99 \times 10^{12}$
Increment	$+5.30 \times 10^{14}$	$+5.55 \times 10^{14}$
Plus basis (1973)	2.39×10^{15}	2.60×10^{15}
Estimated total food available in 2000	2.92×10^{15} kcal/yr	3.15×10^{15} kcal/yr
Calories per person per day	4745 kcal	1991 kcal
Calories per person per day worldwide (assuming equitable distribution)	2762 kcal	

The fact that food production has undergone no significant increase recently, together with the rising cost of fertilizer, argues strongly that it would be irresponsible to suggest that the world population can be fed by achieving spectacular advances in food productivity soon enough to feed the additional 2 billion expected by the year 2000. However, relatively minor modifications of present production techniques might enable us just to squeeze through to that time. Table 24-5 shows how it might be possible to maintain the developing world at present caloric levels (which, we must remember, are not good). This would necessitate a probably acceptable reduction of caloric input in the industrialized nations that would make up the deficit in self-produced food supplies in developing nations by increased levels of exports to them. The concept presented in Table 24-5 might be attainable, although it would be expensive and require some changes in food habits in the industrialized nations. Perhaps the most difficult goal to attain is the proposed 15% increase in agricultural output since, as has been emphasized, most good crop land is already in use. Consequently, achieving the additional increment will cost more in terms of fertilizer, machinery, and labor costs than an equivalent amount of production on currently used land.

Since the developing nations contain most of the tropical forests of the world, perhaps they may increase production by growing selected crops on partially cleared forest. Total clearing presents many problems that would probably make food production on such land uneconomic.

The industrialized nations can achieve a very sizeable increment in available food by stopping the practice of feeding meat-producing animals on grain (Note 24-1). Nutritionally, this would be an improvement since it would result in leaner meat. Techniques are currently available to attain a very substantial reduction of production loss from spoilage and pests, making a 50% reduction of such losses probably readily attainable in both the industrialized and developing worlds. Somewhat more speculative is the proposal to harvest fish less selectively and to prepare fish meal from sterilized whole fish. This practice is in use, but scruples against processing of entire fish generally prevent its use in the western world. General acceptance

Note 24-1. *The Grain Drain*

As Figure 24-18 shows, total grain use per capita rises with income, although the amount used directly for human food actually falls. The difference in the two curves is due to the amount of grain fed to animals for production of meat, milk, butter, and eggs. In the industrialized world a very large proportion of this grain fed to cattle for meat production must be deemed a waste. Until now this waste has been economically acceptable to the meat producer because it has been offset by the ease and rapidity with which grain fed cattle can be finished off for market and because of the consumer preference for tender meat, which requires grain finished, feed-lot animals rather than animals grazed on open range.

This waste of grain seems even more unfortunate when one considers how ingeniously the cow's stomach is constructed for a low-grade diet. The stomachs of ruminants such as the cow, sheep, and goat contain a large compartment in which bacterial and protozoan digestion of food takes place. Carbohydrates, including cellulose, which animals cannot digest without bacterial or protozoan assistance, are rapidly fermented, contributing to microbial growth, which the host's intestine eventually harvests, and supplying some molecules such as propionic and butyric acids directly to the metabolism of the host animal. Since cellulose comprises up to 60% of the range grass diet, the advantage of this symbiotic mechanism is obvious.

Microbial metabolism actually increases the amount of protein available to the host when nonprotein enriched diets are fed. This happens because the microbes of the cow's stomach produce their own protein from nonprotein nitrogen in the diet. Significant increases in protein available to the cow can be achieved by enriching the diet with urea or even with uric acid, which the farmer has in plentiful supply in chicken feces. The cow also contributes to protein synthesis by the stomach microbes by returning some of its own metabolically produced urea to the stomach for further incorporation into microbial protein. When the diet is very rich in protein, as in grain finishing, a larger fraction is lost as ammonia before it can be incorporated into microbial protein. Therefore, there is little net gain in protein available to the host and the use of grain for feeding meat-producing animals would seem doubly wasteful. In the first instance it could be more efficiently consumed directly by humans, and in the second instance its protein content is inefficiently sacrificed in the interest of its carbohydrate content.

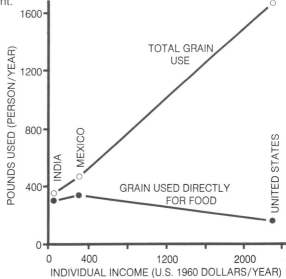

Figure 24-18 Grain consumption rises with income due to its use as animal food in affluent societies. Data from U.S. Department of Agriculture.

of this technique would readily allow at least a 5% increase in the marine catch.

Institution of this relatively modest array of measures to increase food supplies would see us into the year 2000 with a worldwide caloric ration of about 2758 kcal/day per capita. This is enough to live on quite adequately if actually eaten by humans, but it must be remembered that today inequitable distribution of food results in significant hunger in countries with a theoretical daily ration of better than 6000 kcal/person. It is thus obvious that there will be extremely serious starvation at the 2758 kcal level unless there is a marked improvement in food distribution. Under any circumstances this would be difficult to achieve, but at the food prices that most likely will prevail in the year 2000, achievement of equitable food distribution on a worldwide basis will probably be more difficult than attaining the modest increases of production suggested in Table 24-5.

Finally, Table 24-5 indicates that the developing nations in the year 2000 will require substantial food imports from the industrialized world, just as they do now. To bring the developing nations up to a daily ration of 2400 kcal/person would require an annual importation of 6.35×10^{14} kcal/year. This would reduce the daily ration in the industrialized countries to about 3680 kcal/day, but otherwise would not be difficult to accomplish. In fact, this caloric deficit is the equivalent of 186 million metric tons of grain, which is not much more than the 128 million metric tons that the grain producing countries exported in 1973. The rub comes when it is realized that much of the 1973 grain export was to countries like the U.S.S.R., which could readily pay for their imports. Clearly, in the year 2000 the world will face difficult economic, humanitarian, and political problems as it works at feeding its population. The tragedy, of course, is that it is an effort that must ultimately fail unless population growth is soon sharply limited.

AGRICULTURE AND THE ECOSYSTEM

Agriculture is by no means wholly beneficial to the ecosystem because it introduces instability. Since we devote so much effort to converting as much land as possible to agriculture, it is important to assess its effects. In general, agriculture can be said to intrude on ecosystem stability in two ways—by reducing species diversity and by interfering with nutrient cycling.

Reduction of Species Diversity

Modern farming techniques (Figure 24-19) promote cultivating and harvesting efficiency by making the land uniform and by using uniform genetic strains of crop plants selected for high yield. Huge fields consisting of a single high productivity crop are more economical than many small fields with diverse crops. Fence rows, the borders between cultivated areas, are in much of the farmlands of the world the last refuge of many types of organisms and these areas are, of course, reduced as farms grow larger in the interests of efficiency. As might be expected, such large scale simplification of terrain patterns and eradication of elements of food webs produce instability. This is experienced eventually in the agricultural system itself. Thus, reduction of habitat variety greatly disturbs the food web involving insects, small rodents, and their predators such as various birds and mammals—foxes and coyotes. Elimination of cover for insect eating birds, hawks, and mammalian predators means that the farmer is in continual danger of a crop eating insect or rodent outbreak since these organisms are essentially freed from population limitation by predation.

The genetic uniformity of crops also brings danger of instability. With very large areas of agricultural lands planted with virtually identical strains of crops, the possibility of widespread losses due to plant disease is greatly increased. Thus, in 1970, about one fifth of the United States corn crop was lost to the southern corn blight; and, earlier, in 1954, 75% of the Durum (a genetic variety) wheat crop was destroyed by wheat stem rust. Although crops are today bred for resistance to existing diseases, new disease varieties emerge frequently and could conceivably wipe out a significant part of a nation's harvest if one appeared during the beginning of a growing period.

These two consequences of the instability of crop ecosystems have required compensatory action by farmers. Rodenticides, insecticides, and herbicides are in wholesale use, artificially preserving ecosystem stability to insure good crop yields. Such intrusions are, of course, themselves dangerous because they may have unexpected untoward effects. Thus application of a long lasting insecticide may simply insure that

Figure 24-19 A young corn crop. Very large plantings of uniform crop strains are necessary for profits but dangerously reduce habitat variability which necessitates continuous tinkering with the environment to maintain high yields, as in the photograph which shows insecticide spraying in progress. (U.S.D.A. photograph.)

insect eating birds are poisoned by eating poisoned insects, perhaps completely offsetting the beneficial effects of the insecticide by killing insect predators.

Agricultural Interference with Nutrient Cycling

Agricultural use of nitrogen, phosphorous, and potassium containing fertilizers in many agricultural areas introduces extremely heavy loads of these elements into nutrient cycles. Fertilizers applied to soil experience immediate losses into streams and ground water. Fertilizers that contribute to increased plant productivity also return more indirectly to streams and ground water by way of the wastes of animals to which crops are fed for meat production and by way of human wastes.

Modern meat production methods particularly aggravate this situation. Cattle no longer are allowed to range widely under conditions that would allow a more gradual and natural return of their wastes to the ecosystem. Instead the economies of the industry require that cattle being fattened for the market be crowded together by the thousands in feedlots. In states where the cattle industry thrives, the total amount of sewage released by feedlots exceeds the amount of human wastes. This constitutes a great burden on local water resources for it must be remembered that feedlots are not provided with sewage systems. Wastes run off into the nearest streams or soak into the ground water. The result is pronounced eutrophication of lakes and streams and possibly directly toxic effects on humans through their water supplies. In addition, the contribution of such wastes to atmospheric nitrogen oxide enrichment has already been mentioned.

Can Agricultural Practices Be Improved?

Is it possible to continue high agricultural output and, at the same time, reduce agricultural damage to the ecosystem? In heavily urbanized societies, agricultural productivity must be kept very high because so few people actually are engaged in farming. The labor saving advantages of superfarms must be kept while we seek measures to reduce their effects on the ecosystem. Returning feedlot wastes to agricultural land is one obvious way to diminish requirements for fertilizer input and at the same time reduce runoff into streams. Planting without complete tilling is a new technique that promises to reduce labor costs and reduce wastage of soil and nutrients due to erosion. Sparing use of poisons to keep down crop pests can reduce the losses of natural predators.

Perhaps the greatest opportunity for improvement of agricultural practices can be found in undeveloped countries. Since these are labor intensive areas with small farms, it is possible to achieve high productivity in reasonable harmony with natural processes in the ecosystem. Rice culture in Indonesia is a particularly telling example (Figure 24-20). Every inch of favorable land and every drop of available water are utilized. Need for fertilizer is reduced to a minimum by alternating dry crops such as sweet potato with rice and by turning ducks into the rice fields to harvest insects which they return to the rice as fertilizing excrement. Small amounts of supplemental protein are even obtained by catching eels in the rice paddies at night. This type of agriculture supports a very large human population with minimal environmental degradation. The cost, of course, is an extremely large human effort on the part of nearly all members of the agricultural economy.

HUNTING

The list of species that we have selectively eliminated from the ecosphere by hunting is very long. Large and aggressive animals were early victims along with many commercially valuable species. In the United States the bison were reduced to a few remnants by persistent hunting and the passenger pigeon was rendered totally extinct, even though its migrating flocks once literally darkened the skies. Many of these decimations have direct effects on human well-

Figure 24-20 Artist's impression of Balinese rice culture shows how generations of diligent terracing allows use of marginal terrain in high productivity agriculture.

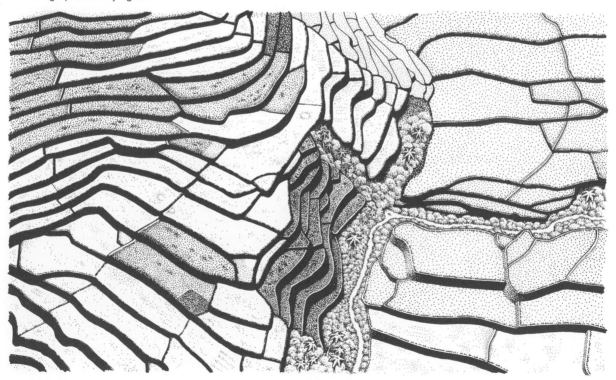

being and are tragic as well. The fate of the great whales is a sad example. A regulated harvest of whales calculated to allow maintenance of their populations would have been one of the most efficient ways to harvest the ocean's great crop of plankton.

Whales held their own during much of the history of whaling. The whaler of the 1800s was a sailing ship, and whales were attacked with hand held lances from small boats at close quarters. Once a whale or two were caught the entire ship's crew had to spend several days of full time work cutting up the carcasses and boiling out the oil. As a result the best whaling ships managed to catch only about one whale every 2 weeks on their long voyages.

With the advent of petroleum products whaling went out of style for a time and then came back into fashion with a vengeance as whales again became valuable, this time as protein sources. The techniques are now completely modernized. The work of catching and working up the carcasses is divided between fast catcher boats and factory ships. Catcher boats are equipped with harpoon guns firing explosive harpoons. In a few years the most favorable species of whales have been reduced to levels making their hunting unprofitable, forcing a switch to smaller or otherwise less desirable species. This chain of events is clearly seen in world catch records (Table 24-6). Despite the activity of the International Whaling Commission, population levels of whales have continued to drop until many of them are virtually extinct. The largest whale of all, the blue whale, is believed down to about 30 individuals worldwide. This is probably not enough for survival considering their low normal reproductive rate.

The moral of the demise of the great whales is clear. Their fate was obvious long before it was inevitable. A moratorium of only 5 years in whale captures during the 1960s would have allowed whale populations to grow and stabilize at levels that would have allowed sizeable harvests into the indefinite future. Yet commercial interests in the several involved countries, particularly Russia and Japan, were unwilling to undergo the immediate financial losses a moratorium would have entailed so they continued whaling. Immediate profits were seized in the face of an unrecognized or ignored ecological imperative; now the whaling industry and humanity have to pay the price.

ECOSYSTEMS MANIPULATION

The migrations and works of man are very likely to introduce new varieties of life into new environments. This can produce profound perturbations in the native populations of such environments for a variety of reasons. The introduced organisms may multiply excessively and at the expense of native organisms, perhaps because of release from pressure from the predators characteristic of their home environments or because the organisms they attack in the new environment are more susceptible to them than their natural prey.

This phenomenon is put to good use in **biological control of pests,** in which possible predator species for certain pests are sought out on a worldwide basis for introduction into the pest's environment; however, the overall result of these transfers is bad. For example, we owe to shipping the worldwide dissemination of cockroaches from their African center of origin. The introduction of the rabbit into Australia is an example known to all. Appropriately, the depredations of the rabbit in Australia were controlled by the deliberate introduction of another organism, a rabbit disease, **myxomatosis,** an initially successful example of biological control.

Canal building is an excellent example of human endeavor that promotes organism transfer on a large scale. The Suez Canal has permitted exchange of marine organisms between the Mediterranean and Red seas without, in this instance, any reported deleterious effects. However, the opening of the St. Law-

Table 24-6 **Antarctic Whale Catch Reflects Switch to Less Desirable Species**

Year	Whales Caught			
	Blue	Fin	Sei	All Species[1]
1940	11,500	—[2]	—	
1950	7,500	20,000	—	
1955	2,000	28,500	500	
1960	1,500	28,000	3,500	
1965	1,000	13,000	19,500	
1970	0	2,700	5,500	14,500
1972	7	1,800	3,800	15,500
1975	0	200	180	11,000

[1] Includes those tabulated plus minke, sperm, and others.
[2] No data.
Data from International Whaling Statistics, 1976.

Figure 24-21 A Great Lakes lamprey and the damage it does demonstrates that caution is necessary with ecosystems manipulation. (Michigan Department of Natural Resources.)

rence Seaway permitted an influx of marine lampreys that caused great destruction of Great Lakes fishes (Figure 24-21). With such facts in mind, many biologists today are concerned over the possible effects of the recurrently proposed sea-level canal across Central America. The Panama Canal has not allowed any great amount of interocean transfer because of its system of locks and the presence of fresh water over much of its length. The situation might prove to be much different with a sea-level canal, because it would have no locks, and tide-facilitated passage of organisms between oceans would be feasible.

Even when we act in the greatest good faith, lack of knowledge of ecological principles may result in our efforts doing more harm than good. A currently debated example is fire prevention in the National Forests of the western United States. When one sees a forest fire in progress or walks through the destruction that such a fire leaves behind, the first thought that comes to mind is that a forest fire is a good thing to prevent. For years this has been philosophy of the Forest Service. Actually, it is not at all certain that it is a wise measure to prevent all forest fires. Fires are a natural process in the forests. Some plants are so adjusted to fire that fire is required if they are to complete their life cycles; others recover very quickly after fires, putting out new vegetation from protected underground roots. It must also be remembered that if fires are relatively frequent, the amount of dead wood on the ground to add to the conflagration is small and the fire sweeps by without becoming very intense. Indeed, it is thought that one of the reasons the giant redwoods survived over several thousands of years is because their thick bark and high placed green vegetation allowed them to withstand moderate intensity fires. The well intentioned policy of fighting every fire in such forests has led to the buildup of more and more down wood so that now, when a fire occurs and gets out of control, it is so intense that these giants are in serious danger and many are killed.

25 Populations

Throughout this book, and particularly in our study of the biosphere, the thought has emerged that the central biological problem of civilization is to limit human population growth so that we do not finally exceed the reasonable carrying capacity of the planet. This subject has many aspects that go beyond biology and into areas of religion, politics, economics, and human customs. Some authorities believe that continually growing populations are essential to a healthy economy. Some racial minorities see the regulation of population growth as attempts at extermination of their races. Some find population limitation a violation of religious beliefs, or of customs long held. Some see a growing population as a source of political and military power. These ideas are not for us to debate here. Our task in this chapter is to provide the scientific background for evaluating such ideas, to discuss how populations grow, to demonstrate general principles, to illustrate the factors increasing or decreasing population growth, and to estimate the consequences of unregulated growth of any population—microbe, mouse, or human.

PLAYING GAMES WITH COMPOUND INTEREST

If you put $100 on deposit at 5% interest compounded annually for 5 years you expect to receive at the end of that time $134.01 (Figure 25-1). To find out how much money you will have at the end of any annual interest period, you use a simple equation:

$$\text{money at end of next interest period} = \text{interest rate} \times \text{money at end of current interest period}$$

Or, the equation can be condensed by letting letters stand for the phrases, as follows: N = money, t = one interest period (1 year in this instance), R = interest rate. Then the interest equation becomes:

$$N_{t+1} = RN_t$$

So, to figure how much you will have at the end of 5 years you apply this equation repetitively five times and the final RN_t that you get is how much the account is worth (Figure 25-1). It is important to note that each successive RN_t is larger by the initial $100 plus all the interest accumulated in the previous interest periods. Once a dollar is earned in interest it also earns interest in subsequent periods. Thus the account is said to earn *compound interest*.

Figure 25-1 shows how such an account starting with $100 would change depending on the size of R. If R is 1, this is the equivalent of what the banker would call 0.0% interest, and the account would not change in value. If R is 1.05, this is the equivalent to banker's interest of 0.05%. As R increases, the growth

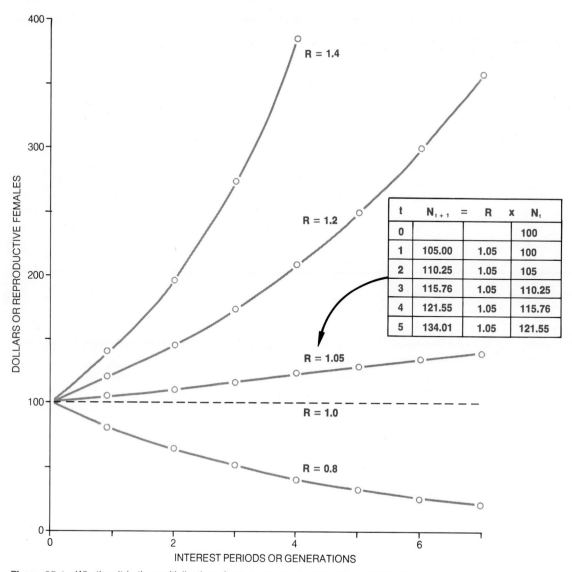

Figure 25-1 Whether it is the multiplication of money or people, the same mathematics applies.

curve for the account curves upwards more steeply. If R is smaller than 1 the growth curve slopes downwards until the account is gone. These results are inevitable as long as the conditions of the equation do not change. If R is 1, the account never changes in value. If R is smaller than 1, the account eventually disappears. If R is larger than 1, the account eventually becomes infinitely large.

These same terms may be used to describe the growth of biological populations in an approximate way. One simply redefines each dollar as a progeny bearing individual. Thus the $100 might become a population of 100 female animals, plus enough males to go around, which have a cycle from reproductive individual to reproductive individual of 1 year and which each have a 5% chance of reproducing in a given year. The curve for the growth of this population would look just like the curve in the bank account graph.

You immediately see that this description of the growth of biological populations cannot be the whole story. There are many populations in existence in which R is greater than 1. Since the equation indicates no upper limit, such populations ought to grow indefinitely. The result would be preposterous. This can be shown by calculations to determine just how large a particular population might grow, according to the compound interest law, if growth could continue indefinitely at existing rates. Figure 25-2 illustrates such a calculation based on the current human population of the world. There were estimated to be 4019×10^6 humans on earth in mid-1976 with an annual rate of increase of 1.8%. Applying an equation similar to the simple interest equation, one can show that the human population would be expected to undergo some startling increments over long periods of time.

To make these more graphic in Figure 25-2 we assume that each human gets 1 m² of space to stand on and that, as the population grows, all of the meter squares required are packed together on the surface of a sphere. The radius of this sphere would therefore be increasing in proportion to the growth of the population. Given these conditions, the compound interest law says that in 658 years the entire surface of the earth (land and water) would be covered with people standing on 1-m squares, some $51,137 \times 10^{10}$ of them. Then, to accommodate continued growth, the human sphere would have to expand off the surface of the planet and move into space. Within 2000 years, the human sphere would have expanded by more than the distance from the earth to the sun. Before 4000 years had passed, the sphere would be expanding outwards at a speed in excess of the speed of light. This absurdity illustrates what we have just said, that when R is larger than 1 the population tends to grow to infinite size. Obviously, there is more to population growth than the compound interest equation has to tell us.

REALITY: MEASURING THE GROWTH OF REAL POPULATIONS

If the compound interest equation does not describe real populations, perhaps we should study data

Figure 25-2 A silly demonstration of the power of unchecked compound interest as applied to human population growth.

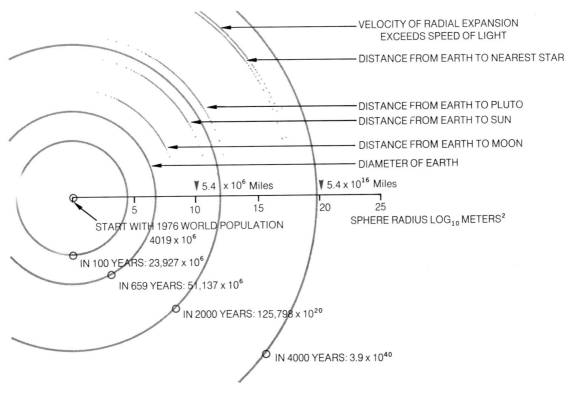

from life and see how populations actually do increase. Then an equation might be found to fit the data and be useful for predicting what will happen in the future.

The Logistic Equation

The American biologist, Raymond Pearl, in 1927 reported experiments on growth of organisms like yeast and fruit flies in the laboratory under conditions of constant food supply. This kind of experiment provides a manageable model of the real world, where the necessities of life are *not* infinite. Therefore this model might provide information as to what really happens in nature. His analysis of an experiment on yeast is shown in Figure 25-3. Instead of growing infinitely according to the compound interest curve, the yeast grew along an S-shaped curve and finally leveled off at a maximum population size. This curve was already known as the **logistic curve,** and Pearl and others found that it applied rather well to many kinds of population growth. The reason for this is that the formula for the logistic growth curve includes a factor

Figure 25-3 Application of the logistic growth equation to the growth of a yeast culture. After R. Pearl, 1927, Quarterly Rev. Biol. **2**:532 and T. Carlson, 1913, Biochem. Zeitschrift **57**:313.

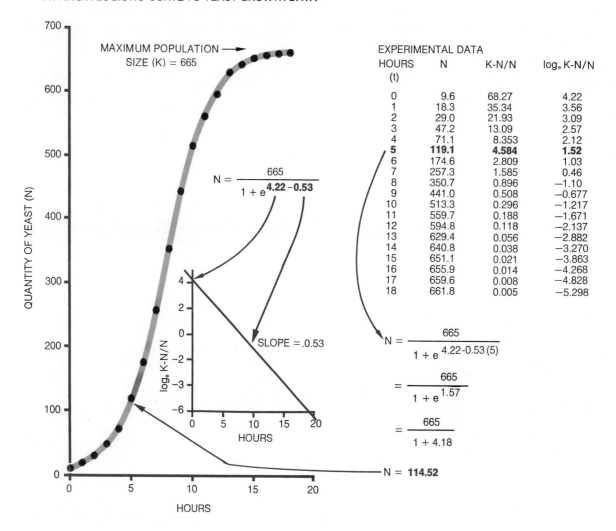

that takes environmental limitations into account. The formula can be approximated by simply adding an environmental limitation term, K, to the compound interest equation, thus

$$N_{t+1} = RN_t K$$

The actual formula and the way it is applied is set forth in Figure 25-3, but the approximation just stated illustrates the important advantage of the logistic equation over the compound interest equation.

What is the meaning of the K term? It represents the remaining capacity of the environment to support further growth of the population. At the beginning of growth K will be large and push the curve up rapidly. As the population utilizes the supportive capacity of the environment, K gets smaller and smaller until it approaches zero. This makes the top part of the curve flatten out instead of continuing upwards, as in the compound interest equation.

This seems fine. The logistic curve at least describes the growth of laboratory populations. Perhaps it can be used to predict the growth of a population in nature before it reaches its maximum. The value of being able to do this would be great. For example, it would allow us to decide if there is really a human population problem by telling us how large the world human population will be when it finally undergoes a flattening out of its growth curve. Unfortunately there are two difficulties standing in the way of doing this. First, if you can't determine K by actual measurement in experiments, as in Figure 25-3, then you have to estimate it from the available part of the growth curve; and, second, not all populations behave like yeast cells as they approach the K limit, or the carrying capacity of the environment.

A Poor Estimate of United States Population Growth

Pearl and his associates ran into the first problem when they attempted to fit a logistic curve to the United States population. In 1920 and again in 1940 they developed a logistic curve to fit the census data, which is available from 1790 in 10-year intervals. As Figure 25-4 shows, they had a rather good fit up through the 1940 census. The logistic curve that they calculated in 1920 predicted a United States population of 136.32×10^6 in 1940. This is only 4% off the actual value of 131.41×10^6. In 1940 they recomputed their logistic equation and used it to formulate a prediction of the United States population to the year 2100. This indicated that the population would then stabilize at 183.59×10^6. Pearl died shortly after his 1940 prediction so he never knew how far off the estimate was, as shown in Figure 25-4.

From the data at hand we cannot determine why the actual census data are so widely divergent from the logistic curve calculated by Pearl. It must be obvious, as Pearl himself noted, that such calculations can only be accurate if there are no changes in the basic conditions affecting population growth during the entire period covered by the calculations. This is easily enough assured in laboratory populations growing under controlled conditions over periods of a few hours to weeks. It is a different matter with a population of humans spreading out across a continent, assimilating immigrants and undergoing innumerable changes in the conditions of life over several centuries. So it is possible that the United States population, responding to changing conditions, is growing along a new logistic and will eventually stabilize at a new and higher K value.

A Boom and Bust Reindeer Herd

We must also consider the second possibility, namely that some kinds of populations are unable to "anticipate" their K value, grow right past them, and then suffer a radical die-off as the essentials of life suddenly become limiting. There are examples of populations behaving in this way. One of the most accurately observed was a population of reindeer established on one of the Pribilof Islands to provide meat for the natives. The subsequent fate of the herd has been studied by Victor Scheffer and is described in the graph in Figure 25-5. Initiated with 4 males and 21 females, the herd multiplied rather gradually until about 1930, when the population began to increase very rapidly until peaking in 1938. At that point a precipitous decline ensued, leaving only 8 individuals alive in 1950. What happened? Scheffer found that the reindeer had almost completely eradicated the lichens (reindeer moss) which are essential winter forage. This overconsumption, coupled with several winters that were more severe than usual, appears to have caused the disaster.

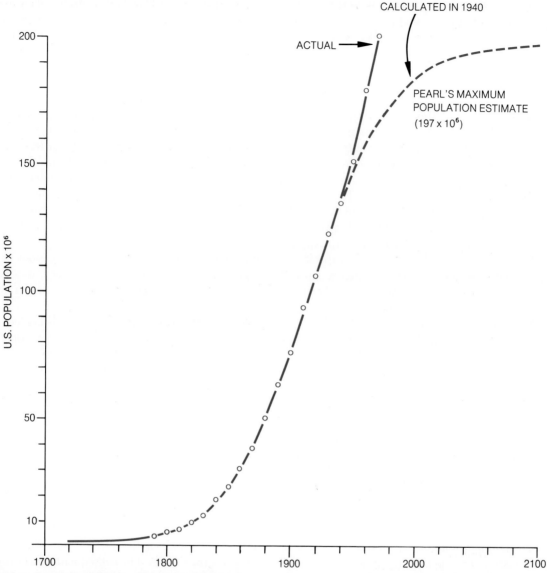

Figure 25-4 A world population estimate by Raymond Pearl, pioneer student of population growth. The estimate is grossly in error. Pearl's data are from *Science* **92**:486 (1940).

If food limitation was the cause of the reindeer die-off, why didn't its approach result in a diminished rate of population increase during the several years preceding the crash? There are several reasons why this may not have happened. One is simply the fact that the increase in numbers of individuals is so large between generations when a population is on the steep part of a logistics curve that it is literally possible for there to be enough food for one generation and not enough for the next. Growth of the reindeer herd came close to that. It is likely that the carrying capacity as far as grazing was concerned was passed only about 3 years before the peak.

The second factor that could have prevented a slowing of the growth of the population, even after the grazing capacity had been exceeded, is the fact that reindeer are reproductive for several years. Females alive in the several years before the crash would have continued reproducing until starvation became severe. The resulting *momentum* of population growth would have carried the herd towards a more severe disaster by continuing to produce new mouths to feed.

The final factor that contributed to the disaster was that grazing land takes a number of years to restore after being completely grazed. Thus, even though the population had commenced to decline, the survivors would have not been significantly helped because the population was still, for a year or two after the peak, above the original optimum carrying capacity and, of course, far above the actual carrying capacity after overgrazing. Hence the fall to nearly zero is not surprising. We learn from this example that a population may not necessarily stabilize near the carrying capac-

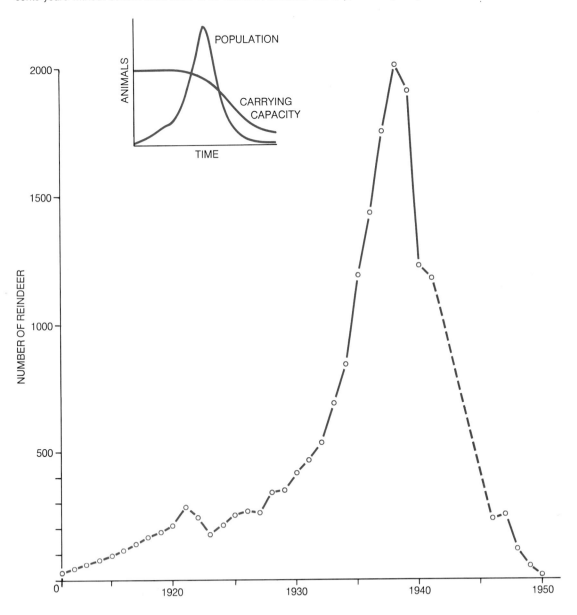

Figure 25-5 Populational history of a reindeer herd transplanted to one of the Pribilof Islands. Dotted line represents years without census data. After V. B. Scheffer, *Scientific Monthly,* **44**:356 (1951).

ity of the environment if its members have long reproductive lives relative to the time course of depletion of essential resources and if the resources themselves are not rapidly restored as soon as population levels are diminished. In Figure 25-5 the inset diagram of population versus carrying capacity suggests what may have happened.

FACTORS INFLUENCING POPULATION GROWTH

Since it is clear that there is no obvious way to predict the future of a "real-life" population simply from its previous history of growth, an estimate of the future growth of a population must come from a painstaking analysis of all factors that contribute to the rate of population growth. In general they may be divided into two classes, which we will call R ("interest rate") factors and K (carrying capacity) factors, reflecting the terms of the logistic equation of growth. K factors are those that determine the environmental carrying capacity for the population—factors like food, shelter, and weather. R factors cannot be completely separated from K factors, since ultimately K factors have their effect on population growth through the R factors, which act directly on birth and death rates. However, R factors include the age structure of the population, the reproductive history of members of the population, and those factors that act directly on birth and death rates such as diseases (Figure 25-6). As the Figure indicates, population size obviously reduces to a matter of the relative magnitudes of birth and death rates. A population grows when it has an excess of births over deaths, is stable when births and deaths are equal, and gets smaller when deaths exceed births. *A population can go from stability to growth either by increasing the birth rate or by decreasing the death rate.*

Human Populations

In regard to human populations, three general patterns of birth and death rates are described. During most of the long history of humanity, growth rates were very low due to a condition of high birth rates and high death rates (Figure 25-7). As populations enter the phase of high social and economic development, death rates fall and are followed by a lagging drop in birth rates. Population growth is high during this period because the fall in death rates is not yet matched by falling birth rates. Once the phase of social and economic development is matured, both death rates and birth rates are low and population growth stabilizes. This shift between two low growth states through a high growth state is thought by many investigators to occur whenever human populations move from an undeveloped to developed state and is called the **demographic transition** (Figure 25-8).

Obviously it is most important to evaluate the possibility that the demographic transition does or will occur for all human populations and whether it

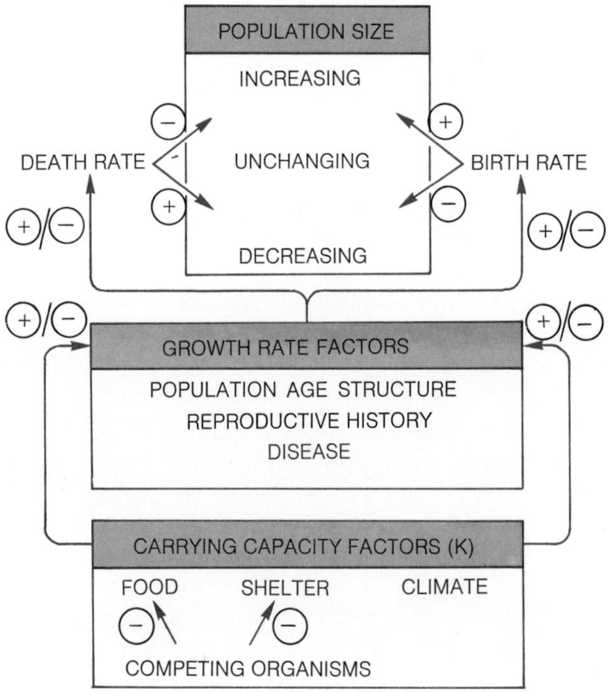

Figure 25-6 The determinants of population size.

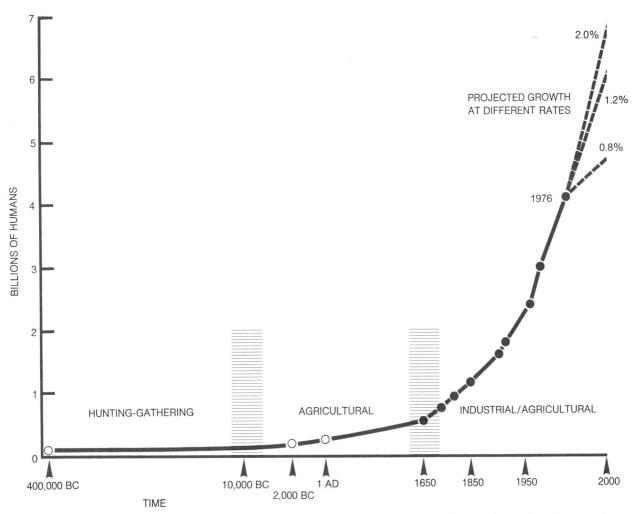

Figure 25-7 Human populations through time with estimates for the year 2000 at three plausible growth rates from the present. Reliable data exist only for the time from A.D. 1650.

will be complete before our species will have exceeded the environmental carrying capacity. It is also of great importance to attempt to identify the factors controlling the reduction in birth rates, since this decline can be seen to be essential to attaining stability after death rates have been reduced in developed populations. This, of course, is the key to population stability in developed societies, otherwise death rates will ultimately have to increase again as carrying capacity is exceeded (Figure 25-8B and C).

The History of Factors Affecting Human Population

Viewing present population growth in the light of population trends in the past emphasizes the unusual and dangerous state of human population growth at the present and illustrates some of the important factors that are currently driving the human population to such high levels. A first impression from a graph of human populations over time (Figure 25-7) is that our numbers have been small until quite recently. Nathan Keyfitz calculates that the total number of people who ever lived is about $70,000 \times 10^6$. It is, of course, impossible to know the time of initiation of this great cohort, but a reasonable estimate is about 600,000 years ago.

Given this, we arrive at the remarkable fact that the average human population size through all of human history has been less than 2 million people. Obviously something unusual has been going on since the 1600s

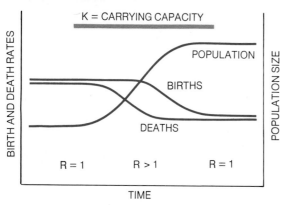
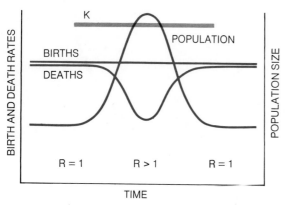
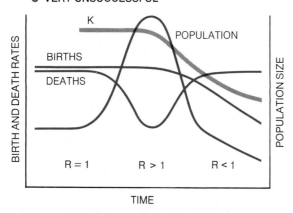

Figure 25-8 Possible results of population growth. A. Successful demographic transition with population growth triggered by declining death rate stabilized below environmental carrying capacity (K) by declining birth rate. **B.** Unsuccessful demographic transition in which inbalance of birth and death rates causes K to be exceeded with the population regressing to the pretransition level. **C.** A very unsuccessful transition episode in which the effect of the population increase is to permanently reduce K, forcing the population to lower than previous levels. R is the "interest rate" factor.

to carry the human population to its present size of 4 billion, which, after all, is 6% of all people who have ever lived, or some 2000 times the average population size through the whole of human history! Identification of the changes associated with this great leap in numbers must necessarily tell us much about what controls human population levels.

Over the greatest part of human history we have no direct knowledge of birth and death rates and must judge population sizes on the scantiest of data. Evidence of age at death, from dental and bone structure, shows that life was short; 30 years was old for those early hunter-gatherers. It seems a certainty that high death rates kept populations small and this, indeed, is the strongest evidence for high birth rates, although it is likely that these never exceed about 50 per 1000 population per year (50.2 was reported for Costa Rica in 1963).

Indeed, in a modern hunter-gatherer society, the !Kung of the Kalahari Desert of Africa, the average birth interval is 4 years. This is thought to be due largely to inhibition of ovulation caused by the 3 to 4 years of breast feeding provided each child (see page 465). Since the prolonged period of breast feeding is necessitated by the absence of other appropriate food for the infant, it is a circumstance that may have also limited birth rates during early human history.

Thus, low population levels persisted for so long because death rates were pushing at the upper ceiling of birth rates. Causes of early death in hunter-gatherer society are easily conjectured: accident, disease, famine due to dependence on erratic food supplies, and inability to build up stores against hard times.

Population Growth in Western Europe

Populations enjoyed their first modest rise with the establishment of farming and village life about 6000 years ago. We presume this must have been due almost wholly to an increased and more predictable food supply and secondarily to the reductions in accidental death and disease associated with the rigors of a migratory life. These modest gains in population continued until the beginnings of the industrial revolution in the early 1700s. The latter part of this period, from about 1100 A.D. to 1700 A.D., is very instructive because quite a bit is known about birth and death rates in this age, and it is one in which the human population was not yet large enough to be impervious to large scale fluctuations from the kinds of limiting factors that must have been of overwhelming significance in earlier times.

Some data on birth rates and longevity are available from marriage, baptismal and burial records between 1560 and 1646 for English villages. These show high birth rates of six or more children per marriage and it seems probable that these rates were typical of earlier times. The high birth rate was substantially offset by a short "normal" life expectancy of about 40 years. In bad times mortalities were far higher.

In the mid 1300s Europe suffered under the Black Death. By then town life, which had obvious advantages, had become thoroughly established in Europe. It facilitated such societal gains as division of labor and the advancement of mercantilism, but the rise of town life occurred long before society was ready to cope with its effects on health. Crowded together in towns without the most rudimentary sanitary facilities, people lacked the protection of low population densities. Farm and small village life at least slowed the spread of communicable diseases.

As might be expected, diseases such as smallpox and influenza became extremely serious; but the greatest killer was the Black Death, or bubonic plague. Particularly in its pneumonic form, in which it spreads from person to person without need for rodent and flea, this disease did frightful damage. The two **pandemics** of the 1300s spread death throughout Europe. The first probably killed 25 million people, about a quarter of the population of Europe, with the majority of deaths occurring in cities. The population of England and Wales actually fell from 4 million in 1350 down to 2 million in 1400 and did not recover to 4 million until 1560. This was the story throughout Europe. The plague returned periodically during the next 300 years, but without the tremendous mortality of the 1384 visitation. It was, however, still a significant force, particularly in the cities. Thus London, in the bad year 1665 lost 68,600 people to plague, or 14.9% of its population. The great fire of the next year may have spared the city a worse visitation.

There is ample evidence of the limiting effects of inadequate food as well as uncontrolled disease on populations in the preindustrial era. Individual communities had to fend for themselves during years of bad harvests, and during such times the new classes of artisans, living in the cities and working at tasks far removed from farming, fared badly. In those days, when times were hard, if the breadwinner of a family died, it was not unusual for the remaining dependents to die in quick succession.

Clearly, until the beginning of the industrial revolution, European society was living about as any primitive society might, maintaining precarious population levels, with birth rates barely staying ahead of high death rates. The major population limiting factors were starvation and disease, serving to hold European society in the first stage of the demographic transition (Figure 25-8).

Mortality rates began to decline with the approach of the industrial revolution, and the great, brief surges of mortality associated with mass starvation and pandemics became more rare. Primary causes lay in agricultural improvements and in development of more efficient means of communication and transport. These made possible wider distribution of harvests, smoothing out the effects of poor crops locally. Agricultural improvements included broadening of the crop base by introduction of new crops such as the potato and changes in farming practice, which increased yield per acre and made less necessary the old practice of allowing land to lie fallow to recover fertility.

Curiously, well into the modern era there was one last, major famine in Western Europe that could be directly associated with one of these improvements. This was the Irish Potato Famine of 1846–1848. It is, perhaps, instructive to the modern agriculturist relying heavily on huge acreages of single crops because the famine was solely due to the virtually total reliance of the Irish on the potato. With nearly 80% of their caloric intake derived from the potato and with small reserves, the Irish immediately began to starve when the potato crop was decimated by the potato blight disease. Two million people migrated out of Ireland or died of starvation, and growth of the Irish population has never been restored to rates maintained before the Famine.

Further reductions of mortality came from general economic improvements associated with the industrial revolution. These brought about some reduction in the extreme poverty of the earlier times. The common citizen could, therefore, afford a warm, dry home in winter. Cleanliness improved, water supplies were cleaned up, and a more nutritious diet became relatively commonplace. The result was a general increase in life expectancy, except in the poor areas of the large cities, which were, sadly, the price paid for the general improvements brought about by industrialization and mercantilism. By 1910 life expectancy in Western Europe had risen to 53 years for males and 56 years for females, an increase of about 13 years for both sexes since 1840.

The final great reductions in mortality awaited developments in medicine and public health. Early infant mortality remained high until scientific ways of fighting infectious diseases were developed. Such landmarks as the advent of vaccination for smallpox (page 503), the recognition of bacterial causes of many diseases, and the discovery of the value of environmental sanitation in blunting the effects of epidemics brought death rates to the lowest levels yet achieved. This trend has continued until now life expectancy has risen to as much as 75 years (Sweden), which means that death from invasive disease is virtually eliminated. Further reductions in death rates in the developed countries await medical solutions to the problems of cardiovascular disease, cancer, and perhaps aging itself.

The inevitable result of falling death rates in the early phases of industrialization was rapid population growth because birth rates were still quite high. The population of Western Europe was entering the demographic transition (Figure 25-8).

Demographic Transition

A simplistic but useful way to consider the effects of the transition from agricultural to industrial societies is shown in the feedback diagrams in Figure 25-9. For both types of societies it is instructive to examine the effects of population size on what we may call individual (or family) and community resources. *Individual resources* are the real wealth available to the individual or family unit, particularly food, clothing, shelter, and access to life supporting services such as medicine. *Community resources* in this context are resources that improve the condition of life of the individual or famiy unit. They include public food reserves and public health and medical facilities.

In the preindustrial society, community resources were essentially nonexistent. The peasant's "excess" production went to the estate owner who provided virtually nothing in return except the right to tenancy of a plot of land. Family resources were improved not at all by additions to the family beyond those necessary to till the restricted alloted acreage. Every additional mouth to feed deprived other family members.

The result was population control by negative feedback. For example, a bad harvest would delay marriages and immediately increase deaths, thus reducing population size on two counts.

Population effects upon both family and community resources switched from negative to positive feedback in industrial society, at least initially. The basic causes were increased agricultural efficiency and the generation of nonagricultural employment. Relatively early in life every addition to the family could go to work, increasing family resources. Equally importantly, the associated improvements in standard in living reduced mortality, allowing the high potential fertility to be expressed more and more in gainfully

Figure 25-9 Factors regulating human populations in pre-industrial and industrial society. Red arrows indicate negative feedback; white arrows indicate positive feedback.

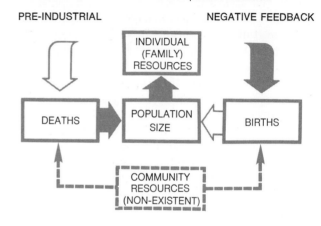

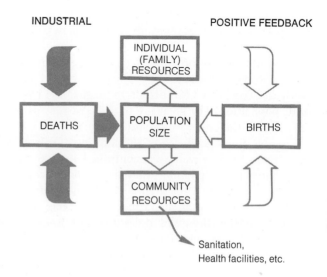

employed adults who both increased the public wealth and added babies to the next generation. Similarly, the diversion of some public wealth to community reserves both assisted in reducing the death rates and lessened the burden of excess births on the poor. Consequently, the demographic transition is characterized by a substantial period of maintained high birth rates in spite of lowered death rates.

How is the Demographic Transition Completed? In general the demographic transition is completed by a decline in birth rates occasioned by the ability to maintain a high standard of living without numerous working children and, indeed, by the realization that, under existing economic conditions, large numbers of children actually reduce the family's standard of living. The result is a breaking of the positive feedback loop between individual resources and births, while even reducing the death rate somewhat. That reduction in the birth rate is associated with an increased standard of living is attested by the fact that historically the birth rate declines first among the most wealthy. As income levels are raised at lower social levels, and as children become more and more an impediment to comfortable living, due for example to restrictions on child labor, birth rates commence to fall at these levels as well.

Various devices were used to attain lowered birth rates. The age of marriage frequently increased and techniques of contraception came into more widespread use. However, in the industrialized countries the fall in birth rates experienced since the 1870s owed little until the 1930s to modern contraceptive techniques. Coitus interruptus and illegal abortions were, until then, the primary means of birth control. Today in the developed countries the availability of sophisticated mechanical and chemical methods of avoiding conception, coupled with prevailing severe economic pressures, have combined to produce low birth rates. Indeed, birth rates are now below replacement level in the Democratic Republic of Germany and in Portugal.

POPULATION GROWTH IN DEVELOPING NATIONS

This sketch of the history of the demographic transition is based on Western Europe since it has the longest continuous population history among those nations that have undergone this process. What about the rest of the world? Canada and the United States have similar histories, although they are not as clear cut examples because migration has played an important role in establishing their population trends. As for the rest of the world—Africa, Asia, Latin America—which we cannot forget includes 75% of the human population, growth rates are, with notable exceptions, generally in a phase of reduced mortality without significant reductions in births. In a sense these countries are undergoing what appears to be a great prolongation of the transition, caused by providing them with technology that reduces the death rate prior to development of standards of living that are sufficiently high to favor lowered birth rates. The governments of virtually all countries in this category have recognized what this arrested demographic transition is costing in terms of providing for swelling populations. They are attempting to correct matters with intensive national family planning (Figure 25-10). These efforts are experiencing various degrees of success as the following examples demonstrate.

China and India

These countries rank first and second in world population. The People's Republic of China with 836.8 million and India with 620.7 million people in mid-1976 have more people than all of the developed nations put together. The extent to which these two nations are able to cope with population growth will to a very large degree determine whether the world will or will not become overpopulated.

China is making a most heartening effort towards population control. In the 1950s the government first endorsed birth control with the goals of improving the health of mothers and children and to give mothers more time to work outside the home. Although there has for many years been a law fixing the minimum marriage at 18 for women and 21 for men, Chairman Mao's request that women defer marriage to age 23 and men to age 26 seems to have been generally respected. Throughout China contraceptive advice and materials are readily available. Once the birth rate target has been set nationally each year, local communes call fertile couples together to decide jointly how many births locally will meet the target and then allocate the quota according to a system of priorities designed to favor families with two children roughly 5 years apart.

Figure 25-10 Family planning. Examples of advertising used to promote family planning in the West Indies and Pakistan. The Pakistani poster reads, "Few children, prosperous family."

reproductive age. If any Asian country can approach population stability through birth rate reduction, it is China.

India represents a more typical Asian situation. The central government does not have the tremendous power of the Chinese government, and there is more widespread poverty. India's population is still growing at 2% per year and it will double in 35 years unless death rates increase or birth rates decrease. Since 1952, national policy has favored slowing population growth, and there have been some results from family planning efforts. The birth rate was down to 35 per 1000 in 1976, but this is offset somewhat by continued slight declines in the death rate.

India is currently mounting a major national program aimed at reducing the birth rate to 1.4% by 1984. If this goal is to be met, it is believed that 33% of reproductively capable couples would have to be protected against conception by 1979 and 45% by 1984. The tremendous effort that this entails is shown by the fact that by 1979 it is proposed to accomplish 18 million sterilizations, supply IUDs to 5.9 million women, and provide conventional contraceptives to 8.8 million persons. Although 2.6 million vasectomies were accomplished in mass vasectomy camps in 1972–1973, achievements are currently projected at about half the target number. Even so, the Department of Family Planning has been forced to make severe budget cuts in recent years due to the severe inflation India is experiencing. Aggressive pursuit of sterilization goals by some regional governments resulted in sporadic sterilization riots and in 1977 became an issue in the national elections. Curiously, sterilization has been a main effort of the program. This is a time consuming procedure for use on the large numbers of women who must be included if the program is to be effective.

Mexico

Populations throughout Latin America are growing at rates of nearly 3% in all countries of any size, except for Argentina, Chile, and Uruguay. Mexico is among the fastest growing with a rate of increase of 3.2% in mid-1977, down from a recent maximum of 3.6%. The significance of such a rate of increase is made apparent in Figure 25-11, which shows projected population growth in the United States, Canada, and Mexico at current rates for the next 70 years. At a sustained rate of 3.2% Mexico will equal the current United States

Abortion is accepted, as is tubal ligation by women who have had two children. In fact, the widely used vacuum-aspiration method of early abortion is a Chinese invention. The result is that birth rates are reported to be as low as 10 per 1000 in major population centers and the national rate of growth is *believed* to be down to about 1.7%. China's progress towards a stable population must, of course, not be taken as an example of what may be easily accomplished in other Asian countries. China is well developed industrially and agriculturally, has a high standard of literacy. It has an all-powerful central government that is widely and enthusiastically supported by young people of

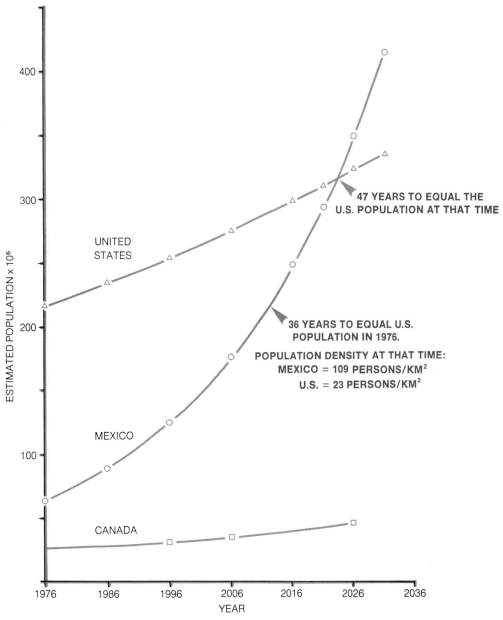

Figure 25-11 A population projection from 1976 for the United States, Canada, and Mexico.

population (at 4.7 times the United States population density) in 36 years. In about 50 years Mexico would exceed the projected United States population for then.

Until recently Mexico has had no national policy of growth limitation. In fact, as recently as 1970 the President favored population growth, although by the time President Echeverria left office he had initiated a birth control policy. In 1974 a new General Law of Population was passed. This law established a National Population Council and favors population stabilization in order to achieve the best possible utilization of human and natural resources in the country. As a result, in 1974 the Ministry of Health began offering family planning at 298 clinics and about 2000 more will be established in rural areas. By the end of 1974 about one million women, about 8% of the reproductive age group, had adopted a family planning method.

In mid-1977 the number of women using some form of birth control was unofficially estimated at 2 million. Using the slogan, "The small family lives better," a sophisticated advertising campaign has been mounted and free contraceptives and birth control pills are issued at government medical clinics. Unfortunately the program has yet to make much impact in poor rural areas. The Roman Catholic Church, typically a potent force in such matters, has not made significant objections. Nonetheless, the situation in Mexico remains extremely serious because effective population limitation efforts have only just commenced and inertia will carry the population to much higher levels, even if highly effective family planning efforts were to be established immediately.

THE OUTLOOK FOR WORLD POPULATION STABILIZATION

A nation entering an era of low mortality with a growing population may not necessarily accomplish the parallel reduction in birth rates that is necessary to achieve population stability, as Figure 25-8 shows. If birth rates are not sufficiently reduced, the environmental carrying capacity (K) will sooner or later be exceeded and death rates must rise again. Due to the lags often experienced in population reduction (recall the case of the reindeer), carrying capacity may be so diminished by a population overshoot that the population may finally recede to a far lower level than before. When translated into human terms, on a national or continental level, this is a horrifying prospect.

Figure 25-12 Doubling times for the populations of selected countries.

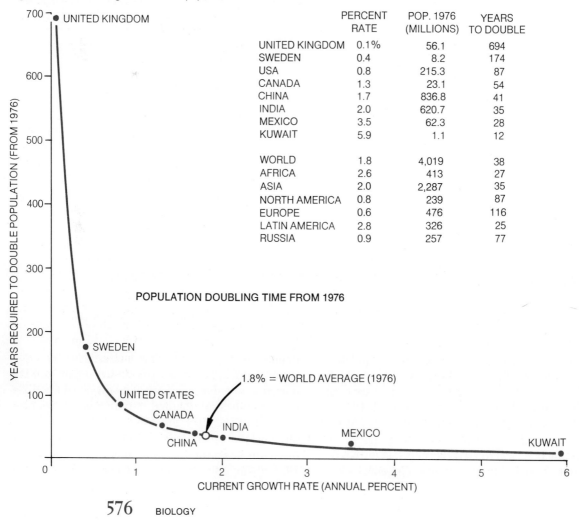

	PERCENT RATE	POP. 1976 (MILLIONS)	YEARS TO DOUBLE
UNITED KINGDOM	0.1%	56.1	694
SWEDEN	0.4	8.2	174
USA	0.8	215.3	87
CANADA	1.3	23.1	54
CHINA	1.7	836.8	41
INDIA	2.0	620.7	35
MEXICO	3.5	62.3	28
KUWAIT	5.9	1.1	12
WORLD	1.8	4,019	38
AFRICA	2.6	413	27
ASIA	2.0	2,287	35
NORTH AMERICA	0.8	239	87
EUROPE	0.6	476	116
LATIN AMERICA	2.8	326	25
RUSSIA	0.9	257	77

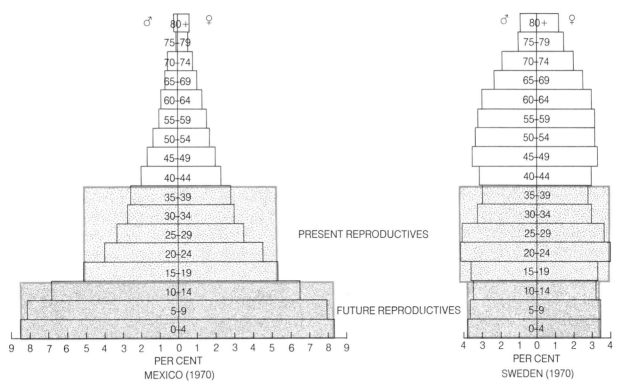

Figure 25-13 The age structure of a population profoundly affects the results of birth limitation programs. As shown for Mexico, the large base of the age pyramid (future reproductives) will continue to force the birth rate up even if substantial reductions in births of currently reproductive individuals are attained. Data from United Nations Demographic Yearbook, 1970.

Imagine what might happen if India, or Mexico, failed to attain population stabilization before exceeding the carrying capacity. The tragedies within a country under such conditions would be terrible enough, and there is every possibility that they would spread to neighboring countries as the desperate populace and its government sought food by any means available. It is thus absolutely essential for world governments to evaluate the barriers to a successful worldwide demographic transition and to take steps to insure that failure does not occur. Unfortunately such estimates are terribly difficult to make, and there appears to be little clear thinking so far about the steps that should be taken.

Population Age Structure and Growth Momentum

We have already seen that many nations are experiencing great difficulty in attaining even modest reductions in birth rates. As Figure 25-12 clearly shows only an insignificant fraction of the world population is down to replacement-level birth rates. Almost all populations in the world will at least double in 50 years.

In order to obtain an optimum estimate of how much family planning efforts might reduce population growth, let us consider what would happen if family planning is so successful that births are reduced to replacement levels. *When* zero population growth is achieved would depend on the *age structure* of the populations involved. To see why this is so examine Figure 25-13, which shows the age structure of the populations of two countries, one that has gone through the demographic transition and one that has not. Replacement-level birth rates in the population that has undergone the transition would produce very little if any change in population size in the future because females in the reproductive years are being replaced by approximately the same number of females maturing from the younger age groups. On the other hand, in the growing population, replacement-level birth rates would definitely not limit population growth for a long time because each age group is followed by a still larger one. Thus as today's reproductive females grow old and become nonrereproductive, they are replaced by still more newly reproductive females from younger age groups. Therefore, replacement-level birth rates would be increased by the larger numbers of marriages that would occur as these younger groups become of reproductive age. Consequently the population of such a country, even if it instantly attains replacement-level births, must increase in response to the maturing wave of 0 to 15 year-olds.

Tomas Frejka has developed a series of population projections that include age-group distributions. They show that, if replacement birth rates are attained in 1980, world populations will still increase 1.7 times over the 1970 population by 2050, when stability would be attained. The developing world, which needs it the least, will experience most of the growth, owing to its populational age structure. If, as seems a more attainable goal, replacement birth rates are not attained until the year 2000, stability could not occur before about 2100 with the developing countries undergoing an astounding 5.5-fold increase as compared with the developed world's 1.8-fold increase. Thus, the tragedy of age structure effects on population momentum is that, no matter how well-intentioned and diligent the developing nations now are about reaching the goal of population limitation, they cannot avoid bearing the brunt of large population increments still to come.

The Dangers of Short Term Population Forecasting

If we look at the great sweep upwards of the human population curve over all time (Figure 25-7) common sense, if nothing else, indicates that the general tendency of that curve is to continue upwards until something dramatic happens. However, the precise position of that curve next year, or in a few years, is open to debate. In general, our personal time scale is so minute, as compared with the entire sweep of the human population curve, that we are likely to assign far too much importance to what are actually minor squiggles in its progress. The rate of rise may increase a little this year or fall a little next year, but such changes cannot be very reliable predictors of the future general progress of the curve. A principal reason for this short term unreliability is that the actual changes recorded over the short term are so small that they must be exceedingly accurately documented if they are to be assigned much significance. Unfortunately, census taking in many parts of the world is so inaccurate that short term reliability is highly doubtful. This is particularly true because the very places in the world where things are probably happening to affect the population curve most are just the places where data taking is the poorest. For example, there probably has not been any good data collected within the last 5 years for much of Africa and for mainland China. Indeed, for the rest of the world the United Nations considers a country to have "complete" birth registration if it is believed that it registers 90% or more of births. Thus it is probably not prudent to pronounce the population bomb defused on the grounds of a few years of fractional declines in the growth rate.

PLANNING FOR THE UNTHINKABLE: UNCHECKED HUMAN POPULATION GROWTH

In order to see the importance of limiting population growth at a level within the carrying capacity of the environment, it is useful to reflect on what may happen in the world if population growth continues unabated. It is, first of all, most unlikely that our just-starting attempts to preserve natural resources and environmental quality will succeed, so that we may look forward to a crowded, unhealthful, and generally uncomfortable planet at best. The conditions of life for our own descendants, on these grounds alone, will be worse than ours, and this is even more true for the rest of the world populace. However, we shall reserve discussion of these growth-dependent environmental problems for the next chapter and concentrate here on a rather clear problem that we have already explored in the last chapter, the problem of food.

We tend to be eternally optimistic and face the future with faith in *technological fixes* of present and future problems. The argument for such optimism is historical. Since so many of our major problems in the past have had technological fixes, why not one or two more? The refutation of the argument comes from two considerations: First, the problem we face is qualitatively totally different from any we have ever experienced. Just look at the upward sweep of the population curve in Figure 25-14. Second, the problem we face finds us sooner or later confronted with clearly

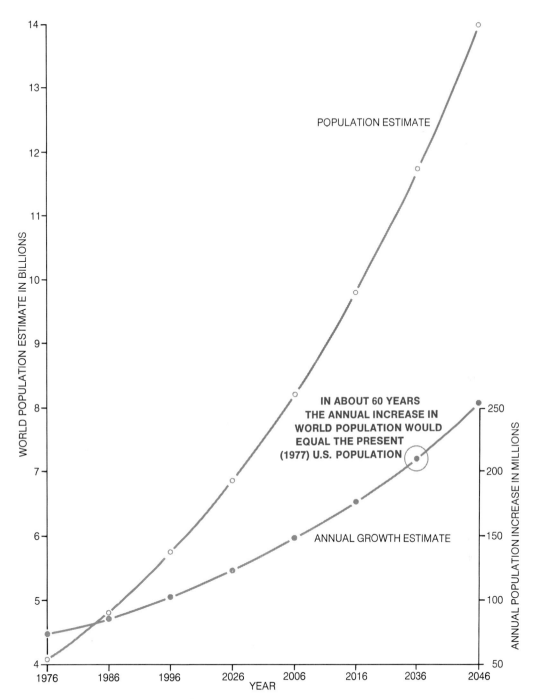

Figure 25-14 World population estimates through A.D. 2046 plotted as total population and annual increment.

impossible tasks in production and fair distribution of food. Even if ways can be found to produce enough food for 11 or 20 or 30 billion people, it is unlikely that costs of production and transportation will allow its equitable distribution, barring a total restructuring of human social order.

One World or None

At least within the industrialized world there is hope of stabilizing populations within the local carrying capacity. One then might assume a calloused attitude towards human misery as long as it occurs outside the industrialized nations and just let nature take its course. In the nineteenth century this might have been possible, but not today. Nonetheless, several eminent thinkers on such matters say that this is essentially what we must do. Proponents of the widely discussed *lifeboat ethics* say that nothing is accomplished if the lifeboat (industrialized nations) is swamped by being filled with improvident swimmers (the starving masses of the developing nations). Indeed, it is argued that aid to the developing nations may merely prolong their collective agony, or even make it worse by providing just enough aid to encourage further breeding, which would then trigger another round of even greater human misery.

The concept of *national triage* differs from lifeboat ethics in that it proposes selective support for those developing nations which, by reason of that support, may be able to make a safe transition to a stable population and economy. Other nations deemed unable to attain stability by receiving aid would be left to their own devices. Thus, according to the triage concept, the limited amount of available aid would be applied so as to have the best possible chance of producing a long term stabilization at least in some aid-receiving countries. The term triage comes from the analogous situation in medical treatment in war or disaster when the number of injured exceeds the available medical resources. Patients then must be sorted into groups, according to whether or not medical attention would be likely to affect their survival.

These are brutal concepts that we instinctively abhor, even if they could be carried out with impunity. On an international scale, with the victims personally unknown and in a far country, the brutality of these concepts is attenuated. However, carry the situation to a personal level—two persons, one strong and one weak, and a loaf of bread that is not enough to keep both alive until rescue—and then the reality of such ideas begins to sink in. Yet, the facts that such concepts are broadcast by well-meaning and humanitarian individuals at least has the good effect of dramatizing the severity of the oncoming crisis.

One certain fact makes it clear that the industrialized world cannot afford to practice such survival ethics. This is the fact that no nation is going to starve by itself, quietly and peaceably and without disturbing world tranquility. If the industrialized world does not assist its neighbors to survive their population and food crises, it seems only a matter of time until one or another hungry nation embarks militarily on an international search for food. An entirely plausible result is generalized war and nuclear destruction on a worldwide basis. Whether civilization could survive this form of solution to the population problem is an open question.

How can the dwindling aid that the industrialized nations can provide—dwindling because their own populations are by no means yet at the zero growth level—be put to sufficiently effective use to stave off violent conclusions? At this point we leave the domain of the science of biology. That is, the biological answer to what must be done is clear, but the means of achievement is not a matter of biology (at least in the academic sense) for it requires nothing more than convincing the great majority of humanity to reproduce at replacement level.

Even though the *means* of persuasion is not a biological problem the *reason* that persuasion is necessary lies, of course, in the biological realm. The great evolutionary drive of every organism is to employ every means available to insure that its genes are maximally represented in the next generation. Being animals that have evolved by just such efforts, it is certain that our reproductive drive is basic and it is hard to restrain. Even so, recall that there are in general two strategies insuring maximal genetic representation in future generations. Organisms either produce large numbers of offspring and pay them little attention, letting sheer numbers insure a sufficient survival rate, or they produce fewer offspring and in various ways improve their chances of survival. The latter tactic is used in a few lower organisms and, of course, is maximized in the primates, as we have seen, page 315.

Since we are so close to the optimum as far as providing insurance of survival to small numbers of offspring, it would seem to take only a little more education to show reproductive age individuals how not only the quality of individual life but how the chances of genetic representation in the next generation may be improved by birth limitation. At the same time it is necessary to move as rapidly as possible to raise the standard of living in the developing countries. The importance of coupling assistance with the standard of living cannot be overemphasized since, as we have seen argued for the demographic transition in the industrialized world, population limitation seems almost an automatic consequence.

Population Stability in Lower Animals

In this discussion of the human population problem we began with some examples concerning population growth in lower organisms and then became immersed in the human population problem, which we found to be complexly enmeshed in political and economic problems that lie outside the realm of biology. These complexities are so overriding that they tend to obscure the underlying biological principles, making it worthwhile, before leaving the subject, to look again at the matter of population growth in lower animals to see if there might still be things to learn with respect to our own problems.

In our previous consideration of population growth and limitation in lower animals, we were primarily concerned with only two matters—the tendency to reproduce and the effect of what we may call extrinsic factors, the necessities of life such as food, and certain limiting factors such as disease. Let us examine animal populations a little more closely and see if they reveal other factors limiting or modifying population growth.

The Balance Between Predator and Prey

If we seek to see why any animal population does not exceed the carrying capacity of its environment, the first thought is that it undoubtedly has predators that limit growth. In nature many such predator-prey systems have been studied, as for example, the classical lemming-fox relationship (page 547). Precisely why this system cycles is not entirely clear. More commonly predator-prey relationships among mammals are rather stable. They appear in the long run to be beneficial even to prey because the predator assists in maintaining the genetic fitness of the prey by culling out the weak and because the predator prevents the prey from overgrowing the carrying capacity of the environment. All this seems quite evident and obviously not of much significance to the human dilemma, except to suggest one of the many reasons that human populations are now so huge.

Intrinsic Limitations of Populations

Our age-old companion, the dog, shows us a vestige of one intrinsic, or within the species, means of population limitation every time it lifts a leg to spray a drop or two of urine on a tree trunk. Like many mammals, it establishes territories. In the case of dogs and their relatives, territory is marked by the scent of urine, which to the sensitive canine nose is an individually specific marker. Other mammals have specialized scent glands for this purpose or conduct ritual combats (Figure 25-15). Birds establish territories by singing or by other displays that have plain meaning to intruders. The value of territory establishment is to ensure a sufficient food supply so that the territory establisher and its mate will be able to rear their offspring. Thus individuals of species become optimally spaced with reference to the productivity of the environment. The more evolutionarily fit individuals occupy territories in the most favorable parts of the environment, and less competitive individuals occupy less favorable territories or perhaps establish no territories at all and fail to reproduce.

Territorial behavior can be seen to be particularly important in preventing a predator from totally overrunning its prey and then dying out. For example, for millenia the immense herds of caribou in Canada and Alaska have been in stable equilibrium with their paramount predator, the artic wolf. At any one time up until the recent past, caribou herds could have readily fed a far larger wolf population. The risk that this might happen, with eventual disastrous effects both on predator and prey, appears to have largely been eliminated by territoriality as well as by the social structure of the wolf pack.

It is difficult to trace territorial behavior into the rich complexity of human social behavior. Only in the most primitive societies does it exist in the primeval form of one family group staking out a territory upon which it exclusively subsists. In the remainder of human societies territoriality is expressed largely in

Figure 25-15 Ritualized combat is often used by male ungulates to establish territories.

terms of possessions that do not interfere much with crowding together of individuals or with their multiplication. Our social behavior seems to have obliterated this particular limitation to population growth, and necessarily so since civilization would hardly have gotten a start if each human family kept to its geographically isolated subsistence territory.

Perhaps for similar reasons another phenomenon that is seen in some lower animals as a population limiting factor does not occur in human societies. This is the physiological and social pathology that appears when some mammal populations grow beyond normal levels. The first indications of the existence of this phenomenon came from observations of inexplicable mass mortalities of animals such as jackrabbits. In several instances of cataclysmic mass mortality of jackrabbits, there was sufficient food available and the deaths were not related to predation. Rabbits were actually observed dying in convulsive seizures. Internal examination revealed manifestations such as adrenal gland deterioration. These signs indicate that the jackrabbits had died of a well known pathology, overactivity of the pituitary-adrenal system. The precipitating cause is thought to have been population dependent in the sense that, in a very large population, upon the advent of warm weather, more frequent than usual reproductive and other interactions of individuals produced an unnatural endocrine stress.

In rats under laboratory conditions similar effects can be produced by crowding. An interesting fact is that laboratory rats maintained their populations levels far below the experimentally established carrying capacity of the environment. When population levels approached the carrying capacity, a variety of physiological and social pathologies greatly reduced births and early survival. Females frequently aborted or gave birth prematurely and were unable to nurse. Abnormal behavior in males served to reduce mating success.

Observation of life in any crowded city readily confirms that this form of population limitation does not affect human populations. Birth rates remain high and infant development is normal, irrespective of crowding, as long as there is adequate food. Once again we see what appears to be suppression of a possibly inherent population limiting mechanism in the face of the advantages of social life.

This brief glance at population limiting mechanisms in lower animals reinforces the idea that the nature of human social life makes us particularly vulnerable to catastrophic overpopulation. We seem to lack, because of our way of life, any of the intrinsic mechanisms that limit population size. If we do not apply our foresightful intelligence to the problem, the only limits that will have effect are the ultimate extrinsic limits of food and environmental destruction. Our wits alone can save us.

26 Degradation of the Environment

Human ecologist Garrett Hardin related a widely discussed and compelling parable when he illustrated a critical failing in our tenancy of the earth in terms of what he called the "Tragedy of the Commons." As he observed, in early times a village often had a *commons*, a pasture for general use. As long as herdsmen did not burden the pasturage with too many cattle the system worked. Inevitably though, a time came when the grazing population approached the maximum supportable by the commons. What then was the most advantageous behavior for the individual herdsman? If he should put another cow out to graze on the commons he would profit exclusively from the gains that cow made. On the other hand, his losses, originating from the damage done by the additional cow to the commons, would be spread among all herdsman. Thus, in terms of that particular herdsman and time, the most profitable behavior would have been to add another cow. This conclusion holds right up to the time when the gain from putting additional cattle on the common would be matched by the losses suffered by the herdsman's entire herd. One can imagine the deteriorated state of the commons at that time.

The Tragedy of the Commons is a valuable tale because it provides, on a comprehensible scale, a plausible example of human activities exceeding the carrying capacity of the environment. Hardin's story is eminently believable. But let us carry it farther and, as is Hardin's intention, scale it up to world dimensions: let the commons become the biosphere, the herdsmen the world's human population and the cattle all the various devices, both industrial and agricultural, by which we wrest a living from the earth. Is the parable still applicable? Will the outcome be the same on a world scale?

The size of the world and the unknown possibilities of the future are great comforts to those who do not choose to believe that the Tragedy of the Commons is applicable on a world scale. They find hope in the apparent immensity of the planet and hold that its resources can never be expended, or at least not expended before it makes no difference to them. Even if they grant that certain resources of the planet are exhaustible, they believe that unknown possibilities for future technical advances (the technological "fix") can be counted on to make good or even improve upon the loss.

We have seen the inevitable outcome of the Malthusian argument as far as human populations and food are concerned. Perhaps this jolt alone is enough to generate respect for the parable of the commons. Even so, it is important to explore other aspects of the commons for, after all, on a world scale the commons is more than just land for food production. All aspects of the world commons must be considered. These include, besides agricultural land, the air and water of the earth and all its natural resources. We will now discuss such matters and show some of the principal ways humanity over-taxes its world commons.

Depletion, Contamination, and Destruction

These three activities are the principal ways humanity jeopardizes the biosphere. We use up nonrenewable resources, such as oil; we contaminate air, water, and land with wastes from home, agriculture, and industry; and we destroy life and degrade land. *These actions are directly attributable to population growth and secondarily to misuse of technology.* Thus, if one agrees that such effects are occurring and that they are not desirable, it might be said that to worry about them is much like a physician worrying about the fever while ignoring the causal bacteria. However, we argue that it is important to study the "symptoms" of human population overgrowth because certain of them are, in themselves, serious enough to precipitate directly or indirectly some final catastrophy. Hence, we should understand these assaults on the environment, and from this understanding proceed to alleviate them as much as possible. *At the least these efforts may buy time to find an equitable solution to the ultimate problem, namely holding human populations at an optimum level.*

RESOURCE DEPLETION

The central component in resource depletion is energy. Indeed, energy depletion is pivotal to the whole spectrum of insults upon the environment, since our efforts to maintain adequate energy supplies cause a spreading and self-feeding sequence of resource depletion, environmental contamination, and destruction. Reciprocally, as resources become depleted, we must increase energy consumption in order to utilize less energetically favorable sources of essential materials. In more general terms, readily available energy is the major underpinning of the economies of all nations that are in even the most modest state of development. In the event that present energy sources are not adequately conserved, or replaced with others, we may eventually expect severe international unrest. To forestall this it is probable that governments lacking adequate energy policies will, at the least, progressively relax hard-won environmental protection standards in efforts to glean the last bit of energy from their own national supplies, hoping to hold off as long as possible total dependency on external supplies. Without adequate energy policies in most nations of the world we face grim prospects of progressive environmental degradation, severely lowered standards of living—particularly in the developed nations—and finally international aggression in the search for energy.

Energy

In the final analysis all available energy originates from nuclear or gravitational forces (Figure 26-1). As we have seen earlier (Chapter 24) the nuclear fires of the sun drive the winds, evaporate water, produce thermal gradients in the oceans, and power photosynthesis. Gravitation contributes to heating the interior of the earth and causes the tides. Spontaneous nuclear fission within the earth helps sustain its internal heat. Finally, our recently developed control of the atom allows the release of nuclear energy in a new and disturbing form.

In the early history of human life, photosynthesis was our sole energy source. The energy in plants was converted into animal power, ours and that of our domestic animals, or provided heat through burning wood. Later, stored photosynthetic energy in the form of peat and coal and still later petroleum and natural gas came into use along with wind and water power.

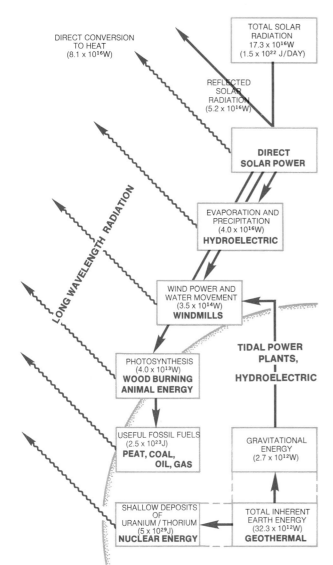

Figure 26-1 Energy sources. See Table 26-1 for explanation of energy terms.

In the most recent phase of human history, energy from controlled nuclear reactions (fission and perhaps ultimately nuclear fusion) has started to contribute to the evergrowing demands of civilization for more and more energy. Geothermal, tidal, and directly harvested solar energy have recently come into use in relatively trivial amounts. However, at present the dominant energy sources are the dwindling fossil fuels, and to a lesser extent, water and nuclear power.

The hard fact is that the developed nations are almost wholly geared to use energy provided by petroleum and natural gas and, within a few years, these will be essentially expended as practical energy sources. The world seems finally to have come to understand this fact and is preparing for transitions to other energy sources. Although petroleum and natural gas have deleterious effects in the form of pollution, the alternative fuel source most readily available for heavy usage is even more polluting. This is coal, whose early heavy use diminished rapidly in the face of plentiful supplies of petroleum. Nuclear power plants are projected to carry a heavier burden of supplying energy in the years ahead, but the future of nuclear energy is in doubt because of fears about safety, particularly with regard to breeder technology which appears to be necessary if nuclear fuels are to last for very long.

Nonpolluting energy sources in the present state of technology are incapable of making a large contribution. Most falling-water power sites have been occupied and are generally deteriorating due to silting of dams. There are only two tidal power sites in the entire world producing power commercially, and none are under construction. Geothermal energy is being used in a few places, but, in the present state of technology, this must be considered a limited and nonrenewable resource. However, geothermal power would last indefinitely if economical methods could be devised to extract the earth's heat from depths below 10,000 ft.

The greatest potential source of energy available to us is solar power and it, of course, does not contribute to our problems of environmental protection. The difficulty is that solar technology most easily produces low grade heat. The heat from the sun is not concentrated enough per unit area of the earth to operate, in a straightforward way, the kinds of machinery that are most efficient for the conversion of heat into electricity. Until this problem is solved, there is not much hope of solar power supplying the immense amounts of energy needed by industry. However, it is already becoming useful for residential and building heating, and in this way reduces the demand on high grade heat sources such as petroleum. Thus, although we may earnestly hope that solar and other nonpolluting power sources will be perfected in time to provide a complete solution to the energy deficit, we are, for the forseeable future, stuck with diminishing petroleum supplies, with about a 200-year supply of unwieldy and heavily polluting coal, and with a new nuclear power industry about which there are many questions, both of safety and power producing capacity.

THE WORLD ENERGY STATUS AND PREDICTIONS OF FUTURE REQUIREMENTS

Current worldwide human use of energy amounts to about 2.9×10^{17} Btu/yr. Examination of Table 26-1 and Figure 26-2 will help you understand this huge number quantitatively, but a glance at Figure 26-3 immediately drives home its meaning in an intuitive way. The figure is a satellite photograph of the United States at night, showing that most significant population centers are readily detectable from hundreds of miles out in space by the light they emit. Since the cumulative efficiency of converting energy into light is low, at best slightly under 7%, one immediately senses that this photic display can only be staged by a country that uses a lot of energy. That is indeed true since only about 2% of the energy used in the United States goes into residential and commercial lighting, only a fraction of which would be visible to the satellite.

Over 95% of the world energy supply comes from fossil fuels. In modern times the principal fossil fuel in use till about 1910 was coal. In Europe coal replaced wood as fuel, and the smog from its burning became one of the curses of city life. Subsequently nearly all of the fourfold increase in energy consumption to present levels has been made up by oil and natural gas. As might be expected, most of the increment in energy requirements has been generated in the developed nations (Figure 26-4) and, indeed, energy consumption is directly related to per capita income (Figure 26-5) in a mutually catalytic way: Increased energy availability increases income, which increases the de-

Table 26-1 Energy Terminology and Conversion Factors

	Unit	Definition
Basic Energy Units		
	calorie (cal)	Heat energy necessary to raise temperature of 1 g distilled water by 1 °C, from 14.5 to 15.5°C.
	kilocalorie (kcal)	1000 cal
	joule (J)	Electrical energy necessary to maintain an electrical current of 1 ampere (amp) for 1 sec at a potential of 1 volt (v). Note that an electrical current due to electron flow in a metallic conductor is measured as the amount of charge flowing per unit time. The unit of charge is the coulomb and 1 coulomb/sec is 1 amp. The charge on the electron is -1.602×10^{-19} coulomb, making 6.24×10^{20} electrons/coulomb.
	British thermal unit (Btu)	252 cal
Measures of energy consumption		
Power (energy used per unit time)	watt (W)	1 J/sec
	kilowatt (kW)	10^3 W
	megawatt (Mw)	10^6 W
Work (power × time)	kilowatt hour (kWh)	1 kW × 1 hr
Conversion factors		
	1 cal	4.186 J
	1 kcal	3.97 Btu
	1 horsepower	746 W, or 0.178 kcal/sec

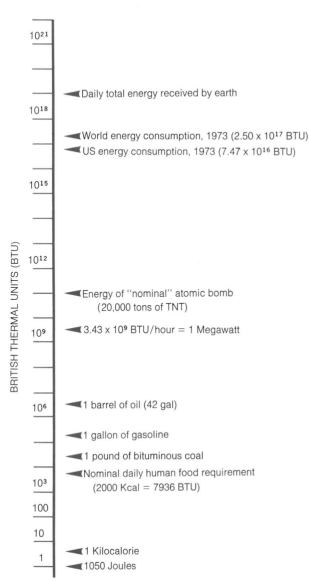

Figure 26-2 Landmarks in energy production and consumption plotted on the Btu scale.

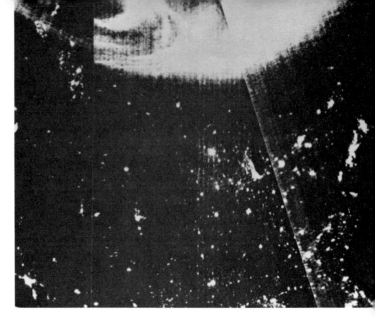

Figure 26-3 Population centers in the United States photographed by their own light from a satellite. The bright swirls at the top are the Northern Lights. (U. S. Air Force photograph.)

mand for materials and services that consume energy. Since in the developing nations increases in energy consumption must be related to the struggle to increase the standard of living and since the developed nations are committed to expanding economies, we can expect no slackening in energy demands in the

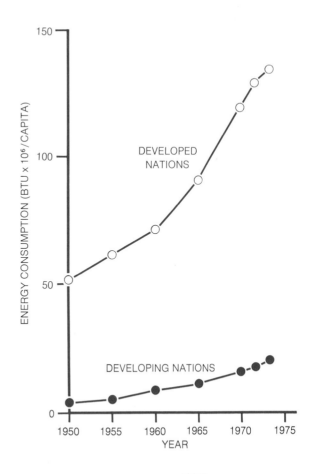

Figure 26-4 Growth in energy consumption is markedly greater in developed than in developing nations. Data from United Nations (1976) *World Energy Supplies*.

DEGRADATION OF THE ENVIRONMENT   587

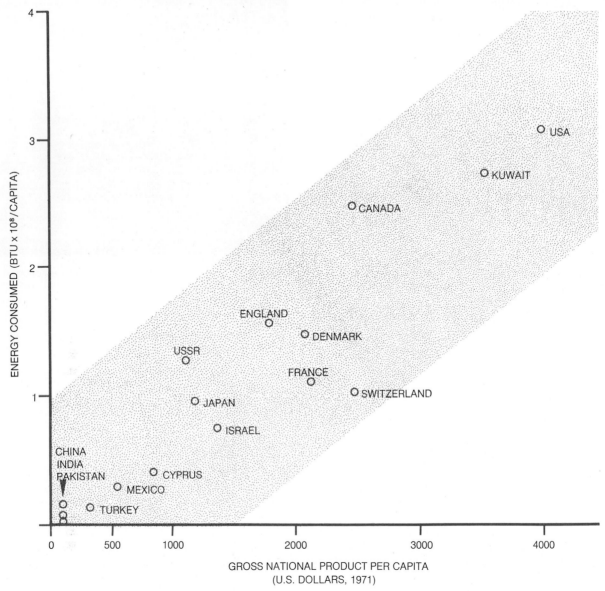

Figure 26-5 Increasing personal and national wealth evokes increasing energy requirements. Data from United Nations (1976) *World Energy Supplies* and Zero Population Growth.

immediate future. One generally accepted estimate of the rate of growth of these energy requirements produces a curve which has a familiar look. You have seen a highly similar one for human populations. It is in fact keyed to the human population growth curve as well as to the spread of industrialization.

As with the human population curve, we may expect the energy curve to have upper limits. If we assume that energy supplies are not limiting, it is possible to assign an ultimate limit to the growth of energy consumption simply in terms of the additional heat load on the ecosphere. Thus L. C. Cole estimates that within less than 800 years, at the present annual increment in power utilization, the additional thermal load would increase the average surface temperature of the earth by 3°C, enough to produce disastrous ecological changes, not the least of which would be melting of the polar ice caps with a resultant rise in sea level of 100 m. However, examination of the world energy suppply shows that it is unlikely that we shall have to worry about this ultimate thermal pollution problem.

PRESENT ENERGY RESOURCES

Fossil Fuels

Peat, coal, oil, and natural gas represent energy capital accumulated by photosynthetic processes, primarily in the distant past, 250 to 350 million years ago. Oil and natural gas were formed when immense quantities of marine plant and animal material were covered by further layers of organic remains or marine sediments. As oxygen was depleted in the covered layers, the ordinary processes of bacterial decomposition were arrested and the remaining organic material was further transformed by pressure and heating as the sediments became more deeply buried. Coal was similarly formed, largely from the remains of freshwater plants. Peat represents an early stage of this process. Further burial and compressional heating of peat over geological stretches of time transformed it through a series of stage culminating in bituminous and finally in the highest quality, cleanest burning coal, anthracite. Peat is still being formed in bogs such as the Great Dismal Swamp of Virginia, and in many parts of the world it is cut and dried to form a useful but low grade fuel, principally for domestic purposes.

Coal reserves are immense. Conservative estimates place world reserves at about 7000 billion metric tons, or about 2×10^{20} Btu. Nearly all of these deposits are in Asia, North America, and Europe. It is difficult to predict how long this supply will last. Its rate of consumption will depend on the availability of other preferred energy sources because coal is cumbersome and environmentally destructive to use (see Environmental Contamination, Sources and Effects). Perhaps the best estimate is that we will not reach 90 percent depletion of coal until 2400.

Petroleum and natural gas reserves are assuredly more limited than coal reserves, and there is controversy over how large the reserves actually are because oil prospecting is a difficult art. Although commercial oil production began only in 1857 we already face imminent exhaustion of both oil and gas stocks. As

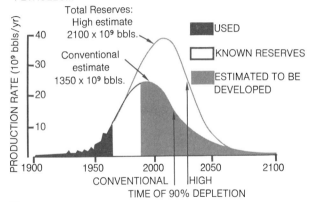
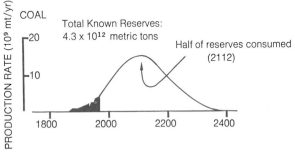

Figure 26-6 Estimated world oil and coal reserves. Large variations in estimated petroleum reserves are seen to have little effect on depletion time. After E. Cook, 1972, Energy for Millenium Three. *Technology Review,* December.

Figure 26-6 shows, rather large uncertainties about the amounts of oil remaining to be discovered have small effect on the time of depletion of useful supplies. The most optimistic estimate forecasts effective depletion of world petroleum supplies within 50 years and of our domestic supplies somewhat earlier, perhaps by the year 2000. Natural gas supplies will have a similar depletion history.

As petroleum and gas supplies diminish, other factors will hasten their departure from the general heating and transportation market. First, since the world's military forces depend on petroleum to operate high performance aircraft, ship and vehicle engines, they will continue to amass reserves of petroleum and thus hasten depletion of civilian supplies. Secondly, both petroleum and natural gas are used as raw materials in synthesis of a huge spectrum of essential products. These include the fertilizers, pesticides, and herbicides that are essential to modern high productivity agriculture, as well as synthetic fiber, rubber, paint, plastics, and a host of essential organic chemicals.

There will obviously come a time, well before total depletion, when petroleum will be too valuable to be used as fuel. Future generations may well question our sanity if we persist much longer in using oil for fuel rather than preserving it for chemical syntheses.

In addition to coal and petroleum, there are very large fossil fuel reserves in oil shales. These are deposits of sedimentary rocks containing a tarlike hydrocarbon, kerogen, which is too viscous to be pumped by conventional petroleum production methods. The best known of the oil shale deposits in the United States is the huge Green River Formation of Wyoming, Utah, and Western Colorado. In the aggregate the United States has sufficient oil shale to supply oil to support our current consumption for nearly 75 years. The problem is that even rich oil shales, those with more than 25 gal of kerogen per ton, are energy intensive and difficult to mine. Energy must be used in such amounts to mine and convert the kerogen to usable oil, that the net energy gain is greatly reduced. However, as oil reserves are depleted, the oil shales will become more economically feasible and perhaps more efficient production techniques will be forthcoming.

Nuclear Energy

The triggering event of the nuclear era was the discovery of radioactivity in 1896 by Henri Becquerel. Although the great British physicist, Lord Ernest Rutherford, within ten years after this event wrote enthusiastically about the enormous energy that might be obtained from nuclear forces, nuclear energy did not begin to enter the realm of practicality until the German physicists Otto Hahn and Fritz Strassman, in 1938, somewhat inadvertently observed the fission of uranium into two lighter weight atoms upon bombardment with neutrons. They actually had hoped this technique would create transuranium elements, elements heavier than uranium. The significance of their discovery was ultimately appreciated by scientists of the Allied Powers and communicated to President Roosevelt, supported by a famous letter from Albert Einstein. This letter pointed out the possibility of producing a tremendously powerful weapon if a nuclear chain reaction could be sustained. Beginning with an appropriation of a few thousand dollars, which grew to over $2 billion within just a few years, the first self-sustaining nuclear chain reaction was demonstrated in 1942 and this led to the first explosion of an atomic bomb in 1945 in the New Mexico desert. The entire world soon thereafter was made aware of the military power of nuclear energy by the bombs dropped on Hiroshima and Nagasaki.

During the war, nuclear reactors were built to produce fissionable elements for use in weapons. These "atomic piles" provided the initial experience that led to the first United States commercial nuclear power plant, which began generating electrical power at Shippingport, Pennsylvania, in 1958, at its design output level of 90,000 kW. Since then all over the world nuclear power plants have grown in number and power capacity. Thus in 1977 about 55 nuclear generating plants in the United States were producing about 41,000,000 kW, and the worldwide nuclear generating output amounted to about 70,000,000 kW. This is an impressive rate of growth for an entirely new and extremely complex technology, yet nuclear generating capacity is still only a small fraction of the

Table 26-2 Development of Nuclear-Electric Generating Capacity

	World		United States	
Year	Total Capacity ($\times$ 1000 KW)	Total Capacity As Fraction of Total Generating Capacity	Total Capacity ($\times$ 1000 kW)	Total Capacity As Fraction of Total Generating Capacity
1954	5	0.00003	—	—
1960	859	0.0016	297	0.0016
1965	6456	0.0083	926	0.0036
1970	18927	0.0170	6493	0.0180
1972	36790	0.0286	15301	0.0366
1974	60568	0.0409	31652	0.0639

Data from *World Energy Supplies,* 1974. U. N. Statistical Office

total electrical generating capacity from conventional fossil fuel or hydropowered power plants (Table 26-2).

The development of the nuclear power industry is a classical example, not only of the conflict that may occur between industrial development and preservation of the environment but of what is perhaps the larger problem, the appearance of technological advances before society is able to cope with them. With respect to the first problem, preservation of the environment, there are strong arguments that nuclear power plants, *when they operate normally*, are less damaging to the environment than, let us say, coal fired plants. Opponents of nuclear power reply by pointing to the unusual hazards posed by liberation of nuclear fuels and waste products into the environment in case of accident or sabotage.

This last possibility is only one of several indications that society is not yet stable enough to embark on a massive nuclear power program without the most elaborate safeguards. Nuclear plants, their fuel, and waste products are uniquely dangerous in the hands of terrorists willing to use them to gain their own ends. Even in the absence of willful dispersal of nuclear materials, adequate safeguards against their accidental dispersal are necessary for periods of time longer than the duration of survival of any society to date. All these facts, together arguing that nuclear power development is uniquely significant to the integrity of the ecosphere, require that we examine nuclear energy carefully.

The Source of Nuclear Power

According to the well-known Einstein equation relating mass to energy,

$$E = mc^2$$

where E is energy, m is mass, and c is the speed of light, we may calculate the energy that will appear whenever there is a reaction in which there is a change in mass. To illustrate, consider what happens when a hydrogen atom absorbs a neutron to become a deuterium atom. It has been determined that this reaction results in the loss of 3.98×10^{-27} g/atom of deuterium formed. Substituting this figure for m in the Einstein equation and inserting the speed of light, 2.998×10^{10} cm/sec, for c, we obtain a value for E of 35.8×10^{-7} ergs, or 8.55×10^{-14} cal. If we divide this energy yield by the mass change that liberated it, namely 3.98×10^{-27} g, we find that the energy equivalent of 1 g is a very large figure, 2.15×10^{13} cal. When it is realized that ordinary chemical combustion of a good fuel such as coal liberates only about 5700 cal/g, making the mass-energy equivalent of 1 g equal to the energy obtained by chemical burning of 3800 metric tons of coal, the incredible energy of the atomic nucleus is apparent.

Fission, Fusion, and Binding Energy

To explain the source of this immense power it must first be realized that, when all atomic nuclei were originally formed, all protons and neutrons contributing to their formation lost mass by conversion to energy, analogously to the events in the example of deuterium formation. The total mass lost on formation of each kind of atom is calculated from the difference in the uncombined mass of each proton and neutron and the mass of the assembled nucleus. This difference, the mass that disappeared by conversion to energy when the nucleus formed, is the **binding energy**. When the binding energies of all stable kinds of atoms are plotted against their mass number (total number of protons and neutrons in the nucleus) a curve is obtained (Figure 26-7) that rises steeply to a maximum at about mass number 60 (nickel) and then declines slowly to mass number 280 (uranium). The significance of the binding energy curve is that moving an atom, by effecting a change in its nuclear structure, from a low point on the curve to a higher point

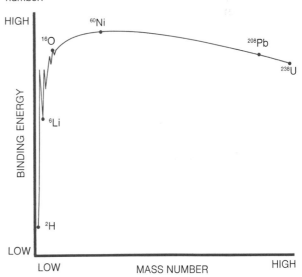

Figure 26-7 Nuclear binding energy as a function of mass number.

increases binding energy. When this happens the nucleus must lose mass and consequently a mass to energy conversion takes place.

The mass-energy conversion happens in different ways on the two arms of the binding energy curve. At the high end, heavy atoms are split into lighter ones having higher binding energies (lying higher on the curve), and the result is mass-energy conversion. This process is called **nuclear fission**. At the low end of the curve, two light atoms may fuse to give rise to one atom having a higher average binding energy than the initial reactants. The result is energy liberation. Called **nuclear fusion**, this process is the fundamental energy releasing process in the sun and other stars. Nuclear fusion can occur only at temperatures of millions of degrees, which are high enough to overcome the repulsive forces between atoms. Our only success in producing nuclear fusion to date has been with the thermonuclear (or hydrogen) bomb in which the requisite conditions for fusion of atomic nuclei are obtained by a preliminary fission explosion.

Nuclear Fission and Nuclear Power Reactors

Atomic nuclei contain nucleons—protons (positive charge) and neutrons (electrically neutral)—held together by nuclear forces which, although thousands of times as strong as the forces holding an electron in orbit around the nucleus, act over only a very short distance, approximately only the diameter of the nucleus. In large nuclei, such as in uranium, the electrical repulsion between protons closely matches the short range nuclear forces that hold the nucleus together. Under certain conditions these forces become sufficiently out of balance for the nucleus either to lose a few nucleons, which is termed **radioactive decay**, or to split into two lighter nuclei, which is nuclear fission. For example, an atom of uranium-238 (^{238}U) may spontaneously lose two protons and two neutrons (an alpha particle, which is the same as a helium nucleus) and become an atom of thorium-234 (^{234}Th),* thus:

$$^{238}_{92}U_{146} \rightarrow {}^{234}_{90}Th_{144} + {}^{4}_{2}He_2$$

---

*In nuclear terminology a nucleus is described as $^{A}_{Z}X_n$, where X is the chemical symbol for the atom and Z is the number of protons, n is the number of neutrons and A is the number of nucleons ($n + Z$). Atoms with the same number of protons but a different number of neutrons are isotopes of each other. Thus there are three isotopes of hydrogen each with one proton but with 0 neutrons (hydrogen), 1 neutron (deuterium), and 2 neutrons (tritium).

Neutrons are excellent projectiles with which to hit an atomic nucleus because their lack of electrical charge does not lead to electrical repulsion. When a neutron of appropriate energy comes near enough to a nucleus, it may be captured. If the nucleus is of the proper type, the resultant instability may cause fission. If the capturing atom is uranium-235 (^{235}U) there are several possible fission fragments, or lighter weight nuclei that may be formed. For example, after absorption of a neutron ($^{1}_{0}n_1$) uranium may undergo fission to form barium-141 (^{141}Ba) and krypton-92 (^{92}Kr), as follows:

$$^{235}_{92}U_{143} + {}^{1}_{0}n_1 \rightarrow$$

$$^{141}_{56}Ba_{85} + {}^{92}_{36}Kr_{56} + 3\,n_1 + \text{ENERGY}$$

The principal source of the very large amount of energy released in this reaction is the kinetic energy of the fission fragments as they push each other apart. In nuclear power reactors this kinetic energy, representing only about 1.5 percent of the potential mass-energy equivalent of uranium-235, is the source of the heat used to produce electricity.

The reason that a nuclear power reactor is able to produce heat in a sustained manner lies in the fact that fission of uranium-235 in the reactor core releases more neutrons than the one required to initiate fission. Consequently, a chain reaction becomes possible if enough uranium-235 nuclei are arranged so as to insure capture of a sufficient number of neutrons, producing still more fissions of uranium-235 nuclei (Figure 26-8). This is not a trivial thing to do for several reasons: First, uranium as it is mined contains only about 0.7% of the fissionable isotope uranium-235. The remainder is uranium-238. Secondly, the neutrons released by fission are initially too energetic to be effective in causing further fissions. Their energy must be reduced by interaction with a moderator in the reactor core before they escape. Water enriched with heavy water (D_2O) serves as a good moderator. D_2O also serves as a coolant for the core and as a means of transferring heat from the core to the electrical generating machinery. In addition, some control mechanism is required to keep the fission process from running away. This is accomplished by control rods of cadmium or boron, which absorb neutrons more efficiently than uranium. Moving the control rods in and out of the core decreases or increases the availability

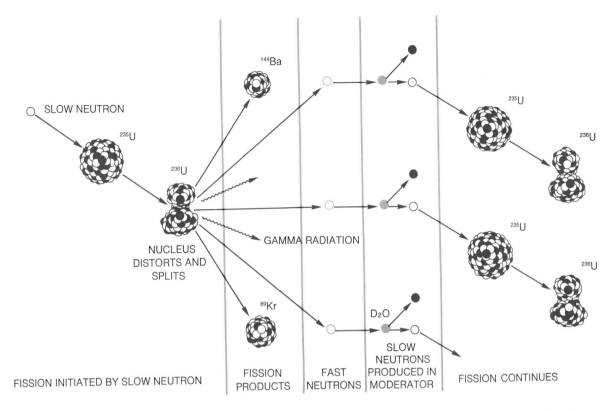

Figure 26-8 Nuclear fission. Slow neutron absorption by an atom of uranium-235 produces an unstable atom of uranium-236. This is shown undergoing fission to give rise to two of several possible fission products and fast neutrons. If sufficient fast neutrons are converted to slow neutrons by collision with atoms such as the hydrogen in heavy water (D_2O) and then interact with more atoms of uranium-235, the fission process becomes self-sustaining.

of neutrons for producing uranium fission. Finally, just the right amount of enriched uranium must be present in the proper geometrical arrangement to insure efficient neutron capture. The arrangement of a nuclear reactor based on these requirements is shown in Figure 26-9.

The Difference Between a Fission Bomb and a Fission Reactor

In order to produce an explosion it is necessary to create a **critical mass** of fissionable material (plutonium-239 or uranium-235) suddenly enough to overcome the tendency of the mass to expand as it heats, which would slow the chain reaction enough to prevent an explosion. This is done either by shooting one subcritical mass into another or by compressing a single, spherical subcritical mass by setting off a surrounding mantle of explosives. This is important to remember because it emphasizes that it takes a great deal of effort to hold a critical mass together for the instant necessary to build up the chain reaction to the point of producing a nuclear explosion. Thus while there are all kinds of ways to make a nuclear reactor explode, the explosion would be a conventional chemical explosion of nowhere near the energy of a nuclear explosion. Naturally, a reactor explosion would be far more serious than a conventional chemical explosion because it carries with it the possibility of releasing intense radioactivity into the environment.

In a fission reactor the uranium fuel is placed in relation to the moderator so that a runaway nuclear reaction can easily be prevented by manipulation of the control rods. The principal problem remaining in normal operation is to insure that the tremendous heat produced is dissipated by the coolant. Nuclear reactors must be constructed with elaborate safeguards to insure against failure of the cooling system since the effects of such a failure are awesome to contemplate. In the event of loss of coolant, and

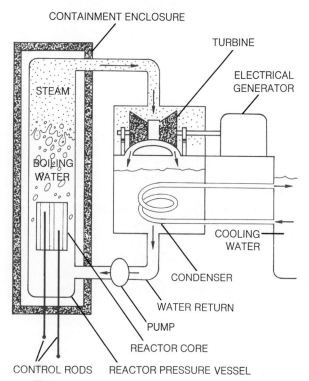

Figure 26-9 A boiling-water type nuclear power system. Steam generated by heat from the reactor core operates a turbine which drives an electrical generator. This conceptual diagram does not show the heat exchanger which may be used to isolate the intensely radioactive water surrounding the core from a steam generator in an independent circuit which may drive the turbine with less radioactive water.

assuming that the nuclear reaction is shut down by insertion of control rods or even dumping the moderator, the residual heat of a conventional power reactor would still be enough to melt some 40 tons of steel per hour!

Although nuclear power plants are constructed with massive containment systems, it is generally conceded that such an accident, termed a melt-down for obvious reasons, would result in the containment being swiftly breached. Since in a melt-down large quantities of gases are formed, including hydrogen, there is a probability of a chemical explosion that would breach the containment and result in venting of radioactive materials into the atmosphere. In addition, the melted core, intensely hot and self-heating, would proceed downwards through the thick steel and concrete containment structure and continue down into the earth for several hundred feet. A maximal accident of this sort, although it would lack the destructive force of a nuclear bomb explosion, would release radioactivity of a similar magnitude to that generated by one of the early fission bombs.

Safety and the Nuclear Fuel Cycle

The aim of the nuclear energy industry is to keep the risk of a major accident down to about 1 per 10^6 years of reactor operation. At this risk level by the end of the century, if the nuclear energy industry develops as planned, a sufficient number of reactors will be in operation to make the probability of a major accident about 1 per hundred years. This is a very low probability and suggests, for example, that major reactor accidents are likely to kill far fewer people than major earthquakes. However, there are problems with such assessments because there is no assurance that nuclear reactors will be operated at rigorous safety levels in all countries and because there has been little experience with the problems of aging and deteriorating reactors. Moreover, the hazard from reactor accidents, since it involves radioactivity, creates problems that are quite different from those arising from other types of accidents. These problems will be discussed later; for now let us continue and examine the nuclear energy industry as a whole for other possible hazards besides those stemming from a major reactor accident.

Radioactive hazards are generated in all phases of the nuclear industry. Uranium mining is dangerous for the miners and atmospheric exposure of uranium mill wastes, containing radium among other radioactive substances, adds to the burden of radioactivity in the biosphere. Reactor fuel production releases small amounts of radioactive gases into the atmosphere, but the principal wastes of the nuclear cycle are produced in the reactor itself. As the nuclear reactions proceed, plutonium-239 and a large array of radioactive fission products are formed. Eventually, the reactor process becomes inefficient. Uranium-235 is consumed, and accumulated fission products absorb neutrons necessary for sustaining the reaction. At this juncture the fuel must be replaced and the spent fuel reprocessed. In this procedure unconsumed uranium-235 is recovered and plutonium may or may not be. The remaining wastes are intensely radioactive and some will remain so for thousands of years. In addition, these wastes generate so much heat that it is difficult to contain them without corrosion of containers and loss to the environment.

Dealing with reactor wastes is one of the most difficult problems in achieving safety in the nuclear industry. In the United States these wastes cannot now be said to be handled in a satisfactory manner. They are usually kept in buried tanks and many of these are known to have leaked. Current thinking on the problem is that the best way to dispose of reactor wastes is to solidify them into a glasslike mass in large cylinders which would then be buried in an extremely stable geological formation. Fortunately, the problem is somewhat diminished by the fact that reactor wastes are of relatively small volume. It is estimated for the United States that the total volume of highly radioactive wastes that will be produced by the year 2000 will amount to about 870 m^3.

Normal operation of a reactor results in the loss of small amounts of radioactive gases into the atmosphere. Most of these gases have short half-lives except for krypton-85 with a half-life (see page 597) of 10.4 years, and tritium with a half-life of 12 years. Their principal route of egress is as stack gas to be diluted in the atmosphere. These emissions are limited by Federal regulations, but the permissible levels have been somewhat variable in accordance with changing evaluations of their hazard. Typically the release rate of radioactive gases has been set at not more than 0.05 curies/sec* (1.5 million curies/yr). Atmospheric dilution apparently occurs effectively as the hot stack gases carry the radioactive load to high altitudes. As a result persons living near reactors gain no more than a few percent of the natural radioactivity background either from stack gases or from plant cooling water.

Breeder Reactors

Uranium-235 actually is not a particularly plentiful element. In fact its price is already rapidly being driven up by scarcity, increasing demand, and, of course, the inevitable manipulation by speculators. The best estimates are that the nuclear industry can survive on uranium-235 alone for only about another 25 years, the estimated economically recoverable world reserves being only about 1 million metric tons (1973). However, the available supply of nuclear fuels can be extended almost indefinitely by a particular type of reactor which actually produces more fissionable fuel than it uses. This is the breeder reactor. A breeder reactor operates much like a conventional, or "burner," reactor in providing thermal energy from fissionable fuel. The difference is that, while it does this, it also produces more fissionable fuel than it consumes. Although this might seem to be a violation of the basic laws of nature, it is not for the following reason. It will be recalled that when a neutron causes fission of uranium-235 up to three neutrons are produced. One of these must be used to sustain the reaction and the remainder often serve no useful purpose. However, it is possible to construct a reactor containing other nuclei, "fertile nuclei," that may be converted to fissionable isotopes by the excess neutrons. Either uranium-238, which is 140 times as plentiful as uranium-235, or another abundant material, thorium-232 may be used. Thorium is converted into uranium-233 and uranium-238 is converted into plutonium-239. Both of these are fissionable and useful as reactor fuel. The net result is that the breeder reactor produces more fuel than it uses by converting otherwise nonfissionable into fissionable material. If breeder technology becomes widespread before uranium-235 supplies are exhausted, our nuclear energy horizons are markedly increased, perhaps creating energy resources with 500 times the energy available in present fossil fuel reserves. We thus face a difficult transition. If conventional burner reactors become too widespread before breeders come into general use, it may be that future nuclear fission developments will be aborted for lack of uranium-235 to effect the conversion of the larger thorium-232 and uranium-238 resources into fissionable material.

The Danger from Breeder Reactors

The principal objection to breeder reactors is that they require establishment of extensive spent fuel repurification facilities, and particularly that they will entail purification and use of plutonium. It should be noted that while plutonium is produced in conventional reactors, the fuel cycle is long enough to reduce substantially the amount of plutonium (by fission) in the spent fuel. Plutonium may be used as well as uranium-235 as a material from which fission bombs can be made, and the large amounts that would be present in a breeder reactor economy raises the specter that enough of it would be diverted to bomb construction by terrorists or even unstable governments very substantially to increase the risk of nuclear explosions.

A bomb can be made from 5 kg of plutonium-239

*See Table 26-8, page 605.

> Radioactive uranium gas leaks from plant in France
>
> 280 pounds of uranium lost at atomic plant, paper says
>
> Uranium losses may be announced

Figure 26-10 Recent newspaper headlines reflect public fears of loss of fissionable material from nuclear fuel reprocessing plants.

or 25 kg of uranium-235 and there will be several hundred thousand tons of these elements circulating about the country at any one time—in reactors, fuel reprocessing plants, or in transit. Substantial amounts are already unaccounted for, and it is generally agreed that overt or covert theft is possible (Figure 26-10).

Plutonium-239 is also one of the most toxic substances known, a hazardous dose being only about 1 µg (microgram) and its half-life is 24,400 years. Consequently, production of this element places the greatest responsibility on society since plutonium contamination of the environment will produce effects long into the future.

Aside from the plutonium problem, another difficulty with breeder reactors as presently envisioned is that they are technically more complex and probably will be more dangerous to operate than conventional burner reactors. This is due to the higher density of fissionable materials that must be maintained in the core.

Nuclear Fission Power in the Balance

Nuclear power in an ideal world could probably be said to be an unmitigated blessing. If nuclear power plants worked without failures and if nuclear fuels are never subverted to ulterior purposes, the large amounts of power available could serve at least as a transition between fossil fuel power and more exotic future power sources; at best nuclear power itself could serve as a major power source into the far distant future. Indeed, under ideal conditions, a nuclear power plant causes less environmental damage than the coal fired power stations that will become more and more common as oil and gas supplies diminish, as shown in Table 26-12, on page 624, which compares the environmental effects of nuclear and coal generating plants of similar generating power.

FUSION POWER

Nuclear fusion reactions may permit harvesting power from the nucleus without most of the radioactive contamination problems of fission reactors and with the promise of essentially limitless fuel supplies. If fusion powered generating plants become feasible, we will have entered a virtual energy millenium. Unfortunately, the problems still to be solved are immense.

If you will look again at the nuclear binding energy curve, Figure 26-7, you will find that the combination of two light atomic nuclei on the steep arm of the curve can produce a single heavy nucleus with a release of energy. Since the binding energy of the heavy nucleus is greater than that of the lighter nuclei that form it, there occurs a decrease in nuclear mass, which appears as energy. To cause two nuclei to fuse it is necessary to overcome the electrical forces of repulsion that keep them apart. At extremely high temperatures, such as occur in the interior of stars, this happens, and, indeed, fusion reactions are the power source of the sun and stars. On earth we have only been able to produce the necessary conditions for fusion in the hydrogen, or thermonuclear, bomb by placing the hydrogen to be fused in the center of a conventional nuclear explosion.

Should it prove possible to make the fusion reaction work in a sustained manner with less heroic measures, the benefits to civilization would be immense. The fusion power reactors envisaged by researchers would be far less hazardous than fission reactors since they produce only one radioactive product, tritium, a heavy hydrogen isotope. Actually, tritium would not be a waste product because it is itself a good fusion fuel. The only other hazards of a fusion reactor would be radioactivity in the reactor structure caused by the neutrons released by the fusion reaction. Thus a fusion power industry would be far less of an environmental hazard than a power industry based on fission.

In addition to the improvement in safety, fusion power would have the advantage of virtually limitless fuel reserves. The basic fuel component that will be used is the heavy hydrogen isotope deuterium. Although deuterium represents only 1/6000th of the hydrogen atoms in water, the vast supply of water on the planet that can be harmlessly relieved of its deuterium represents a fusion fuel supply equal to perhaps 100 thousand times the total amount of energy available from uranium and thorium, even with the most efficient possible breeder technology. Thus fusion power represents a safe energy haven to which, unfortunately, it seems we must cross by a makeshift and dangerous bridge of coal, fission, and other less important power sources from our present but vanishing gas- and oil-based energy system.

Biological Effects of Radioactivity

Now that we have explored the technology of nuclear fission power and have seen why fission power may prove to be essential to the world energy budget in the near future, it is essential to consider the biological effects that may result from release of radioactive wastes into the environment.

WHAT MAKES RADIOACTIVITY DANGEROUS?

The ultimate answer is that radioactive isotopes are capable of promoting destructive chemical reactions in living tissue for variable periods of time. This period is described in terms of what is called the **half-life** of a particular isotope. The half-life is a fundamental and unchangeable characteristic of an isotope. Depending on the isotope, half-lives range from fractions of a second to thousands of years. Each nucleus of a radioactive isotope has a characteristic probability of undergoing radioactive disintegration during which it emits energy and is converted into another isotope. The average frequency of these disintegrations determines the half-life.

For example, let us say a sample of a radioactive isotope is found to have its disintegration rate fall to half the initial value in 15 minutes. Then we speak of its half-life as 15 minutes. This halving continues on in the same way. Thus the second 15 minute interval would reduce the rate existing at the end of the first 15 minutes to half, and so on with halvings every 15 minutes until the disintegration rate becomes immeasurably small (Figure 26-11). Ten half-lives re-

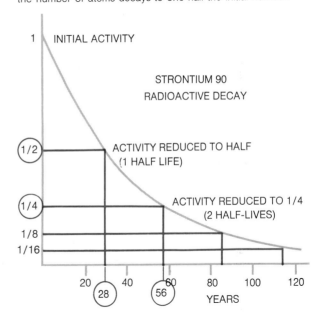

Figure 26-11 The half-life of a radioisotope, the time in which the number of atoms decays to one-half the initial number.

HALF-LIVES OF SOME OTHER IMPORTANT ISOTOPES

SODIUM	^{24}Na	15 HOURS
IODINE	^{131}I	8.05 DAYS
PHOSPHORUS	^{32}P	14.3 DAYS
RADIUM	^{226}Ra	1622 YEARS
CARBON	^{14}C	5770 YEARS
URANIUM	^{338}U	4.51×10^9 YEARS

DEGRADATION OF THE ENVIRONMENT

duces the radioactivity of a substance to 1/1000 the starting level. Since the way that a radioactive isotope produces destructive changes in living tissue is by means of the energy released by radioactive disintegration, it is obvious that the half-life of an isotope is one factor that must be considered in determining how long it will be dangerous.

The other factors in assessing danger relate to the *nature of the energy* released by disintegration and to the *biological affinity* of the isotope, that is, the extent to which the isotope is accumulated in biological systems (as, for example, radioactive iodine, which the thyroid accumulates, or radioactive strontium, which the body treats like calcium, thereby accumulating it in bone).

HOW ENERGY IS RELEASED IN RADIOACTIVE DISINTEGRATION

Energy is released in several forms when radioactive disintegration occurs. As shown in Table 26-3 energy may be carried by different kinds of particles or by electromagnetic radiation (refer to Figure 6-3), which, as you recall from the discussion of photosynthesis on page 83 may be thought of either as wavelike radiation or as streams of photons. Whether or not these disintegration products interact with matter and the nature of the interactions depends on certain of their properties. For example, a neutron, since it bears no electrical charge, is able to strike into an atomic nucleus, whereas an alpha particle, with a positive charge, interacts only with the electrons outside the nucleus.

RADIATION EFFECTS ON ATOMS AND MOLECULES

The types of radiation produced by radioactive disintegration may collectively be called **ionizing radiation** because the packets of energy that form them are large enough to cause ionization, that is, to dislocate electrons and break chemical bonds. The importance of the size of the energy packet is emphasized by the fact that a dose of x rays that would be lethal might represent in total only about 100 cal, which is about the energy one would receive from a few minutes in the sun. The difference is that the 100 cal of sunlight is partitioned among innumerable low-energy, visible light and ultraviolet photons, which do not carry enough energy to break chemical bonds.

Ionization may be caused directly or indirectly. A charged particle with sufficient energy may directly disrupt the atomic structure of the radiated tissue causing chemical changes. X rays, gamma rays, and neutrons act indirectly. An x-ray or gamma-ray photon may collide with an electron, transfer energy to it, and

Table 26-3 Some Important Types of Ionizing Radiation and High Energy Particles

	Properties	Relative Biological Effect (RBE)*
Radiation		
X rays	0 charge; electromagnetic radiation or streams of photons; gamma rays of shorter wavelength than x rays, have more carrying and penetrating power than x rays	1
Gamma rays (γ)		
Particles		
Electrons (beta particles)	− charge; wide range of energies	1
Protons	+ charge; mass about 2000 × mass of electron	1
Alpha particle	+ charge; 2 protons + 2 neutrons (= nucleus of helium atom)	10–20
Neutrons	0 charge; mass about that of proton	1–20

*See page 599.

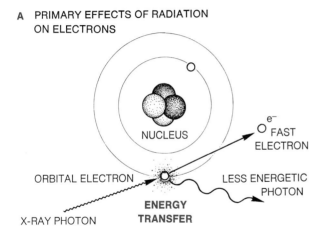

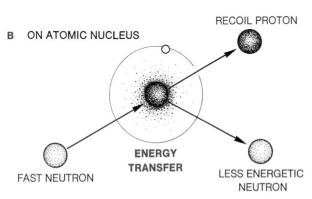

Figure 26-12 Primary effects of radiation. A. An x-ray photon collides with and transfers energy to an electron. The resulting fast electron is able to break chemical bonds. **B.** A neutron collides with a hydrogen nucleus, ejecting the proton as a recoil proton capable of producing a path of ionization and resultant chemical reactions.

cause it to be accelerated out of orbit as a fast electron capable of breaking chemical bonds. The radiation photon, having lost some of its energy, progresses onward until further collisions completely absorb its energy. Similarly, a neutron colliding with an atomic nucleus may eject a proton, which will generate a path of ionization (Figure 26-12).

The effect of ionization is the production of chemically reactive **free radicals**. Thus, if an atom has an electron knocked out of orbit, the atom has a net positive charge while the dislocated electron may join another atom giving it a negative charge. Such atoms are highly reactive and produce a large variety of chemical changes in tissues leading to biological effects.

RADIATION DOSAGE

The effect that each type of radiation may have on living tissues is not a straightforward property of the energy that it carries because of the variation in characteristics among types of radiation. Since it is important to have a system of quantification of radiation effects that applies to all forms of radiation, the following scheme has been developed:

The fundamental measure of radiation intensity is called the **rad.** The rad is defined as the amount of radiation that deposits 100 ergs of energy per gram of target tissue. Owing to the properties just described for the different types of radiation, 1 rad may have markedly different effects on biological systems, depending on the type of radiation. To correct for this effect, the **relative biological effect (RBE)** for each type of radiation has been empirically determined in comparison with the effects of x rays. Thus we see in Table 26-3 that the RBE of beta particles (electrons) is 1, which means that 1 rad of beta particles has the same biological effect as 1 rad of x rays; or we see that neutrons may be far more active in biological systems than x rays because their RBE, depending on their energy, ranges from 1 to 20. Finally, to allow ready intercomparison between radiation types, the dose in rads is multiplied by the RBE to give a new measure, the **rem***, thus,

$$\text{dose in rems} = \text{dose in rads} \times \text{RBE}$$

THE EFFECTS OF RADIOACTIVITY ON ORGANISMS

When we consider the fact that radiation may penetrate deeply into the organism to produce chemical changes in any molecule, it must be obvious that radioactivity has many effects. It is useful to divide these into acute and long term classes, although they merge into one another. There is no difference in the fundamental mechanisms of acute and long term effects since both are caused by randomly induced chemical changes.

---

*Rem stands for roentgen equivalent man. One rem is that dose in rads that produces a biological effect equivalent to that produced by 1 rad of x rays. The millirem (mrem) is 0.001 rem. Wilhelm Roentgen was the discoverer of x rays.

Acute Effects of Radioactivity

In acute radiation exposure, the chemical changes are so numerous that they produce failure in critical organ systems and lead, directly or indirectly, to death before the effects classified as long term have time to be expressed. Although radiation has effects on all types of molecules, it has the most biologically deleterious effects on DNA because of the multiplicative effects of DNA damage. Thus if a particle of radiation hits and puts out of action an enzyme molecule, the effect will go unnoticed since the affected cell probably has a substantial number of molecules of the enzyme and may produce more from the appropriate DNA template (review page 112). However, if the radiation particle happens to hit that DNA segment responsible for production of an enzyme, there is a chance that the DNA will be mutated and produce a malfunctioning enzyme or be unable to produce any enzyme at all. Depending on the half-life of the enzyme, the stricken cell will eventually run out of the

Note 26-1 *Acute Effects of Radioactivity on the Human*

The following quotation is from an account* of one of the relatively few instances in which acute radiation cases have been under close medical observation from soon after the incident until death. The victim received a total body dose of about 8800 rads (neutrons and gamma rays).

Late in the afternoon of July 24, 1964, a 38-year-old married father of 9 was pouring a "dirty" mixture containing ^{235}U from a polyethylene cylinder 12.5 cm in diameter and 120 cm high into a tank 63 cm in diameter containing sodium carbonate. A critical volume was attained by the new geometry near the completion of this operation, and a nuclear excursion occurred. The patient recalled a flash of light and was hurled backward and stunned, but did not lose consciousness. He immediately ran from the building to an emergency shack 200 yards away, discarding his clothing as he ran. There he was joined by 4 other occupants of the plant who had been alerted by the radiation alarm system. Almost at once the patient complained of abdominal cramps and headache, vomited, and was incontinent of diarrheal stool, which according to his colleagues was bloody. He was wrapped in warm blankets and taken to a nearby hospital (but not admitted) and transferred at once to the Rhode Island Hospital. He arrived at 7:49 p.m., 1 hour and 43 minutes after the accident.

He was taken at once to an isolated section of the emergency receiving service. He was complaining of severe abdominal cramps, headache, thirst and chilliness, and was perspiring profusely. He was incontinent of brownish but nonbloody diarrheal stool . . . The patient had transient difficulty in enunciating words. The neck was supple, and the lungs were clear to auscultation. The heart was not enlarged, there were no murmurs, and the quality of the sounds was good. The abdomen was rigid . . .

By 4 hours the blood pressure had dropped to 85/40, and the pulse had risen to 110 . . . By 8 to 10 hours after admission the patient reported that he felt well. There was no other evidence of adrenal failure. The temperature had risen to 102°F. The left hand and forearm, which had held the container, and were nearest the reaction, became edematous and red. Conjunctivitis and periorbital edema appeared on the left. He was alert and cooperative, and spent the time reading and talking. Visual acuity for newsprint seemed normal.

The patient was quite comfortable but slightly restless on the morning after exposure. The edema of the hand and forearm was increasing, and the fingers were moved with difficulty . . . The condition of the patient had deteriorated by the morning of the 2d day. He was restless, fatigued and apprehensive and had become more dyspneic. The left hand and forearm were badly swollen and livid, and there was massive edema of the upper arm . . . Vision had diminished to a point where he was unable to read 1-inch type, but he could still distinguish faces.

From this point the blood pressure could be maintained only with increasing difficulty. The heart increased in rate, and the sounds became tic-tac in quality. Six hours before death the patient became extremely restless and disoriented, the urinary output ceased, and the blood pressure could no longer be obtained. He died 49 hours after the accident.

* J. S. Karas and J. B. Stanbury (1965). Fatal radiation syndrome from an accidental nuclear excursion. *New England Journal of Medicine*, **272**:755–761.

Table 26-4 Acute Radiation Effects (Whole Body Dosage)

Dosage (rems)	Outcome	Primary Effects
10,000 and above	Death within 1–2 days	Neural and cardiovascular breakdown; kidney failure
1,000	Death incidence 90% in 1–2 weeks	Depletion of formative cells of rapid turnover tissues, gastrointestinal mucosa and blood; hemorrhage
500	Death 50–100% probable in 1–2 months	Depletion of blood forming cells; infections made severe by lack of cells conferring immunity; hemorrhage
250		
200	No mortality	Moderate depletion of blood white cells
100		

enzyme; this may cause its death, or, if it survives and reproduces, all its descendants will have an identically malfunctioning enzyme. When the dose of radioactivity is high enough, cells become unable to divide and produce viable daughter cells, presumably by a multiplicity of defects in the genetic apparatus. The results of this are most immediately felt in tissues where there is rapid and continual cell renewal. These include the circulatory system, the lining of the digestive tract, and the skin.

Table 26-4 lists some of the acute effects of high doses of radioactivity delivered over very short time intervals. Doses of about 10,000 rem cause death within 1 or 2 days. A primary cause is neurological deterioration, which is thought to be a secondary consequence of brain swelling brought about by sudden kidney failure. A medically well described example of such a fatality occurring after an accident in a nuclear fuel reprocessing plant is described in Note 26-1.

Lower but still highly lethal dosages lack the neurological effects, and death ensues from more gradually established effects resulting from disruption of important cell systems that require continual renewal. The intestinal mucosa fails to renew itself, resulting in hemorrhage, impairment of digestive processes, and infection by invasion of intestinal bacteria. Impairment of blood cell formation makes infection particularly hazardous because the immune system is inactive in the face of bacterial invasion.

Long Term Effects of Radioactivity

Radiation doses below 200 rem seldom have significant acute effects but have long term effects that may not be expressed for many years or, in the case of genetic effects, may not appear until subsequent generations. Nonspecific life shortening, that is life-shortening without an obvious cause, is one of the long term effects identified in animal experiments. Thus it is found that acute exposure to 100 rads reduces the life expectancy of mice by a little over 5 percent. There has been no clear cut demonstration of this effect in the human, since it is impossible to measure small variation in the human life span without unacceptable controls over the conditions of life. Statistical studies comparing life spans of medical radiologists (who until recently tended to receive large doses of radiation during their work) and Japanese nuclear bomb survivors with normal populations have shown no evidence of nonspecific aging. However, since radiation effects on experimental animals are generally quite comparable to those on humans where they are measurable, it is likely that the effect exists.

Cancer is a well recognized long term effect of radiation. Results of a typical experiment demon-

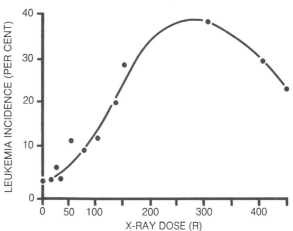

Figure 26-13 The relationship between radiation dose and incidence of leukemia in mice.

Data from Upton, A., *Cancer Res.* 21:717 (1961).

strating this fact on mice are shown in Figure 26-13, where it is seen that the frequency of occurrence of a form of leukemia steadily increases from barely detectable at 25 rad to about 35 percent at 300 rad. The decline in frequency above 300 rad is thought to be due to direct killing of cells at higher dosages, thereby reducing the number of viable cells at risk of becoming cancerous. Data also exists for human groups exposed to higher than ordinary radiation levels. Leukemia incidence is demonstrably elevated in the 24,000 Japanese nuclear bomb survivors who received doses greater than 10 rads. Japanese fishermen and Marshall Islanders exposed to fallout from the Bikini bomb tests have higher than normal incidence of cancer. Uranium miners suffer greater than normal incidence of lung cancer, probably from the radioactive dust they inhale. At maximal exposure rates in the Colorado uranium mines, the lung cancer incidence is over 20 times the normal expectancy. Bone cancer was a common fate of persons employed in radium painting the dials of watches. These unfortunates ingested large amounts of radioactive paint owing to the practice of shaping paintbrush tips by licking them. In more recent times it was a popular medical practice to treat "enlarged" thymus in children with x rays. Follow-up studies on these now mature persons show an increased incidence of thyroid cancer, which follows from the close proximity of the thymus and thyroid.

The many uncertainties of the data on humans make it difficult to estimate the actual cancer risk per rem of exposure. Two uncertainties in particular are difficult to evaluate. One is whether or not the *rate* of irradiation (whether the dose is received over short or long spans of time) influences the outcome; the second has to do with whether there is a *threshold* of radiation effect. That is, is there a radiation dose below which there is no effect, or does any amount, however small, have an effect? As far as cancer induction is concerned, there appears to be a dose-rate effect. The same dosage delivered over a shorter period of time has a larger effect than the same dosage over a longer period. There is much uncertainty as to whether there is a threshold for cancer induction. It appears that there may be and that it may be accounted for in terms of cellular repair processes that are effective if not swamped by large amounts of radiation damage.

Considering all these uncertainties, perhaps the best estimate of the chances of developing a radiation induced cancer within 20 years after exposure is 5.9 cases per 100,000 persons exposed per rem of exposure. If it is true that there is no threshold and if the dose-rate effect is small for low radiation dosages, then this cancer incidence becomes, for the 4.2 billion present world population, a rate of 250,000 cases per rem increment in the normal environmental radiation level, assuming uniform worldwide exposure of all persons. Thus the dose-rate and threshold problems are by no means simply games scientists play with vanishingly small effects. On these terms every slight increment in the world's radiation load has a price in human suffering.

Hotspots

Many isotopes are selectively assimilated by parts of the body and thus have a far more severe effect than would be indicated by the measured whole body dose. Thus, iodine-131 is accumulated by the thyroid and strontium-90 by bone. In these organs, hotspots are formed in which the radiation intensity is far higher than indicated by the whole body dosage. For the same reasons the danger from plutonium has recently been evaluated upwards. Workers with refined plutonium tend to be exposed to very fine airborne particles of the metal. If ingested, the particles pass through the gut relatively harmlessly, but if lodged in the lungs they become extremely hazardous. It has been calculated that a particle of plutonium lodged in the lung might deliver only 0.3 mrem (millirem) averaged over the entire mass of the lung, but actually be delivering 4000 rems to the tissue in the immediate vicinity. If so, the industrial safety limits for plutonium exposure might not be exceeded even though the subject has received a fatal, if highly localized, radiation dose.

The Genetic Effects of Radiation

The genetic effects of ionizing radiation have long been known. Nearly 50 years ago, geneticist H. J. Muller inaugurated this area of study by producing gene mutations in the fruit fly, *Drosophila*, with x rays. It is now known that ionizing radiation may produce both chromosomal and gene mutations on the basis of a large variety of experiments on plants, insects, mammals, and even human chromosomes in tissue culture. Chromosomal mutations have been observed in persons exposed to nuclear fall-out. As with radiation-induced cancer, there probably is no

Table 26-5 Natural Radioactive Isotopes

Isotopes Decaying to a Stable Form in One Step	Half-life (years)	Isotopes Decaying Through a Radioactive Series to a Stable Form	Half-life
^{40}K (potassium)	1.2×10^9	^{232}Th (thorium) to ^{208}Pb (lead), as follows:	
^{50}V (vanadium)	4.0×10^{14}	^{228}Ra (radium)	1.39×10^{10} yr
^{87}Rb (rubidium)	6.2×10^{10}	^{228}Ac (actinium)	6.7 yr
^{115}In (indium)	6.0×10^{14}	^{228}Th (thorium)	6.1 hr
^{138}La (lanthanum)	1.0×10^{11}	^{224}Ra (radium)	1.9 yr
^{142}Ce (cerium)	5.0×10^{15}	^{220}Rn (radon)	3.6 day
^{144}Nb (niobium)	3.0×10^{15}	^{216}Po (polonium)	52 sec
^{147}Sm (samarium)	1.2×10^{11}	^{212}Pb (lead)	0.16 sec
^{176}Lu (lutetium)	5.0×10^{10}	^{212}Bi (bismuth)	10.6 hr
^{187}Re (rhenium)	4.0×10^{12}	^{208}Tl (thallium)	60.5 min
^{192}Pt (platinum)	1.0×10^{15}	^{208}Pb (lead)	3.1 min
		^{235}U (uranium) to Pb (lead) in 11 steps	7.1×10^9 yr
		^{238}U (uranium) to ^{206}Pb (lead) in 14 steps	4.5×10^9 yr

threshold for mutational effects. The best estimate of the amount of radiation required to double the spontaneous mutation rate in the human is about 30 rem for acute exposure and about 100 rem for chronic exposure.

RADIATION LEVELS IN THE ENVIRONMENT

Having seen the effects of ionizing radiation and the doses required to produce them, we are now able to evaluate the biological risks of radiation from natural and man made sources.

Natural environmental radioactivity has two origins: cosmic rays and the natural radioactivity of the earth. Cosmic rays are high energy particles emitted by the sun and other astronomical sources. They bombard the earth continuously, and their interaction with matter at high altitudes results in the arrival at the earth's surface of a complex mixture of particles and the radioactive isotopes tritium and carbon-14. Cosmic radiation intensity is higher at high altitudes and is also influenced by the earth's magnetic field. A person living in Denver receives about a one third larger cosmic radiation dose than someone living at sea level.

The natural radioactivity of the earth's crust has many sources. Some comes from potassium-40, which has a half-life so long that it persists from the formative events of the elements. The radioactive heavy elements such as uranium and thorium, as well as many others, contribute substantial amounts of radiation to the environment (Table 26-5). Many of these show sufficient local variation in distribution to produce significant variations in surface radioactivity in

Table 26-6 Annual Exposure to Radioactivity in the United States

Source	Millirems per person per year
Natural sources: cosmic rays, earth radioactivity	70–200
X-ray diagnosis	75–100
Chest x ray = 200 mrems	
Intestinal tract series = 20 rems	
Minor sources: TV (x rays), watch dials, high altitude flight	2
Nuclear plants (1975)	
Average dose apportioned to entire United States population	0.002
Estimate for persons living within a few miles of a nuclear plant	5

various parts of the earth. Within the United States, the average annual exposure to natural radiation from all sources amounts to from 70 to 200 rems per person, whereas in parts of Brazil the natural radiation level is 1600 rems per person per year. For comparison Table 26-6 lists other low level sources of radiation exposure and shows that the principal man-made low level source is the medical x ray. One chest x ray delivers about 200 rems.

Medical X Rays

Medical x rays and radiotherapy are currently the major man-made causes of exposure to ionizing radiation. In 1970, 76.4 million persons had medical x rays and 59.2 million had dental examinations in the United States. Dosages used are not high enough to cause a measurable incidence of cancer but there is considerable interest in their genetic effects because of the very large numbers of exposures of reproductive age individuals and because of the general consensus that there is no threshold for radiation induced genetic effects. Consequently, the principal consideration in evaluating the risks from medical x rays for adults is the dose that the various types of procedures delivers to the gonads. Table 26-7 shows, as would be expected, that the dose to the gonads varies with the region of the body that is under examination. The gonad dose can be kept low since in most procedures, except those involving the midregions of the body, it is possible to screen the gonads with lead. Recognition of the genetic risk with resultant precautions and limitation of nonessential use of ionizing radiation has resulted in holding the dose received by reproductively capable persons to about 36 mrads/yr. This is thought to be much more than 1000 times smaller than the dose required to double the natural mutation rate. Even so, the world-wide additional mutation burden from medical x radiation is far from insignificant when one considers the immense number of exposures. The benefits of medical x rays to the patient must be always carefully balanced against the genetic risks.

Radiation of the fetus is to be avoided except in the severest emergencies. One authority maintains that any fetus that receives more than 10 rads during the first 6 weeks of pregnancy is a candidate for therapeutic abortion because of the strong likelihood of congenital abnormalities. Studies of pregnancies during which x rays were used, as well as evidence from surviving pregnant Japanese nuclear bomb survivors, show a high incidence of children born with abnormalities which included microcephaly, hydrocephaly, spina bifida, and clubfoot. The most sensitive period appears to be during the phase of organogenesis between the fourth and twentieth weeks.

RADIOACTIVITY HAZARD FROM A MAXIMAL NUCLEAR REACTOR ACCIDENT OR NUCLEAR BOMB EXPLOSION

In evaluating the risks from a possible hazard, it is useful to begin by considering the worst possible outcome. As far as nuclear bomb explosions are concerned the exercise is perhaps unnecessary since, except for the possibility of an isolated (nontest) explosion by a terrorist group, we may most probably expect bomb explosions to occur in very large numbers, if they occur at all, with consequences of a global nature. Events on such a scale lend themselves poorly to comprehensible quantification, so let us begin by considering the effects of a more imaginable but still extremely severe nuclear hazard, a complete meltdown and explosive core release from a nuclear power reactor. The chances of such an occurrence are deemed very small by many experts.

Should there be a maximal accident, the chances are that the radioactive release will occur within a day of the initial event. Since many power reactors are in or near populated regions, this means that a hasty and probably partially ineffectual evacuation of nearby populations will occur. There will be uncertainties as to wind direction changes during release of radioactiv-

Table 26-7 Radiation Dose to Gonads from X-ray Examinations

Examination Type (may involve more than 1 film)	Dose (mrads)	
	Testes	Ovaries
Skull	1	4
Teeth	8	1
Chest	5	8
Upper gastroinintestinal	137	558
Barium enema	1585	805
Hip	1064	309
Lower leg	2	1

Table 26-8 Examples of Radioactive Materials in Standard Reactor (1000 MW)

Isotope	Half-life	Amount (10^6 Ci)*
Noble gases		
^{85}Kr (krypton)	10.76 yr	0.6
^{133}Xe (xenon)	5.3 days	170
Iodines		
^{131}I	8.05 day	85
^{133}I	0.875 day	170
Telluriums		
^{132}Te	3.25 day	120
Cesiums		
^{137}Cs	30 yr	5.8
Alkaline earths		
^{89}Sr (strontium)	50.6 day	110
^{90}Sr	27.7 yr	5.2
Others		
^{91}Y (yttrium)	59 day	140
^{95}Zr (zirconium)	65.5 day	160
^{140}La (lanthanum)	1.66 day	160
^{143}Pr (praseodymium)	13.6 day	150
^{238}Pm (promethium)	86.4 yr	0.1
^{239}Pm	24,390 yr	0.01
Total inventory (43 isotopes)		3675×10^6 Ci

* 1 curie is an amount of material in which there are 3.7×10^{10} radioactive disintegrations per second. One Ci represents a large amount of radiation. For example, the Marshall Islanders described on page 602 are estimated to have had 1 day after exposure to fallout a body burden of radioisotopes amounting to about 10.5 microcuries (3.7×10^4 disintegrations/sec), principally in the form of ^{131}I, ^{89}Sr, and ^{140}Ba.

Data from H. W. Lewis, et al. (1975). Report to the American Physical society by the study group on light-water reactor safety. Revs. Mod. Physics, **47**:S1–S121.

ity into the air, and these will be expected to result in further uncertainties in the evacuation process. Nonetheless evacuation of nearby populations is absolutely essential to reduce casualties. Duration of evacuation may range from years to days, depending on distance and initial weather conditions and the extent to which decontamination is feasible and financially justifiable. The reactor site itself will remain dangerous and unuseable for many years.

According to one estimate, a 1000 MW reactor would contain 3675×10^6 curies of radioactive isotopes that might be released (Table 26-8). This represents an immense amount of radiation. If it were possible to distribute it evenly through the human population of the world, the dose per person would be about 0.9 curie per person, an abundantly lethal dose. To illustrate this, consider that the Marshall Islanders, who were as a group quite severely affected by bomb test fallout, are estimated to have received only about 10.5 microcuries per person. Thus the *distributed* world dose from one power reactor is more than 80,000 times the dose received by those unfortunate people.

Although such an alarmist statement is useful in emphasizing that reactors deserve respect, it must be realized that much of the radioactive load of a ruptured reactor is expected to remain in the immediate vicinity (by melting into the earth), and most of the remainder will be distributed in a downwind plume, in which the danger zone for acute effects will depend on variables such as wind velocity, surrounded by a larger zone where there would be elevated cancer and genetic risks. Even so the Atomic Energy Commission estimated in 1973 that the worst possible accident might kill 45,000 persons and injure 100,000. Estimates from other sources suggest that there would be on the order of an additional 50,000 cases of delayed cancer and developmental and genetic defects in immediate ensuing generations. The AEC report indicated that contamination might force land use restrictions for 500 years in an area the size of Pennsylvania.

As the preceding paragraphs demonstrate, should a nuclear power plant behave badly it will be very bad, indeed. It is also true that when it behaves well and all its wastes are safely stored—except for the relatively modest amounts of radioactive gases that are continuously vented into the air—a nuclear power plant is certainly far less immediately damaging to the environment than a coal or oil fired power plant. Table 26-12 lists the environmental costs of conventional and nuclear power plants. From these figures it is clear that the conventional power plant is the more damaging, beginning with the environmental damages of mining and continuing through to the emission of air pollutants. Thus, if the dangers of nuclear plant accidents or of dispersion of reactor wastes could be held to an acceptable minimum, then it would clearly be desirable to make as complete a switch to nuclear power as we can.

The current public controversy over nuclear power makes two points abundantly clear: (1) there is no agreement as to the definition of acceptable risk and, in fact, in the eyes of the public acceptable risks vary in a remarkable way with the *type* of risk; and (2) there is great public suspicion of the reliability figures stated by the nuclear industry.

With respect to the problem of acceptable risk it seems that the acceptability of a risk depends markedly on its familiarity and on whether or not it is undertaken voluntarily. Thus, by far the greatest individual risk that a person can undertake in the normal events of life is to submit to travel in an automobile (3 deaths per 10,000 population per year). If the automobile fatality rate were experienced in the nuclear industry (where the maximum probable mortality is said to be about 2 deaths per 100,000), or probably any other industry, that industry would undoubtedly be speedily shut down. The familiarity of the automobile and the fact that we use it voluntarily and for our immediate personal convenience renders its risks acceptable. The suspicion of the nuclear industries' reliability figures raises questions that are, of course, not entirely of a scientific nature. But at least we should note that when these suspicions are tranlated into the context of the worldwide nuclear industry, they take on even more substance because there is no guarantee that all nuclear reactors in the world will be operated at the safety standards that maintain in the United States and other major technological nations. Finally, it must be remembered that our safety estimates have not benefitted from direct experience with aging nuclear plants.

Nuclear Bomb Effects

Undoubtedly the most acute environmental hazard that civilization has produced, aside from the fundamental one of an over sufficiency of people, is the gigantic stockpile of nuclear weapons. It is now sufficient to kill many times over the entire populations of those nations likely to become involved in nuclear war. All out use of this arsenal would, in all likelihood, represent the final destructive act of our civilization.

We seem to have become dangerously callous about the power of nuclear weapons as directly evidenced by Hiroshima and Nagasaki. Those tragedies were a long time ago, and far away and, despite all, life has gone on. It may, therefore, be of some educational value to consider some Atomic Energy Commission estimates of what nuclear weapons might do in North America. First of all, we must note that today's nuclear bombs and missiles have their destructive energy described in terms of megatons (millions of tons) of TNT rather than the kilotons of the World War II weapons. Secondly, we must consider whether the weapon is fired at ground level or high enough in the air so that the fireball does not reach the ground. A ground burst produces far more "local" and immediate radioactive fallout than an air burst. The fireball of an airburst rises swiftly into the stratosphere. It carries with it much of the explosion-generated radioactivity, and this remains aloft for a very long period of time, providing opportunity for decay of short lived isotopes before the fallout returns to earth. To produce maximum damage and injury, a low altitude or surface burst would probably be used.

A low altitude explosion centered on Washington, D.C., in addition to prodigious local destruction, would, if weather and winds were right, generate severely injurious fallout in the major cities of the Northeast—the metropolitan complex stretching from Philadelphia to Boston. Or, as the AEC has estimated, one 20 megaton weapon exploded at 20,000 ft over Manhattan Island (without, let us add several *days* warning) would kill 4,000,000 outright and leave another 4,000,000 injured. The damage would be so widespread that there would probably be no way to provide significant help to the survivors.

The events described are for one weapon delivered to one major city. It must be remembered, however, that a nuclear attack—unless it is a terrorist operation—will involve as many weapons as the attacker thinks will be necessary to overwhelm defenses and produce enough destruction to prevent a retaliatory attack. Thus the festivities, if they ever begin, will involve a mass attack with over a thousand warheads delivered over the space of perhaps an hour and for which the recipients will have only the briefest warning. Such an attack could easily kill 60 million people immediately and cause another 60 million deaths in the following month. That would be half of the present United States population. In all likelihood identical reciprocal effects would be felt by the attacking nation since the brief warning of incoming weapons would provide time for a return strike. The two nations involved would have destroyed themselves. This fact is, of course, the source of our greatest hope that such an exchange will never occur.

Although the participants in a nuclear war will obviously have dealt themselves mortal damage, it remains to explore the effects of the radioactivity released upon the nonparticipating remainder of the world. If we assume that the war does not spread beyond the territories of the initiators and if the participants are both located in the same hemisphere, in all probability the northern hemisphere, then the equatorial separation of atmospheric and marine circulation (page 527) will isolate the fallout in the hemisphere of origin. Thus, there is relatively little fear of genetic damage in the nonparticipating hemisphere. In the half of the earth where fallout would be occurring, the survivors of the war would have far more demanding problems than genetic ones.

Nonparticipating territories in the same hemisphere would receive varying amounts of fallout depending on their proximity to the bursts and whether or not the explosions were predominantly aerial or ground bursts. Individuals receiving large amounts of radiation would be likely to die before reproducing, and that would serve to reduce the genetic load of radiation-induced mutations. As for the remainder of the population of the hemisphere, the most accepted estimate is that radiation would not more than double the natural mutation rate for more than a year or two. Probably the germ plasm of the species could readily survive this insult, but mutations will represent a final, long term cost of a nuclear war. It is a sobering footnote to history that consideration of genetic effects evidently did not enter into the decision at the highest level to use nuclear bombs on Japan. Clement Attlee, Prime Minister of Great Britain at the time the decision was made by President Truman, wrote in *Twilight of Empire* that, as far as he knew, neither Churchill nor Truman knew anything of the genetic effects of an atomic explosion.

Are There Energy Alternatives?

Plainly we are between a rock and a hard place when it comes to energy. Faced with growing energy demands and rapidly diminishing supplies of the currently most acceptable fuels (oil and gas), we seem to have only the choice of coal and nuclear power and, possibly in the indefinite future, fusion power, or of doing without. Solar and other less environmentally destructive energy sources promise useful contributions, but it appears that none of them can provide the backbone of projected energy requirements.

However, even the seemingly workable alternatives, coal and nuclear, plus wringing out the last of the earth's available oil, present a difficulty that has not been mentioned yet, namely, the growing cost of new energy sources. We have, of course, personally experienced the recent rises in heating oil and gasoline prices, but these are probably nothing compared with what we may expect to pay for energy if supply attempts to keep up with currently projected demand. A. B. Lovins, a physicist working on energy problems, illustrates this by estimating the costs of increasing a consumer's energy supply by the equivalent of 1 barrel of oil per day. He found that in the 1960s the cost was only a few thousand dollars, principally spent in the areas of transportation, marketing, and distribution. Today, the increased costs of obtaining oil from more and more forbidding places is illustrated by the estimated costs of Alaskan North Slope oil. If we remember that the Alaska Pipeline was the most expensive venture ever undertaken with private capital, it is not surprising that the 1 barrel/day increment from that source is estimated to cost up to $25,000. The spiral continues upwards with the estimate for the 1 barrel increment (electrical energy equivalent) from a new coal-fired generating plant placed at about $150,000 or about $200,000 to $300,000 for the 1 barrel equivalent from a new nuclear plant. The conclusion is that in addition to all other problems, environmental and otherwise, energy costs may rise so high in the near future as to prevent, for purely economic reasons, the currently projected rates of growth of the energy supply.

DOING MORE WITH LESS ENERGY

Many observers believe that the United States, the most profligate energy consumer in the world, could easily improve its energy efficiency by a factor of three or four. To do this would require a combination of technical "fixes" and reduction of consumption. The former involves the use of more energy-saving technology to produce results that otherwise would have a

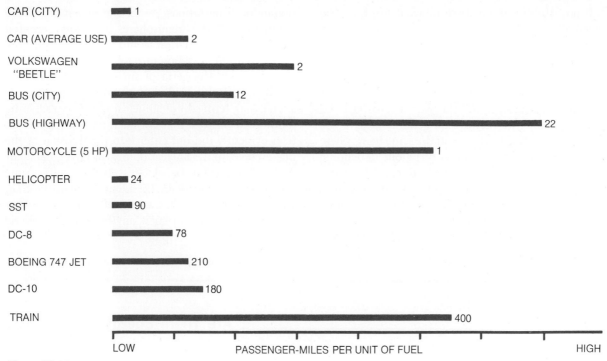

Figure 26-14 **Energy efficiency of various modes of transportation.** Numbers indicate typical number of passengers. Data from Rice, R. A. System Energy and Future Transportation. *Technol. Rev.,* Jan., 1972, p. 31.

higher energy cost. Improved insulation and design of homes to reduce heat loss is an important example. Reduction of consumption is clearly possible in transportation. A shift to smaller automobiles and to mass transportation are two obvious examples. Such efforts are not always easily accomplished, as evidenced by the continued excellent sales of large American cars and by the persistence of the SST, an extravagance whose energy efficiency is less than half that of the 747 (Figure 26-14). In addition to saving energy, the general tendency of economic systems that accomplish their ends with diminished energy consumption is that they tend to be labor-intensive. This has the great advantage of reducing the social problems stemming from unemployment as high energy industries are phased out.

Recycling of materials is also an important aspect of improving our energy balance (Note 26-2).

A shift to what he terms "soft technologies" is recommended by Lovins to take up the slack of energy requirements not fully mitigated by reduced consumption and technological fixes. The soft technologies would reduce the necessity of immense future investments in more huge power technologies, particularly nuclear. He points out that it is not sensible to use premium fuels for tasks for which high energy quality is unnecessary. Perhaps the most obvious example is electrical house heating. Electricity is inefficiently produced from premium fuels and thus should not be wasted on tasks that require only low grade heat. Such tasks—house heating, operation of hot water heaters, and even residential lighting—could be performed by soft energy technology and thus save much premium energy for tasks for which it alone is well suited. Soft energy technologies are those for which the energy source is local—wind and sunlight—and which may be harvested by local, even household systems of modest complexity. These solutions also have the advantage of being labor intensive and thus of value in maintaining employment as the national energy base shifts.

Such concepts are well worth pursuing, for it must be remembered that even a 1 percent reduction in conventional energy use will save the United States the equivalent of 100 million barrels of petroleum per year.

Note 26-2 *Fuel From Wastes*

One of the contributory causes of our rapid depletion of resources is waste. Each American generates about 3 kg/day of solid wastes. It is bad enough to plunder the earth as we do, but it is unconscionable not to utilize that plunder thoroughly before going back for more. Only recently have attempts been made to recycle wastes on a large scale. Ferrous metals, principally from automobiles, are extensively reclaimed, as are aluminum and glass. Such recycling, in addition to preserving natural resources, conserves the energy that would otherwise be used to produce these materials from scratch and reduces somewhat the waste disposal problem. There remains a vast bulk of carbon-containing wastes—paper, cloth, food scraps, plastics, rubber, and so on. These wastes may be converted to useful products by pyrolysis. This is destructive distillation at high temperatures and in the absence of air. The process produces combustible gases (principally hydrogen, methane, and ethylene), light oil, and tar. The remainder, called char, is also combustible. The energy budget is favorable, requiring about 1 Btu input to yield 4 Btu in useable fuels, and the process is less polluting than ordinary incineration of wastes, because the pyrolysis is conducted in closed vessels.

These wastes may be converted less expensively into a powdery dry fuel by heating and grinding without pyrolysis. About 1 ton of trash is converted, with modest energy cost, into half a ton of a coal substitute fuel. It has about the same energy content as intermediate grade coal but has a lower sulfur content.

Environmental Contamination and Destruction

A 50-year-old fisherman from a village about 8 kilometers from Minamata City developed progressive ataxia, dysarthria, and intention tremor during January, 1960. He had no fever, headache, or stiff neck. Within a week he was unable to walk; he became mute, violent, and had to be restrained. When observed in February, 1960, he was grossly demented; he would growl, stare wildly, and bare his teeth when disturbed. His pupils were dilated and equal and his eye movements appeared normal. There were few or no spontaneous movements of the extremities, although when he was approached by an examiner he would claw his hands in a threatening attitude. Stimulation of the plantar aspect of the foot resulted in violent withdrawal. No detailed examination was possible.

L. T. Kurland, *et al.* 1960
Minamata Disease. World Neurology, 1:370–395

This fisherman was one of countless victims of one of the ways in which we contaminate the environment. He and others, similarly afflicted down through the years, were victims of only one of the innumerable kinds of chemical wastes that find their way into the environment—into air, water, soil, and food—and into our persons. The culprit in this instance was mercury. This liquid metal has been with us as an industrial poison for a very long time. Undoubtedly you were at an early age introduced to one type of mercurial poisoning by way of the Mad Hatter in *Through the Looking Glass*. The name is apt. Until fairly recently hatters were rather commonly mad, or at least afflicted with severe neurological disorders, if they worked with felt because felt preparation required the use of mercury compounds. It was not until 1941 that the use of mercury nitrate was outlawed in the hat industry in the United States.

By no means did that stop outbreaks of mercury poisoning, which continue to this day. Some, in a sense, were deliberate, as when methylmercury dicyanodiamide is used as a fungicide on grain and then

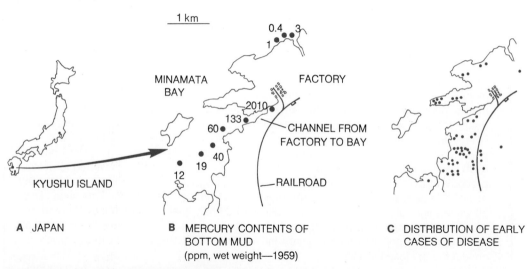

Figure 26-15 Mercury and Minimata Disease. A. Location of Minimata Bay. **B.** Mercury content of bottom mud in the bay in ppm. The highest concentration was 2010 ppm in the channel leading from factory into upper bay. **C.** Location of residences of early victims of the disease. After Kurland, L. T., et al., (1960) Minimata Disease. *World Neurology* 1:370.

accidentally finds its way into human food. So it was that, in 1969, seven members of a New Mexico family ate meat from animals fed methylmercury contaminated grain. Severe mental retardation of four children resulted. Or, in 1972, there were 6530 cases and 459 fatalities in Iraq caused by eating grain treated with organic mercury compounds. The grain had been treated because it was intended for use as seed rather than for direct human consumption.

In such cases the means of poisoning is direct and obvious. But what about the fisherman? His dose of mercury came from eating the seafood that he caught every day in Minamata Bay, a small bay at the southern end of one of the islands of Japan, Kyushu (Figure 26-15). Determining what disabled him and the many others similarly afflicted at about the same time and then tracing the origins of the contaminating mercury required persistent medical detective work.

Initially the disease was called *kibyo*, Japanese for "mysterious illness." At first an encephalitis was suspected; however, when bacteriological and viral studies proved negative, attention switched to the possibility that the disease might be a poisoning from eating fish and shellfish. Nearly all of the immediate victims, about 83 children and adults, lived in villages around Minamata Bay and relied heavily on fishing in the bay for food (Figure 26-15). Suspicion was focused even more on this possibility by the discovery of cats suffering the same symptoms. Cats with the disease staggered, had convulsions, or whirled violently in circles. Cats are, of course, partial to fish. An experiment was performed in which normal cats were fed marine life from the bay. They, too, came down with the disease.

Certain features that appeared at autopsies of human victims suggested that the marine organisms were not transmitting a biological poison, as can readily happen, but that they might have their effect by carrying a metal poison such as manganese, selenium, or thallium. After nearly 2 years of studies it was realized that the symptoms were very similar to those already well known for poisoning by organic compounds of mercury (Figure 26-16). At this point the mercury content was measured in the silt of Minamata Bay, in fish and shellfish from the bay, and in tissues and urine of patients. As Figure 26-15 shows, there was a great deal of mercury in the Bay. There were also large amounts in the disease victims. In normal individuals mercury in the urine was below 16 mg/liter while patients in the first few months after appearance of symptoms had from 40 to 116 mg/liter mercury in the urine.

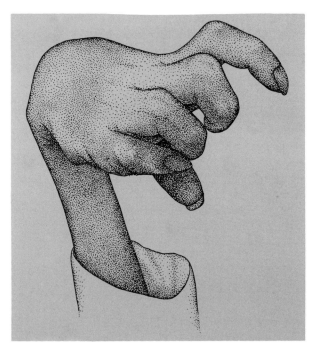

Figure 26-16 The hand of a victim of Minimata Disease states the tragic neurological effects of mercury poisoning.

What was the mercury source? Suspicion finally came to rest on a chemical plant a short distance from the Bay. This plant had produced a variety of chemicals for many years but started production of vinyl chloride in 1949. This is significant because the process uses mercury as a catalyst. Moreover, it developed that, in 1950, a year before the disease appeared, wastes from this plant, which had previously been dumped into the Shiranuhi Sea, were switched by a direct channel into Minamata Bay. Methyl mercury was identified in the waste water of this factory and was found to be discharged into the Bay in amounts ranging from 500 to 2000 g/day. Eventually the operators of the plant, after a court decision against them, apologized to the victims and made large reparations. By early 1975 the responsible corporation had paid damages to patients in excess of 80 million dollars.

The outbreak ultimately involved far more persons than those in the immediate vicinity of Minamata Bay. By late 1974 there were 798 "officially verified" patients living around the shores of the Shiranuhi Sea. Further casualties of this episode of environmental contamination may be expected since organic mercury compounds are persistent in the environment.

Biological Effects of Mercury

When taken by mouth, inorganic forms of mercury are not very quickly absorbed into the body and that which is absorbed is excreted rather rapidly. Organic mercury compounds such as methyl mercury (CH_3Hg^+) are a different matter. They are absorbed rapidly and are lost slowly from the body with a half-life in excess of 70 days. Mercury vapor is also particularly hazardous, since absorption through the lungs is more efficient than the gastrointestinal route. Mercury in the body is accumulated particularly by kidneys and brain. In the latter it concentrates in regions such as the cerebellum, which fact is reflected in the severe locomotor disturbances in mercury poisoning. Rarely is there significant recovery.

The long half-life of mercury in living tissues causes its accumulation in higher and higher concentrations as it progresses through food webs to larger organisms that are ultimately used for human food. Thus, in the Minamata Bay disaster the concentrations of mercury in shellfish, fish, crabs, and lobsters caught in the bay also ranged to 30 ppm or slightly higher. These levels probably represent nearly lethal amounts. Accumulation is more striking in studies conducted in regions where the initial mercury concentrations are not high. For example, in a freshwater food chain, the alga eaters at the bottom of the chain were found to have maximum mercury concentrations of only 0.18 ppm, whereas the value for top predators of the system was 5.82 ppm, a better than 30-fold increase in concentration. It is thus little wonder that persons eating seafood from Minamata Bay were poisoned. They were at the receiving end of a chain of what has been called **biological magnification**.

Another characteristic of the way biological systems react to mercury also renders it more dangerous than it otherwise might be. This is the fact that inorganic mercury that enters the environment can be converted to the far more toxic organic forms by a variety of microorganisms. Not only is it then more toxic, but it is also concentrated in many aquatic organisms by direct uptake from the water as well as by eating contaminated food. The fact that mercury in the environment is readily convertible from the inorganic form means that its toxicity will persist in a region of contamination for a very long time. It is not like an entirely organic molecule that may be rendered permanently harmless when broken down. Any compound of mercury is toxic, and even if the contaminating mercury should be somehow converted to the elemental form, it can be incorporated into organic compounds again by bacterial action. The result is that once a site like Minamata Bay is contaminated with mercury, the mercury remains there in a dangerous form until it diffuses away or is buried below the level of organisms that contribute to the food chains that humans harvest.

Ignorance and Economic Expediency

Why did the Minamata tragedy happen? In many instances of environmental contamination with harmful results to human life the principal contributory cause is ignorance. When DDT was discovered to be such a wonderful insecticide, it came into virtually universal use before anyone realized that it was accumulated by biological systems or fully understood its toxic effects on vertebrates. We can no longer afford this type of ignorance. What makes the Minamata affair particularly bitter is the fact that the toxic effects of mercury have been well known for years. Since the factory operators must have known that their wastes contained mercury, their only excuse, if it can be called that, is that they thought the mercury would be sufficiently diluted to prevent clinical symptoms appearing. Having seen the suffering and paid the financial costs engendered by their action, the operators probably wish that they had gone to the effort of cleaning up their plant effluents in the first place.

They were, as a matter of fact, behaving in a relatively common way for the industrial world. The operator of any industry is in business to make money and quite sensibly is not going to make any changes in his operations that reduce profits unless they are absolutely necessary. To clean up a plant's effluent costs money, and any sensible industrialist will not embark on such a course unless the law requires such action or unless it is likely that more money will be lost by not cleaning up the effluent than by doing so. We are back once more to the Tragedy of the Commons. This time the Commons is the air and water that is vital to life.

But perhaps Minamata was the exception and things are not as bad as they seem. Unfortunately, this is not so. Even as the Minamata episode was being investigated, there was another virtually identical mercury poisoning episode in Japan. Even in the United States, although there have been no major mercury poisonings, other examples of poisoning abound. As recently as 1975 production of a highly toxic insecticide, kepone (Chlordecone), in a small plant under the most rudimentary safety conditions resulted in severe poisoning of several workers and contamination of the commercial fisheries of the James River in Virginia. We must conclude that ignorance and the use of technology without sufficient regard for human welfare are continually resulting in great damage to the ecosphere and ultimately to our own persons. The following section outlines the principal ways these damages are caused and indicates the nature of their biological costs.

ENVIRONMENTAL CONTAMINATION: SOURCES AND EFFECTS

There are now so many humans that environmental contamination is unavoidable. Even with the best of intentions, human demands on the resources of the planet are, owing simply to our numbers, so great and the wastes from human activities are so plentiful that there is simply no economically realistic total technological solution to the problem. As we have already seen for the energy problem, the basic factor is human population growth. Limitation of population growth and reduction of the material demands of humanity are the primary factors that can establish limits to the extent of contamination of the environment. However, within these limits, more widespread understanding of the means and risks of contamination together with an increasingly humane use of technology may serve to limit the impact of the inevitable sufficiently to permit the world to remain a reasonably healthful and esthetically tolerable place in which to live.

The task is immense. Civilization pumps more than 800 million tons of pollutants into the air yearly, enough not only to produce severe direct effects on life but probably also enough to cause modifications in climate and in the amount of ultraviolet radiation that reaches the earth's surface. Pollution of marine and fresh waters is equally severe, as evidenced by the 66 billion dollar estimated cost of reducing water pollution within the continental United States alone for the 5-year period ending in 1975. The sources of all these contaminants are virtually impossible to cata-

Table 26-9 Air Pollutants

Pollutant	Sources	Typical Effects
carbon monoxide (CO)	Natural decay (93% of total) Internal combustion engines, 25–120 ppm* in city traffic Cigarette smoking, 300 ppm in smoke	O_2 deprivation by inactivation of hemoglobin.
carbon dioxide (CO_2)	Low concentration normal in atmosphere, elevated levels due to increased burning of fossil fuels.	Possible increase in global temperature.

sulfur oxides (SO_x)
  sulfur dioxide (SO_2)
  sulfure trioxide (SO_3)
  sulfuric acid (H_2SO_4)
	Sulfur containing fuel (coal, oil). SO_2 converts to SO_3 and sulfuric acid in air.	Respiratory tract irritation, damage to vegetation, destruction of metals, paint, plastics.

nitrogen oxides
  nitric oxide (NO)
  nitrogen dioxide (NO_2)
	Natural processes (1 billion metric tons/yr world)
	Internal combustion engines, industrial boilers (50 billion metric tons/yr world)
		Respiratory tract disease, emphysema
		Damage to vegetation
		Metal corrosion

photochemical oxidants
  ozone (O_3)
  peroxyacetyl nitrate (PAN)
    $CH_3CO_3NO_2$
	Reaction of NO_2 with vaporized hydrocarbons in sunlight	Reacts with hydrocarbons to form toxic compounds producing eye and respiratory tract irritation, damage to vegetation, deterioration of fabrics, rubber.

hydrocarbons
  methane (CH_3)
  benzene (C_6H_6)
  (many others)
	Partially burned petroleum fuels
	Car and truck refueling
	Industrial processes
	Forest fires and agricultural burning
	Natural decay
		Contribute to formation of photochemical smog, some are carcinogenic

particulates (aggregates of solids or liquids up to 500 μ dia.) asbestos	Building and ship insulation Automobile brake lining wear Certain types of rock used in road surfacing	asbestosis (emphysema-like effects) lung cancer, mesothelioma
coal dust	coal mining	black lung disease of miners
metals lead	Fuel anti-knock compounds (use undergoing reduction), coal burning	neurological disorders
mercury	Industrial processes, agricultural chemicals	neurological disorders
allergins	Primarily pollen, especially ragweed Fungi from moldy hay Chemicals used in polyurethane paint	hay fever, asthma farmer's lung asthma
miscellaneous fluorine compounds	Phosphate fertilizer manufacture Aerosol propellants (chlorofluorocarbons such as $CFCl_3$)	Damage to vegetation Possible reduction of ozone layer
industrial and agricultural chemicals ammonia	Fertilizer	Corrosive and lethal in high concentrations
chlorine	Industrial processes	Corrosive and lethal in high concentrations, skin disorders in chronic exposure.

*Parts per million (volume)

log, and their number constantly increases at a rate limited only by the ingenuity of technology to add new products to the list of combustion products, home and agricultural wastes, fertilizers, pesticides, and industrial materials that already flow through the environment.

AIR AND WATER POLLUTION

Essentially all pollutants appear sooner or later in air or water. There they may exert direct effects on living systems or undergo secondary conversion into substances that do. They may have a long dwell-time, as we have seen in the case of mercury in water, or they may be rapidly broken down or otherwise rendered harmless. The complexity of biological systems and the tremendous variety of contaminants renders it virtually impossible to predict their effects in any detail. There are no general rules except the general one that any new intrusion on the environment more than likely will be harmful. The following examples are illustrative.

Air Pollution

Air pollutants, both gaseous and particulate, are produced by natural sources, such as wind-blown dust or volcanic activity, and by human activities, particularly transportation, power plants, industry, heating, and waste disposal. Many of these pollutants are hazardous to human health, especially as causes of respiratory diseases and cancer, and also are damaging to crops and domestic animals and even to property. Some of the principal kinds of air pollutants are listed in Table 26-9. We will consider in detail some of those that illustrate particularly important aspects of air pollution.

Ozone. In the stratosphere ozone (O_3) is formed by the action of high energy ultraviolet light on oxygen to make a critically important screen to filter out biologically damaging ultraviolet light. However, in the lower atmosphere, ozone can be an important pollutant. High concentrations of ozone are developed by the photolytic breakdown of NO_2 (nitrogen dioxide), a prevalent air pollutant produced in large amounts by internal combustion engines. Solar energy converts NO_2 into NO (nitric oxide) and O (monatomic oxygen), an extremely reactive oxidant. This combines with molecular oxygen to form ozone. Since the reaction is readily reversible, very high concentrations of ozone would not build up except when there are other molecules present that can also react with NO. Thus O_3 is to a large extent prevented from breaking down, and its concentration increases in the air when hydrocarbons are present. Hydrocarbons in the air become oxygenated, and in this state are highly reactive with NO. Consequently, when hydrocarbons are present in quantity, as in areas with heavy automobile traffic, O_3 builds up to high concentrations. This interrelationship between O_3, NO, and NO_2 is demonstrated by the sequence of buildup of these pollutants during the daylight hours. During the night, levels of all three pollutants are low. As automobile traffic increases in the morning and the sun rises, both NO and NO_2 begin to increase in concentration. When O_3 begins to peak, NO actually decreases as hydrocarbons promote its conversion to NO_2 (Figure 26-17).

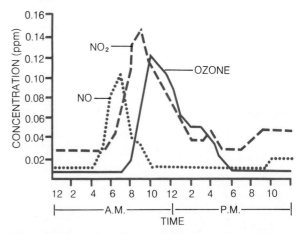

Figure 26-17 Daily variation in pollutant levels shows the interaction of sunlight and automobile exhaust. U. S. Dept. H.E.W., Air Quality Criteria. No. AP-63, 1970.

In animals O_3 enters the body through the respiratory system (Figure 26-18), although it may also produce eye irritation directly. In high concentrations it causes inflammation of the respiratory tract, pulmonary edema, and difficulty with respiration. As with any irritant of the respiratory tract, its effects are far more serious in individuals already suffering from respiratory difficulties such as asthma or emphysema, or who smoke (Figure 26-19). The fundamental actions of O_3 are undoubtedly related to its oxidizing ability since biological antioxidants (page 98) such as vitamin E prolong the survival of experimental animals exposed to high levels of ozone.

Carbon Monoxide. This product of incomplete combustion is equal in concentration to all the other pollutants characteristic of the urban environment. The principal sources are vehicular and stationary gasoline engines. Cigarette smoking provides the smoker with a personal carbon monoxide (CO) enriched environment. Even in the absence of pollution large amounts of CO enter the air from biochemical processes in ocean and marshlands as well as from volcanoes and forest fires. Processes that are not well understood serve to limit the general increase of CO in the atmosphere, although very high local concen-

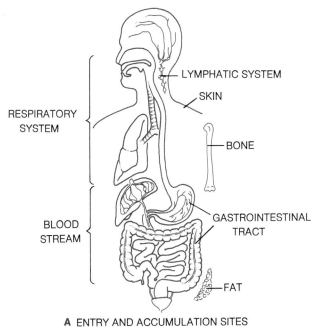

A ENTRY AND ACCUMULATION SITES

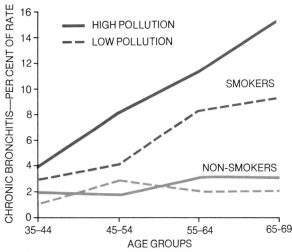

Figure 26-19 Atmospheric pollutants are more likely to be associated with lung disease in persons already prone to respiratory difficulty, as shown by this comparison of the incidence of chronic bronchitis in smokers and non-smokers. After Lambert, P. M. and D. D. Reid (1970) *Lancet*.

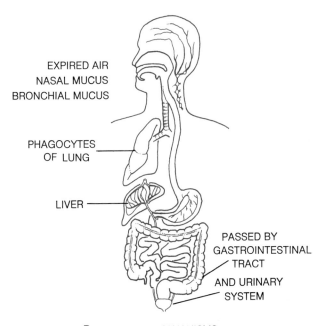

B CLEARING MECHANISMS

Figure 26-18 Sites of access, accumulation and clearance of pollutants in the body. A. Skin, respiratory and gastrointestinal systems are sites of entry with distribution and accumulation taking place in blood and lymphatic systems and accumulation taking place in bone, fat and other tissues.
B. The primary locations of clearance mechanisms are respiratory, gastrointestinal and urinary systems. Many important detoxifying enzymes are found in the liver.

trations are attained under heavy traffic conditions, particularly when atmospheric conditions prevent air circulation.

In high concentrations CO is, of course, lethal because it reacts with hemoglobin to form methemoglobin, which is ineffective in respiration. Lethal concentrations are attained only in enclosed spaces, usually when there is incomplete combustion. Faulty gas heaters or charcoal fires used for space heating are notorious sources of lethal concentrations of CO. Concentrations attainable under open air conditions have relatively minor effects on human behavior, resulting in somewhat increased reaction times and impairment of judgement. This might, of course, be contributory to accidents, since the principal times of exposure to CO are during heavy traffic. There are also indications that chronic exposure to CO may worsen heart disease.

Carbon Dioxide. Carbon dioxide (CO_2) is a normal constituent of the atmosphere and an essential component of the world respiratory cycle. It is directly harmful to organisms only in concentrations that are virtually impossible to attain in the open air. It is nonetheless a cause for worry because the increasing amounts of CO_2 that are entering the air, principally owing to the burning of fossil fuels (Figure 26-20), may cause a worldwide temperature elevation. The

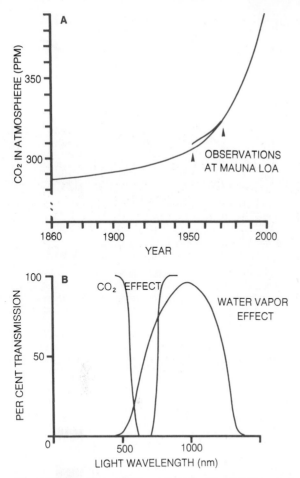

Figure 26-20 Atmospheric carbon dioxide and the possibility of world heating. A. Estimate of the amounts of carbon dioxide in the atmosphere through the year 2000. Measurements at Mauna Loa Observatory between 1958 and 1972 support this estimate. **B.** Both carbon dioxide and water vapor are barriers to loss of long wavelength thermal energy from the earth but admit high energy, short wavelength solar radiation. Based on Wang, W. C., et al., (1976) *Science* 194: 685, and Dyssen, D. and D. Jagner (1972) *Changing Chemistry of the Oceans*, p. 48. Wiley, New York.

causal mechanism is called the "greenhouse effect"* because one of the factors that results in heating greenhouses is the greater transparency of glass to short wavelength than to long wavelength radiation. Visible light enters easily and interacts with matter in the greenhouse producing long wavelength infrared radiation. This is trapped by the glass and thus con-

---

*Although commonly used, the greenhouse example is not very good because heating of a sunlit greenhouse is facilitated by reduction of air flow as well as by entrapment of infrared by glass.

tributes to heating the enclosed space. As Figure 26-20 shows, the contents of the earth's atmosphere work similarly. Water vapor, for example, does not transmit light well below about 500 nm (violet) but does let longer wavelengths through. Thus there is an atmospheric window for heat loss. CO_2 "closes" a major part of that window by absorbing heat rays in the 600–700 nm range. One would thus expect that, if the atmospheric CO_2 levels rose enough, the world would become warmer.

As the figure shows, atmospheric CO_2 does seem to be rising. Is there a comparable increase in world temperatures? The question is not simple to answer because, as CO_2 rises, other things occur in the atmosphere that might obscure the CO_2 effect. The most important of these is the accumulation of particulate matter in the air. This would be expected to parallel the rise in CO_2 since both are produced by burning fossil fuels. One group of scientists appears to have gotten around this problem by analyzing temperature records in the Southern Hemisphere. This approach was used because fossil fuel pollution occurs mostly in the northern Hemisphere. Particulates do not stay aloft very long and this fact along with the equatorial division of atmospheric circulation (see page 527) confines the particulates to the Northern Hemisphere while the CO_2 freely diffuses into the Southern Hemisphere. Thus it is believed that the Southern Hemisphere provides a relatively uncomplicated measure of the extent of the CO_2 greenhouse effect. The investigators believe that there is a temperature rise occurring, less than 0.1°C over the several years that data are available. However, since the projection in Figure 26-20 indicates a continued marked increase in atmospheric CO_2, there is cause for worry. Some authorities even believe that the CO_2 burden from burning coal will prevent its full exploitation as an energy source. Finally, the problem of atmospheric CO_2 is further complicated by worldwide progressive destruction of forest vegetation. It may prove necessary actively to reforest since plants trap CO_2 in photosynthesis.

Sulfur Compounds. Most sulfur compounds in the atmosphere originate from burning fossil fuels, principally coal. On burning of fossil fuels, sulfur compounds enter the atmosphere as SO_2 and SO_3. The SO_3 immediately combines with water vapor to form sulfuric acid (H_2SO_4) and the SO_2 is slowly converted into SO_3. The result can be an extremely corrosive environment in which metals, limestone building materials, and many fabrics are attacked. Even brief exposures can be extremely damaging to vegetation.

Effects on humans may be devastating. The worst recorded air pollution episodes, "killer smogs," have been characterized by the presence of high concentrations of SO_2 emissions combined with high levels of particulates and water vapor during stagnant atmospheric conditions. The highest per capita death rate ever recorded for an air pollution episode was recorded in 1948 in Donora, Pennsylvania. Smog was responsible, over a 4-day period, for 20 deaths and 5900 illnesses out of a population of only about 14,000.

Nitrogen Compounds. Nitric oxide (NO) and nitrogen dioxide (NO_2) are produced in high temperature internal combustion engines by the reaction of nitrogen from air or fuel with oxygen. The two compounds are often referred to as NO_x, which denotes the total of the two oxides without reference to relative concentrations of each. In the discussion of ozone we have already seen how NO_2 reacts with organic molecules to facilitate the buildup of ozone. One of the commonest compounds thus formed is PAN, peroxyacetyl nitrate.

$$CH_3-\underset{\underset{O}{\|}}{C}-O-O-NO_2$$

All of these nitrogen compounds are noxious but do not appear to have any serious effects on humans at the concentrations that currently appear in the environment. However, PAN in particular is toxic to plants.

Photochemical smog is a mixture of these nitrogen compounds with various hydrocarbons together with the reaction products occurring in the presence of sunlight, for example, ozone. Photochemical smog is a particular plague of cities such as Los Angeles, where bright sunlight, heavy vehicular traffic, and terrain and atmospheric conditions combine to produce a particularly obnoxious brownish orange haze (Figure 26-21). The irritating effects of PAN are the source of most public complaints about photochemical smog.

Hydrocarbons

More than 150 gaseous hydrocarbons are emitted into the atmosphere by incomplete combustion of

Figure 26-21 Los Angeles smog, before and during a severe episode.

gasoline and diesel fuel used in automotive transport. A second very large source is evaporation of fuel during refining, transport of fuel, and filling vehicle fuel tanks. The effects of most of these hydrocarbons are unknown. A few aromatic compounds are known carcinogens, and many are known to directly contribute to the buildup of photochemical smog.

Fluorocarbons ($CFCl_3$, CF_2Cl_2, and so on) are used for refrigeration, as propellants in aerosol sprays, and as foam plastic expanders. These have no significant ground-level effects, but strong evidence has developed that these stable compounds are carried into the upper atmosphere where they participate in chemical reactions that would be expected to reduce the protective layer of ozone. High energy ultraviolet light decomposes them, releasing chlorine, which reacts with ozone as follows:

$$Cl + O_3 \rightarrow ClO + O_2$$
$$ClO + O \rightarrow Cl + O_2$$

Thus, a relatively small amount of chlorine reaching the stratosphere has a very large effect because it is regenerated by the reactions that break down ozone.

Particulates. These are an extremely heterogeneous collection of pollutants that form particles or droplets small enough to remain airborne for a long period of time. Many are carbonaceous, sooty, and contribute to the heath hazards of killer smogs. Some carry aromatic molecules such as benzypyrine, which, when breathed, may cause lung cancer, or when swallowed with respiratory tract mucus (Figure 26-18), produce stomach cancer, which reminds us that the first known occupationally related cancer was in chimney sweeps.

Particulates constitute acute health hazards in certain industries. The coal dust of coal mines accumulates in the lungs to produce a progressive, extremely debilitating and frequently fatal disease, black lung. Similarly, quarry workers often suffer from silicosis. Asbestos has relatively recently come to light as the cause of a silicosis-like disease, asbestosis, and as an extremely hazardous cancer causing agent. Its use in the building trades has been sharply curtailed.

The common factor in these conditions is the action of various forms of silica (SiO_2) on the lungs. Silica induces extensive phagocyte activity in the lung and appears to produce aberrant responses in the phagocytic cells so that large amounts of fibrous tissues and nodules are formed. The result is seriously impaired pulmonary function and greatly increased susceptibility to bacterial invasion, particularly tuberculosis. Asbestos produces similar effects and, in addition, lung cancer, beginning on the order of 20 years after beginning of exposure. A peculiarity of asbestos exposure is a rapidly fatal form of cancer, mesothelioma, which involves the pleural lining of the lungs or the peritoneum. Although the disease is rare in the general population, it accounts for nearly 10% of the deaths of long term asbestos workers.

Lead is a highly unusual particulate because, if it were not for human activities, it would never appear as an air pollutant. Human activities tremendously increase the flow of lead in the environment. Natural release of lead by weathering of rock is estimated to be about 380×10^9 g/yr worldwide, whereas about 400×10^9 g/yr enters the atmosphere as gasoline additives. Lead is added to gasoline usually as tetraethyl lead, $(C_2H_5)_4Pb$, to improve the antiknock properties of gasoline, that is, to make it burn more smoothly. The result has been a great increase in the lead content of the surface waters of the oceans and very substantial concentration in plants and animals. Along highways plants commonly have nearly 100 times the normal levels of lead. Similarly, lead poisoning is reported in children living in areas of traffic congestion.

Lead is absorbed through the lungs or digestive tract and is only slowly excreted. It tends to accumulate in bone. Acute lead poisoning is rare, but it is fairly common as a chronic disease, which is characterized by gastrointestinal, kidney, and central nervous system involvement. The effects are not reversible. Fortunately, lead is now rapidly being removed from gasoline. Curiously, the primary instigation is not its toxicity but the fact that it interferes with the catalytic converters now required on new cars to eliminate other forms of air pollution.

Water Pollution

In 1974 the Congress passed the Safe Drinking Water Act prompted in part by studies showing the presence of trace amounts of many dangerous chemicals, including known carcinogens, in drinking water from the lower Mississippi River. This action recognized the obvious, that the rivers of the United States are serving a dangerous multifunctional role as sources of drinking water, as sewers for human and animal wastes, and as dumps for soluble industrial wastes (Figure 26-22). In the United States the Federal Water Quality Administration estimates that there has been a sixfold increment in water pollution since 1910. In many parts of the country primitive water purification plants and equally primitive, or sometimes nonexistant, sewage treatment plants provide only a thin line of protection against this onslaught. As one student of the problem observed, the Mississippi is the colon of America. Of course, the problem is not America's alone, as evidenced by the Rhine, which suffers even more than the Mississippi since it flows through industrial zones of more than one country, making pollution control legislation difficult.

Figure 26-22 A British river with a detergent "head." Such scenes are now rarer due to the increased use of biodegradable detergents.

Contamination of the lakes and rivers of the nation has a compounding effect, which makes the hazard more serious than simply the direct effects of the contaminants that get through the purification systems and into our drinking water. Many pollutants also act to change the ecology of these bodies of water in such a way that their natural detoxification effects are diminished. A healthy body of water is able to cope with a certain amount of contamination and, in the case of many substances, render them harmless. But once this capacity has been exceeded, the body of water suffers a degenerative positive feedback and becomes nothing less than a sewer conveying contaminants from one place to another and even adding noxious substances emanating from the degenerative changes in the stream itself.

Before 1900 the major problem in water pollution in the United States came from viral and bacterial disease. Typhoid, cholera, bacillary dysentery, and a host of other agents of disease entered streams and lakes as raw sewage. Today, although there are millions of gallons of raw sewage still entering the waterways, the input has been drastically curtailed and drinking water is almost universally treated by filtration and chlorination. This has produced a great improvement, but there is still room for progress as evidenced by the U.S. Public Health Service estimate that about 40,000 cases of disease per year are caused in the United States by bacterial or viral water contamination. Bacterial counts continue to reach levels high enough regularly to force curtailment of swimming and fishing or shellfish gathering in many areas along the ocean shore and on major lakes and rivers. Despite the improvements in sewage treatment, the steadily increasing total load of wastes entering bodies of water together with heating (thermal pollution) from power plant cooling water has, in many areas, increased the danger of contracting diseases from swimming or other use of untreated water. In addition to favoring the traditional diseases contracted from contaminated water, conditions of high organic content and warmth are thought to favor bacterial growth that is necessary for multiplication of an ameba (*Naegleria*), which is the causal agent of a dangerous human disease, amebic meningoencephalitis.

The traditional sand filtration followed by chlorine treatment of water for human consumption may itself produce hazardous chemicals because of the large amount of organic matter in many water sources. Sand filters do not remove all of this organic matter, and, when it undergoes chlorination, a number of compounds such as chloroform are produced. Chloroform,

Table 26-10 Pollutants of Freshwater

Pollutant	Typical Effects	Sources
DISEASE AGENTS		
viruses		Raw human sewage contaminating food and drinking water. Rate in developed countries due to chlorination of water supplies and sewage treatment.
infectious hepatitis	Various acute or chronic diseases	
poliovirus	Dysentery	
bacteria		
cholera		
typhoid fever		
bacterial dysentery		
protozoa		
Entameoba histolytica	Dysentery	
Naegleria	Amebic meningoencephalitis	Thermally polluted water
OXYGEN USING WASTES (biodegradable)	Reduction of available oxygen causes loss of desirable organisms and replacement by anaerobes, with resultant production of odorous, toxic substances	Sewage Paper mill wastes Meat-processing wastes Feed-lot runoff (animal manure aggregates 10 times the BOD as human wastes in U.S.A.)
PETROLEUM	Direct toxicity to all forms of life, physical smothering effects	Waste from transportation, heating, industry: accidents. Used motor and industrial oil is the major source (3 million metric tons/yr. worldwide) About 200 million metric tons/yr. produced in U.S.A.
SYNTHETIC ORGANIC COMPOUNDS		
detergents		
bio-degradable	Phosphate eutrophication	Household, commercial washing
non-bio-degradable	Long acting surfactant effects—foaming rivers.	Household, commercial washing
pesticides	See Table 26-12	Agricultural, industrial wastes
herbicides	See Table 26-12	Agricultural, industrial wastes
industrial wastes	At least 2000 substances are involved in an industrialized economy, producing a wide gamut of effects.	Industrial accidents, waste disposal
PLANT NUTRIENTS		
nitrates, nitrites	In drinking water are dangerous to infants by inactivating hemoglobin; may form carcinogenic N-nitrosoamines	Fertilizer runoff; manure; natural weathering; nitrogen fixation by microorganisms
phosphates	Eutrophication	Agriculture
potassium compounds	Eutrophication	Agriculture
INORGANIC SUBSTANCES		
arsenic	Skin and liver cancer, a cumulative lethal substance	Arsenic based pesticides and herbicides account for most of 13,000 metric tons/yr. worldwide production.
cadmium	Circulatory and kidney disease	Mine wastes, battery manufacture, electroplating
chlorine	Generally lethal to aquatic life; possible carcinogenic effects by formation of trihalomethanes in reaction with organic substances in water	Industrial; water and sewage purification
mercury	Central nervous system and kidney effects	Industrial wastes; agriculture (fungicides) coal burning
sulfuric acid	Direct lethality to aquatic organisms	Coal mine wastes

and perhaps other compounds so produced, are carcinogenic. A possible solution to the problem is to equip treatment plants with charcoal filters which are able to remove dissolved organic matter from water. The costs of charcoal filtration are much higher than sand filtration.

Today, however, the major worries about water pollution center on the tremendously increased chemical input from sewage and manure, household wastes, and from industrial and agricultural activities. The general categories of wastes that these sources produce are listed in Table 26-10. Substances in two of these categories damage the environment by what might be called forced feeding of ecosystems. These substances are wastes that increase oxygen consumption and thereby promote unbalanced overgrowth, principally of bacteria.

Adequate amounts of dissolved oxygen in water are extremely important to water quality. When excessive amounts of wastes such as sewage, farm manure, and any of a number of other biological products assimilable by bacteria enter bodies of water, the bacteria that are present multiply and carry out aerobic decomposition, which reduces the wastes ultimately to CO_2 and water. The oxygen required by the bacteria to carry this process to completion is measured as the **biological oxygen demand (BOD)** and is an indicator of the water pollution potential of wastes. The higher the BOD, the greater the pollution potential. Ordinary domestic sewage has a BOD of about 200 mg/liter, although certain types of sewage may have BODs of more than 2000 mg/liter. The solubility of oxygen in fresh water at 15°C is about 10 mg/liter. This means that sewage must be diluted rapidly to avoid total depletion of available oxygen. When dilution is insufficient and oxygen falls to very low levels, the stream or lake ecosystem undergoes drastic deteriorative changes. Bacterial decomposition processes switch to anaerobic mechanisms, and these produce many toxic and offensive products since the decomposition process cannot be carried all the way to CO_2 and water (see page 73). These products plus the lack of oxygen will kill or drive off many forms of animal life, which will tend to be replaced by overgrowth of a few kinds of organisms that are able to survive anaerobically or with very low O_2. The result is a body of water that has lost its ability to purify itself, a harborage of pathogenic bacteria esthetically offensive to nose and eye.

Another way that imbalance is produced in bodies of water by the forced feeding mechanism is by excessive supply of nutrients that normally are the limiting factors in growth of organisms. In fresh water the principal nutrient limits on algal growth are nitrogen and phosphate. Much phosphorous comes from sewage treatment plants and a lesser amount from fertilizers. A very large nitrogen enrichment results from agriculture. The amounts of fertilizers used in countries able to afford them can only increase since modern, high intensity agriculture depends on heavy fertilization (see page 538).

In fresh water the result of this great enrichment from runoff is **eutrophication**. This is a natural process, for example, in the evolution of a lake which might have begun owing to glacial action with low nutrient content and consequently a low density of plant life. This is the oligotrophic (few foods) state. As the lake accumulates nutrients by drainage from the surrounding watershed, plant life increases and the lake becomes eutrophic. Since during this process organic materials accumulate in the lake, bottom sediments build up, plant life increases, and the lake ultimately may evolve into a marsh. When this natural sequence is tremendously accelerated by addition of inorganic nutrients, events are rather more ugly. They include a tendency to a great overgrowth of algae. These algal blooms produce excess organic matter resulting in elevated BODs, which may render the body of water anaerobic over at least part of the year. The result is that the natural cycle of self-purification of the body of water is abolished and another sewer has been created.

Many authorities believe that nutrient runoff from land into the ocean produces an analogous condition in which marine dinoflagellates, single celled algae, undergo rapid growth to produce the phenomenon known as **red tide**. Fish mortality is high during a red tide owing to the production of a toxin by the dinoflagellates.

Nitrites and nitrates also have direct effects on human health. Use of fertilizers increases their concentrations sufficiently in water and food sometimes to cause direct toxic effects. Nitrites react with hemoglobin and render it incapable of carrying oxygen. Poisoning may occur by direct consumption of nitrites or by the action of bacteria that form nitrites from nitrates that are common in food. This can be an especially serious problem in very young infants be-

Table 26-11 Some Hazardous Chemicals of Environmental Concern

Chemical	Typical Effects	Sources
INDUSTRIAL CHEMICALS		
PVC (polyvinyl chloride) $R-(CH_2CHCl)_N-R'$	Liver cancer	About 10 million metric tons/yr. worldwide production, a basic component in plastics industry
PCB (polychlorinated biphenyl) [structure with X-sites where Cl may be inserted]	Long-term effects similar to DDT	Used for heat transfer and insulation in power transformers, lubricants. Highly persistent in environment, accumulated by body fat
PBB (polybrominated biphenyl)	Persistent skin disorders, joint disease, neurological disorders	From plants manufacturing fire retardants. One notable accident contaminated cattle feed in Michigan in 1974.
PESTICIDES		
chlorinated hydrocarbon insecticides		
DDT [structure]	Kidney and liver damage, endocrine disorders, impaired neural function	Used worldwide as insecticide since 1945. Use declining but still highly persistent in environment. Other chlorinated insecticides include aldrin, dieldrin, chlordane, heptachlor, endrin, methoxychlor
kepone (chlordecone) [structure]	Serious neurological effects, possible liver cancer	Insecticide manufacture: one serious contamination caused closing of the James River in Virginia to fishing in 1975. Closely related Mirex has been used for fire ant control in the South.
organophosphate insecticides		
phosvel	Acute effects due to anticholinesterase action in CNS; chronic neurological effects due to myeline degeneration	Not used in U.S.A. but has caused poisoning of manufacturing workers in Bayport, Texas, in 1973–75. Still used in other parts of world.
Malathion [structure]	Actue effects due to anticholinesterase action in nervous system	General purpose insecticide, the least toxic of the organophosphates. Not persistent in environment.
herbicides		
2,4,5-T [structure]	Chloracne (persistent skin disorder), liver and kidney failure, developmental abnormalities	Used as weed and vegetation killers owing to plant hormone mimicry.
2,4-D [structure]		
dioxin (TCDD) [structure]		Major contamination in 1976 due to manufacturing accident in Northern Italy

cause their digestive systems are often not acidic enough to prevent growth of nitrite forming bacteria.

SPECIFIC HAZARDOUS POLLUTANTS

Finally we come to a huge and varied collection of pollutants that are directly toxic. Some are the same substances that we discussed as air pollutants, for example asbestos (Table 26-9). Many others are predominantly water pollutants owing to their chemical properties and manner of entry into the environment. Some of these are listed in Table 26-10.

Inorganic toxicants we are already amply acquainted with from the discussion of mercury and Minamata disease. There are several others of great importance (Table 26-11). All are characterized by a long half-life in the human body and great permanence once established in the environment. Consequently, even slight repetitive contaminations with these substances can be expected eventually to cause serious troubles.

The organic toxicants that appear in water are primarily pesticides and certain industrial products. The classical example is DDT, which starkly delineates the problems that can occur with indiscriminate introduction of such substances into the environment. DDT also serves to illustrate the often delicate balance that may occur between the beneficial and detrimental effects of such compounds.

DDT was undoubtedly the most effective weapon ever developed against insects. It was a powerful tool against insect-borne diseases, particularly malaria and typhus, and even today there are those who believe that we may expect agricultural disaster if it is not reintroduced into general use. The trouble with DDT was that, in the first instance, it was rather stupidly employed. It was sprayed about the world with great abandon and, since it is extremely persistent in the environment, those insects that survived the initial poisonings were continually in its presence. The result was development of many DDT resistant genetic strains of insects.

The second difficulty was that, although the toxic dose of DDT is rather high for vertebrates as compared with insects, DDT proved to be toxic to vertebrates because it is highly bioaccumulative. For example, George Woodwell and Charles Wuster found in a Long Island, N.Y., salt marsh that, although the water had only 5×10^{-5} ppm of DDT, it built up through the food chain of marsh inhabitants to more than 70 ppm in fish-eating birds. The result was great mortality among fish- and insect-eating birds through direct action on the nervous system and through interference with reproductive success, principally by effects on egg shell production. Such studies were only a few of the many that showed that DDT was accumulating everywhere, even in human milk (Figure 2-5). The primary impetus to this work was Rachael Carson's call-to-arms, *Silent Spring*.

Further investigations showed that DDT was even affecting marine life by killing the delicate larval stages of crabs and shrimps. These live in estuaries, which are the nurseries of many forms of marine life. Unfortunately, estuaries are the direct recipients of large quantities of runoff from land and, hence, tend to build up high concentrations of whatever is being broadcast upon the watershed. A good example is San Francisco Bay, which receives runoff from a large agricultural area drained by the San Joaquin River. That river was estimated to deposit about 4000 lb/yr of pesticides in the bay until control efforts began.

Since DDT is also found in organisms far out at sea, there is the possibility that damage may be done in offshore waters. There is some evidence, for example, that low concentrations of DDT can produce a marked reduction in the rate of photosynthesis of phytoplankton. If this reduction is a significant effect, there could be global perturbations because the phytoplankton are the underpinning of all marine food chains and are essential oxygen producers.

DDT continues to be used in large amounts in many parts of the world although it has been banned in the United States since 1972, except for certain very restricted and critical uses. Unfortunately, its very slow rate of degradation means that it will be with us for a very long time even if its further use were to be abolished immediately.

Another group of chlorinated aromatic compounds, the polychlorinated biphenyls, or PCBs are chemically similar to DDT and have similar biological actions. Stability under very high temperatures and other properties have made them extremely useful as cooling fluids in electrical transformers, lubricants, hydraulic fluids, and as ingredients in plastics. Their danger is now so evident that the only manufacturer in the United States has voluntarily restricted its PCB

Table 26-12. A Comparison of Some Social and Environmental costs of Coal and Nuclear Power Plants.
(Costs related to a 1000 MW plant unless otherwise stated)

Cost Category	Coal Plant (bituminous)	Nuclear Plant (light water reactor)
Fuel consumption	3×10^6 metric tons/yr	0.9 metric ton/yr
Fuel production and transport		
accidents	$60/10^9$ MWh	$5/10^9$ MWh
disease	$1000/10^9$ MWh (mostly black-lung)	$30/10^9$ MWh (cancer estimate)
Land use		
power plant and ancillary services (fuel preparation, storage, transportation)	4000 acres	500 acres
mining	35,000 acres/yr (strip mining)	20,000 acres/yr
waste storage and reprocessing (radioactive)	—	100 acres/yr
Water consumption		
cooling and wash water	6×10^6 metric tons of "black water"	13,000 gal (radioactive)
	2×10^7 metric tons evaporated	2.5×10^7 metric tons evaporated
Air pollution		
deaths from air pollution	0.02/yr	2×10^{-5}/yr
pollutant volume	1.6×10^5 metric tons/yr, mostly sulfur and nitrogen oxides	5×10^5 Ci (principally Kr^{85})

Sources: The Council on Environmental Quality (1973) Energy and the Environment-Electric Power, U.S. Gov. Printing Office, Washington, D.C.; Wilson, R. and W. J. Jones (1974) Energy, Ecology and the Environment. Academic Press, New York; Cook, E. (1976) Man, Energy, Society. W. H. Freeman and Co., San Francisco.

sales to uses in confined systems. However, this altruistic act does not prevent importation of PCBs and, as recently as 1975, it was estimated by the Environmental Protection Agency that about 4.5 million kg of PCBs are lost into the environment each year. One result is that fish in certain parts of the Hudson River have up to 350 ppm of PCBs, rendering them unfit for food since the Food and Drug Administration has set the safe level of PCBs in food at 5 ppm.

Thermal Pollution

Heating of rivers, lakes, and marine shore waters by industrial activities is a serious threat to the aquatic environment. Because most of the life of these waters—plants, invertebrates, and fish—are unable to thermoregulate, their metabolism is at the mercy of environmental temperature. In fact, summer temperatures of typical American rivers normally come within 5 to 10°F of the lethal temperature for most of the fish that inhabit them. Such streams cannot be used extensively for industrial cooling without serious risk, because a single large electrical power plant is capable of increasing the temperature of its cooling river by upwards of 10°F for a substantial distance downstream. Nuclear generating plants considerably worsen matters, as they require about 50 times as much cooling as generating plants that utilize fossil fuels.

The solution to this dilemma is by no means clear, because there is simply not enough water to satisfy future demands without drastic thermal effects upon stream ecology. Thus, of the estimated 1,200 billion gal of water per day that is available as surface and subsurface water in the United States, one quarter of this amount will have to be used within the next 10 years for generator plant cooling. Obviously, more expensive cooling procedures will have to be put to use and we shall therefore have to pay, in increased costs of electrical power, the price of environmental protection.

Thermal pollution is not totally bad. It appears that ocean water warmed by power plants may be useful in accelerating the growth of certain valuable marine organisms. A pilot plant using sea water from an electrical generating plant in Los Angeles is rearing Eastern lobsters to edible size in about half the time it takes in their natural habitat. As the Eastern lobster fishery becomes depleted, there may soon come a time when such operations will be profitable. Probably, however, they can be considered only as a very minor rectification of the general damage that is being done nationwide by thermal pollution.

APPENDIXES

1 Numbers and Measurement

SCIENTIFIC NOTATION: HANDLING LARGE AND SMALL NUMBERS

For convenience, use powers of 10 notation for handling large and small numbers. For example:

5,550,000,000	is written	5.55×10^9
5,550,000	is written	5.55×10^6
0.000 000 555	is written	5.55×10^{-7}

Addition and subtraction of numbers in scientific notation: Convert numbers to same power of 10 and add or subtract first terms, thus:

$$\begin{array}{r} 5.55 \times 10^6 \\ + 5.55 \times 10^5 \end{array} \quad \text{becomes} \quad \begin{array}{r} 5.55 \times 10^6 \\ + 0.55 \times 10^6 \\ \hline 6.10 \times 10^6 \end{array}$$

To multiply: multiply first terms together and add the powers of 10, thus:

$$(5.55 \times 10^5)(5.55 \times 10^6) = 30.8 \times 10^{11}$$

To divide: divide the first terms and subtract the powers of 10, thus:

$$\frac{5.55 \times 10^5}{5.55 \times 10^6} = 1 \times 10^{-1} \text{ (or 0.10)}$$

List of powers of 10

1×10^{12}	1,000,000,000,000.	1 billion (English), 1 million million
1×10^9	1,000,000,000.	1 billion (U. S.), 1000 million
1×10^6	1,000,000.	1 million

1×10^3	1,000.	
1×10^2	100.	
1×10^1	10.	
1×10^{-1}	0.1	
1×10^{-2}	0.01	1 hundredth
1×10^{-3}	0.001	1 thousandth
1×10^{-4}	0.000 1	1 ten thousandth
1×10^{-6}	0.000 001	
1×10^{-9}	0.000 000 001	
1×10^{-12}	0.000 000 000 001	

THE METRIC SYSTEM AND EQUIVALENTS

Length

1 kilometer (km) = 1000 meters (m) = 0.62 statute miles
1 m = 10^{-3} km = 100 centimeters (cm) = 39.37 inches (2.2 cubits)
1 cm = 10^{-2} m = 10 millimeters (mm)
1 mm = 10^{-3} m = 1000 microns (μ)
1 μ = 10^{-6} m = 1000 nanometers (nm) = 1 micrometer (μm)
1 nm = 10^{-9} m = 1 millimicron (mμ) = 10 Angstroms (Å)
1 Å = 10^{-10} m = 0.1 nm

1 statute mile = 1.61 km
1 foot = 0.30 m
1 inch = 24.4 mm

Area

1 hectare (ha; square hectometer) = 0.01 km² = 10,000 m² = 2.47 U. S. acres

1 km² = 100 ha = 0.386 square miles (mi²)

1 m² = 10,000 cm² = 10.76 square feet (ft²)

Volume

1 barrel (bbl) = 42 U. S. gallons = 158.76 liters (l)
1 liter = 1000 milliliters (ml) = 1.06 U. S. quarts
1 ml = 10^{-3} l = 1 cubic centimeter (cm²) = 1000 microliters (μl)
1 μl = 10^{-6} l

Mass

1 metric ton (mt) = 1000 kilograms (kg) = 2204.6 pounds avoirdupoise (lbs)
1 kg = 1000 grams (g) = 2.204 lbs.
1 g = 10^{-3} kg = 1000 milligrams (mg)
1 mg = 10^{-3} g = 1000 micrograms (μg)
1 μg = 10^{-6} g = 1000 nanograms (ng)
1 ng = 10^{-9} g

ENERGY AND WORK

$$\text{work} = \text{force} \times \text{distance}$$

1 joule (J) = 1 kg × 1 m
1 calorie (cal) = heat to raise 1 g distilled water 1°C, from 14.5°C to 15.5°C
1 kilocalorie (Kcal)   = 1000 cal = 4190 J
1 electron volt (eV)   = 3.8×10^{-23} Kcal
1 kilowatt hour (kWh) = 3.6×10^6 J = 3412 British thermal units (Btu)
1 Btu = 0.252 Kcal   = 1055 J

Energy equivalents

$$1 \text{ bbl oil} = 5.8 \times 10^6 \text{ Btu} = 1700 \text{kWh}$$
$$1 \text{ mt coal} = 27 \times 10^6 \text{ Btu} = 8100 \text{ kWh}$$

Temperature

freezing point of water =   0°C (Celsius or centigrade) = 32°F (Fahrenheit)
boiling point of water = 100°C   = 212°F

Conversion between temperature scales

$$\text{degrees C} = 5/9(F - 32)$$
$$\text{degrees F} = 9/5(C + 32)$$

Atmospheric pressure

Atmospheric pressure, average sea level = 1.03 kg/cm² (14.7 lb/in²)

2 Time Scale of Earth Evolution

Years from Present (in millions of years)	Period Designations	Principal Events	
		Geological	Biological
4600		Origin of the protoplanet, gravitational heating	
4000		Prebiological synthesis (?)	
3500	PRECAMBRIAN		First organisms (?)
3000			Stromatolites
2000		Atmospheric oxygen buildup	
700–230	PALEOZOIC ERA		
475	Cambrian		Algae, fungi, bacteria, most marine invertebrates protochordates
425	Ordovician		
413	Silurian		Vascular land plants
355	Devonian		First seed plants, fish abundant
355–280	Carboniferous { Mississippian	Extensive seas and limestone deposits form in interior U. S.	Amphibians
	Pennsylvanian	Coal bed formation	Forest trees, reptiles
230	Permian	Appalachian chain uplift	Coniferous trees
230–65	MESOZOIC		
185	Triassic	Dominance of savanna habitats	Mammals
130	Jurassic		Angiosperms, dinosaurs dominant
65	Cretaceous	Rise of Rocky Mountains	Dominance of angiosperms, end of dinosaurs
65–present	CENOZOIC		
53	Paleocene		
37	Eocene	Submergence of Gulf states	Extensive forests, first horses
26	Oligocene	Dry climate in southwest North America	
12	Miocene	Alps rise, climate cools	
2	Pliocene		Grasslands spread, mastodons
1.5	Pleistocene	Several glaciations	Emergence of modern man
	Recent		

3 A Classification of Organisms

NONCELLULAR

    Viruses (status uncertain, but probably derived from cellular organisms)

CELLULAR

MONERA (prokaryotes)
    Phylum Schizophyta (bacteria)
        Rickettsias (require an insect to complete life cycle, smallest of the bacteria)
        Eubacteria (true bacteria)
    Phylum Cyanophyta (blue green algae)

EUKARYOTA
KINGDOM PROTISTA
    Phylum Mastigophora (flagellates)
        Class Phytomastigina (photosynthetic)
        Class Zoomastigina (animal flagellates)
    Phylum Sarcodina (amebae)
    Phylum Sporozoa (parasitic forms such as *Plasmodium*, malarial parasite)
    Phylum Ciliophora (ciliates)
KINGDOM FUNGI
    Phylum Ascomycota (sac fungi and yeasts)
    Phylum Zygomycota (plant rusts and bread molds)
    Phylum Basidiomycota (mushrooms and bracket fungi)
    Phylum Myxomycota (slime molds)

KINGDOM PLANTAE
    Phylum Pyrrophyta (dinoflagellates)
    Phylum Chrysophyta (diatoms)
    Phylum Phaeophyta (brown algae)
    Phylum Rhodophyta (red algae)
    Phylum Chlorophyta (green algae)
    Phylum Bryophyta (mosses, liverworts)
    Phylum Tracheophyta (vascular plants)
        Subphylum Psilophyta (mostly extinct, primitive land plants)
        Subphylum Lycophyta (club mosses)
        Subphylum Sphenophyta (horsetails)
        Subphylum Pteropsida (ferns, seed plants)
            Class Filicinae (ferns)
            Class Gymnospermae (cone-bearing plants)
                Order Cycadales (tree ferns)
                Order Coniferales (pines, redwoods, fir)
            Class Angiospermae (covered-seed, flowering plants)
                Subclass Monocotyledonae (grasses, palms)
                Subclass Dicotyledonae (trees, most other flowering plants)
KINGDOM ANIMALIA
SUBKINGDOM PARAZOA (sponges)
SUBKINGDOM EUMETAZOA (true metazoans)
    Phylum Coelenterata (coelenterates)
        Class Hydrozoa (hydroids, primarily stalked forms)
        Class Scyphozoa (true jellyfish)
        Class Anthozoa (corals, sea anemones)

Phylum Ctenophora (comb jellies)
Phylum Platyhelminthes (flatworms)
    Class Turbellaria (free-living planarians)
    Class Trematoda (parasitic flukes)
    Class Cestoda (parasitic tapeworms)
Phylum Mesozoa (parasitic on invertebrates)
Phylum Nemertia (ribbon worms)
Phylum Rotifera (wheel animals)
Phylum Nematoda (roundworms)
Phylum Entoprocta (moss animals)
Phylum Sipunculida (marine worms)
Phylum Echiura (marine worms)
Phylum Annelida (segmented worms)
    Class Hirudinea (leeches)
    Class Polychaeta (segmented marine worms)
    Class Oligochaeta (earthworms)
Phylum Mollusca (molluscs)
    Class Monoplacophora (primitive marine molluscs)
    Class Amphineura (chitons)
    Class Gastropoda (snails)
    Class Bivalvia (clams)
    Class Scaphopoda (tusk shells)
    Class Cephalopoda (octopods, squids)
Phylum Onychophora (segmented worms with characteristics suggestive of arthropods)
Phylum Arthropoda
    Class Chelicerata (horseshoe crabs, spiders, scorpions)
    Class Trilobita (extinct marine forms)
    Class Crustacea (crabs, shrimps, lobsters)
    Class Insecta (insects)
    Class Diplopoda (millipedes)
    Class Chilopoda (centipedes)
Phylum Ectoprocta (bryozoans)
Phylum Brachiopoda (lamp shells)
Phylum Chaetognatha (arrow worms)
Phylum Echinodermata (echinoderms)
    Class Crinoidea (stalked sea lilies, feather stars)
    Class Asteroidea (starfish)
    Class Ophiuroidea (brittle stars)
    Class Echinoidea (sea urchins, sand dollars)
    Class Holothuroidea (sea cucumbers)
Phylum Hemichordata (acorn worms)
Phylum Chordata (chordates)
  Subphylum Urochordata (sea squirts)
  Subphylum Cephalochordata (lancelets)
  Subphylum Vertebrata (vertebrates)
    Class Agnatha (lampreys)
    Class Chondrichthyes (sharks, rays)
    Class Osteichthyes (bony fish)
    Class Amphibia (frogs, toads, salamanders)
    Class Reptilia (snakes, lizards, turtles, crocodiles, alligators)
    Class Aves (birds)
    Class Mammalia (mammals)
      Subclass Prototheria
        Order Monotremata (egg-laying mammals)
      Subclass Metatheria
        Order Marsupalia (pouched mammals)
      Subclass Eutheria (placental mammals)
        Order Insectivora (shrews, moles)
        Order Chiroptera (bats)
        Order Carnivora (cats, dogs, bears, seals)
        Order Rodentia (rats, mice, squirrels, porcupines, beavers)
        Order Lagomorpha (hares and rabbits)
        Order Primates (lemurs, lorises, tarsiers, monkeys, apes, human)
        Order Artiodactyla (even-toed ungulates: cattle, deer)
        Order Perissodactyla (odd-toed ungulates: horses, rhinoceruses)
        Order Proboscidea (elephants)
        Order Cetacea (whales, porpoises)
        Order Sirenia (sea cows)

4 The Darwin and Wallace Publications

The following two papers were communicated to the Linnean Society on behalf of Darwin and Wallace by Charles Lyell and Joseph Hooker and were originally published in the *Journal of the Linnean Society* in 1858. In their letter of transmittal, Lyell and Hooker wrote

These gentlemen having, independently and unknown to one another, conceived the same very ingenious theory to account for the appearance and perpetuation of varieties and of specific forms on our planet, may both fairly claim the merit of being original thinkers in this important line of inquiry; but neither of them having published his views, though Mr. Darwin has for many years past been repeatedly urged by us to do so, and both authors having now unreservedly placed their papers in our hands, we think it would best promote the interests of science that a selection from them should be laid before the Linnean Society.

DARWIN'S PAPER

Extract from an unpublished work on species, by C. Darwin, Esq., consisting of a portion of a chapter entitled, "On the Variation of Organic Beings in a state of Nature; on the Natural Means of Selection; on the Comparison of Domestic Races and true Species."

De Candolle, in an eloquent passage, has declared that all nature is at war, one organism with another, or with external nature. Seeing the contented face of nature, this may at first well be doubted; but reflection will inevitably prove it to be true. The war, however, is not constant, but recurrent in a slight degree at short periods, and more severely at occasional more distant periods; and hence its effects are easily overlooked. It is the doctrine of Malthus applied in most cases with tenfold force. As in every climate there are seasons, for each of its inhabitants, of greater and less abundance, so all annually breed; and the moral restraint which in some small degree checks the increase of mankind is entirely lost. Even slow-breeding mankind has doubled in twenty-five years; and if he could increase his food with greater ease, he would double in less time. But for animals without artificial means, the amount of food for each species must, on an average, be constant, whereas the increase of all organisms tends to be geometrical, and in a vast majority of cases at an enormous ratio. Suppose in a certain spot there are eight pairs of birds, and that only four pairs of them annually (including double hatches) rear only four young, and that these go on rearing their young at the same rate, then at the end of seven years (a short life, excluding violent deaths, for any bird) there will be 2048 birds, instead of the original sixteen. As this increase is quite impossible, we must conclude either that birds do not rear nearly half their young, or that the average life of a bird is, from accident, not nearly seven years. Both checks probably concur. The same kind of calculation applied to all plants and animals affords results more or less striking, but in very few instances more striking than in man.

Many practical illustrations of this rapid tendency to increase are on record, among which, during peculiar seasons, are the extraordinary numbers of certain animals; for instance, during the years 1826 to 1828, in La Plata, when from drought some millions of cattle perished, the whole country actually swarmed with mice. Now I think it cannot be doubted that during the breeding-season all the mice (with the exception of a few males or females in excess) ordinarily pair, and therefore that this astounding increase during three years must be attributed to a greater number than usual surviving the first year, and then breeding, and so on till the third year, when their numbers were brought down to their usual limits on the return of wet weather. Where man has introduced plants and animals into a new and favourable country, there are many accounts in how surprisingly few years the whole country has become stocked with them. This increase would necessarily stop as soon as the country was fully stocked; and yet we have every reason to believe, from what is known of wild animals, that all would pair in the spring. In the majority of cases it is most difficult to imagine where the checks fall—though generally, no doubt, on the seeds, eggs, and young; but when we remember how impossible, even in mankind (so much better known than any other animal), it is to infer from repeated casual observations what the average duration of life is, or to discover the different percentage of deaths to births in different countries, we ought to feel no surprise at our being unable to discover where the check falls in any animal or plant. It should always be remembered, that in most cases the checks are recurrent yearly in a small, regular degree, and in an extreme degree during unusually cold, hot, dry, or wet years, according to the constitution of the being in question. Lighten any check in the least degree, and the geometrical powers of increase in every organism will almost instantly increase the average number of the favoured species. Nature may be compared to a surface on which rest ten thousand sharp wedges touching each other and driven inwards by incessant blows. Fully to realize these views much reflection is requisite. Malthus on man should be studied; and all such cases as those of the mice in La Plata, of the cattle and horses when first turned out in South America, of the birds by our calculation, &c., should be well considered. Reflect on the enormous multiplying power inherent and annually in action in all animals; reflect on the countless seeds scattered by a hundred ingenious contrivances, year after year, over the whole face of the land; and yet we have every reason to suppose that the average percentage of each of the inhabitants of a country usually remains constant. Finally, let it be borne in mind that this average number of individuals (the external conditions remaining the same) in each country is kept up by recurrent struggles against other species or against external nature (as on the borders of the Arctic regions, where the cold checks life), and that ordinarily each individual of every species holds its place, either by its own struggle and capacity of acquiring nourishment in some period of its life, from the egg upwards; or by the struggle of its parents (in short-lived organisms, when the main check occurs at longer intervals) with other individuals of the same or different species.

But let the external conditions of a country alter. If in a small degree, the relative proportions of the inhabitants will in most cases simply be slightly changed; but let the number of inhabitants be small, as on an island, and free access to it from other countries be circumscribed, and let the change of conditions continue progressing (forming new stations), in such a case the original inhabitants must cease to be as perfectly adapted to the changed conditions as they were originally. It has been shown in a former part of this work, that such changes of external conditions would, from their acting on the reproductive system, probably cause the organization of those beings which were most affected to become, as under domestication, plastic. Now, can it be doubted, from the struggle each individual has to obtain subsistence, that any minute variation in structure, habits, or instincts, adapting that individual better to the new conditions, would tell upon its vigour and health? In the struggle it would have a better chance of surviving; and those of its offspring which inherited the variation, be it ever so slight, would also have a better chance. Yearly more are bred than can survive; the smallest grain in the balance, in the long run, must tell on which death shall fall, and which shall survive. Let this work of selection on the one hand, and death on the other, go on for a thousand generations, who will pretend to affirm that it would produce no effect, when we remember what, in a few years, Bakewell effected in cattle, and Western in sheep, by this identical principle of selection?

To give an imaginary example from changes in progress on an island:—let the organization of a canine animal which preyed chiefly on rabbits, but sometimes on hares, become slightly plastic; let these same changes cause the number of rabbits very slowly to decrease, and the number of hares to increase; the effect of this would be that the fox or dog would be driven to try to catch more hares: his organization, however, being slightly plastic, those individuals with the lightest forms, longest limbs, and best eyesight, let the difference be ever so small, would be slightly favoured, and would tend to live longer, and to survive during that time of the year when food was scarcest; they would also rear more young, which would tend to inherit these slight peculiarities. The less fleet ones would be rigidly destroyed. I can see no more reason to doubt that these causes in a thousand generations would produce a marked effect, and adapt the form of the fox or dog to the catching of hares instead of rabbits, than that greyhounds can be improved by selection and careful breeding. So would it be with plants under similar circumstances. If the number of individuals of a species with plumed seeds could be increased by greater powers of dissemination within its own area (that is, if the check to increase fell chiefly on the seeds), those seeds which were provided with ever so little more down, would in the long run be most disseminated; hence a greater number of seeds thus formed would germinate, and would tend to produce plants inheriting the slightly better-adapted down.*

Besides this natural means of selection, by which those individuals are preserved, whether in their egg, or larval, or mature state, which are best adapted to the place they fill in nature, there is a second agency at work in most unisexual animals, tending to produce the same effect, namely, the struggle of the males for the females. These struggles are generally decided by the law of battle, but in the case of birds, apparently, by the charms of their song, by their beauty or their power of courtship, as in the dancing rock-thrush of Guiana. The most vigorous and healthy males, implying perfect adaptation, must generally gain the victory in their contests. This kind of selection, however, is less rigorous than the other; it does not require the death of the less successful, but gives to them fewer descendants. The struggle falls, moreover, at a time of year when food is generally abundant, and perhaps the effect chiefly produced would be the modification of the secondary sexual characters, which are not related to the power of obtaining food, or to defence from enemies, but to fighting with or rivalling other males. The result of this struggle amongst the males may be compared in some respects to that produced by those agriculturists who pay less attention to the careful selection of all their young animals, and more to the occasional use of a choice mate.

WALLACE'S PAPER

From "On the Tendency of Varieties to Depart Indefinitely from the Original Type," by Alfred Russel Wallace

One of the strongest arguments which have been adduced to prove the original and permanent distinctness of species is, that varieties produced in a state of domesticity are more or less unstable, and often have a tendency, if left to themselves, to return to the normal form of the parent species; and this instability is considered to be a distinctive peculiarity of all varieties, even of those occurring among wild animals in a state of nature, and to constitute a provision for preserving unchanged the originally created distinct species.

In the absence or scarcity of facts and observations as to varieties occurring among wild animals, this argument has had great weight with naturalists, and has led to a very general and somewhat prejudiced belief in the stability of species. Equally general, however, is the belief in what are called "permanent or true varieties,"—races of animals which continually propagate their like, but which differ so slightly (although constantly) from some other race, that the one is considered to be a variety of the other. Which is the variety and which the original species, there is generally no means of determining, except in those rare cases in which the one race has been known to produce an offspring unlike itself and resembling the other. This, however, would seem quite incompatible with the "permanent invariability of species," but the difficulty is overcome by assuming that such varieties have strict limits, and can never again vary further from the original type, although they may return to it, which, from the analogy of the domesticated animals, is considered to be highly probable, if not certainly proved.

It will be observed that this argument rests entirely on the assumption, that varieties occurring in a state of nature are in all respects analogous to or even identical with those of domestic animals, and are governed by the same laws as regards their permanence or further variation. But it is the object of the present paper to show that this assumption is altogether false, that there is a general principle in nature which will cause many varieties to survive the parent

---

*I can see no more difficulty in this, than in the planter improving his varieties of the cotton plant.—C.D. 1858.

species, and to give rise to successive variations departing further and further from the original type, and which also produces, in domesticated animals, the tendency of varieties to return to the parent form.

The life of wild animals is a struggle for existence. The full exertion of all their faculties and all their energies is required to preserve their own existence and provide for that of their infant offspring. The possibility of procuring food during the least favourable seasons, and of escaping the attacks of their most dangerous enemies, are the primary conditions which determine the existence both of individuals and of entire species. These conditions will also determine the population of a species; and by a careful consideration of all the circumstances we may be enabled to comprehend, and in some degree to explain, what at first sight appears so inexplicable—the excessive abundance of some species, while others closely allied to them are very rare.

The general proportion that must obtain between certain groups of animals is readily seen. Large animals cannot be so abundant as small ones; the carnivora must be less numerous than the herbivora; eagles and lions can never be so plentiful as pigeons and antelopes; the wild asses of the Tartarian deserts cannot equal in numbers the horses of the more luxuriant prairies and pampas of America. The greater or less fecundity of an animal is often considered to be one of the chief causes of its abundance or scarcity; but a consideration of the facts will show us that it really has little or nothing to do with the matter. Even the least prolific of animals would increase rapidly if unchecked, whereas it is evident that the animal population of the globe must be stationary, or perhaps, through the influence of man, decreasing. Fluctuations there may be; but permanent increase, except in restricted localities, is almost impossible. For example, our own observation must convince us that birds do not go on increasing every year in a geometrical ratio, as they would do, were there not some powerful check to their natural increase. Very few birds produce less than two young ones each year, while many have six, eight, or ten; four will certainly be below the average; and if we suppose that each pair produce young only four times in their life, that will also be below the average, supposing them not to die either by violence or want of food. Yet at this rate how tremendous would be the increase in a few years from a single pair! A simple calculation will show that in fifteen years each pair of birds would have increased to nearly ten millions! whereas we have no reason to believe that the number of the birds of any country increases at all in fifteen or in one hundred and fifty years. With such powers of increase the population must have reached its limits, and have become stationary, in a very few years after the origin of each species. It is evident, therefore, that each year an immense number of birds must perish—as many in fact as are born; and as on the lowest calculation the progeny are each year twice as numerous as their parents, it follows that, whatever be the average number of individuals existing in any given country, twice that number must perish annually,—a striking result, but one which seems at least highly probable, and is perhaps under rather than over the truth. It would therefore appear that, as far as the continuance of the species and the keeping up the average number of individuals are concerned, large broods are superfluous. On the average all above one become food for hawks and kites, wild cats and weasels, or perish of cold and hunger as winter comes on. This is strikingly proved by the case of particular species; for we find that their abundance in individuals bears no relation whatever to their fertility in producing offspring. Perhaps the most remarkable instance of an immense bird population is that of the passenger pigeon of the United States, which lays only one, or at most two eggs, and is said to rear generally but one young one. Why is this bird so extraordinarily abundant, while others producing two or three times as many young are much less plentiful? The explanation is not difficult. The food most congenial to this species, and on which it thrives best, is abundantly distributed over a very extensive region, offering such difference of soil and climate, that in one part or another of the area the supply never fails. The bird is capable of a very rapid and long-continued flight, so that it can pass without fatigue over the whole of the district it inhabits, and as soon as the supply of food begins to fail in one place is able to discover a fresh feeding-ground. This example strikingly shows us that the procuring a constant supply of wholesome food is almost the sole condition requisite for ensuring the rapid increase of a given species, since neither the limited fecundity, nor the unrestrained attacks of birds of prey and of man are here sufficient to check it. In no other birds are these peculiar circumstances so strikingly combined. Either their food is more liable to failure, or they have not sufficient power of wing to search for it over an extensive area, or during some season of the year it becomes very scarce, and less wholesome substitutes have to be found; and thus, though more fertile in offspring, they can never increase beyond the supply of food in the least favourable seasons. Many birds can only exist by migrating, when their food becomes scarce, to regions possessing a milder, or at least a different climate, though, as these migrating birds are seldom excessively abundant, it is evident that the countries they visit are still deficient in a constant and abundant supply of wholesome food. Those whose organization does not permit them to migrate when their food becomes periodically scarce, can never attain a large population. This is probably the reason why woodpeckers are scarce with us, while in the tropics they are among the most abundant of solitary birds. Thus the house sparrow is more abundant than the redbreast, because its food is more constant and plentiful,—seeds of grasses being preserved during the winter, and our farm-yards and stubble-fields furnishing an almost inexhaustible supply. Why, as a general rule, are aquatic, and especially sea birds, very numerous in individuals? Not because they are more prolific than others, generally the contrary; but because their food never fails, the sea-shores and river-banks daily swarming with a fresh supply of small mollusca and crustacea. Exactly the same laws will apply to mammals. Wild cats are prolific and have few enemies; why then are they never as abundant as rabbits? The only intelligible answer is, that their supply of food is more precarious. It appears evident, therefore, that so long as a country remains physically unchanged, the numbers of its animal population cannot materially increase. If one species does so, some others requiring the same kind of food must diminish in proportion. The numbers that die annually must be immense; and as the individual existence of each animal depends upon itself, those that die must be the weakest—the very young, the aged, and the diseased,—while those that prolong their existence can only be the most perfect in health and vigour—those who are best able to obtain food regularly, and avoid their numerous enemies. It is, as we commenced by remarking, "a struggle for existence," in which the weakest and least perfectly organized must always succumb.

Now it is clear that what takes place among the individuals of a species must also occur among the several allied species of a group,—viz. that those which are best adapted to obtain a regular supply of food, and to defend themselves against the attacks of their enemies and the vicissitudes of the seasons, must necessarily obtain and preserve a superiority in population; while those species which from some defect of power or organization are the least capable of counteracting the vicissitudes of food, supply, &c., must diminish in numbers, and, in extreme cases, become altogether extinct. Between these extremes the species will present various degrees of capacity for ensuring the means of preserving life; and it is thus we account for the abundance or rarity of species. Our ignorance will generally prevent us from accurately tracing the effects to their causes; but could we become perfectly acquainted with the organization and habits of the various species of animals, and could we measure the capacity of each for performing the different acts necessary to its safety and existence under all the varying circumstances by which it is surrounded, we might be able even to calculate the proportionate abundance of individuals which is the necessary result.

If now we have succeeded in establishing these two points—1st, that the animal population of a country is generally stationary, being kept down by a periodical deficiency of food, and other checks; and, 2nd, that the comparative abundance or scarcity of the individuals of the several species is entirely due to their organization and resulting habits, which, rendering it more difficult to procure a regular supply of food and to provide for their personal safety in some cases than in others, can only be balanced by a difference in the population which have to exist in a given area—we shall be in a condition to proceed to the consideration of varieties, to which the preceding remarks have a direct and very important application.

Most or perhaps all the variations from the typical form of a species must have some definite effect, however slight, on the habits or capacities of the individuals. Even a change of colour might, by rendering them more or less distinguishable, affect their safety; a greater or less development of hair might modify their habits. More important changes, such as an increase in the power or dimensions of the limbs or any of the external organs, would more or less affect their mode of procuring food or the range of country which they inhabit. It is also evident that most changes would affect, either favourably or adversely, the powers of prolonging existence. An antelope with shorter or weaker legs must necessarily suffer more from the attacks of the feline carnivora; the passenger pigeon with less powerful wings would sooner or later be affected in its powers of procuring a regular supply of food; and in both cases the result must necessarily be a diminution of the population of the modified species. If, on the other hand, any species should produce a variety having slightly increased powers of preserving existence, that variety must inevitably in time acquire a superiority in numbers. These results must follow as surely as old age, intemperance, or scarcity of food produce an increased mortality. In both cases there may be many individual exceptions; but on the average the rule will invariably be found to hold good. All varieties will therefore fall into two classes—those which under the same conditions would never reach the population of the parent species, and those which would in time obtain and keep a numerical superiority. Now, let some alteration of physical conditions occur in the district—a long period of drought, a destruction of vegetation by locusts, the irruption of some new carnivorous animal seeking "pastures new"—any change in fact tending to render existence more difficult to the species in question, and tasking its utmost powers to avoid complete extermination; it is evident that, of all the individuals composing the species, those forming the least numerous and most feebly organized variety would suffer first, and, were the pressure severe, must soon become extinct. The same causes continuing in action, the parent species would next suffer, would gradually diminish in numbers, and with a recurrence of similar unfavourable conditions might also become extinct. The superior variety would then alone remain, and on a return to favourable circumstances would rapidly increase in numbers and occupy the place of the extinct species and variety.

The variety would now have replaced the species, of which it would be a more perfectly developed and more highly organized form. It would be in all respects better adapted to secure its safety, and to prolong its individual existence and that of the race. Such a variety could not return to the original form; for that form is an inferior one, and could never compete with it for existence. Granted, therefore, a "tendency" to reproduce the original type of the species, still the variety must ever remain preponderant in numbers, and under adverse physical conditions again alone survive. But this new, improved, and populous race might itself, in course of time, give rise to new varieties, exhibiting several diverging modifications of form, any of which, tending to increase the facilities for preserving existence, must, by the same general law, in their turn become predominant. Here, then, we have progression and continued divergence deduced from the general laws which regulate the existence of animals in a state of nature, and from the undisputed fact that varieties do frequently occur. It is not, however, contended that this result would be invariable; a change of physical conditions in the district might at times materially modify it, rendering the race which had been the most capable of supporting existence under the former conditions now the least so, and even causing the extinction of the newer and, for a time, superior race, while the old or parent species and its first inferior varieties continued to flourish. Variations in unimportant parts might also occur, having no perceptible effect on the life-preserving powers; and the varieties so furnished might run a course parallel with the parent species, either giving rise to further variations or returning to the former type. All we argue for is, that certain varieties have a tendency to maintain their existence longer than the original species, and this tendency must make itself felt; for though the doctrine of chances or averages can never be trusted to on a limited scale, yet, if applied to high numbers, the results come nearer to what theory demands, and, as we approach to an infinity of examples, become strictly accurate. Now the scale on which nature works is so vast—the numbers of individuals and periods of time with which she deals approach so near to infinity, that any cause, however slight, and however liable to be veiled and counteracted by accidental circumstances, must in the end produce its full legitimate results.

Let us now turn to domesticated animals, and inquire how varieties produced among them are affected by the principles here enunciated. The essential difference in the condition of wild and domestic animals is this,—that among the former, their well-being and very existence depend upon the full exercise and healthy condition of all their senses and physical powers, whereas, among the latter, these are only partially exercised, and in some cases are absolutely unused. A wild animal has to search, and often to labour, for every mouthful of food—to exercise sight, hearing, and smell in seeking it, and in avoiding dangers, in procuring shelter from the inclemency of the seasons, and in providing for the subsistence and safety of its offspring. There is no muscle of its body that is not called into daily and hourly activity; there is no sense or faculty that is not strengthened by continual exercise. The domestic animal, on the other hand, has food provided for it, is sheltered, and often confined, to guard it against the vicissitudes of the seasons, is carefully secured from the attacks of its natural enemies, and seldom even rears its young without human assistance. Half of its senses and faculties are quite useless; and the other half are but occasionally called into feeble exercise, while even its muscular system is only irregularly called into action.

Now when a variety of such an animal occurs, having increased power or capacity in any organ or sense, such increase is totally useless, is never called into action, and may even exist without the animal ever becoming aware of it. In the wild animal, on the contrary, all its faculties and powers being brought into full action for the necessities of existence, any increase becomes immediately available, is strengthened by exercise, and must even slightly modify the food, the habits, and the whole economy of the race. It creates as it were a new animal, one of superior powers, and which will necessarily increase in numbers and outlive those inferior to it.

Again, in the domesticated animal all variations have an equal chance of continuance; and those which would decidedly render a wild animal unable to compete with its fellows and continue its existence are no disadvantage whatever in a state of domesticity. Our quickly fattening pigs, short-legged sheep, pouter pigeons, and poodle dogs could never have come into existence in a state of nature, because the very first step towards such inferior forms would have led to the rapid extinction of the race; still less could they now exist in competition with their wild allies. The great speed but slight endurance of the race horse, the unwieldly strength of the ploughman's team, would both be useless in a state of nature. If turned wild on the pampas, such animals would probably soon become extinct, or under favourable circumstances might each lose those extreme qualities which would never be called into action, and in a few generations would revert to a common type, which must be that in which the various powers and faculties are so proportioned to each other as to be best adapted to procure food and secure safety,—that in which by the full exercise of every part of his organization the animal can alone continue to live. Domestic varieties, when turned wild, must return to something near the type of the original wild stock, or become altogether extinct.

We see, then, that no inferences as to varieties in a state of nature can be deduced from the observation of those occurring among domestic animals. The two are so much opposed to each other in every circumstance of their existence, that what applies to the one is almost sure not to apply to the other. Domestic animals are abnormal, irregular, artificial; they are subject to varieties which never occur and never can occur in a state of nature; their very existence depends altogether on human care: so far are many of them removed from that just proportion of faculties, that true balance of organization, by means of which alone an animal left to its own resources can preserve its existence and continue its race.

The hypothesis of Lamarck—that progressive changes in species have been produced by the attempts of animals to increase the development of their own organs, and thus modify their structure and habits—has been repeatedly and easily refuted by all writers on the subject of varieties and species, and it seems to have been considered that when this was done the whole question has been finally settled; but the view here developed renders such an hypothesis quite unnecessary, by showing that similar results must be produced by the action of principles constantly at work in nature. The powerful retractile talons of the falcon- and the cat-tribes have not been produced or increased by the volition of those animals; but among the different varieties which occurred in the earlier and less highly organized forms of these groups, those always survived longest which had the greatest facilities for seizing their prey. *Neither did the giraffe acquire its long neck by desiring to reach the foliage of the more lofty shrubs, and constantly stretching its neck for the purpose, but because any varieties which occurred among its antitypes with a longer neck than usual* at once secured a fresh range of pasture over the same ground as their shorter-necked companions, and on the first scarcity of food were thereby enabled to outlive them. *Even the peculiar colours of many animals, especially insects, so closely resembling the soil or the leaves or the trunks on which they habitually reside, are explained on the same principle; for though in the course of ages varieties of many tints may have occurred, yet those races having colours best adapted to concealment from their enemies would inevitably survive the longest. We have also here an acting cause to account for that balance so often observed in nature,—a deficiency in one set of organs always being compensated by an increased development of some others—powerful wings accompanying weak feet, or great velocity making up for the absence of defensive weapons;* for it has been shown that all varieties in which an unbalanced deficiency occurred could not long continue their existence. The action of this principle is exactly like that of the centrifugal governor of the steam engine, which checks and corrects any irregularities almost before they become evident; and in like manner no unbalanced deficiency in the animal kingdom can ever reach any conspicuous magnitude, because it would make itself felt at the very first step, by rendering existence difficult and extinction almost sure to follow. An origin such as is here advocated will also agree with the peculiar character of the modifications of form and structure which obtain in organized beings—the many lines of divergence from a central type, the increasing efficiency and power of a particular organ through a succession of allied species, and the remarkable persistence of unimportant parts such as colour, texture of plumage and hair, form of horns or crests, through a series of species differing considerably in more essential characters. It also furnishes us with a reason for that "more specialized structure" which Professor Owen states to be a characteristic of recent compared with extinct forms, and which would evidently be the result of the progressive modification of any organ applied to a special purpose in the animal economy.

We believe we have now shown that there is a tendency in nature to the continued progression of certain classes of varieties further and further from the original type—a progression to which there appears no reason to assign any definite limits—and that the same principle which produces this result in a state of nature will also explain why domestic varieties have a tendency to revert to the original type. This progression, by minute steps, in various directions, but always checked and balanced by the necessary conditions, subject to which alone existence can be preserved, may, it is believed, be followed out so as to agree with all the phenomena presented by organized beings, their extinction and succession in past ages, and all the extraordinary modifications of form, instinct, and habits which they exhibit.

Ternate, February, 1858.

INDEX-GLOSSARY

Note to user:
If you need a definition to a word, look for italicized page numbers if the word is not actually defined in the Index-Glossary. Letters following page numbers have the following meaning: f = figure; n = note; t = table.

A

α amylose, breakdown in human digestion, 374f
  structure 374f
A site, see Aminoacyl site of ribosome
aa-tRNA, see Aminoacyl tRNA
Abalone, reproductive strategy, 281f
ABO blood groups
  in blood transfusion, 486f
  and fetal health, 485, 486f
Aboral pole, of coelenterates, 260, 261f
Aboreal life, effects on primate evolution, 312, 313
Abortion
  causes of spontaneous, 492
  critical periods for, 470
  methods, 469
  moral aspects, 470
  numbers of U.S., 469
  U. S. Supreme Court ruling on, 469
Absorption, of food in human small intestine, 376, 377f
Absorption spectrum, of chlorophyll, 88f, 89
Acellular organisms. Organisms bounded by a single cell membrane and containing organelles accomplishing functions analogous to the parts of multicellular organisms. 183
Acetabularia, 181f
Acetycholine, secretion by parasympathetic nervous system, 349
Acetyl coenzyme A
  molecular importance of, 101
  structure, 101
  and tricarboxylic acid cycle, 100f, 101
Acetyl groups, and tricarboxylic acid cycle, 99, 101
Acetylcholine
  as neuromuscular transmitter, 345, 346f
  synaptic agonism and antagonism, 442, 443f, 444
  as synaptic transmitter, 344, 344f
Acetyl-CoA, see Acetyl coenzyme A
Acid
  organic molecules as, 42
  secretion by human stomach, 375f
  weak organic, 72f, n
Acoeles, primitive flatworms, 263, 264f
Acquired characteristics
  inheritance of, 285
  theories of, 123
Acridine, as mutagen, 125f, 126
Acrosomal cap. Part of sperm head which reacts with egg to cause fusion of egg and sperm. 449
  of human sperm, 449
ACTH, see Adrenocorticotropic hormone
Actin, role in muscle contraction, 346f
Action potential
  of muscle, 346f
  of plants, 222
  see also Nerve impulse
Action spectrum, of photosynthesis, 88f, 89
Active transport, 70
  by kidney, 407, 408
  and transport of materials in plants, 216, 216f, 217, 218f
Acupuncture, 439
Adaptation
  effects on evolution of species, 297, 298
  of sensory receptors, 351
Adaptive convergence. The assumption of similar appearance and function by diverse forms of life, usually attributable to restrictive selective forces. 240
  of aquatic animals, 240, 241f
  of desert plants, 240, 241f
Adaptive radiation. The evolution of a related, homogeneous group of organisms into an array of forms specialized for diverse modes of life.
  mammals, 281
  marsupials, 296f
Adenine
  in plant cytokinins, 221, 221f
  structure, 55f
Adenine dinucleotide, structure, 75f
Adenosine monophosphate
  cyclic, 137, 138f, 139
  as second messenger, 139
Adenosine triphosphate
  and glycolysis, 72, 73f
  and hexose monophosphate pathway, 80, 81f
  high energy bonds of, 72f
  and photosynthesis, 92f, 93f
  structure, 72f
Adenyl cyclase, role in second messenger system, 363, 364f
ADH, see antidiuretic hormone
Adhesiveness, of xylem tubes for water, 217

641

Adjustment to environment, as function of organ systems, 192
Adrenal cortex, structure, 359f, 360
Adrenal glands
  cellular structure, 359f
  compensatory hypertrophy of, 357
  and control of thermoregulation, 389f, 390
  effects on embryonic development, 481
  hormones secreted by, 359f, 360, 365, 366
  interactions with organs of the human body, 359f
  interaction with pituitary, 359, 365, 366f
  location of, 207f, 408f
  see also Adrenal cortex and Adrenal medulla
Adrenal medulla
  control of blood glucose by, 365, 366
  hormones secreted by, 365
  neurosecretory function, 365
  release of epinephrine by, 359f, 360, 365, 366
  responses to stress, 366
  role in regulating blood glucose, 382, 383f, 385
  structure, 359f, 360
Adrenaline, see Epinephrine
Adrenergic synapse. Synapse in which the transmitter is noradrenaline. 442
  agonism and antagonism by drugs, 443f, 444
Adrenocortical hormones, 360
  release by pituitary, 357, 359f
  role in regulating glucose, 383f, 385
Adrenocorticotropin, see Adrenocorticotropic hormone
Adrenogenital syndrome, and development of adrenal glands, 481
Aerosol propellants, as pollutants, 617
Africans, origin and distribution of early races, 329, 330f
Agassiz, Louis, 290
Age
  effect on population growth, 577, 577f
  and necessity for cell division, 142
Aging
  biological significance of, 514
  mechanisms of, 514
  process in humans, 516
Agnathous fishes, as primitive vertebrates, 275, 275f
Agonist. An entity opposed by another, as a muscle or drug, the opposing entity being an antagonist. 414
  muscles as, 414
Agriculture
  current productivity of, 548, 549t
  development of by early humans, 325, 325f, 326
  and ecosystem stability, 548, 556
  effects on human populations, 569f, 570
  improvement of methods, 558
  interference with nutrient cycling, 557
  and reduction of genetic diversity of crops, 556
  and supply of human food, 548

unsuitability of rain forest for, 544
  see also Food production
Ainu, human racial variety, 329f
Air, restoration of oxygen in by plants, 211
Air pollution
  sources of, 613t, 614
  types of, 614
Alanine
  stereoisomers, 43f
  structure, 43f
Albinism, human genetic disorder, 493t
Alchemy, 3, 3f
Alcohol, effects on fetal health, 483
Aldehyde group, 53f
Aldosterone
  release by adrenal glands, 359f
  release by testes, 361
Algae
  blue-green, 245
  fossils of, 63f
  gamete cilia of, 243f
  golden, 247f, 248, 249
  types of, 250
Allantois, of reptile egg, 280f
Alleles. A pair of genes influencing the same process and occurring at the same site on homologous chromosomes. 157
  crossing over of, 174, 175, 176f
  and incomplete dominance of genes, 170
Allergic reaction. Reaction of the organism to foreign substances in which the normal resistance mechanism (antibody formation) becomes modified to produce harmful effects, including hay fever, hives, and asthma. 504
  in immunization, 504
Allergins, as air pollutants, 613
Allosteric enzymes, 140
Aloplex lagopus, Arctic Fox, 547
Alpha carbon, of amino acids, 46, 47f
Alpha control system, of muscle, 415
Alpha helix, of proteins, 48, 48f
Alpha rhythm, 430d
  in electroencephalogram, 424n
Alpha-tocopherol, see Vitamin E
Alternation of generations, in coelenterates, 260f, 261
Alveoli. The small air pockets or sacs that make up lung tissue and serve as the sites of gas exchange between the air in the lung and the capillaries lining its walls. 391
  role in human ventilation, 391
  structure, 392f
Alveolus, see Alveoli
Amebas
  of slime mold, 185, 185f
  structure, 247f
American Indian, human racial variety, 329f
Amino acid analyser, 50
Amino acids. Organic acids containing at least one basic amino group (—NH$_2$ group). Most of the important amino acids have the general formula R—CH(NH$_2$) COOH and are the essential building blocks of proteins. 41

absorption in human small intestine, 376, 377f
  active transport of by kidney, 407, 408
  analysis in determining phylogeny, 239f
  attachment to tRNA, 132
  as part of glucose cycle, 383f, 385
  in proteins, 41
  RNA codons for, 116f
  structures, 43f
  variations of in cytochrome c, 239f
Amino group, 41
  of amino acids, 41, 42f, 43f
Amino peptidase, role in human digestion, 378f
Aminoacyl site, of ribosome, 134
Aminoacyl tRNA, 133f, 134
Ammonia
  as air pollutant, 613t
  in nitrogen cycle, 538
  structure, 202f
Ammonium, use of by plants, 219
Amnesia, 432
Amniocentesis, 487
  information gained from, 487
Amnion. One of the membranes enclosing the embryo in reptiles, birds and mammals: formed of ectoderm and mesoderm from the embryo.
  of human placenta, 461
  of reptile egg, 280f
Amniotic cavity, in human development, 482f
Amniotic fluid, 461
  of human placenta, 461
Amphetamine, neural effect of, 443f
Amphibians
  characteristics of, 278, 278f
  heart of primitive, 305f
  reliance on water for reproduction, 278, 278f
  types of, 278, 278f
Ampulla, of echinoderm tube foot, 272f
Amygdalin, see Laetrile
Amylopectin, role in human digestion, 374f
Anabolic steroids, use by athletes, 454
Analogous structures. Similarity in appearance and function resulting from convergent evolution instead of common ancestry. 240
Analogy, of crab and vertebrate limbs, 242
Anaphase
  of meiosis, 160f
  of mitosis, 148, 148f
Anatomy
  comparative, 234, 238
  use in classifying organisms, 234, 238
Anchoveta, fishery for, 542
Androgametophytes, of gymnosperms, 255f
Androgen. Male sex hormone, 19-carbon steroid, for example testosterone. 361
  release by adrenal cortex, 359f
  released by testes, 361
  role in human reproductive cycle, 452f
Androsporangia, of gymnosperms, 255f
Androspores, of gymnosperms, 255f

Androstrobilii, of gymnosperms, 255f
Aneuploid chromosome number, 176
Angiosperms, characteristics of, 256
Ångstrom, 36f, 37
Animalia, Kingdom, see Specific type of animal
Ankylosing spondylitis, 502
Annelida, Phylum, see Annelids
Annelids
　coelom of, 265, 265f
　comparative embryology of, 238, 238f
　locomotion of, 265f
　nephridia of, 201f
　segmentation of, 265f, 266
　structure of, 265, 265f
Annual plants, 222
　and success of angiosperms, 256
Anoxia, cause of death, 489
Antagonist, 414
　muscles as, 414
　see Agonist
Antelope, 282f
Anterior lobe
　of pituitary gland, 356, 356f
　see also Anterior pituitary
Anterior pituitary
　and endocrine system control, 207
　control of gonads by hormones from, 360, 361
　control of thyroid by hormones from, 358f
　control of thryoid by, 358f
　in human development, 481
　role in regulating blood glucose, 382, 383f
Anterior surface, of bilateral animals, 261f
Anther, of flowers, 225f
Anther, pollen production by, 226f
Antheridia, of moss, 252f
Anthropoidea, Suborder, see Anthropoids
Anthropoids
　classification of, 311, 311t
　comparison of hands and feet of, 312, 313f
　prehensile tail of, 312
　see also Humans; Apes
Antibiotics
　and treatment of disease, 504, 505f
　transference of resistance to in bacteria, 155, 156n
　types of resistant bacteria, 504
Antibodies. Proteins normally present in the body or produced in response to infection or introduction of foreign substances into the body. Antibodies serve to agglutinate and/or precipitate foreign particles, including bacteria, and to neutralize *toxins*, or poisonous substances, produced by invading organisms. 500
　chemical nature of, 501f
　clonal selection theory of formation, 501, 502f
　direct absorption by infants, 376
　as proteins, 38
　used in determining primate phylogeny, 315
Anticoagulants, and circulatory disease, 495
Anticodon, of tRNA, 132, 132f

Antidiuretic hormone
　function in osmoregulation, 409, 410
　release by pituitary, 356, 356f
　release governed by osmotic receptors, 357
Antigen. A substance, usually protein, which when introduced into an organism to which it is foreign, results in antibody formation.
Antigen-antibody reaction, and immunity, 500
Antioxidants, biomolecules as, 98, 99f
Antiparallel, chains of DNA, 117, 118f
Antipyretic. Something that relieves fever. 390
　use in controlling fever, 390
Antisera, use in comparative serology, 238, 239f
Anus, in human digestive system, 373f
Anvil, of human ear, 354f
Aorta
　role in heart circulation, 401f, 402
　as site of blood chemoreceptors, 395, 396f
Aortic arches, in human development, 476, 477f, 478
Aortic bodies, 395
　function in ventilation control, 395, 396f
Apes, classification of, 311t
　close relationship to humans, 316
Apical meristem, 212
　and growth of plants, 212, 213f
Apparent intelligence, 433
Appendix, location in human digestive system, 373f
Aquarium, as an ecosystem, 524
Archaeoptryx, fossil bird, 278, 280f
Archegonia, of moss, 252f
Archegonium, of gymnosperms, 255f
Arctic Fox, population cycles of, 547, 547f
Arenicola, in detritus food chain, 545, 546f
Arginine, structure, 43f
Argon-40, use in dating fossils, 236f
Aristotle, classification scheme of, 230n
Army ant, collectives of, 189, 189f
Aromatic rings, of amino acids, 301f
Arrhenius, S., 61
Arteries, characteristics of human, 404, 405
　location of human, 401f
Arteriole. Arterial vessels of approximately 30 um internal diameter with walls containing adrenergically controlled smooth muscle but with less elastic tissue than in arteries. 405
　structure, 404
Arteriosclerosis. A disease condition in which blood flow through arteries is limited by the development of thickenings of their linings.
　causes of, 495
Arteriovenous shunts, 406
Arthritis, association with HLA genes, 502
Arthropoda, Phylum, see Arthropods
Arthropods
　characteristics of, 266, 267, 267f, 268
　effects of exoskeleton on, 267
　exoskeleton of, 202, 203f

　molting of, 203, 208f, 209
　nervous system limitations in, 267
　nervous system of, 206f, 207
　polyphyletic origin of, 245
　respiration of insects, 267
　roles in human disease, 499
　segmentation of, 266, 267
　stereotyped behavior of, 269
　types of, 266, 267f
Artificial respiration, as positive pressure breathing, 394f, 395
Asbestos
　as air pollutant, 613t, 618
　as carcinogen, 512
　effect on lungs, 618
Asbestosis, cause of, 618
Ascaris
　as a roundworm, 265
　coelom of, 265
　cross section of, 265
　use in determining zygote chromosome number, 172
Ascending aorta, and lung circulation, 392f
Ascending reticular system, and wakefulness, 438
Ascorbic acid
　structure, 99f
　see Vitamin C
Asexual reproduction, by budding in jellyfish, 260f
Asparagine, structure, 43f
Aspartic acid, structure, 43f
Aspirin, as antipyretic, 390
Aster, 147
　function in mitosis, 147, 148f
Asthma, and pollution, 614
Asymetric carbon, of amino acids, 46d, 47f
Athlete's foot, fungal animal parasite, 256
Atmosphere
　early accumulation of oxygen in, 96
　effects on solar radiation, 527, 528f
Atom. The smallest particle of an element that can exist and show the properties of that element. 38, 40d
　structural model of, 84f, 85
Atomic bomb, see Nuclear bomb
Atomic number, 85
"Atomic pile," 590
Atoms, and organization of organisms, 191f
ATP, see Adenosine triphosphate
ATP synthetase, and oxidative phosphorylation, 105
Atria
　function of human, 401, 401f, 402, 403f
　pressure receptors of, 402
　see also Left atrium; Right atrium
Atrial siphon, of sea squirts, 272, 273f
Atrioventricular node, 403f
Attlee, C., 607
Auditory nerve, human, 354f
Australopithecines, see *Australopithecus*
Australopithecus
　characteristics of early ape-human, 318t, 319
　life-style of, 320

INDEX-GLOSSARY　**643**

Australopithecus (cont.)
  skull of, 12f
  use of weapons by, 319
Autecology, 522
Autonomic ganglia, see Sympathetic ganglia
Autonomic nervous system. The involuntary system of nerves controlling certain glands, the smooth muscle of gut and blood vessels, the heart and other organs. It has two major divisions, the sympathetic and parasympathetic systems.
  components of, 349
  control of human digestion, 379, 382
  functions of, 349
  transmitter chemicals of, 349
Autosomes, 174
Autozooid, of sea pansy Renilla, 188f
Auxin. A plant hormone secreted by *meristematic* tissue, which causes growth in a longitudinal direction. 220
  and control of plant growth, 220, 220f
  effects on plant buds and fruit, 221
  effects on plant cell walls, 220
  structure, 220f
A.V. node, see Atrioventricular node
Avery, O. T., 54, 113
Axillary artery, 401f
Axillary vein, 401f
Axon. An extension of a nerve cell that normally conducts nerve impulses away from the cell body. 341
  conduction velocities determined by diameter, 268
  of motor neuron, 340f, 341
  of neuron, 205f, 206
Axon hillock, of neuron, 205f

B

B lymphocytes, role in combating disease, 501, 501f
Backbone
  as maladapted for human needs, 299
  see also Spinal cord; Vertebrae
Bacteria
  antibiotic resistant strains of, 504
  as causes of human disease, 497, 497f
  characteristics of, 245, 246, 246f, 246t
  colonies of, 182f
  colonies as multicellular organisms, 182f, 184
  DNA replication rate in, 123
  general structure of, 246f
  kinds of, 245, 246, 246t
  metabolic diversity of, 246t
  symbiotic luminous, 546f
  transformation of, 153, 154f
  true, 245, 246, 245t
Bacteriophage T1, and *E. coli* replica plating experiments, 127, 127f
Bacteriophage T4, use in estimating gene size, 124
Balancing organs, of comb jellies, 262f
Balanoglossus
  feeding in, 272
  as protochordate, 272
Balme, D. M., 164
Basal body, of cilia, 243f
Basal metabolic rate. The rate of metabolism of an organism in a fasting and resting state. 367
  energy requirement for, 367
Bases, pairing by in DNA, 117
Basidiocarps, mushrooms as, 257f
Basidiomycetes, see Fungi
Basidiomycota, subphylum, see Mushrooms; Puffballs; Truffles
Bats, and mammalian diversity, 282f
  as pollinators, 227f
Battery, of nerve cells, 341, 342f
Bayliss, William, 380
Beadle, G., 112
Beak, of squid, 268, 270f
Bears, example of speciation of, 231, 231f
Beche de mer, oriental sea cucumber dish, 270
Becquerel, H., 590
Beef, use of grain in production of, 555n
Behavior, affecting population size, 581
  and conditioning of octopuses, 270
Behavior
  drugs and, 441
  effects of drugs on, 442
  human sex specific, 454
  instinctive insect, 268
  stereotyped insect, 268
Benzene
  as air pollutant, 613t
  structure, 41f
Benzer, S., 124
Benzypyrine, as pollutant, 618
Bergson, H., 285
Bering Straits, migration of humans across, 324
Bernard, Claude, 197
  description of milieu interieur, 197
Beta helix, of proteins, 48, 48f
Bikini, fallout effects of bomb tests on, 602
Bilateral symmetry. Symmetry in which one unique plane divides the body into two mirror image halves. 261
  evolution of, 263
Bile duct
  in human development, 481
  role in human digestive system, 382
Bile, role in human digestive system, 382, 382f
  see also Bile salts
Bile salts, hormonal control for release of, 380
Binding energy, nuclear, 591, 591f
Binocular vision, development of in primates, 313
Binomial nomenclature, 229, 230n, 231
Biochemical morphology, 238
Biochemical pathways, evidence for evolution in, 300, 301f
Biofeedback
  and alpha rhythm, 430
  medical use, 429, 430
Biogenic amines
  as brain transmitters, 442
  see also Catecholamines
Biogeochemical cycles, 531
Biogeographer, study of fossils by, 303
Biological clocks, 221
  demonstration of in plants, 222
  and environmental coordination of plants, 221
Biological evolution
  as biological history, 20
  early stages of, 77, 108, 108t
Biological memory, and evolution, 21
Biological oxygen demand, 621
  and water pollution, 621
Biological pest control, 559
Biology
  as governed by history, 19
  and human affairs, 25
  levels of organization, 521
  major themes, 19, 21
Biome, 542
  major types of, 542, 543f
Biosphere. That part of the earth and atmosphere in which life exists.
  human jeopardization of, 584
Biparental inheritance, 164n
Bipedal locomotion, development by primates, 314, 316, 317f, 318
Biphenyl, see Polybrominated; Polychlorinated biphenyl
Birds
  adaptations for flight in, 278, 280f
  advances over reptiles, 278, 280
  characteristics of, 278, 280f
  thermoregulation by, 387, 387f
Birth
  events of human, 461, 464f
  fetal adaptation during, 464
Birth control
  by Australian aborigines, 465
  effectiveness of contraceptive methods, 467t
  types of, 465
Birth interval, among !Kung people of Africa, 570
Birth rate
  in China, 573
  in Costa Rica, 570
  in Democratic Republic of Germany, 573
  in early English villages, 570
  in early humans, 570
  historical means of decreasing, 573
  in India, 574
  in Mexico, 574
  in Portugal, 573
  variations in human, 570–576
Bison, reduction of by hunting, 558
Biston betularia, moth, 293f
"Black box" concept, in ecology, 521
Black Death, see Plague
Black lung disease, cause of, 613t
Black widow spider venom, neural effect of, 443f, 444

Blastocoel. The central cavity of the blastula. 471
  of human embryo, 471
Blastocyst. The blastula stage in mammalian development, characterized by a thin sphere of cells containing a localized cell mass that will become the embryo. 460d
  formation of human, 460, 460f
Bleeding, control of, 406
Blind experiment. An experiment in which the experimenter lacks foreknowledge of certain critical details as a precaution against unconscious or other bias. 8
Blind spot, of human eye, 352f, 353
Blood
  composition of, 399
  dialysis of human, 496, 497f
  route through heart, 401f
  types of human, 485, 485f, 486f
  volume in humans, 399
  see also ABO blood groups
Blood clots, formation liked with oral contraceptive, 468
  formation of, 400, 400f
  structure of, 400f
Blood vessels
  role in controlling blood pressure, 406
  and surface area, 198, 198f
  see also Arteries and veins
Blue whale
  probable extinction of, 559t
  size of, 299f
Blue-green algae, characteristics of, 245
BMR, see Basal metabolic rate
BOD, see Biological oxygen demand
Body fluids, composition in marine animals, 68t
Bohr effect, of pH on oxygen transport, 397, 398f
Bohr planetary model, of atom, 84f, 85
Bomb
  atomic, see Bomb, nuclear
  hydrogen, 604
  nuclear, 593
  thermonuclear, 604
Bond
  carbon, 46
  covalent, 84f
  high energy, 72f, 74
  hydrogen, 48, 58f
  ionic, 84f
  peptide, 44d, 44f
  phosphodiester, 55f
  sulfur, 48, 48f
  thiol, 101
Bony fish, diversity of, 275, 275f, 276
Borlag, Norman, 552
Botulinum toxin, neural effect of, 444
  mode of action, 490
Boveri, T., 172
Bowman's capsule, of nephron, 409f
Brachial artery, 401f
Brachial vein, 401f
Brachiation, 314d
  effects on primate evolution, 314

Brachiocephalic vein, 401f
Brain
  control of blood pressure by, 406
  and control of thermoregulation, 389f, 390
  crucial period in development, 483
  development of human, 474, 474f
  evolution of, 306, 306f
  evolution of in hominids, 319, 320
  evolution of large primate, whale and sea lion, 313, 314f
  heart-inhibiting center of, 402
  of insects, 208f
  interpretation of sensory information by, 344
  olfactory lobe of, 372f
  parts of human, 347, 348f
  as part of nervous system, 208
  ways to study human, 411
  see also Cerebellum; Cerebral cortex
Brain size
  related to cultural development in humans, 326
  related to usage of hands by primates, 316, 317
Bread mold, mutants of, 114n
Breast feeding, effect on birth interval of !Kung people, 570
  inhibitory effect on ovulation, 465
Breathing, see Ventilation
Breeder reactor, see Nuclear power reactor, breeder type
Bridges, C. B., 174
Briggs, R., 17
British thermal unit, 568t
Bronchi, 391
  structure, 392f
Bronchioles, 391
  role in human ventilation, 391
Bronchitis, and smoking, 510f
Brown algae, characteristics of, 250
  organizational similarity to terrestrial plants, 250
  regional specialization in, 250
  reproduction in, 250, 251f
Bryan, William Jennings, 290
Brontosaurus, size of, 299f
Bryophyta, Phylum, see Bryophytes
Bryophytes
  characteristics of, 252, 253f
  life cycle of moss, 252f
  reproductive structures of moss, 252f
Btu, see British thermal unit
Buccal siphon, of sea squirts, 272, 273f
Budding, by jellyfish strobilus, 260f
Bulbourethral glands. Glands which produce a secretion that lubricates the male urethra. Also known as Cowper's glands. 448
  human, 448
Burgess Shale, importance to fossil record, 237
Burkitt's lymphoma, 511
Burnett, Sir MacFarlane, 520
Burning, as oxidation-reduction reaction, 71

Bushmen, human racial variety, 329f, 330

C

Cactus, adaptations to control water loss in, 219
Caffeine, neural effects of, 445
Calcitonin
  function of, 357
  release by parathyroid glands, 357
Calcium carbonate
  hormonal control of, 357, 358
  residence time in oceans, 532
  role in muscle contraction, 346f
Caloric intake, by humans, 549t
Calorie 71, Appendix 2
Calorimetry. Measurement of the heat generated by a system. 540
Calvin cycle, efficiency of, 93
  of photosynthesis, 87, 93, 93f
Cambrian Age, and evolution of the atmosphere, 96
  rapid speciation during, 306
cAMP, see Cyclic AMP; Adenosine monophosphate, cyclic
Canals, and ecosystem manipulation, 559
Cancer
  action of interferon, 507
  current U.S. death rate from, 507
  environmental factors and, 509, 512
  Hodgkin's disease as, 502
  induced by radiation, 509
  of lungs in uranium miners, 594
  prevention of, 511
  quack cures for, 4
  radiation caused, 509
  and rights of the terminally ill, 513
  theories of causes, 509
  treatment for, 511
  types of, 508, 509t
Canine teeth, of chimpanzees, 314f
Canines, reduction of in early hominids, 319
Cannon, W. B., 333
Capacitation period, 449
Capillaries. Small fine blood vessels having thin walls (one cell thick) across which materials can pass to or from tissues. 405
  characteristics of human, 405
  of human digestive system, 381f
  of human skin, 386f
  of human small intestine, 377f
  of kidney, 409f
  role in thermoregulation, 387, 389f
Carbohydrate structure, 53f, 54
Carbohydrates
  breakdown in tricarboxylic acid cycle, 99, 100f
  functions in cells, 38, 53
  and regulation of blood glucose, 383f
Carbon
  bonding, 45f, 46
  compounds of, 40, 41t
  description, 40
  long term storage in carbon cycle, 534

Carbon (cont.)
  and organic molecules, 40
  tetrahedron structure of, 46
Carbon cycle, 534, 534f
  linkage with phosphorus cycle, 536
Carbon dioxide, early observation on use by plants, 211
  effect on earth's temperature, 525, 616
Carbon dioxide fixation, 93
  generated in tricarboxylic acid cycle, 101
  receptors, and ventilation control, 395, 396f
  role in carbon cycle, 534, 534f, 535
  role in respiratory control, 337, 338f
  role in ventilation control, 395, 396f
Carbon monoxide, as poison, 26, 490
  role in carbon cycle, 534, 534f, 535
  sources as pollutant, 614
Carbon tetrachloride, as poison, 490
Carbon-14, use in dating fossils, 236f, 237
Carbonic acid, role in respiratory control, 337, 338f
Carbonic anhydrase, role in control of respiration, 338f, 339
Carboxypeptidase, role in human digestion, 378f
Carboxyl group. An aggregation of carbon, oxygen and hydrogen atoms having the formula —COOH. Carboxyl groups are the acid groups of organic molecules. 41
  acidity of, 42, 43f
  of amino acids, 41, 42f
Carcinogen. Substance capable of producing cancer. 509
Carcinoma, 508
Cardiac pacemaker, 495
  and treatment of circulatory disease, 495
Cardiovascular accident, *see* Circulatory system failure
Carotid artery, site for blood chemoreceptors, 395, 396f
Carotid bodies, 395
  function in ventilation control, 395, 396f
Carpel, of flowers, 224, 225f
Carrier molecules
  of respiratory chain, 102f, 103
  spectropotometry of reactions, 106f, 107
Carson, Rachael, 623
Cartilagenous fishes, 275, 275f, 276
Case, S., 487f
Casparian strips, of plant roots, 216f
Castration, as form of birth control, 465
Catalase, function of, 99, 99f
Catalyst, 49
Catastrophism, 286
  as explanation for creation of fossils, 287
Catecholamines. Biologically active compounds formed of a benzene nucleus with two adjacent hydroxyl groups (catechol) and an amine, compounds such as epinephrine, norephinephrine, serotonin and dopamine. 365
  as brain transmitters, 442
  release by adrenal medulla, 365
Catecholamine-O-methyl transferase, role in transmitter breakdown, 444
Caucasians, human racial variety, 330
Caucasoids, evolution and distribution of early, 329, 330f
CCK-PZ, *see* Cholecystokinin-pancreozymin
Cebiodea, Superfamily, *see* Cebiods
Cebiods
  classification of new world monkeys, 311t
  development of prehensile tail by new world monkeys, 312
Celibacy, effectiveness in birth control, 465
Cell. A small mass of protoplasm bounded by a semipermeable membrane, containing at least one nucleus and capable alone or with other cells of performing the fundamental functions of life. 33
  of motor neuron, 340f, 341
  of neuron, 205f
Cell division
  in eukaryotes, 145
  necessity for, 142
  in prokaryotes, 144
  synchrony with nuclear division, 149
Cell membrane. The living boundary between the cell and its surroundings which regulates inflow and outflow of many substances. 34
  chemical analysis of, 70
  pores through, 70f, 71
  structure of, 70f
  as target of hormones, 363, 364f
Cell plate, formation during mitosis, 147f
Cell theory, 31, 33
Cell wall. The rigid, non-living, permeable wall surrounding a plant cell outside the plasma membrane; usually composed largely of cellulose (with other materials added according to the specialization of the particular cell). 34
  effects of auxin on, 220
  formation of in plant cells, 213
  plasmodesmata through, 213
Cells
  composing tissues, 191f
  differentiation of in plants, 212, 213f
  and organization of organisms, 191f
  sensory, 350f, 351, 353, 353f
Cellular differentiation, 181
Cellular respiration. The oxygen-requiring metabolic pathway used to obtain energy from food. 99
  energetic advantages of, 107
  metabolic pathways of, 99, 100f
Cellular slime mold, as multicellular organism, 185, 185f
Cellulose
  in sea squirt tunic, 272
  structural characteristics of, 54
Central nervous system, 207
Central response selector, of homeostatic system, 334, 334f
Centrioles, 147
  function in mitosis, 147
Centromere, 146
Cephalization. Concentration of exteroreceptors and central nervous system at one end of the body axis. 261
  in flatworms, 264
  lack of in coelenterates, 261
Ceratium, structure, 247f
Cerebellum
  general function of human, 348f
  location of human, 348, 348f
  role in muscle control, 415
Cerebral cortex
  effects of damage on higher behavior, 421
  evolutionary development of, 417
  higher functions of, 420
  increasing dominance of in vertebrates, 306f
  location of functional areas, 418, 419f
  motor areas of, 418
  prefrontal lobes of, 420
  primary sensory areas of, 419f
  sensory association areas, 420
  size of, 418
  see also Cerebrum
Cerebral hemispheres, lateralization of functions, 420
  see Cerebrum
Cerebrum
  general function of human, 348
  involvement in ventilation control, 396f
  location of human, 347, 348f
  see also Cerebral cortex
Cervix, human, 450f
Cestodes, *see* Tapeworms
Changeaux, J. P., 140
Chardin, Tielhard de, 285
Chemical evolution, of earth's environment, 20, 60
Chemical reactions
  covalent, 84f, 86
  ionic, 84f, 86
  types of, 84f, 85
Chemical transmitter. Chemical released by nerve impulses at a nerve ending and serving to initiate action in an adjacent nerve or effector cell. 343
  transfer of information across synapse by, 343, 344f
Chemiosmotic theory, of oxidative phosphorylation, 105, 105f
Chemoautotroph. Organism able to obtain carbon from carbon dioxide and energy by oxidation of certain reduced inorganic compounds; chemolithotroph. 59
  and energy input to biosphere, 540
  Nitrobacter as, 537
  Nitrosomas as, 537
Chemoreceptor
  ciliary nature of, 243f
  for control of ventilation, 395, 396f
  types of, 350f, 351
Chewing, role in human food processing, 374
Chiasma. A connection between homologous chromosomes providing visible evidence of crossing over. 159
  of crossing-over chromosomes, 159, 161f

Chimaera. Organism formed of tissues of several genetic origins. 472
  human, 472
  of mouse embryos, 472
Chimney sweeps, scrotal cancer of, 509
Chimpanzee
  language learning in, 326, 327f
  morphology compared to other primates, 312f
  teeth of, 314
China, regulation of population growth in, 573
Chitons, as molluscs, 268
Chlordecone, *see* Kepone
Chlorine, and problems in water purification, 621
  as air pollutant, 613t
Chloroform, as carcinogen, 621
Chlorophyll
  absorption spectrum of, 88f
  in chloroplasts, 87, 88f, 89, 90f
  and photosynthesis, 87, 88f, 90f
  structure, 87, 90f
Chlorophyll a, structure, 90f, 242f
Chlorophyta, Phylum *see* Green algae
Chloroplasts
  commensal theory of origin, 150
  description of, 37, 87, 89f, 242f
  as episomes, 149
  and photosynthesis, 88
Chloropromazine, neural effect of, 443f, 444
Choanocytes, of sponges, 257, 258f
Choanoflagellata, and sponge evolution, 257
Cholecystokinin-pancreozymin, role in human digestion, 380
Cholesterol, similarity of hormones to, 360
Cholinergic synapse. Synapse in which acetylcholine is the transmitter. 442
  effects of drugs, 442, 443f, 444
Chordates
  characteristics of, 273, 274
  derivation of from protochordates, 273
  notochord of, 273
  similarities to tunicate larva, 273
  skeleton of, 202, 203f
Chorion, of reptile egg, 280f
Chorionic gonadotropin, role in pregnancy, 455, 456f
Chromatids, 146
  crossing over of, 174, 175, 176f
  replication of, 146, 146f
Chromatography, 50
Chromoplasts, of bacteria, 144
Chromosomal mutation, 124
  types of, 157, 158f
Chromosome. Nuclear structure carrying part of the cell's DNA in a matrix of protein and histones. The naked DNA strand of a prokaryote is loosely termed a chromosome. 35
  bacterial, 145f
  banding patterns of, 178, 178f
  as carriers of genetic traits, 172, 174
  centromeres of, 146, 146f
  chiasmata of, 159

  chromatids of, 146
  crossing over in, 159, 161f
  changes affecting meiosis, 178
  continuity through life of cell, 172
  effects of deletions, 178
  effects of number changes, 176, 177
  effects of structure changes, 177
  experimental variation in numbers of, 172
  of eukaryotes, 144, 145
  gene alleles of, 157
  haploid or diploid numbers of, 146
  homologous, 157
  human, 177f, 178f
  independent segregation of, 169
  kinetochores of, 146
  linkage groups of genes in, 174
  linkage maps of, 174, 175f
  maps of, 174, 175f, 176f
  methods of mapping human, 178
  polyploid numbers of, 146
  and polyploidy, 176
  replication and division of, 145
  sex, 174
  structure, 145
  types of mutations affecting, 157, 158f
Chrysophyta, Phylum *see* Algae, golden and diatoms
Churchill, Sir Winston, 607
Chymotrypsin, role in human digestion, 378f
Cilia, 37
  arrangement of fibrils in, 243f
  cross section of, 243f
  in human ventilating system, 393
  in sensory receptors, 350f, 353f
  structural similarities throughout biosphere, 242f, 243
  structure of, 243f
Ciliates
  characteristics of, 248
  specialized types of, 183f
Cilium, *see* Cilia
Circadian rhythms
  demonstration of in plants, 222
  in human activity, 437
Circulation, in humans, 399-407
Circulatory disease, prevention and treatment, 495
Circulatory system, 193
  of chordates, 274f
  control of human, 403, 403f, 406
  function in animals, 194, 197, 199f
  Harvey's studies of, 405n
  in human development, 476, 477f
  of molluscs, 268, 270
  of plants, 214, 215f, 218f
  parts of human, 392f, 401f, 401-406, 406f, 408f
Circulatory system failure, types of human diseases involving, 494, 494f, 495
Citrate, and tricarboxylic acid cycle, 100f, 101
Citric acid, and tricarboxylic acid cycle, 100f, 101
Clams, gill structure of, 269f

Classification
  criteria used in, 234
  evolutionary significance of schemes, 303
  of flowers, 233
  of mammals, 233
  mono- verus polyphyletic, 245
  of organisms, 229, 230n
Claude-Haber process, and fertilizer production, 538
Cleavage
  and comparative morphology of organisms, 238
  of human zygote, 471
Cleft palate, human congenital malformation, 493t
"Clever Hans" phenomenon, 431
Clitoris, human, 450f, 451
Clone. A population of organisms descended from a common ancestor by mitosis. 151
  difficulty of inheriting changes in, 151
Clostridium botulinum, toxin from, 490
Clostridium tetani, toxin from, 490
Clothing, and human thermoregulation, 387, 388f
Club moss, *see* Lycopods
CO, *see* Carbon monoxide
CoA, *see* Coenzyme A
Coal, reserves of, 589, 589f
Cocaine, neural effect of, 443f
Coccyx, rudimentary organ in humans, 310
Cochlea, of human ear, 354, 354f
Codon. Set of three nucleotides in mRNA that govern addition of a particular amino acid to a polypeptide undergoing synthesis. 132
  amino acids specified by, 116f
  of mRNA, 132
  stop, 135
Coelenterata, Phylum *see* Coelenterates
Coelenterates
  asexual reproduction in, 260f
  characteristics of, 260f, 261
  colonies of, 188, 188f
  colonies of as collectives, 189
  Hydra, 186, 187f
  life history of, 260f, 261
  nematocysts of, 262
  nervous system of, 206f, 207
  sea pansy, *Renilla*, 188, 188f
  symmetry of, 261, 261f
  types of, 260f
Coelocanths
  and evolution of fossil limb, 276
  as living fossils, 237, 237f
Coelom. The mesoderm lined body cavity of a member of the higher metazoa.
  of earthworm, 265, 265f
  effects of segmentation on, 266
  of eumetazoans, 261
  evolution of, 265
  of human embryo, 482f
  structural types of, 265
Coenzyme, *101*
Coenzyme A

Coenzyme A (cont.)
  acetylation of, 101
  structure, 101f
Coenzyme Q
  function in respiratory chain, 103, 104f
  structure, 102f
Cofactor, 101
Cohesiveness, of water molecules, 217
Coitus interruptus, form of birth control, 465
Colchicine
  effects on mitosis, 148
  microtubule disruption by, 147
Coleridge, Samuel Taylor, 437
Collar cells, of sponges, 258f
Collared lizards, 279f
Collection duct, of kidney, 409f
Collectives, 189
  colonial coelenterates as, 189
  of social insects, 189
Collenchyma, of plants, 213f
Colloidal suspension, 34
Colon
  in human development, 482
  role in human digestion, 379
Colony
  of bacteria, 182f
  as a collective, 189
  division of labor in, 181
  evolution of, 182
  formation of in sea pansy *Renilla*, 188, 188f
  of hydrozoans, 260f
  of insects, 189
  of *Volvox*, 186, 187f
  polymorphism in coelenterate, 260f, 262
Color vision
  evolution of by primates, 313
  of humans, 353
Colorado River, soil discharge from, 531f
Coloration, protective, 295
Comb jellies
  control of comb plates in, 262
  embryonic differentiation of, 472
  feeding in, 262
  structure, 262, 262f
Comb plates, of comb jellies, 262, 262f
Commensalism. A relationship between organisms that is beneficial to one and not harmful to the other. 150
Common carotid artery, 401f
Common iliac artery, 401f
Common iliac vein, 401f
Communication
  in chimpanzees, 326
  neuroid, 222
  as transmitter for human cultural development, 326
Community, characteristics of, 523
Community ecology, 523
Companion cells, of phloem cells, 214
Comparative anatomy, of embryonic stages, 238, 238f
Comparative anatomy, use in classification of organisms, 234

Comparative biochemistry, use in determining phylogeny, 238, 239f, 241, 242
Comparative serology, use in classifying organisms, 238, 239f
Compound interest, and population growth, 561, 562f
Compound microscope, 32, 33f
COMT, see Catecholamine-O-methyltransferase
Concentration gradients, importance in plant transport system, 216, 217, 218f
Concept, as unit of cultural evolution, 326
Conditioning
  behavioral, 427, 428f
  in octopus, 270
  Pavlovian, 429
  and sympathetic nervous system, 429
Condom, form of birth control, 465
Conduction velocity, of nerves as function of diameter, 268
Cone cells, of human retina, 352, 353, 353f
Cones
  of gymnosperms, 255f
  types of pine, 256
Congenital malformations, causes of, 490, 491
Conifers
  cones of, 255f
  fertilization in, 255f
  leaves of, 255f, 256
  life history of, 255f
Conjugation. Reciprocal exchange of genetic material by two cells, especially ciliated protozoa. 248
  in *Paramecium*, 248, 249f
Conservative replication, of DNA, 121, 122f
Consumers, in ecosystems, 544
Continental drift, as isolating mechanism in speciation, 294, 295f
Continents, separation of, 295f
Contraception, effectiveness of methods, 467t
Contractile proteins, of motile cells, 37
Contractile vacuoles, and osmotic equilibrium, 69, 69f
Controlled experiments, 8
Controls. Experiments performed, in conjunction with experimental tests of an hypothesis, to detect the intrusion of extraneous factors. 8
Convergence
  adaptive, 240
  molecular, 241, 242f, 243, 244
  of mollusc and vertebrate eyes during evolution, 270, 270f
Convulsion, 424f
Copepods, in marine food chain, 545, 546f
Copulins, 457
  hormonal regulation of, 457, 458f
CoQ, see Coenzyme Q
CoQH2, see Reduced coenzyme Q
Coral reef, 23f
Corey, R. B., 48
Cornea, of human eye, 352f
Coronary arteries, and heart disease, 495
Corpus allatum, of insect brain, 208f

Corpus callosum, 423
Corpus cardiacum, of insect brain, 208f
Corpus luteum
  endocrine role of, 452f, 453
  formation in ovary, 450f
  function in pregnancy, 455
Cortex
  of plant roots, 216f
  of plants, 212, 213f
Cosmic rays, 83, 783f
Cosmozoa theory, of life's origin, 61
Coughing, as cleansing mechanism, 393
Countercurrent exchange system, thermoregulation as, 387
Covalent bond, 84f, 85
Cranial nerves, origin of, 348
Creation Society, opposition to evolution, 290
Cretinism, 435t
Crick, Frances, 115
Crinoids, as echinoderms, 270, 271f
Cristae, of mitochondria. Infoldings of the mitochondrial inner membrane bearing the enzymes of oxidative phosphorylation. 35
Critical mass. The least amount of a fissionable element sufficient to sustain a chain reaction. 593
  of nuclear reaction, 593
Crocodiles, as reptiles, 279, 279f
Cro-Magnon Man
  art of, 322, 322f, 323
  characteristics of, 318f, 322
  life-style of, 322, 323
Crossing over. Mutual exchanges of like parts between homologous chromosomes, occurring during chromosome pairing at meiosis.
  frequency of, 175, 176f
  of homologous chromosomes, 159, 161, 161f
  and linkage groups, 176f
  use in constructing linkage map, 176
Cross-over zone, of homologous chromosomes, 161f
Crown-of-thorns, predatory sea star, 270, 271
Crustaceans, as arthropods, 267f
  green glands of, 201f
Ctenophora, Phylum see Comb jellies
Ctenophores, see Comb jellies
Curare, neural effect of, 443f, 444
Currents
  driven by solar radiation, 527
  principal ocean, 529f
Cuvier, Georges, 286
CVA, see Circulatory system failure
CVM, see Cytomegalovirus
Cyanide, as poison, 490
Cybernetics, applicability to biological systems, 334
Cyclic AMP
  and drug effects on target cells, 445
  and gene regulation in *E. coli*, 137, 138f
  second messenger, 139, 363, 364, 364f, 384

Cyclic phosphorylation, in photosynthesis, 92f
Cysteine
  structure, 43f
  and sulfur bonding, 48, 48f
Cystic fibrosis, human genetic disorder, 493t
Cytochrome a
  as respiratory carrier molecule, 103, 104f
  structure, 102f
Cytochrome b, as respiratory carrier molecule, 103, 104f
Cytochrome c
  as respiratory carrier molecule, 103, 104f
  comparative structure of among organisms, 239f
Cytochromes, structure, 102f
Cytokinins, 221
  effects on plants, 221
  similarity to tRNA bases, 221
  structure, 221f
Cytomegalovirus, and fetal health, 484
Cytoplasm. That part of the protoplasm of a cell lying externally to the nuclear membrane. 34
  controlling influence on cell division, 149
Cytosine, structure, 57f

D

2,4-D, as hazardous chemical, 622t
Dalkon Shield, health problems of, 468
Dark reactions
  efficiency of in photosynthesis, 93
  of photosynthesis, 87, 93, 93f
Darwin, Charles, 285, 287f
  belief in Pangenesis, 123, 163
  evolutionary theory of, 285, 291, 292t, Appendix 4
  observation of Galapagos finches, 287, 287f
  observations on movements of plants, 219, 220f
  publications of, 287, 288, Appendix 4
  views on first humans, 317
  views on human evolution, 309, 310, 317
DDT, see 1,1,1-trichloro-2,2-bis(p-chlorophenyl)ethane
de Saussure, N. T., 211
Death
  causes of in U. S., 494t
  legal certification of, 496
Decussation, in nervous system, 416
Defecation, 379
Dehydration synthesis, 44, 44f
Dehydrogenase, of NAD and NADP, 75f
Deletion, of genes, 157, 158f
Demographic transition. The switch from a state of stable population sustained by high birth and death rates to one with low birth and death rates. 568
Demographic transition, in human population growth, 568, 570f, 572
Dendrites. Processes of nerve cells that receive information from other neurons or from extra-neuronal sources in the instance of certain sensory cells. 341
  of motor neuron, 340f, 341
  of neuron, 205f
Deoxyribonucleic acid
  amount in a gene, 124
  antiparallel chains of, 117, 118f
  and bacterial transformation, 153
  compared with RNA, 55f, 56
  complementary messages of, 120
  and control of bacterial sex by F factor, 155
  and control of protein synthesis, 111
  double helix structure of, 115, 118f
  of eukaryotes compared to prokaryotes, 144, 145, 146
  and evolution of replication, 140
  information storage by, 113
  mutation of, 124, 125, 125f
  pairing rule, 117
  polymerase for, 120, 121
  polynucleotide chains of, 117
  recombinant experiments with, 141
  recombination of in prokaryotes, 153, 154f
  repair mechanisms for, 128
  repeating bases coding for histones, 146
  replication of, 115, 118, 119f, 121, 122f
  replication of in eukaryotes, 145, 146f
  replication of in prokaryotes, 144, 145f
  replication rate of, 121
  spacer, 131
  structural model of, 111f, 118f
  as transforming principle, 112
  triplet code of, 113, 116f
Deoxyribonucleic acid structure, 55f
  primary, 115
  secondary, 115
  tertiary, 115
Deoxyribose phosphate, structure, 55f
Depolarization, of neurons, 341, 342f, 350
Derepression, of genes, 137, 138
Dermal ossicles, of sea cucumbers, 272f
Dermis. The mesodermic inner layer of the skin, made up chiefly of connective tissue with some smooth muscle, nervous tissue, and sensory receptors. 386
  developmental origin of human, 479
  of human skin, 386, 386f
DES, see Diethylstilbesterol
Desynchronized sleep, 437
Detritus food chain, 545
Deuterostomes
  characteristic development of, 266
  phylogeny of, 244f
Deuterostomia, see Deuterostomes
Development
  complexity cost of multicellularity, 181
  embryonic human, 447
  nuclear regulation of, 17
DHAP, see Dihydroxyacetone phosphate
Di- and tri-peptidases, roles in human digestion, 378f
Diabetes
  causes in adults, 492, 493, 493t
  role of pancreas in, 360
Dialysis, of human blood, 496, 497f
Diamond, Jared, 234
Diaphragm, role in human ventilation, 393, 393f
Diastole, of human heart, 401f
Diatomaceous earth, formation of, 248
Diatoms, 180f
  and marine photosynthesis, 249
  skeletons of, 247f, 248, 249f
  structure, 247f, 248
Dicrostonyx, lemming, 547
Diet
  and regulation of food intake in humans, 369, 370t, 371t
  effects on evolution of primate teeth, 314, 314f
Diethylstilbesterol, 483
Differential growth, of plants, 220, 220f
Differentiation. The modification of different parts of a cell or an organism for specific functions as, for example, during the development of an embryo.
  causes in development, 471
  of cells in colony, 181
  cellular, 181
  of ctenophore embryo, 472
  environmental controls of, 472
  of meristem cells in plants, 212, 213f
Diffusion. The process whereby particles of liquids, gases, or solids spread from regions of higher concentration to regions of lesser concentration as a result of random molecular movement. 65
  importance in human ventilation, 391, 392
  importance in respiration, 200, 200f
Digestion
  control of, 383f
  in humans, 374–380
Digestive hormone system, control of human digestion by, 379
Digestive nerve plexus, control of human digestion by, 379
Digestive system, 193
  development of human, 481, 482f
  function of human, 373
  structure of human, 373f
Digitalis, and circulatory disease, 495
Dihydroxyacetone phosphate, structure, 73f
Dimer, DNA as, 130, 130f
Dinoflagellates
  and marine photosynthesis, 249
  in red tide, 621
Dinosaurs
  prevention of adaptation by size, 298, 299f
  reasons for extinction of, 280
  as reptiles, 279f
  size compared to whale, 299f
Dioxin
  as hazardous chemical, 622t
  structure, 622t
Dipeptide, 44f
1,3-diphosphoglycerate, structure, 73f

Diploblastic organisms, 258
  types of, 258f
Diplococcus, extinct reptile, 279f
Diploid chromosome number, *146,* 176
Diploidy
  genetic effects of, 157, 158
  significance of, 157
Dire wolf, extinction caused by early man, 324
Disease
  categories of human, 491
  genetic susceptibility to, 502
  human genetic, 493t
  prevention of spread, 507
Distal convoluted tubule, of nephron, 409f
DNA, *see* Deoxyribonucleic acid
DNA polymerase, function in replication, 120, 121
Dobzhansky, Theodosius, 300
Dolphin, 282f
Domesticated animals, representing controlled speciation, 288, 291, 292
Dominance. The quality of a gene that allows it to produce the same phenotypic effect when present singly or doubly. 166
  as exception to rule, 171
Dominant genes, 166, 171
Donora smog episode, 617
Dopamine, as neural transmitter, 444
Dopaminergic synapse, 442
Dorsal aorta, in human development, 476, 477f
Dorsal root, of spinal cord, 348, 348f
Dorsal surface, of bilateral animals, 261f
Double helix, of DNA, 115, 118f
Doubling time, for human populations, 576f
Down's disease
  karyotype of, 177f
  detection by amniocentesis, 488
  effects on brain functions, 435t
Dreaming
  creativity during, 437
  mental activity during, 437
Driesch, H., 15, 472
Drosophila
  chromosomes of, 174, 175f
  crossing over of chromatids in, 174, 176f
DNA replication rate in, 123
  genetic traits of, 174, 175f
  linkage map of, 174, 175f
  study of linkage groups using, 174, 175f
  salivary chromosome endomitosis of, 148f, 149
  x-ray mutations of, 602
Drugs
  and hospitalization of mental patients, 441
  effects on fetal development, 483
  psychoactive, 441
Dryopithecus, early hominid, 318f, 319
Ductus arteriosus, congenital malformation of heart, 492f
Duodenum
  in human development, 481

  location in human digestive system, 373f
Dynamic equilibrium, across membranes, 67

E

Ear
  developmental origin of human, 478
  inner, 354, 354f
  middle, 354, 354f
  outer, 354, 354f
  sensitivity of human, 355
  sensory receptors of, 350f, 354f
  structure of human, 354, 354f
Eardrum, of human ear, 354, 354f
Early Pleistocene Age, *see* Pleistocene Age
Earth
  effects of pollutants on temperature of, 524
  formation of, 57
  history of, 59f
  history of temperature of, 525f
  primitive environment, 59
Earthworm
  locomotion of, 265f
  nephridium of, 201f
  structure of, 265, 265f
EBV, *see* Epstein-Barr virus
Ecdysone, and control of molting, 208f
Echidnas, as primitive mammals, 280, 281f
Echinodermata, Phylum *see* Echinoderms
Echinoderms
  characteristics of, 270, 271f
  economic impact of, 270
  as evolutionary dead end, 272
  locomotion of, 272, 272f
  skeletons of, 272, 272f
  structure, 272f
  types of, 270, 271f
Ecology. The study of the interactions between and among living things and the physical environment. 521
  levels of study, 522–524
Ecosphere, 523d
  characteristics of, 523, 524
  see Biosphere
Ecosystem. Any interacting system of organisms and environment that is largely independent of any other. 22
  human effects on stability of, 548
  manipulation of by humans, 559, 560
  stability factors, 547, 548
Ecosystems ecology, 523
Ection, army ant, 189f
Ectoderm. One of the three primary cell layers of the animal embryo (*see* Endoderm; Mesoderm) from which forms the integument and nervous system. 473
  of coelenterates, 260f, 261
  formation of in human embryo, 473
Ectopic implantation, of embryo, 460
Ectothermic animals, 385f
  disadvantages of ectothermy in, 385
Edelman, G., 501
EEG, *see* Electroencephalography

Effector. A cell or organ that performs actions under direction of the nervous system, usually a muscle or gland.
  of homeostatic systems, 334, 334f
  role in homeostasis, 336
EF-G, *see* Elongation factors
EF-T, *see* Elongation factors
Egg. A female germ cell often characterized by large amounts of food material stored in the cytoplasm.
  description of human, 162
  formation of, 157f, 162
  of reptiles, 278, 280f
  see also Ovum and oogenesis
Einstein, Albert, 590
Ejaculate, *see* Semen
Ejaculation, in humans, 458
Ejaculatory duct, *448*
  human, 448
Elan vital, 285
Electrical currents, associated with nerve impulse, 341, 342, 343, 343f
Electrical synapse, 343
Electroencephalogram, as evidence of death, 496
Electroencephalography, 424n
Electrolytes
  control of by human kidney, 409
  in solution, 58f
Electromagnetic spectrum, 83, 83f
Electromotive force, and oxidation-reduction reactions, 91n
Electron, 85
  energy levels of, 84f
  excited state of, 86
  ground state of, 86
  ionization of, 86f
  valence, 85
Electron acceptor, in photosynthesis, 92f
Electron carrier molecules, of respiratory chain, 102f, 103
Electron microscope, 32, 33f
Electron shells, 84f, 85
Electron transport, 90
  in photosynthesis, 92
  and photosynthesis evolution, 94
Electron volt, 83
Elements
  of living systems, 39t, 40
  trace, 39t
Elongation factors, in protein synthesis, 134, 135f
Embryo
  formation of ancestral structures by, 300
  of reptiles, 280f
Embryology
  and homology of vertebrate structures, 310, 310f
  of human fetus, 462f, 463f
  of humans, 473–482
  importance to phylogeny, 238, 238f
  of protostomes versus deuterostomes, 266
Emotions, neurophysiological basis for, 439
Emphysema
  and pollution, 614

and smoking, 510f
Emulsification, of fats by bile salts, 382
Encasement, theory of, 164n
End product repression, of gene activity, 139
Endoarterioectomy, 495
Endocrine system, 193
  and control of human reproduction, 451–456
  control of temperature regulation by, 386, 388, 389f
  controlling processes of, 207
  glands of, 207, 355–362
  influence on human digestion, 379
  interaction with nervous system, 207
  location of glands of, 207f
  overlap with nervous system, 205
  role in homeostasis, 336, 336f
Endoderm. The embryonic germ layer that gives rise to the digestive tract and associated organs. 473
  formation of in human embryo, 473
  of coelenterates, 260f, 261
Endodermal cells, of plant roots, 216, 216f
Endodermis, of plant roots, 216f
Endometrium, development of, 455
Endomitosis, of *Drosophila* salivary chromosomes, 149
Endoplasmic reticulum. A series of channels through the cytoplasm often bordered by ribosomes and important in manufacturing and transporting enzymes, etc. 35, 35f
Endorphins, as brain transmitter, 442, 445
Endoskeleton, 202, 203f
  compared to exoskeleton, 202, 203f
Endosperm mother cell, 227
  and fertilization in flowers, 227
Endosperm nucleus, of flower egg cell, 226f
Endostyle, of sea squirts, 272
Endothelium, of capillaries, 40
Endothermic animals. Animals that maintain a constant or relatively constant body temperature regardless of external conditions or environmental temperature. 385
  thermoregulation of, 385–390
Energy and environmental degradation, 584
  conservation of, 609
  consumption related to income, 588f
  contents of foods, 368t
  effect of usage on earth's temperature, 588
  equilibrium in ecosphere, 526
  flow in biosphere, 540
  future prospects for, 586
  historical development of sources, 584
  means to reduce demand for, 607
  need for national policies for, 584
  nuclear, 590
  present sources of, 585
  recovery from wastes, 609n
  required for human activities, 368t
  trophic levels of, 540
  types available to ecosphere, 526
  ultimate sources of, 584, 585f

yield from fossil fuels, 586
Energy of activation, 7
Energy levels, of electrons, 84f
Enkephalins, as brain transmitters, 442, 445
Entamoeba histolytica, amebic dysentery protist, 248
Entelechy, 16
Environmental contamination, effects on fetal health, 483
Environmental exchange, as function of organ systems, 192
Enzyme induction, 136, 138f
Enzymes
  activation by second messenger, 363, 364, 364f
  allosteric, 140
  for blood clotting, 400f
  control of activity in, 139, 140
  effects of temperature on reaction rates of, 385
  human digestive, 376, 378f
  linkage with genes, 112
  modulator molecules for, 139
  operon control of, 137
  protective, 98
  protein catalysts as, 38, 52d
  zymogens of, 139
Eocene Age, appearance of first modern humans during, 324
Eohippus
  value of fossils of in phylogeny of horse, 236
  and evolution of modern horse, 304f
Epidermal cells, of plants, 213, 213f
Epidermis. An outer covering; in animals the outer layer of the skin that overlies the layers containing blood vessels, nerves, etc.; in plants, the layer of tissue one-cell thick on the surfaces of the leaves. 386
  human, 386, 386f
  of plant roots, 216f
  of plants, 213f, 218
Epididymus. Convoluted tube connecting sperm-forming tubules of testis with the vas deferens. 448d
  human, 448
Epigenesis, 164n
Epinephrine
  effects on temperature regulation, 386
  release by adrenal medulla, 359f, 360, 365
  role in regulating blood glucose, 383f, 385
  role in stress response, 366
  secretion by sympathetic nervous system, 349
Episomal factors, 149
  genetic independence of, 149
  of eukaryotes, 149
Epstein-Barr virus, and cancer, 511
Equilibrium, human sensory structures of, 350f, 354, 354f
  osmotic, 67, 68f
Equus, evolution of modern horse, 304f
Erection, of human male, 458
Ergotropic nervous system, 442

Error catastrophy theory, of aging, 514, 515
Erythroblastosis fetalis. Disease of newborn characterized by destruction of red blood cells usually due to immune reaction with mother. 484
  causes of, 484
  from ABO compatibility, 485
Erythrocyte, *see* Red blood cell
Erythropoietin, stimulating red blood cell production, 399
Erythrose-4-phosphoric acid, as amino acid precursor, 301f
Escherichia coli
  demonstration of recombinants of, 153, 154f
  and gene regulation, 136
  genetic transfers by viruses in, 153
  radioautography of replicating DNA in, 144, 145f
  and recombinant DNA research, 141
  replica plating of, 127, 127f
  sexual types of, 155
Esophagus, in human development, 481
Estradiol-17-beta, structure, 452f
Estrogens
  developmental effects of, 360, 361
  effects on emotions, 457
  release by adrenal cortex, 359f
  release by ovary, 360
  role in human reproductive cycle, 452f
  structure, 452f
Estrone
  role in human reproductive cycle, 452f
  structure, 452f
Ethane, structure, 41f
Ethanol, neural effect of, 443f
Ethiopians, human racial variety, 330
Euglena
  plant and animal characteristics of, 247, 247f, 249
  structure, 247f, 249, 250
Euglenoids, characteristics of, 249
Eukaryotes, 143
  cell division in, 143
  characteristics of, 143, 143f, 144
  classification of, 246, 256, 257
  sexual reproduction in, 155
Eumetazoa, true Metazoa, 257
Eumetazoans
  characteristics of, 258
  phylogeny of, 259f
  tissue types of, 258, 261
Euploid chromosome number, 176
Eurypterids, as reason for evolution of armor by fish, 275
Eustacean tubes
  developmental origin, 478
  of human ear, 354, 354f
Eutrophication. Artificial increase in nutrients in an ecosystem, usually resulting in marked overgrowth of certain constituent organisms and disruption of natural relationships between organisms. 539
  contribution from feedlots, 557
  and phosphorus enrichment, 539

INDEX-GLOSSARY 651

Eutrophication (cont.)
  and waste enrichment of water, 621
eV, see Electron volt units
Evolution. Genetic change in organisms over time leading to the appearance of new species. 293
  acceleration of in early Cambrian, 96
  and adaptive change in species, 298
  biological, 20, 108, 108t, 77
  biological beginnings, 77
  chemical, 20, 60
  as cumulative process, 300
  early opposition to theory of, 289, 290
  early stages of biological, 108, 108t
  effects of medicine on human, 332
  effects of neoteny on, 263
  effects of selective reproduction on, 293
  effects of specialization on, 263
  evidence for, 287f, 288, 291, 293, 302, 307
  human control of, 332
  of humans, 308, 312, 319, 322, 325, 326
  hypotheses of, 291, 292t
  importance of mutation in, 126
  lengthy verification process, 300
  mechanisms of, 294, 295, 297
  of nucleic acids, 140
  osmotic, 69
  of populations rather than species, 292
  of proteins, 140
  of protostomes versus deuterostomes, 266
  of replication, 140
  representation of by systematics, 303
  social, 22
Evolvability, 298
  universal characteristic of species, 298
Excitable, biological meaning, 31
Excitation, role in nervous system, 341, 342f, 344
Excited state, of electrons, 86, 86f
Excretion
  by plants, 219
  types of in animals, 202f
  water loss associated with, 202, 202f
Excretory system, 193
  control of water balance, 201f
  development of human, 479
  first animals to develop, 264, 264f
  of humans, 406–411
  link with circulatory system, 201f
  osmoregulatory function of, 201
  types of wastes produced by, 202f
  waste disposal function of, 202
Exercise
  energy requirements of, 368t
  effects on heart rate, 402, 403
Exobiology. The study of the possibility and probable nature of life in extraterrestrial settings. 21
Exoskeleton, 202, 203f
  advantages and limitations of in arthropods, 267
  compared to endoskeleton, 202, 203f
  of echinoderms, 271f, 272, 272f
Exteroreceptors. Sensory receptors monitoring the surface and surroundings of the organism, especially visual and auditory receptors. 351
  types of, 351
Extinction, of species caused by early humans, 324, 325
Extrachromosomal inheritance, 149
Eye
  development of human, 476
  sensitivity of human, 355
  of squid compared to vertebrates, 270, 270f
  structure of human, 351, 352f, 353f
Eye spot, of *Volvox*, 186, 187f
Eye vesicle, in human development, 474f

F

F factor, of bacteria, 155
F replication site, and initiation of bacterial replication, 155
F1 generation, 166, 167f
F2 generation, 166, 167f
F-6-P, see Fructose-6-phosphate
Facilitated diffusion, and absorption in small intestine, 376
Fallopian tube. Tube with opening beside ovary, leading to cavity of uterus. 450
Fallopian tube, human female, 450f
Family Hominidae, see Hominids
Family Pongidae, see Pongids and apes
Faraday, M., 12
Farm community, ecology of, 522
Fat
  emulsification by bile salts, 382
  absorption in human small intestine, 376, 377f
Fatigue, neurophysiological, 427
Fatty acids
  conversion in small intestine, 376
  in lipids, 54f
  structure, 54f
FDP, see Fructose-1,6-diphosphate
Feedback controls, of neuroendocrine system, 357
Feedback loops, role in homeostasis, 334, 335f
Feeding
  control of in humans, 369, 383f
  and food processing in humans, 373
  mechanisms used by marine organisms, 546f
Feedlots, effects on ecosystem, 557
Felis concolor, mountain lion, 231
Felis leo, African lion, 231
Femoral artery, 401f
Fermentation. Energy production by anaerobic metabolism. 107
Ferns
  characteristics of filicinophytes, 254, 256f
  life history of, 254f
Ferredoxin, in nitrogen fixation, 537
Fertility, effects of oral contraceptive on, 468
Fertilization
  of flowers, 226f
  of flowers by wind and animals, 227, 227f
  human, 459
  in gymnosperms, 255f
  on land versus in sea, 209
Fertilizers, effect on nitrogen cycle, 538
Fetus
  drug effects on, 483
  effects of maternal disease on, 483
  effects of rubella on, 483
  effects of undernourishment, 483
  hemolytic disease of, 484
  immune reactions with mother, 484
  means of monitoring health of, 487, 487f, 488
  mortality of, 482
  nutrition and health, 482
  sex determination of, 486
Fever, and control of disease, 390
Fibrinogen, formation of blood clots by, 400, 400f
Fight or flight, see Stress
Filicinophytes, see Ferns
Filter feeders, in marine food chain, 545
Filter feeding
  in molluscs, 268, 269f
  in protochordates, 272, 272f
Filtration, function of kidney, 407, 408f
Finches, of Galapagos, 287, 287f
Fins, evolution into legs in vertebrates, 276, 277f
Fire, prevention as manipulation of ecosystems, 560
First filial generation, see F1 generation
Fish
  bony, 275, 275f, 276
  cartilaginous, 275, 275f, 276
  evolution of, 275, 276
  fossils of, 235f
  heart of primitive, 305f
  lobe-finned, 237, 237f, 276
  transition to land of, 276
  as vertebrates, 274, 275, 275f
Fisher, R. A., 167
Fishing tentacles, of comb jellies, 262f
Fission. Form of cell division in simple organisms. 144
  in prokaryotes, 144
  nuclear, see Nuclear fission
Five Kingdom System, for classifying organisms, 244, Appendix 3
Flagella, 37
Flagellae, of *Volvox*, 186, 187f
  structural similarities among living organisms, 242, 243
Flagellates, and origin of higher animals, 248
Flagellum, see Flagella
Flame cells, of flatworms, 264, 264f
Flatworms
  advances over coelentetrates, 264
  characteristics of, 264, 264f, 265
  cross section of, 195f
  distribution limitation by water, 195, 196
  organ systems of, 195f

parasitic types, 264, 264f
surface area/volume ratio of, 195, 195f, 200f
types of, 263, 264, 264f
Flavin mononucleotide
structure, 102f
as respiratory carrier molecule, 102f, 103
Flemming, W., 172
Flowers
fertilization of, 225, 226f, 227, 227f
parts of, 223, 225f
Fluorocarbons, as pollutants, 617
FMN, see Flavin mononucleotide
FMNH2, see Reduced flavin mononucleotide
Folacin, see Folic acid
Folic acid
deficiency in pregnancy, 483
destruction of by ultraviolet light, 297
Follicle stimulating hormone
and initiation of female hormonal cycles, 360
reproductive cycle trigger, 453
role in human reproductive cycle, 452f, 453, 454
Follicles, see thyroid follicles and ovarian follicles
Food chain. A sequence of organisms through which energy and material flow by grazing and predation. 81
detritus, 545
marine, 545
origin, 81
terrestrial, 547
Food production
estimates of human, 550, 550t
FAO estimate in 1973, 551t
possible improvements in, 552, 553
related to cost of production, 553
related to population growth, 553f
see also Agriculture and productivity
Food webs, and ecosystem stability, 547
Foods, energy content of, 368t
Foramen magnum, 319
of early ape-humans, 319
Foramen ovale, congenital malformation of heart, 464, 492f
Foraminiferans, skeletal deposits of, 248
Forebrain
general function of human, 347
of human embryo, 474f
location of human, 347
Formaldehyde, structure, 39f
Formic acid, 40
Formyl group, 133, 133f
Formylmethionine, as protein synthesis initiator, 133
Fossil fuels
carbon storage by, 534
origin, 589
reserves of, 589, 589f
Fossil record
fragmentary nature of, 236
gaps in, 306
Fossils

dating using radioisotope decay, 236f, 237
as evidence for evolution, 20, 291, 303
"living," 237, 238
types of information preserved in, 235, 235f
use in classification of organisms, 234, 235f
use in determining human phylogeny, 317, 318f, 319, 321f
Fovea, of human eye, 352f, 353
Fox, C. A., 459
Fox, C. B., 459
Fraternal twins. Twins resulting from the independent fertilization of two eggs. 460
Free radical, from radioactive decay, 599
Freemartin, 481
hormonal causes of, 457, 481
Frejka, T., 578
Frequency coding system, nervous system as, 344
Frogs
as amphibians, 278, 278f
brain of, 306f
control of metamorphosis in, 363, 363f
positive pressure breathing in, 394f, 395
Fructose, and regulation of blood glucose, 383f
Fructose-1,6-diphosphate, structure, 73f
Fructose-6-phosphate, structure, 73f
Fruit, and facilitated distribution of angiosperms, 256
Fruiting body, of mushrooms, 256, 257f
FSH, see Follicle-stimulating hormone
Fucus
brown alga, 250
environmental control of development of, 472
Fungi, Kingdom, see Fungi
Fungi
cellular slime mold, 185, 185f
characteristics of, 256
life history of, 256, 257f
mushrooms, truffles and puffballs, 256, 257f
origin and evolution of, 256
sac types, 256
as saprophytes, 256
Fusion
atomic, 83
nuclear, see Nuclear fusion
Fustule, of diatoms, 249f

G

G-6-P, see Glucose-6-phosphate
Gage, Phineas, 422
Galactose, and regulation of blood glucose, 383f
Galactose transferase, viral transmission of gene for, 153
Galactosemia
human disease, 493t
and mental retardation, 435t
prevention by genetic engineering, 153

Galapagos tortoise, 279f
Gall bladder
function in human digestion, 380, 382
location in human digestive system, 373f
structure, 380f
Gamete. A mature germ or reproductive cell, containing the haploid number of chromosomes, ready to fuse with another gamete to initiate a new individual. 156
formation of, 157f, 162
Gametes
as carriers of genetic traits, 168f, 169
derivation from specific embryonic cells, 209
fertilization of on land or in sea, 209
origin in human development, 480
production of by brown algae, 250, 251
Gametophytes. Sex-cell producing generation in plant life cycle, made up of haploid cells. 224d
of brown algae, 250, 251f
of ferns, 254f
of flowering plants, 224
of gymnosperms, 255f
of moss, 252f
of psilophytes, 253
of seed plants, 254, 255f
Gamma control system, of muscle, 414
Ganglia. Localized groups of nerve cells with synaptic interconnections, usually devoted to a limited array of functions. 207
as part of nervous system, 207
Garden pea, usefulness in genetics studies, 165f, 166
Gas exchange
importance in human ventilation, 391
in plants compared to animals, 218
Gastrin, role in human digestion, 379
Gastropods, as molluscs, 269
Gastrovascular cavity, of colenterates, 260f, 261
Gastrulation. Cell movements occurring in embryonic development, after blastulation, in which mesoderm and endoderm move from the outside to the inside of the embryo. 473
in human development, 473
Gemmules, of heredity, 123, 163
Gene. The unit of heredity; in molecular terms that segment of DNA coding for a single polypeptide.
amount of DNA in, 124
chemical nature of, 112, 113
frequency, 293
histocompatability, 502
hybridization experiments with, 141
linkage with enzymes, 112
operator, 137d, 138f
promoter, 138f
regulation of action, 136, 137, 138f, 139
regulator, 137d, 138f
structural, 137d, 138f
Gene frequency. The proportion of a gene to its alleles in an interbreeding popula-

Gene frequency (cont.)
tion. Gene frequencies change the least in evolutionary stable populations. 293
effects of isolating mechanisms on, 294
Gene pool. The total genetic material freely circulated by reproductive processes in a population. 293
Gene mutation. Any change in the nucleotide sequence of a segment of DNA coding for a single polypeptide. 124
masking effects of diploidy on, 157
Generative cell, of pollen grain, 226f
Genes, alleles for, 157
and continuous variation of genetic traits, 171, 172, 173f
determination of cross over frequencies of, 175, 176f
duplicate, 178
incomplete dominance of, 169, 170f
independent assortment of chromosomes carrying, 168f, 169
mapping human, 178
and phenotype, 169
similarity among primate, 314
variable penetrance of, 170f, 171
Genetic code, 113, 116f
degeneracy of, 116f
mutation of, 124, 125f, 126
universality of, 141
Genetic disease, types of human, 493t
Genetic diversity, importance of sex in creating, 151, 152f
Genetic engineering, and prevention of disease, 153
Genetic recombination, bacteria, 153, 154f
Genetic traits
and chromosomes, 172
continuously varying, 171, 172, 173f
incomplete dominance of, 169, 170f
independent assortment in gametes, 169, 174
linkage of, 174
transmission of, 167, 167f, 168
variable penetrance of, 170f, 171
Genetics
history of, 163, 172
use in determining phylogeny, 240
Genital ridges, 480
hormonal influences in development, 480
in human development, 480, 481f
Genitalia, development of human external, 480, 481, 481f
Genome. The total complement of genes of an organism. 149
Genotype. A description of the genetic composition of an individual with reference to one or a few characters.
of garden peas, 167, 167f, 168
Genus, binomial naming of, 231
Geological cycles, 530
Geothermal energy, 585
Germ cells, origin in human development, 480
Germ disc, in human development, 473

Giant fibers, and improved reactions in predators, 270
Gibberellins, 221
effects on plants, 221
Gills
of mushroom, 256, 257f
respiratory function of, 200
structure of in fish, 200f
Gill slits, of vertebrates, 273, 274f
Ginko biloba, maidenhair tree, 238, 238f
Glaciation, historical pattern of, 524
Glands, *see* Pineal gland; Thymus
Glia, 340
effects on conduction velocity of nerves, 340
Globulins, 49
Glomerulus
development of (renal), 480f
of kidney, 409f
Glucagon
release by pancreas, 360
role in regulating blood glucose, 383f, 384, 385
Glucocorticoids
and glucose regulation, 383f, 385
release by adrenal cortex, 359f, 360
Gluconeogenesis, control of by adrenal gland, 359f, 360
Glucose
active transport of by kidney, 407, 408
breakdown in glycolysis, 72, 73f
control of blood levels of, 365, 382, 383f, 384
and control of human food intake, 369
cyclic stereoisomers of, 73f
effects of stress on levels of, 385
formation of glycogen from, 384f
and gene regulation in *E. coli*, 137, 138f
potential oxidation energy from, 100
structure, 53f, 73f
Glucose metabolism, and control of human respiration, 338f
Glucose-6-phosphate
and glycolysis, 72
and hexose monophosphate pathway, 81, 81f
and regulation of blood glucose, 383f
structure, 73f
Glutamic acid
as amino group donor, 301f
as brain neurotransmitter, 442
structure, 43f
Glutamine, structure, 43f
Glyceraldehyde-3-phosphate
and glycolysis, 72, 74
structure, 73f
Glycerol
conversion in small intestine, 376
in lipids, 54
structure, 54f
Glycine
as brain neurotransmitter, 442
dehydration synthesis of, 44
structure, 43f

Glycogen
and regulation of blood glucose levels, 382, 383f, 384
structure, 384f
Glycogen storage disease, 493t
Glycolysis, 72d, 73f
efficiency compared with aerobic oxidation, 107
summary, 75f
Glycosidic linkage, 53f
Glycylglycine, structure, 44f
GnRF, *see* Gonadotropin releasing factor
Goiter, 491
Golden algae, as plant-like protists, 247f, 248, 249, 249f
Golgi apparatus, 35d, 35f
Gonadal hormones, environmental controls for release of, 357
Gonadotropin releasing factor, role in human reproduction, 453
Gonads
effects of hormones on, 360, 361
endocrine functions of, 360
hormones released by, 360
Gonium, structure, 247f
Gonorrhoea, penicillin resistance of, 156n
Goose bumps, and thermoregulation in humans, 387
Gorilla
hands and feet of, 313f
morphology compared to other primates, 312f
skull of, 312f
Gossypium, cotton, genetics and phylogeny, 240
Grain drain, in meat production, 555n
Grana, of chloroplasts, 87
Grasses, adaptive characteristics of, 298
Grasses, and evolution of grazers, 298
Gravity, orientation of plants to, 220
Grazers, evolution of teeth of, 298
Great Dismal Swamp, peat from, 589
Great saphenous vein, 401f
Green algae, as plants, 250
Green glands. A pair of large glands in lobsters, crayfish, and related crustaceans, which act as organs of *excretion*; they have outlets at the bases of the antennae to the exterior.
excretory function of, 201f
structure, 201f
Green revolution, and improved food production, 552
Green River Formation, oil reserves of, 590
Greenhouse effect. Entrapment of long wave radiation by a barrier transparent to short wave radiation. 525
causes of, 525
and CO_2 pollution, 616
Grey matter, of spinal cord, 348
Griffith, F., 112
Ground state, of electrons, 86, 86f
Growing tip, control of growth by in plants, 220, 220f

Growth
  control of in plants, 220, 220f
  early stages of in plants, 212
  of living versus nonliving, 18, 29
  by molting in insects, 208f, 209
Growth hormone
  of humans, 356, 356f
  of insects, 208, 209
Growth rate, of humans, 453f
GTP, see Guanosine triphosphate
Guanine, structure, 55f, 202f
Guano, as phosphate source, 539
Guanosine triphosphate
  role in protein synthesis initiation, 134
  and tricarboxylic acid cycle, 101
Guard cells, function in plant gas exchange, 218, 219f
Gulf Stream, 530
Gymnosperms
  fertilization in, 255f, 256
  life history of Jeffrey pine, 255f
  reproductive structures of, 255f
  types of, 255
Gynogametophytes, of gymnosperms, 255f
Gynosporangia, of gymnosperm gametophytes, 255f
Gynosporocytes, of gymnosperms, 255f
Gynostrobili, of gymnosperms, 255f

H

Habituation, behavioral, 427, 428f
Hadley cells, and vertical mixing of air, 529f
Haeckel, E., 521
Hagfish, as primitive vertebrate, 275, 275f
Hahn, O., 590
Hair, effects on thermoregulation, 387
Hair, of human skin, 386f, 387
Haldane, J. B. S., 59
Half-life, 597
  and age determination of fossils, 236f
  of radioactive elements, 597f
Hallucinogens, neural effects of, 445
Haploid chromosome number, 146, 176
Hardin, Garrett, 583
Hare lip, human genetic disease, 493t
Harvey, Sir William, 405n
  discovery of human circulation by, 405n
Hayflick limit, 515
  and aging, 515
HCG, see Chorionic gonadotropin
HCN, see Cyanide
Heart
  congenital malformation of, 492, 492f
  control of human, 403, 403f
  development of human, 476, 477f, 478f, 482f
  electrical excitation of, 403f
  evolution of vertebrate, 305, 305f
  muscles of, 402, 403f
  structure of human, 401, 401f, 402, 403f
  structure of vertebrate, 305f

Heart attack, causes, 400, 401
Heartbeat, cycle of, 402, 403f
Heat, from atomic motion, 87
Heavy water, 592
Heliozoans
  pseudopods of sun animals, 247f, 248
  structure, 247
Helium, atomic structure of, 84f
Hematite, 98
Heme, structure, 242
Hemicholinium, neural effect of, 443f, 444
Hemodialysis, see Blood; Dialysis
Hemoglobin
  comparison of amino acid sequences in vertebrate, 315t
  fetal, 476
  function in oxygen transport, 397, 398f
  in sickle-cell anemia, 110
  role in respiratory control, 338f, 339
  structure, 398f
Hemophelia, human disease, 493t
Henslow, John Stevens, 285
Hepatic artery, role in hepatic portal system, 382
Hepatic portal system
  connection with small intestine, 377f
  role in human digestion, 381, 381f
Hepatic portal vein, 401f
  embryonic origin, 476
Herbicides, as hazardous chemicals, 622t
Hereditary traits, transmission of, 166, 167f, 168f
Heterogametic organisms, 162
Heterokaryon. A cell with multiple nuclei of varied genetic composition. 178
Heterozygous organisms. Diploid organisms with different alleles of the same gene. 157
  incomplete dominance of genes in, 170
Hexosaminidase A, and Tay-Sach's disease, 488
Hexokinase, and glycolysis, 73f, 74
Hexose monophosphate pathway, 80, 81f
High energy bond. Chemical bond which releases a large amount of free energy upon hydrolysis, as found in ATP.
  of ATP, 72f, 74
  of CoA, 101
Hindbrain
  general function of human, 347
  in human development, 474, 474f
  location of human, 347
Hindgut, in human development, 481, 482f
Hippocampus, 412f
Hiroshima, 590
Histidine, structure, 43f
Histocompatibility genes, and disease susceptibility, 502
HLA system, see Human leukocyte antigen system
Hodgkin's disease, association with HLA system, 502
Homeostasis. Maintenance of stability in the face of variation. 333

characteristics of regulating systems for, 334, 334f
  history of concept, 333
  human mechanisms of, 336, 336f, 337
Hominidae, Family see Hominids
Hominids
  classification of human family, 311, 311t
  time of evolutionary separation from pongids, 317
  types of, 318f, 319, 320, 322, 323
  see also Humans
Hominoidea, Superfamily see Hominoids
Hominoids
  characteristics of, 312, 312f
  classification of, 311, 311t
  see also Humans; Apes
Homo erectus
  advances over *Australopithecus*, 320
  brain size of, 320
  characteristics of early hominid, 318f, 319
  Java Man, 320
  life-style of, 320
  Peking Man, 320
  radiation (evolutionary) of, 320
  skull of, 321f
Homo sapiens, see Humans and Cro-Magnon Man
Homo sapiens neanderthalensis, see Neanderthal Man
Homo sapiens sapiens, see Humans; Cro-Magnon Man
Homologous chromosomes. A pair of chromosomes carrying a largely similar sequence of genes. 157
  alleles of, 157
  crossing over in, 159, 161f
  pairing during meiosis, 159, 161f
  pairing patterns used in determining phylogeny, 240
  segregation difficulties of, 178
Homologous structures. Structures with common ancestry. 240
Homonuculus, 164f
Homozygous organisms. Diploid organisms carrying the same alleles of particular genes. 157
  produced by selfing, 165, 166
Hooke, R., 33
Hormone. A substance produced by one part of the body and transported by the blood, sap, or other fluids to another part of the body upon which it has a specific effect beyond that of a simple nutrient or metabolite. 207
  age dependent effects of, 363
  cellular effects of, 363, 364
  controlling human digestion, 379, 379f, 380
  fast versus slow effects of, 364
  of plants, 220, 221
  second messengers as, 363, 364, 364f
  specificity of, 363, 364
  stimulation of transcription by, 364
  human, 356–366

Horowitz, W., 79
Horse
  brain of, 306f
  fossil record of, 303, 304f
Horsetail, see Sphenophytes
Hottentots, human racial variety, 330
Hudson River, contamination with PCBs, 624
Human development, embryonic stages of, 462f, 463f
Human leukocyte antigen system, and disease, 502
Human populations
  growth of, 519
  necessity of limitation, 519
Human reproduction, hormonal control of, 451, 452f
Human survival
  adult similarities with other vertebrates, 310
  aging of, 516
  birth rates of, 570–576
  brain of, 411–425
  circulatory system of, 399–406
  classification of, 311t
  cultural evolution of, 326
  desirability of racial uniformity of, 331, 332
  development of agriculture by, 325
  development of reproductive system, 453, 453f, 454
  diseases of, 499–507
  embryonic development of, 473–482
  embryonic homologies with other vertebrates, 310, 310f
  endocrine system of, 355–362
  energy requirements of, 367, 368t
  evidence for evolution from lower forms, 310
  evolution of, 308, 312, 319, 322, 325, 326
  excretory system of, 407–411
  food digestion in, 374–380
  food intake requirements of, 367, 368t, 370t, 371
  hands and feet compared to other primates, 313f
  homeostatic mechanisms of, 336, 336f
  hormones secreted by glands of, 355–362
  imperfections in adaptations of, 299
  intelligence of, 433–436
  learning in, 426–433
  life expectancy of, 516
  morphology compared to other primates, 312f
  nervous system of, 339, 340, 347
  osmoregulation by kidneys of, 409, 410
  population growth of, 561–572
  problems of, 520
  races of, 328, 329, 330
  reproduction in, 447–470
  rudimentary organs of, 310
  sense organs of, 351, 352f, 353, 353f
  sexual behavior of, 457
  sexual development of, 479
  skull compared to other primates, 321f

  thermoregulation in, 385–389
  see also Primates
Hunger, and regulation of food intake in humans, 369
Hunting, by early humans as cause of animal extinction, 324
  effects on ecosystems, 558
Hutterites, reproductive rate of, 465
Huxley, T. H., 34, 285, 289, 290
  evolution debate with Bishop Wilberforce, 289, 290
Hybrid vigor. Markedly improved viability over the parental types as shown by the progeny of two inbred varieties. 158
  production by diploidy, 158
Hybridization
  of garden peas, 168f, 169
  of somatic cells, 178
Hydra
  asexual budding of, 186, 187f
  feeding in, 186, 187f
  lack of medusa stage in, 261
  as multicellular organism, 186, 187f
  nematocysts of, 186, 187f
  nerve net of, 206f, 207
  structure of, 186, 187f
  tissues of, 190
Hydraulic skeleton, and use of coelome for locomotion, 265f
Hydrocarbons, as pollutants, 617
Hydrochloric acid
  as poison, 490
  secretion by human stomach, 375f
Hydrogen
  carrier molecules for, 103
  generated in tricarboxylic acid cycle, 101
Hydrogen bonding, of proteins, 48, 48f
Hydrogen bonds, of water, 58f
Hydrogen bomb, 592
Hydrogen ion, and chemiosmiotic theory of oxidative phosphorylation, 105
Hydrogen peroxide, breakdown by catalase, 99, 99f
Hydrolysis. Disruption of a chemical bond by the insertion of water. 44
Hydrostatic skeleton, 202, 203f
Hydroxyl ions, and acidity, 42
Hydroxyl ion, and chemiosmotic theory of oxidative phosphorylation, 105
5-Hydroxytryptamine, see Serotonin
Hydrozoans
  alternation of generation in, 260f
  colony of, 260f
  medusae of, 260f, 261
  polyps of, 260f, 261
  type of coelenterates, 260f
Hymen, human, 451
Hyperglycemia, see Diabetes
Hypertension. Greater than normal pressure of blood against the arterial walls. 495
  causes of, 495
Hypertonic medium, 199f
Hyphae, of fungi, 256, 257f
Hypogastric vein, 401f
Hypophyseal portal system, role in human

reproduction, 452f
Hypothalmus
  and control of hunger in humans, 369
  and control of thermoregulation, 389f, 390
  evolutionary development of, 417
  general function of human, 347
  and human reproductive cycles, 453
  interactions with pituitary gland, 356, 356f, 357
  location of human, 347, 361f
  and pituitary control, 207
  and regulation of osmolarity, 409
  role in regulating glucose, 384
  role in stress responses, 366
  secretion of releasing factor by, 357
Hypothesis. A tentative explanation of fact, of lesser certainty than a theory and superior to a hunch. 5
  testing and development, 5
Hyracotherium, see *Eohippus*

I

Identical twins. Twins derived from the same egg. 460
  use in intelligence studies, 434
IF-1, see Initiation factors
IF-2, see Initiation factors
IF-3, see Initiation factors
Immune system. The organs and tissues of an animal responsible for the production and distribution of antibodies that react with foreign protein of bacterial or other origin.
  (and microbial disease, 500)
  and organ transplantation, 496
Immunity. Resistance to a disease agent engendered by antibodies. 500
  mechanisms of, 500
Immunization, mechanisms of, 504
Immunoglobin, 501
  and antibodies, 501
Implantation
  ectopic, 460
  in human pregnancy, 460
Inbred organisms, sexual selection against, 297
Incisors, of chimpanzee, 314f
Incomplete dominance, 170
  of genes, 169, 170f
Independent assortment, of genetic traits, 168f, 169, 174
India, regulation of population growth in, 574
Inducer, in gene regulation, 138f
Induction
  embryonic, 474
  role of mesoderm in, 475, 475f
Industrial melanism. The predominance of darker color phases of an organism in an environment where the general background coloration is darker than normal

owing to soot and other wastes of industry.
    as proof of survival of the fittest, 293, 293f, 294
Industrial revolution, effects on human population, 571
Infants, absorption of protein by, 376
Infectious mononucleosis, viral cause of, 511
Inferior mesenteric artery, 401f
Inferior oblique muscle, of human eye, 352f
Inferior rectus muscle, of human eye, 352f
Inferior vena cava, 401f
"Inflammatory" disease, and HLA genes, 502
Ingen-Housz, J., 211
Inhalation, muscular control of, 393
Inheritance
    biparental, 164n
    extrachromosomal, 149
    of genetic traits, 165
    theories of, 163
Inheritance of acquired characteristics, 123
    Darwin's belief in, 289
    theory of, 285
Inheritibility, and origin of life, 64
Inhibition, role in nervous system, 345
Initiation factors, in protein synthesis, 134
Initiator complex, for protein synthesis, 134
Initiator tRNA, 134
Ink sac, of squid, 270f
Inner cell mass, of human embryo, 471
Inner ear, human, 354, 354f
Inner membrane, function in mitochondria, 105f, 107
Inner segment, of human photoreceptors, 353f
Insecta, Class see Insects
Insecticides, as hazardous chemicals, 622t, 623
Insectivore, brain of, 306f
Insects
    advantages and limitations of exoskeleton to, 266
    as arthropods, 267f
    castes of, 189, 189f
    chemoreceptor structure in, 243f
    collectives of, 189
    colonies of, 189
    control of molting in, 208f
    growth hormones of, 208f
    Malpighian tubules of, 201f
    neurohormones of, 208f
    as pollinators, 227f
    roles in human disease, 499f
    social, 189, 189f
    trachael respiration in, 266
Insightful learning, 431
Instinctive behavior. Behavior that is innate and inherited, rather than learned.
    of arthropods, 268
    and sexual selection behavior, 297
Insulin
    and diabetes, 492
    effects on glucose levels, 360
    release by pancreas, 360
    role in regulating blood glucose, 383f, 384, 385
    structure, 50n
Integument, 193
    functions of in animals, 198, 199f
    of marine organisms, 198, 199f
    of plants, 213f, 218, 219f
Intelligence
    apparent, 433
    as continuously varying genetic trait, 172
    environmental effects on, 434, 434f
    of human races, 331, 332
    potential, 433
    as product of memory, 433
Intelligence quotient, 434
    genetic effects on, 435
    social differences in, 435
Interest, compound, 561, 562f
Interferon, and disease resistance, 507
Intermediate lobe, of pituitary gland, 356, 356f
Internal jugular vein, 401f
Internal transport, as function of organ system, 192
Internuncial neuron. A neuron of the central nervous system in the pathway between sensory input and motor output.
    of reflex arc, 365
Interoreceptors. Receptors monitoring the internal environment. 351
    types of, 351
Interphase, of mitosis, 147, 147f
Intestinal capillary system, 381f
Intestinal villi, of small intestine, 376, 377f
Intrauterine device
    effectiveness of, 467t
    health problems of, 468
Intrinsic factor, role in intestinal transport of vitamin B12, 376
Iodine
    accumulation by tunicate endostyle, 272
    and synthesis of thyroxin, 357, 358f
Ion pump, and osmotic equilibrium, 68
Ionic bond, 84f, 85
Ionization, 86, 86f
Ionizing radiation. Radiation with sufficient energy to break chemical bonds, resulting in chemically reactive ionized molecules. 598t, 599
    from radioactive decay, 592
I.Q., see Intelligence quotient
Iris, of human eye, 351, 352f
Irish potato famine, effect on human population, 571
Iron ore, formation of deposits, 98
Isaacs, A., 507
Ischemic phase of female reproductive cycle, 455, 455d, 456f
Islets of Langerhans
    of pancreas, 360
    role in regulating blood glucose, 382, 383f
Isogametic organisms, 162d
Isoleucine, structure, 43f
Isomerase, and glycolysis, 73f
Isotope, 85

IUD, see Intrauterine device

J

Jacob, F., 137, 140
James River, pollution of, 612
Japan Current, 530
Java Man, see Homo erectus
Jaws, evolution of by fish, 275, 276
Jeffrey pine, life history of, 255f
Jellyfish
    strobilus of, 260f
    type of coelenterates, 260f
Jenner, William, 503n
Johnson, V. E., 459
Juvenile hormone, and insect molting, 208f, 209
Juvenility, extension of in primates, 315

K

K factor, in population growth, 569, 569f
Kammerer, P., and midwife toad evolution hoax, 285
Kangaroo rat, 282f
Karyotype, for Down's disease, 177f
Kelp
    life history of, 251f
    skeleton of, 203f
Kepone
    effects of, 622t
    as hazardous chemical, 622t
    structure, 622t
Kerogen, from oil shales, 590
Ketone group, 53f
Keyfitz, Nathan, 569
Khorana, H., 113
Kidney
    development of human, 479, 480f
    excretory function of, 201f
    function of human, 407, 408, 408f, 410
    location of human, 408f
    nephron of, 201f
    structure of, 201f
    transplantation of human, 496
    triggering red blood cell production, 399
"Killer smog," 617
Kinase, activation by second messenger, 363, 364, 364f
Kinetochore, 146
    and effects of mutation on segregation, 178
    function during cell division, 146
King, R., 17
Kingdom Animalia, see Specific types of animals
Kingdom Fungi, see Fungi
Kingdom Monera, see Blue-green algae; Bacteria
Kingdom Plantae
    diversity of, 250
    see Plants and specific types of plants

Kingdom Protista, *see* Protists
Kingdoms, of organisms, 244, 244f
Klinefelter's disease, effects on brain function, 435f
Knee jerk, as a reflex, 365
Kodiak bear, 231f
Koestler, Arthur, 285
Krebiozen, proposed cancer cure, 513

L

Labia majora, human, 451
Labia minora, human, 451
Labor, division of, 180
Labrador Current, 530
Labyrinth, of human ear, 354f
Lac operon, *see* Lactose operon
Lactate
   accumulation in active muscles, 76
   formation in glycolysis, 74
   structure, 72fn, 73f
Lactation, initiation in humans, 464
Lacteal vessels
   function in lymphatic system, 406
   of human small intestine, 376, 377f
Lactic acid
   formation in glycolysis, 72, 73f
   as liver glucose source, 382, 383
Lactose, and regulation of blood glucose, 383f
Lactose operon, of *E. coli*, 137, 138f
Laetrile, proposed cancer cure, 513
Lamarck, Jean Baptiste, 285
   theory of acquired characteristics, 285
Lamellae, of chloroplasts, 87
Lampreys
   effects on Great Lakes, 560
   as fish, 275f
   as primitive vertebrates, 275, 275f
Lancelets
   feeding in, 274f
   structure, 274f
   as vertebrates, 274, 275f
Land, food chain of, 547
   trophic levels of, 547
Language, use of by primates, 326
Large intestine
   function in human digestion, 377, 379
   location in human digestive system, 373f
Large ribosomal unit, 134
Larvacea, Class *see* Larvaceans
Larvaceans
   as protochordates, 272, 273
   neoteny in, 273
Larvae
   of sea squirts, 273
   tadpole, 273
Larynx, location of, 358f
Lascaux, caves containing art of early humans, 322f, 323, 323f
Lateral nuclei, role in controlling human food intake, 369
Lateral rectus muscle, of human eye, 352f
Laterite soil, of rain forest, 544

Latimeria
   and evolution of vertebrate limbs, 276
   as living fossil, 237, 237f
Latin, use in classifying organisms, 231
Latitude, effect on solar radiation, 526, 526f
Lead
   as air pollutant, 613t
   as poison, 618
Learning
   human, 426
   insightful, 431
   perceptual, 431
Leaves
   development of in plants, 214
   mesophyll cells of, 214
Lederberg, J., 153
Leeuwenhoek, A., 32, 33
Left atrium, function of human, 401, 401f, 402, 403f
Left auricle, *see* Left atrium
Left ventricle, function of human, 401f, 402, 403f
Legs, evolution of by vertebrates, 276, 277f
Lehninger, A., 17
Lemming, population cycles of, 547, 547f
Lens
   of human eye, 351, 352f
   of squid eye, 270f
Lenticels, 218
Leopard frogs
   as amphibians, 278, 278f
   speciation of, 302
Leucine, structure, 43f
Leucocytes, *see* White blood cells
Leucotomy, 422
Leukemia. A commonly fatal form of cancer involving the blood-forming tissues, usually characterized by anemia and greatly increased numbers of white blood cells in circulation. 508
   radiation induction of, 601, 601f
LH, *see* Luteinizing hormone
Life
   early stages in evolution of, 108t
   energy sources for, 71
   on Mars, 62
   most essential attributes of, 64
   properties, 27, 27t, 28f
   time of origin, 58
   and water, 57, 58
Life expectancy, of humans, 516, 572
"Lifeboat ethics," and human survival, 580
Ligase, in DNA replication, 119f, 121
Light
   detection by human eye, 351, 353, 355
   orientation of plants to, 220, 220f
   penetration of in the sea, 250
   speed of, 83, 83f
   spectrophotometric measurements of, 106f, 107
   visible, 83, 83f
   wavelengths of, 83
Light cycle, of photosynthesis, 87, 93, 93f
Light reactions, of photosynthesis, 87, 93
Lignin, in plant cell walls, 213

Lindeman, J., 507
Linkage group, 174
   and crossing over of homologous chromosomes, 174
   and segregation of genetic traits, 174
   use in determining cross over frequencies, 175, 176f
Linkage map, construction using cross over frequencies, 174, 175f, 176
Linking forms, and incompleteness of fossil record, 237
Linnaeus, Carolus, 229, 230n, 230f
Linnaeus, classification scheme of, 230t
Lion, 282f
Lipid structure, 54, 54f
Lipids, functions in cells, 38, 54
Liver
   function in human digestive system, 381–385
   function in regulating blood glucose, 383, 383f
   in human development, 481, 482f
   location in human digestive system, 373f
   nervous control of, 382
   portal circulatory system of, 381, 381f
   structure, 381f
Liver capillary system, 381f
Liverworts, *see* Bryophytes
Living systems, characteristics, 27, 27t
Lizards, as reptiles, 279f
Lobe fins, and evolution of vertebrate limbs, 276
Lobsters, culture in power plant cooling water, 624
Lobule, of human liver, 381, 381f
Locomotion, evolution of primate, 314, 318
Logistic equation, of population growth, 564, 564f
Long term memory
   of humans, 411, 432
   of octopus, 270
Long-day plants, 222
Lovins, A. B., 607
Lowei, O., 10
LRU, *see* Large ribosomal unit
LSD, *see* Lysergic acid diethylamide, 445
Lung cancer
   increasing incidence of, 512
   in uranium miners, 602
Lungfish, and evolution of land vertebrates, 276, 277f
Lungs
   effects of pollutants on, 614
   evolution of, 276, 277f
   in human development, 481, 482f
   oxygen loading of, 339
   respiratory function of, 201f
   structure of human, 391, 392f
   structure of in animals, 200f
Luteinizing hormone
   and initiation of female hormonal cycles, 360
   role in human reproductive cycle, 452f, 453, 454
Lycophyta, Subphylum *see* Lycopods

Lycopods
  characteristics of, 253, 253f
  primitive vascular plants, 252, 252f
Lye, as poison, 490
Lyell, Charles, 285, 286
Lymph. The pale coagulable fluid bathing the tissues of the body and consisting of the liquid plasma portion of the blood. 405
  function of, 405, 406
Lymphatic system, function of, 405–406
Lymphocytes, role in combating disease, 500
Lysergic acid diethylamide, neural effects of, 445
Lysine, structure, 43f
Lysosomes. Membrane bounded cytoplasmic bodies containing digestive enzymes able to break down worn out cellular components. 35

M

Macaque, 282f
Macrocystis
  life history of kelp, 251f
  and ocean productivity, 542
Macromeres, of annelid and mollusc embryos, 238f
Mad Hatter, disease of, 609
Maidenhair tree, as living fossil, 238, 238f
Maintenance, costs of in multicellular organisms, 181
Malaria, life cycle of, 499f
Malathion, as hazardous chemical, 622t
Malnutrition, effects on fetal development, 483
Malpighian tubules
  excretory function of, 201f
  structure of, 201f
Malthus, R., 288, 519
  principles controlling human populations, 288
Mammal
  advances over reptiles in, 280
  characteristics of, 280
  evolution of limbs, 276, 277f
  heart of, 305f
  maternal care of young by, 281
  time scale of evolution of, 317
  types of, 280, 281f, 282f
Mammary glands
  of mammals, 280, 281f
  role in human reproduction, 452f
Mantle (terrestrial), contribution to earth's geological cycles, 530
Mantle, of molluscs, 269f, 270f
Mantle cavity, of molluscs, 268, 269f, 270f
Marsupials, adaptive radiation of, 296f
Mass number, 85
Mass-energy conversion, 592
  equation for, 591

Mastadon, extinction caused by early man, 324
Masters, W. H., 459
Maternal care, in primates, 315
Maupertius, 164
Mayr, Ernst, 234
Mechanistic theories of life, 17
Mechanoreceptors, types of, 350f, 351
Medawar, P., 8
Medial nuclei, role in controlling human food intake, 369
Medical research, tactics of, 517
Medulla oblongata
  control of ventilation by, 395, 396f
  control of blood pressure by, 406
  general function in humans, 347
  location of human, 347, 348f, 361f
Medusae, of coelenterates, 260f, 261
Medvedev, Z., 515
Megaspore mother cell, 225
  egg cell production from in flowers, 225, 226f
Megaspores of seed plants, 254
Meiosis. The two successive cell divisions by which a diploid cell reduces its chromosome complement to the haploid state. 157
  and behavior of hereditary traits, 174
  in brown algae, 250, 251f
  first division of, 161f
  in flowers, 224, 225, 226f
  genetic consequences of, 159
  reduction divisions of, 157
  second division of, 161f
  stages of, 159, 160f
  summary of divisions, 159
Melanism, advantages to humans, 295
Melatonin
  functions of, 361
  release by pineal gland, 361
  structure, 361f
Membrane
  cell, 34d, 70f
  chemical analysis of, 70
  dynamic equilibrium across, 67
  inner mitochondrial, 107
  nuclear, 34
  outer mitochondrial, 107
  pores through, 70f, 71
  semipermeable, 66
  respiratory functions in mitochondria, 104
Memory
  chemical coding of, 433
  experiments on temporal lobe of brain, 432
  human, 426
  long term, 411, 432
  needed for homeostatic system function, 334
  of octopus, 270
  possible role of RNA coding in, 433
  role of protein synthesis in, 433
  short term, 411, 432
Mendel, G. J., 165

  experiments of, 165
  proof of independent assortment of genetic traits, 168f, 169
Mendelian traits, behavior of, 169, 172
Meningoencephalitis, and water pollution, 620
Menarche, 453f
Menstruation
  and female reproductive cycle, 458f
  uterine stages of, 456f
Mercury
  accumulation in food webs, 611
  as poison, 609
Merychippus, ancestral horse, 303, 304f
Mesenteries, supporting organs in coelom, 265, 265f
Mesentery, of human embryo, 482f
Mesoderm. The embryonic germ layer that forms blood, heart, muscle and connective tissue. 473
  embryonic induction by, 475, 475f
  first formation of by flatworms, 264
  formation of in human embryo, 473
Mesohippus, ancestral horse, 304f
Mesophyll cells, of plant leaves, 214
Mesosome, of prokaryotes, 144
Messenger RNA
  codons of, 132
  ribosomal binding of, 132
  transcription of, 130
Metabolic pathways, origin of, 79, 80f
Metabolic pump, role in nerve impulse generation, 341
Metabolism
  early observations of plant, 211, 211f
  production of body heat by, 385
Metamorphosis, hormonal control of frog, 363, 363f
Metaphase
  of meiosis, 161f
  of mitosis, 147, 148f
Metastasis. The spread of cancer cells from the primary site of origin to other growing sites within the body, often by way of the circulatory system. 508
Metazoa, true, 257
Methane
  as air pollutant, 613t
  recovery from sewage, 545
  role in carbon cycle, 534, 534f, 535
  structure, 39f, 40
Methionine, structure, 43f
Methyl alcohol
  from waste fermentation, 545
  structure, 39f, 40
Methylmercury, *see* Mercury
Metric system, 37, Appendix 1
Mexico, regulation of population growth in, 575
Microbial disease
  demonstration of causal agents, 497, 497f
  immune system defense against, 500
  natural human defenses against, 500
  stages of human, 500

INDEX-GLOSSARY 659

Micromeres, of annelid and mollusc embryos, 238f
Micron, 36f, 37
Micropyle, of flower ovule, 226f
Microscope, structure, 32f
Microscopy, 32n
Microspores, of seed plants, 254
Microtubules
 disruption of colchicine, 147, 148
 function in mitosis, 147, 148
 involvement in cell division, 146, 147, 148
Microvilli, of small intestine, 376, 378f
Microwaves, 83, 83f
Midbrain, in human development, 474, 474f
Middle ear
 development of bones of, 478
 human, 354, 354f
Midgut, in human development, 481, 482f
Midwife toad, inheritance of acquired characteristics by, 285
Milieu exterieur, 197
 provision of, 197
Milieu interieur, 197
 and concept of homeostasis, 333
Miller, Neal, 429
Minamata disease, 609
Mineralocorticoids, release by adrenal cortex, 360
Miocene Age, and separation of apes and humans, 317, 319
Miscarriage, in humans, 461
Mitchell, P., 105
Mitochondria, 35d, 35f
 commensal theory of origin, 150
 and control of human respiration, 338f
 as episomes, 149
 as site of oxidative phosphorylation, 104
Mitosis. Eukaryote cell division resulting in provision of an identical chromosome complement to each daughter cell.
 stages of, 147, 148, 148f
MOA, see Monoamine oxidase
Modulator molecules, 139
 and control of enzyme activity, 139, 140
Molars, of chimpanzee, 314f
Mole, 282f
Molecular hybridization, of DNA and RNA, 131
Molecule. The smallest particle into which a chemical compound may be broken down without losing the properties of that compound. 40
 organic, 40
 polar, 58f
Molecules, and organization of organisms, 191f
Mollusca, Phylum see Molluscs
Molluscs
 characteristics of, 268, 269f, 270f
 comparative embryology of, 238, 238f
 eyes of, 270, 270f
 giant neurons of, 270
 organ systems of, 268, 269f, 270f
 segmentation of, 268
 structure, 269f, 270f
 theoretical ancestor of, 269f
 types of, 268, 269f, 270f
Molting, 203
 and control of growth in arthropods, 203, 208f, 209
Monera, Kingdom, 245
Mongoloid idiocy, see Down's disease
Mongoloids, evolution and distribution of early, 329, 330f
Monoamine oxidase, and neural effects of drugs, 443f, 444
Monod, J., 137, 140
Mononucleosis, see Infectious mononucleosis
Monophyletic organisms, 245
Monosaccharides, structure, 53f
Monotremes, primitive egg-laying mammals, 280, 281f
Mortality ratio, 510f
Mosses, true, see Bryophytes
Motoneurons, see Motor neurons
Motor axon, 347
 see also Motor Neuron
Motor cells, of plants, 222
Motor end plate, 206
Motor neuron
 control of muscle contraction by, 345, 346f
 emergence from spinal cord, 348f, 349
 as part of reflex arc, 365
 structure of, 340f
Motor unit. The group of muscle cells directly controlled by one motor neuron. 347
 and control of muscular contraction, 347
Movement, by differential growth in plants, 220, 220f
Movement receptors, role in ventilation control, 395, 396f
mRNA, see Messenger RNA; Ribonucleic acid, messenger
Mucosal lining, of human stomach, 376
Mucus
 in human ventilating system, 393
 role in human digestion, 376
Mudskippers, and evolution of land vertebrates, 276, 277f
Muller, H. J., 63, 602
Multicellular organism. An organism composed of many cells that are interdependent on each other and cannot sustain living processes without each other. 184d
 characteristics of, 184
 colonies of, 189
 evolution of, 179
Multicellularity, 184
 advantages of, 180
 derived by cell compartmentalization, 183
 evolutionary advantages of, 182
 in plants, 212, 227
 price of, 181
Multiple sclerosis, and HLA genes, 502
Muscle fibers, nerve end plates on, 340f
Muscles
 control of contraction of, 345, 346, 346f, 347
 control of skeletal, 413, 415
 controlling human eye, 352f, 353
 development of human, 479
 innervation of, 346f
 motor end plate of, 206
 nervous control of, 345, 346f
 oxygen transport in, 398f
 stretch receptors of, 344, 345f
 structure of, 346f
 tetanized, 346
Muscle contraction, control of, 346, 346f, 347
Muscular system, 193
Mushrooms
 life history of, 256, 257f
 rapid uptake of water by, 256
Mutagenic chemicals, 125, 125f
Mutants, identification of in bread mold, 114n, 115f
Mutation
 caused by radiation, 602
 by chemical agents, 125, 125f
 chromosomal, 124, 157, 158f
 effects of diploidy on, 158, 178
 environmental influences and, 123
 and evolution, 126
 gene, 124
 of vertebrate hemoglobin gene, 315, 316
 randomness of, 126, 127f, 128
 rate of, 128
 role in aging, 514
 single word, 125f, 126
 somatic, 158
 by wording change in genetic code, 124
Mycelium, of fungi, 256, 257f
Myelin
 effects on nerve conduction velocity, 268
 and electrical currents of nerve, 341, 343f
 formation of, 343f
Myoglobin, role in oxygen transport, 398f
Myoneme, of Vorticella, 183f
Myosin, role in muscle contraction, 346f
Myxomatosis, and ecosystem manipulation, 559

N

NA, see Noradrenaline; Epinephrine
NAD, see Nicotinamide adenine dinucleotide
NAD-NADP transhydrogenase, 103
NADH, see Reduced nicotinamide adenine dinucleotide
NADP, see Nicotinamide adenine dinucleotide phosphate
Naegleria, ameba, in polluted water, 620
Nagasaki, 590
Nanometer, 36f, 37, Appendix 1
Narcotics, effects on death rate, 441f
"National triage," and human survival, 580
Natural gas, reserves of, 589
Natural resources, depletion of, 584

Natural selection
  elimination of effects on humans, 332
  theory of, 291, 292t
Neanderthal Man
  characteristics of, 318f, 322
  cultural advancement of, 320
Negative feedback. In a dynamic system, any feedback process that tends to oppose change and returns the system to its steady state. 334
  role in homeostasis, 334, 335f
Negative feedback system
  human thermoregulation as, 389f
  muscle control as, 414
  ventilation system as, 395, 396f
Negritos, evolution and distribution of early, 329, 330f
Negroes, evolution and distribution of early, 329, 330f
Negroids, evolution and distribution of early, 329, 330f
Nematoblasts, nematocyst forming cells, 262
Nematocysts
  function of in coelenterates, 262
  of *Hydra*, 186, 187f
Nematoda, Phylum, see Nematodes
Nematodes
  structure of, 265, 265f, 266
  see also Ascaris
Neocortex, of brain, 418
Neopilina, living "fossil" mollusc, 268
Neotenin, insect juvenile hormone, 208f
Neoteny. Sexual reproduction by a larval organism. 263
  of ancestral bilateral metazoan, 263
  in protochordates, 273
Nephric tubule, role in excretion, 407, 409f
Nephridia. Excretory organs characteristic of many invertebrates. The general form is a tube opening at one end into the *coelom* (body cavity) and discharging at the other through a pore in the body wall.
  excretory function of, 201f
  of flatworms, 264, 264f
  structure, 201f
Nephron. One of the excretory units that make up the kidney, particularly in vertebrates. 407
  function in human kidney, 407, 409f
  structure in human kidney, 201f, 407, 409f
Nerve impulse
  conduction of, 341, 343, 343f
  currents associated with, 341, 343f
  frequency coding of, 344, 345
  generation of, 341, 342f
Nerve net, structure of, 206f, 207
Nerves, see Neuron; Nervous system
Nervous system. The system made up of nerves, sensory receptors, ganglia, and usually a brain, that receives sensory impulses and integrates and coordinates the responses, as well as directs the general functioning of the organism. 193
  absence in plants, 223, 224n
  of chordates, 274f
  components of, 205, 206, 206f, 207
  components of human, 347, 348, 349
  control of adrenal medulla by, 359f
  control of human digestion by, 370
  and control of human reproduction, 453, 457, 458, 459
  control of temperature regulation by, 386, 388, 389f
  development of human, 474, 474f
  effects of segmentation on, 266
  ergotropic, 442
  function compared with endocrine system, 355
  function of neurons of, 339, 340, 341, 342, 343, 344
  of humans, 441–446
  influence on pituitary gland, 357
  integration of information by, 345
  language of, 344
  limitations of in invertebrates, 267, 268
  overlap with endocrine system, 205
  role in homeostasis, 336, 336f
  of sea squirts, 272, 273, 273f
  of squids, 268, 270, 270f
  trophotropic, 442
  types of in animals, 206f, 207
  autonomic, see Autonomic nervous system
  see also Neuroendocrine system
  parasympathetic, see Parasympathetic nervous system
  sympathetic, see Sympathetic nervous system
Neural crest, 474
  in human development, 474f
Neural groove, 474
  in human development, 474f
Neural plate, 474d
  in human development, 474
Neural transmitters, types of, 206, 442, 445
Neural tube, 474
  in human development, 474f
Neuroendocrine system, 205
  effects on temperature regulation, 386, 388, 389f
  pituitary involvement in, 357
Neuroid conducting system. Non-neural tissues able to conduct action potentials, 222d
  plant coordination by, 222
Neuron, 205
  conduction velocity of, 268, 340, 341
  functions of, 205, 206
  integration of information by, 345
  language of, 344
  motor end plate, 206, 340f
  neurosecretory, 356
  number in human nervous system, 340
  sensory types of, 206, 350f
  signal transmission in, 341, 342f, 343f
  structure of, 205f, 340
  synapse of, 206
  transmitter chemicals of, 206
  see also Motor neuron; Sensory neuron
Neurosecretory nerves, of pituitary, 356, 356f
Neurospora crassa
  mitochondria as episomes of, 149
  mutation research using, 112, 114n, 115f
  poky mutant of, 149
  as sac fungus, 256
Neutron, 85
New Guinea, classification of bird species of, 234
Newts, as amphibians, 278f
Nicotinamide adenine dinucleotide
  function in glycolysis, 73f, 75
  from hexose monophosphate pathway, 80, 81f
  role in hydrogen transport, 103
  structure, 75f
  and tricarboxylic acid cycle, 101, 103
Nicotinamide adenine dinucleotide phosphate
  and photosynthesis, 92, 92f, 93f
  role in biosynthesis, 103
  structure, 75f
Nicotine, neural effect of, 443f
Night vision, function of rhodopsin in, 353
Nitrate
  use of by plants, 219
  production by *Nitrobacter*, 537
Nitric oxide, as pollutant, 613t, 614
Nitrification, 536f, 537, 538
Nitrite, production by *Nitrosomas*, 537
Nitrobacter, in nitrification, 536f, 537
Nitrogen
  forms in atmosphere, 538
  oxides effects on ozone layer, 538
  total amount on earth, 536
Nitrogen cycle, 536, 536f
  biological phase of, 537
Nitrogen compounds, as pollutants, 617
Nitrogen dioxide, as pollutant, 617
Nitrogen fixation, biological mechanism of, 537
Nitrogen scavenging, in nitrogen cycle, 536f, 537
Nitroglycerine, and circulatory disease, 495
Nitrosomas, in nitrification, 536f, 537, 538
Nitrous acid, as mutagen, 125, 125f
Node of Ranvier. The myelin-lacking junction between two adjacent sheath cells wrapped around a vertebrate nerve cell axon. 341
  role in nerve impulse transmission, 341, 343, 343f
Nonconservative replication, of DNA, 121, 122f
Nondisjunction, 177
Noradrenaline
  release by adrenal medulla, 360
  synaptic agonism and antagonism, 443f, 444
Norepinephrine, see Noradrenaline
North America, human migration to, 324
Nose, role in ventilation, 393
Notochord, 273
  of chordates, 273, 274f

Notochord (cont.)
  formation in human development, 473, 474f
  of sea squirts, 273f
  types of animals having, 274
Novocaine, neural effect of, 443f
Nuclear bomb, first test, 590
Nuclear fission, 591
  discovery, 590
Nuclear fusion, 591
Nuclear power plants
  environmental hazards of, 624t
  first U.S. commercial, 590
Nuclear power reactor
  breeder type, 595
  mechanism of operation, 594f
  safety mechanisms of, 594
  wastes from, 595
Nuclear power, world wide production of, 590t
Nuclear reactor, 590
Nuclear-cytoplasmic ratio, in human cleavage, 471
Nucleic acid. Complex acid occurring in all living cells, composed of purine and pyrimidine bases, a five-carbon sugar (*ribose* in the case of the ribonucleic acids, and *deoxyribose* in the case of the deoxyribonucleic acids) and phosphoric acid. Nucleic acids carry the genetic information that directs the functioning of the cell. 56
  structure, 55f
  structure, RNA compared to DNA, 130, 130f
Nucleic acids
  and control of protein synthesis, 111
  in DNA, 55f
  and evolution of replication, 140
  functions in cells, 38
  in RNA, 55f
Nucleolus. An organized body of predominantly protein structure found in the nucleus and variously regarded as a center of synthetic activity or a storage organelle. The nucleoli usually disappear during mitosis but are reformed after each division. 35
Nucleoplasm, 34
Nucleoside. A product of the hydrolysis of a *nucleotide*; consisting of a sugar amine joined to one of the purine bases (adenine or guanine) or one of the pyrimidine bases (cytosine, uracil, or thymine). 56
Nucleotide. A product of the hydrolysis of a nucleic acid, consisting of a *nucleoside* joined to a phosphate group ($-PO_4\equiv$). 55f, 56
Nucleotide bases, as genetic code, 116f, 117
Nucleus. An organized protoplasmic structure found in most plant and animal cells, which contains the nucleic acid material responsible for directing the function and reproduction of the cell. The nucleus is usually surrounded by a double membrane perforated by pores.
  of cell, 34d, 35, 35f
  controlling influence on cell division, 149
  division synchronization with cell division, 149
Nullosomy, 177
Nyctea, snowy owl, 547

O

Ocean
  photosynthesis in, 545
  principal currents of, 529f
Oceans
  food chain of, 545
  primary production of, 541
  thermocline of, 541
  tropic levels of, 547
Ochromonas, 247f
Octopus
  brooding of eggs by, 281f
  characteristics of, 268, 269f
  learning and memory in, 270
  as predator, 546f
Oil shales, as energy reserve, 590
Olfaction, role in controlling human food intake, 370, 372f
Olfactory lobe, of human brain, 372f
Olfactory nerve, connection with human tongue, 372f
One gene/one enzyme hypothesis, 112
Ontogeny
  and formation of ancestral structures by embryos, 300
  recapitulation of phylogeny by, 263
Oocyte. A cell of the female germ line that, undergoing meiosis, forms the egg cell (ovum). 360
  development of in ovary, 360
Oogenesis, in humans, 451f
Oparin, A. I., 59
Oparin-Haldane Theory
  of life's origin, 59
  and origin of metabolic pathways, 79
Open systems, 27
Operator gene, 137d, 138
Operon, 137
Operon, genes in, 137
Operon, and gene regulation, 137, 138f
Operon, lactose, 137, 138f
Opiates, self-produced, 445
Opposable thumbs, of primates, 313f
Opsin, visual protein, 352, 353
Optic nerve, exit from human eye, 352f, 353, 353f
Optic nerve, of squid eye, 270f
Optical isomers, 46
Oral cavity, in human development, 481, 482f
Oral contraceptive
  and calcium metabolism, 468
  effectiveness of, 467t
  effects on later fertility, 468
  health problems of, 468
Oral pole, of coelenterates, 260, 261f
Orangutan, hands and feet of, 313f
Orangutan, morphology compared to other primates, 312f
Order Primates, *see* Primates
Organ systems
  development of human, 474–482
  of echinoderms, 272, 272f
  effects of coelom on, 266
  effects of segmentation of, 266
  and evolution of land vertebrates, 277
  functional classification of, 192
  functions of in animals, 194t
  making up organisms, 191f, 192
  of molluscs, 269f, 270f
  of plants, 214, 228t
  of plants compared to animals, 228t
  of sea squirts, 273, 273f
  ten types found in animals, 193
Organ transplantation, moral aspects, 496
Organelle. A specialized part of a cell (or of a one-celled organism) performing a specific function or set of functions often analogous to those performed by an organ of a higher animal or plant. 34, 35f
Organic molecule. Molecules originally thought to be formed only by biological processes, molecules containing carbon and hydrogen. 40
Organisms
  classification of, 229, 230, 231
  as composed of organ systems, 191f
  five kingdoms of, 244, 244f
  levels of organization in, 191f
  numbers of kinds of, 229
Organophosphate insecticides, as hazardous chemicals, 622t
Organ. A differentiated structure (such as a heart or kidney or, in a plant, leaf or flower) made up of various cells and tissues and adapted for performing a specific function. (Organs concerned with similar functions are grouped into systems.) 190
  and cellular differentiation, 181
  complimentary functions of in organ systems, 192
  composition of, 191f
  human sensory, 349, 351, 352f, 353, 353f
  of plants, 214, 228t
  substitutes for human, 496
  supported by mesenteries in coelom, 265, 265f
  transplantation of human, 496
Orgasm, in humans, 458
Orgel, L. E., 77
Osmolarity. The "strength" of a solution in terms of the amount of dissolved substances in proportion to the volume of solvent. 409
Osmolarity, control of by human kidney, 409
Osmoregulation, 201
Osmoregulation, and water control, 201
Osmoregulation, by humans, 409, 410
Osmosis. Flow or diffusion of water through a *semipermeable membrane* separating

two solutions of unequal concentration. 66, 66f
Osmosis, serving transport of materials in plants, 216, 216f, 217, 218f
Osmotic equilibrium, of body fluids, 67, 68f
Osmotic evolution, of cells, 69
Osmotic receptors, in hypothalamus, 357
Osmotic stress
  adaptations to by land vertebrates, 277
  and evolution of respiratory organs, 198
  and evolution of terrestrial animals, 198
  offset by turgor pressure in plant cells, 212
Outbred organisms, sexual selection of, 297
Outer ear, human, 354, 354f
Outer membrane, function in mitochondria, 105f, 107
Outer segment, of human photoreceptors, 353f
Oval window, of human ear, 354f
Ovarian follicle. The cellular sac in which the ovum matures in the ovary. 360, 449
  development of, 450f, 456f
  production of hormones by, 360
Ovary
  endocrine function of, 360
  follicles of, 449, 450f
  germ cell content of, 449
  hormones released by, 360
  human female, 449, 450f
  initiation of hormonal cycles during development, 360
  location of, 207f
  of flower, 224, 225f
  structure of, 360
Ovists, 164n
Ovulation
  environmental control of, 357
  and female reproductive cycle, 458f
Ovules, of flower, 224, 225f
Ovum, and events of human fertilization, 459
Ovum, development of human, 449, 450, 450f
Oxalic acid, as poison, 490
Oxaloacetate, and tricarboxylic acid cycle, 100f, 101
Oxidant, 71, 91
Oxidation, of glucose in cellular respiration, 99
  of wood, 6
Oxidation-reduction potentials, 91n
  of photosynthesis, 91, 92f
Oxidation-reduction reaction, 71
  electron transfer in, 91n
  of respiratory chain, 104f
Oxidative metabolism, respiratory chain in, 91
Oxidative phosphorylation
  chemiosmotic theory of, 105, 105f
  in cellular respiration, 104
Oxides, of wood, 7
Oxygen
  buildup in early atmosphere, 96
  carrying capacity of blood for, 338f, 339
  early observations on use by plants, 211
  effects on metabolic evolution, 98
  and origin of life, 59
  and photosynthesis evolution, 94
  role in ventilation control, 395, 396f
  transport by hemoglobin, 398f
Oxygen debt, of active muscles, 76
Oxygen receptors, and ventilation control, 395, 396f
Oxytocin
  release by pituitary, 356, 356f
  role in human birth, 461
  triggers for release of, 357
Ozone
  and photosynthesis evolution, 94
  as pollutant, 614
  as ultraviolet shield, 59
Ozone layer
  early development of, 97
  effect of nitrogen oxides on, 538
  effect on solar radiation, 527
  effects of fluorocarbons on, 617

P

P gene, *see* Promoter gene
P site, *see* Peptidyl site of ribosome
Pacemaker, cardiac, 495
Pain
  gate theory of, 440
  neurophysiological basis of, 439
Pairing rule, of DNA polynucleotides, 117
Paleozoic age, rapid speciation by, 306
Palsy, *see* Parkinson's disease
PAN, *see* Peroxyacetyl nitrate
Pancreas
  and diabetes, 360
  endocrine function of, 360
  islets of Langerhans in, 360
  location in human digestive system, 373f
  location of, 207f
  release of hormones by, 360
  role in human digestion, 380
  role in regulating blood glucose, 382, 383
  structure of human, 380f
Pancreatic juice, role in human digestion, 380
Pangenesis, theory of heredity, 123d, 163
Panting, function in thermoregulation, 387, 389f
Pantothenic acid, structure, 101f
Papules, of echinoderms, 272f
Paramecium
  characteristics of, 247, 247f, 248
  conjugation in, 248, 249f
  contractile vacuoles of, 69, 69f
  locomotion in, 248
  structure, 247f, 248
Parasites, flatworms as, 264, 264f
Parasympathetic nervous system, 347
  connections with sympathetic, 349, 349f
  control of heart by, 349
  control of human digestion, 379, 382
  and control of human thermoregulation, 388, 389f, 390
  nerves of, 349, 349f
  transmitter chemicals of, 349
Parathion, as poison, 491
Parathyroid gland
  developmental origin of human, 470
  function of, 357
  hormones of, 357
  location of, 207f, 358f
Parazoa, Kingdom or Subkingdom, *see* sponges
Parenchyma cells, of plants, 212, 213f, 214
Parenchyma, of plants, 213f
Parkinson's disease, 435t
Particulates, as pollutants, 618
Passenger pigeon, reduction by hunting, 558
Pasteur, L., 30
Pauling, L., 48
Pavlov, I. N., 429
PBB's, *see* Polybrominated biphenyl
PCBs, *see* Polychlorinated biphenyl
Pearl, R., 564
Peat, as energy source, 589
Pedicellaria, of echinoderms, 272f
Peduncle, of sea pansy *Renilla*, 188f
Peking Man, *see* Homo erectus
Pelecypoda, Class *see* Clams
Penicillin
  bacterial resistance to, 156n
  as disease combatant, 505f
Penicillium, as fungus, 256
Penis, human, 448f
Peony, learned chimpanzee, 327
PEP, *see* Phosphoenolpyruvate
Pepsin
  role in human digestion, 378f
  secretion during digestion, 376
Pepsinogen, secretion during digestion, 376
Peptidase, role in human digestion, 378f
Peptide bond. The chemical bond joining one peptide with another by reaction of the amino group of one with the carboxyl of the other, forming the bond R—NH—CO—R. 44, 44f
  formation of in protein synthesis, 134, 135, 135f
Peptidyl site, of ribosome, 134
Peptidyl transferase, function in protein synthesis, 134
Perennial plants, 222
Pericycle, of plant roots, 216f
Peripheral nervous system, 347
Peristalsis. Successive waves of contraction along a hollow cylinder of muscle (as an intestine or the body of a worm) that force the contents onward. 375
  role in human digestion, 375, 376
Peritoneum. A mesodermal sheet lining the body cavity. 265
  and evolution of coelom, 265, 265f
Pernicious anemia, and vitamin B12, 376
Peroxyacetyl nitrate, effects of, 617
Peruvian upwelling, 542
Petit mal convulsion, 424f

Petroleum
  current supplies of, 589, 589f
  reserves of, 589
1,3-PGA, *see* 1,3-diphosphoglycerate
2-PGA, *see* 2-phosphoglycerate
3-PGA, *see* 3-phosphoglycerate
pH, effect on oxygen transport in blood, 397, 398f
pH receptors, role in ventilation control, 395, 396f
Phaeophyta, Phylum, *see* Brown algae
Phagocyte. Cell found in blood and tissues capable of taking up foreign material by engulfing it (phagocytosis) and hence important as a bodily defense mechanism. 81
  response to silica, 618
Pharyngeal basket, of sea squirts, 272, 273f
Pharyngeal gill slits, of chordates, 274f
Phenotype. The term applied to a characteristic evident or expressed in an organism. 163
  of garden peas, 167, 167f, 168
  gene interactions affecting, 169, 170f
Phenylalanine
  biosynthesis of, 301f
  and phenylketonuria, 171, 171f
  structure, 301f
Phenylalanine hydroxylase, and phenylketonuria, 171, 171f
Phenylketonuria
  as example of single gene mutation, 171, 171f
  human disease, 493t
Pheromone. A chemical used in communication among organisms of the same species.
  human sexual, 457
Phloem
  flow of materials in, 217, 218f
  of plants, 213f, 214, 215f, 217, 218f
  pressure generation in, 217, 218f
  primary, 213f, 216f
  of roots, 216, 216f
  transport functions of, 216, 216f, 217, 218f
Phloem cells
  companion cells of, 214
  cytoplasmic retention of, 214
  of plants, 212, 213f, 214
  sieve plates between, 214
  tissue formed by, 214
  vacuoles of, 214
Phosphodiester linkage, of nucleotides, 55f
Phosphoenolpyruvate
  as CO_2 acceptor in dry climate plants, 543
  and glycolysis, 75
  structure, 73f
Phosphoenolpyruvic acid
  as amino acid precursor, 301f
  biosynthesis, 301f
6-phosphofructokinase, and glycolysis, 73f
6-phosphoglucolactone, structure, 81f
2-phosphoglycerate, structure, 73f

3-phosphoglycerate
  and glycolysis, 74
  and photosynthesis, 93, 93f
  structure, 73f
3-phosphoglycerol phosphate, and glycolysis, 74
Phosphorus cycle, 538, 539f
Phosphorylation
  as means of preventing diffusion, 81
  of glucose, 72
Phosvel, as hazardous chemical, 622t
Photoautotroph. An organism using light energy to generate energy and material by reduction of carbon dioxide. 340d
  and energy input into biosphere, 540
Photoblepharon, flashlight fish, 546f
Photochemical smog
  composition of, 617
  contribution of nitrogen oxides, 538
Photon, 83
  interactions with atoms, 86, 86f
Photoperiodism, 222
  effects on plant growth cycles, 222
Photophosphorylation, 82
  origin, 82
  and photosynthesis evolution, 94
Photoreactivation, of damaged DNA, 128
Photoreceptors, types of human, 351, 353f
Photosynthesis, 82
  action spectrum of, 88f, 89
  and control of gas exchange in plants, 218, 219f
  efficiency of, 93
  electron transport chains in, 91
  equations for, 90, 94
  evolution of, 94
  global factors affecting, 525
  marine organisms responsible for, 249
  in ocean, 545
  origin, 82
  and pigments of algae, 250
  plant adaptations to reduce water loss, 543
  process in modern plants, 87
Phycobillins, of algae, 250
Phylogeny
  of flowers, 233
  of major types of organisms, 244, 244f
  of mammals, 232
  methods of determining, 234, 238, 239, 240
  recapitulation of during development, 263
  reflected in classification schemes, 303
Phylum Annelida, *see* Annelids
Phylum Arthropoda, *see* Arthropods
Phylum Coelenterata, *see* Coelenterates
Phylum Ctenophora, *see* Comb jellies
Phylum Echinodermata, *see* Echinoderms
Phylum, Mollusca, *see* Molluscs
Phylum Nematoda, *see* Nematodes
Phylum Platyhelminthes, *see* Flatworms
Physiology
  of human food intake, 369
  and human homeostasis, 336, 337

Phytol
  of chlorophyll, 89, 90f
  structure, 90f
Phytoplankton, 180f
Pigment layer, of human eye, 352f
Pigments
  control of by melatonin, 361
  of human photoreceptors, 353
  and photosynthesis in algae, 250
Pigmies, human racial variety, 330
Pill, contraceptive, *see* Oral contraceptive
Pineal gland
  day-night cycle control by, 361f, 362
  functions of in vertebrates, 356
  hormones released by, 361
  and initiation of vertebrate reproductive cycles, 453
  location of, 207f, 361, 363f
  visual function of, 361
Pinocytosis, 81
Pinocytosis, role in human digestion, 376
Pith parenchyma, of plants, 212, 213f
Pithecanthropine Man, *see Homo erectus*
Pithocanthropines, *see Homo erectus*
Pituitary gland
  development of, 356
  hormones released by, 356, 356f
  interactions of with hypothalamus, 356, 356f
  lobes of, 356, 356f
  location of, 207f
  as master gland, 355, 358f, 359f, 360, 361
  neuroendocrine function of, 357
  role in human reproduction, 452f, 453
  see also Anterior pituitary; Posterior pituitary
Pituitary portal system, role in human reproduction, 453
Placebo, 8
Placenta
  formation of human, 455, 461, 462f
  of mammals, 281
Placoderms, as primitive vertebrates, 275, 275f
Plague, 489, 499f
  effect on human population, 571
Planaria, similarity to coelenterate planula, 263
Planarians, as flatworms, 264, 264f
Plantae, Kingdom, *see* Plants
Plants
  adaptations to control water loss in, 219
  annual, 222
  carnivorous, 222, 223f, 224n
  cellular differentiation of, 212, 213f
  characteristics of cells of, 212, 213f
  coordination in, 219, 221
  differential growth of, 220, 220f
  early observations on metabolism of, 210, 211f
  fertilization in flowers of, 226f, 227, 227f
  flowering, 255, 256
  kingdom of, 250
  neuroid communication in, 222
  organ systems of, 214, 228t

664   INDEX-GLOSSARY

perennial, 222
pressures generated in, 217
primitive vascular, 252, 252f
psychology of, 224n
reproductive parts of, 223, 225f
seed-bearing, 254, 255
specialized regions of, 211
tissues of, 212, 213f, 215f, 216f
transition from sea to land, 251
transport system of, 214, 215f, 216, 217, 218f
types of seed-bearing, 254, 255
see also Specific plant type
Planula
coelenterate larva, 260f, 261, 26
resemblance to bilateral metazoan ancestor, 263
Planuloid, theoretical bilateral metazoan ancestor, 263
Plasmodesmata, 34, 213
through plant cell walls, 213
Plasmodium, malaria protist, 248
Plasmolysis. The process of causing living cytoplasm to shrink away from the wall (or original perimeter) of a cell by causing the cell to lose water by osmosis (as by placing it in a concentrated salt solution). 69
Plastids, 37
of typical plant cell, 35f
Platelets
functions of, 399, 400
source of, 399, 400
Platyhelminthes, Phylum, see Flatworms
Platypus, primitive mammal, 280, 281f
Pleistocene Age, and evolution of humans, 317, 319
Pleura, covering of the lung, 396f
Plutonium-239, 596
Plutonium-239
from breeder reactor, 595
toxicity of, 602
Phenylketonuria, biochemical pathway affected by, 170, 171f
Phenylpyruvic acid
and phenylkatonuria, 171, 171f
structure of, 171f
Plasmid, 156
and bacterial resistance to antibiotics, 156n, 255
Pneumonic plague, 499f
Poisons
modes of action, 490
of respiration, 490
Poky, mutant of *Neurospora*, 149
Polar bear, 231f
Polar body, 162
formation in humans, 450, 451f
formation of, 157f, 162
Polar molecule, 58f
Polio, see Poliomyelitis
Poliomyelitis, costs of eradication, 517
Politics, and human survival, 521
Pollen grain, formation of in flowers, 224, 226f

Pollen mother cell, and pollen grain formation, 226f
Pollen tube, formation during fertilization, 226f, 227
Pollution, 613
air, 613
thermal, 624
water, 613
Polybrominated biphenyl, as environmental hazard, 622t
Polychlorinated biphenyl, as environmental hazard, 622t, 623
Polydactyly, 164, 165f
Polydactyly, as example of variable gene penetrance, 170f, 171
Polymerase, function in replication, 120, 121
Polymerization, as step toward evolution of replication, 140
Polymers, of carbohydrates, 54
Polymorphism, 262
of coelenterates, 260f, 262
Polynucleotide, 55f
Polypeptides, as hormones, 356
Polyphyletic organisms, 245
Polyploidy, 146
development of, 176
Polyps
of coelenterates, 260f, 261
of sea pansy *Renilla*, 188, 188f
Polysaccharides, 54
Polyvinyl chloride
as air pollutant, 622t
as environmental hazard, 622t
structure, 622t
Pongidae, Family, see Pongids and apes
Pongids, classification of apes, 311, 311t
Pons, general function of human, 348
location of human, 348
Population, 522
Population growth
compounded interest and, 561, 561f
in developing nations, 573
effect of age structure on human, 577, 577f
effects of unlimited human, 578, 579f
estimates of human, 563, 563f
factors influencing, 569
of humans based on age structure, 578, 579f
of humans through history, 568, 569f
outlook for stabilizing human, 576
pathological manifestations of, 582
Pearl's equation and U.S. growth, 565, 566f
problems in predicting human, 578
regulation of human, 573, 574, 575
of reindeer, 565, 567f
Populations
characteristics of, 522
cycles of, 547, 547f
effect of agriculture on human, 569f, 570
effect of industrial revolution on human, 571
evolution of, 292

growth of, 561–582
of humans through time, 568, 569f
Malthus' principles of human, 288, 289
mechanisms of stability in animals, 581, 582
Pore cells, of sponges, 258f
Porphyrin ring
of chlorophyll a, 90f, 242f
of hemoglobin, 398f
prebiological synthesis, 90f
structure, 90f, 242f
Portuguese Man-O-War, type of hydrozoan, 260f
Positive feedback system, blood clotting as, 400f
Post-anal tail, of chordates, 274f
Posterior lobe, of pituitary gland, 356, 356f
Posterior surface, of bilateral animals, 261f
Post-natal depression, in humans, 464
Post-reproductive mutation theory, of aging, 514, 515
Potassium, role in nerve impulse, 342f
Potassium-40, use in dating fossils, 236f
Potato blight, fungal plant parasite, 256
Potential intelligence, 433
Power, solar, 585
Prairie biome, 544
Preadaptation, of humans for communication, 326
Precambrian Age, 96
Precession, 524
Precession, of earth's axis, 524
Predators
echinoderms as, 270, 271
squids as, 268, 270, 270f
Preformation, 164n
Pregnancy, events of human, 455, 456f, 460, 461
Prehensile tail, of anthropoids, 312
Premack, D., 431
Premolar teeth, of chimpanzee, 314f
Pressure, generation in plant xylem and phloem, 217, 218f
Pressure receptors
of heart, 402
type of mechanoreceptor, 351
Priestly, J., 211
Primary phloem
of plants, 213f
of roots, 216f
Primary producers, role in energy production, 540
Primary production
agricultural effect on terrestrial, 542
in biosphere, 540
conditions affecting terrestrial, 542
geographical distribution of, 542, 543f
and human food production, 550t
limiting factors in oceans, 541
of land, 541t, 542
of major regions of earth, 541f
measurement of, 540
of oceans, 541, 541t

INDEX-GLOSSARY 665

Primary structure
  of DNA, 115
  of proteins, 44d, 45f
Primary xylem, of plants, 213
Primates
  basis for social behavior in, 315
  brain size compared to other vertebrates, 314f
  classification of, 311, 311t
  evolution of communication in, 326, 327
  evolution of longer life span in, 328
  evolution of skulls of, 321f
  hands and feet of, 313f
  maternal care and extended juvenility in, 315
  nervous system of, 206f, 207
Primates, Order, see Primates
Primitive node, in human development, 473f
Primitive seas, carbon content of, 77
Primitive streak, in human development, 473, 473f
Principle of multiple assurance, 337
  in ventilation control, 395
Probability, and statistics, 9
Proconsul, see Dryopithecus
Progesterone
  release by ovary, 360
  role in female reproductive cycle, 455
  structure, 452f
Proglottids, of tapeworms, 264f
Prokaryotes. Organisms without a membrane-bounded nuclear region, the bacteria and blue-green algae. 143d
  cell division in, 144
  characteristics of, 143, 143f, 144t
  classification of, 245
  sex in, 153, 155
Prolactin, in human reproductive cycle, 452f
Proline, structure, 43f
Promoter gene, 138f
Pronucleus, 459
Propane, structure, 41f
Prophase
  of meiosis, 159, 160f
  of mitosis, 147, 147f
Prosimians
  classification of tree shrews, 311, 311t
  see also Tree shrews
Prosimii, Suborder, see Prosimians; Tree shrews
Prostaglandins, 362f
  and abortion induction, 469
  characteristic effects of, 362f
  structure, 322f
  types of, 362, 362f
Prostate gland, 448
  hormones released by, 362
  human, 448, 448f
Protein structure
  analysis of, 50
  and biological function, 44
  control of by DNA, 111
  determining factors, 46
  primary, 44d, 45

  quaternary, 44d, 45f
  secondary, 44d, 45f, 46
  tertiary, 44d, 45f
  types, 48, 48f, 49
  x-ray diffraction studies of, 48
Protein synthesis
  completion phase of, 135
  energy demands of, 136
  initiation factors for, 134
  operon control of, 137, 138f
  peptide bond formation in, 134, 135, 135f
  and regulation of gene action, 136
  ribosomal involvement in, 131
  steps of, 132
  tRNA role in, 132, 133, 133f
proteins
  amino acids of, 41, 41t, 43f
  and acidity, 42
  as catalysts, 49
  as enzymes, 38, 52
  and evolution, 140
  fingerprinting of for determining phylogeny, 239, 239f
  functional types, 42f
  functions in cells, 38, 41t, 42f, 49
  as hormones, 356
  human digestion of, 378f
  molecular weights of, 41t
  repressor, 137
  synthesis in plants, 219
Protista, Kingdom
  monophyletic origin of, 245
  see Protists
Protists, animal-like, 247f, 248
  characteristics of, 246, 247, 247f, 248, 249
  conjugation of ciliate, 248, 248f
  plant-like, 247f, 248, 249
  types of, 247f, 248, 249
Protochordates
  characteristics of, 272
  as evolutionary link with chordates, 273
  types of, 272, 272f
Proton, 85
Proton gradient, across mitochondrial membrane, 105
Protoplasm, 34
Protostomes
  characteristic development of, 266
  phylogeny of, 244f
Protostomia, see Protostomes
Protozoa, 247, 247f, 248
Protozoa, specialization of, 183f
Provirus, 511
Proximal convoluted tubule, of nephron, 409f
Pseudopodia, 37
Pseudoscience, 4
Psilophyta, Subphylum, see Psilophytes
Psilophytes
  characteristics of, 253, 253f
  primitive vascular plants, 252f
  reproduction in, 253, 253f
Psilopsids, see Psilophytes
Psychology, of plants, 224n
Puberty, endocrine initiation of, 451, 452f

Puffballs, as fungi, 256
Pulmonary artery
  and lung circulation, 392f
  role in heart circulation, 402
Pulmonary vein, and lung circulation, 392f
Pulse labeling, of DNA, 146
Pupil, of human eye, 351, 352f
Pure science. Scientific investigations without reference to practical goals. 12
Purines, structures, 55f
Purines, types, 55f, 56
Purkinje fiber. A specialized heart muscle cell that transmits signals to contract from one part of heart to another.
  of human heart, 403f
PVC, see Polyvinyl chloride
Pyrimidines
  structure, 55f
  types, 55f, 56
Pyrogen, release during fever, 390
Pyrolysis, 609n
  and energy recovery, 609n
Pyruvate
  entry into tricarboxylic acid cycle, 100f, 101
  production in glycolysis, 73f, 75
  structure, 73
Pyruvic acid, and glycolysis, 72

Q

Quaternary structure, of proteins, 44d, 45f
Quinone, 103

R

R factor, in population growth, 569, 569f
Rabinowitch, E. I., 82
Races
  desirability of uniformity of human, 331
  intelligence of human, 331, 332
  origin and early distribution of human, 328, 329, 330f
  types of human, 329, 329f, 330, 330f
  and selection pressure for melanism, 295
  viability of human hybrids of, 328
Rad, 599
Radial artery, 401f
Radial symmetry. Structural plan allowing division of organism into two mirror image halves by any longitudinal bisector intercepting the longitudinal axis. 261
  of coelenterates, 260f, 261
  of comb jellies, 262
  of echinoderms, 270
Radiation
  acute biological effect, 600, 600n
  biological threshold for, 602
  and cancer induction, 601, 601t
  danger of fetal exposure to, 604
  dosage terminology for, 599
  genetic effects, 602
  hazard from nuclear bomb, 606

hazard from reactor accident, 604
hotspots for biological action of, 602
ionizing, 598t, 599
long term effect, 601
radiation dose in medical x-rays, 603t, 604t
relative biological effect of, 599
solar, 83, 83f
spectrum of, 83, 83f
see also Radioactivity
Radioactive decay, 592
Radioactivity
  biological hazards of, 599
  discovery of, 590
  effects on matter, 598
  natural levels in environment, 603, 603t
  see also Radiation
Radioautography, 144
  of replicating DNA, 144, 145f
Radioisotopes, use in dating fossils, 236, 237
Radiolarians, skeletal deposits of, 248
Radiowaves, 83, 83f
Radula, of molluscs, 269f
Rain forest
  characteristics of, 543
  distribution of, 543
  unsuitability for agriculture, 544
Rainfall, pattern on earth, 532
Ramapithecus
  characteristics of early hominid, 318f, 319
  skull of, 321f
Rana pipiens, speciation of leopard frog, 302
Rao, P. N., 149
Rapid eye movement (REM) sleep, 437
Rat, kangaroo, 282f
Rays, as fish, 275f, 276
Receptors
  of homeostatic systems, 334, 334f
  sensory types of, 350, 350f
  stretch, 414f
  tendon, 414f
Recessive genes, 166, 171
Rechschaffen, A., 436
Recombinant DNA, research with, 141
Recombination
  of bacteria, 153, 154f
  genetic, 153
Rectum, human, 448f, 450f
Red algae, phycobilins of, 250
Red blood cell
  increase in numbers at high altitudes, 399
  as link between ventilation and circulation 242f, 395, 396, 397, 398f
  proximity to alveoli in lungs, 392f
  role in respiratory control, 337, 338f
Red tide, and water pollution, 621
Redi, F., 30
Reduced coenzyme Q, in respiratory chain, 103, 104f
Reduced flavin mononucleotide, in respiratory chain, 103, 104f
Reduced nicotinamide adenine dinucleotide, in respiratory chain, 103, 104f

Reducing agents, biomolecules as, 98
Reducing power, generation by hexose monophosphate pathway, 79, 81f
Reductant, 71
  in oxidation-reduction reaction, 91n
Reduction divisions, of meiosis, 157
Reflex. A simple neural pathway that is the basis of act that can be caused without conscious involvement. 347
Reflex arc
  components of, 364, 365
  speed of, 364, 365
Reflex ovulation, hypothalamic control of, 357
Regulator gene, 137, 138f
Reindeer, population growth of, 565, 567f
Relative biological effect (of radiation), 599
Relaxin
  effects of, 360
  release by ovary, 360
Release factors, for stopping protein synthesis, 135
Releasing factor, effects on pituitary, 357
REM, *see* Rapid eye movement sleep
Renal artery, 401f
Renal vein, 401f
Renilla
  polymorphism in, 188, 188f
  polyps of sea pansy, 188, 188f
  sea pansy, 188, 188f
Replica plating, of bacteria, 127, 127f
Replication
  conservative, 122f
  of DNA, 111d, 112f, 115, 116f
  evolution of, 140
  nonconservative, 122f
  and origin of life, 64
  semi-conservative, 121, 122f
Replication fork, 121
Repression
  of gene activity by end product, 139
  of gene activity by substrate, 137, 138f
Repressor protein, 137
  and gene regulation, 137, 138f
Reproduction
  in brown algae, 250, 251f
  cellular, 142
  in coelenterates, 260f, 261
  complexity cost in multicellular organisms, 181
  essential accuracy of, 111
  as function of organ systems, 192
  in mushroom fungi, 256, 257f
  in primates, 315
  in primitive vascular plants, 253f, 254f
  in seed plants, 254
  of sponges, 257, 258
  sexual, 151
Reproduction rate, effects on evolution, 297
Reproductive cycle
  cause of cyclic nature, 457
  of human female, 454, 455, 456f
Reproductive drive, of humans, 580
Reproductive system, 193

derivation during early embryonic life, 209
development of human, 453, 453f, 454, 480, 481, 481f
Reproductive system
  evolution of specialized structures in, 209
  functions of, 209
  general types of in animals, 209
  human female, 449, 450, 450f
  human male, 447, 448f
  of plants, 223, 225f
  types of structures in, 209
Reptiles
  advances over amphibians of, 278, 279f
  egg of, 278, 280f
  and evolution of tetrapod limbs, 276, 277f
  extinct forms of, 278, 278f
  heart of primitive, 305f
Reserpine, neural effect of, 443f
Residence time, for components of biogeochemical cycles, 531
Resonance transfer
  and chlorophyll, 90, 90f
  in chloroplasts, 90
Respiration
  in arthropods, 266
  cellular, 99
  in echinoderms, 272f
  homeostatic control of human, 337, 338f, 339
  poisons of, 490
Respiratory carrier molecules, 102f, 103
Respiratory center, of medulla, 395, 396f
Respiratory chain. The linked system of catalysts that facilitate the oxidation-reductions involved in cellular metabolism. 91
  function in cellular respiration, 103
  oxidation-reduction reactions of, 104f
Respiratory system, 193
  function in animals, 200
  in human development, 481
  types of in animals, 200f
Results (scientific), verification of by scientific method, 8
Reticular formation, 417f
Retina
  of squid eye, 270f
  structure of human, 351, 353, 353f
Retrieval, as role of kidney, 407, 408f
Rh factor, as cause of fetal disease, 484, 484f
Rhesus factor, *see* Rh factor
Rheumatoid arthritis, and HLA genes, 502
Rhizobium, in nitrogen fixation, 537
Rhizoids, of psilophytes, 253, 253f
Rhizomes, of ferns, 254f
Rhodinius prolixus, blood-sucking bug, 208f
Rhodophyta, Phylum *see* Red algae
Rhodopsin, and human vision, 353
Rhue family, polydactyly in, 164, 165f
Rhythm method, of birth control, 468
Rib cage, role in human ventilation, 393, 393f
Riboflavin phosphate, *see* Flavin mononucleotide

Ribonucleic acid
  codons for amino acids, 116f
  compared with DNA, 55f, 56
  and control of protein synthesis, 111
  messenger, 130
  ribosomal, 131
  synthesis stimulated by hormones, 364
  transfer, 131, 132f, 133f
  types of, 129
Ribonucleic acid structure, 55f, 130f
  compared to DNA, 130, 130f
Ribose, as nucleic acid building block, 55f, 56
Ribose-1,5-diphosphate, and photosynthesis, 93, 93f, 94
Ribose phosphate, 55f
Ribose-5-phosphate, and hexose monophosphate pathway, 81f
Ribose-5-phosphate, structure, 81f
Ribosomal RNA, function of, 131
Ribosomes, 35
  aminoacyl site of, 134
  composition of, 131
  large unit of, 134
  peptidyl site of, 134
  and rough endoplasmic reticulum, 35f
  small unit of, 134
  subunits of, 131
Ribulose-1,5-diphosphate, structure, 93f
Ribulose-5-phosphate, structure, 81f
Rice
  culture in Indonesia, 558, 558f
  high yield strains of, 552
Right atrium, function of human, 401, 401f, 402, 403f
Right auricle, see Right atrium
Right ventricle, function of human, 401f, 402, 403f
Rickettsias, description of, 245, 246
Risk, definition of acceptable, 605
Rivers, contribution to geological cycles, 530, 531f
RNA, see Ribonucleic acid
Rod cells, of human retina, 352, 353, 353f
Roman Catholic Church, and population control, 576
Root hairs
  development of in plants, 214
  function in pressure generation, 216
  of plant roots, 216f
Roots
  cells of, 216f
  development of in plants, 214
  pressure generation in, 216, 216f
Roundworms, see Nematodes
Rous, Peyton, 509
Roux, W., 172
rRNA, see Ribosomal RNA; Ribonucleic acid, ribosomal
Rubella, and fetal malformation, 483
Rudimentary organs, of humans, 310
Rutherford, Lord Ernest, 590

S

S phase, of mitosis, 147
S.A. node, see Sino-atrial node
Sabretooth tiger, extinction caused by early man, 324
Sacrum, human, 448f, 450f
Sagittal plane, through bilateral animals, 261f
Salamander, as amphibian, 278, 278f
Salivary glands
  function in human food processing, 374
  location in human digestive system, 373f
Salps, as protochordates, 272
Salt, taste cells of human tongue, 372f
Saprobes, 545
  in detritus food chain, 545
Saprophyte, fungi as, 256
  psilophyte gametophyte as, 253
Sarah, learned chimpanzee, 431
Sarcodina, types of, 248
Sarcoma, 508d
Sarcoplasmic reticulum, role in muscle contraction, 346f
Saunders, John, 515
Scheffer, V., 565, 567f
Schleiden, M. J., 33
Schrodinger, E., 11
Schwann cell, formation of myelin by, 343f
Schwann, T., 33
Science, 2
  applied, 12
  importance, 1
  and morality, 13
  pure, 12
Scientific method
  procedures, 4
  and science, 2
  verification of results, 8
Scientists, 1
Scopes, see Monkey trial
Scyphozoans, see Jellyfish
Sea anemones, as coelenterates, 261
Sea cucumbers, as echinoderms, 270, 271f
Sea lilies, see Crinoids
Sea pansy, structure of, 188, 188f
Sea squirts
  feeding in, 272, 273
  larval forms of, 273f
  metamorphosis in, 273f
  structure, 272, 273f
Sea stars, as echinoderms, 270, 271f
Sea urchins
  as echinoderms, 270, 271f
  genetic experiments with, 172
  skeleton of, 203f
Sebaceous glands, see Sweat glands
Second filial generation, see F2 generation
Second messenger, 139
  cyclic AMP as, 139
Secondary structure
  determining factors in proteins, 46
  of DNA, 115
  of proteins, 44d, 45f, 46, 48, 48f

protein types of, 48, 48f
Secretin, role in human digestion, 380
Sedimentary rock, carbon storage by, 534
Seeds
  advantages of in plants, 254
  production of in gymnosperms, 255f
Segmentation
  of annelids, 265f, 266
  of arthropods, 266, 267f
  of molluscs, 269
Segmented organisms, higher eumetazoans as, 261
Selection pressure, 297
  due to isolating mechanisms, 294
  due to sexual selection, 297
Self fertilization, of garden peas, 166
Selfing, see Self fertilization
Semen, 449
  composition of human, 448, 449
Semicircular canals
  analogy between vertebrates and molluscs, 270
  receptors of, 350f
Semi-conservative replication, of DNA, 121, 122f
Seminal vesicle. Storage organ for mature sperm in the male. 448
  human, 448f
Seminiferous tubule, 448
  as source of male hormones, 361
Semipermeable membranes, 66
Sense organs, types of human, 349
Sensitive plants, movement in, 222
Sensitization, behavioral, 427, 428f
Sensory neuron, 206
  entrance to spinal cord, 348f, 349
  generation of impulses in, 344
  see also Sensory receptors
Sensory receptors
  adaptation of, 351
  for carbon dioxide in blood, 395, 396f
  human smell, 370, 372f
  human taste, 371, 372f
  olfactory, 370, 371, 372f
  for oxygen level in blood, 395, 396f
  specificity of, 350f, 351
  transduction by, 351
  types of, 350, 350f, 351
  for ventilatory movement, 396f
Sequoia, xylem pressure in, 217
Serine, structure, 43f
Serotonin
  as precursor for melatonin, 361f
  structure, 361f
Serotonin releasing synapses, 442d
Sertoli cell, 449
Sessile organisms, 273
  sea squirts as, 273
Sewage
  typical BOD of, 621
  as water pollutant, 621
Sex
  determination by types of chromosomes, 174
  evolutionary value of, 151

origins of, 152
Sex hormones, release by adrenal cortex, 359f
Sex specific behavior, in humans, 454
Sexual development, of humans, 478
Sexual reproduction
    of bacteria, 153, 155
    of eukaryotes, 155
    role of in rapid Cambrian speciation, 306
Sexual selection. A mechanism of evolution in which the female is said to choose among various possible mates on the basis of certain characteristics and thus insure preservation of these characteristics in future generations. 297
    effects on evolution, 297
Shadow reflex, habituation of, 427
Sharks
    brain of, 306f
    as fish, 275f, 276
Shell, of molluscs, 269f
    of reptile egg, 280f
Shivering, role in thermoregulation, 386, 389f, 390
Shope's papilloma, vial cause, 509
Short term memory
    of humans, 411, 432
    of octopuses, 270
Short-day plants, 222
Sickle cell anemia
    incomplete gene dominance in, 170
    human genetic disease, 110, 493t
Silent Spring, 623
Silica, effect on lungs, 618
Silicosis, cause of, 618
Sino-atrial node, 403f
Sinuses, as means of increasing surface area, 198, 198f
Sinusoids, of liver capillary system, 382, 383
Siphon, of squid, 268, 270f
Siphonozooid, of sea pansy, *Renilla*, 188f
Siphons, of sea squirts, 272, 273f
Size
    and adaptability of species, 298, 299
    as advantage of multicellularity, 181
Skeletal system, 193
    endoskeletal, 202, 203f
    exoskeletal, 202, 203f
    hydrostatic, 202, 203f
    types of in animals, 202, 203f
Skeleton
    characteristics of in early hominids, 319
    development of human bones, 479
    of echinoderms, 271f, 272, 272f
    formation of deposits of protistan, 248
    of squid, 270f
    types of protistan, 247f, 248, 249f
Skin color, as continuously varying genetic trait, 172
Skulls, of hominids, 321f
Sleep
    circadian periodicity of, 437
    role in brain function, 436
    types of, 437
Slime mold

cellular, 185, 185f, 256
characteristics of, 256
doubtful classification of, 256
life history of, 185
two types of, 256
Slow wave sleep, 437
Small intestine
    function in human digestion, 376, 377f
    in human development, 482
    location in human digestive system, 373f
    structure, 377f
Small ribosomal unit, 134
Smallpox, history of eradication, 502n
Smell
    decreased importance in primates, 313
    role in controlling human food intake, 369, 370, 372f
    sensory receptors for, 370
Smog. Air pollution sufficient to be visible as a haze, brought about by volatile wastes of human activity, often with interaction with sunlight to produce photochemical smog.
    causes of, 617
Smoking, and cancer, 509, 510f
Snakes, as reptiles, 279f
Snowy owl, population cycle of, 547
Social behavior, linked with estrous in primates, 314
Social insects, collectives of, 189
Sodium, role in nerve impulse, 342f
"Soft" technology, 608
Solar constant, 526
Solar power, 585
Solar radiation
    amount received by earth, 526
    and circulation of air and water, 527, 529f
    effect of atmosphere on, 527, 528f
    effect of latitude on, 526, 526f
    effect of ozone layer on, 527
    seasonal effects and, 528f
    spectrum of, 83, 83f
Somatic cell hybridization, use in gene mapping, 178
Somatic mutations, 158
    protection from effects by diploidy, 158
Somatosensory map, of cerebral cortex, 419f
Somatostatin, synthesis of gene for, 141
Somite, 479
    of chordates, 274f
    developmental origin of human, 479
Sori, of ferns, 254f
Sound, detection by human ear, 354, 355
Sour, taste cells of human tongue, 372f
Southern corn blight, 556
Spacer DNA, 131
Special creation, 284
    experimental verification of, 285
Specialization, as advantage of multicellularity, 180
Speciation, 285
    by evolution, 285
    by inheritance of acquired characteristics, 285
    of leopard frog, 302

    by special creation, 285
    theory of, 300
Species
    adaptation of, 297, 298, 299
    binomial naming of, 231
    creation of, 285
    distribution of as evidence for evolution, 291
    objective reality of classification of, 234
    stability of throughout time, 285
    survival of the fittest, 291
Species ecology, *see* Autecology
Species diversity, agricultural reduction of in ecosystems, 556
Species Plantarium (book), 231
Spectrophotometry, 106f, 107
Speech
    as by product of ventilation, 393
    evolution of, 327
Speed, of light, 83
Sperm
    capacitation period of, 449
    description of, 162
    development in testes, 361
    and events of human fertilization, 459
    formation of, 157f, 162
    structure of human, 449, 449f
Sperm count, and fertility, 449
Spermatid, 448
Spermatogenesis, 447
    in humans, 447, 448, 451
Spermatozoa, *see* Sperm
Spermists, 164n
Sperry, R. W., 423
Sphenophyta, Subphylum, *see* Sphenophytes
Sphenophytes
    characteristics of, 253, 253f
    primitive vascular plants, 252, 252f
Sphincter. A ring-shaped muscle surrounding a body opening or channel that it is capable of closing by contraction. 379d
    nervous control of human digestive system, 379
Spicules, of sponges, 258f
Spiders, as arthropods, 267f
Spiegelman, S., 64
Spinal cord
    as part of nervous system, 207
    as part of reflex arc, 365
    cross section, 348f
    structure of human, 348, 348f
Spinal nerves, origin in spinal cord, 348
Spindle apparatus, 147
    function in mitosis, 147
Spindle fibers, 147
    disruption by colchicine, 148
    function in mitosis, 147, 148f
Spiny echidnas, primitive mammals, 280, 281f
Spirochaetes, description of, 245
Split brain, 422, 423f
Sponges
    characteristics of, 257, 258f
    evolution of, 257, 258

Sponges (*cont.*)
    as an evolutionary dead end, 258
    simple and complex types of, 258f
    water flow in, 258f
Spontaneous generation, 30n
Spores
    of basidiomycetes, 256, 257f
    of brown algae, 250, 251f
    of ferns, 254f
    formation of in protists, 248
    of moss, 252f
    production of by slime molds, 185, 256
    production of in flowers, 224, 226f
    of seed plants, 254
Sporophyte. The diploid phase of a plant life cycle, produces spores. 223
    of brown algae, 250, 251f
    of ferns, 254f
    of flowering plants, 223
    of gymnosperms, 255f
    of moss, 252f
    of psilophytes, 253, 253f
    of seed plants, 155f
Squids
    characteristics of, 268, 270, 270f
    locomotion in, 268, 270f
    as predators, 268, 270, 270f
SRU, *see* Small ribosomal unit
Stamen, of flowers, 224, 225f
Standard of living, effects on human population growth, 580
Staphylococcus, antibiotic resistant strains of, 504
Starch, 54
    breakdown during digestion in humans, 347f, 375
    forms of, 374f
    and regulation of blood glucose, 383f
Starling, E. H., 380
Statistics, and verification of experimental results, 9
Stentor, structure, 247f
Stereoisomerism, 46, 47f
Stereotyped behavior, of arthropods, 268
Sterilization
    and birth control, 465
    of human female, 465
    by sectioning vas deferens, 465
Steroids, hormones as, 360
Stevenson, Robert Louis, 437
Stigma, of flower, 224, 225f
Stirrup, of human ear, 354f
Stoma, *see* Stomata
Stomach
    as example of organ, 190, 192f
    function in human digestive system, 190, 192f, 373, 373f, 376
    in human development, 481
    location in human digestive system, 373f
    structure of human, 375f
    tissues of, 190
Stomach ulcers, and smoking, 510f
Stomata, and gas exchange in plants, 218, 219f
Stomatal cells, and gas exchange in plants, 218, 219f
Stop codon, function of, 135
Strassman, F., 590
Stratigraphic method, of dating fossils, 237
Streptococcus pneumoniae, 112
Streptomycin
    as bacterial replication blocker, 155
    as disease combatant, 505f
Stress
    effects on blood glucose levels, 385
    effects on heart activity, 402
    neuroendocrine responses to, 366
    ventilative, 397
Stretch receptor
    functioning of, 344, 345f
    in muscles, 414f
Strobilus, and budding of jellyfish, 260f
Stroke
    causes of, 400, 401
    see also Circulatory system failure
Stromatolites, 97, 98f
Structural gene, 137, 138
Strychnine, as poison, 490
Style, of flower, 224, 225f
Subclavian artery, 401f
Subclavian vein, role in lymphatic system, 406
Suborder Anthropoidea *see* Anthropoidea
Suborder Prosimii, *see* Prosimians; Tree shrews
Substrate repression, of gene activity, 137, 138f
Sucrose, and regulation of blood glucose, 383f
Suez Canal, and exchange of aquatic organisms, 559
Sugar
    absorption in human small intestine, 376, 377f
    facilitated diffusion in small intestine, 376
    fermentation of, 107
    structure, 53f, 54
Sulfanilamide, as disease combatant, 505f
Sulfur bonds, of proteins, 48, 48f
Sulfur compounds, as pollutants, 616
Sulfur dioxide, as pollutants, 613t, 616
Sulfur trioxide, as pollutant, 613t, 616
Sulfuric acid, as poison, 490
Super molecules, of water, 58f
Superfamily Cebiodea, *see* Cebiods
Superfamily Hominoidea, *see* Hominoids
Superior cervical ganglion, location of, 361f
Superior rectus muscle, of human eye, 352f
Superior vena cava, 401f
    and lung circulation, 392f
Superorganisms, 189
Superorganisms, *see* Collectives
Superoxide anion, 99
Superoxide dismutase, function, 99, 99f
Surface area, importance in respiration, 200, 200f
Surface area/volume ratio
    of flatworm, 195, 195f
    limiting sizes of cells and organisms, 195
    and necessity for cell division, 142
    and need for circulation, 194, 195, 197, 198f
    ways to increase, 195, 196f, 197f
Survival of the fittest, 291
Sutherland, E. W., 363
Swallowing, role in human food processing, 375
Sweat glands, role in human thermoregulation, 386f, 387, 389f
Sweet, taste cells of human tongue, 372f
Swim bladders, of fish, 276
Swivelase, in DNA replication, 123
Symmetry, types of animal, 261, 261f
Sympathetic ganglia, of sympathetic nervous system, 348f, 349, 349f
Sympathetic nervous system, 347
    control of heart by, 349, 366
    control of human digestion, 379, 382
    control of human thermoregulation, 388, 389f, 390
    nerves of, 349, 349f
    role in stress responses, 366
    transmitter chemicals of, 349
Synapse, 206, 343
    in brain, 442
    conduction of information across, 343, 344f
    inhibitory, 446
    types of, 343
Synaptic cleft, 344f
Synaptic layer, of human retina, 353f
Synaptic vesicles, 343
    containing acetylcholine, 343, 344f
Syncytial organisms, and evolution of multicellularity, 184
Synthetic phase, *see* S phase
Syphilis, effects on human fetus, 484
Systema Naturae, 230n, 231
Systematics, 229
    evolutionary significant information of, 303
Systole, of human heart, 401f

T

2,4,5-T
    as hazardous chemical, 622t
    structure, 622t
T lymphocytes, role in combating disease, 501, 501f
Tadpole, larval amphibian, 278, 278f
Tadpole larva, of sea squirts, 273, 273f
Tapeworms, as flatworms, 264, 264f
Target tissues, examples of effects of hormone on, 363f
Taste
    and regulation of human food intake, 371, 372f
    sensory receptors for, 371
    sensor receptors of in fly, 350f
Taste bud, human, 371, 372f
Tatum, E. W., 153
Taxonomy, 229
    of flowers, 233

of mammals, 232
Tay-Sach's disease, 435t
  detection by amniocentesis, 488
TCCD, see Dioxin
Technologists, 1
Tectoral membrane, of human ear, 354f
Teeth
  evolution of high crowns in grazers, 298
  evolution of in primates, 314, 314f, 319, 320
Telophase
  of meiosis, 160f
  of mitosis, 148, 148f
Temperature
  CO_2 effect on earth's, 616
  of earth through time, 525f
  effect of energy consumption on earth's, 588
  pollutant effects on earth's, 525
Temperature regulation, see Thermoregulation
Temporal lobe, 348f
  role in human memory, 432
Tendon receptor, of muscles, 414f
Tentacle, of squid, 270f
Teratoma, 509
Teredo, molluscan shipworm, 268
Tennessee "monkey" trial, of Scopes for teaching evolution, 290
Territoriality, 297
  and population limitation, 581
  and sexual selection pressure, 297
Tertiary structure, of DNA, 115
  of protein, 44, 45f
Testes
  control by pituitary hormones, 361
  endocrine function of, 360
  hormones released by, 360, 361
  location of, 207f
Testis
  human, 447, 448f
  interstitial cells, 448f
  seminiferous tubule of, 448
  Sertoli cells of, 448f, 449
Testosterone
  in human development, 453
  release by testes, 361
  role in human reproductive cycle, 452f
  structure, 452f
Tetanized muscle. A muscle maintained in the maximally contracted stage for periods substantially longer than the single contraction duration, by repetitive stimulation. 346
  causes of, 346
Tetanus, see *Clostridium tetani*
Tetracycline, as disease combatant, 505f
Tetraethyl lead, pollution from, 618
Tetrahedron, 46
Tetrapods, as vertebrates, 274, 275
Tetrapods, evolution of, 276, 277f
Tetrapyrrole group
  evolution of, 243
  ubiquity throughout biosphere, 242f, 243
Tetrapyrrole ring

and chlorophyll, 87
  in cytochromes, 102f, 103
  structure, 90f
Tetrapyrroles
  linear, 250
  of algae, 250
Thalmus
  functions of human, 416
  general function of human, 347
  location of human, 347
Thalassemia, human genetic disorder, 493t
Thalidomide, 14
  effects on fetal limb formation, 483
Thermal pollution, from power plants, 624
Thermocline, 541
Thermonuclear bomb, 592
Thermoreceptors, types of, 350f, 351
Thermoregulation
  controls of, 388, 389f, 390
  mechanisms of, 385-390
Thiol bond, of CoA, 101
Thirst, hypothalmic control of, 410
Thompson, Sir W., 34
Thorax, role in human ventilation, 391, 393, 393f
Threonine, structure, 43f
Thrombin, effects on blood clotting, 400f
Thromboplastin, effects on blood clotting, 400, 400f
Thrombosis. Clot formation within heart or blood vessels. 400
  linked with birth control pill, 468
Thrombus, 400
Thymine, structure, 55f
Thymus
  developmental origin of human, 479
  location of, 207f
  possible functions of, 356
Thyroglobin, in thyroid follicle, 358f
Thyroid follicles, synthesis and storage of thyroxin by, 357, 358f
Thyroid gland
  cellular structure of, 358f
  and control of thermoregulation, 289f, 390
  and endocrine system control, 207
  feedback control involving pituitary, 357
  homology in sea squirts, 272
  hormones of, 357, 358f
  in human development, 481
  location of, 207f, 358f
Thyroid hormone
  age dependency of effects of, 363, 363f
  effects on metamorphosis in frogs, 363, 363f
  effects on pituitary, 357
  effects on temperature regulation, 386
Thyroid stimulating hormone
  release of by pituitary, 357, 358f
  specificity of, 363
Thyrotropic hormone, see Thyroid stimulating hormone
Thyroxin, endocrine function of, 207
Tissues, 190
  and cellular differentiation, 181

composition of, 191f
  layers of in eumetazoans, 258, 159f, 260
  and organization of organisms, 191f
  of plants, 212, 213, 213f, 214
Toads, as amphibians, 278, 278f
Tobacco
  effects on fetal health, 483
  relationship to human health, 510f
Tongue, distribution of taste cells in humans, 372f
Tonsils, developmental origin of human, 479
Toxins, of coelenterates, 262
Trachea
  location of, 358f
  respiratory function of in insects, 200
  role in human ventilation, 391, 393, 393f
  size limitations imposed by, 266
  structure of human, 392f
  structure of insect, 200f
Tracheophyta, Phylum, see Tracheophytes
Tracheophytes
  primitive types of vascular, 252, 253, 253f
  reasons for success of, 252
  species diversity of, 252
  vascular plants as, 251, 252, 254, 256
Trade winds, 529f, 530
Tragedy of the Commons
  applied to pollution, 612
  concept of, 583
Tranquilizers, neural effects of, 443f
Transducers, sensory receptors as, 351
Transformation, of bacteria, 153, 154f
Transforming principle, 112
Transcription, 112
  control of by gene repression, 136, 137, 138f, 139
  of DNA into RNA, 112, 112f, 116f
  description of, 130
  effects of hormones on, 364
  exactness of, 131
Transfer RNA
  aminoacyl, 134
  attachment of amino acids to, 132
  function of, 132, 133, 133f
  initiator, 134, 134f
  structure, 132, 132f
  unusual bases in, 132
Translation, 112
  control of, 136, 139
  of RNA into protein, 112d, 112f, 116
Translocation, of genes, 157, 158f
Transmitter chemical, 206
Transpiration. In a plant, the process of water loss from the ventilating surfaces in the leaves. 217
  and adaptations to control water loss in plants, 219
  importance in generating xylem pressure, 217, 218f
Transplants, of human organs, 496
Transverse tubule system, and control of muscle contraction, 346f
Tree frogs, as amphibians, 278, 278f
Tree shrews, characteristics of, 313f

Tree shrews (cont.)
  hands and feet of, 313f
Trematodes, as flatworms, 264, 264f
Treponema pallida, syphillis spirochaete, 242
Tricarboxylic acid cycle, 99, 100f
  net energy yield, 101
1,1,1-trichloro-2,2-bis(p-chlorophenyl)ethane, 623
  accumulation in ecosystem, 623
  as hazardous chemical, 622t
  as poison, 490, 491
  structure, 622t
Trilobite
  fossils of, 235f
  and Down's disease, 177
Triplet code, of DNA, 113, 116f
Triploblastic organisms, 261
  types of, 259f
Trisomy, 177
Tristearin, structure, 54f
tRNA, *see* Transfer RNA; Ribonucleic acid, transfer
Trophic levels, 540, 540f
  marine and terrestrial compared, 547
  transfer costs between, 544
  transfer strategies of, 544
Trophoblast, 460
  of human embryo, 471
Trophotropic nervous system, 442
True mosses, *see* Bryophytes
Truffles, as fungi, 256
Truman, Harry, 607
Trypanosomes, structure, 247f
Trypsin, role in human digestion, 378f
Tryptophan
  as precursor for melatonin, 361, 361f
  structure, 43f
TSH, *see* Thyroid stimulating hormone
TTS, *see* Transverse tubule system
Tube cell nucleus, of pollen grain, 226f
Tube foot, of echinoderms, 272, 272f
Tunic, of sea squirts, 272, 273f
Tunicates, *see* Sea squirts
Turbellarians, as flatworms, 264, 264f
Turgor, offsetting osmosis, 67
Turgor pressure, 212
  and function of guard cells, 218, 219f
  offsetting osmosis in plant cells, 212
Turtles, as reptiles, 279, 279f
Twinning, in humans, 459
Twins
  fraternal, 459, 460f
  identical, 459, 460f
  use in intelligence studies, 434
Tyrosine
  biosynthesis of, 301f
  and phenylketonuria, 171, 171f
  structure, 43f, 301f
  and synthesis of thyroxin, 357, 358f

U

Ulcers, formation by human stomach, 376

Ulnar artery, 401f
Ultrasound, use in imaging fetus, 487, 487f
Ultraviolet light
  effect of ozone layer on, 527
  as mutagen, 128
  as part of electromagnetic spectrum, 83, 83f
  reduction by ozone, 97
  resistance to in dark races, 297
Umbilical cord, of human, 461
Undernutrition, effects on fetal health, 483
Upwelling
  and anchoveta fishery, 542
  causes of, 541, 542
Uracil, structure, 55f
Uranium, and discovery of nuclear fission, 590
Uranium-235, nuclear decay of, 592
Uranium-238, nuclear decay of, 592
Urea
  elimination by human kidney, 409
  structure, 202f
Urethra
  human, 448, 448f, 450f
  location of human, 408f
Uric acid
  storage by reptilian embryos, 280f
  structure of, 202f
Urinary bladder, human, 408f, 448f, 450f
Urinary system, parts of human, 407, 408f
Uterine aspiration, as method of abortion, 469
Uterus, human, 450f

V

Vaccination, history of, 503n
Vacuole. Membrane-bordered, usually fluid-filled, space within cytoplasm of a cell performing a variety of functions, as water-balance in contractile vacuoles, digestion in the food vacuoles of protozoa. 35, 35f
Vagina, human, 450f
Vagus nerve
  control of heart by, 349, 349f
  effects on human heart, 403f
  and regulation of vagus nerve plexus, 379
  role in regulating hunger, 369
Valence electrons, 85
Valence shell, of electrons, 85
Valine, structure, 43f
Valinomycin, as disease combatant, 505f
van Helmont, Jean Baptiste, 210
Variable penetrance, 171d
  of genes, 170f, 171
Variable expression, of genes, 170f, 171
Variolation, and treatment of smallpox, 503n
Vas deferens, 448
  human, 448
Vas deferens section, use in regulating population growth, 575
Vascular cambium

  of plants, 212, 213f
  of plant roots, 216f
Vasectomy, *see* Vas deferens section
Veins
  characteristics of human, 404, 405
  location of human, 401f
  one-way valves of, 402, 404
Venereal disease, bacterial resistance to, 156n
Ventilation, 390
  adaptation to stress, 397, 398
  control of, 395, 396f
  homeostatic control of human, 337, 338f
  in humans, 390-399
  medullary control of, 395, 396f
  by negative pressure, 393, 393f
  by positive pressure, 393, 394f, 395
Ventral root, of spinal cord, 348, 348f
Ventricles, function of human, 401f, 402, 403f
Venus's fly trap, feeding in, 222, 223f
Verification, of results by scientific method, 8
Vertebrae, 274
  protecting spinal cord, 348
  of vertebrates, 274
Vertebrates
  brain sizes of, 314f
  characteristics of, 274, 275f
  evolution of brain in, 306, 306f
  evolution of heart in, 305, 305f
  evolution of limbs in, 276, 277f
  homology of limbs of, 240, 240f
  kidney of, 201f
  major divisions of types, 274
  nervous system of, 206f, 207
  skeleton of, 274
  transition from sea to land, 276
  types of, 275, 275f, 276, 277f, 278, 279, 280
Vertical migrators, 545
  and marine food chain, 545, 546f
Vine snake, 279f
Violence, and organization of human brain, 438
Virus
  and cancer, 509
  control of human, 506f, 507
  effect on cell multiplication in tissue culture, 511
  life history of, 596f
  structure, 506f
Viruses
  DNA replication in, 120
  as genetic carriers between hosts, 153
Viscera. General term for the organs lying within the abdomen and thorax. 349
  of humans, 349
  innervation of, 349
Visible light, 83, 83f
Vision
  causes of poor, 351
  color, 353
  emphasis on in primates, 313
  evolution of binocular primate, 313

night, 353
Visual phosphene, 420d
Vitalism, 15
Vitamin A, required for vision, 351, 352
Vitamin B, absorption in human large intestine, 377
Vitamin B12, intestinal transport mechanism for, 376
Vitamin B17, see Laetrile
Vitamin C
   as reducing agent, 98
   structure, 99f
Vitamin E
   as reducing agent, 98
   structure, 99f
Vitamins
   destruction of by ultraviolet light, 297
   treatment of during human digestion, 376
Vitelline blood vessels, 476, 477f
Volcanoes, contribution to geological cycles, 530
Voluntary muscle, see also Skeletal muscle
Volvox
   cellular differentiation in, 186, 187f
   daughter colonies of, 187f
   as multicellular organisms, 186, 187f
   structure of, 186, 187f
   vegetative cells of, 187f
von Baer, Karl Ernst, 310
Vorticella, 183f

W

Wald, G., 12, 238
Wall, cell, 34, 35f
Wallace, Alfred Russel, 285, 289f
   evolutionary theories of, 288
   publications of, 288, Appendix 4
   views on human evolution, 309
Warburg, O., 74
Wastes
   conversion to fuel, 609n
   elimination by kidney, 408, 409
Water
   and cellular osmotic balance, 69
   control of by human kidney, 409
   distribution on earth, 532
   effects of chlorine use, 620t
   hydrogen bonding, 58f
   influence of sun's distance on state of, 524
   movement from soil into plants, 216, 216f, 217, 218
   and origin of life, 57
   and osmotic instability, 201
   as photosynthesis substrate, 87
   plant adaptations to limitations of, 542
   polarity of, 58f
   production in cellular respiration, 104
   properties of, 58f
   reabsorption in human large intestine, 377
   release in dehydration synthesis, 44
   structure, 58f
Water cycle, 532, 534f
Water pollution
   from nutrient enrichment, 621
   from sewage, 618
   from wastes, 621
Water vascular system, of echinoderms, 272, 272f
Watson, James, 115
Wavelength, of light, 83
Waxy cuticle, of plants, 213f, 218
Weiner, N., 333
Weismann, A., 172
Whales
   and ocean productivity, 542
   effects of hunting on, 558t, 559
Wheat stem rust, 556
White blood cells
   functions of, 339, 400
   in lymph, 405, 406
   source of, 399, 400
White matter, of spinal cord, 348f
Whittaker, R. H., 283
Wilberforce, Bishop Samuel, 289, 290
Will-o'-the-wisp, 534
Wind
   driven by solar radiation, 527
   and pollen transport, 227f
   principle, 529f
Woodwell, G., 623
Wuster, C., 623

X

X chromosome, in insects, 174
X-rays
   and cancer, 509
   as part of electromagnetic spectrum, 83, 83f
   radiation dose of medical, 603t
X-ray diffraction, 48
Xylem
   of plant roots, 216, 216f
   of plants, 213f, 214, 215f, 216, 218f
   pressure generation in, 217
   primary, 213f
   transport functions of, 216, 216f, 217, 218f
Xylem cells
   of plants, 212, 213f, 214
   tissue formed by, 214

Y

Y chromosome, in insects, 174
Yeast, fermentation by, 107
Yolk sac, of reptile egg, 280f

Z

Z bands, of muscle, 346f
Zoospores, of brown algae, 250, 251f
Zwitterion, 42
   structure, 42f
Zygote, 156
   demonstration of chromosome number of, 172
   formation of, 156
   formation of human, 459
Zymogen, 139